Handbook of Research Methods in Health Social Sciences

Pranee Liamputtong
Editor

Handbook of Research Methods in Health Social Sciences

Volume 3

With 192 Figures and 81 Tables

Editor
Pranee Liamputtong
School of Science and Health
Western Sydney University
Penrith, NSW, Australia

ISBN 978-981-10-5250-7   ISBN 978-981-10-5251-4 (eBook)
ISBN 978-981-10-5252-1 (print and electronic bundle)
https://doi.org/10.1007/978-981-10-5251-4

Library of Congress Control Number: 2018960888

© Springer Nature Singapore Pte Ltd. 2019

This work is subject to copyright. All rights are reserved by the Publisher, whether the whole or part of the material is concerned, specifically the rights of translation, reprinting, reuse of illustrations, recitation, broadcasting, reproduction on microfilms or in any other physical way, and transmission or information storage and retrieval, electronic adaptation, computer software, or by similar or dissimilar methodology now known or hereafter developed.

The use of general descriptive names, registered names, trademarks, service marks, etc. in this publication does not imply, even in the absence of a specific statement, that such names are exempt from the relevant protective laws and regulations and therefore free for general use.

The publisher, the authors, and the editors are safe to assume that the advice and information in this book are believed to be true and accurate at the date of publication. Neither the publisher nor the authors or the editors give a warranty, express or implied, with respect to the material contained herein or for any errors or omissions that may have been made. The publisher remains neutral with regard to jurisdictional claims in published maps and institutional affiliations.

This Springer imprint is published by the registered company Springer Nature Singapore Pte Ltd.
The registered company address is: 152 Beach Road, #21-01/04 Gateway East, Singapore 189721, Singapore

To my mother:
Yindee Liamputtong
and
To my children:
Zoe Sanipreeya Rice and Emma Inturatana
Rice

Preface

Research is defined by the Australian Research Council as "the creation of new knowledge and/or the use of existing knowledge in a new and creative way so as to generate new concepts, methodologies, inventions and understandings." Research is thus the foundation for knowledge. It produces evidence and informs actions that can provide wider benefit to a society. The knowledge that researchers cultivate from a piece of research can be adopted for social and health programs that can improve the health and well-being of the individuals, their communities, and the societies in which they live. As we have witnessed, in all corners of the globe, research has become an endeavor that most of us in the health and social sciences cannot avoid. This Handbook is conceived to provide the foundation to readers who wish to embark on a research project in order to form knowledge that they need. The Handbook comprises four main parts: Traditional Research Methods in Health and Social Sciences, Innovative Research Methods in Health Social Sciences, Doing Cross-Cultural Research in Health Social Sciences, and Sensitive Research Methodology and Approach. This Handbook attests to the diversity and richness of research methods in the health and social sciences. It will benefit many readers, particularly students and researchers who undertake research in health and social science areas. It is also valuable for the training needs of postgraduate students who wish to undertake research in cross-cultural settings, with special groups of people, as it provides essential knowledge not only on the methods of data collection but also salient issues that they need to know if they wish to succeed in their research endeavors.

Traditionally, there are several research approaches and practices that researchers in the health social sciences have adopted. These include qualitative, quantitative, and mixed-methods approaches. Each approach has its own philosophical foundations, and the ways researchers go about to form knowledge can be different. But all approaches do share the same goal: to acquire knowledge that can benefit the world. This Handbook includes many chapters that dedicate to the traditional ways of conducting research. These chapters provide the "traditional ways of knowing" that many readers will need.

As health and social science researchers, we are now living in a moment that needs our imagination and creativity when we carry out our research. Indeed, we are now living "in the new age" where we will see more and more "new experimental works" being invented by researchers. And in this new age, we have witnessed many innovative and creative forms of research in the health and social sciences. In this Handbook, I also bring together a unique group of health and social science researchers to present their innovative and creative research methods that readers can adopt in their own research. The Handbook introduces many new ways of doing research. It embraces "methodological diversity," and this methodological diversity will bring "new ways of knowing" in the health and social sciences. Chapters in this Handbook will help to open up our ideas about doing research differently from the orthodox research methods that we have been using or have been taught to do.

Despite the increased demands on cross-cultural research, discussions on "culturally sensitive methodologies" are still largely neglected in the literature on research methods. As a result, researchers who are working in cross-cultural settings often confront many challenges with very little information on how to deal with these difficulties. Performing cross-cultural research is exciting, but it is also full of ethical and methodological challenges. This Handbook includes a number of chapters written by researchers who have undertaken their research in cross-cultural settings. They are valuable to many readers who wish to embark on doing a cross-cultural research in the future.

Globally too, we have witnessed many people become vulnerable to health and social issues. It will be difficult, or even impossible, for health and social science researchers to avoid carrying out research regarding vulnerable and marginalized populations within the "moral discourse" of the postmodern world, as it is likely that these population groups will be confronted with more and more problems in their private and public lives as well as in their health and well-being. Similar to undertaking cross-cultural research, the task of conducting research with the "vulnerable" and/or the "marginalized" presents researchers with unique opportunities and yet dilemmas. The Handbook also includes chapters that discuss research that involves sensitive and vulnerable/marginalized people.

This Handbook cannot be born without the help of others. I would like to express my gratitude to many people who helped to make this book possible. I am grateful to all the contributors who worked very hard to get their chapter done timely and comprehensively. I hope that the process of writing your chapter has been a rewarding endeavor to you as well. My sincere appreciation is given to Mokshika Gaur who believes in the value of the volume on research methods in health social sciences and has given me a contract to edit the Handbook. I also thank Tina Shelton, Vasowati Shome, and Ilaria Walker of Springer who helped to bring this book to life.

I dedicate this book to my mother Yindee Liamputtong, who has been a key person in my life. It was my mother who made it possible for me to continue my education amidst poverty. Without her, I would not have been where I am now. I also dedicate the book to my two daughters, Zoe Sanipreeya Rice and Emma Inturatana Rice, who have formed an important part of my personal and professional lives in Australia.

Sydney, Australia   Pranee Liamputtong

Contents

Volume 1

Section I   Traditional Research Methods in Health and Social Sciences 1

1 **Traditional Research Methods in Health and Social Sciences: An Introduction** 3
   Pranee Liamputtong

2 **Qualitative Inquiry** 9
   Pranee Liamputtong

3 **Quantitative Research** 27
   Leigh A. Wilson

4 **The Nature of Mixed Methods Research** 51
   Cara Meixner and John D. Hathcoat

5 **Recruitment of Research Participants** 71
   Narendar Manohar, Freya MacMillan, Genevieve Z. Steiner, and Amit Arora

6 **Ontology and Epistemology** 99
   John D. Hathcoat, Cara Meixner, and Mark C. Nicholas

7 **Social Constructionism** 117
   Viv Burr

8 **Critical Theory: Epistemological Content and Method** 133
   Anastasia Marinopoulou

9 **Positivism and Realism** 151
   Priya Khanna

10 **Symbolic Interactionism as a Methodological Framework** 169
   Michael J. Carter and Andrea Montes Alvarado

ix

11	Hermeneutics: A Boon for Cross-Disciplinary Research Suzanne D'Souza	189
12	Feminism and Healthcare: Toward a Feminist Pragmatist Model of Healthcare Provision Claudia Gillberg and Geoffrey Jones	205
13	Critical Ethnography in Public Health: Politicizing Culture and Politicizing Methodology Patti Shih	223
14	Empathy as Research Methodology Eric Leake	237
15	Indigenist and Decolonizing Research Methodology Elizabeth F. Rix, Shawn Wilson, Norm Sheehan, and Nicole Tujague	253
16	Ethnomethodology Rona Pillay	269
17	Community-Based Participatory Action Research Elena Wilson	285
18	Grounded Theory Methodology: Principles and Practices Linda Liska Belgrave and Kapriskie Seide	299
19	Case Study Research Pota Forrest-Lawrence	317
20	Evaluation Research in Public Health Angela J. Dawson	333
21	Methods for Evaluating Online Health Information Systems Gary L. Kreps and Jordan Alpert	355
22	Translational Research: Bridging the Chasm Between New Knowledge and Useful Knowledge Lynn Kemp	367
23	Qualitative Interviewing Sally Nathan, Christy Newman, and Kari Lancaster	391
24	Narrative Research Kayi Ntinda	411
25	The Life History Interview Erin Jessee	425
26	Ethnographic Method Bonnie Pang	443
27	Institutional Ethnography Michelle LaFrance	457

28	**Conversation Analysis: An Introduction to Methodology, Data Collection, and Analysis** Sarah J. White	471
29	**Unobtrusive Methods** Raymond M. Lee	491
30	**Autoethnography** Anne Bunde-Birouste, Fiona Byrne, and Lynn Kemp	509
31	**Memory Work** Lia Bryant and Katerina Bryant	527
32	**Traditional Survey and Questionnaire Platforms** Magen Mhaka Mutepfa and Roy Tapera	541
33	**Epidemiology** Kate A. McBride, Felix Ogbo, and Andrew Page	559
34	**Single-Case Designs** Breanne Byiers	581
35	**Longitudinal Study Designs** Stewart J. Anderson	603
36	**Eliciting Preferences from Choices: Discrete Choice Experiments** Martin Howell and Kirsten Howard	623
37	**Randomized Controlled Trials** Mike Armour, Carolyn Ee, and Genevieve Z. Steiner	645
38	**Measurement Issues in Quantitative Research** Dafna Merom and James Rufus John	663
39	**Integrated Methods in Research** Graciela Tonon	681
40	**The Use of Mixed Methods in Research** Kate A. McBride, Freya MacMillan, Emma S. George, and Genevieve Z. Steiner	695

Volume 2

41	**The Delphi Technique** Jane Chalmers and Mike Armour	715
42	**Consensus Methods: Nominal Group Technique** Karine Manera, Camilla S. Hanson, Talia Gutman, and Allison Tong	737

| 43 | **Jumping the Methodological Fence: Q Methodology** | 751 |

Tinashe Dune, Zelalem Mengesha, Valentina Buscemi, and Janette Perz

| 44 | **Social Network Research** | 769 |

Janet C. Long and Simon Bishop

| 45 | **Meta-synthesis of Qualitative Research** | 785 |

Angela J. Dawson

| 46 | **Conducting a Systematic Review: A Practical Guide** | 805 |

Freya MacMillan, Kate A. McBride, Emma S. George, and Genevieve Z. Steiner

| 47 | **Content Analysis: Using Critical Realism to Extend Its Utility** | 827 |

Doris Y. Leung and Betty P. M. Chung

| 48 | **Thematic Analysis** | 843 |

Virginia Braun, Victoria Clarke, Nikki Hayfield, and Gareth Terry

| 49 | **Narrative Analysis** | 861 |

Nicole L. Sharp, Rosalind A. Bye, and Anne Cusick

| 50 | **Critical Discourse/Discourse Analysis** | 881 |

Jane M. Ussher and Janette Perz

| 51 | **Schema Analysis of Qualitative Data: A Team-Based Approach** | 897 |

Frances Rapport, Patti Shih, Mia Bierbaum, and Anne Hogden

| 52 | **Using Qualitative Data Analysis Software (QDAS) to Assist Data Analyses** | 917 |

Pat Bazeley

| 53 | **Sequence Analysis of Life History Data** | 935 |

Bram Vanhoutte, Morten Wahrendorf, and Jennifer Prattley

| 54 | **Data Analysis in Quantitative Research** | 955 |

Yong Moon Jung

| 55 | **Reporting of Qualitative Health Research** | 971 |

Allison Tong and Jonathan C. Craig

| 56 | **Writing Quantitative Research Studies** | 985 |

Ankur Singh, Adyya Gupta, and Karen G. Peres

| 57 | **Traditional Academic Presentation of Research Findings and Public Policies** | 999 |

Graciela Tonon

| 58 | **Appraisal of Qualitative Studies** | 1013 |

Camilla S. Hanson, Angela Ju, and Allison Tong

59 **Critical Appraisal of Quantitative Research** 1027
    Rocco Cavaleri, Sameer Bhole, and Amit Arora

60 **Appraising Mixed Methods Research** 1051
    Elizabeth J. Halcomb

Section II   Innovative Research Methods in Health Social Sciences **1069**

61 **Innovative Research Methods in Health Social Sciences: An Introduction** 1071
    Pranee Liamputtong

62 **Personal Construct Qualitative Methods** 1095
    Viv Burr, Angela McGrane, and Nigel King

63 **Mind Maps in Qualitative Research** 1113
    Johannes Wheeldon and Mauri Ahlberg

64 **Creative Insight Method Through Arts-Based Research** 1131
    Jane Marie Edwards

65 **Understanding Health Through a Different Lens: Photovoice Method** 1147
    Michelle Teti, Wilson Majee, Nancy Cheak-Zamora, and Anna Maurer-Batjer

66 **IMAGINE: A Card-Based Discussion Method** 1167
    Ulrike Felt, Simone Schumann, and Claudia G. Schwarz-Plaschg

67 **Timeline Drawing Methods** 1183
    E. Anne Marshall

68 **Semistructured Life History Calendar Method** 1201
    Ingrid A. Nelson

69 **Calendar and Time Diary Methods** 1219
    Ana Lucía Córdova-Cazar and Robert F. Belli

70 **Body Mapping in Research** 1237
    Bronwyne Coetzee, Rizwana Roomaney, Nicola Willis, and Ashraf Kagee

71 **Self-portraits and Maps as a Window on Participants' Worlds** .. 1255
    Anna Bagnoli

72 **Walking Interviews** 1269
    Alexandra C. King and Jessica Woodroffe

73 **Participant-Guided Mobile Methods** 1291
    Karen Block, Lisa Gibbs, and Colin MacDougall

74	**Digital Storytelling Method**	1303

Brenda M. Gladstone and Elaine Stasiulis

75	**Netnography: Researching Online Populations**	1321

Stephanie T. Jong

76	**Web-Based Survey Methodology**	1339

Kevin B. Wright

77	**Blogs in Social Research**	1353

Nicholas Hookway and Helene Snee

78	**Synchronous Text-Based Instant Messaging: Online Interviewing Tool**	1369

Gemma Pearce, Cecilie Thøgersen-Ntoumani, and Joan L. Duda

79	**Asynchronous Email Interviewing Method**	1385

Mario Brondani and Rodrigo Mariño

80	**Cell Phone Survey**	1403

Lilian A. Ghandour, Ghinwa Y. El Hayek, and Abla Mehio Sibai

81	**Phone Surveys: Introductions and Response Rates**	1417

Jessica Broome

82	**The Freelisting Method**	1431

Marsha B. Quinlan

83	**Solicited Diary Methods**	1447

Christine Milligan and Ruth Bartlett

84	**Teddy Diaries: Exploring Social Topics Through Socially Saturated Data**	1465

Marit Haldar and Randi Wærdahl

85	**Qualitative Story Completion: A Method with Exciting Promise**	1479

Virginia Braun, Victoria Clarke, Nikki Hayfield, Naomi Moller, and Irmgard Tischner

Volume 3

Section III Doing Cross-Cultural Research in Health Social Sciences ... 1497

86	**Doing Cross-Cultural Research: An Introduction**	1499

Pranee Liamputtong

87	**Kaupapa Māori Health Research**	1507

Fiona Cram

88	**Culturally Safe Research with Vulnerable Populations (Māori)** .... Denise Wilson	1525
89	**Using an Indigenist Framework for Decolonizing Health Promotion Research** .... Karen McPhail-Bell, Alison Nelson, Ian Lacey, Bronwyn Fredericks, Chelsea Bond, and Mark Brough	1543
90	**Engaging Aboriginal People in Research: Taking a Decolonizing Gaze** .... Emma Webster, Craig Johnson, Monica Johnson, Bernie Kemp, Valerie Smith, and Billie Townsend	1563
91	**Space, Place, Common Wounds and Boundaries: Insider/Outsider Debates in Research with Black Women and Deaf Women** .... Chijioke Obasi	1579
92	**Researcher Positionality in Cross-Cultural and Sensitive Research** .... Narendar Manohar, Pranee Liamputtong, Sameer Bhole, and Amit Arora	1601
93	**Considerations About Translation: Strategies About Frontiers** ... Lía Rodriguez de la Vega	1617
94	**Finding Meaning: A Cross-Language Mixed-Methods Research Strategy** .... Catrina A. MacKenzie	1639
95	**An Approach to Conducting Cross-Language Qualitative Research with People from Multiple Language Groups** .... Caroline Elizabeth Fryer	1653
96	**The Role of Research Assistants in Qualitative and Cross-Cultural Social Science Research** .... Sara Stevano and Kevin Deane	1675
97	**Indigenous Statistics** .... Tahu Kukutai and Maggie Walter	1691
98	**A Culturally Competent Approach to Suicide Research with Aboriginal and Torres Strait Islander Peoples** .... Monika Ferguson, Amy Baker, and Nicholas Procter	1707
99	**Visual Methods in Research with Migrant and Refugee Children and Young People** .... Marta Moskal	1723
100	**Participatory and Visual Research with Roma Youth** .... Oana Marcu	1739

| 101 | Drawing Method and Infant Feeding Practices Among Refugee Women 1757
June Joseph, Pranee Liamputtong, and Wendy Brodribb

| 102 | Understanding Refugee Children's Perceptions of Their Well-Being in Australia Using Computer-Assisted Interviews 1777
Jeanette A. Lawrence, Ida Kaplan, and Agnes E. Dodds

| 103 | Conducting Focus Groups in Terms of an Appreciation of Indigenous Ways of Knowing 1795
Norma R. A. Romm

| 104 | Visual Depictions of Refugee Camps: (De)constructing Notions of Refugee-ness? 1811
Caroline Lenette

| 105 | Autoethnography as a Phenomenological Tool: Connecting the Personal to the Cultural 1829
Jayne Pitard

| 106 | Ethics and Research with Indigenous Peoples 1847
Noreen D. Willows

| 107 | Conducting Ethical Research with People from Asylum Seeker and Refugee Backgrounds 1871
Anna Ziersch, Clemence Due, Kathy Arthurson, and Nicole Loehr

| 108 | Ethical Issues in Cultural Research on Human Development ... 1891
Namrata Goyal, Matthew Wice, and Joan G. Miller

Section IV   Sensitive Research Methodology and Approach: Researching with Particular Groups in Health Social Sciences **1905**

| 109 | Sensitive Research Methodology and Approach: An Introduction 1907
Pranee Liamputtong

| 110 | "With Us and About Us": Participatory Methods in Research with "Vulnerable" or Marginalized Groups 1919
Jo Aldridge

| 111 | Inclusive Disability Research 1935
Jennifer Smith-Merry

| 112 | Understanding Sexuality and Disability: Using Interpretive Hermeneutic Phenomenological Approaches 1953
Tinashe Dune and Elias Mpofu

113	**Ethics and Practice of Research with People Who Use Drugs** ... Julaine Allan	1973
114	**Researching with People with Dementia** ... Jane McKeown	1991
115	**Researching with Children** ... Graciela Tonon, Lia Rodriguez de la Vega, and Denise Benatuil	2007
116	**Optimizing Interviews with Children and Youth with Disability** ... Gail Teachman	2023
117	**Participant-Generated Visual Timelines and Street-Involved Youth Who Have Experienced Violent Victimization** ... Kat Kolar and Farah Ahmad	2041
118	**Capturing the Research Journey: A Feminist Application of Bakhtin to Examine Eating Disorders and Child Sexual Abuse** ... Lisa Hodge	2061
119	**Feminist Dilemmas in Researching Women's Violence: Issues of Allegiance, Representation, Ambivalence, and Compromise** ... Lizzie Seal	2079
120	**Animating Like Crazy: Researching in the Animated Visual Arts and Mental Welfare Fields** ... Andi Spark	2093
121	**Researching Underage Sex Work: Dynamic Risk, Responding Sensitively, and Protecting Participants and Researchers** ... Natalie Thorburn	2111
122	**The Internet and Research Methods in the Study of Sex Research: Investigating the Good, the Bad, and the (Un)ethical** ... Lauren Rosewarne	2127
123	**Emotion and Sensitive Research** ... Virginia Dickson-Swift	2145
124	**Doing Reflectively Engaged, Face-to-Face Research in Prisons: Contexts and Sensitivities** ... James E. Sutton	2163
125	**Police Research and Public Health** ... Jyoti Belur	2179

| 126 | **Researching Among Elites** | 2197 |

Neil Stephens and Rebecca Dimond

| 127 | **Eliciting Expert Practitioner Knowledge Through Pedagogy and Infographics** | 2213 |

Robert H. Campbell

Index 2225

About the Editor

Pranee Liamputtong is a medical anthropologist and Professor of Public Health at the School of Science and Health, Western Sydney University, Australia. Previously, Pranee held a Personal Chair in Public Health at the School of Psychology and Public Health, College of Science, Health and Engineering, La Trobe University, Melbourne, Australia, until January 2016. She has also previously taught in the School of Sociology and Anthropology and worked as a Public Health Research Fellow at the Centre for the Study of Mothers' and Children's Health, La Trobe University. Pranee has a particular interest in issues related to cultural and social influences on childbearing, childrearing, and women's reproductive and sexual health. She works mainly with refugee and migrant women in Melbourne and with women in Asia (mainly in Thailand, Malaysia, and Vietnam). She has published several books and a large number of papers in these areas.

Some of her books in the health and social sciences include *The Journey of Becoming a Mother Among Women in Northern Thailand* (Lexington Books, 2007); *Population Health, Communities and Health Promotion* (with Sansnee Jirojwong, Oxford University Press, 2009); *Infant Feeding Practices: A Cross-Cultural Perspective* (Springer, 2011); *Motherhood and Postnatal Depression: Narratives of Women and Their Partners* (with Carolyn Westall, Springer, 2011); *Health, Illness and Wellbeing: Perspectives and Social Determinants* (with Rebecca Fanany and Glenda Verrinder, Oxford University Press, 2012); *Women, Motherhood and Living with HIV/AIDS: A Cross-Cultural Perspective* (Springer, 2013); *Stigma, Discrimination and Living with HIV/AIDS: A Cross-Cultural*

Perspective (Springer, 2013); *Contemporary Socio-Cultural and Political Perspectives in Thailand* (Springer, 2014; *Children and Young People Living with HIV/AIDS: A Cross-Cultural Perspective* (Springer, 2016); *Public Health: Local and Global Perspectives* (Cambridge University Press, 2016, 2019); and *Social Determinants of Health: Individuals, Communities and Healthcare* (Oxford University Press, 2019).

Pranee is a Qualitative Researcher and has also published several method books. Her most recent method books include *Researching the Vulnerable: A Guide to Sensitive Research Methods* (Sage, 2007); *Performing Qualitative Cross-Cultural Research* (Cambridge University Press, 2010); *Focus Group Methodology: Principle and Practice* (Sage, 2011); *Qualitative Research Methods, 4th Edition* (Oxford University Press, 2013); *Participatory Qualitative Research Methodologies in Health* (with Gina Higginbottom, Sage, 2015); and *Research Methods in Health: Foundations for Evidence-Based Practice*, which is now in its third edition (Oxford University Press, 2017).

About the Contributors

Mauri K. Ahlberg was Professor of Biology and Sustainability Education at the University of Helsinki (2004–2013). On his 69th birthday, January 01, 2014, he had to retire. He has since the 1980s studied research methodology. He is interested in theory building and its continual testing as a core of scientific research. He has developed an integrating approach to research. He is an expert in improved concept mapping.

Farah Ahmad is an Associate Professor at the School of Health Policy and Management, Faculty of Health, York University. Applying equity perspective, she conducts mixed-method research to examine and improve the healthcare system for psychosocial health, vulnerable communities, access to primary care, and integration of eHealth innovations. The key foci in her research are mental health, partner violence, and cancer screening.

Jo Aldridge is a Professor of Social Policy and Criminology at Loughborough University, UK. She specializes in developing and using participatory research methods with "vulnerable" or marginalized people, including children (young carers), people with mental health problems and learning difficulties and women victims/survivors of domestic violence.

Julaine Allan is a Substance Use Researcher and Practitioner with over 30 years' experience in social work. Julaine's research is grounded in a human rights approach to working with stigmatized and marginalized groups. Julaine is also Senior Research Fellow and Deputy CEO at Lyndon, a substance treatment, research, and training organization. She holds conjoint positions at the National Drug and Alcohol Research Centre at UNSW and Charles Sturt University.

Jordan M. Alpert received his Ph.D. in Communication from George Mason University and is currently an Assistant Professor in the Department of Advertising at the University of Florida.

Andrea Montes Alvarado is an M.A. candidate in Sociology at California State University, Northridge. Her main research interests are in education, immigration, demography, and social inequality. Her current research explores the interpersonal relations and lived experiences of drag queens, specifically how drag queens construct and maintain families of choice within the drag community.

Stewart Anderson is a Professor of Biostatistics and Clinical and Translational Medicine at the University of Pittsburgh, Graduate School of Public Health.

His areas of current methodological research interests include (1) general methodology in longitudinal and survival data analysis, (2) modern regression techniques, and (3) methods in the design and analysis of clinical trials. He has over 25 years of experience in cancer clinical trial research in the treatment and prevention of breast cancer and in mental health in mid- to late-life adults.

Mike Armour is a Postdoctoral Research Fellow at NICM, Western Sydney University. Mike's research focus is on implementing experimental designs that can replicate complex clinical interventions that are often seen in the community. Mike has extensive experience in the design and conduct of clinical trials using a mixed-methods approach to help shape trial design, often including clinicians and community groups.

Amit Arora is a Senior Lecturer and National Health and Medical Research Council (NHMRC) Research Fellow in Public Health at Western Sydney University, Australia, where he teaches public health to undergraduate and postgraduate students. His research focuses on developing interventions to improve maternal and child health and oral health. Amit's research expertise includes mixed-methods research, health promotion, and life course approach in health research.

Kathy Arthurson is the Director of Neighborhoods, Housing and Health at Flinders Research Unit, Flinders University of South Australia. Her past experiences as a Senior Policy Analyst in a range of positions including public health, housing, and urban policy are reflected in the nature of her research, which is applied research grounded in broader concepts concerning social inclusion, inequality, and social justice.

Anna Bagnoli is an Associate Researcher in the Sociology Department of the University of Cambridge. Anna has a distinctive interest in methodological innovation and in visual, arts-based, and other creative and participatory approaches. She teaches postgraduates on qualitative analysis and CAQDAS and supervises postgraduates engaged with qualitative research projects. Her current work looks at the identity processes of migrants, with a focus on the internal migrations of Europeans, particularly young Italians.

Amy Baker is an academic in the Occupational Therapy Program at the University of South Australia. Her teaching and research focuses on mental health and suicide prevention, particularly working with people of culturally and linguistically diverse backgrounds and research approaches that are qualitative and participatory in nature.

Ruth Bartlett is an Associate Professor based in the Faculty of Health Sciences, University of Southampton, with a special interest in people with dementia and participatory research methods. Ruth has designed and conducted several funded research projects using innovative qualitative techniques, including diary and visual methods. Ruth has published widely in the health and social sciences and teaches and supervises postgraduate students.

Pat Bazeley has 25 years' experience in exploring, teaching, and writing about the use of software for qualitative and mixed-methods analysis for social and health research. Having previously provided training, consulting, and retreat facilities to researchers through Research Support P/L, she is now focusing on writing and researching in association with an Adjunct Professorial appointment at Western Sydney University.

Linda Liska Belgrave is an Associate Professor of Sociology in the Department of Sociology at the University of Miami. Her scholarly interests include grounded theory, medical sociology, social psychology, and social justice. She has pursued research on the daily lives of African-American caregivers of family with Alzheimer's disease, the conceptualization of successful aging, and academic freedom. She is currently working on research into the social meanings of infectious disease.

Robert F. Belli is a Professor of Psychology at the University of Nebraska–Lincoln. He received his Ph.D. in Experimental Psychology from the University of New Hampshire. Robert's research interests focus on the role of memory in applied settings, and his published work includes research on autobiographical memory, eyewitness memory, and the role of memory processes in survey response. The content of his work in surveys focuses on methods that can improve retrospective reporting accuracy.

Jyoti Belur is a Lecturer in the UCL Department of Security and Crime Science. She worked as a Lecturer before joining the Indian Police Service as a Senior Police Officer. She has undertaken research for the UK Home Office, College of Policing, ESRC, and the Met. Police. Her research interests include countering terrorism, violence against women and children, crime prevention, and police-related topics such as ethics and misconduct, police deviance, use of force, and investigations.

Denise Benatuil obtained her Ph.D. in Psychology from Universidad de Palermo, Argentina. She has a Licenciatura in Psychology from UBA, Argentina. She is the Director of the Psychology Program and Professor of Professional Practices in Psychology at Universidad de Palermo, Argentina. She is member of the Research Center in Social Sciences (CICS) at Universidad de Palermo, Argentina.

Sameer Bhole is the Clinical Director of Sydney Dental Hospital and Sydney Local Health District Oral Health Services and is also attached to the Faculty of Dentistry, the University of Sydney, as the Clinical Associate Professor. He has dedicated his career to address improvement of oral health for the disadvantaged populations with specific focus on health inequities, access barriers, and social determinants of health.

Mia Bierbaum is a Research Assistant at Macquarie University's Centre for Healthcare Resilience and Implementation Science, within the Australian Institute of Health Innovation. Mia had worked on the Healthy Living after Cancer project, providing a coaching service for cancer survivors, and conducted population monitoring research to examine the perceptions of cancer risk factors and behavioral change for patients. Her interests include health systems enhancement, behavioral research, and the prevention of chronic disease.

Simon Bishop is an Associate Professor of Organizational Behavior at Nottingham University Business School and a founding member of the Centre for Health Innovation and Learning. His research focuses around the relationship between public policy, organizational arrangements, and frontline practice in healthcare. His work has been published in a number of leading organization and policy and health sociology journals including *Human Relations*, *Journal of Public Administration Research and Theory*, and *Social Science & Medicine*.

Karen Block is a Research Fellow in the Jack Brockhoff Child Health and Wellbeing Program in the Centre for Health Equity at the University of Melbourne. Karen's research interests are social inclusion and health inequalities with a focus on children, young people, and families and working in collaborative partnerships with the community.

Chelsea Bond is a Munanjali and South Sea Islander Australian woman and Health Researcher with extensive experience in Indigenous primary healthcare, health promotion, and community development. Chelsea is interested in the emancipatory possibilities of health research for Indigenous peoples by examining the capabilities rather than deficiencies of Indigenous peoples, cultures, and communities.

Virginia Braun is a Professor in the School of Psychology at the University of Auckland, Aotearoa/New Zealand. She is a feminist and Critical Psychologist whose research has explored the intersecting areas of gender, bodies, sex/sexuality, and health/well-being across multiple topics and the possibilities, politics, and ethics of lives lived in neoliberal times.

Wendy Brodribb is an Honorary Associate Professor in the Primary Care Clinical Unit at the University of Queensland in Brisbane, Australia. She has a background as a medical practitioner focusing on women's health. Her Ph.D. investigated the breastfeeding knowledge of health professionals. Wendy's research interests are focused on infant feeding, especially breastfeeding, and postpartum care of mother and infant in the community following hospital discharge.

Mario Brondani is an Associate Professor and Director of Dental Public Health, University of British Columbia. Mario has developed a graduated program in dental public health (as a combined M.P.H. with a Diploma) in 2014 in which he currently directs and teaches. His areas of research include dental public health, access to care, dental geriatrics, psychometric measures, policy development, and dental education.

Jessica Broome has designed and applied qualitative and quantitative research programs for clients including top academic institutions, Fortune 500 companies, and grassroots community groups, since 2000. Jessica holds a Ph.D. in Survey Methodology from the University of Michigan.

Mark Brough is a Social Anthropologist with extensive experience in social research and teaching. Mark specializes in the application of qualitative methodologies to a wide range of health issues in diverse social contexts. He has a particular focus on strength-based approaches to community health.

Katerina Bryant completed a Bachelor of Laws and Bachelor of Arts degree in 2016. She is currently completing an honors degree in the Department of English and Creative Writing at Flinders University. Her nonfiction has been published widely, including in the *Griffith Review*, *Overland*, *Southerly*, and *The Lifted Brow*.

Lia Bryant is an Associate Professor in the School of Psychology, Social Work and Social Policy, University of South Australia, and is Director of the Centre for Social Change. She has vast experiences in working with qualitative methods and has published several books and refereed journal articles. She has authored two books and edited books. One of these relates specifically to research methodologies and methods. Lia also teaches innovation in research to social science students.

Anne Bunde-Birouste is Director of Yunus Social Business for Health Hub and Convener of the Health Promotion Program at the Graduate School of Public Health and Community Medicine, University of New South Wales in Sydney Australia. She specializes in innovative health promotion approaches for working with disadvantaged groups, particularly community-based research in sport for development and social change, working with vulnerable populations particularly youth.

Vivien Burr is a Professor of Critical Psychology at the University of Huddersfield. Her publications include *Invitation to Personal Construct Psychology* (2nd edition 2004, with Trevor Butt) and *Social Constructionism* (3rd edition 2015). Her research is predominantly qualitative, and she has a particular interest in innovative qualitative research methods arising from Personal Construct Psychology.

Valentina Buscemi is an Italian Physiotherapist with a strong interest in understanding and managing chronic pain conditions. She completed her M.Sc. in Neurorehabilitation at Brunel University London, UK, and is now a Ph.D. student at the Brain Rehabilitation and Neuroplasticity Unit, Western Sydney University. Her research aims to investigate the role of psychological stress in the development of chronic low back pain.

Rosalind Bye is the Director of Academic Program for Occupational Therapy in the School of Science and Health at Western Sydney University. Rosalind's research interests include adaptation to acquired disability, family caregiving experiences, and palliative care. Rosalind employs qualitative research methods that allow people to tell their story in an in-depth and meaningful way.

Breanne Byiers is a Researcher in the Department of Educational Psychology at the University of Minnesota. Her research interests include the development of measurement strategies to document changes in cognitive, health, and behavioral function among individuals with severe disabilities, assessment and treatment of challenging behavior, and single-subject experimental design methodology.

Fiona Byrne is a Research Officer in the Translational Research and Social Innovation (TReSI) Group, part of the School of Nursing and Midwifery at Western Sydney University. She works in the international support team for the Maternal Early Childhood Sustained Home-visiting (MECSH) program.

Robert Campbell is a Reader in Information Systems and Research Coordinator for the School of Creative Technologies at the University of Bolton. He holds a Ph.D. in this area and is on the board of the UK Academy for Information Systems. He has been professionally recognized as a Fellow of the British Computer Society and as a Senior Fellow of the Higher Education Academy. For many years, he was an Information Technology Specialist in the UK banking sector.

Michael J. Carter is an Associate Professor of Sociology at California State University, Northridge. His main research interests are in social psychology and microsociological theory, specifically the areas of self and identity. His current work examines how identities motivate behavior and emotions in face-to-face versus virtual environments. His research has appeared in a variety of academic journals, including *Social Psychology Quarterly* and *American Sociological Review*.

Rocco Cavaleri is a Ph.D. candidate and Associate Lecturer at Western Sydney University. He has been the recipient of the Australian Physiotherapy Association (APA) Board of Director's Student Prize and the New South Wales APA Award for academic excellence. Rocco has been conducting his research with the Brain Rehabilitation and Neuroplasticity Unit at Western Sydney since 2014. His research interests include chronic illnesses, continuing education, and exploring the nervous system using noninvasive technologies.

Jane Chalmers is a Lecturer in Physiotherapy at Western Sydney University and is also undertaking her Ph.D. through the University of South Australia. Jane's research focus has been on the assessment, management, and pathology of pelvic pain and in particular vulvodynia. She has extensive research experience using a range of methodologies and has recently used the Delphi technique in a novel way to create a new tool for assessing pelvic pain in women by using a patient-as-expert approach.

Nancy Cheak-Zamora is an Assistant Professor in the Department of Health Sciences at the University of Missouri. She conducts innovative research to inform policy-making, advocacy, service delivery, and research for youth with special healthcare needs and adults with chronic medical conditions.

Betty P.M. Chung, a graduate from the University of Sydney, Australia, has worked in the field of palliative and end-of-life care as practicing nurse and researcher. Her focus is particularly in psychosocial care for the dying and their families, family involvement to care delivery, good "quality of death," and well-being at old age. She has methodological expertise in qualitative interpretation and close-to-practice approaches in health sciences research.

Victoria Clarke is an Associate Professor in Sexuality Studies at the University of the West of England, Bristol, UK. She has published three prize-winning books, including most recently *Successful Qualitative Research: A Practical Guide for Beginners* with Virginia Braun (Sage), and over 70 papers in the areas of LGBTQ and feminist psychology and relationships, appearance psychology, human sexuality, and qualitative methods.

Bronwyne Coetzee is a Lecturer in the Department of Psychology at Stellenbosch University. She teaches introductory psychology and statistics to undergraduate students and neuropsychology to postgraduate students. Bronwyne's main interests are in health psychology, specifically in behavioral aspects of pediatric HIV and strengthening caregiver and child relationships to improve adherence to chronic medications.

Ana Lucía Córdova-Cazar holds a Ph.D. in Survey Research and Methodology from the University of Nebraska–Lincoln and is a Professor of Quantitative Methods and Measurement at the School of Social and Human Sciences of Universidad San Francisco de Quito, Ecuador. She has particular interests in the use of calendar and time diary data collection methods and the statistical analysis of complex survey data through multi-level modeling and structural equation modeling.

Jonathan C. Craig is a Professor of Clinical Epidemiology at the Sydney School of Public Health, the University of Sydney. His research aims to improve healthcare and clinical outcomes particularly in the areas of chronic kidney disease (CKD) and more broadly in child health through rigorous analysis of the evidence for commonly used and novel interventions in CKD, identifying gaps/inconsistency in the evidence, conducting methodologically sound clinical trials, and application of the research findings to clinical practice and policy.

Fiona Cram has tribal affiliations with Ngāti Pahauwera (Indigenous Māori, Aotearoa New Zealand) and is the mother of one son. She has over 20 years of Kaupapa Māori (by Māori, for Māori) research and evaluation experience with Māori (Indigenous peoples of Aotearoa New Zealand) tribes, organizations, and communities, as well as with government agencies, district health boards, and philanthropic organizations.

Anne Cusick is Professor Emeritus at Western Sydney University, Australia, Professor and Chair of Occupational Therapy at the University of Sydney, Editor in Chief of the *Australian Occupational Therapy Journal*, and inaugural Fellow of the Occupational Therapy Australia Research Academy. She has widely published in social, medical, health, and rehabilitation sciences.

Angela Dawson is a Public Health Social Scientist with expertise in maternal and reproductive health service delivery to priority populations in Australian and international settings. Angela has also undertaken research into innovative approaches to deliver drug and alcohol services to Aboriginal people. Angela is nationally recognized by the association for women in science, technology, engineering, mathematics, and medicine and is a recipient of the 2016 Sax Institute's Research Action Awards.

Kevin Deane is a Senior Lecturer in International Development at the University of Northampton, UK. His educational background is in development economics, but his research draws on a range of disciplines including political economy, development studies, economics, public health, and epidemiology. His research interests continue to focus on mobility and HIV risk, local value chains, transactional sex, and female economic empowerment in relation to HIV prevention.

Virginia Dickson-Swift was a Senior Lecturer in the La Trobe Rural Health School, La Trobe University, Bendigo, Australia. She has a wealth of experience in teaching research methods to undergraduate and postgraduate students throughout Australia. Her research interests lie in the practical, ethical, and methodological challenges of undertaking qualitative research on sensitive topics, and she has published widely in this area.

Rebecca Dimond is a Medical Sociologist at Cardiff University. Her research interests are patient experiences, clinical work, the classification of genetic syndromes and their consequences, and reproductive technologies. Her work is currently funded by an ESRC Future Research Leaders Award.

Agnes Dodds is an Educator and Evaluator and Associate Professor in the Department of Medical Education, the University of Melbourne. Agnes' current research interests are the developmental experiences of young people. Her current projects include studies of the development of medical students, selection into medical school, and school-related experiences of refugee children.

Suzanne D'Souza is a literacy tutor with the School of Nursing and Midwifery at Western Sydney University. Suzanne is keenly interested in the academic writing process and is actively involved with designing literacy strategies and resources to strengthen students' writing. Suzanne is also a doctoral candidate researching the hybrid writing practices of nursing students and their implications for written communicative competence.

Joan L. Duda is a Professor of Sport and Exercise Psychology in the School of Sport, Exercise and Rehabilitation Sciences at the University of Birmingham, UK. Joan is one of the most cited researchers in her discipline and is internationally known for her expertise on motivational processes and determinants of adherence and optimal functioning within physical and performance-related activities such as sport, exercise, and dance.

Clemence Due is a Postdoctoral Research Fellow in the Southgate Institute for Health, Society and Equity at Flinders University. She is the author of more than 50 peer-reviewed academic articles or book chapters, with a primary focus on research with adults and children with refugee backgrounds. Her research has focused on trauma, child well-being, housing and health, access to primary healthcare, and oral health.

Tinashe Dune is a Clinical Psychologist and Health Sociologist with significant expertise in sexualities and sexual and reproductive health. Her work focuses on the experiences of the marginalized and hidden populations using mixed methods and participatory action research frameworks. She is a recipient of a Freilich Foundation Award (2015) and one of Western's *Women Who Inspire* (2016) for her work on improving diversity and inclusivity in health and education.

Jane Edwards is a Creative Arts Therapy Researcher and Practitioner and is currently Associate Professor for Mental Health in the Faculty of Health at Deakin University. She has conducted research into the uses of the arts in healthcare in a range of areas. She is the Editor in Chief for *The Arts in Psychotherapy* and edited *The Oxford Handbook of Music Therapy* (2016). She is the inaugural President of the International Association for Music & Medicine.

Carolyn Ee is a GP and Research Fellow at NICM, Western Sydney University. Carolyn has significant experience in randomized controlled trials having conducted a large NHMRC-funded trial on acupuncture for menopausal hot flushes for her Ph. D. Carolyn combines clinical practice as a GP and acupuncturist with a broad range of research skills including health services and translational research as well as mixed-methods approaches to evaluating the effectiveness of interventions in the field of women's health.

Ulrike Felt is a Professor in the Department of Science and Technology Studies and Dean of the Faculty of Social Sciences at the University of Vienna. Her research focuses on public engagement and science communication; science, democracy, and governance; changing knowledge cultures and their institutional dimensions; as well as development in qualitative methods. Her work is often comparative between national context and technoscientific fields (especially life sciences, biomedicine, sustainability research, and nanotechnologies).

Monika Ferguson is a Research Associate in the Mental Health and Substance Use Research Group at the University of South Australia. Her current program of research focuses on educational interventions to reduce stigma and improve support for people at risk of suicide, both in the health sector and in the community.

Pota Forrest-Lawrence is a University Lecturer in the School of Social Science and Psychology, Western Sydney University. Pota holds a Ph.D. in Criminology from the University of Sydney Law School. Her research interests include drug law and policy, criminological and social theory, criminal law, and people and risk. She has published in the areas of cyber security, illicit drugs, and media.

Bronwyn Fredericks is an Indigenous Australian woman from Southeast Queensland (Ipswich/Brisbane) and is Professor and Pro Vice-Chancellor (Indigenous Engagement) and BHP Billiton Mitsubishi Alliance Chair in Indigenous Engagement at Central Queensland University, Australia. Bronwyn is a member of the National Indigenous Research and Knowledges Network and the Australian Institute of Aboriginal and Torres Strait Islander Studies.

Caroline Fryer is a Physiotherapist and a Lecturer in the School of Health Sciences at the University of South Australia. Her interest in equity in healthcare for people from culturally diverse backgrounds began as a clinician and translated to a doctoral study investigating the experience of healthcare after stroke for older people with limited English proficiency.

Emma George is a Lecturer in Health and Physical Education at Western Sydney University. Emma teaches across a range of health science subjects with a focus on physical activity, nutrition, health promotion, and evidence-based research methodology. Her research aims to promote lifelong physical activity and improve health outcomes. Emma's research expertise includes men's health, intervention design, implementation and evaluation, mixed-methods research, and community engagement.

Lilian A. Ghandour is an Associate Professor at the Faculty of Health Sciences, American University of Beirut (AUB). She holds a Ph.D. from the Johns Hopkins Bloomberg School of Public Health and a Master of Public Health (M.P.H.) from the American University of Beirut. Her research focuses on youth mental health and substance use, and Dr. Ghandour has been involved in the implementation and analyses of several local and international surveys. She has over 40 publications in high-tier peer-reviewed international journals.

Lisa Gibbs is an Associate Professor and Director of the Jack Brockhoff Child Health and Wellbeing Program at the University of Melbourne. She leads a range of complex community-based public health studies exploring sociocultural and environmental influences on health and well-being.

Claudia Gillberg is a Research Associate with the Swedish National Centre for Lifelong Learning (ENCELL) at Jonkoping University, Sweden. Her research interests include feminist philosophy, social citizenship, lifelong learning, participation, methodology, and ethics. Claudia divides her time between the UK and Sweden, closely following developments in health politics in both countries and worldwide. In her work, she has expressed concerns about the rise in human rights violations of the sick and disabled.

Brenda Gladstone is Associate Director of the Centre for Critical Qualitative Health Research and Assistant Professor at Dalla Lana School of Public Health,

University of Toronto. Brenda's research uniquely examines intergenerational experiences and effects of mental health and illness, focusing on parental mental illnesses. She teaches graduate-level courses on qualitative methodologies and uses innovative visual and participatory research methods to bring young people's voices into debates about their mental/health and social care needs.

Namrata Goyal is a Postdoctoral Fellow at the New School for Social Research. She received her Ph.D. in Psychology from the New School for Social Research and completed her B.A. at York University. Her research interests include cultural influences on reciprocity, gratitude, and social support norms, as well as developmental perspectives on social expectations and motivation.

Adyya Gupta is a Research Scholar in the School of Public Health at the University of Adelaide, Australia. She has an expertise in applying mixed methods. Her interests are in social determinants of health, oral health, and behavioral epidemiology.

Talia Gutman is a Research Officer at the Sydney School of Public Health, the University of Sydney. She has an interest in patient and caregiver involvement in research in chronic kidney disease and predominantly uses qualitative methods to elicit stakeholder perspectives with the goal of informing patient-centered programs and interventions. She has conducted focus groups with nominal group technique both around Australia and overseas as part of global studies.

Elizabeth Halcomb is a Professor of Primary Healthcare Nursing at the University of Wollongong. Elizabeth has taught research methods at an undergraduate and postgraduate level and has been an active supervisor of doctoral candidates. Her research interests include nursing workforce in primary care, chronic disease management, and lifestyle risk factor reduction, as well as mixed-methods research.

Marit Haldar is a Professor of Sociology. Her studies are predominantly on childhood, gender, and families. Her most recent focus is on the vulnerable subjects of the welfare state and inequalities in healthcare. Her perspectives are to understand the culturally and socially acceptable in order to understand what is unique, different, or deviant. She has a special interest in the development of new methodologies and analytical strategies.

Camilla S. Hanson is a Research Officer at the Sydney School of Public Health, University of Sydney. Her research interest is in psychosocial outcomes for patients with chronic disease. She has extensive experience in conducting systematic reviews and qualitative and mixed-methods studies involving pediatric and adult patients, caregivers, and health professionals worldwide.

John D. Hathcoat is an Assistant Professor of Graduate Psychology and Associate Director of University Learning Outcomes Assessment in the Center for Assessment and Research Studies at James Madison University. John has taught graduate-level courses in educational statistics, research methods, measurement theory, and performance assessment. His research focuses on instrument development, validity theory, and measurement issues related to "authentic" assessment practices in higher education.

Nikki Hayfield is a Senior Lecturer in Social Psychology in the Department of Health and Social Sciences at the University of the West of England, Bristol,

UK. Her Ph.D. was an exploration of bisexual women's visual identities and bisexual marginalization. In her research she uses qualitative methodologies to explore LGB and heterosexual sexualities, relationships, and (alternative) families.

Lisa Hodge is a Lecturer in Social Work in the College of Health and Biomedicine, Victoria University, and an Adjunct Research Fellow in the School of Nursing and Midwifery, University of South Australia. Her primary research interests include eating disorders and sexual trauma in particular and mental health more broadly, as well as self-harm, the sociology of emotions, and the use of creative expression in research methodologies.

Anne Hogden is a Research Fellow at Macquarie University's Centre for Healthcare Resilience and Implementation Science, within the Australian Institute of Health Innovation, and a Visiting Fellow at the Centre for Health Stewardship, Australian National University (ANU). Her expertise is in healthcare practice and research. Her research uses qualitative methodology, and she is currently focusing on patient-centered care, patient decision-making, and healthcare communication in Motor Neurone Disease.

Nicholas Hookway is a Lecturer in Sociology in the School of Social Sciences at the University of Tasmania, Australia. Nick's principle research interests are morality and social change, social theory, and online research methods. He has published recently in *Sociology* and *The British Journal of Sociology*, and his book *Everyday Moralities: Doing It Ourselves in an Age of Uncertainty* (Routledge) is forthcoming in 2017. Nick is current Co-convener of the Australian Sociological Association Cultural Sociology group.

Kirsten Howard is a Professor of Health Economics in the Sydney School of Public Health at the University of Sydney. Her research focuses on methodological and applied health economics research predominantly in the areas of the assessment of patient and consumer preferences using discrete choice experiment (DCE) methods as well as in economic evaluation and modeling. She has conducted many discrete choice experiments of patient and consumer preferences in diverse areas.

Martin Howell is a Research Fellow in Health Economics in the Sydney School of Public Health at the University of Sydney. His research focuses on applied health economics predominantly in the areas of assessment of preferences using discrete choice experiment (DCE) methods applied to complex health questions including kidney transplant and the trade-offs to avoid adverse outcomes of immunosuppression, preferences, and priorities for the allocation of deceased donor organs and relative importance of outcomes in nephrology.

Erin Jessee is a Lord Kelvin Adam Smith Research Fellow on Armed Conflict and Trauma in the Department of Modern History at the University of Glasgow. She works primarily in the fields of oral history and genocide studies and has extensive experience conducting fieldwork in conflict-affected settings, including Rwanda, Bosnia-Hercegovina, and Uganda. Her research interests also include the ethical and methodological challenges that surround conducting qualitative fieldwork in highly politicized research settings.

James Rufus John is a Research Assistant and Tutor in Epidemiology at Western Sydney University. He has published quantitative research articles on community perception and attitude toward oral health, utilization of dental health services, and public health perspectives on epidemic diseases and occupational health.

Craig Johnson is from the Ngiyampaa tribe in western NSW. Craig did over 30 years of outdoor and trade-based work; he moved to Dubbo in 1996 and became a trainee health worker in 2010. Craig is now a registered Aboriginal Health Practitioner and qualified diabetes educator and valued member of the Dubbo Diabetes Unit at Dubbo Base Hospital. Craig is dedicated to lifelong learning and improving healthcare for Aboriginal people.

Monica Johnson is from the Ngiyampaa tribe in western NSW. Monica has always been interested in health and caring for people and has worked in everything from childhood immunization to dementia. Monica works for Marathon Health (previously Western NSW Medicare Local) in an audiology screening program in western NSW. Monica has qualifications in nursing and has been an Aboriginal Health Worker for the past 8 years.

Stephanie Jong is a Ph.D. candidate and a sessional academic staff member in the School of Education at Flinders University, South Australia. Her primary research interests are in social media, online culture, and health. Stephanie's current work adopts a sociocultural perspective using netnography to understand how online interactions influence health beliefs and practices.

June Mabel Joseph is a Ph.D. candidate in the University of Queensland. She has a keen interest in researching vulnerable groups of populations – with a keen interest on mothers from refugee background in relation to their experiences of displacement and identity. She is also passionate in the field of breastfeeding, maternal and infant health, sociology, and Christian theology.

Angela Ju is a Research Officer at the Sydney School of Public Health, the University of Sydney. Her research interest is in the development and validation of patient-reported outcome measures. She has experience in conducting primary qualitative studies with patients and health professionals and in systematic review of qualitative health research in the areas of chronic disease including chronic kidney disease and cardiovascular disease.

Yong Moon Jung, with a long engagement in quantitative research, has strong expertise and extensive experience in statistical analysis and modeling. He has employed quantitative research skills for the development and evaluation purposes at either a policy or program level. He has also been teaching quantitative methods and data analysis for the undergraduate and postgraduate courses. Currently, he is working as part of the Quality and Analytics Group of the University of Sydney.

Ashraf Kagee is a Distinguished Professor of Psychology at Stellenbosch University. His main interests are in health psychology, especially HIV and mental health, health behavior theory, and stress and trauma. He has an interest in global mental health and lectures on research methods, cognitive psychotherapy, and psychopathology.

Ida Kaplan is a Clinical Psychologist and Director of Direct Services at the Victorian Foundation for Survivors of Torture. Ida is a specialist on research and practice in the area of trauma, especially for refugees and asylum seekers.

Bernie Kemp was born in Wilcannia and is a member of the Barkindji people. Bernie has spent most of the past 25 years working in Aboriginal health in far western NSW providing health checks and preventing and treating chronic disease. Bernie has qualifications in nursing and diabetes education and is a registered Aboriginal Health Practitioner. Bernie recently relocated to Dubbo to be closer to family and works at the Dubbo Regional Aboriginal Health Service.

Lynn Kemp is a Professor and Director of the Translational Research and Social Innovation (TReSI), School of Nursing and Midwifery, Western Sydney University. Originally trained as a Registered Nurse, Lynn has developed a significant program of community-based children and young people's research that includes world and Australian-first intervention trials. She is now leading an international program of translational research into the implementation of effective interventions at population scale.

Priya Khanna holds a Ph.D. in Science Education and Master's degrees in Education and Zoology. She has been working as a Researcher in medical education for more than 10 years. Presently, she is working as an Associate Lecturer, Assessment at the Sydney Medical School, University of Sydney, New South Wales.

Alexandra King graduated from the University of Tasmania in December 2014 with a Ph.D. in Rural Health. Her doctoral thesis is entitled "Food Security and Insecurity in Older Adults: A Phenomenological Ethnographic Study." Alexandra's research interests include social gerontology, social determinants of health, ethnographic phenomenology, and qualitative research methods.

Kat Kolar is a Ph.D. candidate, Teaching Assistant, and Research Analyst in the Department of Sociology, University of Toronto. She is also completing the Collaborative Program in Addiction Studies at the University of Toronto, a program held in collaboration with the Centre for Addiction and Mental Health, the Canadian Centre on Substance Abuse, and the Ontario Tobacco Research Unit.

Gary L. Kreps received his Ph.D. in Communication from the University of Southern California. He currently serves as a University Distinguished Professor and Director of the Center for Health and Risk Communication at George Mason University.

Tahu Kukutai belongs to the Waikato, Ngāti Maniapoto, and Te Aupouri tribes and is an Associate Professor at the National Institute of Demographic and Economic Analysis, University of Waikato. Tahu specializes in Māori and Indigenous demographic research and has written extensively on issues of Māori and tribal population change, identity, and inequality. She also has an ongoing interest in how governments around the world count and classify populations by ethnic-racial and citizenship criteria.

Ian Lacey comes from a professional sporting background and was contracted to the Brisbane Broncos from 2002 to 2007. Ian has expertise in smoking cessation, completing the University of Sydney Nicotine Addiction and Smoking Cessation

Training Course and the Smoke Check Brief Intervention Training. He has recently completed a Certificate IV in Frontline Management and Diploma in Management.

Michelle LaFrance teaches graduate and undergraduate courses in writing course pedagogy, ethnography, cultural materialist, and qualitative research methodologies. She has published on peer review, preparing students to write across the curriculum, e-portfolios, e-research, writing center and WAC pedagogy, and institutional ethnography. She has written several texts and her upcoming book, *Institutional Ethnography: A Theory of Practice for Writing Studies Researchers*, will be published by Utah State Press in 2018.

Kari Lancaster is a Scientia Fellow at the Centre for Social Research in Health, UNSW, Sydney. Kari is a Qualitative Researcher who uses critical policy study approaches to contribute to contemporary discussions about issues of political and policy significance in the fields of drugs and viral hepatitis. She has examined how policy problems and policy knowledges are constituted and the dynamics of "evidence-based" policy.

Jeanette Lawrence is a Developmental Psychologist and Honorary Associate Professor of the Melbourne School of Psychological Sciences. Jeanette's current research specializes in developmental applications to cultural and refugee studies. With Ida Kaplan, she is developing age- and culture-appropriate computer-assisted interviews to assist refugee and disadvantaged children to express their thoughts, feelings, and activities.

Eric Leake is an Assistant Professor of English at Texas State University. His areas of research include the intersection of rhetoric and psychology as well as civic literacies and writing pedagogy.

Raymond M. Lee is Emeritus Professor of Social Research Methods at Royal Holloway University of London. He has written extensively about a range of methodological topics including the problems and issues involved in research on "sensitive" topics, research in physically dangerous environments, and the use of unobtrusive measures. He also provides support and advice to the in-house researchers at Missing People, a UK charity that works with young runaways, missing and unidentified people, and their families.

Caroline Lenette is a Lecturer of Social Research and Policy at the University of New South Wales, Australia. Caroline's research focuses on refugee and asylum seeker mental health and well-being, forced migration and resettlement, and arts-based research in health, particularly visual ethnography and community music.

Doris Leung graduated from the University of Toronto and worked there as a Mental Health Nurse and Researcher focused on palliative end-of-life care. Since moving to Hong Kong, she works at the School of Nursing, the Hong Kong Polytechnic University. Her expertise is in interpretive and post-positivistic qualitative research approaches.

Nicole Loehr is a Ph.D. candidate in the School of Social and Policy Studies at Flinders University. Her doctoral work examines the integration of the not-for-profit, governmental, and commercial sectors in meeting the long-term housing needs of resettled refugees and asylum seekers in the private rental market. She has worked as a Child and Family Therapist and in community development and youth work roles.

Janet Long is a Health Services Researcher at the Australian Institute of Health Innovation, Macquarie University, Australia. She has a background in nursing and biological science. She has published a number of studies using social network analysis of various health and medical research settings to demonstrate silos, key players, brokerage and leadership, and strategic network building. Her other research interests are in behavior change, complexity, and implementation science.

Colin MacDougall is concerned with equity, ecology, and healthy public policy. He is currently Professor of Public Health at Flinders University and Executive member of the Southgate Institute of Health, Society and Equity. He is a Principal Fellow (Honorary) at the Jack Brockhoff Child Health and Wellbeing Program at the University of Melbourne. He studies how children experience/act on their worlds.

Catrina A. MacKenzie is a Mixed-Methods Researcher investigating how incentives, socioeconomic conditions, and household well-being influence conservation attitudes and behaviors, with a particular focus on tourism revenue sharing, loss compensation, and resource extraction. For the last 8 years, she has worked in Uganda studying the spatial distribution of perceived and realized benefits and losses accrued by communities as a result of the existence of protected areas.

Freya MacMillan is a Lecturer in Health Science at Western Sydney University, where she teaches in health promotion and interprofessional health science. Her research focuses on the development and evaluation of lifestyle interventions for the prevention and management of diabetes in those most at risk. Particularly relevant to this chapter, she has expertise in working with community to develop appropriate and appealing community-based interventions.

Wilson Majee is an Assistant Professor in the Master of Public Health program and Health Sciences Department at the University of Missouri. He is a Community Development Practitioner and Researcher whose primary research focuses on community leadership development in the context of engaging and empowering local residents in resource-limited communities in improving their health and well-being.

Karine E. Manera is a Research Officer at the Sydney School of Public Health, the University of Sydney. She uses qualitative and quantitative research methods to generate evidence for improving shared decision-making in the area of chronic kidney disease. She has experience in conducting focus groups with nominal group technique and has applied this approach in global and multi-language studies.

Narendar Manohar is a National Health and Medical Research Council Postgraduate Research candidate and a Research Assistant at Western Sydney University, Australia. He teaches public health and evidence-based practice to health science students and has published several quantitative research studies on several areas of public health. His interests include evidence-based practice, quantitative research, and health promotion.

Oana Marcu is a Researcher at the Faculty of Political and Social Sciences of the Catholic University in Milan. Her interests include qualitative methods (the ethnographic method, visual and participatory action research). She has extensively worked with Roma in Europe on migration, transnationalism, intersectionality, and health. She teaches social research methodology and sociology of migration.

Anastasia Marinopoulou is teaching political philosophy, political theory, and epistemology at the Hellenic Open University and is the coeditor of the international edition *Philosophical Inquiry*. Her publications include the recent monograph *Critical Theory and Epistemology: The Politics of Modern Thought and Science* (Manchester University Press, 2017). Her latest research award was the Research Fellowship at the University of Texas at Austin (2017).

Kate McBride is a Lecturer in Population Health in the School of Medicine, Western Sydney University. Kate teaches population health, basic and intermediate epidemiology, and evidence-based medicine to undergraduate and postgraduate students and has also taught at Sydney University. Kate's research expertise is in epidemiology, public health, and the use of mixed methods to improve health at a population level through the prevention of and reduction of chronic and non-communicable disease prevalence.

Angela McGrane is Director of Review and Approvals at Newcastle Business School, Northumbria University. She is currently investigating the effect of placement experience on student perceptions of themselves in a work role through a longitudinal study following participants from first year to graduation.

Jane McKeown is a Mental Health Nurse specializing in dementia research, practice, and education, working for the University of Sheffield and Sheffield Health and Social Care NHS Foundation Trust. Jane has a special interest in developing and implementing approaches that enable people with dementia to have their views and experiences heard. She also has an interest in the ethical aspects of research with people with dementia.

Karen McPhail-Bell is a Qualitative Researcher whose interest lies in the operation of power in relation to people's health. Karen facilitates strength-based and reciprocal processes in support of community-controlled and Aboriginal and Torres Strait Islander-led agendas. She has significant policy and program experience across academic, government and non-government roles in health equity, health promotion, and international development.

Joan Miller is Professor of Psychology at the New School for Social Research and Director of Undergraduate Studies in Psychology at Lang College. Her research interests center on culture and basic psychological theory, with a focus on interpersonal motivation, theory of mind, social support, moral development, as well as family and friend relationships.

Rodrigo Mariño is a Public Health Dentist and a Principal Research Fellow at the Oral Health Cooperative Research Centre (OH-CRC), the University of Melbourne. Rodrigo has an excellent publication profile and research expertise in social epidemiology, dental workforce issues, public health, migrant health, information and communication technology, gerontology, and population oral health. Rodrigo has also been a consultant to the Pan American Health Organization/World Health Organization in Washington, DC.

Anna Maurer-Batjer is a Master of Social Work and Master of Public Health student at the University of Missouri. She has a special interest in maternal and child health, particularly in terms of neurocognitive development. Through her graduate research assistantship, she has explored the use of photovoice and photo-stories among youth with Autism Spectrum Disorder.

Cara Meixner is an Associate Professor of Psychology at James Madison University, where she also directs the Center for Faculty Innovation. A scholar in brain injury advocacy, Cara has published mixed-methods studies and contributed to methodological research on this genre of inquiry. Also, Cara teaches a doctoral-level course on mixed-methods research at JMU.

Zelalem Mengesha has Bachelor's and Master's qualifications in Public Health with a special interest in sexual and reproductive health. He is passionate about researching the sexual and reproductive health of culturally and linguistically diverse communities in Australia using qualitative and mixed-methods approaches.

Dafna Merom is a Professor in the School of Science and Health, Western Sydney University, Australia. She is an expert in the area of physical activity measurement, epidemiology, and promotion. She has recently led several clinical trials comparing the incidence of falls, heart health, and executive functions of older adults participating in complex motor skills such as dancing and swimming versus simple functional physical activity such as walking.

Magen Mhaka-Mutepfa is a Senior Lecturer in the Department of Psychology, Faculty of Social Sciences, the University of Botswana. Previously, she was a Teaching Assistant in the School of Public Health at the University of Sydney (NSW) and was a recipient of Australia International Postgraduate Research Scholarship. Her main research interests are on health and well-being, HIV and AIDS, and ageing.

Christine Milligan is a Professor of Health and Social Geography and Director of the Centre for Ageing Research at Lancaster University, UK. With a keen interest in innovative qualitative techniques, Christine is an active researcher who has led on both national and international research projects. She has published over a hundred books, journal articles, and book chapters – including a recently published book (2015) with Ruth Bartlett on *What Is Diary Method?*

Naomi Moller is a Chartered Psychologist and Lecturer in Psychology at the Open University. Trained as a Counseling Psychologist, she recently coedited with Andreas Vossler *The Counselling and Psychotherapy Research Handbook* (Sage). Her research interests include perceptions and understandings of counselors and counseling, relationship, and family research including infidelity.

Marta Moskal is a Sociologist and Human Geographer based at the University of Glasgow, UK. Her research lies within the interdisciplinary area of migration and mobility studies, with a particular interest in family, children, and young people's experiences and in using visual research methods.

Elias Mpofu is a Rehabilitation Counseling Professional with a primary research focus in community health services intervention design, implementation, and evaluation applying mixed-methods approaches. His specific qualitative inquiry orientation is interpretive phenomenological analysis to understand meanings around and actions toward health-related quality of community living with chronic illness or disability.

Sally Nathan is a Senior Lecturer at the School of Public Health and Community Medicine, UNSW, Sydney. She has undertaken research into consumer and community participation in health as well as research approaches which engage and partner directly with vulnerable and marginalized communities and the organizations that represent and advocate for them, including a focus on adolescent drug and alcohol treatment and Aboriginal health and well-being.

Alison Nelson is an occupational therapist with extensive research, teaching, and practice experience working alongside urban Aboriginal and Torres Strait Islander people. Alison has a particular interest in developing practical strategies which enable non-Indigenous students, researchers, and practitioners to understand effective ways of working alongside Aboriginal and Torres Strait Islander Australians.

Ingrid A. Nelson is an Assistant Professor of Sociology at Bowdoin College, in Brunswick, Maine, USA. Her research examines the ways that families, schools, communities, and community-based organizations support educational attainment among marginalized youth. Her current work examines the educational experiences of rural college graduates, through the lens of social capital theory.

Christy E. Newman is an Associate Professor and Social Researcher of health, sexuality, and relationships at the Centre for Social Research in Health, UNSW, Sydney. Her research investigates social aspects of sexual health, infectious disease, and chronic illness across diverse contexts and communities. She has a particular passion for promoting sociological and qualitative approaches to understanding these often culturally and politically sensitive areas of health and social policy.

Mark C. Nicholas is Executive Director of Institutional Assessment at Framingham State University. He has published qualitative research on program review, classroom applications of critical thinking, and its assessment.

Kayi Ntinda is a Lecturer of Educational Counseling and Mixed-Methods Inquiry Approaches at the University of Swaziland. Her research interests are in assessment, learner support services service learning, and social justice research among vulnerable populations. She has authored and coauthored several research articles and book chapters that address assessment, inclusive education, and counseling practices in the diverse contexts.

Chijioke Obasi is a Principal Lecturer in Equality and Diversity at the University of Wolverhampton, England. She is a qualified British Sign Language/English Interpreter of Nigerian Origin and was brought up in a bilingual and bicultural household.

Felix Ogbo is a Lecturer in the School of Medicine, Western Sydney University. Felix studied at the University of Benin (M.B.B.S.), Benue State University (M.H.M.), and Western Sydney University – M.P.H. (Hons.) and Ph.D. He is an active collaborator in the landmark Global Burden of Disease Study – the world's largest scientific effort to quantify the magnitude of health loss from all major diseases, injuries, and risk factors by year-age-sex-geography and cause, which are essential to health surveillance and policy decision-making.

Andrew Page is a Professor of Epidemiology in the School of Medicine, Western Sydney University. He has been teaching basic and intermediate epidemiology and population health courses to health sciences students for 10 years and has published over 150 research articles and reports across a diverse range of population health topics. He has been a Research Associate at the University of Bristol and has also worked at the University of Queensland and University of Sydney in Australia.

Bonnie Pang is a Lecturer at the Western Sydney University and a school-based member of the Institute for Culture and Society. As a Sociocultural Researcher, she specializes in ethnographic methods, social theories, and youth health and physical activity. She has over 10 years of research experiences in exploring Chinese youth communities in school, familial, and neighborhood environments in Australia and Hong Kong.

Gemma Pearce is a Chartered Psychologist currently working in the Health Behaviour and Interventions Research (BIR) group at Coventry University, UK. Her research focuses on women's health, public health, and self-management support of people with long-term conditions. She specializes in systematic reviews and pragmatic research methods (including qualitative, quantitative, and mixed-methods research). She developed a synchronous method of online interviewing when conducting her Ph.D.

Karen Peres is an Associate Professor of Population Oral Health and Director of the Dental Practice Education Research Unit (DPERU) at the University of Adelaide. Her main research interests are in the epidemiology of oral diseases, particularly in social inequality in oral health, mother and child oral health-related issues, and life course and with over 100 publications in peer-reviewed journals.

Janette Perz is a Professor of Health Psychology and Director of the Translational Health Research Institute, Western Sydney University. She researches in the field of reproductive and sexual health with a particular focus on gendered experiences, subjectivity, and identity. She has demonstrated expertise in research design and analysis and mixed-methods research.

Rona Pillay (nee' Tranberg) is a Lecturer and Academic Course Advisor BN Undergraduate. She has unit coordinated research units and teaches primarily in research. Her research interests are in health communication, teams and teamwork, decision-making, and patient safety based on ethnomethodology and conversational analysis. Her additional interests lie in communication of cancer diagnosis in Aboriginal people and treatment decision-making choices.

Jayne Pitard is a Researcher in education at Victoria University, Melbourne, Australia. She has completed a thesis focused on her teaching of a group of students from Timor-Leste. She has held various teaching and research positions within Victoria University for the last 27 years.

Jennifer Prattley is a Research Associate at the Cathie Marsh Institute for Social Research, University of Manchester, UK. She works across a variety of ageing and life course studies, with research questions relating to women's retirement, frailty, and social exclusion. Jennifer has expertise in the analysis of longitudinal data and an interest in applying and appraising new and innovative quantitative methods. She also maintains an interest in statistics and mathematics education and the teaching of advanced methods to nonspecialists.

Nicholas Procter is a Professor and Chair of Mental Health Nursing and Convener of the Mental Health and Substance Use Research Group at the University of

South Australia. Nicholas has over 30 years' experience as a mental health clinician and academic. He works with various organizations within the mental health sector providing education, consultation, and research services in the suicide prevention and trauma-informed practice areas.

Marsha B. Quinlan is a Medical Anthropologist focusing on the intersection of health with ethnobiology and an Associate Professor in the Department of Anthropology and affiliate of the School for Global Animal Health at Washington State University, Pullman, WA, USA. Her research concentrates on ethnomedical concepts including ethnophysiology, ethnobotany, and ethnozoology, along with health behavior in families, and psychological anthropology.

Frances Rapport is a Professor of Health Implementation Science at Macquarie University's Centre for Healthcare Resilience and Implementation Science within the Australian Institute of Health Innovation. She is a social scientist with a background in the arts. Frances has won grants within Australia and the United Kingdom, including the Welsh Assembly Government, the National Institute for Health Research in England, and Cochlear Ltd. to examine the role of qualitative and mixed methods in medical and health services research.

Liz Rix is a Researcher and Academic working with Gnibi College of Indigenous Australian Peoples at Southern Cross University, Lismore. She is also a practicing Registered Nurse, with clinical expertise in renal and aged care nursing. Liz currently teaches Indigenous health to undergraduate and postgraduate health and social work professionals. Her research interests include Indigenous health and chronic disease, reflexive practice, and improving the cultural competence of non-Indigenous clinicians.

Norma R. A. Romm is a Research Professor in the Department of Adult Education and Youth Development at the University of South Africa. She is author of *The Methodologies of Positivism and Marxism: A Sociological Debate* (1991); *Accountability in Social Research: Issues and Debates* (2001); *New Racism* (2010); *People's Education in Theoretical Perspective: Towards the Development* (with V. McKay 1992); *Diversity Management: Triple Loop Learning* (with R. Flood 1996); and *Assessment of the Impact of HIV and AIDS in the Informal Economy of Zambia* (with V. McKay 2006).

Rizwana Roomaney is a Research Psychologist, Registered Counselor, and Lecturer in the Psychology Department at Stellenbosch University. She teaches research methods and quantitative data analysis to undergraduate and postgraduate students. Rizwana's main interests are health psychology and research methodology. She is particularly interested in women's health and reproductive health.

Lauren Rosewarne is a Senior Lecturer at the University of Melbourne, Australia, specializing in gender, sexuality, media, and popular culture. Her most recent books include *American Taboo: The Forbidden Words, Unspoken Rules, and Secret Morality of Popular Culture* (2013); *Masturbation in Pop Culture: Screen, Society, Self* (2014); *Cyberbullies, Cyberactivists, Cyberpredators: Film, TV, and Internet Stereotypes* (2016); and *Intimacy on the Internet: Media Representations of Online Connections* (2016).

Simone Schumann is a doctoral candidate and former Lecturer and Researcher in the Department of Science and Technology Studies, University of Vienna. Her main research interest is in public engagement with emerging technologies, especially on questions of collective sense-making and the construction of expertise/power relations within group settings. In her dissertation project, she focuses on the case of nano-food to understand how citizens encounter and negotiate an emerging food technology in Austria.

Claudia G. Schwarz is a Postdoc Researcher at the University of Vienna and Lecturer and former Researcher in the Department of Science and Technology Studies. Her current research interests lie mainly in the public understanding of and engagement with science and technology. In her dissertation project, she examined how laypeople use analogies as discursive devices when talking about nanotechnology.

Lizzie Seal is a Senior Lecturer in Criminology at the University of Sussex. Her research interests are in the areas of feminist, historical, and cultural criminology. She is the author of *Women, Murder and Femininity: Gender Representations of Women Who Kill* (Palgrave, 2010); *Transgressive Imaginations: Crime, Deviance and Culture* (with Maggie O'Neill, Palgrave, 2012); and *Capital Punishment in the Twentieth-Century Britain: Audience, Justice, Memory* (Routledge, 2014).

Kapriskie Seide is a doctoral student in the Department of Sociology at the University of Miami with concentrations in Medical Sociology and Race, Ethnicity, and Immigration. Her current research explores the impacts of infectious disease epidemics on everyday life and human health and their overlap with lay epidemiology, social justice, and grounded theory.

Nicole Sharp is a Lecturer in the School of Science and Health at Western Sydney University. Nicole's research interests include the in-depth experiences of people with disability, particularly at times of significant life transition. Nicole is focused on using inclusive research methods that give voice to people who may otherwise be excluded from participation in research.

Norm Sheehan is a Wiradjuri man and a Professor. He is currently Director of Gnibi College SCU. Basing on his expertise in Indigenous Knowledge and Education that employs visual and narrative principles to activate existing strengths within Indigenous education contexts for all individuals and learning communities, he has led the development of two new degrees: the Bachelor of Indigenous Knowledge and the Bachelor of Aboriginal Health and Wellbeing.

Patti Shih is a Postdoctoral Research Fellow at Macquarie University's Centre for Healthcare Resilience and Implementation Science, within the Australian Institute of Health Innovation. She teaches qualitative research methods in public health and has conducted qualitative and mixed-methods projects across a number of healthcare settings including HIV prevention in Papua New Guinea, e-learning among medical students, and professional training in the aged care sector.

Abla Mehio Sibai is a Professor of Epidemiology at the Faculty of Health Sciences, American University of Beirut (AUB). She has led a number of population-based national surveys, including the WHO National Burden of Disease

Study and the Nutrition and Non-communicable Disease Risk Factor (NNCD-BRF) STEPWise study. She is the author of over 150 scholarly articles and book chapters. Abla holds a degree in Pharmacy from AUB and a Ph.D. in Epidemiology from the London School of Hygiene and Tropical Medicine.

Ankur Singh is a Research Fellow in Social Epidemiology at the University of Melbourne, Australia. Ankur has published in different areas of public health including oral epidemiology, tobacco control, and health promotion. He is an invited reviewer for multiple peer-reviewed journals.

Valerie Smith was born in Dubbo and is a Wiradjuri woman. Val started in the healthcare industry in 2012, becoming an Aboriginal Health Worker with the Dubbo Regional Aboriginal Health Service in 2014. Val has recently moved to Port Macquarie and is currently taking a career break to spend time with her family.

Jennifer Smith-Merry is an Associate Professor of Qualitative Health Research in the Faculty of Health Sciences at the University of Sydney. She has methodological expertise in a range of qualitative methods including inclusive research, process evaluation of policy and services, narrative analysis, and critical discourse analysis. Her research focuses on mental health service and policy and consumer experiences of health and healthcare.

Helene Snee is a Senior Lecturer in Sociology at Manchester Metropolitan University, UK. Her research explores stratification with a particular focus on youth and class. Helene is the author of *A Cosmopolitan Journey? Difference, Distinction and Identity Work in Gap Year Travel* (Ashgate, 2014), which was short-listed for the BSA's Philip Abrams Memorial Prize for the best first and sole-authored book within the discipline of Sociology.

Andi Spark led the animation program at the Griffith Film School for 10 years following two decades in various animation industry roles. Working as an animation director and producer, she has been involved with projects ranging from large-scale outdoor projection installations to micro-looping animations for gallery exhibitions and small-screen devices and to children's television series and feature-length films and supervised the production of more than 200 short animated projects.

Elaine Stasiulis is a community-based Research Project Manager and Research Fellow at the Hospital for Sick Children and a doctoral candidate at the University of Toronto. Her work has involved an extensive range of qualitative and participatory arts-based health research projects with children and young people experiencing mental health difficulties and other health challenges.

Genevieve Steiner is a multi-award-winning NHMRC-ARC Dementia Research Development Fellow at NICM, Western Sydney University. Gen uses functional neuroimaging and physiological research methods to investigate the biological bases of learning and memory processes that will inform the prevention, diagnosis, and treatment of cognitive decline in older age. She also conducts many rigorous clinical trials including herbal and lifestyle medicines that may reduce the risk of dementia.

Neil Stephens is a sociologist and Science and Technology Studies (STS) scholar based at Brunel University London. He uses qualitative methods, including ethnography, to research innovation in biomedical contexts and explore the cultural and political aspects of the setting. He has researched topics including human embryonic stem cell research, biobanking, robotic surgery, and cultured meat.

Sara Stevano is a Postdoctoral Fellow in Economics at SOAS. She has a background in development economics, with interdisciplinary skills in the field of political economy and anthropology. Her research interests include the political economy of food and nutrition, agrarian change, and labor markets, with a focus on sub-Saharan Africa (Mozambique and Ghana in particular).

James Sutton is an Associate Professor of Anthropology and Sociology at Hobart and William Smith Colleges in Geneva, New York, USA, where he serves as a member of the Institutional Review Board. His research is focused in the areas of criminology and criminal justice, and his substantive research emphases include prisons, gangs, sexual assault, white-collar crime, and criminological research methods.

Roy Tapera is an Epidemiologist and Medical Informatician with more than 10 years of experience in the field of public health. He is currently a Lecturer of Epidemiology and Health Informatics at the University of Botswana, School of Public Health.

Gail Teachman is a Researcher and an Occupational Therapist. She is currently completing postdoctoral studies at McGill University, Montreal, Canada, and was a trainee with the Critical Disability and Rehabilitation Studies Lab at Bloorview Research Institute in Toronto, Canada. Gail's interdisciplinary research draws on critical qualitative inquiry, rehabilitation science, social theory, and social studies of childhood.

Gareth Terry is a Senior Research Fellow at the Centre for Person Centred Research at the Auckland University of Technology. He comes from a background in critical health and critical social psychologies and currently works in critical rehabilitation studies. His research interests are in men's health, gendered bodies, chronic health conditions, disability and accessibility, and reproductive decision-making.

Cecilie Thøgersen-Ntoumani is an Associate Professor in the School of Psychology and Speech Pathology at Curtin University, Western Australia. She conducts mixed-methods research with a particular emphasis on the development, implementation, and evaluation of theory-based health behavior interventions.

Natalie Thorburn is a Ph.D. candidate at the University of Auckland, New Zealand. Natalie works as Policy Advisor in the community sector, and her background is in sexual violence and intimate partner violence. Her research interests focus on sexual violence, sexual exploitation, and trafficking, and she sits on the board of Child Alert, an organization committed to ending sexual exploitation of children through child trafficking, child prostitution, and child pornography.

Irmgard Tischner is a Senior Lecturer in Social Psychology at the University of the West of England, Bristol, UK, and member of the Centre for Appearance Research. Focusing on poststructuralist, feminist, and critical psychological approaches, her research interests include issues around embodiment and subjectivity, particularly in relation to (gendered) discourses of body size, health, and physical activity in contemporary western industrialized societies.

Allison Tong is an Associate Professor at the Sydney School of Public Health, the University of Sydney. She developed the consolidated criteria for reporting qualitative health research (COREQ) and the enhancing transparency in reporting the synthesis of qualitative health research [ENTREQ], which are both endorsed as key reporting guidelines by leading journals and by the EQUATOR Network for promoting the transparency of health research.

Graciela Tonon is the Director of the Master Program in Social Sciences and the Research Center in Social Sciences (CICS) at Universidad de Palermo, as well as the Director of UNICOM, Faculty of Social Sciences, Universidad Nacional de Lomas de Zamora, Argentina. She is Vice-President of External Affairs of the International Society for Quality of Life Studies (ISQOLS) and Secretary of the Human Development and Capability Association (HDCA).

Billie Townsend is a non-Aboriginal woman born in Dubbo with family connections to the local area. Billie graduated with a double degree in International Studies and History in 2015 from the University of Wollongong. Billie worked on the Aboriginal people's stories of diabetes care study as a Volunteer Research Assistant for the University of Sydney, School of Rural Health, prior to returning to Wollongong to complete her honors year.

Nicole Tujague is a descendant of the Gubbi Gubbi nation from Mt Bauple, Queensland, and the South Sea Islander people from Vanuatu and the Loyalty Islands. Nicole has extensive experience interpreting intercultural issues for both Indigenous and non-Indigenous stakeholders. She has the ability to engage with Aboriginal and Torres Strait Islanders effectively to enable and support changes on an individual and family level.

Jane M. Ussher is Professor of Women's Health Psychology, in the Centre for Health Research at the Western Sydney University. Her research focuses on examining subjectivity in relation to the reproductive body and sexuality and the gendered experience of cancer and cancer care. Her current research include sexual health in CALD refugee and migrant women, sexuality and fertility in the context of cancer, young women's experiences of smoking, and LGBTI experiences of cancer.

Bram Vanhoutte is Simon Research Fellow in the Department of Sociology of the University of Manchester, UK. His research covers a wide field both in terms of topics and methods used. He has investigated aspects of the political socialization of youth and network influences on community-level social cohesion to more recently life course perspectives on ageing. Methodologically, he is interested in how to improve measurement and modeling using survey and administrative data and has recently developed an interest in life history data.

Lía Rodriguez de la Vega is a Professor and Researcher of the University of Palermo and Lomas de Zamora National University (Argentina). She obtained her doctoral degree in International Relations from El Salvador University, Argentina, in 2006. She is also a postdoctoral stay at the Psychology Faculty, Universidade Federal de Rio Grande do Sul (UFRGS, Porto Alegre, Brazil, in the context of the Bilateral Program Scholarship MINCYT-CAPES, 2009).

Randi Wærdahl is an Associate Professor of Sociology. Her work has an everyday perspective on transitions and trajectories in childhood and for families, in consumer societies, in times of social and economic change, and in changing contexts due to migration. She combines traditional methods of interviews and surveys with participatory methods such as photography or diary writing.

Morten Wahrendorf is a Senior Researcher at the Centre for Health and Society, University of Duesseldorf, Germany, with substantial expertise in sociology and research methodology. His main research interest are health inequalities in ageing populations and underlying pathways, with a particular focus on psychosocial working conditions, patterns of participation in paid employment and social activities in later life, and life course influences.

Maggie Walter is a member of the Palawa Briggs/Johnson Tasmanian Aboriginal family. She is Professor of Sociology and Pro Vice-Chancellor of Aboriginal Research and Leadership at the University of Tasmania. She has published extensively in the field of race relations and inequality and is passionate about Indigenous statistical engagement.

Emma Webster is a non-Aboriginal woman who has lived in Dubbo over 20 years and has family connections to the area. Emma joined the University of Sydney, School of Rural Health, in 2015 after working in the NSW public health system for 21 years. She has extensive experience in guiding novice researchers through their first research study having worked in the NSW Health Education and Training Institute's Rural Research Capacity Building Program.

Johannes Wheeldon has more than 15 years' experience managing evaluation and juvenile justice projects. He has worked with the American Bar Association, the Open Society Foundations, and the World Bank. He has published 4 books and more than 25 peer-reviewed papers on aspects of criminal justice, restorative justice, organizational change, and evaluation. He is an Adjunct Professor at Norwich University.

Sarah J. White is a Qualitative Health Researcher and Linguist with a particular interest in using conversation analysis to understand communication in surgical practice. She is a Senior Lecturer at the Faculty of Medicine and Health Sciences at Macquarie University, Sydney.

Matthew Wice is a doctoral student at the New School for Social Research where he studies Developmental and Social Psychology. His research interests include the influence of culture on moral reasoning, motivation, and the development of social cognition.

Nicola Willis is a Pediatric HIV Nurse Specialist. She is the Founder and Director of AfricAid, Zvandiri, and has spent 14 years developing and scaling up differentiated HIV service delivery models for children and adolescents in Zimbabwe. Nicola's work has been extensively influenced by the engagement of HIV-positive young people in creative, participatory approaches as a means of identifying and informing their policy and service delivery needs.

Noreen Willows uses anthropological, qualitative, and quantitative methodologies for exploring the relationships between food and health, cultural meanings of food and health, how food beliefs and dietary practices affect the well-being of communities, and how sociocultural factors influence food intake and food selection.

Noreen aims to foster an understanding among academics and nonacademics of the value of community-based participatory research.

Denise Wilson is a Professor in Māori Health and the Director of Taupua Waiora Centre for Māori Health Research at Auckland University of Technology. Her research and publications focus on Māori health, health services access and use, family violence, cultural responsiveness, and workforce development.

Elena Wilson is a Ph.D. candidate in the rural health research program, "Improving the Health of Communities through Participation," in the Rural Health School at La Trobe University, Bendigo. Elena's research interests are research methods, research ethics, and community participation with a focus on rural health and well-being.

Leigh Wilson is a Senior Lecturer in the Ageing Work and Health Research Group at the University of Sydney. Leigh comes from a public health and behavioral science background and has worked as an Epidemiologist and Health Service Manager in the NSW Health system. Her key research interests are the epidemiology of ageing, the impact of climate change (particularly heatwaves) on the aged, and the impact of behavior and perceptions of age on health.

Shawn Wilson has worked with Indigenous people internationally to apply Indigenist philosophy within the contexts of Indigenous education, health, and counselor education. His research focuses on the interrelated concepts of identity, health and healing, culture, and well-being. His book, *Research is Ceremony: Indigenous Research Methods*, has been cited as bridging understanding between mainstream and Indigenist research and is used as a text in many universities.

Jess Woodroffe works in research, academic supervision, teaching, and community engagement. Her research interests include community partnerships and engagement, social inclusion, health promotion, inter-professional learning and education, health sociology, qualitative research, and evaluation.

Kevin B. Wright is a Professor of Health Communication in the Department of Communication at George Mason University. His research focuses on social support processes and health outcomes in both face-to-face and online support groups/communities, online health information seeking, and the use of technology in provider-patient relationships.

Anna Ziersch is an Australian Research Council Future Fellow based at the Southgate Institute for Health, Society and Equity at Flinders University. She has an overarching interest in health inequities, in particular multidisciplinary and multimethod approaches to understanding the social determinants of health, and in particular for refugees and asylum seekers.

Contributors

Mauri Ahlberg Department of Teacher Education, University of Helsinki, Helsinki, Finland

Farah Ahmad School of Health Policy and Management, York University, Toronto, ON, Canada

Jo Aldridge Department of Social Sciences, Loughborough University, Leicestershire, UK

Julaine Allan Lyndon, Orange, NSW, Australia

Jordan Alpert Department of Advertising, University of Florida, Gainesville, FL, USA

Stewart J. Anderson Department of Biostatistics, University of Pittsburgh Graduate School of Public Health, Pittsburgh, PA, USA

E. Anne Marshall Educational Psychology and Leadership Studies, Centre for Youth and Society, University of Victoria, Victoria, BC, Canada

Mike Armour NICM, Western Sydney University (Campbelltown Campus), Penrith, NSW, Australia

Amit Arora School of Science and Health, Western Sydney University, Sydney, NSW, Australia

Discipline of Paediatrics and Child Health, Sydney Medical School, Sydney, NSW, Australia

Oral Health Services, Sydney Local Health District and Sydney Dental Hospital, NSW Health, Sydney, NSW, Australia

COHORTE Research Group, Ingham Institute of Applied Medical Research, Liverpool, NSW, Australia

Kathy Arthurson Southgate Institute for Health, Society and Equity, Flinders University, Adelaide, SA, Australia

Anna Bagnoli Department of Sociology, Wolfson College, University of Cambridge, Cambridge, UK

Amy Baker School of Health Sciences, University of South Australia, Adelaide, SA, Australia

Ruth Bartlett Centre for Innovation and Leadership in Health Sciences, Faculty of Health Sciences, University of Southampton, Southampton, UK

Pat Bazeley Translational Research and Social Innovation Group, Western Sydney University, Liverpool, NSW, Australia

Linda Liska Belgrave Department of Sociology, University of Miami, Coral Gables, FL, USA

Robert F. Belli Department of Psychology, University of Nebraska-Lincoln, Lincoln, NE, USA

Jyoti Belur Department of Security and Crime Science, University College London, London, UK

Denise Benatuil Master Program in Social Sciences and CICS, Universidad de Palermo, Buenos Aires, Argentina

Sameer Bhole Sydney Dental School, Faculty of Medicine and Health, The University of Sydney, Surry Hills, NSW, Australia

Oral Health Services, Sydney Local Health District and Sydney Dental Hospital, NSW Health, Surry Hills, NSW, Australia

Mia Bierbaum Centre for Healthcare Resilience and Implementation Science, Australian Institute of Health Innovation (AIHI), Macquarie University, Sydney, NSW, Australia

Simon Bishop Centre for Health Innovation, Leadership and Learning, Nottingham University Business School, Nottingham, UK

Karen Block Melbourne School of Population and Global Health, The University of Melbourne, Melbourne, VIC, Australia

Chelsea Bond Aboriginal and Torres Strait Islander Studies Unit, The University of Queensland, St Lucia, QLD, Australia

Virginia Braun School of Psychology, The University of Auckland, Auckland, New Zealand

Wendy Brodribb Primary Care Clinical Unit, Faculty of Medicine, The University of Queensland, Herston, QLD, Australia

Mario Brondani University of British Columbia, Vancouver, BC, Canada

Jessica Broome University of Michigan, Ann Arbor, MI, USA
Sanford, NC, USA

Mark Brough School of Public Health and Social Work, Queensland University of Technology, Kelvin Grove, Australia

Katerina Bryant Department of English and Creative Writing, Flinders University, Bedford Park, SA, Australia

Lia Bryant School of Psychology, Social Work and Social Policy, Centre for Social Change, University of South Australia, Magill, Australia

Anne Bunde-Birouste School of Public Health and Community Medicine, UNSW, Sydney, NSW, Australia

Viv Burr Department of Psychology, School of Human and Health Sciences, University of Huddersfield, Huddersfield, UK

Valentina Buscemi Western Sydney University, Sydney, NSW, Australia

Rosalind A. Bye School of Science and Health, Western Sydney University, Campbelltown, NSW, Australia

Breanne Byiers Department of Educational Psychology, University of Minnesota, Minneapolis, MN, USA

Fiona Byrne Translational Research and Social Innovation (TReSI) Group, School of Nursing and Midwifery, Ingham Institute for Applied Medical Research, Western Sydney University, Liverpool, NSW, Australia

Robert H. Campbell The University of Bolton, Bolton, UK

Michael J. Carter Sociology Department, California State University, Northridge, Northridge, CA, USA

Rocco Cavaleri School of Science and Health, Western Sydney University, Campbelltown, NSW, Australia

Jane Chalmers Western Sydney University (Campbelltown Campus), Penrith, NSW, Australia

Nancy Cheak-Zamora Department of Health Sciences, University of Missouri, Columbia, MO, USA

Betty P. M. Chung School of Nursing, The Hong Kong Polytechnic University, Hong Kong, SAR, China

Victoria Clarke Department of Health and Social Sciences, Faculty of Health and Applied Sciences, University of the West of England (UWE), Bristol, UK

Bronwyne Coetzee Department of Psychology, Stellenbosch University, Matieland, South Africa

Ana Lucía Córdova-Cazar Colegio de Ciencias Sociales y Humanidades, Universidad San Francisco de Quito, Diego de Robles y Vía Interoceánica, Quito, Cumbayá, Ecuador

Jonathan C. Craig Sydney School of Public Health, The University of Sydney, Sydney, NSW, Australia
Centre for Kidney Research, The Children's Hospital at Westmead, Sydney, NSW, Australia

Fiona Cram Katoa Ltd, Auckland, Aotearoa, New Zealand

Anne Cusick Faculty of Health Sciences, Sydney University, Sydney, Australia

Angela J. Dawson Australian Centre for Public and Population Health Research, University of Technology Sydney, Sydney, NSW, Australia

Kevin Deane Department of Economics, International Development and International Relations, The University of Northampton, Northampton, UK

Lía Rodriguez de la Vega Ciudad Autónoma de Buenos Aires, University of Palermo, Buenos Aires, Argentina

Virginia Dickson-Swift LaTrobe Rural Health School, College of Science, Health and Engineering, LaTrobe University, Bendigo, VIC, Australia

Rebecca Dimond School of Social Sciences, Cardiff University, Cardiff, UK

Agnes E. Dodds Melbourne Medical School, The University of Melbourne, Melbourne, VIC, Australia

Suzanne D'Souza School of Nursing and Midwifery, Western Sydney University, Sydney, NSW, Australia

Joan L. Duda School of Sport and Exercise Sciences, University of Birmingham, Birmingham, West Midlands, UK

Clemence Due Southgate Institute for Health, Society and Equity, Flinders University, Adelaide, SA, Australia

Tinashe Dune Western Sydney University, Sydney, NSW, Australia

Jane Marie Edwards Deakin University, School of Health and Social Development, Geelong, VIC, Australia

Carolyn Ee NICM, Western Sydney University (Campbelltown Campus), Penrith, NSW, Australia

Ghinwa Y. El Hayek Department of Epidemiology and Population Health, Faculty of Health Sciences, American University of Beirut, Beirut, Lebanon

Ulrike Felt Department of Science and Technology Studies, Research Platform Responsible Research and Innovation in Academic Practice, University of Vienna, Vienna, Austria

Monika Ferguson School of Nursing and Midwifery, University of South Australia, Adelaide, SA, Australia

Pota Forrest-Lawrence School of Social Sciences and Psychology, Western Sydney University, Milperra, NSW, Australia

Bronwyn Fredericks Office of Indigenous Engagement, Central Queensland University, Rockhampton, Australia

Caroline Elizabeth Fryer Sansom Institute for Health Research, University of South Australia, Adelaide, SA, Australia

Emma S. George School of Science and Health, Western Sydney University, Sydney, NSW, Australia

Lilian A. Ghandour Department of Epidemiology and Population Health, Faculty of Health Sciences, American University of Beirut, Beirut, Lebanon

Lisa Gibbs Melbourne School of Population and Global Health, The University of Melbourne, Melbourne, VIC, Australia

Claudia Gillberg Swedish National Centre for Lifelong Learning (ENCELL), Jonkoping University, Jönköping, Sweden

Geoffrey Jones Centre for Welfare Reform, Sheffield, UK

Brenda M. Gladstone Dalla Lana School of Public Health, Centre for Critical Qualitative Health Research, University of Toronto, Toronto, ON, Canada

Namrata Goyal Department of Psychology, New School for Social Research, New York, NY, USA

Adyya Gupta School of Public Health, The University of Adelaide, Adelaide, SA, Australia

Talia Gutman Sydney School of Public Health, The University of Sydney, Sydney, NSW, Australia

Centre for Kidney Research, The Children's Hospital at Westmead, Westmead, NSW, Australia

Elizabeth J. Halcomb School of Nursing, University of Wollongong, Wollongong, NSW, Australia

Marit Haldar Department of Social Work, Child Welfare and Social Policy, Oslo and Akershus University College of Applied Sciences, Oslo, Norway

Camilla S. Hanson Sydney School of Public Health, The University of Sydney, Sydney, NSW, Australia

Centre for Kidney Research, The Children's Hospital at Westmead, Westmead, NSW, Australia

John D. Hathcoat Department of Graduate Psychology, Center for Assessment and Research Studies, James Madison University, Harrisonburg, VA, USA

Nikki Hayfield Department of Health and Social Sciences, Faculty of Health and Applied Sciences, University of the West of England (UWE), Bristol, UK

Lisa Hodge College of Health and Biomedicine, Victoria University, Melbourne, VIC, Australia

Anne Hogden Centre for Healthcare Resilience and Implementation Science, Australian Institute of Health Innovation (AIHI), Macquarie University, Sydney, NSW, Australia

Nicholas Hookway University of Tasmania, Launceston, Tasmania, Australia

Kirsten Howard School of Public Health, University of Sydney, Sydney, NSW, Australia

Martin Howell School of Public Health, University of Sydney, Sydney, NSW, Australia

Erin Jessee Modern History, University of Glasgow, Glasgow, UK

James Rufus John Translational Health Research Institute, School of Medicine, Western Sydney University, Penrith, NSW, Australia

Capital Markets Cooperative Research Centre, Sydney, NSW, Australia

Craig Johnson Dubbo Diabetes Unit, Dubbo, NSW, Australia

Monica Johnson Marathon Health, Dubbo, NSW, Australia

Stephanie T. Jong School of Education, Flinders University, Adelaide, SA, Australia

June Joseph Primary Care Clinical Unit, Faculty of Medicine, The University of Queensland, Herston, QLD, Australia

Angela Ju Sydney School of Public Health, The University of Sydney, Sydney, NSW, Australia

Centre for Kidney Research, The Children's Hospital at Westmead, Westmead, NSW, Australia

Yong Moon Jung Centre for Business and Social Innovation, University of Technology Sydney, Ultimo, NSW, Australia

Ashraf Kagee Department of Psychology, Stellenbosch University, Matieland, South Africa

Ida Kaplan The Victorian Foundation for Survivors of Torture Inc, Brunswick, VIC, Australia

Bernie Kemp Dubbo Regional Aboriginal Health Service, Dubbo, NSW, Australia

Lynn Kemp Translational Research and Social Innovation (TReSI) Group, School of Nursing and Midwifery, Ingham Institute for Applied Medical Research, Western Sydney University, Liverpool, NSW, Australia

Priya Khanna Sydney Medical Program, University of Sydney, Camperdown, NSW, Australia

Alexandra C. King Rural Clinical School, Faculty of Health, University of Tasmania, Burnie, Tasmania, Australia

School of Pharmacy, Faculty of Health, University of Tasmania, Hobart, Tasmania, Australia

Nigel King Department of Psychology, University of Huddersfield, Huddersfield, UK

Kat Kolar Department of Sociology, University of Toronto, Toronto, ON, Canada

Gary L. Kreps Department of Communication, George Mason University, Fairfax, VA, USA

Tahu Kukutai University of Waikato, Hamilton, New Zealand

Ian Lacey Deadly Choices, Institute for Urban Indigenous Health, Brisbane, Australia

Michelle LaFrance George Mason University, Fairfax, VA, USA

Kari Lancaster Centre for Social Research in Health, Faculty of Arts and Social Sciences, UNSW, Sydney, NSW, Australia

Jeanette A. Lawrence Melbourne School of Psychological Science, The University of Melbourne, Melbourne, VIC, Australia

Eric Leake Department of English, Texas State University, San Marcos, TX, USA

Raymond M. Lee Royal Holloway University of London, Egham, UK

Caroline Lenette Forced Migration Research Network, School of Social Sciences, University of New South Wales, Kensington, NSW, Australia

Doris Y. Leung School of Nursing, The Hong Kong Polytechnic University, Hong Kong, SAR, China

The Lawrence S. Bloomberg Faculty of Nursing, University of Toronto, Toronto, ON, Canada

Pranee Liamputtong School of Science and Health, Western Sydney University, Penrith, NSW, Australia

Nicole Loehr School of Social and Policy Studies, Flinders University, Adelaide, SA, Australia

Janet C. Long Australian Institute of Health Innovation, Macquarie University, Sydney, NSW, Australia

Colin MacDougall Health Sciences Building, Flinders University, Bedford Park, SA, Australia

Catrina A. MacKenzie Department of Geography, McGill University, Montreal, QC, Canada

Department of Geography, University of Vermont, Burlington, VT, USA

Freya MacMillan School of Science and Health and Translational Health Research Institute (THRI), Western Sydney University, Penrith, NSW, Australia

Wilson Majee Department of Health Sciences, University of Missouri, Columbia, MO, USA

Karine Manera Sydney School of Public Health, The University of Sydney, Sydney, NSW, Australia

Centre for Kidney Research, The Children's Hospital at Westmead, Westmead, NSW, Australia

Narendar Manohar School of Science and Health, Western Sydney University, Sydney, NSW, Australia

Oana Marcu Università Cattolica del Sacro Cuore, Milan, Italy

Rodrigo Mariño University of Melbourne, Melbourne, VIC, Australia

Anastasia Marinopoulou Department of European Studies, Hellenic Open University, Patra, Greece

Anna Maurer-Batjer Department of Health Sciences, University of Missouri, Columbia, MO, USA

Kate A. McBride School of Medicine and Translational Health Research Institute, Western Sydney University, Sydney, NSW, Australia

Angela McGrane Newcastle Business School, Northumbria University, Newcastle-upon-Tyne, UK

Jane McKeown School of Nursing and Midwifery, The University of Sheffield, Sheffield, UK

Karen McPhail-Bell University Centre for Rural Health, University of Sydney, Camperdown, NSW, Australia

Poche Centre for Indigenous Health, Sydney Medical School, The University of Sydney, Camperdown, NSW, Australia

Abla Mehio Sibai Department of Epidemiology and Population Health, Faculty of Health Sciences, American University of Beirut, Beirut, Lebanon

Cara Meixner Department of Graduate Psychology, Center for Faculty Innovation, James Madison University, Harrisonburg, VA, USA

Zelalem Mengesha Western Sydney University, Sydney, NSW, Australia

Dafna Merom School of Science and Health, Western Sydney University, Penrith, Sydney, NSW, Australia

Translational Health Research Institute, School of Medicine, Western Sydney University, Penrith, NSW, Australia

Joan G. Miller Department of Psychology, New School for Social Research, New York, NY, USA

Christine Milligan Division of Health Research, Lancaster University, Lancaster, UK

Naomi Moller School of Psychology, Faculty of Social Sciences, The Open University, Milton Keynes, UK

Andrea Montes Alvarado Sociology Department, California State University, Northridge, Northridge, CA, USA

Marta Moskal Durham Univeristy, Durham, UK

Elias Mpofu University of Sydney, Lidcombe, NSW, Australia

Educational Psychology and Inclusive Education, University of Johannesburg, Johannesburg, South Africa

Magen Mhaka Mutepfa Department of Psychology, University of Botswana, Gaborone, Botswana

Sally Nathan School of Public Health and Community Medicine, Faculty of Medicine, UNSW, Sydney, NSW, Australia

Alison Nelson Allied Health and Workforce Development, Institute for Urban Indigenous Health, Brisbane, Australia

Ingrid A. Nelson Sociology and Anthropology Department, Bowdoin College, Brunswick, ME, USA

Christy Newman Centre for Social Research in Health, Faculty of Arts and Social Sciences, UNSW, Sydney, NSW, Australia

Mark C. Nicholas Framingham State University, Framingham, MA, USA

Kayi Ntinda Discipline of Educational Counselling and Mixed-Methods Inquiry Approaches, Faculty of Education, Office C.3.5, University of Swaziland, Kwaluseni Campus, Manzini, Swaziland

Chijioke Obasi University of Wolverhampton, Wolverhampton, UK

Felix Ogbo School of Medicine and Translational Health Research Institute, Western Sydney University, Campbelltown, NSW, Australia

Andrew Page School of Medicine and Translational Health Research Institute, Western Sydney University, Campbelltown, NSW, Australia

Bonnie Pang School of Science and Health and Institute for Culture and Society, University of Western Sydney, Penrith, NSW, Australia

Gemma Pearce Centre for Advances in Behavioural Science, Coventry University, Coventry, West Midlands, UK

Karen G. Peres Australian Research Centre for Population Oral Health (ARCPOH), Adelaide Dental School, The University of Adelaide, Adelaide, SA, Australia

Janette Perz Translational Health Research Institute, School of Medicine, Western Sydney University, Sydney, NSW, Australia

Rona Pillay School of Nursing and Midwifery, Western Sydney University, Sydney, NSW, Australia

Jayne Pitard College of Arts and Education, Victoria University, Melbourne, VIC, Australia

Jennifer Prattley Department of Social Statistics, University of Manchester, Manchester, UK

Nicholas Procter School of Nursing and Midwifery, University of South Australia, Adelaide, SA, Australia

Marsha B. Quinlan Department of Anthropology, Washington State University, Pullman, WA, USA

Frances Rapport Centre for Healthcare Resilience and Implementation Science, Australian Institute of Health Innovation (AIHI), Macquarie University, Sydney, NSW, Australia

Elizabeth F. Rix Gnibi Wandarahn School of Indigenous Knowledge, Southern Cross University, Lismore, NSW, Australia

Lia Rodriguez de la Vega Ciudad Autónoma de Buenos Aires, Buenos Aires, Argentina

Norma R. A. Romm Department of Adult Education and Youth Development, University of South Africa, Pretoria, South Africa

Rizwana Roomaney Department of Psychology, Stellenbosch University, Matieland, South Africa

Lauren Rosewarne School of Social and Political Sciences, University of Melbourne, Melbourne, VIC, Australia

Simone Schumann University of Vienna, Vienna, Austria

Claudia G. Schwarz-Plaschg Research Platform Nano-Norms-Nature, University of Vienna, Vienna, Austria

Lizzie Seal University of Sussex, Brighton, UK

Kapriskie Seide Department of Sociology, University of Miami, Coral Gables, FL, USA

Nicole L. Sharp School of Science and Health, Western Sydney University, Campbelltown, NSW, Australia

Norm Sheehan Gnibi Wandarahn School of Indigenous Knowledge, Southern Cross University, Lismore, NSW, Australia

Patti Shih Centre for Healthcare Resilience and Implementation Science, Australian Institute of Health and Innovation (AIHI), Macquarie University, Sydney, NSW, Australia

Ankur Singh Centre for Health Equity, Melbourne School of Population and Global Health, The University of Melbourne, Melbourne, VIC, Australia

Valerie Smith Formerly with Dubbo Regional Aboriginal Health Service, Dubbo, NSW, Australia

Jennifer Smith-Merry Faculty of Health Sciences, The University of Sydney, Sydney, Australia

Helene Snee Manchester Metropolitan University, Manchester, UK

Andi Spark Griffith Film School, Queensland College of Art, Griffith University, Brisbane, QLD, Australia

Elaine Stasiulis Child and Youth Mental Health Research Unit, SickKids, Toronto, ON, Canada

Institute of Medical Science, University of Toronto, Toronto, ON, Canada

Genevieve Z. Steiner NICM and Translational Health Research Institute (THRI), Western Sydney University, Penrith, NSW, Australia

Neil Stephens Social Science, Media and Communication, Brunel University London, Uxbridge, UK

Sara Stevano Department of Economics, University of the West of England (UWE) Bristol, Bristol, UK

James E. Sutton Department of Anthropology and Sociology, Hobart and William Smith Colleges, Geneva, NY, USA

Roy Tapera Department of Environmental Health, University of Botswana, Gaborone, Botswana

Gail Teachman McGill University, Montreal, QC, Canada

Gareth Terry Centre for Person Centred Research, School of Clinical Sciences, Auckland University of Technology, Auckland, New Zealand

Michelle Teti Department of Health Sciences, University of Missouri, Columbia, MO, USA

Cecilie Thøgersen-Ntoumani Health Psychology and Behavioural Medicine Research Group, School of Psychology and Speech Pathology, Curtin University, Perth, WA, Australia

Natalie Thorburn The University of Auckland, Auckland, New Zealand

Irmgard Tischner Faculty of Sport and Health Sciences, Technische Universität München, Lehrstuhl Diversitätssoziologie, Munich, Germany

Allison Tong Sydney School of Public Health, The University of Sydney, Sydney, NSW, Australia

Centre for Kidney Research, The Children's Hospital at Westmead, Westmead, Australia

Graciela Tonon Master Program in Social Sciences and CICS-UP, Universidad de Palermo, Buenos Aires, Argentina

UNICOM- Universidad Nacional de Lomas de Zamora, Buenos Aires, Argentina

Billie Townsend School of Rural Health, Sydney Medical School, University of Sydney, Dubbo, NSW, Australia

Nicole Tujague Gnibi Wandarahn School of Indigenous Knowledge, Southern Cross University, Lismore, NSW, Australia

Jane M. Ussher Translational Health Research Institute, School of Medicine, Western Sydney University, Sydney, NSW, Australia

Bram Vanhoutte Department of Sociology, University of Manchester, Manchester, UK

Randi Wærdahl Department of Social Work, Child Welfare and Social Policy, Oslo and Akershus University College of Applied Sciences, Oslo, Norway

Morten Wahrendorf Institute of Medical Sociology, Centre of Health and Society (CHS), Heinrich-Heine-University Düsseldorf, Medical Faculty, Düsseldorf, Germany

Maggie Walter University of Tasmania, Hobart, Tasmania, Australia

Emma Webster School of Rural Health, Sydney Medical School, University of Sydney, Dubbo, NSW, Australia

Johannes Wheeldon School of Sociology and Justice Studies, Norwich University, Northfield, VT, USA

Sarah J. White Macquarie University, Sydney, NSW, Australia

Matthew Wice Department of Psychology, New School for Social Research, New York, NY, USA

Nicola Willis Zvandiri House, Harare, Zimbabwe

Noreen D. Willows Faculty of Agricultural, Life and Environmental Sciences, University of Alberta, Edmonton, AB, Canada

Denise Wilson Auckland University of Technology, Auckland, New Zealand

Elena Wilson Rural Health School, College of Science, Health and Engineering, La Trobe University, Melbourne, VIC, Australia

Leigh A. Wilson School of Science and Health, Western Sydney University, Penrith, NSW, Australia

Faculty of Health Science, Discipline of Behavioural and Social Sciences in Health, University of Sydney, Lidcombe, NSW, Australia

Shawn Wilson Gnibi Wandarahn School of Indigenous Knowledge, Southern Cross University, Lismore, NSW, Australia

Jessica Woodroffe Access, Participation, and Partnerships, Academic Division, University of Tasmania, Launceston, Australia

Kevin B. Wright Department of Communication, George Mason University, Fairfax, VA, USA

Anna Ziersch Southgate Institute for Health, Society and Equity, Flinders University, Adelaide, SA, Australia

Section III

Doing Cross-Cultural Research in Health Social Sciences

Doing Cross-Cultural Research: An Introduction

86

Pranee Liamputtong

Contents

1 Introduction 1499
2 About the Section 1500
References 1505

Keywords

Cross-cultural research · Multiculural societies · Indigenous peoples · Marginalized people · Ethical issues · Methodological issues · Moral challenges

1 Introduction

Globally, cross-cultural research has become increasingly essential. In multicultural societies like the UK, USA, Canada, New Zealand, and Australia, there has been an increasing number of people from different cultural and linguistic backgrounds. Meeting the needs of our multicultural society requires a cultural awareness of the diversity and commonality of people's beliefs and practices. It is argued that this can be obtained by research, particularly with culturally sensitive approach. Therefore, cross-cultural research is a valuable tool for advancing an awareness of belief systems and practices among diverse cultural groups. The need for culturally competent research is now urgent in view of current social and health policies in many developed countries and their attempts to address the needs of multiethnic populations.

The presence of indigenous populations in countries such as Canada, the USA, New Zealand, and Australia has a great ramification for social science researchers. These indigenous people have been colonized, damaged, and become marginalized

P. Liamputtong (✉)
School of Science and Health, Western Sydney University, Penrith, NSW, Australia
e-mail: p.liamputtong@westernsydney.edu.au

in their own native lands. Due to a concern about reducing inequalities between the indigenous people and the "white" populations, there have been attempts to include these marginalized people in the research arenas. Again, this has barely been discussed in the literature. In this Handbook, I will include some examples on researching with the indigenous population.

Conducting research with marginalized people in cross-cultural settings is rife with methodological, ethical, and moral challenges (Liamputtong 2010, 2013). Researchers are challenged by a vast array of issues for carrying out their research with people in cross-cultural arenas. In this section, I bring together salient issues for the conduct of culturally appropriate research. The task of undertaking cross-cultural research can present researchers with unique opportunities, but also provide dilemmas. The section will give some thought-provoking points so that our research may proceed relatively well and yet remains ethical in our approach. To make the section more comprehensive, I am covering both methodology and procedural sensibilities (including ethics) in this Handbook.

2 About the Section

The section on doing cross-cultural research in health social sciences comprises four parts. Section 1 focuses on methodologies and research processes in cross-cultural research. In Section 2, language issues are included. Section 3 embraces chapters that discuss culturally sensitive research methods and processes, and the last part dedicates to ethics in cross-cultural research in health social sciences.

In ▶ Chap. 87, "Kaupapa Māori Health Research," Fiona Cram writes about ▶ "Kaupapa Māori Health Research." She argues that Kaupapa Māori is "literally a Māori way" which has been theorized in response to the colonization in Aotearoa New Zealand. Māori (Indigenous peoples) have been marginalized in their own lands, as evidenced by widespread social disparities. Kaupapa Māori health research, according to Cram, "promotes a structural analysis of Māori health disparities that moves the discourse away from victim-blaming and personal deficits to more fully understanding people's lives and the systemic determinants of their health and wellness." In this chapter, Cram uses Kaupapa Māori health research to illustrate the nature of the research paradigm as well as discusses what it means for Māori researchers undertaking Māori health research.

Denise Wilson, in ▶ Chap. 88, "Culturally Safe Research with Vulnerable Populations (Māori)," writes about culturally safe research with vulnerable populations but particularly with Māori. Wilson contends that often, vulnerable populations are subjected to some forms of social marginalization. Researchers' decisions and the research processes they employ can further increase the risk of vulnerability and marginalization among these people. She contends that "creating culturally responsive and safe spaces and research contexts with Maori, and others vulnerable within research settings, are needed to minimize participants' vulnerability and marginalization and counter unhelpful constructions about them." In this chapter, she offers some strategies that aim at minimizing the vulnerability of individuals participating in research. She provides a framework based on the concepts of partnership, participations, protection, and power to help researchers' cultural responsiveness in order to get the research story right, and importantly, to improve the utility of their research.

In ▶ Chap. 89, "Using an Indigenist Framework for Decolonizing Health Promotion Research," Karen McPhail-Bell, Alison Nelson, Ian Lacey, Bronwyn Fredericks, Chelsea Bond, Mark Brough write an interesting chapter on using an Indigenist framework for decolonizing health promotion research. The chapter explores the way the principles of Indigenist research informed the study, as a critical reflection of the methodology's achievement of a decolonizing research agenda. The flow of Maiwah (the Brisbane River in Australia) is used as a metaphor for the diverse authorship. The flow of Maiwah signifies the dialogical approach of the research; what Linda Smith (2005) refers to as a "tricky ground." The flow of Maiwah also shows us the possibilities of research where researcher and participants co-create new knowledge in support of their own agendas.

In ▶ Chap. 90, "Engaging Aboriginal People in Research: Taking a Decolonizing Gaze," the Dubbo Aboriginal Research Team (Craig Johnson, Monica Johnson, Bernie Kemp, Valerie Smith, Emma Webster and Billie Townsend) write about engaging Aboriginal people in research. They contend that much research in Australia has been done on Aboriginal people, but Aboriginal people themselves have received little benefit from it. This has added to distrust between Aboriginal and non-Aboriginal people over many years. In this chapter, they share aspects of their research that value Aboriginal people. The authors also discuss tensions which occur between the "scientific way" and the "culturally appropriate way" and offer how they resolve this.

Chijioke Obasi discusses issues of identity in research in ▶ Chap. 91, "Space, Place, Common Wounds and Boundaries: Insider/Outsider Debates in Research with Black Women and Deaf Women." She examines the impacts of the identity of the researcher, participants, and the various identity interchanges that take place in her research with five culturally Deaf (white) women and 25 Black (hearing) women. The chapter offers a reflexive account of the research; however, it is done in a way that centralizes the perspectives of the participants. In this chapter, Obsi raises a number of important questions, for example, should researchers seek out participant perspectives on the insider/outsider debates in research? and in what ways does the identity interchange between the researcher and the researched have an impact on the research process?

"Researcher Positionality in Sensitive Qualitative Research" is about researcher positionality in cross-cultural and sensitive research and is written by Narendar Manohar, Pranee Liamputtong, Sameer Bhole, and Amit Arora (▶ Chap. 92, "Researcher Positionality in Cross-Cultural and Sensitive Research"). The authors argue that the status of the insider and outsider is an important concept for cross-cultural and sensitive research and in recent years, the concept of placement of the researcher has received much attention. The way research participants "place" the researchers, and vice versa, is crucial for the success of any research. In this chapter, the authors introduce the concept of researcher positionality, examine the debates on researcher positionality in cross-cultural and sensitive research, and discuss "placing" issues including gender, age, culture and ethnicity, social class, and shared experiences.

In Section 2, in the language issues in cross-cultural research section, three chapters are included. In ▶ Chap. 93, "Considerations About Translation: Strategies About Frontiers," considerations about translation are discussed by Lía Rodriguez de la Vega. She argues that the translation of any text entails different methodological issues that range from linguistic treatment, grammatical issues of the languages considered, lexical issues and how to approach them, space and time considerations in a given textual construction, implied ethical questions, and so on. This chapter draws on a bibliographic review and the experiences of researchers who translate or have translated different types of texts in a nonprofessional manner.

In ▶ Chap. 94, "Finding Meaning: A Cross-Language Mixed-Methods Research Strategy," Catrina A. Mackenzie writes about finding meaning in a cross-language mixed-methods research strategy. She argues that the literature that devoted to methodological issues arising from working through an interpreter is sparse. Interpreters are crucial to the research process when a foreign researcher conducts research with an indigenous culture and when the researcher is not fluent in the local language. In this chapter, the author discusses an experientially developed cross-language research strategy.

Caroline Elizabeth Fryer, in ▶ Chap. 95, "An Approach to Conducting Cross-Language Qualitative Research with People from Multiple Language Groups," writes about an approach to conducting cross-language qualitative research with people from multiple language groups. She writes that a lack of shared preferred language between researcher and participant creates complexity and additional challenges in the research process. This is particular so when participants are from multiple language groups. In this chapter, Fryer introduces a research approach and methods which have been successfully used to conduct in-depth interviews with people from multiple language groups in a constructivist grounded theory study. She offers key strategies for conducting culturally competent and rigorous research at modest cost. She contends that this approach can enable health researchers to take "able to speak English" out of the inclusion criteria of studies and hence those who have limited English proficiency can be more included in health research.

▶ Chapter 96, "The Role of Research Assistants in Qualitative and Cross-Cultural Social Science Research," written by Sara Stevano and Kevin Deane, is on the role of research assistants in cross-cultural social science research. As we have seen, cross-cultural research often involves working with research assistants who assist with the data collection activities. Their participation in the research project has ramifications for the quality of the research. This chapter discusses a set of key practical decisions that researchers should make when planning their research fieldwork. The authors also explore how the triangular power dynamics between research participants, research assistants, and researchers impact on the research process and outcomes.

The next section is on culturally sensitive research methods in cross-cultural research and there are nine chapters. In ▶ Chap. 97, "Indigenous Statistics," Tahu Kukutai and Maggie Walter discuss Indigenous Statistics. The authors contend that in Anglo-colonizing nation states such as Canada, Australia, Aotearoa New Zealand,

and the United States (CANZUS), statistics about Indigenous peoples are a common feature. In this chapter, they "contrast these statistics with those from statistical research using processes and practices that are shaped by Indigenous methodologies." Indigenous methodologies are characterized by their "prioritization of Indigenous methods, protocols, values, and epistemologies." They conclude the chapter with two examples of what Indigenous quantitative methodologies look like in practice from Aotearoa NZ and Australia.

▶ Chapter 98, "A Culturally Competent Approach to Suicide Research with Aboriginal and Torres Strait Islander Peoples" is about a culturally competent approach to suicide research with Aboriginal and Torres Strait Islander Peoples and is written by Monika Ferguson, Amy Baker, and Nicholas Procter. Suicide has profound, and ongoing, impacts for Aboriginal and Torres Strait Islander Peoples and has been identified as an area requiring further research. In this chapter, the authors outline a culturally competent approach for conducting social and emotional well-being research, from the perspective of non-Aboriginal researchers. They outline sensitivities associated with conducting research as non-Aboriginal researchers and then introduce important ethical principles, which can be used to guide culturally competent practice throughout the research journey. Specific methodological approaches with an emphasis on those that are participatory in nature are also outlined.

In ▶ Chap. 99, "Visual Methods in Research with Migrant and Refugee Children and Young People," visual methods in research with migrant and refugee children and young people were presented by Marta Moskal. In this chapter, Moskal examines how visual methods can be utilized in understanding and interpreting migrant and refugee children's worlds. She argues that visual methods can secure the engagement and reflexivity among children who may not feel comfortable with a traditional interview, focus group, or survey methods. Several visual methods including drawing, maps, photographs, and videos are discussed in the chapter.

Oana Marcu, in ▶ Chap. 100, "Participatory and Visual Research with Roma Youth," writes about participatory and visual research with Roma youth. Oana discusses methods, tools, and strategies which can be used in peer-research with young people belonging to minorities, from migrant backgrounds or marginalized ethnic groups. In this chapter, research strategies including the participatory design process and the selection of specific levels of participation in all stages are discussed. Visual and participatory methods are illustrated with examples from two research projects: the representation of drugs and the migratory experience from a gendered perspective.

In ▶ Chap. 101, "Drawing Method and Infant Feeding Practices Among Refugee Women," the drawing method and infant feeding practices among refugee women is presented by June Joseph, Pranee Liamputtong, and Wendy Brodribb. This chapter adopts the postmodern methodological framework to unravel the multiple truths that drive the perceptions and perspectives of infant feeding among mothers from refugee backgrounds (Myanmarese and Vietnamese) in Brisbane. Since the research trend of gaining visual access to the lives of mothers from refugee backgrounds is new, the authors also outline some tips and tricks that steered their initially rocky data collection journey. The chapter illustrates ways in which women from refugee backgrounds conceptualize motherhood and infant feeding. The authors also delineate the usefulness of using drawing as a research method for researchers who work with refugee women and/or in a similar research domain.

In ▶ Chap. 102, "Understanding Refugee Children's Perceptions of Their Well-Being in Australia Using Computer-Assisted Interviews," Jeanette A. Lawrence, Ida Kaplan and Agnes E. Dodds write about understanding refugee children's perceptions of their well-being using computer-assisted interviews. They suggest that children from refugee backgrounds have the ability and right to contribute to research knowledge. However, they need methods that enact respect and are theoretically appropriate for them. Two computer-assisted interviews (CAIs) were developed as research tools in their research with refugee children. The usefulness of the methodology is discussed in relation to the need to understand the perspectives of refugee children and other children about their well-being.

Norma Romm, in ▶ Chap. 103, "Conducting Focus Groups in Terms of an Appreciation of Indigenous Ways of Knowing," writes about conducting focus groups in terms of an appreciation of Indigenous ways of knowing. In this chapter, Romm offers deliberations around the facilitation of focus groups in a manner which embraces Indigenous ways of knowing. Indigenous knowing, in this chapter, is "defined as linked to processes of people collectively constructing their understandings by experiencing their social being in relation to others." The chapter illustrates how the conduct of focus groups can be geared towards this epistemology.

▶ Chapter 104, "Visual Depictions of Refugee Camps: (De)Constructing Notions of Refugee-ness?" is about visual depictions of refugee camps which can deconstruct notions of refugee-ness written by Caroline Lenette. The author argues that visual representations of asylum seekers and refugees can have a marked impact on how these individuals are perceived in politically stable contexts, especially in western nations. Using refugee camps as the example, the author applies the framework of "humanitarian sentimentalism" which describes four typifications (or tropes) associated with images of humanitarian crises (Personification, Massification, Care, and Rescue) to visual depictions of refugee camps. The author contends that visual methodologies can provide a rich dimension to critical discussions on complex and multifaceted issues.

Jayne Pitard writes about autoethnography as a phenomenological tool that connects the personal to the cultural in ▶ Chap. 105, "Autoethnography as a Phenomenological Tool: Connecting the Personal to the Cultural." Autoethnography, Pitard writes "retrospectively and selectively writes about experiences that have their basis in, or are made possible by, being part of a culture and/or owning a specific cultural identity." In researching her role as the teacher of a group of vocational education professionals from Timor-Leste, she conducted a phenomenological study using autoethnography to portray the existential shifts in her cultural understanding. I also utilized vignettes to firstly place her within the social context and then to explore her positionality as a researcher. In this chapter, the author presents the framework that others can adopt in a cross-cultural setting.

The last section is on ethical issues in cross-cultural research and comprises three chapters. In ▶ Chap. 106, "Ethics and Research with Indigenous Peoples," Noreen Willow writes about ethical health research involving Indigenous peoples. Willow contends that states and academic institutions have an obligation to support ethical research with Indigenous peoples which would result in the elimination of health disparities among Indigenous peoples and others. Health research that respects Indigenous self-determination, and is safe, ethical, and useful for participants, necessitates increased capacity among Indigenous and non-Indigenous peoples alike. She argues that non-Indigenous researchers need appropriate ethical guidelines to follow and training opportunities that offer guidance on Indigenous ways of knowing, the social determinants of health, strength-based research approaches, community-based participatory research, and how to engage in culturally appropriate ways with Indigenous peoples.

Conducting ethical research with people from asylum seeker and refugee backgrounds is written by Anna Ziersch, Clemence Due, Kathy Arthurson, and Nicole Loehr in ▶ Chap. 107, "Conducting Ethical Research with People from Asylum Seeker and Refugee Backgrounds." In this chapter, the authors outline issues that need to be considered when working on health and other research with people with asylum seeker or refugee backgrounds in countries of resettlement. The chapter not only highlights the utility of a Social Determinants of Health framework, but also outlines the importance of ethical research which balances the considerations of formal ethics committees with the need for the voices of the most vulnerable people within this population to be heard. The chapter also offers some appropriate methodologies, including emerging and innovative research methods such as visual scales, photovoice, photolanguage, and digital storytelling, and discusses the ways in which these data collection methods contribute to high quality quantitative and qualitative data. In the last section of the chapter, the authors cover the challenges of working cross-culturally and the need to make sure that research is "culturally appropriate, consultative and meaningful."

In ▶ Chap. 108, "Ethical Issues in Cultural Research on Human Development," the last chapter in this section, ethical issues in cultural research on human development were discussed by Namrata Goyal, Matthew Wice, and Joan G. Miller (▶ Chap. 108, "Ethical Issues in Cultural Research on Human Development"). The authors argue for the importance of attending to culture in all phases of the research process. They also highlight ways that promote the ethical sensitivity of cultural research which enhances its explanatory force. They address ethical aspects of study design and data collection and point out ways that harm, coercion, and invasion of privacy may result from inadequate attention to cultural meanings and practices. Importantly, they discuss the impact of drawing unsound or stereotypical conclusions about culture and human development. They conclude their chapter by outlining ways in which culturally sensitive research can enhance both ethics and research quality.

References

Liamputtong P. Performing qualitative cross-cultural research. Cambridge: Cambridge University Press; 2010.
Liamputtong P. Qualitative research methods, 4th edn. Melbourne: Oxford University Press; 2013.
Smith LT. On tricky ground: Researching the native on the age of uncertainty. In Denzin N, Lincoln YS, editors. The Sage handbook of qualitative research, 3rd edn. Thousand Oaks, CA: Sage; 2005. p. 85–107.

Kaupapa Māori Health Research

87

Fiona Cram

Contents

1	Introduction	1508
2	A Kaupapa Māori Research Paradigm	1511
	2.1 Ontology: The Nature of Reality	1512
	2.2 Epistemology: Relationship between the Knower and What would be Known	1514
	2.3 Methodology: Appropriate Approaches to Systemic Inquiry	1515
	2.4 Axiology: Nature of Ethics	1516
3	Conclusion and Future Directions	1520
	References	1522

Abstract

Kaupapa Māori is literally a Māori way. It is a response to the colonization in Aotearoa New Zealand that has seen Māori (Indigenous peoples) marginalized in our own lands, as evidenced by widespread health, education, socioeconomic, and other Māori-non-Māori disparities. What began in the late 1980s as Kaupapa Māori research within Māori education has spread to other disciplines, including Māori health. Kaupapa Māori health research promotes a structural analysis of Māori health disparities that moves the discourse away from victim-blaming and personal deficits to more fully understanding people's lives and the systemic determinants of their health and wellness. Describing this work as occurring within a Kaupapa Māori inquiry paradigm enables the exploration of its axiological (i.e., ethical), ontological (i.e., theory about the nature of reality), epistemological (i.e., theory of knowledge), and methodological (i.e., theory about how to find out things) assumptions. Kaupapa Māori health research is called upon to illustrate the nature of the paradigm as well as what it means practically for Māori researchers undertaking Māori health research. As part of transdisciplinary

F. Cram (✉)
Katoa Ltd, Auckland, Aotearoa, New Zealand
e-mail: fionac@katoa.net.nz

research teams with Māori colleagues, Pākehā (non-Māori) researchers also have important roles to play in this research. The mission of Kaupapa Māori health research is ensuring that Māori health research informs an agenda of Māori being Māori, being fully human, and living in health and prosperity.

Keywords

Kaupapa Māori · Māori · Indigenous · Paradigm · Decolonization · Collaboration · New Zealand

1 Introduction

Colonized peoples have been compelled to define what it means to be human because there is a deep understanding of what it has meant to be considered not fully human, to be savage. (Smith 2012, p. 28)

Camara Jones (2000), a public health practitioner, tells a story about flower seeds growing in different types of flower boxes. Red flower seeds are planted in a flower box that has rich soil, full of nutrients. Pink flower seeds are planted in a flower box that has poor soil, full of clay and rocks. The red flower seeds flourish in their soil: they germinate, grow, and bloom. The pink flower seeds do not fare so well: few germinate, they grow spindly, and the blooms are few and far between. A person looking at just the mature plants might easily come to the conclusion that the red flower seeds are much better than the pink flower seeds. However, an examination of the soil will soon show them that something else has affected the "success" of the seeds. Jones uses the soil as a metaphor for understanding racism, that is, "behavior that stems from a belief that people can be differentiated mainly or entirely on the basis of their ancestral lineage" (Cochrane 1991, p. 127). This chapter is akin to the story of the flower seeds, with a slight but important twist. As a metaphor for colonization, this story begins in a previous era, when the pink flower seeds were sown in soil that nurtured them and they too thrived. The inquiry then becomes about how these flower seeds – Indigenous peoples – were exiled from the best soil or land and found themselves in marginal places that did not nourish them. And how the red flower seeds – colonists – came into possession of the best soil and what this has meant for the flourishing of all seeds.

The process of colonization depends upon the dehumanization of Indigenous peoples (Smith 2012). In other words, the belief that Indigenous peoples do not have human rights is foundational to the transfer of Indigenous resources and territories out of their hands and into the hands of newcomers. Through colonization, Indigenous peoples are driven to a low point and colonizers rise on a wave of artificial prestige. It should not be surprising, therefore, that the health and well-being of Indigenous peoples suffers under the burden of colonization. Even though Aotearoa New Zealand was one of the last places where the British Crown made a treaty with Indigenous peoples, with the 1840 Treaty of Waitangi, and the impacts of colonization upon Māori (Indigenous peoples of Aotearoa New Zealand) were foreseen, they were not forestalled (Orange 1987; Walker 2004).

The relatively short history of the colonization of Aotearoa me te Waipounamu (the Māori names for these lands, shortened here as Aotearoa) saw Māori very quickly go from considering themselves ordinary (Orange 1987; Salmond 1991) to being seen by the British newcomers as different and deficient compared to their White ethnocentric norms. This helped justify the redistribution of Māori land and resources to the newcomers, with the outcome that Māori were pushed to the margins of a renamed, colonized New Zealand (Walker 2004). This marginalization and the vulnerability of Māori children and whānau (families) is reflected in lower life expectancy (some 7.3 years lower for Māori compared to the NZ European population) and the segregation of two-thirds of Māori households within the most deprived neighborhoods (Ministry of Health 2015).

The role of research in the colonization of Māori is not often considered. Rather descriptions tend to focus on broken treaty promises, land wars, and subsequent land confiscations, disease, and overwhelming settler numbers. However, in his comparison of political and scientific colonialism, Nobles (1991) places research alongside these other forces as central to colonization. He describes scientists as believing they have the right to access any knowledge, and then export this raw material from communities so that it might be "processed" into books, patents, new drugs, and wealth. The result is that a people's knowledge is often relocated outside their community or tribal boundaries and beyond their control. This description can be applied to how non-Indigenous researchers journeyed until very recently within Māori communities.

Some of the first non-Indigenous scientists who wrote about Māori were aboard the Endeavor when it visited Aotearoa in 1769–1970. Captain James Cook wrote: "The Natives of this Country are strong, raw boned, and well-made" (Beaglehole 1968, p. 278), while Joseph Banks (1896, p. 240) reported: "A further proof, and not a weak one, of the sound health that these people enjoy, may be taken from the number of old people that we saw." Banks felt that the health of Māori was due to a sound health philosophy system rather than good luck. He wrote: "Such health drawn from so sound principles must make physicians almost useless" (Salmond 1991, p. 279). In Jones' (2000) gardener's tale speak, the soil for Māori at this precolonial time was rich and nurturing.

When Darwin visited some 60 years later in 1835, he described Māori as "...fearsome people...a more warlike race of inhabitants could be found in no part of the world...[whose] shifty looks betrayed a fierce cunning, and tattooed face revealed a base nature" (Desmond and Moore 1991, p. 174–5). At this time, missionaries in this country held views that denied the humanity of Māori, setting the scene for the subordination of Māori and the theft of Māori land and resources.

> Māori were, according to Henry Williams, 'governed by the Prince of Darkness'. Robert Maunsell thought Māori songs were 'filthy and debasing'. Even the Catholic Bishop Pompallier thought of the Māori as 'infidel New Zealanders'. (Walker 1994, p. 102)

From the 1830s, Māori health was declining in response to contact with newcomers. At the beginning of the nineteenth century, the Māori population ranged from 200,000 to half a million (Durie 1998). By 1856, when a census recorded 56,049 Māori and 59,413 Pākehā (non-Māori), politician Isaac Featherston speculated, "[t]he Maoris are dying out, and nothing can save them. Our plain duty as good, compassionate colonists is to smooth down their dying pillow" (Dow 1999, p. 48). By 1887, the Pākehā newcomer population was more than 700,000 while the Māori population had further declined to just under 40,000 (Pool 1991). While the Māori population started to recover in 1890s, the imposition of colonial health, economic, education, and land policies continued to marginalize Māori. The rapid decline in Māori health has been attributed to land loss that led to the loss of an economic base and a source of identity, the undermining of cultural knowledge and connectedness (Kunitz 1994), and the demoralization of Māori (Orange 1987). In places were land seizure was most rapid, for example, the ratio of Māori children to women was lowest (indicating high child mortality) (Pool 1991; Kunitz 1994). At the heart of Māori land loss, in turn, are the broken promises of the 1840 Treaty of Waitangi that guaranteed Māori continued possession of their resources, rangatiratanga (sovereignty), and citizenship rights (Orange 1987).

Fast forward to the latter part of the twentieth century, when Māori were increasingly migrating to urban centers in the 1950s and 1960s, the colonial agenda intensified for all New Zealanders to mix and integrate to become one culture – largely by turning Māori into British New Zealanders (Walker 2004). Pākehā researchers largely undertook research on Māori during this time. Te Awekotuku (1991, p. 12) describes this as a continuation of "...many decades – even centuries – of thoughtless, exploitative, mercenary academic objectification." In a similar vein, Linda Smith (1992, p. 7) calls these researchers "willing bedfellows of assimilationist, victim-blaming policies." These researchers lacked an understanding that the knowledge system and worldview of newcomers to this land was being privileged (Smith 2012a). They also failed to recognize that research is about power, and power commands the (re)distribution of resources (Te Awekotuku 1991).

When Indigenous attendees at a 1992 meeting held by the Royal Commission on Aboriginal Peoples in Canada talked about research as a colonizing tool, they described Aboriginal communities as having "been researched to death" (Brant Castellano 2004, p. 98). While they may have been meaning some metaphorical death, they could just as easily have been referring to research being implicated in the deaths of Aboriginal people and of researchers failing to address this in their various studies within First Nations communities. The reply of an Aboriginal elder to this talk was that "[i]f we have been researched to death, maybe it's time we started researching ourselves back to life" (Brant Castellano 2004, p. 98). For Māori, this life has to be a fully human life, a Māori life – as emphasized by Linda Smith (2012) in this chapter's opening quote.

As an intervention strategy, Kaupapa Māori rests upon the principles that inform this life (see below, Ontology). A Kaupapa Māori inquiry paradigm speaks to how research can support this decolonization agenda. Here this paradigm is described with a particular focus on Māori health that, in turn, requires an expansion of understandings of health to reflect Māori culture, traditions, and beliefs (see below, Epistemology). The writing of other Indigenous peoples is included to support and expand on points made. The result is not a checklist or a how-to guide. Rather, this chapter is a request for health research that advocates for a world that includes a Māori world. This re-inclusion of Māori as ordinary, as normal, is a Treaty of Waitangi obligation as well as a cultural survival necessity.

2   A Kaupapa Māori Research Paradigm

When seeking ways to eliminate widespread health, education, socioeconomic, and other Māori-non-Māori disparities, Māori leaders at a 1984 hui (conference) rejected assimilation or integration with newcomers and asserted that solutions lay in Māori being Māori. Kaupapa Māori is literally a Māori way. It is a response to the history of colonization in Aotearoa New Zealand that has seen Māori go from being ordinary to being the "other" in our own lands. Henry and Pene (2001, p. 237) describe Kaupapa Māori as "both...a resistance and reconstruction strategy and a culturally appropriate approach for Māori." What then began in the late 1980s as Kaupapa Māori (by Māori, with Māori), research within Māori education has spread to other disciplines including health, where Kaupapa Māori health research has infiltrated quantitative and qualitative Māori health research and evaluation since the early 1990s. This research promotes a structural analysis of Māori health disparities that moves the discourse away from victim blaming to more fully understanding people's lives and the determinants of Māori health disparities. Kaupapa Māori research has been described as a methodology (Smith 2012), an approach and a framework (Edwards et al. 2005). It has also been described as an inquiry paradigm (Cram et al. 2015; Cram and Mertens 2015), and it is this description that is pursued here.

In the 1980s, Guba and Lincoln (2005) drew on Thomas Kuhn's work in the physical sciences to articulate inquiry paradigms within the social sciences. In their work, different paradigms are able to co-exist within the social sciences, and the different worldviews these paradigms represent are characterized by their philosophical assumptions about the nature of reality (ontology), the relationship between the knower and what would be known (epistemology), and what are considered to be appropriate approaches to systemic inquiry (methodology) (see also ▶ Chaps. 6, "Ontology and Epistemology," and ▶ 90, "Engaging Aboriginal People in Research: Taking a Decolonizing Gaze"). Calling upon the language of paradigms in support of a Māori responsive research agenda echoes Linda Smith's (2012) description of key concepts that are embedded within a culture that provide rallying points; drawing people together on cultural terms for dialogue about a purposeful dream. For Māori, these concepts include tino rangatiratanga (sovereignty) and whānau. Recently, a Whānau Ora (Māori family wellness) initiative has provided a key cultural concept that enables Māori to come together to work toward a common vision. If the language of paradigms can also be a way to draw peoples together for some sense of solidarity, purposeful dreaming, and transformation, then it may also serve a decolonization agenda (Cram and Mertens 2015). The use of the language of inquiry paradigms builds upon the foundation laid by Chilisa (2012) and others of using this language to deconstruct White research paradigms and to pose an alternative Indigenous research paradigm. Describing Kaupapa Māori as a paradigm has also enabled a dialogue about decolonization to occur with other theorists who are committed to seeing research and evaluation contribute to social justice and equity around the world (Cram and Mertens 2015). This connection has firmly embedded axiology (ethics) as a key component of any inquiry paradigm (Mertens 2009). In the next part of this chapter, the assumptions of the four components of a Kaupapa Māori research paradigm (i.e., ontology, epistemology, methodology, and axiology) are defined. The potential role of Pākehā researchers within a Kaupapa Māori inquiry paradigm is touched upon in each section, with the proviso that the biggest thing a Pākehā researcher can bring is an open, reflexive mind and the knowledge that their world is just one of many.

2.1 Ontology: The Nature of Reality

Te ao Māori – the Māori world – is whakapapa – the genealogical ties that bind people with people, with the environment, and with the cosmos. Māori know this world through whakawhanaungatanga – the processes of understanding connectedness and relationships (Barlow 1991). Whakapapa means "to lay one thing upon another," as in one generation upon the next. Barlow (1991, p.173) writes "[W]hakapapa is a basis for the organization of knowledge in respect of the creation and the development of all things." This worldview emphasizes spirituality and spirit and the familial ties that bind people to one another. This is a reason for reciprocity as it is a way of honoring these relationships, with this described for Māori as an economy of affection (in contrast to a colonial economy of exploitation) (Henry and Pene 2001).

The nature of reality within Kaupapa Māori is represented by principles of Tino rangatiratanga (Self-determination), He taonga tuku iho (Cultural aspirations), Ako (Culturally preferred pedagogy), Kia piki ake i ngā raruraru o te kainga (Socioeconomic mediation), Whānau (Extended family structure), and Kaupapa (Collective philosophy). An overview of these principles is given below (see also ▶ Chap. 88, "Culturally Safe Research with Vulnerable Populations (Māori)").

2.1.1 Tino rangatiratanga (Self-Determination)

Māori sovereignty or self-determination is guaranteed in Article 2 of the Treaty of Waitangi. The importance of tino rangatiratanga has been emphasized at hui around the country since 1840. In the 1980s, tino rangatiratanga became part of the new Māori health movement where Māori claimed health initiatives as their own (Durie 1998). For example, at the 1994 Māori Health Decade gathering, Te Ara Ahu Whakamua, "[t]he most powerful and insistent message was the repeated call for Māori control and Māori management of Māori resources...'By Māori, for Māori'" (Te Puni Kōkiri 1994, p. 7).

2.1.2 He taonga tuku iho (Cultural Aspirations)

The assertion within Kaupapa Māori that being Māori is valid and legitimate, and to be taken for granted draws tikanga Māori (custom), te reo Māori (language), and mātauranga Māori (knowledge) into the contemporary environment as invaluable taonga (treasures) passed down from past generations (Smith 2012a). Within health services delivery, they are also seen as components of health practitioners' cultural competency to provide health care to Māori patients (Cram 2014).

2.1.3 Ako (Culturally Preferred Pedagogy)

Ako means both learner and teacher, with this principle describing culturally preferred pedagogies (both traditional and contemporary) that recognize that everyone has something to share and something to learn. Ako reinforces the importance of learning within tuakana-teina (older-younger) pairings and in groups. This might be, for example, learning to tie shoelaces or learning skills for the self-management of long-term health conditions.

2.1.4 Kia piki ake i nga raruraru o te kainga (Socioeconomic Mediation)

This principle recognizes that Kaupapa Māori mediation practices acknowledge and can successfully intervene in the socioeconomic disadvantage and negative pressures that are often experienced by whānau.

2.1.5 Whānau (Extended Family Structure)

As a way for Māori to organize their social world, the whānau has been a persistent structure in the face of colonization (Smith 2012). Whānau can still be defined as a collective concept that comprises three or more generations of the descendants of a significant marriage. Even when whānau are not contained within the same household they will still be whānau.

2.1.6 Kaupapa (Collective Philosophy)

A collective vision and commitment binds people together, whether they be kin whānau or kaupapa (agenda) whānau. It is kaupapa that connects whānau, hapū (subtribe), Iwi (tribe), and Māori community collectives to their aspirations for positive cultural, economic, social, and political well-being.

In summary, Kaupapa Māori ontology articulates what it means to be Māori, to be fully human within a Māori world. This world is about relationships and connectedness to other people, to the environment, and to the cosmos. It is a world that thrives within an economy of affection. While Māori researchers may be immersed in this world and be able to speak as part of a transdisciplinary team undertaking Kaupapa Māori health research, it is important that Pākehā researchers on the team also undertake to become familiar with this world – not so they can fully represent it but rather so they can support their Māori colleagues and advocate for their right to live, breathe, and be Māori within a research context.

2.2 Epistemology: Relationship between the Knower and What would be Known

Mātauranga Māori, often translated as knowledge, is a Māori tool for organizing and thinking about knowledge, including about our place in the world. As Tau (2001, p. 73) writes: "[M]atauranga Māori is simply the epistemology of Māori – it is what underpins and gives point and meaning to Māori knowledge." There is a sacredness about mātauranga Māori and, therefore, also the cultural practices tied to knowledge and learning (Smith 2012). The purpose of knowledge is to serve and uphold the mana (status) of the community. Traditionally, not all knowledge was available to everyone. Rather, some knowledge was available to all peoples to enable day-to-day living, while other tapu (sacred) knowledge was entrusted to only a few who would protect it and use it appropriately (Smith 2012). The transmission of this knowledge was, therefore, within Māori schools of learning, under the tutelage and guidance of tribal experts (Te Awekotuku 1991).

The cultural practices that link knowers to what would be known invariably acknowledge and work within a web of relationships that is the Māori world (see above, Ontology). The meandering (rather than linear) pathways of these connections and relationships are embedded within Indigenous languages, with knowledge coming from keen observation and interpretative messages (Deloria 1999) and reflexive learning cycles (Henry and Pene 2001). As Brant Castellano (2004, p. 98) writes: "Aboriginal knowledge has always been informed by research, the purposeful gathering of information and the thoughtful distillation of meaning." Thus, research becomes "**re-search**," as explained by Aboriginal educator Bob Morgan; it is about searching familiar pathways with a new inquisitiveness, new questions, new tools, and new companions. In addition to formal inquiry, epistemology is also an everyday occurrence, where events and objects can be looked at anew.

A key challenge to mātauranga Māori is epistemological racism. Scheurich and Young (1997) write that "White racism or White supremacy became interlaced or interwoven into the founding fabric of modernist western civilization" (p. 7) through "racially biased ways of knowing" (p. 4). The result has been the exclusion of other non-White epistemologies including mātauranga Māori, and the consequential negative impacts on Indigenous and minoritized peoples around the world (Smith 2012). The dual task of Kaupapa Māori research is, therefore, to engage with mātauranga Māori as well as with the deconstruction and challenging of White epistemology (Smith 2012a).

Critically, for Māori health, the emphasis on illness within the health sector needs to continue to be challenged so that other knowing about health and well-being can be fully recognized. Māori cultural aspirations are for holistic health and well-being that includes physical, mental, and spiritual health and extends to economic security and whānau support (Te Puni Kōkiri 1994), a good education and a healthy home free from violence, and political representation and environmental protection (Pomare et al. 1995). Holistic models of Māori health, including Te Whare Tapa Wha (Durie 1985) and Te Wheke (Pere 1988), speak to these aspirations. They also align with other Indigenous peoples where health is equivalent to the "harmonious coexistence of human beings with nature, with themselves, and with others, aimed at integral well-being, in spiritual, individual, and social wholeness and tranquility" (UN Permanent Forum on Indigenous Issues 2009, p. 157). Likewise, the recent Whānau Ora (Māori family wellness) initiative acknowledges and aims to strengthen the connectedness of whānau members, as well as increase the inclusion of whānau within society. Within this initiative, the well-being of whānau is measured through a holistic framework that acknowledges Māori conceptions of health and well-being. The six major whānau goals developed by the Taskforce on Whānau-Centered Initiatives (2010, p. 43) are whānau self-management, healthy whānau lifestyles, full whānau participation in society, confident whānau participation in te ao Māori (the Māori world), economic security and successful involvement in wealth creation, and whānau cohesion.

In summary, a Kaupapa Māori research epistemology is mātauranga Māori – Māori knowledge. In order to make space for mātauranga Māori the dual task of Kaupapa Māori research is to explore mātauranga hauora – knowledge of wellness – and critique other knowledge systems that challenge and undermine that knowledge. Pākehā researchers can support a Māori kaupapa by enquiring after the knowledge systems that challenge mātauranga Māori. For example, in a project on Māori patients' interactions with Pākehā general practitioners, the Māori researchers talked with patients while the Pākehā researchers talked with general practitioners in a parallel research process. This gave insight into mātauranga hauora as well as Pākehā discourses that supported or undermined it within primary health care (Cram et al. 2006).

2.3 Methodology: Appropriate Approaches to Systemic Inquiry

Methodology is about how to find things out, how to gain knowledge (Guba and Lincoln 2005; see ▶ Chap. 97, "Indigenous Statistics"). Kaupapa Māori methodology informs the selection of health research projects that have the potential to make a difference for Māori. This makes research both culturally prescribed and culturally acceptable. Linda Smith (2012) lists a series of questions that can be asked about research that can support the decision-making of both researchers and Māori communities about whether a project should proceed. These include questions about who defined the research question or problem, who will benefit or possibly be negatively impacted by the research, and who will gain knowledge from the research. The time has passed when researchers are in sole control of a research agenda. Now the expectation is that Māori health research should respond to Māori needs, priorities, and aspirations (Cram 2015).

Methodological considerations, especially the questions that research is aiming to answer, also inform the choice of research methods. Paipa et al. (2015) recommend that methods be selected for Kaupapa Māori evaluations using the principles of whakapapa (kinship connections), whakawhanaungatanga (making connections), whakawātea (a cleansing approach), whakaae (agreement), and whakamana (enhancement of authority). This will help ensure that the chosen methods will align with Māori values and make cultural sense. For example, researchers at Te Rōpū Rangahau Hauora a Eru Pōmare and Ngāti Kahungunu Iwi Incorporated (Ngāti Kahungunu tribal authority) used multi-method research to examine the health effects on Māori of employment loss due to factory closure. These included Kaupapa Māori epidemiology (Keefe et al. 2002), focus groups and interviews (Cram et al. 1997), and document review (during an evaluation) (see Cram and Mertens 2015 for an overview of this research).

In summary, methodology is responsive to the questions asked by Māori about the mission and vision of the research they are being asked to be involved in. Both Māori and Pākehā researchers need to be skilled enough to rise to this challenge – bending method rules where they need to be bent in order for research to be culturally relevant while at the same time ensuring that the research is also scientifically credible and valid.

2.4 Axiology: Nature of Ethics

Guba and Lincoln (2005) did not explicitly refer to axiology in their description and subsequent discussion of paradigms – confining themselves primarily to ontology, epistemology, and methodology. Axiology – what is done to gain knowledge, and what that knowledge will be used for – has been brought to the fore in Mertens' (2009) transformative paradigm.

For Māori, ethical obligations within research are about the maintenance and strengthening of relationships, and the cultural processes that help ensure this. "Concern for people's own well-being and prosperity ensured the observation of ethical practice, because a scholar was accountable to [their learning] community" (Te Awekotuku 1991, p. 8). This is similar to the operation of a communal ethic within Canadian Aboriginal communities that is "intimately related to who you are, the deep values you subscribe to, and your understanding of your place in the spiritual order of reality" (Brant Castellano 2004, p. 103). A key element of the Kaupapa Māori axiological assumption is attention to the cultural norms and protocols of the community in which research takes place. Seven "community-up" values describe ethical protocols that guide research practice (Pipi et al. 2004; Cram 2009; Smith 2012) (see Table 1 for an overview).

2.4.1 Aroha ki te tāngata: A Respect for People

- *About allowing people to define their own space and to meet on their own terms*

The essential characteristics associated with relationships are that they are respectful and trusting. Rituals of first encounter enable researchers to negotiate the space between themselves and those they want to involve in their research (Irwin 1994). Kaumātua (Māori elders) can assist researchers to make appropriate judgments and decisions within cultural spaces, including leading customary practices. For example, the E Hine project – a Kaupapa Māori health research on young Māori women (less than 20 years of age) having babies – has been guided by a Kāhui Kaumātua

Table 1 "Community-Up" approach to defining research conduct

Cultural values (Smith 1999)	Researcher guidelines (Cram 2001)	Kaupapa Māori health research practices
1. Aroha ki te tāngata	Be respectful – Allow people to define their own space and meet on their own terms	• Meet at places convenient to communities • Use cultural protocols to bridge the space between researchers and community • Engage elders to facilitate rituals of encounter • Acknowledge and strengthen connections
2. He kanohi kitea	Meet people face to face, and to also be a face that is known to and seen within a community	• Be known to communities • Be respectful and professional when encountering people known in other roles • Allow Māori research team members to take a lead Budget for regular meetings in community
3. Titiro, whakarongo... kōrero	Look and listen (and then maybe speak) – Develop understanding in order to find a place from which to speak	• Use all your senses during encounters • Encourage all peoples to participate • Know the value of silence • Feedback on the key points you are taking away
4. Manaaki ki te tangata	Share, host, and be generous	• Look after visitors, especially those uncomfortable in research contexts • Ensure needs and concerns are canvassed and addressed • Offer appropriate support Ensure that attendance is cost neutral for community members
5. Kia tūpato	Be cautious – Be politically astute, culturally safe, and reflective about insider/outsider status	• Know a community's history and current political standing • A research role is an "outsider" role, even when researchers are from a community • Be guided by community leaders about the appropriateness of methods • Ensure that participants and researchers are kept safe
6. Kaua e takahia te mana o te tāngata	Do not trample on the "mana" or dignity of a person	• Choose research methods that authentically represent lived realities • Allow time for people to participate in genuine ways • Ensure researchers have the skills to engage with community members • Maintain the integrity of research

(continued)

Table 1 (continued)

Cultural values (Smith 1999)	Researcher guidelines (Cram 2001)	Kaupapa Māori health research practices
7. Kia māhaki	Be humble – Do not flaunt your knowledge; find ways of sharing it	• Be honest about what the researchers do not know • Return information and knowledge to communities • Support communities to utilize research knowledge • Be responsive to community health priorities, and work to advance research in those areas

(Elders Group) who provide cultural advice and also accompany the researchers during their consultation and feedback meetings. These elders hold knowledge of relationships among community members, and among tribal groups. This project also sought input from a Rōpū Māmā – a young Māori mothers advisory group – to ensure that the project was responsive to participants' lived realities (Lawton et al. 2013).

Research teams for other Kaupapa Māori research projects have been guided by team members who have the necessary cultural expertise (Pipi et al. 2004). For example, many of the researchers on projects looking at the health impacts of Māori job loss had genealogical connections with this tribe as well. The researchers noted that their whakapapa (genealogical connections) only got them "in the door"; it was their research skills and professionalism that prevented them from being shown out that door by Māori communities still wary of research (Cram et al. 1997).

2.4.2 He kanohi kitea: A Face That Is Known
- *About the importance of researchers being known to the people*

Researchers are expected to be clear and honest about their research, including how the research findings will be used and how participants and other community stakeholders will be kept informed about the findings. Maori communities call upon researchers who intend to conduct studies in their communities to come face-to-face with them (Cram 2009), and to explain the nature of the research and the relationships between the researchers and tribe or community. This allows potential research participants to use all their senses to assess the advantages and potential disadvantages of their involvement. These meetings may also signal the beginning of a relationship that will extend beyond the end of the research project itself. The research on factory closure described above, for example, is a series of research projects conducted within a 25-year research relationship with Ngāti Kahungunu.

2.4.3 Titiro, whakarongo… kōrero – Look, Listen… Speak
- *About taking the time to become accustomed to a setting in order to find your voice*

This practice is about allowing time for people to discover each other's intricacies, so that mutual thinking can emerge between the parties involved in a research project.

Researchers should use all their senses to engage with communities, and also when undertaking research with Māori participants. There is value in silence and also in being able to offer appropriate responses and support. For example, evaluation encounters with Māori organizations in the 1990s were sometimes difficult because of an organization's past experiences with evaluators. Time taken to sit and to listen to the concerns of these organizations may have delayed the start of an evaluation, but was time well spent getting to know one another so that when the evaluators spoke they did so having gained some insight into the organization's reality. Evaluation relationships around issues of healthy activity and eating in one case, and issues of intimate partner violence in another case then grew into three to 6 year collaborations to explore and revise what works for Māori (Cargo and Cram 2003).

2.4.4 Manaaki ki te tangata: Host People
- *About having a collaborative approach to research, research training and reciprocity*

In research settings researchers may have something meaningful to contribute, as information flows in both directions. This occurred in Kaupapa Māori research with whānau about their children's asthma. The interviewing researcher was a nurse who could answer health questions asked by whānau, and who could connect whānau with a Māori asthma health organization that was under the leadership of another member of the research team (Jones et al. 2010). This enabled the research team to be responsive to whānau, rather than just one-way recipients of whānau information.

Manaaki is also expressed through koha or an offering to research participants to thank them for taking the time to be involved in the research project. This offering might be money or vouchers (e.g., petrol or grocery vouchers), books, or food. It is not an incentive to be involved and, in many cases, participants will not be aware they will receive such a thank you until after their research participation has ended.

2.4.5 Kia Tūpato: Be Careful
- *About being politically astute, culturally safe and reflexive about our insider/outsider status*

Kia tūpato is a caution to researchers that they need to take care, especially when negotiating spiritual spaces within their research. Research on Māori sudden infant death syndrome that involved interviewing whānau, especially Māori men who had lost a child, allowed participating whānau to follow their preferred rituals of whakawhanaungatanga (establishment of relationships) and karakia (prayer) so that the research space would be safe for them. Case workers were also present during these interviews to provide additional support to whānau (Edwards et al. 2005). One of the projects looking at the impact of employment loss on Māori involved accessing ex-workers' personnel records for an epidemiological study. The researchers appreciated that some workers had passed away and some were ill so they invited an elder to bless the records before any were looked at (Keefe et al. 1999).

2.4.6 Kaua e takahia te mana o te tangata: Do Not Trample on the Status of the People

- *Take care to acknowledge and respect people's roles and responsibilities*

Respect for people's status must filter through all stages of a research project, beginning with Kaupapa Māori research being responsive to Māori needs, priorities, and aspirations (Cram 2015). For example, research relationships should first be established with the tribal authorities for the regions that researchers want to work in. People also need to be able to take the time necessary for consultation and decision-making, with researchers ensuring they have the skills on their team to ensure this happens. This includes being flexible about the choice of research methods, or about the inclusion of additional methods that will give a community confidence its views will be well represented.

2.4.7 Kia māhaki: Be Humble

- *Do not flaunt your knowledge; share it and use your qualifications to benefit a community*

This final practice is about sounding out ideas with people, disseminating research findings, providing community feedback that keeps people informed about the research process and the findings. When research requests are outside researchers' discipline and expertise, they need to be honest about this, and then work with communities to connect them with researchers who can help. When research is conducted with a community there should be clear communication back to that community about the findings, along with support for any advocacy the community wishes to then take to improve their health and well-being. Communities must decide what is empowering, with researchers supporting when they can.

> Shared knowledge...is one of the key tools for empowering the people... the results of research [should be] made available to form part of the knowledge base of the people and to help them make decisions. (Mutu 1998, p. 51)

In summary, the community-up research practices provide guidance to researchers about engaging with communities and participants to both design and deliver Māori health research. They were first proposed as a guide for both Māori and Pākehā researchers. They have since become an example of how Māori might describe their views of good research practices.

3 Conclusion and Future Directions

A Kaupapa Māori inquiry paradigm is transformation space as it asserts the validity and legitimacy of being Māori and the right of Māori to live lives and undertake research that facilitate Māori health and well-being. At a high level, Kaupapa Māori has much in common with the theories of other groups who are seeking to enlarge their own worlds within an oppressive and marginalizing White "mainstream," for example, feminist, African-American, and deaf communities (Mertens 2009). The calling into existence of these worlds remakes these peoples as fully human, valuable contributing members of society, and helps ensure their cultural survival. At a local level, Kaupapa Māori addresses the oppression of Māori within their own lands, and the loss of those lands due to broken treaty guarantees. In this way it is unique. The justice Kaupapa Māori seeks is decolonization, including the return of stolen lands and resources as essential to Māori health and well-being. It is a decolonization project that seeks the return of Aotearoa.

This chapter has used the language of inquiry paradigms to distil Kaupapa Māori into component parts of epistemology, ontology, methodology, and axiology. The language of paradigms allows for this segmentation, but this is a device only rather than a truth about our world. It has allowed the examination of how Māori and (briefly) Pākehā researchers might be involved in a Māori agenda of health research that desires Māori health in its fullness, as it can be within a Māori world. In addition to the return of Māori land, this world is about the strengthening of Māori connectedness to one another, to the environment, and to the cosmos. It is about the revitalization of an economy of affection that was displaced during colonization by a capitalist economy (Henry and Pene 2001).

It is essential that social change is facilitated in a way that is commensurate with the day-to-day realities of the people affected (Cram 2015). This cannot be an exercise that is undertaken in, and confined to, the academy. Such moves turn the Indigenous transformative goal of sovereignty through decolonization into a metaphor and "kills the very possibility of decolonization; it recenters whiteness, it resettles theory, it extends innocence to the settler, it entertains a settler future" (Tuck and Yang 2012, p. 3). A Kaupapa Māori inquiry paradigm must, therefore, always be vigilant about the appropriation of decolonization and the distraction of health researchers away from a mission of eliminating Māori health disparities and enabling Māori to live in health and wellness as it is defined within a Māori cultural view. Health disparities will continue until there is also social, economic, and political equity for Māori (Durie 1998).

The future direction for Kaupapa Māori health research is, therefore, challenging disparities and seeking to fully represent Māori realities and aspirations. While the capacity of Māori health researchers has grown over the past 25 years, there is still a need for Pākehā health researchers who are willing to contribute their skills and expertise to this mission. The chapter did not begin from the point of view of Pākehā researchers because this is not the right starting point. Cross-cultural research is not about non-Indigenous researchers learning the passwords and byways that allow them to access Indigenous communities. Rather, it is about these colleagues starting from a place of commitment, combined with willingness to learn that can only come about when they realize that the privileged world they inhabit has been built on broken promises and Māori loss and marginalization. This may prompt guilt or anger on their part but, more helpfully, it should prompt commitment and questions about what they might do to facilitate the re-inclusion of Māori within Aotearoa New Zealand.

Waiho i te toipoto, kaua i te toiroa.
Let us keep close together, not far apart.

References

Banks J. Journal of the Right Hon. Sir Joseph Banks during Captain Cook's first voyage in H.M.S. Endeavour in 1768–71 to Terra del Fuego, Otahite, New Zealand, Australia, the Dutch East Indies, etc. New York: The Macmillan Company; 1896.
Barlow C. Tikanga whakaaro: key concepts in Māori culture. Auckland: Oxford University Press; 1991.
Beaglehole JC. The journals of captain James cook, VI. Cambridge: Cambridge University Press; 1968.
Brant Castellano M. Ethics of aboriginal research. J Aborig Health. 2004;1:98–114.
Cargo T, Cram F. Evaluation of the Atawhainga Te pa harakeke programme. Auckland: International Research Institute for Maori and Indigenous Education; 2003.
Chilisa B. Indigenous research methodologies. Thousand Oaks: SAGE; 2012.
Cochrane R. Racial prejudice. In: Cochrane R, Carroll D, editors. Psychology and social issues. A tutorial text. London: Falmer Press; 1991.
Cram F. Rangahau Māori: Tona tika, tona pono. In: Tolich M, editor. Research ethics in Aotearoa. Auckland: Longman; 2001. p. 35–52.
Cram F. Maintaining indigenous voices. In: Mertens D, Ginsberg P, editors. SAGE handbook of social science research ethics. Thousand Oaks: SAGE; 2009. p. 308–22.
Cram F. Improving Māori access to health care: integrated research report. Auckland: Katoa Ltd.; 2014.
Cram F. Harnessing global social justice and social change with multiple and mixed method research. In: Hesse-Biber SN, Johnson RB, editors. The Oxford handbook of mixed and multiple methods research. New York: Oxford University Press; 2015. p. 667–87.
Cram F, Mertens D. Transformative and indigenous frameworks for mixed and multi method research. In: Hesse-Biber SN, Johnson BB, editors. The Oxford handbook of mixed and multiple methods research. New York: Oxford University Press; 2015. p. 99–109.
Cram F, Keefe V, Ormsby C, Ormsby W, Ngāti Kahungunu Iwi Incorporated. Memorywork and Māori health research: discussion of a qualitative method. He Pukenga Kōrero. 1997;3(1):37–45.
Cram F, Kennedy V, Paipa K, Pipi K, Wehipeihana N. Being culturally responsive through Kaupapa Mȧori evaluation. In: Hood S, Hopson R, Frierson H, editors. Continuing the journey to reposition culture and cultural context in evaluation theory and practice. Charlotte, NC: Information Age Publishing; 2015. p. 289–311.
Cram F, McCreanor T, Smith L, Nairn R, Johnstone W. Kaupapa Māori research and Pākehā social science: epistemological tensions in a study of Māori health. Hūlili. 2006;3:41–68.
Deloria V Jr. Spirit and reason: the Vine Deloria, Jr., reader. Golden: Fulcrum Publishing; 1999.
Desmond A, Moore J. Darwin. London: Penguin; 1991.
Dow D. Māori health & government policy 1840–1940. Wellington: Victoria University Press; 1999.
Durie M. A Maori perspective of health. Soc Sci Med. 1985;20(3):483–6.
Durie M. Whaiora: Maori health development. 2nd ed. Auckland: Oxford University Press; 1998.
Edwards S, McManus V, McCreanor T. Collaborative research with Māori on sensitive issues: the application of tikanga and kaupapa in research on Māori sudden infant death syndrome. Soc Policy J N Z. 2005;25:88–104.

Guba EG, Lincoln YS. Paradigmatic controversies, contradictions, and emerging confluences. In: Denzin NK, Lincoln YS, editors. The SAGE handbook of qualitative research. 3rd ed. Thousand Oaks: SAGE; 2005. p. 191–215.

Henry E, Pene H. Kaupapa Māori: locating indigenous ontology, epistemology and methodology within the academy. Organ. 2001;8:234–42.

Irwin K. Māori research methods and processes: an exploration. SITES. 1994;28:25–43.

Jones CP. Levels of racism: a theoretic framework and a gardner's tale. Am J Public Health. 2000;90(8):1212–5.

Jones B, Ingham T, Davies C, Cram F. Whānau Tuatahi: Māori community partnership research using a Kaupapa Māori methodology. MAI Rev. 2010;3:1.

Keefe V, Ormsby C, Robson B, Reid P, Cram F, Purdie G, et al. Kaupapa Māori meets retrospective cohort. He Pukenga Kōrero. 1999;5(1):12–7. Koanga (Spring).

Keefe V, Reid P, Ormsby C, Robson B, Purdie G, Baxter J, et al. Serious health events following involuntary job loss in New Zealand meat processing workers. Int J Epidemiol. 2002;31:1155–61.

Te Puni Kōkiri. Te Ara Ahu Whakamua: Proceedings of the Māori Health Decade Hui, March 1994. Wellington: Whaia te Whanaukataka: Oraka Whanau: The wellbeing of Māori whanau. A discussion document. 1994.

Kunitz S. Disease and social diversity: the European impact on the health of non-Europeans. New York: Oxford University Press; 1994.

Lawton B, Cram F, Makowharemahihi C, Ngata T, Robson B, Brown S, et al. Developing a Kaupapa Māori research project to help reduce health disparities experienced by young Māori women and their babies. AlterNative Int J Indigen Peoples. 2013;9(3):246–61.

Mertens DM. Transformative research and evaluation. New York: Guildford Press; 2009.

Ministry of Health. Ministry of health. (2015, October 8). Retrieved 2 Mar 2016, from Neighbourhood deprivation: http://www.health.govt.nz/our-work/populations/maori-health/tatau-kahuku ra-maori-health-statistics/nga-awe-o-te-hauora-socioeconomic-determinants-health/neighbourh ood-deprivation.

Mutu M. Barriers to research: the constraints of imposed frameworks. A keynote address to Te Oru Rangahau Māori Research and Development conference. Palmerston North: Massey University; 1998. 7–9 July.

Nobles W. Extended self: Rethinking the so-called negro self-concept. In: RL. Jones, editor. Black psychology (3rd Edition). Berkley: Cobb & Henry; 1991. p. 295–304.

Orange C. The treaty of waitangi. Wellington: Allen and Unwin; 1987.

Paipa K, Cram F, Kennedy V, Pipi K. Culturally responsive methods for family centered evaluation. In: Hood S, Hopson R, Frierson H, editors. Continuing the journey to reposition culture and cultural context in evaluation theory and practice. Charlotte: Information Age Publishing; 2015. p. 151–78.

Pere RR. Te Wheke: Whaia te maramatanga me te aroha. In: Middleton S, editor. Women and education in Aotearoa, vol. 1. Wellington: Unwin & Irwin; 1988. p. 6–19.

Pipi K, Cram F, Hawke R, Hawke S, Huriwai T, Mataki T, et al. A research ethic for studying Māori and iwi provider success. Soc Policy J N Z. 2004;23(December):141–53.

Pomare E, Keefe-Ormsby V, Ormsby C, Pearce N, Reid P, Robson B, et al. Hauora III: Māori standards of health. Wellington: Te Rōpū Rangahau Hauora a Eru Pomare; 1995.

Pool I. Te Iwi Maori: a New Zealand population, past present and projected. Auckland: Auckland University Press; 1991.

Salmond A. Two worlds: first meeting between Māori and Europeans 1642–1772. Auckland: Viking; 1991.

Scheurich I, Young M. Coloring epistemologies: are our research epistemologies racially biased? Educ Res. 1997;26(4):4–16.

Smith LT. Te Rapunga i te Ao Marama: the search for the World of Light. In: The issue of research and Māori, Monograph, vol. 9. Auckland: Research Unit for Māori Education, University of Auckland; 1992.

Smith LT. Decolonizing methodologies – research and indigenous peoples. London: Zed Books; 1999.

Smith GH. Kaupapa Māori: the dangers of domestication. Interview with Te Kawehau Hoskins and Alison Jones. N Z J Educ Stud. 2012;47(2):10–20.

Smith Linda T. Decolonizing methodologies – research and indigenous peoples. 2nd ed. London: Zed Books; 2012.

Taskforce on Whānau-Centred Initiatives. Whānau Ora: report of the taskforce on Whānau-Centred initiatives, to hon. Tariana Turia, minister for the community and voluntary sector. Wellington: Taskforce on Whānau-Centred Initiatives; 2010.

Tau T. Mātauranga Māori as an epistemology. In: Sharp A, McHugh P, editors. Histories, power and loss: uses of the past – a New Zealand commentary. Wellington: Bridget Williams Books; 2001. p. 64–81.

Te Awekotuku N. He tikanga whakaaro: research ethics in the Māori community. Wellington: Manatu Māori; 1991.

Tuck E, Yang KW. Decolonization is not a metaphor. Decol Indig Educ Soc. 2012;1(1):1–40.

UN Permanent Forum on Indigenous Issues. State of the world's indigenous peoples. New York: United Nations; 2009.

Walker R. Māori resistance to state domination. Kia Pūmau Tonu. In: Proceedings of the Hui Whakapūmau Māori Development Conference. Palmerston North: Massey Univeristy Māori Studies Department; 1994.

Walker R. Ka whawhai tonu: struggle without end. Revised ed. Auckland: Penguin; 2004.

Culturally Safe Research with Vulnerable Populations (Māori)

88

Denise Wilson

Contents

1 Introduction	1526
2 The Significance of Culture	1527
3 Māori and Research	1529
4 Research at the Interface	1531
5 Culturally Responsive and Safe Research	1535
6 A Framework for Culturally Safe Research	1535
7 Conclusion and Future Directions	1540
References	1541

Abstract

Vulnerable populations are often subjected to some form of social marginalization. This contributes to persistent inequities in their social and health outcomes, and differences in their access to and use of necessary services. Researchers' decisions and the research processes they utilize can further increase their risk of vulnerability and marginalization. Historically, Māori (indigenous peoples of Aotearoa New Zealand) experiences with research often yielded little benefit for them, instead frequently reinforcing negative stereotypes and perpetuating deficit explanations and inaccuracies. Today, many Maori remain suspicious of researchers and their agendas and are reluctant to engage in research. Yet, quality evidence and generating accurate "stories" are crucial to inform optimal strategies to resolve persistent social and health inequities. Nonetheless, evidence founded on dominant cultural research paradigms and sociocultural realities and interpretations can worsen people's vulnerability and marginalization within social and health research contexts. Creating culturally responsive and safe spaces and research contexts with Maori, and others vulnerable within research settings,

D. Wilson (✉)
Auckland University of Technology, Auckland, New Zealand
e-mail: dlwilson@aut.ac.nz

© Springer Nature Singapore Pte Ltd. 2019
P. Liamputtong (ed.), *Handbook of Research Methods in Health Social Sciences*,
https://doi.org/10.1007/978-981-10-5251-4_31

are needed to minimize participants' vulnerability and marginalization and counter unhelpful constructions about them. In this chapter, the importance of understanding the impact differing worldviews can have on researchers, research methodology, and research conduct with vulnerable populations will be discussed. Strategies will be presented aimed at minimizing the vulnerability of those participating in or targeted for research. A framework based on the concepts of partnership, participation, protection, and power is provided to assist researchers' cultural responsiveness, getting the research story right, and importantly, to improve the utility of their research.

Keywords

Indigenous research · Cultural responsiveness · Cultural safety · Māori · Māori-centered · Kaupapa Māori research

1 Introduction

Undertaking research that accurately portrays "the story" of those participating in the research, and which is conducted and interpreted in socially and culturally responsible ways, is a fundamental requirement for undertaking cross-cultural research, particularly with those considered vulnerable (Liamputtong 2010). For the purposes of this chapter, vulnerable groups are those peoples whose strengths and positive attributes are generally overlooked, and who are confronted with differential risks and health burdens in comparison to others living in their community or country. Within the context of research, this means those considered vulnerable, marginalized or belonging to minority groups are potentially at risk of further burden by research that is conducted on them (not with them) and interpretations that bear little or no resemblance to their realities. Oftentimes, those who belong to "vulnerable" groups live with ongoing health and social inequities (Ruger 2008). Addressing these inequities requires understanding the people belonging to these groups, their life circumstances, and the sometimes multitude of factors impacting their daily lives. In order to understand and then produce useful knowledge and interventions requires culturally responsive and safe research – researchers' decisions and actions can inadvertently increase people's vulnerability and serve to marginalize them within and throughout the research process.

Indigenous peoples globally can be considered vulnerable within a background of persistent social and health inequities they face, and their subjection to discriminatory research processes that have overlooked the significance of their culture (Smith 2012; see also ▶ Chap. 15, "Indigenist and Decolonizing Research Methodology," ▶ 87, "Kaupapa Māori Health Research," ▶ 89, "Using an Indigenist Framework for Decolonizing Health Promotion Research," ▶ 90, "Engaging Aboriginal People in Research: Taking a Decolonizing Gaze," ▶ 97, "Indigenous Statistics," and ▶ 106, "Ethics and Research with Indigenous Peoples"). Moreover, researchers have often disregarded and disrespected people's important cultural traditions and practices. There are numerous accounts of research undertaken in disrespectful and demeaning ways that have not only negated the importance of the people but "trampled" on their cultural ways and their status as indigenous peoples (Chilisa 2012; Smith 2012; Walter and Andersen 2013). Linda Smith (2012, p. 1) notes the impact of researchers on Māori as a process that "...stirs up silence, it conjures up bad memories, it raises a smile that is knowing and distrustful." "Outsiders" have come into indigenous communities as detached observers without establishing meaningful relationships with the people. It is not unusual for indigenous peoples to endure situations but at the same time withhold or filter information they share. Researchers have then gone away and analyzed their observations with no real understanding of their accuracy and the nuances of language and cultural practices. Such approaches to research produce research that lacks cultural responsiveness, and can only be considered culturally unsafe.

The aim of this chapter is to assist researchers developing their cultural responsiveness, getting the research story right, and importantly, to ultimately improve the utility of their research for indigenous peoples. In this chapter, Māori (the indigenous peoples of Aotearoa New Zealand) will be referred to and links with other indigenous peoples who have been colonized made where appropriate. To begin an overview of what culture is presented. Thereafter, the impact of differing worldviews on research, the selection of research methodology, and the way in which research is conducted is discussed. The significance of the role that culture plays when planning and undertaking research is explored briefly, within the context of culturally responsive and culturally safe research practice for Māori, the indigenous peoples of Aotearoa New Zealand. Strategies aimed at minimizing (and ideally eliminating) participant vulnerability will be presented together with a framework based on the concepts of partnership, participation, protection, and power (see also ▶ Chap. 87, "Kaupapa Māori Health Research").

2 The Significance of Culture

Culture refers to the shared beliefs, values, and everyday practices groups of people undertake, influenced by their unique worldview (Wilson and Hickey 2015). It prescribes expectations related to interactions, behaviors, and protocols to guide the way things are done – for instance, processes of engagement with those not belonging to the group under study. It differs from the concepts of ethnicity and race. Ethnicity refers to larger groupings of people belonging to different "cultural" groups based on a common heritage or nations (for example, Pacific peoples, Greek, or Asian). On the other hand race classifies people based on their physical features (such as Caucasian, Polynesian, Asian, and Melanesian) (Bhopal 2004).

Individuals grow up within a cultural milieu created by their family of origin, and influenced by those they interact with, for instance other groups of people with differing cultures (such as those based on age, gender, generation, sexual orientation, occupation, socioeconomic status, ethnic origin, migrant or refugee experiences, spiritual or religious beliefs, and disability (Nursing Council of New Zealand 2011)). The ensuing interactions with differing and similar groups over time shapes and modifies people's culture in addition to the social and technological changes that in turn alter what behaviors are considered acceptable and what is expected. In this way, culture is dynamic and contributes to the growing contemporary diversity that occurs not only between cultural groups but within them. Importantly, it influences how people see and understand the world. If we ignore people's culture when embarking on research, its findings and outcomes may not fully realize the intended aims. The message is that culture counts, and that it can "eat" the best planned research strategy if ignored. Researchers are, therefore, advised to give serious consideration to how they include those people from different cultures to ensure their cultural needs are incorporated into planning, implementing, and reporting their research.

Globally, indigenous peoples share similarities in their worldviews and cultural practices. Generally, they are defined by a holistic and spiritual view of the world, which is defined by their whakapapa (genealogy), relationships, and interactions with not only living people but also their ancestors, living creatures, and the environment (Battiste 2000; Chilisa 2012; Sherwood and Kendall 2013). A defining feature that sets indigenous peoples aside from those belonging to the dominant western cultural groups is their collective orientation and the associated responsibilities and obligations to others. This contrasts with the dominant cultural focus on individualism, first and foremost.

The nuances and subtleties of culture and its concepts are embedded in language, which can be considered the window to people's culture. Thus, how to best access and understand people's culture is an essential consideration when planning and conducting research. Marsden (cited in Royal 2003) highlights the importance of culture, stating:

> Cultures pattern perceptions of reality into conceptualisations of what they perceive reality to be; of what is to be regarded as actual, probable, possible or impossible. These conceptualisations from what is termed the 'worldview' of a culture...lies at the very heart of culture, touching, interacting with and strongly influencing every aspect of the culture. (p. 56)

Vulnerable people are subjected to services that are constructed and delivered based on evidence informed by dominant epistemologies which privilege how information is gathered, analyzed, and interpreted. This is referred to "epistemological domination," whereby one epistemology is exclusively privileged over another – in the case of this chapter non-Māori western epistemologies are promoted over Māori indigenous epistemologies (Moewaka Barnes et al. 2008; see also ▶ Chap. 87, "Kaupapa Māori Health Research"). Ironically, indigenous researchers using indigenous methodologies are often required to still conform to and address dominant Western research requirements.

Ignoring the significance of culture increases the risk of producing research outcomes that lack relevance and utility. Such an approach disregards people's historical and everyday realities and circumstances, risking research outcomes that either lack relevance or are inappropriate. Importantly, researchers need to understand these historical and contemporary dimensions and how they identify and bridge the "gap" existing between the researchers' cultural location and that of those being researched. Researchers can have the best intentions and strategies, but framing and planning research with vulnerable populations from a different worldview is often driven by hegemonic and unconscious biases that promote ignoring the crucial role of culture.

3 Māori and Research

To understand the need for culturally responsive and safe research approaches, there is also a need to understand the historical and contemporary realities of those considered vulnerable, such as indigenous peoples similar to Māori. Without question, colonization has had significant and damaging effects on Māori, comparable to indigenous peoples in Australia, Canada, and the United States. Undoubtedly, colonization has had detrimental effects on indigenous peoples (Sherwood 2013). The long established ways of observing, knowing, and understanding the world Māori possessed was systematically replaced by "new" scientific ways of knowing – ignoring the crucial role of culture. Colonization and imperialism led to historic injustices, which have in turn created the social milieu to perpetuate ongoing contemporary socioeconomic and political disenfranchisement (Smith 2012). Accompanying the colonization of Aotearoa New Zealand was the systematic invalidation of indigenous knowledge and ways of knowing that sustained indigenous communities for hundreds of years, handed down from generation to generation. The dispossession of land (people's tūrangawaewae – a place to stand in the world), language (te reo Māori), disconnection from supports, deculturation of protective cultural practices, along with widespread depopulation through disease and warfare contributed to the loss of Māori ways of knowing and being in the world (Durie 1998). Policies of assimilation further aided this loss, such as banning te reo Māori being spoken in schools, while urbanization in the 1950s and 1960s removed people from their vital whānau (extended family) supports and cultural practices. Legislation, such as the Tohunga Suppression Act 1907, banned the practices of traditional healers which led to the loss of important cultural knowledge. Consequently, mātauranga Māori (traditional Māori knowledge) was invalidated and demeaned, instead replaced by Western scientific paradigms that became the prevailing and accepted way of viewing knowledge (Durie 1998).

It was from dominant Western epistemological stances that researchers repeatedly characterized Māori within deficit constructs, regularly depicting them negatively. Māori have been subjected to generations of research about them but not with them being adequately involved. Subsequently, research has been considered to be of little or no benefit or relevance to Māori, particularly when based on inaccurate interpretations that becomes common understandings and "truths." This most often arises from researchers functioning on unconscious biases and lacking essential cultural insights to accurately collect, analyze, and interpret the data. Such approaches circumvent "getting the story right" (Smith 2012). Moreover of concern are the culturally inappropriate and unacceptable processes researchers often use to engage with Māori and obtain their information. Jahnke and Taiapa (1999) highlight the importance of culture, stating:

Knowledge which makes sense in one particular context cannot always be understood through the tools that govern the understanding of other belief systems and worldviews. (p. 41)

Often, researchers left with inaccurate and incomplete information, all the while having "trampled" on the mana (status and esteem) of Māori communities. Understandably, many Māori to this day remain suspicious of researchers' and research agendas.

Research is important to inform responses to addressing the serious and persistent inequities Māori suffer in their health and social outcomes, as well as their experiences and the quality of health service access and use. Research findings and recommendations based on Western paradigms and research approaches do not always work for Māori, particularly when they promulgate uninformed and unhelpful discourses. This is primarily because such findings can be devoid of any analysis or consideration of the competing intersections of cultural relevance and multiple oppressions. Research undertaken without cultural insights and relevance, and a critical understanding of the historical and contemporary social and health inequities that indigenous peoples are faced with, ignores the ongoing systemic issues they encounter when accessing health or social services.

For instance, while "big picture" research using quantitative methodologies can show us what is happening on a large scale, it is unable to provide the contextual issues that are important for effectively addressing disparities and inequities in social and health well-being. However, smaller qualitative, culture-based research studies using appropriate processes and methods that seek people's actual experiences can provide greater insight and offer counter-understandings than the existing prevailing ones (Liamputtong 2010). For example, the research mentioned in Box 1 demonstrates the systemic and health-provider issues young pregnant Māori women encounter when seeking early maternity care which are often absent in "mainstream" research studies. This study utilized Māori research approaches which enabled researchers to access and gain the trust of this group of young Māori women, who are 'known' to avoid services when they are confronted with or anticipate negative responses from health-care providers.

Box 1 Countering Prevailing Discourses About Inequities

In Aotearoa New Zealand, pregnant Māori women less than 20 years of age are at higher risk of poor maternal and infant outcomes. Negative portrayals are prevalent about young Māori women who are pregnant, particularly because they are seen to present late and "fail" to attend early maternity care, which is believed to contribute to increased risk of poor outcomes for both mother and baby. Such portrayals position "blame" onto vulnerable young women and is frequently supported by evidence generated using a dominant cultural lens. Thus, "common" understandings promulgated are that these young Māori

(continued)

Box 1 (continued)

women are remiss and somewhat neglectful regarding their own health and that of their babies. Makowharemahihi et al. (2014) interviewed 44 young Māori women under 20 years of age who were pregnant or recently pregnant using a kaupapa Māori research approach. They reported a contrary "story" indicating that the majority of these women accessed health care services early to confirm their pregnancy, but they received inadequate information about how to navigate the maternity care system. Also evident were missed opportunities and failure to respond by health-care providers in assisting these women to access maternity care (usually a midwife). This in turn led to fragmented maternity care and inhibited them having a seamless pathway of care. Those health-care providers who provided practical assistance to access a lead maternity carer and women with supportive whānau and friends were better able to access seamless maternity care earlier in their pregnancy. This study using a Māori approach to gather and analyze data shows that contrary to common understandings of healthcare providers, young pregnant Māori women are proactively connecting with primary health care services early – they often encounter barriers difficult to overcome.

Makowharemahihi et al. (2014).

4 Research at the Interface

Research lacking connection to Māori realities will not result in the expected impact, and therefore, researchers must be cognizant of local Māori cultural values and practices (Durie 2004). The research interface enables a place for the best approach to be used to answer a research question(s) being investigated that draws on both indigenous methodologies and knowledge and suitable western research methodologies and/or methods. It is a space in which mutual dialogue and negotiation can occur, and whereby culturally appropriate and safe processes and actions can be established – a place without contest or challenge where indigenous researchers meet with those researchers who work from western methodological epistemologies to ensure research is designed to best reflect indigenous realities (see Fig. 1).

For Māori, the research interface is contingent on healthy bicultural relationships based on Te Tiriti o Waitangi that recognizes and respects Māori culture and peoples. The Treaty of Waitangi in 1840 was signed by representatives of the British Crown and Māori chiefs. This Treaty sets out an agreed relationship with expectations between Māori and the British Crown. Nevertheless, significant differences exist between Te Tiriti o Waitangi (the Māori language version) and the Treaty of Waitangi (the English language version), although the English version is deferred to officially. However, Māori often refer to Te Tiriti o Waitangi, which is the version their ancestors signed. In signing Te Tiriti o Waitangi, they ceded governorship (not

Fig. 1 The research interface (Note: Adapted from Wilson and Neville (2009))

sovereignty) of their lands and were guaranteed the rights of tino rangatiratanga (self-determination), to sell their land, have equitable outcomes, and be protected. Historically, the rights enshrined within the Treaty have not been upheld, and until the 1960s, it was generally not honored breaching the rights of Māori (Hudson and Russell 2009; Wilson and Haretuku 2015). Breaches of the Te Tiriti o Waitangi have compromised the cultural identity of Māori and for many denied the opportunities to access te ao Māori (the Māori world). Cram (2012) claims that breaches of Te Tiriti o Waitangi are the origins of Māori overrepresentation in poor social outcomes (see also ▶ Chap. 87, "Kaupapa Māori Health Research").

Establishing healthy bicultural relationships based on the notion of reciprocity ensures research processes and outcomes are:

- Beneficial to Māori
- Inclusive of Māori values, aspirations, and needs
- Privilege an indigenous Māori worldview
- Māori retain self-determination and control over the research to make sure their interests are protected

Such bicultural relationships are premised on the notion of equitable relationships between indigenous groups, like Māori, and those undertaking the research. The research relationship, according to Hudson and Russell (2009), should be founded on the equal status of all parties, the veracity of the engagement, and equitable outcomes – it can be considered a framework for not only balanced relationships but also for restoring power to Māori.

Keeping central and prominent the indigenous cultural imperatives of those being researched throughout the research process is crucial to honor the rights embedded within Te Tiriti o Waitangi. In Aotearoa, ethical review processes require researchers to demonstrate their responsiveness to Māori based on the principles of the Treaty of Waitangi – that is, partnership, participation, and protection (discussed later in the chapter).

Māori research methodologies, such as Kaupapa Māori Research (see for example, Smith 2012), Māori-centered research (for instance, Durie 1997; Wilson 2004), and their several variations, represent strategic Māori decolonizing and reclaiming the Māori research space and regaining control over research about them. It is about research with, by, and for Māori, particularly kaupapa Māori (Smith 2000). It is fundamentally nested within te ao Māori (Māori worldview) and informed by mātauranga Māori and Māori beliefs, values, and practices. It normalizes te ao Māori (rather than excludes as seen in western scientific approaches) within a research context (Moewaka Barnes 2000). Smith (2000, 2012) identifies the following dimensions of kaupapa Māori research:

- It is about being Māori,
- It is grounded in Māori viewpoints and ideologies,
- It normalizes the validity and legitimacy of Māori language and culture, and
- It is concerned with tino rangatiratanga (self-determination/autonomy) over Māori well-being.

Smith also locates kaupapa Māori within localized critical theory, which critiques the power structures and social inequities within the unique historical and contemporary contexts Māori experience. While the emancipatory goals of kaupapa Māori is somewhat debated (see Bishop 1998, for example) given the failure to meet these goals, it does help to interrogate the unique local contexts and social and health inequities confronting Māori.

Without doubt, there are numerous variations within how kaupapa Māori research is expressed, which is reflective of Māori in general – Māori are not a homogenous group of people, but rather a diverse people with variations on how they see the world from iwi and hapū to iwi and hapū. Added to this diversity is the contemporary reality for many with varying connections culturally to their Māori whakapapa. According to Moewaka Barnes (2000), it is a "distinctive" approach. For example, Irwin (1994) and Bishop (1996) utilized whānau for the supervision and way in which they structured their research, and Bishop's work is based on whakawhanaungatanga. More recently, constructivist forms of kaupapa Māori have also emerged (Eketone 2006; Walker et al. 2006).

Māori-centered research is underpinned by three cornerstones to ensure Māori interests in research are upheld. These are, whakatuia (integration), whakapiki tangata (enablement), and mana Māori (control) (Durie 1997). This means that under the guidance of indigenous researchers, Māori cultural values, practices, and processes are integrated into the research design and processes, Māori participation is ensured, and Māori control over the research and its outcomes is safeguarded. Box 2 shows an example of research where researchers leading a project erroneously interpreted, and then used Māori-centered research without ensuring these three cornerstones informed the development of the research. An example is then presented using the three cornerstones of Māori-centered research to illustrate culturally responsive research practice.

Box 2 Moving from Flawed to Culturally Responsive Practice

Exemplar	Flawed practice	Culturally responsive practice

The research team designed a study building on a Kaupapa Māori research study and aimed to include Māori as participants along with others in this new study controlled by non-Māori research. The only mention of Māori in the substantive research proposal was in the background with a brief reference to the prior study, the inequities for Māori in the area of research, and in the responsiveness to Māori section. The researchers claimed this was a Māori-centered study because Māori researchers were involved and some of the participants would be Māori. Nonetheless, the control of the research was under the guises of researchers who were not Māori. The advisory group challenged this was a Māori-centered study and wanted the anomaly to be resolved. They claimed that the focus was deficit-based rather than strengths-based, the measures were focused on negative behaviors and lacked validation with Māori. Although they acknowledged that there some attempts were made to include Māori cultural values and seek cultural advice, its implementation did not flow through to the research proposed presented to the group	Erroneously thinking that because the research was based on a previous kaupapa Māori study and that Māori would be involved as researchers and participants, it was Māori-centered
	The deficit-based approach adopted did not meet the needs and aspirations of the advisory group
	The absence of a by, with, and for Māori approach – that is, not Māori initiated, inadequate collaboration with Māori, and the participant make-up did not reflect a Māori participant group, but rather a mixture of ethnic and research backgrounds
		Whakapiki tangata (Enablement)
		  Māori researchers leading the prior study would be the lead investigators for this study
		  Māori would be the significant researchers and participants
		  Establishing respectful relationships with relevant Māori whānau (extended family), hapū (sub-tribe), iwi (tribe) and communities should underpin the research
		Whakatuia (Integration)
		  Throughout the proposal, there would be evidence of Māori cultural processes and practices informing the research design, particularly for the engagement, data collection, data analysis, and dissemination and translation of the research findings
		  Culturally appropriate and validated measures would be used in the research
		  Ethical considerations should include culturally based ethics together with mainstream ethical requirements
		Mana Māori (Control)
		  The advisory group would have been involved early in the planning stages of the research so the aspirations and needs of Māori would be met in the current study
		  The research outcomes must be beneficial for improving the well-being of Māori and reducing inequities within the area of study

5 Culturally Responsive and Safe Research

Ko tau hikoi i runga i ōku whāriki
Ko tau noho i tōku whare
E huakina ai tōku tātau tōku matapihi
Your steps on my whariki (mat), your respect for my home, open my doors and windows

This whakataukī (traditional proverb) outlines the importance of culturally responsive and safe research and reinforces the importance of respectful practice. The research interface is a space to establish and develop culturally responsive and safe research practices. Culturally responsive and safe research refers to the researchers' intentions to, and processes of, creating a research environment founded on collaboration with and the inclusion of those people belonging to "vulnerable" groups to ensure their cultural values, beliefs, and practices are not only recognized and respected but are integrated into all facets of the research (Wilson and Neville 2009). This collaboration and inclusion ideally begins from the inception of a research idea to the dissemination and translation of the research findings. Ideally, the research team should include researchers who belong to a "vulnerable" group(s) and informed by an advisory group comprising cultural advisors and community members. Cunningham (2000) provides a framework for researchers to identify the focus of their research with Māori:

- Research not involving Māori – where Māori are not involved. Having said this, consideration should be given to the research's potential benefits for Māori (see, for example, Hudson et al. 2010)
- Research that may include Māori – whereby participants may be Māori but the focus of the study is not specifically Māori focused.
- Māori-centered research – Māori are major participants and researchers, and the data are analyzed using a Māori lens, but the research is situated within a "mainstream" institution.
- Kaupapa Māori research – research by Māori, with Māori, and for Māori, reflected in the catchphrase "nothing about us, without us."

Hudson et al. (2010) outlines the expectations for three levels of practice – that is, research without Māori as a focus, Māori-centered research and Kaupapa Māori Research. Fundamental to creating culturally responsive and safe research spaces is the engagement and formation of relationships between researchers and those being researched. For indigenous people, engagement and relationships are cultural cornerstones which do not alter in the presence of research.

6 A Framework for Culturally Safe Research

Culturally safe research is located within the context of Te Tiriti o Waitangi/the Treaty of Waitangi (Fig. 2). Whether research is culturally safe must be decided by those groups of people being researched (Wilson and Neville 2009). Wilson and Neville (2009) claim that culturally safe research:

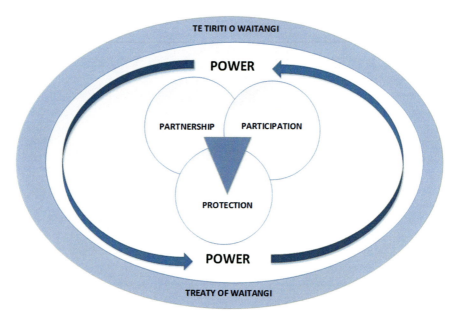

Fig. 2 The four Ps of culturally safe research

...begins from the inception of a research idea when relationships are established with those who belong to the group being researched, and extends to the dissemination of the findings. It is about research participants feeling included, respected, and that they can trust the researchers and what they will do with the information shared about them. (p. 72)

The principles of partnership, participation, and protection, sometimes referred to as the 3Ps, are commonly used to apply the Treaty of Waitangi to research practice, although a fourth 'P' is also presented here – that is, the principle of power which researchers can choose to either use cautiously and responsibly or misuse (Wilson and Neville 2009).

Partnership involves researchers establishing meaningful relationships with the appropriate Māori (tangata whenua (people of the land)/mana whenua (local iwi for whom researchers are intending to research), whānau, hapū, iwi, or Māori communities). Building meaningful and enduring relationships is crucial to the success of the research, and importantly to the credibility of the researchers within Māori communities and the integrity of the research (Smith 2012; Wilson and Neville 2009). Establishing these relationships requires identifying the right people to connect with and ideally meeting face to face. These relationships are not one-off or sporadic interactions, but instead they last throughout, and sometimes beyond, the life of the research. Importantly, established relationships are essential to access and engage participants belonging to vulnerable groups. They know who researchers need to connect with, and who possess the community credibility necessary to involve those people who are suspicious and distrust researchers. Moreover, participating in research is not necessarily a priority for many who are vulnerable, especially when many face daily life challenges.

Māori have culturally-based ethical principles to guide research being undertaken with Māori (Hudson et al. 2010). These ethical principles reflect traditional Māori values associated with tika (being correct and appropriate), pono (being honest and having integrity), aroha (having unconditional respect), manaaki (taking care of others), whānau (collective extended family), and whanaungatanga (establishing relationships and connections), for instance (Mead 2003). Using these ethical principles, in addition to institutional ethical requirements, ensures research participants and researchers alike are kept culturally safe. Therefore, being respectful of the people that you interact with as well as the relevant cultural processes and practices should underpin all actions.

Researchers unfamiliar with Māori tikanga (cultural processes) should undertake early consultation about obtaining the necessary cultural advice and guidance needed to be followed to keep the research team and the research culturally safe. It is crucial that, where possible, researchers arrange to meet face to face with Māori – the known face is important within Māori communities for engagement and data collection. Community-based research assistants who are known within the community can be helpful. It is crucial when engaging with Māori whānau, hapū, iwi, and communities, researchers observe how things are done, observe and listen to what is being said before speaking, acting with humility at all times, and building into research proposals manaakitanga (taking care of research participants) (see Box 3). This may mean that providing kai (food) when going to meetings or interviews as part of the process and offering koha to participants to acknowledge and value participants' contributions and time. It is useful to note that research would not get done without participants' input – being seen to look after and care for participants is noted by Māori, and respected when it is carried out.

Box 3 Ethical Research Values

Values		Application in research
Aroha ki te tangata	Respect people at all times	Always approach Māori participants and people with the utmost respect by establishing mutually beneficial relationships with Māori cultural advisors and act on their advice and guidance
He kanohi kitea	The known face	Researchers who are known to Māori communities and who are respected and seen as credible are more likely to successfully recruit research participants. If researchers are unknown in the community, they then should have people as co-researchers who are known on the research team is essential
Titiro, whakarongo, kōrero	Look, listen then speak	Humility is highly valued in Māori culture, therefore, it is important to observe and listen before speaking. Researchers should avoid assuming they are "the experts" and remember everyone brings expertise to the research partnership

(continued)

Box 3 (continued)

Values		Application in research
Manaaki ki te tangata	Sharing, hosting, and being generous	Taking care of your participants is crucial. Providing kai (food) at meetings or interviews is one way of conveying researchers' understanding of cultural processes and the role of kai and their commitment to looking after them. Also the concept of reciprocity is important, therefore, koha (gifts) are important recognition and the valuing of the time and important knowledge gathered from people for the research
Kia tūpato	Be cautious	To avoid breaching tikanga (cultural processes), it is important to be both politically astute and aware
Kaua e takahia te mana o te tangata	Do not trample on the mana of the people	Upholding the mana of people, no matter how different they are from the researchers, is important. People are born with and acquire mana over their life – it can be diminished by researchers not valuing people and respecting cultural traditions, for instance
Kaua e māhaki	Don't parade your knowledge	As mentioned earlier, researchers having a sense of humility is important – remembering that the people are "experts" on their life and realities they must be treated accordingly and thus respected at all times. Researchers conveying a sense of self-importance and expertise can show they lack the necessary humility to work effectively with the people

Adapted from (Smith 1999)

Researchers need to be mindful and respectful of both individual and collective rights (which may mean, for example, that iwi or hapū consent and support is required for the research to occur in an area in addition to individual participants' consent). It is also important for researchers to recognize the diversity in cultural identity and connection amongst contemporary Māori (as they were prior to colonization). They range from those with strong cultural identities and connections to those who know they are Māori but who are disconnected from their cultural lands and traditions and without the necessary connections to rectify this. The New Zealand Census 2013 shows there are 70,000 more people who claim Māori ancestry than those who identify ethnically as Māori (Statistics New Zealand 2013). This diversity needs to be considered when planning research, and caution must be taken, assuming all participants are culturally connected (know where they are from and who they are) and can speak te reo Māori, for instance.

Box 4 Examples of Ethical Relationship Building

Examples of practice lacking cultural responsiveness	Culturally responsive engagement
Initiating engagement: researchers planned to undertake a study requiring regular input and feedback from a variety of key informants. Because of Māori over-representation in the area of study, it was decided to include Māori key informants. A list of names was compiled and in due course people were contacted by email. A potential Māori key informant expressed dismay and concern about the process used. The research team was distressed by this negative response	Planning to develop research relationships requires researchers to commit to organizing with the appropriate and relevant Māori (be it whānau, hapū, iwi, or community) to initially meet kanohi ki te kanohi (face to face) to discuss planned research and seek their input. Writing letters or sending emails at the beginning is not a culturally appropriate form of engagement. If it is not practical to physically meet with the appropriate people or group, then at a minimum a phone conversation should be had and an explanation provided conveying the researchers' awareness of the preferable need to meet and discuss the request. In similar circumstances, researchers are advised to offer an apology for the process they used to preserve possible future relationships and credibility
Consultation: researchers met with local Māori iwi to discuss their planned research and to obtain evidence of their consultation process required by the funding agency. When they arrived the researchers conveyed an attitude of paternalism and presented themselves as "experts" that Māori should be thankful to have interested. This curtailed the meeting before the researchers expected – they wondered what had happened	A key value for Māori is humility – displaying being knowledgeable about everything and as the experts does not build the confidence in Māori that participants will be looked after during the research or that researchers will listen to what they have to say. Therefore, Māori in this community have decided not to proceed further with the conversation and are highly unlikely to participate in the research
Participation in the research design and process: the principal investigator decided to have preliminary discussions with a local kaumātua (elder) about a research idea. The meeting began with whakawhanaungatanga (connecting by sharing where they were from and who they were). Following this the researcher shared what was intended and invited feedback from the kaumātua. The kaumātua thought the research would be of value but suggested that for it to be of real value to their community, he suggested they slightly altered aspects of what they were planning. He said if these changes were made, their community would support the research and help with the participant recruitment	This example demonstrates the researcher's willingness to engage with the Māori community, establish key relationships, and involve them in the early planning stages of the research. In addition to this, the researcher showed a willingness to respect an important cultural practice – that is, making connections and establishing relationships. This is an illustration of partnership, protection and protection in action, and a willingness of the researcher to use power in a positive and appropriate manner

Participation requires Māori involvement in the entire research process, from the beginning of a research idea to the dissemination and translation of findings for their application. This then optimizes research aligning to the aspirations and needs of vulnerable groups, such as Māori, the outcomes will be meaningful and useful, and Māori are involved in the interpretation of the findings. In addition to consultation and collaboration activities, Māori should also be significant research team members in order to build research capacity and capability.

Protection requires thoughtfulness about how Māori cultural beliefs, values, and practices are identified and understood so that they can be respected and protected during the research process. Furthermore, the benefits researchers reap from undertaking research in the forms of employment, gaining qualifications, the kudos arising from research, and the presentation and publication of the findings in various forums are obvious. It is equally important research is beneficial for those being researched. Thus, researchers need to be clear and transparent about what the actual and potential benefits are for participants and the wider community.

Power is another area which researchers must be mindful about as they possess power that can either be used to benefit participants or be misused. The power researchers wield comes in the form of the decisions they make about planning and designing the research, such as deciding to establish relationships with the appropriate people or to short circuit the process by coercing people who are not necessarily the right people. Moreover, the process for interpreting and reporting research findings and the worldview used to do this are important decisions that can result in beneficial or less than beneficial outcomes. How power impacts the research and the scope for sharing it must be considered by researchers, and ideally is something that is discussed early in the research with Māori.

Using partnership, participation, protection, and power as a framework guides culturally responsive and safe research practice that is situated within the context of Te Tiriti o Waitangi. Such an approach goes someway to promoting research that "gets the story right."

7 Conclusion and Future Directions

The notion of vulnerability is a somewhat contested notion that can imply some form of deficit for those being assigned such a label. In this chapter, vulnerability has been framed as those groups of people being faced with persistent inequities and being subjected to discriminatory research practices that overlook not only their culture but also their everyday realities. High-quality research outcomes are essential if they are to usefully contribute to eliminating inequities and their associated burdens, and importantly achieve health equity. Getting the story right is an essential ingredient for transformative research that makes a difference for indigenous peoples like Māori. Therefore, inclusion of indigenous ways of being that privilege a worldview based on whakapapa (genealogy) which is holistic, spiritual, relational, and ecological in nature is without question. Culturally relevant approaches to research facilitates understanding the sociocultural contexts within which inequities and disadvantage occur, and how these impact on Māori access to social and health services and the achievement of equity. In order to get their story right requires the selection of approaches that accurately reflect the people's voice along with their historical and contemporary realities and that focuses on generating transformational Māori outcomes. Discontent and frustration with research extends to the research methodologies and approaches used to research them as a people (Walker et al. 2006). Kaupapa Māori research, defined as research "by Māori, for Māori, with Māori" (Smith 2012), legitimizes and validates Māori epistemology, ontology, and axiology. Kaupapa Māori Research and Māori-centered research both offer Māori the necessary agency to safeguard the research process and generate knowledge that is valid and useful to Māori Cram (2014).

Culturally responsive and safe research is essential – this can be achieved in part by the inclusion of indigenous researchers and indigenous research methodologies. Nevertheless, this alone is insufficient as research will continue to be done on indigenous peoples by nonindigenous researchers. Culturally responsive and safe research requires researchers to not only consider their personal cultural locations and the imposition this can have on the quality of the research and the methodologies selected but also consider the unconscious biases they may bring to their research practice. It also calls for researchers to make sure that they engage with communities in a culturally appropriate and respectful manner. This requires establishing relationships and involving indigenous peoples early in the research process through to its completion. Within the context of Te Tiriti o Waitangi, the framework of partnership, participation, protection, and power can usefully be used to guide researchers. Culturally responsive and safe indigenous research should primarily be focused on respectful and fruitful research that aims to achieve equity and improve social and health outcomes.

References

Battiste M, editor. Reclaiming indigenous voice and vision. Vancouver: UBC Press; 2000.
Bhopal R. Glossary of terms relating to ethnicity and race: for reflection and debate. J Epidemiol Community Health. 2004;58(6):441–5.
Bishop R. Collaborative research stories: Whakawhanaungatanga. Palmerston North: Dunmore; 1996.
Bishop R. Freeing ourselves from neo-colonial domination in research: a Maori approach to creating knowledge. Int J Qual Stud Educ. 1998;11(2):199–219.
Chilisa B. Indigenous research methodologies. Thousand Oaks: SAGE; 2012.
Cram F. Safety of subsequent children – Māori children and whānau: a review of selected literature for the families commission – Kōmihana ā Whānau research report 2/12. 2012. Retrieved from http://www.superu.govt.nz/sites/default/files/SoSC-Maori-and-Whanau.pdf.
Cram F. Measuring Māori wellbeing: a commentary. MAI J. 2014;3(1):18–32.
Cunningham C. A framework for addressing Maori knowledge in research, science and technology. Pac Health Dialog. 2000;7(1):62–9.
Durie M. Identity, access and Maori advancement. In: The indigenous future: edited proceedings of the New Zealand educational administration society research conference. Auckland/New Zealand: Auckland Institute of Technology/New Zealand Educational Administration Society Research Conference; 1997. p. 1–15.
Durie M. Whaiora: Maori health development. 2nd ed. Auckland: Oxford University Press; 1998.
Durie M. Understanding health and illness: research at the interface between science and indigenous knowledge. Int J Epidemiol. 2004;33(5):1138–43.
Eketone A. Tapuwae: a vehicle for community change. Community Dev J. 2006;41(4):467–80.

Hudson M, Russell K. The treaty of Waitangi and research ethics in Aotearoa. J Bioeth Inq. 2009;6(1):61–8.

Hudson M, Milne M, Reynolds P, Russell K, Smith B. Te ara tika guidelines for Maori research ethics: a framework for researchers and ethics committee members. Auckland: Health Research Council; 2010.

Irwin K. Māori research methods and processes: an exploration. Sites. 1994;28:24–43.

Jahnke HT, Taiapa J. Maori research. In: Davidson C, Tolich M, editors. Social science research in New Zealand: many paths to understanding. Auckland: Longman Pearson Education; 1999. p. 39–50.

Liamputtong P. Research methods i health: foundations of evidence-based practice. South Melbourne: Oxford University Press. 2010.

Makowharemahihi C, Lawton BA, Cram F, Ngata T, Brown S, Robson B. Initiation of maternity care for young Māori women under 20 years of age. N Z Med J. 2014;127(1393):52–61.

Mead HM. Tikanga Māori: living by Māori values. Wellington: Huia; 2003.

Moewaka Barnes H. Kaupapa Māori: explaining the ordinary. 2000. Retrieved from http://www.kaupapamaori.com/assets/explaining_the_ordinary.pdf.

Moewaka Barnes H, McCreanor T, Edwards S, Borell B. Epstemological domination: social science ethics in Aotearoa. In: Mertens DM, Ginsberg P, editors. The Sage handbook of social research ethics. Thousand Oaks: SAGE; 2008. p. 442–57.

Nursing Council of New Zealand. Guidelines for cultural safety, the Treaty of Waitangi, and Maori health in nursing education and health. 2011. Retrieved from http://www.nursingcouncil.org.nz/download/97/cultural-safety11.pdf.

Royal TAC. The woven universe: selected writings of Rev. Māori Marsden. Ōtaki: Te Wānanga-o-Raukawa; 2003.

Ruger JP. Social risk management: reducing disparities in risk, vulnerability and poverty equitability bioethics and human rights. Med Law. 2008;27:109–18.

Sherwood J. Colonisation – it's bad for your health: the context of aboriginal health. Contemp Nurse. 2013;46(1):28–40.

Sherwood J, Kendall S. Reframing spaces by building relationships: community collaborative participatory action research with aboriginal mothers in prison. Contemp Nurse. 2013;46(1):83–94.

Smith LT. Decolonizing methodologies: research and indigenous peoples. Dunedin: University of Otago Press; 1999.

Smith LT. Kaupapa Māori research. In: Battiste M, editor. Reclaiming indigenous voice and vision. Vancouver: UBC Press; 2000. p. 225–47.

Smith LT. Decolonizing methodologies: research and indigenous peoples. 2nd ed. London: Zed Books; 2012.

Statistics New Zealand. 2013 census quick statistics about Māori. 2013. Retrieved from www.stats.govt.nz.

Walker S, Eketone A, Gibbs A. An exploration of kaupapa Maori research, its principles, processes and applications. Int J Soc Res Methodol. 2006;9(4):331–44.

Walter M, Andersen C. Indigenous statistics: a qualitative research methodology. Walnut Creek: Left Coast Press; 2013.

Wilson D. Ngā kairaranga oranga – the weavers of health and wellbeing: a grounded theory study. Wellington: Massey University – Wellington; 2004. Retrieved from http://hdl.handle.net/10179/992.

Wilson D, Haretuku R. Te Tiriti o Waitangi/treaty of Waitangi 1840: its influence on health practice. In: Wepa D, editor. Cultural safety in Aotearoa New Zealand. Melbourne: Cambridge University Press; 2015. p. 79–99.

Wilson D, Hickey H. Māori health: Māori- and whānau-centred practice. In: Wepa D, editor. Cultural safety in Aotearoa New Zealand. Melbourne: Cambridge University Press; 2015. p. 235–51.

Wilson D, Neville S. Culturally safe research with vulnerable populations. Contemp Nurse. 2009;33(1):69–79.

Using an Indigenist Framework for Decolonizing Health Promotion Research

Karen McPhail-Bell, Alison Nelson, Ian Lacey, Bronwyn Fredericks, Chelsea Bond, and Mark Brough

Contents

1 Introduction	1544
2 Maiwah as Place: Place as Ontology	1546
3 Indigenist Research	1548
3.1 Resistance as Its Emancipatory Imperative	1549
3.2 Political Integrity	1552

K. McPhail-Bell (✉)
University Centre for Rural Health, University of Sydney, Camperdown, NSW, Australia

Poche Centre for Indigenous Health, Sydney Medical School, The University of Sydney, Camperdown, NSW, Australia
e-mail: karenmcphailbell@gmail.com

A. Nelson
Allied Health and Workforce Development, Institute for Urban Indigenous Health, Brisbane, Australia
e-mail: alison.nelson@iuih.org.au

I. Lacey
Deadly Choices, Institute for Urban Indigenous Health, Brisbane, Australia
e-mail: Ian.Lacey@iuih.org.au

B. Fredericks
Office of Indigenous Engagement, Central Queensland University, Rockhampton, Australia
e-mail: b.fredericks@cqu.edu.au

C. Bond
Aboriginal and Torres Strait Islander Studies Unit, The University of Queensland, St Lucia, QLD, Australia
e-mail: c.bond3@uq.edu.au

M. Brough
School of Public Health and Social Work, Queensland University of Technology, Kelvin Grove, Australia
e-mail: m.brough@qut.edu.au

© Springer Nature Singapore Pte Ltd. 2019
P. Liamputtong (ed.), *Handbook of Research Methods in Health Social Sciences*,
https://doi.org/10.1007/978-981-10-5251-4_32

   3.3   Privileging Indigenous Voices 1556
4   Conclusion and Future Directions 1559
References 1560

Abstract

This chapter provides a critical reflection on an ethnographic approach led by a non-Indigenous researcher in partnership with an Indigenous community-controlled health organization, and a team of Indigenous and non-Indigenous supervisors, advisors, critical friends, and mentors. The chapter explores the way the three interrelated principles of Indigenist research informed the study, as a critical reflection of the methodology's achievement of a decolonizing research agenda. The flow of Maiwah (the Brisbane River in Australia) provides a metaphor for the chapter's diverse authorship. Maiwah's tributaries, inlets, and banks represent author voices at different points while the one River flowing represents coming together to form a broader collective story of the research that still respects the authors' individual positioning. Maiwah's flow also signifies the dialogical approach of the research – "tricky ground" (Smith 2005) for non-Indigenous researchers seeking to privilege Indigenous voices while remaining accountable to their own White privilege, particularly given that at its most basic level, research requires the "extraction of ideas" from participants. Yet, the flow of Maiwah also shows us the possibilities of research, where in this case, researcher and participants together cocreated new knowledge in support of their agendas. This process enabled both research outcomes and increased research capacity and confidence in the host agency and researcher. On this account, decolonizing research is perhaps more about relationship and devolving control over the process than it is about particular methods, and the respectful negotiation of epistemological meanings and representation of particular knowledges that can result.

Keywords

Indigenist research · Decolonizing methodologies · Health promotion · Indigenous Australians · Non-Indigenous Australians

1   Introduction

Non-Indigenous people largely control health research in Australia. Health research has, therefore, either ignored the presence and needs of Indigenous Australians or has maintained a colonial stance, conducting research on Indigenous people rather than with Indigenous people. While there has been a growth in recognition of this colonial dynamic within health research and a resultant call for a decolonizing of the space (Sherwood and Edwards 2006; Fredericks and Anderson 2013), guidance on how to unfold a postcolonial research practice remains limited. Using a case study of a "decolonizing" health promotion research project conducted by a non-Indigenous researcher in partnership with an Indigenous community-controlled organization, under the supervision of Indigenous and non-Indigenous scholars, this chapter provides a critical reflection of the learning gained from this experience and examples of strategies that may assist others in the decolonizing endeavor.

The argument for improved research processes and ethical frameworks to redress Indigenous health inequality is already well established (Thomas et al. 2014). In Australia, the National Health and Medical Research Council (NHMRC 2003) guidelines emphasize community-driven and community-controlled participatory approaches, while elsewhere the importance of research translation and collaboration has been emphasized (Laycock et al. 2011). Positive change has been taking place in Indigenous research, with governments listening to Indigenous Australians (Fredericks et al. 2011), who are now actively engaged in determining the details of research, including the conditions under which it should take place (Fredericks 2007). Despite this, "Aboriginalism" continues in health research, characterized below:

> ...the story about Aborigines told by whites using only white people's imaginations. Aboriginal voices do not contribute to this story, so in Aboriginalism, the Aborigines always become what the white man imagines them to be. (Walton and Christie 1994, p. 82; cited in Rigney 2006, p. 34)

Decolonization requires a centering of the concerns and worldviews of the colonized "other" (Chilisa 2012; Smith 2012). To do so requires having a critical understanding of the underlying assumptions, motivations, and values informing research practice (Smith 2012). Thus, the decolonization of research involves research reform according to Indigenous peoples' aspirations of empowerment and self-determination (Bainbridge et al. 2015). It involves a process of questioning how the disciplines – in this case, health promotion – "others" Indigenous people by continuing to legitimize the "positioned superiority of Western knowledge" (Chilisa 2012, p. 14; see also Said 1993; see also ▶ Chap. 15, "Indigenist and Decolonizing Research Methodology," ▶ 87, "Kaupapa Māori Health Research," ▶ Chap. 88, "Culturally Safe Research with Vulnerable Populations (Māori)," ▶ Chap. 90, "Engaging Aboriginal People in Research: Taking a Decolonizing Gaze," and ▶ 97, "Indigenous Statistics").

Decolonization is a question for both Indigenous and non-Indigenous people. Decolonizing research involves liberating the "captive minds" of both the colonized and the colonizer from oppressive conditions that silence and marginalize the voices of the colonized (Chilisa 2012, p. 14). It requires a consensus of effort but also recognition of the different voices involved. To provide some practical insight into how this can be approached, this chapter examines a case study of a health promotion research project led by a non-Indigenous researcher in partnership with an Indigenous community-controlled health organization, and a team of Indigenous and non-Indigenous supervisors, advisors, critical friends, and mentors. The reflection provided here explores the decolonizing agenda in health promotion research by drawing on the three principles of Indigenist research (Rigney 1999; see also ▶ Chap. 15, "Indigenist and Decolonizing Research Methodology") using the metaphor of Maiwah, the Brisbane River in Australia.

2   Maiwah as Place: Place as Ontology

A starting point for writing about decolonizing research is to introduce how the lives and worldviews of the authors are intertwined. To do this, we have deliberately chosen to use a metaphor which privileges Indigenous ways of knowing, to introduce each author's position in relation to the Place of this study, the flowing Maiwah. Maiwah meanders through South East Queensland (SEQ), the Place upon which the Institute for Urban Indigenous Health (IUIH) operates (Box 1). IUIH is where the study that has informed this chapter took place. Karen McPhail-Bell was a PhD candidate supervised by Mark Brough, Chelsea Bond, and Bronwyn Fredericks, while Alison Nelson and Ian Lacey were IUIH staff involved in the PhD study. Bronwyn, Chelsea, and Ian identify as Indigenous Australians, while Karen, Mark, and Alison identify as non-Indigenous Australians.

> **Box 1 Overview of IUIH**
> The research site of this study was the Deadly Choices team within IUIH, based in Brisbane, the capital city of the state of Queensland in Australia. In 2009, the four community-controlled health services in SEQ established IUIH, with the leadership and support of the Queensland Aboriginal and Islander Health Council. IUIH's establishment was a strategic response by these bodies to the growth and geographic dispersion of Indigenous populations within SEQ, and their associated under-servicing (IUIH 2011). IUIH maintains that its model for operation is the only one of its kind in urban Australia, providing a way for Indigenous communities to have a voice in improving health outcomes in an urban setting. IUIH has worked with SEQ AMSs to drastically increase the number of clinics and facilities available to Indigenous Australians in the SEQ region. Since 2009, these clinics have grown from 5 to 18, with more planned, and IUIH has expanded from six staff members in 2009 to over 430 employees in 2016 across IUIH and the clinics it established.
>
> IUIH's core business involves coordination and integration of comprehensive primary health care across SEQ, including clinical care, health promotion, and preventative health, all of which work together to achieve its vision of equitable health outcomes for urban Indigenous Australians. IUIH established Deadly Choices in 2010 as part of its response to rising prevalence of chronic disease in Indigenous communities, funded primarily through the Australian Government Closing the Gap initiative. Deadly Choices is IUIH's key health promotion initiative.

The flow of Maiwah provides a metaphor for this chapter's diverse authorship. Maiwah's tributaries, inlets, and banks represent our voices at different points in this chapter (and in the research project), while the one River flowing represents our coming together to form a broader collective story of the research that still respects our individual positioning. As the main watercourse and water source for Brisbane,

Maiwah begins in the Brisbane Range approximately 140 km from central Brisbane. It begins high from water falling from Mount Langley; quietly, gently Maiwah moves downwards, gathering more waters as it travels, growing, and taking different forms and objects on the journey. Maiwah connects with the Stanley River and runs into Lake Wivenhoe, meandering eastward as "a serpentine river which follows you as you drive through the inner suburbs of Brisbane" (Megarrity 2015, p. 101). Maiwah winds and curves, smoothing the rocks it covers, depositing soil, playing with sticks; it widens and deepens, strengthens as a force, as it edges closer to its end. Its character is established, dictating its own path through the Brisbane Maiwah valley and the heart of the city of Brisbane. Maiwah eventually runs into Quandamooka (Moreton Bay) where Stradbroke Island sits in Maiwah's mouth. At this river mouth, salt water and winds wash sand through the channels of Quandamooka into the catchment area of the Brisbane River (Martin 2003). Two bodies of water meet and meld. There are points of difference and points of connectivity, and an energy in the relationship of the waters as they come together.

As Maiwah journeys and connects with land, water, animals, and so on, flowing with various eddies and swirls along its path, so too did this research begin, grow, and connect with people and communities. As is Maiwah's way, this research process built relationships and a project that created a new, shared knowledge. Research always requires negotiation with different stakeholders and their needs. In the case of this chapter, we bring different disciplinary and cultural knowledges, experiences, and viewpoints. Like Maiwah's tributaries, each of us is connected to the other in some way; mutual respect and trust are key as each negotiates the ever-present tensions in trying to engage in decolonizing research. Thus, this chapter is both an expression of different knowledges and viewpoints as well as a demonstration of common themes. Like Maiwah's flowing waters, there were times when we encountered obstacles that required navigating, and this too is part of decolonizing research, which makes the end product broader and deeper (see Box 2).

Box 2 Mark Reflects on His Position as Principal Supervisor in the PhD Research Project

Movement toward postcolonial research requires careful reflection of the "self-other" and an associated questioning of the colonial imaginations of expert/instructor. Here, the neat binary of researcher/researched needs attention too, since the dialogue of this project also involved a team of Indigenous and non-Indigenous advisors as well as interaction with an Indigenous organization which whilst predominantly made up of Indigenous workers, included non-Indigenous workers as well. Thus, the challenge of a postcolonial dialogue criss-crosses multiple overlapping spaces of academia and community. As a non-Indigenous academic in the role of "principal supervisor" a powerful White institutional framework surrounded my own position privileging my authority. I have observed a great deal of "White fragility" (DiAngelo 2011)

(continued)

> **Box 2** (continued)
> within academia when confronted with questions of racism and the "other," hence non-Indigenous researchers attempting to operate cross-culturally within a postcolonial paradigm must be prepared to question both their own fragilities as well the disciplinary and institutional fragilities surrounding them. Indigenous authority is quickly dismantled by White fragility within the dialogic encounter. Maiwah flows best when all tributaries are open and this requires awareness of the ways in which those tributaries can become blocked.

By positioning the research as the flow of Maiwah, we are seeking to engage the principles of Indigenist research (Rigney 1999, 2006) to demonstrate an alternative epistemology and ontology to that more commonly seen in Western research and textbooks. Mary Graham (2006, p. 6) talks about the importance of Place in knowledge for the way it explains "how and why something comes into the world" and provides a balance and rebalance when used like an "ontological compass." Attention to ontology is important for Indigenous and non-Indigenous researchers alike to develop an awareness and sense of self, of belonging and the responsibilities that belonging brings, including how we relate to ourselves and others (Martin 2003).

3 Indigenist Research

Following Maiwah, different authors speak at different times in this chapter. At some points, boxes highlight the voice of a particular author. At other times, the voice of particular authors leads as they write about specific aspects of the research, as tributaries that make the whole river. In this section, the way the three interrelated principles of Indigenist research informed the study is discussed (Rigney 1999). These are:

(a) Resistance as its emancipatory imperative
(b) Political integrity
(c) Privileging Indigenous voices

Together, these principles function to clarify what is and is not meant by Indigenist research; they provide a strategy for research rather than a process (Foley 2003). We recognize that the explicit criteria of Indigenism are the Indigenous identity and colonizing experience of the writer (Rigney 2006). We also recognize that collaboration and participation models involving non-Indigenous researchers in Indigenist research, undertaken with examination of relationships (among the researcher and researched, to knowledge and to self), can lead to greater agency of Indigenous people and thus enact an Indigenist approach (Smith 2001, cited in Martin 2008; see also ▶ Chaps. 15, "Indigenist and Decolonizing Research

Methodology," ▶ 87, "Kaupapa Māori Health Research," ▶ 88, "Culturally Safe Research with Vulnerable Populations (Māori)," ▶ 90, "Engaging Aboriginal People in Research: Taking a Decolonizing Gaze," and ▶ 97, "Indigenous Statistics"). In other words, when based upon self-reflexivity and dialogical characteristics, different types of relationships in collaborations can be meaningful, useful, and powerful, and achieve an Indigenist research agenda (Martin 2008). As an example, the genesis of this research case study involved a discussion between Karen and Bronwyn regarding the prospect of Karen undertaking a PhD in Indigenous health. Karen shared her caution in light of the negative and harmful history of Eurocentric research and practice. Bronwyn responded in a way that for Karen presented an opening that saw the research project beginning 2 years later (McPhail-Bell 2015, p. 8) (see Box 3).

> **Box 3 Bronwyn Emailed Karen Regarding Her Consideration and Caution to Do a PhD in Indigenous Health**
> ...we need good people working with us, not around us and over us...if you do it well then it helps us. If you work with us it helps us. If you look at issues that Aboriginal and Torres Strait Islander people want you to look at then it helps us...To do nothing also maintains the inequities! ... this is what I get upset at some folks for, like if they really wanted to address inequities in Australia they would put a lot more effort into Indigenous health, Indigenous housing, Indigenous employment, Indigenous education etc.... and into their own and other peoples relationships with Indigenous peoples in this country... to leave alone leaves the issues alone and unaddressed... Problem is too many people act in this way and then it doesn't get picked up...and Aboriginal people are left to do it on our own, the sickest are left to do, the poorest are left to do it, the most disadvantaged are left to struggle once again. So go for it Karen, it won't be easy but it will be worth it!!

Following Maiwah's path of connecting and flowing, this section provides a reflection of how this PhD study satisfied the principles of Indigenist research, to contribute to dialogue regarding the partnership of Indigenous and non-Indigenous researchers and practitioners in Indigenist research.

3.1 Resistance as Its Emancipatory Imperative

The first principle of Indigenist research, resistance as its emancipatory imperative, is explicitly concerned with liberation from colonial domination (both in research and society) (Rigney 2006). This principle positions Indigenist research as part of Indigenous peoples' struggle for self-determination. Maiwah's history of flooding shows us the power of resistance in order to travel new territory and bring change over time. Brisbane is built on a flood plain, where flooding is periodic (van den Honert and McAneney 2011). The 2011 Brisbane floods saw heavy, continuous rainfall into the Wivenhoe and Somerset Dams and along the path of Maiwah. When water was released from the dam, Maiwah burst forth, resisting the chartered parameters of the dams, flooding 18,000 properties in metropolitan Brisbane (van den Honert and McAneney 2011). Maiwah's flooding and change over time is like the continued Indigenous resistance to the ongoing colonial practices and attitudes of many non-Indigenous people, including health professionals (Bond et al. 2012; Fredericks et al. 2012; McPhail-Bell et al. 2015). This resistance is about change for the emancipation of Indigenous people. In the research example drawn on for this chapter, there were several ways that resistance was enacted as an emancipatory tool. These included: development of the research question, framing the research outside "typical" health promotion approaches to research, strengths-based orientation, and shared control over the research process and outcomes.

As this research demonstrated, while Indigenous research can be a tool of colonization, it can also be a tool to counter ongoing forms of Indigenous dispossession (Martin 2008). To do this, Indigenist research focuses upon the survival and celebration of Indigenous people's resistance to racist oppression, and to cease continuation of oppression against Indigenous people (Rigney 1999; Foley 2003). An example of this can be found in this study's research question, which the authors argue was emancipatory in nature. The nature of the research question was open-ended to ensure that the PhD research was grounded in the everyday lives of Indigenous people. The research question served to achieve the study's aim to contribute to the decolonization of health promotion practice and accordingly, to recognize Indigenous knowledge, skills, and experience in health promotion. Such an aim was a direct rejection of health promotion's tendency to position Indigenous people as victims requiring outside intervention, and as lacking knowledge and expertise in health promotion. It was also a rejection of health promotion's ahistorical self-positioning regarding colonization and its role in that (McPhail-Bell et al. 2015). To achieve this aim, the research enquired into the daily practice of a cross-section of Indigenous and non-Indigenous health promotion practitioners, in an urban setting of Indigenous-led health promotion. The resulting research question was: *How do health promotion practitioners in an urban Indigenous setting make sense of and navigate the tensions inherent to health promotion in daily practice?*

Note that the research question opened with the presumption that there was something to learn from Indigenous-led health promotion practice for all of health promotion. The presumption of Indigenous expertise also informed a review of the health promotion literature from a decolonizing lens. Researchers are responsible for examining the historical basis of the methodologies and disciplines involved in their research to achieve a decolonizing agenda (Nakata 2007). In this study, the historical basis for health promotion and its relationship to Indigenous peoples was critiqued, including whether it was modeled on learning from Indigenous peoples as equals (McPhail-Bell et al. 2013). Likewise, the basis of the selected methodology (critical ethnography) was reviewed, including the role of non-Indigenous women in its use and in Australia's relations to Indigenous peoples more broadly, given it was a non-Indigenous woman leading the research.

There is an awkward tension for non-Indigenous researchers to negotiate in their efforts to apply decolonizing approaches to research, not only because Indigenist research requires that Indigenous researchers are privileged (Rigney 2006). Nonetheless, this is far from reason to not be involved. Rather, non-Indigenous researchers must come to the research in opposition to the colonial nature of "Aboriginalism" and the unequal power dynamics between Indigenous and non-Indigenous people (Rigney 2006). Thus, as the "research tool" (Madden 2010) it was essential that the lead researcher remain critically reflexive of their own position in the research, given its influence on the research approach and interpretation (see Box 4).

Box 4 Karen's Reflection on Identifying Her Position and Assumptions in the Research

I began the research with particular assumptions, as every researcher does (identified or not). My standpoint is grounded in knowledges of different realities to those of Indigenous people (Moreton-Robinson 2000), so I spent time to articulate my ways of being, doing and knowing (Martin 2003; Moreton-Robinson and Walter 2010). This is because as non-Indigenous researchers, we have a responsibility to investigate our own subjectivities and our society, political and cultural positioning when engaging with Indigenous peoples (Fredericks et al. 2014).

I view Indigenous people not as "other" or "non-other," but as a people whose subject position is "fixed in its inalienable relation to land" (Moreton-Robinson 2003, p. 31). This viewpoint means that I view myself as being in relation to Indigenous sovereignty, which helps me to better understand my complicity with benevolent practices that normalize Whiteness and the construction of people as passive objects (Riggs 2004, p. 8). It also means that I began this study with the assumption that Indigenous-led health promotion carries strengths from which mainstream health promotion could learn. There was no one driver of this assumption but a major reason was my belief in the power of self-determination for people's health. This belief was against a backdrop of mainstream health system failure to effectively act to address poor Indigenous health, and to genuinely partner with Indigenous peoples and communities. My assumption of strength and its associated analysis could allow for possibilities to move beyond colonial control, towards a place of mutual respect (Said 1978).

To be emancipatory, the research required a theoretical framework to guide the work to be process driven, as well as outcome oriented (Rigney 2001). Such a framework supported the research project and its aim by drawing on a decolonizing approach. In practice, this translated to being strengths-based and guided by theoretical tools that enabled the research to move towards decolonizing epistemologies and identify the manifestation of racism in research practices and knowledge production (Rigney 1999). These theoretical tools were: critical race theory and postcolonialism to position the lead researcher, and health promotion as a discipline and practice; and the cultural interface, to move beyond the constraint of a pre-defined and reactive anticolonial agenda (Nakata 2007). As a whole, the framework aligned with a decolonized approach by critiquing the dominance of a Euro-Western paradigm and inquiring into how knowledges converge and evolve through their daily enactment. Together, these tools enabled the decolonizing approach to give voice to the researched and move from a deficit-orientation (Smith 2012), along with expansion of assumptions that underpin existing health promotion theory and practice.

3.2 Political Integrity

Political integrity, the second principle of Indigenist research, highlights the importance of Indigenous ways of being and knowing based on Indigenous philosophies, cultural values, and beliefs (Rigney 2006). This principle is reflected in the use of Maiwah as a tool for describing the ways in which we have merged our different knowledges and experiences to produce this chapter and the research underpinning it, rather than drawing on a more "standard" Western knowledge system to describe this work. The research presented throughout this chapter also highlights the processes through which a non-Indigenous researcher can maintain this political integrity when working in partnership with Indigenous participants and supervisors.

Political integrity in Indigenist research requires that there is a social link between research and the political struggle of Indigenous communities (Rigney 1999). It emphasizes the need for Indigenous communities to build their own capacity mechanisms to gain the benefits of research (Rigney 2006). However, while Rigney (2006) argues that political integrity necessitates Indigenous researchers conduct the research, he also acknowledges that non-Indigenous researchers can draw upon these principles if upholding the struggle for genuine self-determination of Indigenous people. The methodologies chosen thus also need to enable a countering of racism and privileging of Indigenous knowledges and experiences for Indigenous emancipation (Rigney 1999).

Fundamental to political integrity in Indigenist research is the premise that the researcher must negotiate with the participants at all levels of the research design, including data collection, analysis, and reporting of findings (Rigney 1999). To do this, there was a need for the researcher to be vigilant in learning to operate in decolonizing ways through significant self-reflection, listening, and learning. Relationality formed the basis of the research methodology; it is ethical to do so and part of a decolonizing approach (Martin 2008; NHMRC 2003; Nicholls 2009). This included spending approximately 12 months building relationships and informing potential participants about the research, including what a PhD would involve. In practical terms, this meant volunteering, attending team meetings and other work activities, and finding ways to support the work of the team being researched (see Box 5). It also necessitated an investment of time to learn about the priorities and needs of those involved, and embed these into the research. This approach required an open agenda for collaborative research planning to take place, embedding the research within the team so that it could adapt to their needs and requirements as the research progressed (see Box 6). The important factor was time in "being there" with potential participants to build relationships, and collaboratively design the research (see Box 7). This is consistent with building trust and cooperation as core tenets of maintaining political integrity in Indigenist research (Rigney 2006).

> **Box 5 Ian Reflects on the Presence of Karen as a Researcher with Their Team**
> Karen was very helpful; she wasn't a researcher; she was just an extra set of hands, just helping. And when we have our community days, any extra sets of hands that we have are very helpful for the team. So most of the time, I didn't see Karen as a researcher, when she attended community days or programs or events; so, that's a positive thing. We didn't have to segregate her and say, "You can't do this task; you're a researcher." She just jumped in and was part of the team. Once you're part of the team and everyone's helping each other out, it just flows. Karen was helping out on the days and would come back and feed back to us and we weren't even aware that she was observing us most of the time. We knew she was there to do the research but once we started the days, and she wasn't sitting back with a notepad and just doing that sort of thing, it just became easier; Karen was just there and part of the team. So, that was helpful to us and helpful to Karen.

> **Box 6 Alison Nelson Reflects on the Relationship-Building Phase of the Research**
> I think the beginning of that process was very much about relationship-building and trust... it was a pretty vague concept when it first started and I think that's a key thing for research in this area, or service delivery, especially if you're starting something new, is that there will be periods of time when it's really uncertain and there's a lack of clarity and it's a bit vague and a bit unclear... I also think that that's part of the success because I think if Karen had come to these guys and gone, "I'm going to do exactly this, this, this and this," it probably wouldn't have worked as effectively as her saying, "I've got this idea of what I'd like to do and then let's just see how it works. I'm just going to build relationship with you and become part of the team. I don't have any agenda really at the moment, I'm just getting to know you guys." That was then what led to the trust, and also helped inform Karen about what might be useful for her research. And I think if you go into that space with this preconceived idea, you're less likely to get that trust back from the team that you're working with... but Karen approached it from that trust and building relationship.

> **Box 7 Ian Reflects on the Initial Stages of the Research and the Way That Karen Worked with the Team**
> Karen was there, helping us at community events; we weren't thinking, "Oh that's Karen, she's evaluating us here." She was just there as part of the team and would come back and feed us back something about that day and we'd be like, "Oh yeah, that's right, we did do that and that was the methodology behind us doing it that way." So, that was good. It was a learning curve for me to understand fully what the research was going to be about and that probably took me a 6 month period to give me the confidence to then instil that into the team, and then be confident enough to speak up at team meetings to say, "When Karen needs something done you need to put it forward" – you know, the participant diaries and all that sort of stuff. If I didn't have the confidence in what was happening, I wouldn't have been able to give that confidence to the team. So, I think that was part of the learning curve.

The research approached decolonization of knowledge by supporting self-determining processes by Indigenous people – including by working with processes of IUIH, as an organization based upon Indigenous self-determination – and opposing colonial structures of knowledge production (Evans et al. 2014). Critical to this approach was learning about and maintaining Indigenous ideals, values, and philosophies as core to the research agenda (Rigney 2006). This meant working within a framework of an Indigenous-led and community-controlled organization and assisting the team being researched to simultaneously describe and understand their ways of knowing and doing, without predetermined theoretical or methodological constraints (see Box 7).

In practice, informal conversations, presentations, and feedback papers during fieldwork were used to inform the team about research learnings and interpretations. These were also used to inform a dialogue with participants and IUIH to ensure the research findings and interpretations reflected and respected the knowledge and experiences of participants. This approach accorded the participants the right to speak for themselves and engage in their own self-reflection in research (Rigney 2006). Regular communication about the research and PhD processes itself, including seeking permissions from participants and IUIH before including that information in public documents, was also part of working in an ethical, decolonizing way. The principle of political integrity was also enacted by supporting the team involved to develop its own practice framework and research capacity, as requested by IUIH (see Boxes 8 and 9).

> **Box 8 Ian Reflects on the Strengths-Based Nature of the Research**
> I think Karen just helping us believe that what we were doing was true to the research. We weren't doing something that wasn't of value; we weren't doing

(continued)

Box 8 (continued)
something that wasn't for the community. The research helped reinforce that we were on the right path. That was one of the biggest things for me because we weren't very good at research and evaluating programs and evaluating effectiveness in the community. We knew we had a model there that was good for the community but we didn't have any facts to reinforce that. That was one of the biggest things for us: the research helped reinforce that we were on the right path. It gave us the self-belief to go the next step with it, to say, "Righto, you've got a good model, the community's buying into it, now we got the evidence behind it – what's the next step to make it even more powerful?" It just helped reinforce that our model was working; and you need that academic validation when you're talking back to governments and others (funding providers... Broncos, etc).

Box 9 Ian Reflects on the Impact of the Research for Him and His Team
The team members have all really improved. Look at (team member A), (team member A's) doing an MBA now. (Team member B) has enrolled in university. (Team member C) has enquired about university. So, all this sort of stuff on the back of what has happened here in the research has been really beneficial for our team. And I think the research has also given tools: it's helped us as a team to evaluate our strengths and our weaknesses, and what plans we can put in place to minimize risks and make sure we are doing the right stuff in the community. It was really beneficial, I think, the whole experience: it instilled a real self-belief in the team. It did because like me and (team member A) personally, when we first started, we had no idea what research was but you look at us now: we're evaluating everything, we're making sure that research is tied into everything we do because it's all evidence-based, what we do. It has really instilled a real self-belief in us personally and as a team. It has helped us be who we are.

Fundamentally, to maintain political integrity, non-Indigenous researchers must support the work of Indigenous communities and their researchers (Rigney 2006). Decolonizing research might also be understood as anticolonial research, where non-Indigenous researchers work as allies with Indigenous people (Max 2005). Advocates of anticolonial research argue that it must be initiated, directed, and controlled by Indigenous people and be of direct benefit to those involved (Max 2005). The principles of ownership, control, access, and possession are self-determination applied to research (Schnarch 2004). Non-Indigenous people can work in collaborative ways with Indigenous people, which will also result in greater critical self-reflection of one's own privilege and position (Max 2005).

The Indigenist research principle of political integrity also maintains a significant tension for non-Indigenous researchers who continue to have a function in representing and speaking for the participants. In this research, while acknowledging that Karen did "speak for" the participants in her PhD thesis, there was also an effort to "speak to" in dialogue with those involved to be accountable and responsible for what was said (Spivak 1988; Alcoff 1991). Mutual respect and power sharing were enacted as further expressions of political integrity through clear acknowledgement of researcher biases implicit in Karen's researcher standpoint. Personal stories were woven into the academic work, recognizing that narratives enable a deep understanding of what it means to be located at the cultural interface (Young 2001). Being a non-Indigenous researcher seeking to learn from Indigenous peoples is a point of contention and a matter with which all researchers must contend; this is, as Linda Smith says, "tricky ground" (Smith 2005, p. 114).

3.3 Privileging Indigenous Voices

Indigenist research focuses on the lived, historical experiences, ideas, traditions, dreams, interests, aspirations, struggles; in other words, it gives voice to Indigenous people (Rigney 1999). The way that inlets and tributaries assemble to form the flowing Maiwah provides a reminder of the importance of privileging Indigenous voices. When tributaries reach the river, they can be diluted in the larger river. Moving water is a powerful force, as is the power of hegemonic Whiteness, which works to silence Indigenous voices (Moreton-Robinson 2006). Therefore, integral to this research has been the use of different ways of privileging Indigenous voices and knowledges, including through engagement with the supervisory team (see Box 10).

> **Box 10 Chelsea Bond Reflects on Her Role on the Supervisory Team**
> Health research interest in Indigenous Australia, both in its absence and its presence has and continues to serve the colonial project. Control over the lives of Indigenous peoples is maintained through the production of knowledge that evidences Indigenous inferiority even in the most benevolent endeavours of "closing the gap" of Indigenous health inequality. The supposed relinquishment of biological notions of race has not hindered the reproduction of racial hierarchical arrangements via Indigenous deficit; socially, culturally, psychologically and intellectually. Within health research Indigenous peoples and knowledges are frequently silenced, subjugated and exploited and this rings true not just for Indigenous peoples as the subjects of study, but also for Indigenous scholars situated within investigative and supervisory teams.
>
> Our presence and position as Indigenous scholars in and of itself does not halt the reproduction of racialized logics, and in fact Rigney (2001) notes

(continued)

> **Box 10** (continued)
> that race continues to be fundamental to the Indigenous scholar's oppression within the research enterprise. This oppression is evidenced by the relegation of the Indigenous scholar to the role of cultural broker/mentor on supervisory or investigative teams. In this role, the Indigenous scholar is cast as a supporting actor providing auxiliary support to the "real" knowledge producers. Configuring the Indigenous scholar as the cultural steward is an effective strategy for authorizing non-Indigenous participation in and knowing of Indigenous social worlds while also maintaining colonial assumptions, which insist that "Indigenous traditions of intelligentsia equate to 'Intellectual Nullius'" (Rigney 2001, p. 4). Recognition of Indigenous intellectual sovereignty does not require the appointment of an Indigenous only cast of researchers, rather it demands a displacement of non-Indigenous peoples from their position of knowing us exclusively and absolutely. It requires a reconfiguring of research questions away from trying to find what's inherently wrong with us, to what can be learnt about us and from us. It requires an acknowledgment of and engagement with Indigenous people as thinkers, knowers and experts whether they are study participants, community stakeholders, partners, co-investigators or supervisory team members.

The silencing of Indigenous voices is evident in research epistemologies and ways of knowing in Australia, where there is little evidence of learning from Indigenous peoples (Rigney 1999). To counter this trend, this research required continued prioritization of Indigenous autonomy and control over the knowledges, languages, and cultures involved. This autonomy and control was achieved through partnership and participatory action. For example, in addition to the planning phase already described, the study design positioned participants as co-researchers, whereby using multimedia participant diaries they could determine what data the study should examine and why. It also required the lead researcher's attention to language to ensure research terminology was presented in meaningful language to participants involved (see Box 11), and that participants' language was reflected in the research dissemination (for example, in the PhD thesis). The team members involved report this process has increased their research capacity and confidence to now determine their own research priorities and agendas (see Box 8 above). Fieldwork involved an ongoing dialogue regarding the researcher's interpretation and representation, which required her to *listen*: an essential value of ethical and respectful praxis (Sherwood 2010, p. 35). That the research prioritized Indigenous control and autonomy meant that her position as White, non-Indigenous researcher and mainstream practitioner was destabilized, where IUIH and the participants carried power regarding their knowledge and their representation in the research. This destabilization is something with which non-Indigenous researchers must contend (see Box 12).

Box 11 Ian and Alison Reflect on the Way Karen Integrated Academic Language and Concepts into Her Communication with the Team

Ian: From my point of view, it was really beneficial for my team to see it from Karen's perspective. The guys do their work, day in, day out, and they didn't realize; the research was happening but they didn't understand that it was being translated in a way, the way Karen did it. I think it was really empowering for them. I think it made them realize that, "hey I am doing my job," and it gave them the strength to continue on and build on this. And I think that proved in the feedback to Karen also, in the interviews she did with them. It was a really empowering experience and as I said, the words that Karen uses in her thesis, she didn't use those words with our guys. That really helped them – it just helped them understand they are part of a research project that is on their lives, their work lives. That really empowered them to become who they are now.

Alison: That's my observation too. I saw a real shift in people's confidence. And even though Karen didn't use the same academic language, she used enough academic language in the way that she was talking that it then empowered people to know what that language meant. So, it wasn't that she was completely removing the academic from it either, she was bringing it but in a way that was accessible. That's my main observation: that increase in confidence. Ian used the word empowering, which I totally agree with, but I think it has helped the team feel confident about what research is and that they are engaged in research and they are doing it and they're confident in doing it.

Box 12 Mark Reflects Upon the Importance of Dialogue in Decolonizing Research

Methodology textbooks provide valuable guidance on how to undertake research; however, they can struggle to convey the human experience of doing research. I was very aware that any social research has this very personal edge to it and often becomes part of the supervisor–student dialogue as research proceeds, but also aware that it often remains a hidden corner in more formal written discussions of methodology. In this project, the attempt to create a decolonizing approach to research was always going to involve more than a mechanical manual of instruction, and would need ideas, processes, interpretations, and ethical judgements produced from dialogue as the research proceeded. The dialogue needed to be collaborative, equal, and respectful, but critically aware that the backdrop of Whiteness would never be conducive to this.

It was important that this research privileged Indigenous voices not just in data collection or PhD production but also in the ways in which the research was represented and disseminated. This included negotiations with IUIH regarding their representation in the PhD thesis, including whether participants and IUIH chose to remain anonymous (as is standard practice in maintaining the confidentiality of participants) or to be named in order to highlight successful practices. There was a tension at times in navigating the needs and requirements of completing a PhD and its theoretical requirements, while simultaneously conducting research that was of value and represented IUIH in a way which "rang true" to the organization's philosophy and practice. These discussions took time and required an ongoing reexamination of the research and its theoretical underpinnings and practices but resulted in mutual respect and understanding of each stakeholder's position and voice. Ongoing collaboration continues in order to disseminate the research findings, which requires creative attention to privileging Indigenous voices in a range of formats and products.

This privileging of Indigenous voices also extended to the writing of this chapter. Like the tributaries forming Maiwah, we have each provided input based on our own knowledges and positions and written in a way that reflected our experiences rather than needing to conform to a particular "way" of producing an academic work. The use of the boxes throughout the chapter has illustrated one way of alerting the reader to each author's "voice" but so too have the use of Maiwah as a metaphor and Indigenist research principles as a guiding framework.

4 Conclusion and Future Directions

As our journey with Maiwah reaches its end in Quandamooka, a place is reached where salt and fresh water mixes, where different people and knowledges mix, to create something new. Almost in a process of renewal, this story with Maiwah shares how an ethnographic approach led by a non-Indigenous PhD candidate in partnership with an Indigenous community-controlled health organization, and a team of Indigenous and non-Indigenous supervisors, advisors, critical friends, and mentors can be enacted as Indigenist research. From this journey, this chapter proposes a set of principles that may assist others in their decolonizing journey:

1. Know yourself: your assumptions; your way of being, doing, and knowing; the basis of your disciplinary training and socialization; the limitations of your knowing. Find ways to remain accountable to this.
2. Recognize Indigenous people as knowers.
3. Decolonizing research takes time and hard work. Commit to this.
4. Decolonizing research requires relationship with Indigenous peoples and communities involved, not only participants. Build the relationship early. Be accountable to this relationship after the research is completed in terms of how the research is used.

5. Collaborate with potential participants from the beginning to form research priorities, parameters, and processes based upon their strengths and priorities. Pay attention to the research question and process just as much to outcomes.
6. Know the power dynamics of the research methodology and methods you use, particularly in relation to their colonial uses, and how your positioning exacerbates those dynamics. The same applies to your examination of the literature.
7. Negotiate research governance and feedback processes that acknowledge Indigenous sovereignty and ownership of Indigenous knowledge and intellectual property. This includes agreement on expected research products and uses of those products.
8. Ensure the process and the product/s are based on reciprocity; this includes the dissemination process.

Partnership made possible the bringing together of different knowledges and expertise in this research, to accomplish more than what could be accomplished alone (Bainbridge et al. 2015). This research project is one example of how decolonizing research can take place; it is not a template for decolonizing processes. Rather, the example teaches us that decolonizing research is more about relationship and devolving control over the process than it is about particular methods, and the respectful negotiation of epistemological meanings and representation of particular knowledges that can result.

Acknowledgments We acknowledge the support of the Institute for Urban Indigenous Health and Deadly Choices.

References

Alcoff L. The problem of speaking for others. Cult Crit. 1991;20:5–32.

Bainbridge R, Tsey K, McCalman J, Kinchin I, Saunders V, Lui FW, … Lawson K. No one's discussing the elephant in the room: contemplating questions of research impact and benefit in aboriginal and Torres Strait islander Australian health research. BMC Public Health. 2015;15:696.

Bond C, Brough M, Spurling G, Hayman N. 'it had to be my choice' indigenous smoking cessation and negotiations of risk, resistance and resilience. Health Risk Soc. 2012;14(6):565–81.

Chilisa B. Indigenous research methodologies. London: SAGE; 2012.

DiAngelo R. White fragility. Int J Crit Pedagog. 2011;3(3):54–70.

Evans M, Miller A, Hutchinson P, Dingwall C. Decolonizing research practice: indigenous methodologies, aboriginal methods, and knowledge/knowing. In: Leavy P, editor. The Oxford handbook of qualitative research. Oxford: Oxford University Press; 2014. p. 179–91.

Foley D. Indigenous epistemology and indigenous standpoint theory. Soc Altern. 2003;22(1):44–52.

Fredericks B. Utilising the concept of pathway as a framework for indigenous research. Aust J Indigenous Educ. 2007;36(Supplement):15–22.

Fredericks B, Anderson M. Aboriginal and Torres strait islander cookbooks: promoting indigenous foodways or reinforcing Western traditions? In: Paper presented at the Peer reviewed proceedings of the 4th annual conference. Brisbane: Popular Culture Association of Australia and New Zealand (PopCAANZ); 2013.

Fredericks B, Adams K, Edwards R. Aboriginal community control and decolonizing health policy: a yarn from Australia. In: Lofgren H, De Leeuw E, Leahy M, editors. Democratizing health: consumer groups in the policy process. Cheltenham: Edward Elgar Publishing Limited; 2011.

Fredericks B, Lee V, Adams M, Mahoney R. Aboriginal and Torres Strait islander health. In: Fleming ML, Parker E, editors. Introduction to public health. 2nd ed. Sydney: Elsevier Australia; 2012. p. 350–72.

Fredericks B, Maynor P, White N, English F, Ehrich L. Living with the legacy of conquest and culture: social justice leadership in education and the indigenous peoples of Australia and America. In: Bogotch I, Shields CM, editors. International handbook of educational leadership and social (in)justice. New York: Springer; 2014. p. 751–80.

Graham M. Introduction to Kummara conceptual framework on place: a discourse on a proposed aboriginal research methodology. Brisbane: Kummara Association Inc; 2006.

IUIH. Intitute for Urban Indigenous Health: strategic plan 2011–2014. Bowen Hills: Institute for Urban Indigenous Health; 2011.

Laycock A. with Walker D, Harrison N, Brands J. Researching Indigenous health: a practical guide for researchers. Melbourne: Lowitja Institute; 2011.

Madden R. Being ethnographic: a guide to theory and practice of ethnography. London: SAGE; 2010.

Martin K. Ways of knowing, being and doing: a theoretical framework and methods for Indigenous re-search and Indigenist research. J Aust Stud. 2003;27(76):203–14.

Martin K. Please knock before you enter – Aboriginal regulation of outsiders and the implications for researchers. Teneriffe: Post Pressed; 2008.

Max K. Anti-colonial research: working as an ally with Aboriginal peoples. In: Sefa Dei GJ, Johal GS, editors. Critical issues in anti-racist research methodologies. New York: Peter Lang; 2005. p. 79–94.

McPhail-Bell K. "We don't tell people what to do" – an ethnography of health promotion with Indigenous Australians in South East Queensland. Doctoral thesis, Queensland University of Technology, Kelvin Grove. 2015. Retrieved from https://eprints.qut.edu.au/91587/.

McPhail-Bell K, Fredericks B, Brough M. Beyond the accolades: a postcolonial critique of the foundations of the Ottawa charter. Glob Health Promot. 2013;20(2):22–9.

McPhail-Bell K, Bond C, Brough M, Fredericks B. "We don't tell people what to do": ethical practice and Indigenous health promotion. Health Promot J Austr. 2015;26(3):195–9.

Megarrity L. Sport, culture and ideology at the Brisbane regatta (1848-73): Aboriginal-European relations in an emerging colonial city. J Aust Colon Hist. 2015;17:101–14.

Moreton-Robinson A. Talkin' up the white woman: Indigenous women and white feminism. St. Lucia: University of Queensland Press; 2000.

Moreton-Robinson A. I still call Australia home: Indigenous belonging and place in a white postcolonising society. In: Ahmed S, editor. Uprootings/regroundings: questions of home and migration. New York: Berg Publishing; 2003. p. 23–40.

Moreton-Robinson A. Whiteness matters. Aust Fem Stud. 2006;21(50):245–56.

Moreton-Robinson A, Walter M. Indigenous methodologies in social research. In: Walter M, editor. Social research methods. Melbourne: Oxford University Press; 2010. p. 1–18.

Nakata M. Disciplining the savages: savaging the disciplines. Canberra: Aboriginal Studies Press; 2007.

NHMRC. Values and ethics: guidelines for ethical conduct in Aboriginal and Torres Strait Islander health research. Canberra: Australian Government; 2003. Retrieved from http://www.nhmrc.gov.au/_files_nhmrc/file/health_ethics/human/conduct/guidelines/e52.pdf.

Nicholls R. Research and Indigenous participation: critical reflexive methods. Int J Soc Res Methodol. 2009;12(2):117–26. https://doi.org/10.1080/13645570902727698.

Riggs D. Benevolence and the management of stake: on being 'good white people'. Philament(4): Untitled. (2004). Accessed 2015. http://www.philamentjournal.com/issue4/riggs-benevolence/.

Rigney L-I. Internationalization of an Indigenous anticolonial cultural critique of research methodologies: a guide to Indigenist research methodology and its principles. Wicazo Rev. 1999;14(2):109–21.

Rigney L-I. A first perspective of Indigenous Australian participation in science: framing Indigenous research towards Indigenous Australian intellectual sovereignty. In: Paper presented at the chamcool conference. Alberta; 2001.

Rigney L-I. Indigenous Australian views on knowledge production and Indigenist research. In: Kunnie J, Goduka N, editors. Indigenous peoples' wisdom and power: affirming our knowledge. Burlington: Ashgate Publishing Ltd; 2006. p. 32–48.

Said EW. Orientalism. New York: Vintage Books Edition; 1978.

Said EW. Culture and imperialism. London: Vintage; 1993.

Schnarch B. Ownership, control, access, and possession (OCAP) or self-determination applied to research. A critical analysis of contemporary first nations research and some options for first nations communities. J Aborig Health. 2004;1(1):81–95.

Sherwood J. Do no harm: decolonising aboriginal health research. Unpublished Doctor of Philosophy thesis. Sydney: University of New South Wales; 2010.

Sherwood J, Edwards T. Decolonisation: a critical step for improving Aboriginal health. Contemp Nurse. 2006;22(2):178.

Smith LT. On tricky ground: researching the native in the age of uncertainty. In: Denzin N, Lincoln YS, editors. The Sage handbook of qualitative research. 3rd ed. Thousand Oakes: SAGE; 2005. p. 85–107.

Smith LT. Decolonizing methodologies: research and Indigenous peoples. 2nd ed. London: Zed Books; 2012.

Spivak GC. Can the subaltern speak? In: Nelson C, Grossberg L, editors. Marxism and the interpretation of culture. Basingstoke: Macmillan Education; 1988. p. 721–313.

Thomas D, Bainbridge R, Tsey K. Changing discourses in Aboriginal and Torres Strait Islander health research, 1914–2014. Med J Aust. 2014;201(1 Supplement):S15–8.

van den Honert R, McAneney J. The 2011 Brisbane floods: causes, impacts and implications. Water. 2011;3(4):1149–73.

Young R. Postcolonialism: an historical introduction. Carlton: Wiley-Blackwell Publishing; 2001.

Engaging Aboriginal People in Research: Taking a Decolonizing Gaze

90

Emma Webster, Craig Johnson, Monica Johnson, Bernie Kemp, Valerie Smith, and Billie Townsend

Contents

1 Introduction	1564
2 Background	1565
3 Design: "Who Else Should We Be Talking To?"	1566
4 Data Collection: "We Wanted It to Have a Social Feel"	1568
5 Analysis: "All that Work for Just a Few Pages?"	1570
6 Findings: "I Have a Good Relationship with My Patients, But they Have Never Talked to Me About that Before"	1571
7 Research Impact: "How Have Aboriginal People Benefitted from the Research?"	1572
8 Applying an Active Decolonizing Gaze: Conclusion and Future Directions	1575
8.1 Tips to Facilitate Participation of Aboriginal People in Your Research	1577
8.2 Things We Could Have Done Better	1577
References	1578

E. Webster (✉) · B. Townsend
School of Rural Health, Sydney Medical School, University of Sydney, Dubbo, NSW, Australia
e-mail: Emma.Webster@sydney.edu.au; billie.townsend5591@gmail.com

C. Johnson
Dubbo Diabetes Unit, Dubbo, NSW, Australia
e-mail: Craig.Johnson1@health.nsw.gov.au

M. Johnson
Marathon Health, Dubbo, NSW, Australia
e-mail: Monica.Johnson@marathonhealth.com.au

B. Kemp
Dubbo Regional Aboriginal Health Service, Dubbo, NSW, Australia
e-mail: berniek@dubboams.com.au

V. Smith
Formerly with Dubbo Regional Aboriginal Health Service, Dubbo, NSW, Australia
e-mail: valjanesmith@yahoo.com

© Crown Copyright 2019
P. Liamputtong (ed.), *Handbook of Research Methods in Health Social Sciences*,
https://doi.org/10.1007/978-981-10-5251-4_33

Abstract

A criticism of some research involving Aboriginal people is that it is not equitable in its design or application, further disadvantaging the poor and marginalized. In Australia, much research has been done on Aboriginal people, but Aboriginal people themselves have benefited little, adding to distrust between Aboriginal and non-Aboriginal people over many years. Is it possible to take "scientific" research practices and transform them into research that can be done **with** a community rather than **on** a community? How can research findings benefit Aboriginal people? This chapter shares our study of Aboriginal people's stories of diabetes care. It is a collaborative story told by four Aboriginal Health Workers and two non-Aboriginal researchers which focuses on methodology rather than findings. We share aspects of our research which we propose values Aboriginal people and invites participation and reciprocity at design, data collection, and research translation stages. We discuss tensions which occur between the "scientific way" and the "culturally appropriate way" and describe how we resolved this. Imposing research designs and practices on Aboriginal people and communities without consideration that each community is unique has the potential to cause further harm and disempowerment. Valuing an Aboriginal way of knowing influenced all aspects of our study design and procedures. We would like to inspire others to challenge methodological norms to develop research methods with their community to allow the unique voice of their community to be heard and for this to facilitate pragmatic change leading to meaningful improvements in health.

Keywords

Aboriginal people · Australia · Participatory research · Community engagement · Social participation · Qualitative research

1 Introduction

> ... [C]olonisation continues today both politically and through health service provision, research and scholarship. This is because the context of causal agents that have impacted upon the health of Indigenous Australians.... have been maintained through problematic constructions of Aboriginal people that were established when the concept of terra nullius was applied to this continent. If we all take up an informed and active decolonizing gaze we can shift this colonial context. It is time to make the change; the knowledge is out there, to stop blaming Indigenous Australians for their health circumstances, and to contribute to providing the very best health care, research and scholarship to the first Australians. (Sherwood 2013, p. 37)

A criticism of research in Aboriginal health is that it is invasive, inappropriate, and unnecessary and undertaken without community consultation (Aboriginal Health & Medical Research Ethics Committee 2013). In Australia, much research has been done on Aboriginal people, but Aboriginal people themselves have benefited little, adding to distrust between Aboriginal and non-Aboriginal people over many years.

Empirical research has not been able to deliver positive health benefits to Aboriginal communities as research has often focused on Western ways of knowing and doing and ignoring the power imbalance that Western research creates between subject and researcher (Prior 2007; Kendall et al. 2011). Sherwood (2013) explains problematic constructions of Aboriginal people have persisted since terra nullius (land belonging to no one) and urges academics to take an "active decolonizing gaze" to research and scholarship to allow Aboriginal ways of knowing to shape research design and application (see also ▶ Chaps. 15, "Indigenist and Decolonizing Research Methodology," and ▶ 89, "Using an Indigenist Framework for Decolonizing Health Promotion Research").

This chapter shares a study of Aboriginal people's stories of diabetes care. It is a collaborative story told by four Aboriginal Health Workers and two non-Aboriginal researchers which focuses on methodology rather than findings. The "active decolonizing gaze" was taken to the usual "scientific" research practices to transform them into research that can be done **with** a community rather than **on** a community. The chapter outlines aspects of our research which we propose value Aboriginal people and invite participation and reciprocity at design, data collection, and research translation stages. It is designed to be an intensely practical work which describes tensions which occurred between the "scientific approach" and the "culturally sensitive approach" and discussed how we resolved this.

This chapter is written for academics and practitioners who would like to challenge methodological norms to develop research methods with their community to allow the unique voice of their community to be heard. Research findings can then facilitate pragmatic change and meaningful improvements in health.

For the ease of the reader, we have described our research in the sequence one would expect in a research report. However, the stages were generally not as clearly defined and were often concurrent rather than consecutive. Tensions are highlighted where they occurred in the chronology of the research.

2 Background

Dubbo is a city in regional New South Wales with a population of 40,000 people of whom 5,000 are Aboriginal (Australian Bureau of Statistics 2011). Dubbo is part of the Wiradjuri nation. There are many Aboriginal nations in Australia and Aboriginal people from many different nations reside in Dubbo, often as a direct result of colonization and dislocation from their Country in past generations. In Australia, cultural protocols exist relating to who can talk on behalf of or make decisions for Aboriginal people. Elders are highly respected as leaders and teachers and are active in the decision-making process. Becoming an Elder is earned through respect rather than an entitlement associated with age or family history. Consensus decision-making is often preferred and even a respected Elder from one nation does not speak on behalf of a community from another nation.

The opportunity for research came from a whole of locality integrated care pilot project to improve diabetes care for Aboriginal people (NSW Department of Health

2015). Partners included a regional public health service provider, the local Aboriginal Community Controlled Health Service, and the private primary care sector (then called Medicare Locals). A Steering Group with membership from all these organizations determined research to identify the model of care experienced by Aboriginal people in Dubbo should be undertaken so the model could be improved. The Steering Group also determined Intellectual Property and any accolades or benefits arising from the research would belong to all organizations and that all would need to endorse the research at key stages such as ethics and the publication of findings.

Each organization nominated one or more staff members to be part of the research team, and the research was to be conducted alongside usual clinical work. The research team consisted of two male Aboriginal Health workers (Bernie Kemp and Craig Johnson) who were qualified diabetes educators, two female Aboriginal Health Workers (Valerie Smith and Monica Johnson) and two non-Aboriginal women (Emma Webster and Billie Townsend). Between the team, there was over 60 years' experience in health and qualifications in nursing, diabetes education, research, and history. Individually, we were all novices in at least one aspect (research, diabetes, or cultural knowledge). Team members knew only one or two of the other team members, but all had strong community connections to Dubbo. Only one of the Aboriginal team members was Wiradjuri.

3   Design: "Who Else Should We Be Talking To?"

The first research team meeting focused on getting to know each other. Having strong existing community connections assisted in this process as we could identify common interests and connections quickly. Establishing trust was important as this allowed each individual to participate fully with no risk to their personal, professional, or cultural identities.

Tension 1: Risk to Identity The professional identity of the individual is an important element of the "scientific approach" versus a "culturally sensitive approach" where personal, cultural, and professional identities are all important and there is a strong sense of responsibility to other Aboriginal people.

Team members clarified during the initial meeting that they had agreed to participate in the research because they wanted to see an improvement in diabetes care for their patients. The meeting concluded with discussion about research and some brainstorming about data collection methods which would be engaging for Aboriginal participants.

Tension 2: Motivation for Research Producing knowledge is a valid reason to undertake research when coming from a "scientific approach," whereas this is not valued when a "culturally sensitive approach" is taken as it neglects the need for reciprocity. In this case, the research took knowledge, but put something back by improving diabetes care for Aboriginal people.

The philosophical underpinning of research was discussed as a group in that first meeting, not by using the accepted academic terms, but by discussing what was important to allow Aboriginal people to feel comfortable to participate in the research. The research was constructivist, accepting people create and recreate a social world (Crotty 1998), and it was decided to privilege Aboriginal cultural knowledge wherever this would make participants or researchers more comfortable.

Grounded theory (Birks and Mills 2015) was chosen as we were trying to depict the model of care experienced by Aboriginal people, and participatory methodologies (Bowen 2015) reflected well how we wanted to work as a research team. Choosing these approaches had the overall effect of equalizing the power between all researchers, who each took the lead when their various skills were best able to contribute to the research. This approach challenged existing research systems as ethics applications, manuscripts for publication, and conference presentations cannot be submitted without a lead investigator being identified. Yet, we had chosen to work collaboratively and not privilege any individual or assume that one investigator could speak on behalf of all. This was resolved to some extent by creating a collective identity for ourselves, the "Dubbo Aboriginal Research Team," for conference presentations. However, research ethics and publishing systems still required a single contact and this role was fulfilled by the person with most research experience, as dealing with the research process and associated forms was not as alienating as they would have been for other members.

Tension 3: Lead Investigator A lead investigator is expected for research undertaken using a "scientific approach" including ethics applications and publications. A "culturally sensitive approach" required a collective identity which did not privilege any individual.

Cultural protocol is important, and it was respectful that elders and key agencies not only know about the research, but also had the opportunity to shape its design. Regular meetings with the research team addressed the question *who else should we be talking to?* which brought involvement of community into the design of the study. Members of the research team visited the Local Aboriginal Lands Council, Dubbo Aboriginal Community Working Party (both are formal entities with local cultural authority), and the Koori Yarning Group (an informal group where elders come together for social interaction) in the design stages. This helped shape the approach to participants, focus group structure, and incentives for participants. These groups kept the research focused on delivering benefits to Aboriginal people with diabetes.

Tension 4: Design of the Research The researcher with the most research experience would lead research design in a "scientific approach," whereas our "culturally sensitive approach" actively asked community members to direct the design stage. We did this by asking *who else should we be talking to?* and incorporated their advice. In this way, cultural knowledge was valued alongside research knowledge.

4 Data Collection: "We Wanted It to Have a Social Feel"

An example of privileging cultural knowledge can be seen in our approach to inviting participants to be part of the study. Our research team attended community groups popular with Aboriginal people to introduce ourselves and explain the research. Members of the team went back to these groups on two to three occasions to build rapport and answer questions. Names and phone numbers of potential participants were collected and they were contacted when the next focus group was planned. This approach was chosen as potential participants would be more comfortable to decline being part of the research once trust had been established as it reduces the fear and any perceived power relationship.

Tension 5: Approach to Participants The usual "scientific approach" would suggest the least coercive approach to participants is that the approach is made by a person unknown to the participant. In our "culturally sensitive approach," we determined the invitation to participate should come from a known person. This was seen as less coercive as the potential participant feels they can say "no" without penalty when they trust the researcher. The research team sought to establish this trust by meeting potential participants in environments where participants felt a sense of belonging and ownership.

Focus groups were piloted with a group of Aboriginal Health Workers. This enabled the research team to practice the roles of facilitator, organizer, and note taker and test that the questions would generate effective discussion. The pilot ran well, with excellent feedback and encouragement from the pilot "participants" which ensured the research team was comfortable that the approach would be successful and that there was clarity around the roles. This proved valuable as not every research team member could be present for all of the data collection events, and the team was easily able to change roles or help each other with the role. Focus groups were always facilitated by at least one Aboriginal member of the team.

Transport to and from focus groups was offered to all participants as many did not have their own transport and public transport was not readily available. Transport was provided by members of the research team. The focus groups were held in a local sporting facility on the riverbank, which, while central to town, is part of the natural environment. This was considered a "neutral" environment by Aboriginal people and had been the location of an Aboriginal reserve in years gone by (reserves were parcels of land set aside by the government where Aboriginal people were expected to live) (NSW Government Office of Environment and Heritage 2012). Some of the participants recalled growing up on the reserve.

The focus groups themselves had a strong emphasis on social relationships and reciprocity. Sessions started with a light snack and a cup of tea and conversation with all research team members in order for people to get to know each other. The research was explained to each person individually and informed consent given and demographic details collected. Focus groups were followed by a barbeque lunch and a "Feltman" (Australian Indigenous Health Infonet 2015) education session conducted by one of the Aboriginal Health Workers who was a qualified diabetes educator. "Feltman" is a life-size person shaped felt wall hanging designed specifically for diabetes education. "Feltman's" internal organs relevant to diabetes are clearly shown, and symbols for sugar can be moved around the body to give a pictorial description of the way hormones like insulin and glucagon work to store and release sugar in the body. The "Feltman" sessions focused on answering questions which had been raised in the focus groups, so each was customized to the needs of each focus group. Participants received a gift bag at conclusion of the session containing written information about diabetes management and the importance of having regular health checks. A voucher to a local fruit and vegetable shop to the value of $10 was also included.

Tension 6: Reciprocity in Focus Groups The usual "scientific approach" would involve the researcher giving the participant a nominal incentive in return for data from or about the individual. In our "culturally sensitive approach," reciprocity involved the exchange of information and knowledge. This was achieved by ensuring focus groups were followed by a diabetes education session tailored to the needs of each group. Social relationships were valued and enhanced with the opportunity to meet over morning tea and socialize at lunch time.

Five gender-specific focus groups were conducted (3 male, 2 female, total participants $n = 25$, $n = 12$ male, $n = 13$ female) as it was thought that gender-sensitive issues might be raised. It is believed that men's business and women's business should be kept separate in Aboriginal culture, and it can therefore be inappropriate to discuss some of these issues in the presence of people from the other gender.

Focus groups were run by the facilitator, who dealt out six "conversation cards" so a different participant had a turn to start with their story. Each conversation card had a different numbered question written upon it, and the participant would read out the question and answer it before other group members shared their thoughts. For example, the card two question was "Who were the people who helped you understand your diabetes?" The group facilitator had a master list of all the questions and could help read the question where required and prompt further exploration. A visual diabetes education tool called a conversation map (Reaney et al. 2012, 2013) was displayed to give visual prompts of the usual aspects of diabetes care. Visual representation of medications, types of diabetes, support groups, feelings, learning about blood sugar levels, complications, and exercise are just some of the visual prompts on the conversation map. Some of the participants were familiar with the conversation map as they had experienced this as part of previous diabetes education sessions. Both the conversation cards and the conversation map made the focus group very interactive with little direction required by the facilitator. Discussion in the focus groups was very rich, with audio recordings running from 85 to 108 min (see also ▶ Chap. 103, "Conducting Focus Groups in Terms of an Appreciation of Indigenous Ways of Knowing").

5 Analysis: "All that Work for Just a Few Pages?"

All researchers coded the first transcript collectively. This was time intensive but meant that all members had a good understanding of the process and that the initial codes had a high level of interrogation before being labeled. Combinations of the research team coded the remaining transcripts. All transcripts were open coded by hand and then cut up and sorted. This made the coding process visual and tangible and allowed intellectual input from all members of the research team in a way that coding on a computer could not.

In grounded theory, constant comparison of codes and incidents is made as each focus group is conducted and new data are added (Birks and Mills 2015; Wong et al. 2017). The slow process of the group coding the first transcript meant that we had conducted our fifth focus group before finishing coding the first. We incorporated constant comparison of each subsequent focus group transcript, but this is not how grounded theory is meant to be done as it is not possible to test emergent theory in subsequent data collection events. The constant comparison process also drives theoretical sampling, where participants with certain characteristics are intentionally chosen to determine if the emergent theory holds (Birks and Mills 2015; Wong et al. 2017). In our case, we attempted theoretical sampling of two newly diagnosed people with diabetes, but this was unsuccessful. Grounded theory purists might argue that it is no longer grounded theory if these tenets cannot be followed. The decision to continue to collect data before having completed coding the first focus group was balanced against the importance of participatory methodologies and cultural values rather than the absolute primacy of grounded theory. It was decided that going with the community momentum and goodwill for the study, being inclusive of the research team in the coding process, and finally not approaching additional newly diagnosed patients were all justifiable. This last point was imperative as Dubbo's only Aboriginal Health Worker's qualified in diabetes education were members of the research team. Establishing a trusting therapeutic relationship with those newly diagnosed cases was determined to be more important than the research.

Tension 7: Divergence from Accepted Methodology The "scientific approach" to grounded theory privileges the method (constant comparison and theoretical sampling), whereas our "culturally sensitive approach" prioritized the enthusiasm for focus groups in the community and the need to include all research team members in the coding of the first transcript over the formal method.

Memo writing is a key analytic tool in grounded theory (Birks and Mills 2015; Wong et al. 2017). Three types of memos were used in this study. The first documented the group debrief after each data collection event, the second documented group discussion of coding and categories, and the third encapsulated a very brief overview ("top line" memo) of our thinking at occasional intervals. Visual diagrams of codes and categories were constructed during meetings, and these were photographed to document the evolution of the model of care experienced by Aboriginal people. The purpose of memo writing in grounded theory is to document the development of the thoughts of the researcher. Most of the analysis happened in discussion during meetings rather than the writing in between meetings, and the writing and pictures simply documented meeting discussion.

6 Findings: "I Have a Good Relationship with My Patients, But they Have Never Talked to Me About that Before"

We had assumed at the commencement of our study that the model of care delivered by health services and health professionals would be the key factor in Aboriginal people coming to learn and manage their diabetes. Instead, we found that this was one of four factors, with the other three being the continuing effects of colonization on physical activity, nutrition and bush medicine, the power of learning about diabetes directly from family, and positive and negative interactions with the health service. This was an important finding for the research team as a whole, as it was new information not previously heard by members of the research team despite their experience, qualifications, and long and enduring relationships with many Aboriginal patients with diabetes. This was attributed to the difference between research-style questioning which is open-ended and positions the participant as the expert, rather than the clinical-style questioning which positions the health professional as the expert. In addition, the nonclinical environment, the social feel of the focus group, and the conversation map put participants at ease and facilitated the discussion on a broad range of topics.

A brief summary of findings yielded by this participatory and culturally sensitive approach is described below. A more detailed description and analysis can be found elsewhere (Webster et al. 2017).

Colonization was found to have a continued effect on health and lifestyle, with contemporary policies restricting desired hunting and fishing practices and dislocation from country resulting in lost access to bush tucker.

> The government bloke was there trying to tell us that if we wanted to go out and shoot a roo and eat it, we had to apply to Canberra first, and one old fella (said)... 'My kids want it tonight not next week or the week after'. (Travis FG1 920–929)

Seeing family members with diabetes shaped participant's views on diabetes. Participants were well aware of diabetes and associated complications and learned about what diabetes was and how to manage it predominantly from family members through the cultural practice of intergenerational learning. The high prevalence of diabetes in Aboriginal families increased exposure to this experiential learning.

> I saw my mother go through hell ... My mother died and my grandmother, her mother. She was totally blind by the time she was 60 and now it's all reflecting on me. (Bianca FG4 88–92)

While some interactions with health services had been very positive, many participants described feeling stereotyped by non-Indigenous health professionals. Participants perceived they were not being heard and that the lack of cultural understanding by health professionals resulted in their reluctance to seek further help.

> One of the worst things I come across is because I'm dark skinned, they think I drink and smoke, I never drank or smoked in my life. And the first thing they say is, they don't even ask the question of me 'You will have to give up the grog and smokes', and I said, 'Mate I've never tried it in my life'. (Ross FG1 1064–1067)

Where participants described receiving the general practitioner model of care, they found it highly acceptable. Pharmacists and Aboriginal Health Workers were the other health professionals who participants described as helping them learn about diabetes.

> Really, really nice... I go every three months and I have a long appointment because first of all I go to the nurse then I go to (the GP) and the nurse weighs you and does all your sugars tells you everything and that and then from there I go to (GP). (Felicity FG4 223–226)

While many participants described difficulty finding support and sticking with programs, many made lifestyle adjustments after their diabetes diagnosis. All participants had goals relating to diabetes, and while there were various levels of control and understanding, many were extremely motivated to master aspects of the disease.

> In the supermarkets I'm forever reading the packets, what do they call it? Carbohydrates 'cause carbohydrates is sugar as well and then underneath you can see sugar, and I always look at that to see how low it is before I buy it... I'm still doing things wrong, and I'm still learning. (Daniel FG3 217–222)

This study provided an explanation for health practitioners, services, and systems to understand how Aboriginal people learn about and manage diabetes (see Fig. 1). The model of care was a small part of learning to understand and manage diabetes for Aboriginal people. Other influences were historical factors such as colonization and witnessing family members suffering from diabetes and interaction with the health service.

Health system and service improvements which build on to the cultural acceptability of intergenerational learning such as supporting holistic patient centered and family centered models of care were recommended. Improving cultural knowledge of non-Aboriginal health professionals and ensuring Aboriginal voices can be heard in priority setting at a service level will improve service credibility with the Aboriginal community. Employing more Aboriginal people in professional roles and more Aboriginal Health Workers who are qualified diabetes educators helps Aboriginal patients feel they are understood.

7 Research Impact: "How Have Aboriginal People Benefitted from the Research?"

Group writing and presentation of research findings happened throughout the research. The focus on presenting the work from the beginning ensured group writing tasks commenced early in the research process. Group discussion and agreement shaped how presentations would proceed and what would be included. The preferred presentation style was collaborative, with up to five of the research

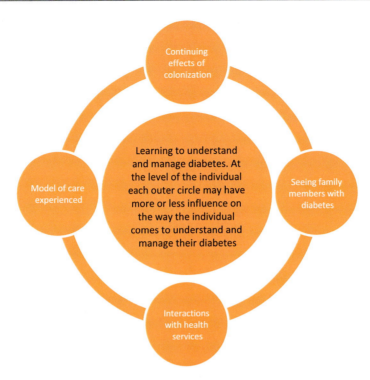

Fig. 1 Theory which explains an Aboriginal perspective of learning to understand and manage diabetes

team involved in a single presentation. All presentations were scripted to ensure multiple presenters did not detract from overall coherency for the audience. All members of the research team were involved with checking and practicing scripts and presentations. This actively involved research team members and contributed to the joint development of ideas and understanding as the research progressed. The presentations were visual and included many photographs and diagrams. Participant quotes read by Aboriginal colleagues were embedded as audio files to bring the cultural tradition of oral knowledge to the presentations.

Regular formal communication with the Dubbo Aboriginal Community Working Party continued throughout the research in addition to informal verbal updates with key individuals. The Working Party kept the research team focused on the importance of improvements in diabetes care for Aboriginal people. The research team presented findings to local hospital and primary health care medical, allied health, and nursing professionals. Presentations were also made to state and national clinical and academic audiences (see Fig. 2 Dissemination summary for detail). A formal report was prepared for partner organizations and a manuscript was prepared for peer review.

For many research projects, achieving the dissemination summary described (see Fig. 2) would be sufficient and signal the final phase of the research. For our study,

Oral presentation of research in progress or research findings or both

- Local Aboriginal Community Working Party
- Local Aboriginal Diabetes Support Group and Carers Support Group
- Local health services (hospital and primary care settings for medical, allied health and nursing staff)
- Diabetes integrated care working party (membership included senior managers, clinicians and patient representatives)
- Local Health District Aboriginal Health Worker Forum
- National Rural Health Conference
- NSW Rural Health and Research Congress

Written publication of research findings

- A full report was circulated to the organizations who formed part of the original working party and the Research Ethics committees who endorsed the research
- Publication in a peer reviewed journal

Fig. 2 Dissemination summary

this did not fulfil the mandate to improve diabetes care for Aboriginal people desired by research team members, the Dubbo Aboriginal Community Working Party, and others.

There have been a number of improvements thus far. An Aboriginal Chronic Disease Support Group has been established, led by Aboriginal Health Workers. Meetings are held monthly in a community setting and a number of agencies work collaboratively to make meetings a success by referring patients, transporting participants, arranging morning tea and lunch and monitoring blood pressure, blood sugar, and weight. Health professionals from different health services attend as guest speakers and stay for the duration of the session to build relationships with participants. A local general practitioner attends regularly, bringing a medical student on clinical placement when available. Support group members report they feel listened to and appreciate the concern shown by the health professionals. They also regard it as a privilege to be involved in the education of future medical personnel. The running of the Support Group represents much stronger collaboration at a service level than had happened previously and has provided a forum for Aboriginal people to learn in a family centered way from each other as well as from health professionals.

Members of the research team have been invited to join a Dubbo-wide leadership group to improve diabetes care. The focus of the leadership group is on improving communication and referral between services, locality-wide workforce education, and service reform. Some small local service changes have taken place, such as Aboriginal Health Worker's being able to contact general practitioners at a single medical practice directly when concerned about the need for medication changes for patients.

Other benefits from the research included the development of research skills for members of the research team and an increased interest in research generally. There has been an increase in the respect for Aboriginal Health Workers by both the community and the services they work for. The Dubbo Aboriginal Community Working Party has suggested their members work with medical staff and students at the local university to develop "yarning style" medical consultations and has invited one of the Aboriginal Health Workers from the research team to continue to provide regular updates on progress. Yarning is a relaxed, sharing conversation between two or more people. Yarning might include finding out more about the person and finding common interests, understanding each other's community or kinship connections and sharing stories.

Tension 8: Value of Published Findings The usual "scientific approach" to research would value publication of research findings in the academic literature as the most important research output. In this way, the "scientific approach" values intellectual property highly. Our "culturally sensitive approach" showed that presentation of research findings back to the community and decision makers and the subsequent changes in practice was most important. The culturally sensitive approach, therefore, valued meaningful improvement in service delivery over publication of findings.

We cannot provide unequivocal evidence to show locality-wide improvements in diabetes care for Aboriginal people as a result of this research. However, the research has raised the profile of diabetes care for Aboriginal people, members of the research team continue to be actively involved in service improvements and the research continues to influence decisions made at an individual and a service level. This extension of involvement beyond publication is part of the expectation of reciprocity and giving something back.

Research impact measurement for studies undertaken with Aboriginal communities continues to develop, with questions about whose right it is to determine research impact and meaningful measures of capacity building, engagement, and reciprocity at the center of the debate (Tsey et al. 2016). This has been discussed by the research team who determined that while good progress has been made, we have not achieved enough improvement in diabetes care to discharge our sense of responsibility to participants or each other.

8 Applying an Active Decolonizing Gaze: Conclusion and Future Directions

This chapter set out to describe how we brought a "decolonizing gaze" to the usual "scientific" research practices. Prior (2007) proposes colonization by research occurs when research restricts cultural representation by dominating and suppressing the other. Colonization by research could, therefore, occur at any or every stage of the research process. Decisions of what can be considered authentic knowledge (epistemology), suitable approaches to inform how the research should proceed (methodology) and ways of collecting data (methods) all have the potential to dominate and suppress. Likewise, explicit or implicit judgments are made about values (axiology) and morals (ethics).

By contrast, decolonized research would be directed and undertaken by Aboriginal people consistent with their world view, specifically to benefit the participants and guided by the values of spirit and integrity, reciprocity, respect, equality, survival and protection, and responsibility (Prior 2007; Aboriginal Health & Medical Research Ethics Committee 2013; Sherwood 2013; Gray and Oprescu 2015). Judged by these criteria, we have fallen short. Honoring the decolonizing gaze was difficult, as many structures such as organizational decision-making (where policy and managers determine priorities), applying for ethics and publication of research findings (requiring a lead researcher) undermined our efforts to work collaboratively and to minimize the power relationships both among the research team and with the community.

Yet, taking a "decolonizing gaze" facilitated participation of Aboriginal people in our research. We were successful in embedding reciprocity in both methodology and method. There was a focus on the value of collective knowledge, the importance of relationships and culturally sensitive data collection methods. Aboriginal researchers developed research skills and non-Aboriginal researchers enhanced their cultural knowledge. It will be possible to build on these successes by continuing long-term research partnerships between health services, academia, and the community. Theoretical models such as engaged scholarship which is driven by principles of reciprocity, mutual benefit, and multidirectional learning to solve social, civic, and ethical problems may prove useful (Bowen 2015). Participatory research has an explicit commitment to collaborative value generation, co-creation of design, and co-production of knowledge (Higginbottom and Liamputtong 2015; Jackson and Greenhalgh 2015). It is assumed strong research transfer occurs when using this approach.

Like others (Kendall et al. 2011), we have found a need to be flexible through the conduct of the research. It was not possible to anticipate all rules for the conduct of the research in advance. Tensions between a "scientific approach" and a "culturally sensitive approach" were faced at each stage of the research process and were resolved by discussion among the research team privileging the cultural comfort of participants, Aboriginal members of the research team, and the Aboriginal community. This approach facilitated involvement of Aboriginal people in research.

Imposing research designs and practices on Aboriginal people and communities without consideration that each community is unique has the potential to cause further harm and disempowerment. Applying the "decolonizing gaze" meant valuing an Aboriginal way of knowing which influenced all aspects of our study design and procedures. Participatory research methods allowed high quality, outcome-driven research to be undertaken with a team of people who have varying levels of cultural understanding and experience with research. Using participatory methods facilitated Aboriginal community involvement in the design and ultimately to benefit from the research.

We would echo Sherwood's call to researchers. There is enough knowledge out there and it is time to apply our scholarship, community links, and good will to bring about change to research processes to drive improvements in health outcomes for Aboriginal people.

We provide below some tips which facilitated participation of Aboriginal people in our research. We have also included reflection on what could have been done better so the reader can consider and incorporate into future research.

8.1 Tips to Facilitate Participation of Aboriginal People in Your Research

- Become familiar with cultural protocols for your community. Ask the question of each other and those you consult with "Who else should we be talking to?"
- Think of experience, knowledge, and qualifications as a collective as this strengthens confidence and trust that you will know how the research should proceed when the time comes
- Commit collectively to a philosophy in the beginning and trust that this will support you all to make the best decision when needed
- Commit collectively to how you want meetings to proceed. In our study, we tried to make meetings suit everyone, but ultimately if someone could not attend, the meeting still went ahead with the absent member catching up with the details at a later stage
- Use visual prompts in data collection, such as the conversation map and conversation cards, to enhance interaction in focus groups
- Present on the research as a group throughout the research process as often as you can as this involves everyone in the analytical process and facilitates the evolution of your thinking
- Involve everyone in the coding and allow more time as it will take longer, but it also maximizes intellectual input from all members
- Use visual methods such as diagrams, white boards, and photographs to document collective progress with analysis
- Develop community connections before you start thinking about research

8.2 Things We Could Have Done Better

- Our approach to reflexivity involved group discussion of what worked well and what could be improved. This could have been enhanced by reflexive practice at an individual level if we had explored what was uncomfortable or comfortable and reflecting on own beliefs, knowledge, and assumptions
- In hindsight, our focus groups did not need to be gender specific as both men and women showed concern for each other and were interested in hearing experiences and opinions of the other gender
- While using Grounded Theory facilitated a broader understanding of diabetes and was an entirely sensible approach, it was a difficult methodology for novice researchers because of the complexity of the terminology and to some extent the rigidity of the form. Ease of methodology for the research novice should be considered when designing the study.

References

Aboriginal Health and Medical Research Council Ethics Committee. AH&MRC guidelines for research into aboriginal health: key principles. 2013. http://www.ahmrc.org.au/media/resources/ethics/ethics-application-resources/271-ah-mrc-guidelines-for-research-into-aboriginal-health-key-principles-1/file.html. Accessed 6 Nov 2016.

Australian Bureau of Statistics. 2011 Census QuickStats. Dubbo LGA (C). 2011. Retrieved from http://www.censusdata.abs.gov.au/census_services/getproduct/census/2011/quickstat/LGA12600?opendocument&navpos=220. Accessed 6 Nov 2016.

Australian Indigenous Health Infonet. Feltman (2010). 2015. http://www.healthinfonet.ecu.edu.au/key-resources/promotion-resources?lid=19500. Accessed 6 Nov 2016.

Birks M, Mills J. Grounded theory: a practical guide. 2nd ed. London: Sage; 2015.

Bowen S. The relationship between engaged scholarship, knowledge translation and participatory research. In: Higginbottom G, Liamputtong P, editors. Participatory qualitative research methodologies in health. London: Sage; 2015. p. 183–99.

Crotty M. The foundations of social research: meaning and perspective in the research process. Crows Nest: Allen & Unwin Australia; 1998.

Gray MA, Oprescu FI. Role of non-indigenous researchers in indigenous health research in Australia: a review of the literature. Aust Health Rev. 2015;40(4):459–65.

Higginbottom G, Liamputtong P, editors. Participatory qualitative research methodologies in health. London: Sage; 2015.

Jackson CL, Greenhalgh T. Co-creation: a new approach to optimizing research impact? Med J Aust. 2015;203(7):283–4.

Kendall E, Sunderland N, Barnett L, Nalder G, Matthews C. Beyond the rhetoric of participatory research in indigenous communities: advances in Australia over the last decade. Qual Health Res. 2011;21(12):1719–28.

NSW Department of Health. NSW integrated care strategy. 2015. http://www.health.nsw.gov.au/integratedcare/Pages/Integrated-Care-Strategy.aspx. Accessed 6 Nov 2016.

NSW Government Office of Environment & Heritage. Living on Aboriginal reserves and stations: research resources for Aboriginal heritage. 2012. http://www.environment.nsw.gov.au/chresearch/ReserveStation.htm. Accessed 6 Nov 2016.

Prior D. Decolonizing research: a shift toward reconciliation. Nurs Inq. 2007;14(2):162–8.

Reaney M, Eichorst B, Gorman P. From acorns to oak trees: the development and theoretical underpinnings of diabetes conversation map education tools. Diabetes Spectr. 2012;25(2):111–6.

Reaney M, Gil Zorzo E, Golay A, Hermanns N, Cleall S, Petzinger U, Koivisto V. Impact of conversation map education tools versus regular care on diabetes-related knowledge of people with type 2 diabetes: a randomized, controlled study. Diabetes Spectr. 2013;26(4):236–45.

Sherwood J. Colonization- It's bad for your health: the context of Aboriginal health. Contemp Nurse. 2013;46(1):28–40.

Tsey K, Lawson K, Kinchin I, Bainbridge R, McCalman J, Watkin F, et al. Evaluating research impact: the development of a research for impact tool. Front Public Health. 2016;4:Article 160. https://doi.org/10.3389/fpubh.2016.00160.

Webster E, Johnson C, Kemp B, Smith V, Johnson M, Townsend B. Theory that explains an Aboriginal perspective of learning to understand and manage diabetes. Aust N Z J Public Health. 2017;41:27–31.

Wong P, Liamputtong P, Rawson H. Grounded theory in health research. In: Liamputtong P, editor. Research methods in health: foundations for evidence-based practice. 3rd ed. Melbourne: Oxford University Press; 2017. p. 138–56.

Space, Place, Common Wounds and Boundaries: Insider/Outsider Debates in Research with Black Women and Deaf Women

91

Chijioke Obasi

Contents

1 Introduction	1580
2 The Research Project	1581
2.1 Research Design	1581
2.2 "Africanist Sista-Hood in Britain": An Evolving Theoretical Framework	1581
2.3 Learning from the Pilot Study	1584
3 Data Collection	1585
3.1 Access	1585
3.2 Interviews	1586
4 Inside Looking Out or Outside Looking In?	1587
4.1 Situating the Researcher in the Research	1587
4.2 Situating Participants on the Insider/Outsider Continuum	1589
5 Connecting to Africanist Sista-Hood in Britain	1592
5.1 Names, Self-Naming, and Naming Identity	1592
5.2 Space and Place in Africanist Sista-Hood	1595
6 Conclusions and Future Directions	1597
References	1598

Abstract

The chapter discusses issues of identity in research. It does this by examining the impacts of the identity of the researcher, participants, and the various identity interchanges that take place. This chapter draws on the perspectives and experiences of participants and researcher in a PhD study with five (Six Deaf women were interviewed but one withdrew due to a conflict of interest.) culturally Deaf (white) women and 25 Black (hearing) women discussing their world of work in UK public sector organizations. The theoretical framework of "Africanist Sistahood in Britain" is that which underpins the positioning of the research and

C. Obasi (✉)
University of Wolverhampton, Wolverhampton, UK
e-mail: Chijiokeobasi@blueyonder.co.uk

researcher. The chapter provides a reflexive account of the research but in a way that centralizes participant perspectives. Two goals have been achieved; firstly, it adds further contribution to the insider/outsider debate by adding participant perspectives on the issue, and secondly, it demonstrates the ways in which the theoretical framework of "Africanist Sista-hood in Britain" can be used in research not just with Black women but also via collaborative approaches with other social groups. In so doing, the chapter raises a number of important questions: Should researchers seek out participant perspectives on the insider/outsider debates in research? In what ways does the identity interchange between researcher and researched have an impact on the research process? What does Africanist Sista-hood in Britain have to offer to Black women and others carrying out research in the field?

Keywords

Insider/outsider research · Black women · Deaf women · Africanist Sista-hood · Black feminism · Womanism · Africana womanism · Deaf research

1 Introduction

In this chapter, I discuss the way a researcher's identity as a Black female, hearing researcher has an impact on the research processes with Black women and culturally Deaf women within the same PhD study. Central to the chapter are discussions of participant perspectives on insider/outsider researcher positioning and "sameness" and "difference" in research. The chapter adds significant contribution to existing discourses by centralizing participant perspectives, as well as adding discussions on Deaf/hearing insider/outsider positionings which have largely been absent from the field. It introduces the framework of "Africanist Sista-hood in Britain" which underpins the study. It uses participant contributions to make links to the central tenets as a source of validation of the developing framework. Alliances are also formed in Deaf cultural discourse in an attempt to foreground cultural and linguistic understandings of Deaf people. In working with these two participant groups, neither Deaf cultural discourse nor Africanist Sista-hood would be adequate in isolation. However, in creating theoretical alliances between the two, there is relevance in the frameworks for analysis for both participant groups.

The Deaf women in the research are culturally Deaf women. Within deaf studies literature, there is much debate about Deaf identity and the ways in which culturally Deaf people form a cultural and linguistic minority group that differs from dominant hearing perceptions of deaf people as disabled. In incorporating cultural Deaf discourse, I write about the distinct Deaf communities who refer to themselves less as disabled as is the dominant hearing articulation, rather their own lived reality is that of members of a cultural and linguistic minority who share a pride in their signed language and cultural norms that are distinct and in some cases in opposition to that of the majority hearing society (Ladd 2003; Lane 2005; Bahan 2008; Padden and Humphries 2005; Bauman and Murray 2014a). Deaf pride in their culture and the efforts made to maintain and nurture it are absent from the dominant constructions of Deaf people as disabled. When writing about culturally Deaf signing individuals or communities, the convention introduced by Woodward (1972) is to write the word Deaf with a capital D and so maintaining a distinction from deaf people outside of this culture who would not describe themselves in this way. This is the convention I will follow in this chapter.

The chapter starts by introducing the research project with discussions on research design, access, pilot study, data collection, and changes to the design post-pilot. Integral to this are discussions of the theoretical framework, its foundations, influences, and scope for research. The chapter draws heavily on participant data to discuss the views that Black women and Deaf women shared on "insider/outsider" positions in research. It does this in a way that incorporates discussions of researcher/researched identity, positive aspects of assumed *common bonds* that can occur between Black female participants and Black female researcher (Johnson-Bailey 1999), and the parameters that should guide hearing with Deaf research.

In the final section of this chapter, participant data is used to validate the theoretical framework of Africanist Sista-hood by demonstrating the ways in which individual identity of Deaf women and Black women is still connected to historical identities of their collectives. It draws on examples from the data about names and self-naming, place and space, and the way this is regulated both physically and professionally both for culturally Deaf women and Black women within their work spaces.

2 The Research Project

2.1 Research Design

The research forms part of a PhD study focussing on perspectives and experiences of equality and diversity in the UK public sector. It seeks to examine whether the rhetoric provided by public sector organizations are validated via the experiences and perspectives of Black women and Deaf women who work within them. It falls within an interpretivist paradigm using a framework of Africanist Sista-hood in Britain, Deaf cultural discourse, and using qualitative research methods.

The primary focus of the study is Black women working in the public sector across a range of organizations. Learning from the work of Patricia Hill Collins, within the framework, is the recognition of the need to acknowledge both individual and collective views on our experiences as Black women while also working in alliance with other marginalized groups for wider social justice (Collins 2000). In this case, the allied social group is culturally Deaf women.

2.2 "Africanist Sista-Hood in Britain": An Evolving Theoretical Framework

I have written in detail about Africanist Sista-hood in Britain (Obasi 2016a) but for ease of reference, I have summarized some of that work within this section.

Africanist Sista-hood in Britain was developed as a result of both top-down and bottom-up approaches to theory. Gibson and Brown (2009) describe top-down theory as any theory that has been formulated prior to the empirical work and bottom-up theory as that which is created through the exploration of data. Working within new terminological frames the framework builds on the work of Black feminists, womanists, and Africana womanists to merge this with organic developments in the process of data collection and analysis in the research.

For many Black women, the search for an analytical framework that centralizes our individual and collective experiences and perspectives has ended with the developments of theories in Black feminism. However, influences of American history and legislation means that direct application to the UK context can be more problematic (Mirza 1997; Young 2000; Reynolds 2002; Dean 2009). Black British feminism does seek to address this issue, but it is the particularities of the British context that brings with it issues of contestation of who is considered Black which are also translated into Black British Feminism. Yet, there are still a significant number of Black women who have declared their dissatisfaction with the theory and more resolutely the terminology of feminism no matter its variant. The history of feminism with the privileged status of white middle class women and the marginalization of Black women makes feminism a bitter pill for many to swallow. Womanism (Walker 1983) has for some provided a useful alternative but in Britain has had much less appeal or recognition (Charles 1997).

"Africanist Sista-hood in Britain" offers original terminology while highlighting points of connection with and divergence from existing theories in an aim to regenerate long-standing debates about epistemological and ontological understandings of Black womanhood. It recognizes the specific location of Black women in Britain and is reflective of the "race," class, and gender relations as well as other intersections that affect us. It does this by drawing on existing frameworks in the field and calling for further contributions to an organic framework.

Crotty (1998), in his much cited work explaining the foundations of social research, describes the way in which researchers can draw on established methodological works to develop one specific to the research in hand. Guest et al. (2012) provide an analogy with the work of Bruce Lee in developing his own fighting style due to his dissatisfaction with existing styles. In so doing, he has not developed a new fighting style but synthesized the most useful techniques from numerous existing ones. In reflecting on this fighting style, Lee describes it as something that is fluid and flexible inviting practitioners to take from it what they choose rather than trying to follow a prescribed process. In Africanist Sista-hood in Britain, a similar fluidity is built in. It is a fluidity that allows for incorporation of the work of our Sistas without being constrained by the frames of feminism.

In moving away from feminist terminology, it is not in an attempt to deny the numerous achievements of Black feminists and womanists to whom we should all remain eternally grateful. Rather, it is an attempt to continue and build on that work but in a way that responds creatively to those both within and outside feminist discourses who have declared their dissatisfaction with the legacy the history and terminology still leaves. Black feminists themselves have recognized the limitations of the terminology of feminism (Hooks 1984; Collins 1996, 2000; Springer 2002). As Davis (2004, p. 95) points out, "we need to find ways to connect with and at the same time be critical of the work of our foremothers." Many Black women have voiced their rejection of feminist terminology, yet this issue has not been adequately addressed by Black feminism. Jain and Turner (2012, p. 76) state: "When we look at the term feminist through the lens of the politics of naming we see that it is not an impartial label and that there are multiple reasons why women are reluctant to identify with it." The dissent that has been voiced for many decades both in Western and so-called Third World Black women's discourses still remains active and unsatisfied.

"Africanist Sista-hood in Britain" is a term that is quite deliberately proposed for a number of reasons and can be broken down into its component parts. The term "Africanist" relates directly to Diaspora, and in doing so, connects us back to the direct or (an)Sista-ral heritage in Africa. An Africanist perspective sets out clear and unambiguous messages about embracing African (an)Sista-ry. In this sense, it differs significantly from the contestations that exist around political use of the term Black in the British context which is incorporated into Black British Feminism.

"Sista" is a term recognized within Black popular culture in the UK and beyond, but is also a term that has historically been articulated within Africa and as part of the migration journey for many Black women in the UK. To be a Sista is different from being a sister as within the term Sista is an implied recognition of a positive association with Black womanhood.

Like womanism and Black feminism, for Africanist Sista-hood, any perspective aimed at ending sexist oppression of all women must also embrace issues of intersectionality in relation, not just to "race" and class, but to issues of oppression facing all social groups (Collins 2000). For many Black women, "race" and gender are aspects of our identity that are recognized as constantly visible, but the intersectional and fluid position of our identities also contributes to the diversity that is Black womanhood. The fight for social justice must remain a central focus in any emancipatory framework. Africanist Sista-hood leaves space for recognition of the shifting nature of oppression which dictates that other forms of oppression may impact on groups and individuals within those groups to a larger extent than sexism and may impact simultaneously as part of their everyday lives.

The "-hood" component of the term is about the collectivity and connectivity which is a driving force behind the concept. It has at the center the values of internal validation, self-definition authentication, creativity, and elevation of subjugated knowledge. Lived experience is a central tenet in which validation and authenticity from within the Sista-hood is gained. Within Africanist Sista-hood it is important to build in the safeguards against this loss of control from those guilty of "knowing without knowing" (Collins 2000, cited in Reynolds 2002). It is recognized that every individual, male or female, Black or white, has a contribution to make, but without the recognition of the importance of both *knowledge* and *experience*, the validity of those contributions will be limited and may need further validation.

Unlike Black British feminism, it is a discourse "in Britain" rather than one centered around a British identity as exists in Black British feminism. It also focuses on the location in Britain rather than restricting it to women with British citizenship. It is inclusive of Africanist women living their life in Britain, but who are either not legally British citizens or do not wish to readily showcase that aspect of their identity given the tumultuous history of British "race" relations (see greater details of this theoretical framework in Obasi 2016a, b).

2.3 Learning from the Pilot Study

Details of the pilot have previously been published (Obasi 2014). Within that publication, I discussed some of the possible benefits of a pilot study, and indeed in this research, the pilot proved very valuable in testing out the theoretical framework, research methods, and analysis. It was also useful to highlight some preliminary findings and use some of these findings and participant perspectives to contribute to further redesign the study as well as improve data collection methods.

In relation to the Deaf interviews post-pilot, in addition to my own reflections, I sought out discussions with other (Deaf and hearing) academics that suggested analysis straight from the data would be a more beneficial approach that might also tackle some of the issues in relation to being *lost in translation* (Stone and West 2012; Obasi 2014) as well as the issues of language and power in research discussed by Temple and Young (2004) and Young and Temple (2014). There are some arguments put forward by Gibbs (2007) that researchers can actually benefit from working directly with the data in any research project rather than working through transcriptions. In this case, some of those advantages were in relation to my need to continually return to the data to check for the signed quotes, which led to continual review of facial expression, intonation, and intensity which all form part of the meaning in British Sign Language (BSL). This all increased authenticity and familiarity with the data.

The decision on how to represent signed data was also a very difficult one, and one in which I have changed my position a number of times. There are a number of equally valid but sometimes opposing perspectives and practices in the area. Signed languages have been willfully suppressed in favor of spoken or written languages (Ladd 2003; Bauman 2008; Ladd 2008; Obasi 2008). In the case of academia, where most of the Deaf women worked, this position is exasperated even further (Bauman 2008; Stone and West 2012; Young and Temple 2014). The decision in moving from signed to a written representation of the language in and of itself can be seen as further contributing to this. Temple and Young (2004) discuss issues of the dynamics of power in relation to minority and majority languages, and the processes that take place during the translation of interviews. They point to the way that transcription into the dominant English language can also have an invisibilizing effect on the source language when translating research interviews. Young and Temple (2014, p. 145) go further and state that "the language that is less powerful is made to disappear and by implication so too do the users....the seemingly straightforward and expedient research practice of transcription is akin to epistemological and ontological vandalism."

Ladd (2003), in his ground-breaking work on Deafhood, also takes a ground-breaking approach to representation, as in some cases not just BSL grammar but also nonmanual representations are written into the quotes of his participants, in way that represents true authenticity. Given the framework, authenticity would be my natural aim. However, there are other issues to consider including issues of authority and voice, raised by Hole (2007a) in a similar endeavor as a hearing woman interviewing Deaf women in sign language. As a nonnative signer I am not confident I could achieve the equivalence of Ladd (2003).

Another consideration raised by Young and Temple (2014) is about the negative impacts that can result from researchers' attempts to follow source language grammatical structure in a written representation as this could reinforce stereotypes of illiteracy, and in the particular case of Deaf people can be inappropriately linked to the "dumb" label inferring stupidity. Najarian (2006) also points to the potential to "trip up the reader" where a signed grammar is followed. For the most part I have opted for a full English translation of the quotes but on occasion left in "a BSL flavor" where the opportunity lends itself.

3 Data Collection

3.1 Access

In terms of access, a form of snowballing was used where participants were asked to recommend others for the research (Liamputtong 2013; Patton 2015). This was a preferred option because of the view that by implication the participant also recommends the researcher as well as endorsing the research. This is particularly important within the Africanist Sista-hood framework as it strengthens the idea of validation from within the collective of the social groups identified.

Linking again to the theoretical framework, as discussed by Browne (2005) in her research with nonheterosexual women, snowball sampling is a method that enables participants to have an ongoing influence on the research beyond their contributions in the interviews. Biernacki and Waldorf (1981) go as far as to say that participants become de facto research assistants. Young and Hunt (2011) write about the way that culturally Deaf perspectives have been excluded from research agendas where Deaf people are seen as objects of data collection rather than shaping the research and what is asked. Snowball sampling, therefore, in some ways provides an opportunity to allow this contribution. This method was particularly attractive to this research because of the opportunities it provides for both validation and monitoring based on the assumption that participants would only recommend others if they had had a positive research experience and felt the research worthwhile (Browne 2005).

In snowball sampling, potential issues of bias and other limitations discussed in the literature (Biernacki and Waldorf 1981; Merton 1972; Browne 2005; Sturgis 2008) have not been overlooked. Snowball sampling does have limitations which it is important to recognize, including the potential for bias that might result from those that are put forward as well as exclusions and boundaries created for those that are not (Biernacki and Waldorf 1981; Browne 2005; Sturgis 2008). The potential exists for participants to put forward only those they feel will support their own position and this cannot be ruled out. However, there were instances of participants making suggestions of other participants who, once interviewed, seemed to represent completely different and sometimes opposing standpoints.

In addressing issues of validity, Faugier and Sargeant (1997) recognize potential bias as "a price which must be paid" in order to gain an understanding of participants and their particular circumstances. In addition to this, there were also limitations in terms of generalizability of findings (see also ▶ Chap. 5, "Recruitment of Research Participants").

3.2 Interviews

Willis (2007) describes research methods as an expression of the research paradigm. I took a qualitative approach to the research, with in-depth interviews for data collection, moving from loosely structured interviews in the pilot study and early fieldwork, to more semi-structured interviews towards the end of the study (see Serry and Liamputtong 2017; see also ▶ Chap. 23, "Qualitative Interviewing"). Given the framework of Africanist Sista-hood in Britain and its recognition of both collective and individual experiences of the participants, focus groups were also planned to provide the opportunity for collective as well as individual responses (Liamputtong 2011). The plan was that the focus groups were to be made up of a sample of those who had already taken part in the interviews.

At the start of the research, my aim was to interview 20 women in total, at least 13 of whom would be Black and at least seven of whom would be Deaf (also recognizing the possibility that some may be both Black and Deaf). The final number of Black participants in the one-to-one interviews was 22. Two focus groups took place with Black women; in one of these focus groups there were an additional three women who did not take part in the one-to-one interviews, bringing the total Black participants to 25.

The snowball started with Black women already known to the researcher and was very successful, with participants generally providing me with the contact details of the next prospective participant after first having sought their permission.

The snowball sampling was not so successful with Deaf participants and the target number of seven was not met. Six Deaf women took part in the one-to-one interviews. Participants provided names but did not generally contact on my behalf first. There was, however, one exception where one Deaf participant went as far as to send out a flyer to her contact lists on my behalf which included my contact details and indeed one participant did contact me for an interview via this method. In addition, one of the participants suggested that I make a signed video of the flyer to be sent out to Deaf participants. However, although one person did respond to this video the interview did not take place in the end because of problems of a lack of availability. In addition, I was not successful in my attempts to hold a focus group with Deaf Women. Participants worked in both professional and nonprofessional grades across the public sector.

In total, 30 participant interviews were used as one of the Deaf participants withdrew due to a conflict of interest that later developed. It is unfortunate that this was the only Deaf participant that was not white as this had added to the diversity and richness of data that was collected. In addition to the participants, I was also supported by a Black Deaf female friend who helped develop some of the signs around the concepts being discussed. As some of the terminology was new, there was no easy equivalent in British Sign Language. Having a Black Deaf perspective on this proved invaluable in terms of dissemination presentations I have done.

As I am a qualified interpreter, I made the decision to conduct the interviews with Deaf women in sign language, bringing me closer to the participants and these were video recorded. Interviews with Black women were in spoken English and voice recorded.

Some preliminary findings from the pilot study were discussed. Participants generally viewed equality and diversity policies within their organization with some level of skepticism with many using words like "tick box" and "tokenistic" to describe their view. There was evidence that participants held cultural and linguistic understandings of Deaf identity rather than disability. The concept of "workplace racial trauma" as well as "chronic racial insults" (Obasi 2013) were discovered where the trauma of racial discrimination leads to an exit from the workplace, and the chronic racial insults being lower level acts of bias, exclusion, and/or stereotyping that participants experienced on a more frequent basis. Chronic racial insults are most similar to the concept of micro invalidations as described by Sue et al. (2007) in their wider discussions of racial microaggressions. The Deaf participant discussions about the additional barriers they face in relation to, for example, the effects of working in English as a second language (British Sign Language being their first language), organization of interpreters, particular barriers in academia and publishing, etc., were interpreted as "the Deaf premium." The discussion about the advantages given to hearing, signing colleagues was interpreted as "occupational circumvention." Issues of "race" were made difficult for Black participants to discuss and as such were similar to the race taboo identified by Gordon (2007).

4 Inside Looking Out or Outside Looking In?

4.1 Situating the Researcher in the Research

Like Maylor (2009), being Black and being female is central to who I am as a researcher. It is also central to who I am as an individual. These aspects of my identity will always be visible and have an impact on how I am perceived by the research participants. However, given my research area, my position as a hearing person, which is often taken for granted outside of my working life, also becomes much more significant in conducting the research with Deaf women. During the research with the two different groups of women, with regards to my own identity and the access that it gave me in terms of the importance of *experience* (Collins 1998, 2000), there were clear shifts in positioning that challenged my own location casting me both as "insider" and "outsider" in the same research study (see also ▶ Chap. 92, "Researcher Positionality in Cross-Cultural and Sensitive Research"). I found, as Johnson-Bailey (1999) states, that the interviewer/interviewee relationships had deeper foundations and were more intimate when there were fewer margins to mitigate (Obasi 2014).

Fawcett and Hearn (2004, p. 203) have described debates about otherness in research as something that is "ongoing, unfinished and probably unfinishable." There has, in recent years, been a growing discourse from academics about insider and outsider debates in research (Mullings 1999; Brayboy 2000; Brayboy and Deyhle 2000; Merriam et al. 2001; Serrant-Green 2002; Innes 2009; Maylor 2009; Ochieng 2010; Gair 2012; Obasi 2014; Suwankhong and Liamputtong 2015). Within these writings, researchers generally provide reflexive accounts of their own position or a summary of the literature, whether or not this is based on the remarks or interactions with participants. However, this chapter takes an original approach in that it centralizes participant perspectives on this issue in a way that adds valuable contributions to this ongoing debate.

It has been argued for some time that it is impossible to separate the researcher from the research or knowledge from the knower (Andersen 1993), and at least in qualitative PhD research, engagement with reflexivity is increasingly becoming a requirement of completion (Obasi 2014).

According to Gair (2012, p. 137) "the notion of insider/outsider status is understood to mean the degree to which a researcher is located either within or outside a group being researched, because of her or his common lived experience or status as a member of that group." Within her paper, Gair talked about "common wounds" experienced both by researcher and participants, which was seen as a positive element to bring to the research relationship. It is clear from some of the responses from the participants that this was a position being afforded to me because of my status as a Black woman. What follows is a reflexive discussion centered on participant responses to insider/outsider perspectives on research.

There have been some examples in the literature where other Black women have discussed the way in which their identity as the researcher has formed part of the participant discussion (Johnson-Bailey 1999: Mullings 1999; Serrant-Green 2002; Maylor 2009; Ochieng 2010). In my own case within this study, similar instances occurred. An example of this is demonstrated by the participant below in her discussions about who should or should not be included within the label BAME (Black Asian and Minority Ethnic):

> It is so broad right it is Black, Asian, minority ethnic so it is very very wide and there are people who use it to play a game to their advantage.........For example there is someone, there is a programme that is run to help BAME people and this guy applied to go and he says he is ethnic minority. He is Jewish; he already has an advantageous position being a white

man. . . . . . . well I shouldn't really say it but there is nowhere for the likes of you and me to hide. If you see what I mean,...We are what it says on the tin! (Devandra, (For reasons of confidentiality and anonymity all participants have been given pseudonyms.) Civil servant).

There were many more similar examples where my identity formed part of the discussions, some of which are provided in other parts of the chapter.

4.2 Situating Participants on the Insider/Outsider Continuum

In addition to incidental references to my identity, and as I continued to think reflexively about my researcher position throughout the research, I felt it important to gain some understanding from participants as to their own perspectives on insider/outsider research. I introduced questions and discussions around sameness and difference of the researcher. Some generic responses were given and two examples are provided below:

> Really it depends on the person. In an ideal world yes signing Deaf person, also their identity their experience growing up. Some might say "oh definitely Deaf" but some Deaf are '**deaf by ear**' ... but are they the best representative for us Deaf signing community? Some hearing that know nothing about Deaf- definitely not!....also they must acknowledge give back something.. ....so hearing not allowed? I won't say that, it depends on the person and the benefit to the signing community.. .. Also it's difficult because Deaf education is poor. Deaf with knowledge and qualification to become researcher are few so we have not much choice. We can't wait fifty years for the education to be sorted out and Deaf become researchers because that would be too late. (Mary, Academic).

Or a Black participant response:

> I would have asked myself the question can this person really get a sense of what it's like for me as a Black female only because it's a different experience although they may empathize it's a totally different experience isn't it? I would have been curious as to why a white person felt the need to explore this. I would have asked the question why is it so important for you to be doing this? So I don't think I would have ruled it out because I would have been curious. Because white people have done research about Black people but it's not always been successful. (Debra, Counsellor).

From these two quotes, it would appear that there is some recognition of trust that can be won or lost by the researcher. For "insiders," the starting position is of trust that can be lost, for example, through the recognition of being "deaf by ear," and therefore, showing a lack of understanding of culturally Deaf perspectives. For "outsiders," there is recognition that trust can be won via willingness to be held to account and explain motivations for the research, or a willingness to give something back to the community the researcher is taking from. There is also recognition of the history of power and exploitation that needs at least to be acknowledged.

While it is important to be able to ask difficult and challenging questions about power dynamics in researcher/researched relationships during the research process,

I would say it is impossible to ask such questions without shining a spotlight on the identity of the researcher themselves. Within my own research, there were many more examples of responses where my own identity was central to the response made. This was regardless of whether I was interviewing Deaf women or Black women as seen in the examples below.

> Chijioke: What do you think of hearing people doing research with Deaf people?
> Louise: Honestly?
> Chijioke: Yes please do.
> Louise: Honestly, would be better Deaf. But you alright you know background Deaf you interpreter yourself that's fine I have no problem with that. Hearing person just interested in learning? I would object to that....better that person can know and empathize (Louise, Administrator).

Similarly but relating to my identity as a Black woman:

> Yes it does help because I know that you've had, well I shouldn't really assume really I shouldn't assume. It is likely that you would have gone through racist or you would have experienced some level of discrimination whereas you know one thing which I have been told over the years since I have been working for this local authority is that "no we don't understand what you're talking about because we're white" and I'm like "you're not even trying".......I have worked with people who are white who put in effort to understand or can identify because of whatever experiences they may go through or in the family or a friend they can draw on that (Tochi, Social worker).

In an earlier paper (Obasi 2014), I discussed a Deaf participant response that was more challenging in questioning the legitimacy and negative impacts of hearing on Deaf research, but in both of these examples above, in addition to highlighting aspects of my own identity, there is also some distinction being made about those in the majority populations being discussed. They are differentiating between those who are part of the majority group (hearing people and white people) but who have some access to "knowledge" of the minority group be that professionally or personally.

The literature on insider/outsider researchers now incorporates the positive elements an "outsider" position can take including objectivity, distance from participants, and theoretical and practical freedom (Merton 1972; Innes 2009). However, this does not address one of the key advantages of "insider" research being that participants are more likely to be more open in what they discuss. In interviews with Black participants where sameness rather than difference was my majority experience, it was clear that this was part of the participant perspectives on the issue. Linking back to the origins of the debate and African-American scholarly views (Innes 2009), some of the Black participants felt that the identity of the researcher would have an impact on how they responded and the subjects they were willing to discuss as demonstrated below.

> Because you are from the same background it makes it easier to say the thing, that like sometimes you have almost hidden racism. Nobody is going to understand that unless you

have been there you know like there are some things that you wouldn't feel open in saying because people would think "oh you have got a chip on your shoulder why would you think that?" Of course yes definitely it makes a difference I would have taken part in the interview but my answers would have been different. It's almost like when you have these diversity groups and ultimately the diversity champion is always normally white or you have these groups and people are asking you things and you are guarded you are guarded against what you say (MJ, Civil Servant).

Or another example:

I think it makes a lot of difference to me because I felt here that I could say exactly what I have said today and not feel bad about that because I think I know you would understand or you know you can empathize and I felt comfortable within that because I know that you wouldn't be judging what I have said or making your own assumptions about it whereas I think if it wasn't a Black woman for me I don't think I would I wouldn't have been as open I would have been more reserved because I probably wouldn't have trusted them as much so yeah for me it makes a lot of difference. (Chemma, Community Worker).

This aspect of the debate was not present in the narratives of Deaf participants, but in considering Deaf cultural discourse, it may be premature to conclude that it was not a perspective that they shared. Hearing research with Deaf people in and of itself can create boundaries and limitations that impact on responses. My position as a Black female researcher and specifically a Black female hearing researcher was an integral part of the research as demonstrated by the participant responses discussed. If Deaf participants held similar perspectives about the *importance of experience* of the researcher in enabling wider and deeper responses, it is more likely that they would share this perspective with a Deaf rather than hearing researcher. Using the theoretical framework allows me to recognize that lack of *experience* may create limitations or require further validation and turning to Deaf cultural discourse and the specific history of hearing on Deaf research raises my awareness of the forms these limitations may take.

Fluidity in insider/outsider experiences is now an established element of academic debates (Mullings 1999; Innes 2009; Ochieng 2010; Obasi 2014). While carrying out the research with Black women, I would describe my position as one which was mostly an "insider" position, but this was not a fixed constant position; it was accompanied by moments and situations of difference. For different reasons to Ochieng (2010), I found that I was cast as both "insider" and "outsider" within the same interview. The participant below was reflecting on bereavement and the inadequacies of the allowances made within her workplace, which she felt did not take account to of her cultural bereavement practices.

But things like in the West Indian community if somebody dies they have a dead house, from the minute they die to the minute they are buried and that can be a week. So when you get special leave if it's the mum and dad or close relative you get three days but in reality well you know you need more days.....
Chijioke: So what happens in that 'dead house' then what does that mean?
Well that's when people are allowed to visit the dead house so when my dad died we had to keep food on the go 24 hours a day, alcohol and hospitality so there's always somebody in

so people can pay their respects and when they come they often leave money or they'll bring some kind of food or something like that, to feed other people so it is from when they die to when they're buried and people come round and tell stories. And people just like for a week or so come to the dead house so you see people "oh I've just heard and I've travelled from..." and its things like the funeral, as I understand it in the British culture, you're invited to a funeral but in the West Indian one you can have 300 people because everybody comes. (Sharon, Social Worker).

My ethnic origin is Nigerian and although there are similar aspects of this in relation to cultural burial processes in my own culture, I was not familiar with the name it was given so felt the need to probe deeper. Similarly, many of the participants dipped in and out of patois on occasion or made gestures that were culturally specific to African Caribbean cultures and in a way that there was an expectation that I would understand. Although I am not part of those cultures and so would probably be classed as an outsider by many academic theorists, I am used to hearing patois from within my social circles. As such, I was able to understand the patois that was spoken to me and most of the cultural references that were made. Young Jr. (2004) talks about the advantages outsider status may bring to the research interview when outsiders probe deeper. Had I been an insider on this occasion there may have been no need to probe for an elaboration on the meaning of the "dead house" because of an assumed knowledge. Indeed in some other interviews, there were probably situations where I did not probe and therefore missed the opportunity for elaboration on a specific point. However, it is also clear from some of the responses that the perception of sameness or insider status did facilitate more freedom of expression, trust, and wider discussion than would have occurred in "outsider" research.

5 Connecting to Africanist Sista-Hood in Britain

5.1 Names, Self-Naming, and Naming Identity

The emancipatory elements of Africanist Sista-hood in Britain have built within them recognition of the need to validate counter constructions to that of the majority. In Deaf cultural discourse, Ladd (2003), Padden and Humphries (2005), Hole (2007a), Bahan (2008), and Young and Hunt (2011) have written about the way that majority hearing constructions of Deaf people as disabled are privileged over those articulated by Deaf people themselves. There is also recognition from Collins and Solomos that we need to acknowledge not just the identity that is imposed on us but also that which we choose. Included within this is often the position of resistance and the use of agency against the dominant constructions which in and of itself can lead to mobilization of collectives (Collins and Solomos 2010). None of the participants in the study described themselves or the wider Deaf community as disabled. Their identity was articulated in relation to their culture, their language, and affiliation with other Deaf people as discussed by this participant during the pilot.

> I'm a Deaf BSL user part of the Deaf world. culture, language all of that is included that is part of it. Hearing? Not much. Most of my hearing friends are interpreters really so I'm not really hearing led I am more Deaf led. ….My needs my access needs are different from disabled people……who for example may use a wheelchair.….. Or blind need Braille and they are visible disabilities my needs are not visible. (Erica, Social Worker).

For both participant groups, discussions on identity were interesting and complex and the value in self-naming and validation came through.

> Chijioke: What is your ethnic origin?
> It depends on the question really; If you mean how I identify or how government label me? If forced to choose I would tick "White British" but if Deaf was there as an option I would choose that. (Mary, Academic).

This response provides a clear link to the work of Lane (2005) and Eckert (2005, 2010), in which they make the case for culturally Deaf people to be recognized as an ethnic group in their own right. Lane provides justification for this perspective by measuring the position of Deaf people or "Deaf-world" against the criteria set out for what constitutes an ethnic group, including issues of language, values, community, history, and art among others (Lane 2005). Interestingly, Mary had no reservations about her nationality being British, which shows there was a clear distinction being made for her own identification as a member of a Deaf ethnic group. Anthias (2010) makes the point that ideational identity can exist alongside legal or judicial identity. Given the power behind anything in the Western world that occupies a legal status, the fight for recognition of Deaf people beyond disability remains an arduous task. Surely, within any emancipatory research, it is perspectives of participants that should be prioritized in how these identities are validated.

Within the interviews and focus groups, there were some discussions of identity and names of some of the Black participants and the way in which the anglicized names they had inherited impacted on their working lives. Within the framework of Africanist Sista-hood, there is recognition of the way in which collective identities of our (an)Sistas can have an impact on our individual identities today. The branding of our enslaved (an)Sistas that took place centuries ago still carries legacies for some Black women in Britain today. This was an issue that was discussed in both focus groups in strikingly similar ways.

Focus group 1

> Participant 1: I occasionally get it "oh you didn't sound Black on the phone".
> Participant 2: So how does Black sound like?
> Participant1: I should have said "a whaa g'on? Ya ready for me come see ya now?" (In patois). (Raucous laughter all round).

Focus Group 2

> I used to be a housing officer ……. they would open the door and they would say "oh you don't sound Black on the phone"… I said oh well what do Black people sound like? You

know, they expect you to have this yardy speech or something or speak like a Jamaican you know.

Other similar examples were given by participants where this was either verbalized by service users or colleagues or where the reception received by the Black staff made them think that similar assumptions had taken place. This also provides validation for the Africanist elements within the framework which links back in some way to African heritage or "African descent" as one participant described her ethnic origin. There were instances like those above where Africanist links were implicit, and others where explicit connections were reflected on.

Chijioke: What is your ethnic origin?

> For a long time I was African Caribbean and then there was, I go through phases and would just tick African and wouldn't tick the Caribbean bit I was kind of like so distant from that bit...... but I have come full circle again and I've gone to African-Caribbean because I think that helps me to understand where I have come from. I think if I just pretend that there isn't a Caribbean element I lose some of my history and my history is that my grandma was dual heritage,... my great grandfather was French and you can see through the ages and you can see through the colouring of some of my family members and all of us are not the same skin colour even my brother and sisters we are all different and I think some time you have to, if you don't grab hold of that history that is Caribbean you forget that you have another side to your history and that's the reason. (Marlene, Social Work Manager).

For Deaf participants too, there was recognition of individual and collective identity that existed simultaneously. The participant below when discussing researcher positioning is talking not just of her own identity, but an identity linking back to a Deaf history of oppression exclusion and discrimination. Again, the framework points the analysis in the direction of historical significances, but it is the theoretical alliance with Deaf cultural discourse that provides the detail and implication of that specific Deaf history and its links to the present day.

> The Deaf community through history experience discrimination, for example, Milan-discrimination (The Milan congress of 1880 is well known in Deaf communities for having had a detrimental impact on the education of Deaf children because of the decision that was made to ban sign language in deaf schools (See Ladd 2003 for comprehensive review).), cochlear implants- discrimination. Because society and governments only respect us for research, research what for? Research **on** me for your benefit then future what? Eradicate Deaf for the philosophy utilitarianism- for the benefit of the greater good compared to the few. The Deaf community is small and hearing society is the majority so that's why. (Mary Academic).

From these very interesting, complex, and history-steeped responses, it becomes clear that, as individuals, it is not always possible to be or be seen as separate from our histories, which is an important factor regularly overshadowed by post-structuralist and post-race perspectives currently flooding academic debates. Although the concept of agency should not be excluded from any academic debates about collectivity and connectivity in identity, those histories will still have an impact on our everyday lives regardless of the extent to which they are acknowledged. Those impacts are not always constant and vary to differing degrees and in differing situations.

5.2 Space and Place in Africanist Sista-Hood

Atewologon and Singh (2010), in their study of black professionals' workplace identity, talk of the way participants recognized a differentiation between a "black British" and "black in Britain" identity. This was also an issue that was echoed by participants in one of the focus groups where the general view was expressed by the participant below:

> I think there is a scale in terms of you can be Black and have an English accent and you can be Black and have an African accent and the English accent will be treated better than the African accent. (Focus Group 2).

Similar issues came out in the analysis of the participants individual interviews, three of which were born in Africa, and of those, two of them were born in countries subject to the apartheid system. When asked about their identity, their view included the base line position of being considered as a human. Though interviewed separately and unknown to each other, part of their responses on the issue provided quite similar perspectives:

> Sometimes there have been times when I am filling forms in I will actually leave it blank or sometimes I have been naughty enough to write human being. (Makosi, Social Worker).

Or another participant

> I just identify myself as a person really, I would say just as a human being who has every right to be here! (Jain, Support Worker).

Within the framework of Africanist Sista-hood **in Britain**, there is the recognition of the need to be flexible enough in moving away from the notion of Britishness which for some is unobtainable and for others is not an aspect of their identity they readily wish to showcase. At the basic level, as Black women, we all have in common the experience of being Black women "in Britain," while recognizing the diversity of experiences this encompasses, the focus becomes one of location rather nationality. This provides a further welcome divergence from the terminology of Black British feminism and the Britishness this encompasses.

The framework recognizes the way in which as Black women our spaces are heavily regulated and policed. Within the narratives, there was evidence of different coping strategies employed to deal with this, some of which were influenced by some of the symbiotic relationships related to "race," gender, and the social spaces occupied by Deaf people.

Three of the five Deaf participants in the study talked in detail about the working relationship with other hearing professionals who can sign. A central theme within the Africanist Sista-hood framework is the concept of *knowing without knowing* taken from Reynold's analysis of Collins' work on Black feminist thought. For the three Deaf participants, this was a significant issue that seemed to cause them problems in the workplace on a regular basis. The hearing signing colleagues who were afforded this position of knowing without knowing were sometimes seen as the instigators as in the first example provided below. But in others examples, the issue was that the hearing signing colleagues were seen by the wider organization to have *knowledge* of Deaf people, their language and/or culture, and that they were the preferable source through which to access this *knowledge* even where the Deaf participant was the obvious person to contact as in the second example provided below.

> Four of them [hearing, signing colleagues] meetings, networking no Deaf involvement. . . .. Them represent us? No thank you! You have to be Deaf. Must Deaf because of experience. We know what we want. Networking is brilliant but bring Deaf or Deaf go by themselves with interpreter. . . . They don't share the information with us because we are second class. . . . I wanted to say if it wasn't for us Deaf you wouldn't have those brilliant jobs. . . .. They are climbing their ladders on our backs! (Janet, Academic).

One of the participants here discusses the way in which her efforts to tackle a personal issue with the human resources section was discussed with another hearing, signing colleague and therefore circumventing the need for direct contact.

> They should involve Deaf directly. For example I went to see Susan [a hearing HR representative] for something. . ..she came back asked Tammy [a hearing signing colleague] about it. I ask her why not ask Susan to ask me herself? She said it was to save time so (Facial expression what can I do?). . .she should ask them to ask me. (Carol, Academic).

Whatever the case, these approaches leave the Deaf person feeling devalued in terms of their professional status but also further strengthens the power relationships that exist in the workplace between signing hearing people and their Deaf colleagues. Hearing signing colleagues by accepting this approach act as enablers to the circumvention then either wittingly or unwittingly become gatekeepers to the information that should rightfully be gained or passed on via direct access to the Deaf person themselves. A further issue is whether these power advantages are acknowledged, challenged, or perpetuated by the hearing colleagues in question.

Issues regarding physical spaces occupied by Black women in the workplace were also discussed in some of the interviews and focus groups. The examples below, however, demonstrate the ways in which the pressure of our hyper-visibility can force us to self-manage it. This participant provides an explanation of the reasons and the manner in which she tries to self-regulate her position of hyper-visibility.

> When Rena (Black participant) comes to my organization and Pat (Black participant) I just shush them straight away. . ..I do it because I don't want to be looked at or be noticed to be honest, and I put myself on mute I say it all the time when I go to work I put myself on mute I can't be who I am. (Focus group 2).

I have written in an earlier paper (Obasi 2013) about the ways in which hyper-visibility and invisibility can exist in the same person, and within the study a participant summarized the issues very well in relation to the collective identity of Black people in her community. Having contributed to discussion about the frustration of trying to progress services for Black community members her conclusion was as follows:

> I always say that as the Black community we are the visible invisibles, we are visible by our skin colour but invisible when it comes to services and that's a real battle (Focus Group 2).

6 Conclusions and Future Directions

Creswell (2013) and Liamputtong (2010) recognize that the position of the researcher has an impact on every stage of the research process, and the same is recognized within this study. My own personal position as a researcher had an impact on what I chose to study, how I chose to study it, who I chose to study it with, and the framework I chose to use. From the contributions of the participants, it seems that my personal and professional identity also had an impact on the interactions that occurred. The framework of Africanist Sista-hood in Britain, while building on existing works, helps create a deep level of analysis of the data when connections are made to collective, historical, and individual experiences. In working in collaboration with Deaf cultural discourse with its attempts to foreground counter constructions of language, culture, and or ethnicity, I also leave space for further contributions from others researching with other social groups. Working with other social groups may necessitate different theoretical alliances and emancipatory approaches.

Johnson-Bailey (1999, p. 669), when carrying out research with other Black women, acknowledges some of the dividing lines between researcher and researched, but concludes that "there were many more times when the experience of a Black woman interviewing a Black woman was advantageous." From the data collected in my own study, it would also appear that this was the majority view both of the participants as well as myself as the researcher.

Deaf women, for the most part, also expressed a preference for Deaf researchers as well as setting parameters within which hearing researchers can work. As outsider hearing researchers, there are mitigations that can be introduced such as interviewing in sign language and incorporating participants in shaping the research, but the position of "partial knowledge" or "professional knowledge" held by hearing signing researchers can be a double-edged sword and one that has been used against Deaf people in the form of oppression and exploitation. It is difficult to see a way in which our identity as hearing researchers can have the same majority advantageous position which would be gained from Deaf with Deaf research.

References

Andersen ML. Studying across difference: race, class and gender in qualitative research. In: Stanfield JH, Dennis RM, editors. Race and ethnicity in research methods. London: SAGE; 1993. p. 39–52.

Anthias F. Nation and post-nation: nationalism, transnationalism and intersections of belonging. In: Collins PH, Solomos J, editors. The Sage handbook of race and ethnic studies. London: SAGE; 2010. p. 221–48.

Atewologun D, Singh V. Challenging ethnic and gender identities: an exploration of UK black professionals' identity construction. Equal Divers Incl Int J. 2010;29:332–47.

Bahan B. On the formation of a visual variety of the human race. In: Bauman H-DL, editor. Open your eyes: deaf studies talking. Minneapolis: University of Minnesota Press; 2008. p. 83–9.

Bauman H-D. Introduction-listening to deaf studies. In: Bauman H-DL, editor. Open your eyes deaf studies talking. Minnesota: University of Minnesota Press; 2008. p. 1–32.

Bauman H-DL, Murray JJ. Deaf gain: an introduction. In: Bauman H-DL, Murray JJ, editors. Deaf gain raising the stakes for human diversity. Minnesota: University of Minnesota Press; 2014a. p. xv–xiii.

Bauman H-DL, Murray JJ. Deaf gain raising the stakes for human diversity. Minnesota: University of Minnesota Press; 2014b.

Biernacki P, Waldorf D. Snowball sapmpling: problems and techniques of chain referral sampling. Soc Methods Res. 1981;10(2):141–63.

Brayboy BM. The Indian and the researcher: tales from the field. Int J Qual Stud Educ. 2000;13(4):415–26.

Brayboy B, Deyhle D. Insider-outsider: researchers in American Indian communities. Theory Pract. 2000;39(3):163–9.

Browne K. Snowball sampling: using social networks to research non-heterosexual women. Int J Soc Res Methodol. 2005;8(1):47–60.

Charles H. The language of womanism: rethinking difference. In: Safia Mirza H, editor. Black British feminism: a reader. London: Routledge; 1997. p. 278–97.

Collins PH. What's in a name? Womanism, black feminism and beyond. Black Sch. 1996;26:9–17.

Collins PH. Fighting words: black women & the search for justice. Minnesota: University of Minnesota Press; 1998.

Collins PH. Black feminist thought: knowledge, consciousness, and the politics of empowerment. London: Routledge; 2000.

Collins PH, Solomos J. Introduction: situating race and ethnic studies. In: Collins PH, Solomos J, editors. The Sage handbook of race and ethnic studies. London: SAGE; 2010. p. 1–21.

Creswell JW. Qualitative inquiry & research design: choosing among five approaches. 3rd ed. London: SAGE; 2013.

Crotty M. The foundations of social research. London: SAGE; 1998.

Davis A. Black women and the academy. In: Bobo J, Hudley C, Michel C, editors. The black studies reader. New York: Routledge; 2004.

Dean J. Whose afraid of third wave feminism? Int Femin J Polit. 2009;11(3):334–52.

Eckert, R. C.. Deafnicity: a study of strategic and adaptive responses to audism by members of the Deaf American community of cultrure. Dissertation, University of Michigan, Ann Arbor. 2005.

Eckert RC. Towards a theory of deaf ethnos: Deafnicity~ D/deaf. J Deaf Stud Deaf Educ. 2010;15(4):317–33.

Faugier J, Sargeant M. Sampling hard to reach populations. J Adv Nurs. 1997;26:790–7.

Fawcett B, Hearn J. Researching others: epistemology, expereience, standpoints and participation. Int J Soc Methodol. 2004;7(3):201–18.

Gair S. Feeling their stories: contemplating empathy, insider/outsider positionings, and enriching qualitative research. Qual Health Res. 2012;22(1):134–43.

Gibbs G. Analyzing qualitative data. London: SAGE; 2007.

Gibson WJ, Brown A. Working with qualitative data. London: Sage; 2009.

Gordon G. Towards bicultural competence- beyond black and white. Stoke on Trent: Trentham Books; 2007.
Guest G, Macqueen KM, Namey EE. Applied thematic analysis. London: SAGE; 2012.
Hole R. Narratives of identity a poststructuralist analysis of three deaf women's life stories. Narrat Inq. 2007a;17:259–78.
Hole R. Working between language and cultures issues of representation, voice and authority intensified. Qual Inq. 2007b;13(5):696–710.
Hooks b. Feminist theory from the margin to the center. Cambridge, MA: South End Press; 1984.
Innes RA. "Wait a second who are you anyways?": the insider/outsider debate and American Indian studies. Am Indian Q. 2009;33(4):440–61.
Jain D, Turner C. Purple is to lavender: womanism, resistance, and the politic of naming. Negro Educ Rev. 2012;62 & 63(1–4):67–88.
Johnson-Bailey J. The ties that bind and the shackles that separate: race, gender, class, and color in a research process. J Q Stud Educ. 1999;12(6):659–70.
Ladd P. Understanding deaf culture: in search of Deafhood. Clevedon: Multilingual Matters; 2003.
Ladd P. Colonialism and resistance: a brief history of deafhood. In: Bauman H-DL, editor. Open your eyes deaf studies talking. Minneapolis: University Of Minnesota; 2008. p. 42–59.
Lane H. Ethnicity, ethics and the deaf-world. J Deaf Stud Deaf Educ. 2005;10(3):291–310.
Liamputtong P. Performing qualitative cross-cultural research. Cambridge: Cambridge University Press; 2010.
Liamputtong P. Focus group methodology: principles and practice. London: Sage; 2011.
Liamputtong P. Qualitative research methods. 4th ed. Melbourne: Oxford University Press; 2013.
Maylor U. What is the meaning of 'black'? Researching 'black' respondents. Ethnic Racial Stud. 2009;32(2):369–87.
Merriam SB, Johnson-Bailey J, Ming-Yeh L, Youngwha K, Ntseane G, Muhamad M. Power and positionality: negotiating insider/outsider status within and across cultures. Int J Lifelong Educ. 2001;20(5):405–16.
Merton R. Insider and outsider: a chapter in sociology of knowledge. Am J Sociol. 1972;78(1): 9–47.
Mirza HS. Introduction: mapping a genealogy of black British feminism. In: Mirza HS, editor. Black British feminism a reader. London: Routledge; 1997.
Mullings B. Insider or outsider, both or neither: some dilemmas of interviewing in a cross-cultural setting. Geoforum. 1999;30:337–50.
Najarian CG. "Between worlds": deaf women, work, and intersections of gender and ability. London: Routledge; 2006.
Obasi C. Seeing the deaf in "deafness". J Deaf Stud Deaf Educ. 2008;13(4):455–65.
Obasi C. Race and ethnicity in sign language interpreter education, training and practice. Race Ethn Educ. 2013;16(1):103–20.
Obasi C. Negotiating the insider/outsider continua: a black female hearing perspective on research with deaf women and black women. Qual Res. 2014;14(1):61–78.
Obasi, C. "The visible invisibles": exploring the perspectives and experiences of Black women and Deaf women of equality and diversity in the public sector through the prism of Africanist Sistahood in Britain. Unpublished PhD Thesis, University of Central Lancashire, England. 2016a.
Obasi, C. 'Africanist sista-hood in Britain': Debating black feminism, womanism and Africana womanism. In: Paper presented at the women of colour conference Black Feiminism, Womanism and the Politics of Women of Colour in Europe. University of Edinburgh 3 Sept 2016. 2016b.
Ochieng B. You know what I mean: the ethical and methodological dilemmas and challenges for black researchers interviewing black families. Qual Health Res. 2010;20(12):1725–35.
Padden C, Humphries T. Inside deaf culture. London: Harvard University Press; 2005.
Patton MQ. 2015. Qualitative research and evaluation methods, 4th edn. Sage: Thousand Oaks, CA.
Reynolds T. Re-thinking a black feminist standpoint. Ethn Racial Stud. 2002;25(4):591–606.
Serrant-Green L. Black on black: methodological issues for black reserearchers working in minority ethnic communities. Nurse Res. 2002;9(4):30–44.

Serry T, Liamputtong P. The in-depth interviewing method in health. In: Liamputtong P, editor. Research methods in health: foundations for evidence-based practice. 3rd ed. Melbourne: Oxford University Press; 2017. p. 67–83.

Springer K. Third wave black feminism? Signs J Women Cult Soc. 2002;27(41):1059–82.

Stone C, West D. Translation, representation and the deaf 'voice'. Qual Res. 2012;12(6):645–65.

Sturgis P. Designing samples. In: Gilbert N, editor. Researching social life. 3rd ed. England: London; 2008. p. 165–81.

Sue DW, Capodilupo CM, Torino GC, Bucceri JM, Holder AMB, Nadal KL, Esquilin M. Racial microaggressions in everyday life: implications for clinical practice. Am Psychol. 2007;62(4):271–86.

Suwankhong D, Liamputtong P. Cultural insiders and research fieldwork: case examples from cross-cultural research with Thai people. Int J Qual Methods. 2015; https://doi.org/10.1177/1609406915621404.

Temple B, Young A. Qualitative research and translation dilemmas. Qual Res. 2004;4(2):161–78.

Walker A. In search of our mothers' gardens. New York: Harcourt, Bruce Jovanovich; 1983.

Willis JW. Foundations of qualitative research, interpretive and critical approaches. London: SAGE; 2007.

Woodward J. Implications for sociolinguistics research among the deaf. Sign Lang Stud. 1972;1:1–7.

Young L. What is black British feminism? Women Cult Rev. 2000;11:45–60.

Young AA Jr. Experiences in ethnographic interviewing about race: the inside and outside of it. In: Bulmer J, Solomos J, editors. Researching race and racism. London: Routledge; 2004. p. 87–202.

Young A, Hunt R. Research with deaf people. London: London School of Economics and Political Science; 2011.

Young A, Temple B. Approaches to social research: the case of deaf studies. New York: Oxford University Press; 2014.

Researcher Positionality in Cross-Cultural and Sensitive Research

92

Narendar Manohar, Pranee Liamputtong, Sameer Bhole, and Amit Arora

Contents

1. Introduction 1602
2. Researcher Positionality 1603
3. Placing Issues in Cross-Cultural and Sensitive Research 1603
   - 3.1 Gender Issues 1604
   - 3.2 Age Issues 1607

N. Manohar
School of Science and Health, Western Sydney University, Sydney, NSW, Australia
e-mail: narendar.manohar@hotmail.com

P. Liamputtong
School of Science and Health, Western Sydney University, Penrith, NSW, Australia
e-mail: P.Liamputttong@westernsydney.edu.au

S. Bhole
Sydney Dental School, Faculty of Medicine and Health, The University of Sydney, Surry Hills, NSW, Australia

Oral Health Services, Sydney Local Health District and Sydney Dental Hospital, NSW Health, Surry Hills, NSW, Australia
e-mail: Sameer.Bhole@sswahs.nsw.gov.au

A. Arora (✉)
School of Science and Health, Western Sydney University, Sydney, NSW, Australia

Discipline of Paediatrics and Child Health, Sydney Medical School, Sydney, NSW, Australia

Oral Health Services, Sydney Local Health District and Sydney Dental Hospital, NSW Health, Sydney, NSW, Australia

COHORTE Research Group, Ingham Institute of Applied Medical Research, Liverpool, NSW, Australia
e-mail: a.arora@westernsydney.edu.au

© Springer Nature Singapore Pte Ltd. 2019
P. Liamputtong (ed.), *Handbook of Research Methods in Health Social Sciences*,
https://doi.org/10.1007/978-981-10-5251-4_35

3.3	Race, Culture, and Ethnicity Issues	1608
3.4	Social Class Issues	1610
3.5	Shared Experiences	1612
4	Conclusion and Future Directions	1613
References		1613

Abstract

The status of the insider and outsider is an important concept for cross-cultural and sensitive research. In recent years, the concept of placement of the researcher has received much attention. Until a few generations ago, researchers who shared the same cultural, social, and linguistic background with those of the research participants mainly conducted research. However, over the last two decades, we have started to witness researchers who have different characteristics to that of the research participants conduct research in health and social sciences. In current times, this has led to the debates of insider versus outsider status of the researchers, as the way research participants "place" the researchers, and vice versa, is vital for the success of any research. In this chapter, we shall introduce the concept of researcher positionality. We will look at the debates on researcher positionality in cross-cultural and sensitive research and discuss "placing" issues such as gender, age, culture and ethnicity, social class, and shared experiences.

Keywords

Sensitive research · Insider · Outsider · Placing issues · Positionality

1 Introduction

In research, the opinions, values, beliefs, and social background of researcher influence him/her in the research process, shaping each methodological and analytical decision he or she makes. This chapter introduces the concept of researcher positionality and its potential influence on the research process making a case for all researchers to consider the importance of researcher positionality in their research.

Research, according to England (1994, p. 82), "is a process, not just a product." Research signifies a mutual space, shaped by both researcher and participants (England 1994). Hence, their identities have the potential to influence the entire research process. Identities are established through our perceptions, not only of others, but also by the ways we expect others to perceive us (Bourke 2014). Kezar (2002, p. 96) states: "Within positionality theory, it is acknowledged that people have multiple overlapping identities. Thus, people make meaning from various aspects of their identity..." Additionally, researchers' biases define the research process. Thus, by identifying their own biases, researchers can gain insight into their approach towards research settings, members of particular groups, and their relationship with research participants. This has led to debates on the insider versus outsider status of the researcher in recent years, especially in cross-cultural and sensitive research (Ergun and Erdemir 2010; Liamputtong 2010; Edmonds-Cady 2012).

In this chapter, we first introduce the concept of researcher positionality and discuss the relationship of researcher positionality with reflexivity. Thereafter, we provide detailed explanation with examples from our research and from other researchers on insider versus outsider perspectives on the impact of gender, age, culture and ethnicity, social class, and shared/unshared experiences on the research process and research outcome(s).

2 Researcher Positionality

The term positionality describes an individual's view and the position he/she has chosen to adopt in relation to a research task (Savin-Baden and Major 2013). These are often shaped by political allegiance, religious faith, gender, sexuality, geographical location, race, culture, ethnicity, social class, age, linguistic tradition, and so on (Sikes 2004). Positionality reflects the position the researcher has taken *within* a given research study (see also ▶ Chaps. 91, "Space, Place, Common Wounds and Boundaries: Insider/Outsider Debates in Research with Black Women and Deaf Women," and ▶ 125, "Police Research and Public Health"). Some aspects of positionality are fixed such as gender and race while others are subjective such as personal experiences.

Positionality requires the researcher to acknowledge and locate their views, values, and beliefs in relation to the research process. Self-reflection is a mandatory ongoing process in any research project as it gives the researcher the ability to identify, construct, and critique their position within the research process. Reflexivity, the concept that researchers should acknowledge and disclose their own selves in the research, seeking to understand their part in it, or influence on the research (Cohen et al. 2011, p. 225), informs positionality. It is a self-reflection on how their views and position might have influenced the research design, the research process, and interpretation of research findings.

Savin-Baden and Major (2013, pp. 71–73) identify three ways of researchers accomplishing positionality. Firstly, researchers locate themselves in relation to the subject, i.e., acknowledging personal positions that have the potential to influence the research. Secondly, they situate themselves in relation to the participants, i.e., how the researchers view themselves, and how others view them. Thirdly, they locate themselves in relation to the research context and the research process, i.e., acknowledging that the research will be influenced by the research context.

3 Placing Issues in Cross-Cultural and Sensitive Research

The way research participant "place" the researchers, and vice versa, is vital for the success of any research (Al-Makhamreh and Lewando-Hundt 2008; Ramji 2008; Wegener 2014; Berger 2015; Hayfield and Huxley 2015). Here, we will discuss the positionality of researcher as perceived by the research participants (Liamputtong 2007, 2010). There are various "placing" issues such as gender, age, culture and ethnicity, social class, and other identities that have a significant impact on the research process and the research outcome(s) (Breen 2007; Al-Makhamreh and Lewando-Hundt 2008; Ramji 2008; Maylor 2009; Al-Natour 2011; Manohar 2013; Wegener 2014; Berger 2015; Hayfield and Huxley 2015; Suwankhong and Liamputtong 2015). Such "placing" issues will be discussed in following sections.

3.1 Gender Issues

Considerations on the influence of gender in fieldwork, particularly by researchers of gender opposite to that of participants, have highlighted the essential role of gender in research depending on the situation (Al-Makhamreh and Lewando-Hundt 2008; Enguix 2012; Takeda 2012; Pante 2014). Such contextual and situational appreciation of the impact of gender in research creates a dynamic perspective of how gender interacts with other social and cultural aspects related to research (Galam 2015).

In our previous research (Arora et al. 2012b, 2014), we noted that migrant mothers whose first language was not English were not comfortable talking about sensitive topics such as breastfeeding and birth experiences to male researchers. We also noted that women from culturally and linguistically diverse backgrounds were uncomfortable talking about breastfeeding in public as they considered the discussion about breast and/or breastfeeding to be of sexual nature. Furthermore, we noted that women were shy talking about breastfeeding issues such as latchment, sore and/or cracked nipples for similar reasons. They particularly felt uncomfortable talking about female issues to a stranger (male researcher) and felt they were only comfortable to their husband and not other men in general. One of the research participants in our study pointed out:

> I just feel uncomfortable talking about breasts to men except if it's my husband. This is something too personal and sexual for me...

In order, to make the interviewees more comfortable, we ensured that both male and female researcher were present at the time of the interview. This helped the interviewee to be at ease, as they could feel better connected to female researchers.

In our research with new mothers, women were comfortable talking about the birth experiences and breastfeeding issues to other women in general. This is because they felt women share the similar body anatomy and could relate to the experiences better. Pingol (2001), in her research on migrant women conducted interviews of their husbands, noted that gender had no effect on her relationship with participants. However, she recognized that female participants were more comfortable in discussing their sexual matters since she was of the same gender.

Interestingly, in our research with English-speaking mothers (Arora et al. 2012a), women were very comfortable talking about breastfeeding and other female matters with male researchers. Our reflections from our previous research highlighted that it is more difficult to conduct cross-gender research with new migrant women. However, if migrants have lived in western countries for long or if they are migrants from English-speaking countries, it is much easier to build rapport and collect good quality data.

Al-Makhamreh and Lewando-Hundt (2008, p. 11) contend that "cultural and social norms construct certain gender expectations that researchers can negotiate and act within." In a sociocultural context, gender is particularly important due to the existence of sexual boundaries and gender domains. Therefore, in all research methods and processes, gender needs consideration primarily because cultural and social traditions require certain expectations from researchers and participants in regard to gender (Järviluoma et al. 2004). Researcher of same gender may encourage the interviewees to be more open about their feelings and thoughts since they may share common assumptions and experiences (Riessman 1994; Liamputtong 2010; Suwankhong and Liamputtong 2015). Therefore, the data gathered by female researchers from female participants may well be different from that collected by male researchers and vice versa. Gill and Maclean (2011) suggest that when the researcher and participant are of same gender and culture, the communication is easier because they can clearly appreciate the mutual aspect of gender and culture, thereby the produced data is more sensibly interpreted. In our research with Mandarin-speaking migrant women we had similar reflections (Arora et al. 2012b). In particular, we noted that Mandarin-speaking women were very appreciative that female Mandarin-speaking researchers were keen to speak to them over the phone, in their own language, give infant feeding advice that was culturally appropriate, and took the time out of their busy schedules for a home-visit. We also noted that the quality of these interview data was excellent as women shared personal insights of their routine lives.

Gender has been the subject of intense debate among feminist researchers. The position and role of men as a researcher has always been a topic of debate in the domain of feminist research. It is due to the fact that men's status in feminism is still marginalized no matter how much they are committed to women's problems and concerns. Many feminists exclude men from feminist research on the grounds that men cannot experience the world in the same way as women. Although it is true that men cannot experience women's problems and concerns in the same way as women, it does not restrict men to make a contribution to feminist research. Feminist researchers feel that female interviewers best interview women respondents. In our breastfeeding research in Sydney, Australia, we noted that female interviewees were willing to discuss their personal lives with female researchers. One of our respondents stated: "Of course I would prefer to talk to a woman because she would understand me. I would not be embarrassed to talk to her about how I feel." Al-Makhamreh and Lewando-Hundt (2008) noted that in spite of being the same-gender, Al-Makhamreh had to consider the respondents' social constructions of gender, employing "informal but respectful nomenclature," for example, referring to respondents as "mother of Sahar" or "father of Jamal" rather than their first names (p. 18). In our research with Arabic migrant women in Australia, we also employed informal respectful nomenclature such as "mother of Mohammed" when male researchers interviewed female Arabic migrants on oral health. Also, we used terms such as "sister" to refer to the respondents since in Arab society it is a culturally acceptable method for "cross-gender interactions" and "a way for a male to frame interaction with a woman in a nonsexual way."

In perspective of a male researcher undertaking research within gender-segregated communities, Abdi Kusow (2003), a Somali male researcher whose research involved Somali migrant community in Canada, experienced difficulty in accessing as well as meaningfully interacting with the female participants. He contends that "in such social arrangements, cultural or racial differences or similarities do not determine insider or outsider status; the social organisation of gender does. What this situation suggests is that a Western, outsider, female ethnographer may have better access than would I as a native male ethnographer" (p. 597). He summarizes his dilemma as follows:

> I was able to talk to several Somali female students at the local universities, but beyond them, finding access to female participants remained a daunting experience. In the Somali social context, one cannot simply call a female participant for an appointment or go to her house without the assistance of a male relative, for one must avoid any suggestion of impropriety or other misunderstandings. If the woman is married, the situation is even more complicated. Married women cannot, at least officially, associate with men other than their husbands or relatives regardless of the circumstances. (p. 597)

Similarly, Brandes (2008) in his cross-gender research in Andalusia, where strict gender domains existed, had to gain access to female participants through their husbands. Apart from gaining access to the participants, the issue of building rapport is critical in cross-gender research as it is noted that often men are more reluctant to open up to a female researcher because they are less used to being questioned by women, and they do not expect women to understand their experiences.

It is believed that sharing the same gender can have multiple advantages for both the interviewer and interviewee, such as reducing the social distance, facilitating communication, and providing a positive and beneficial experience (Finch 1993). In her research on breast cancer conducted by Dusanee Suwankhong and Pranee Liamputtong in Australia (Suwankhong and Liamputtong 2015), they were recognized as an insider by the research participants as they shared the same gender. One of her research participant said:

> I sometimes check my breasts like this [her hand press on her breasts when showering]. I want to check if there is no lump or anything like that . . . abnormal thing. If there is something wrong I can go to the doctor early. Mine is possibly small size, hah (her face turned red and still holding her breasts). See, if you are a male interviewer, I would not touch mine and show you like this. I would feel too embarrassed! But you are a woman. I am not too shy to talk about this thing and can show my breast to you as I should, why not.

On the other hand, Kusek and Smiley (2014, p. 160) noted that in spite of their cross-gender apprehensions, the male participants were "approachable, friendly, and eager to share their stories." In their opinion, being in a position of female researcher might be beneficial, as male participants were not only boasting about their professional successes but also shared personal problems probably to gain empathy, a behavior which would have been less verbalized in case of male researcher. However, Kusek and Smiley also noticed that their gender made the female respondents more comfortable and acquire easy access to their homes which they felt male researchers would not have necessarily gained. They certainly experienced reluctance to research participation, but gender had no link to it.

3.2 Age Issues

Besides gender, age of researcher also influences the research process, particularly in establishing relationship and trust. In several cultures across the world, older age implies respect. Therefore, older participants might not extend due respect to younger researchers which can jeopardize the quality of research findings. For example, particularly in a South Asian rural society perspective, a young unmarried female researcher would find it difficult to engage and establish rapport with older married female and male participants. Stiedenroth (2014, p. 84) in her research in Pakistani rural communities writes:

> I experienced that my age (22 and 24 years old during each fieldwork period respectively) and my status (unmarried and childless) were more central to my positionality than being a western foreigner, at least in respect to contact with men in the field.

On the other hand, Suwankhong and Liamputtong (2015) highlight the importance of being recognized as an insider in research rather than the influence of age. They highlight that a young researcher would be seen as a family member and the access to participants and building trust with the participants becomes easier if accepted as an insider.

The age of researcher is crucial in a sense that it creates a learning sensitivity to difference in expression of thoughts based on the age of participants. Generally speaking, younger participants share a wealth of lived experiences with young researchers. However, older participants often do not share a great deal of information with younger researchers. This may be due to the fact that older research participants feel a lack of similar shared experiences with younger researchers. Underwood et al. (2010) noted that a young, novice researcher (aged between 28 and 30 years at the time of research) observed that younger participants provided detailed information during interviews whereas older interviewees responded with short answers and made comments that suggested either the questions were not relevant to them or they did not understand the interviewer's questions and perspective. In our breastfeeding research in Sydney, Australia, we had a similar experience. One of the interviewer/researcher said:

> Most study participants were 25–35 years old, (and) I am in that age range, so I felt I could relate to them from that standpoint.

In one of our current research projects with older Australians (70+ years), we faced many issues with recruitment of study participants. We believe that one of the key reasons for difficulties in recruitment is that the researcher recruiting the study participants is only a 30-year-old young woman. Further, we noted that the study participants, particularly men, did not feel connected to the younger researcher as they believed that the researcher would not understand their perspective on health issues. We noted that not only were the study participants harder to recruit but they did not participate well in the research process. The interview responses were relatively short, and it made the researcher feel that she had not explained herself well or she was not a good researcher. One of our research participants said: "I feel I haven't explained myself very well. Maybe I'm not grasping what you want."

Interestingly, older migrant men in our current work in Sydney, Australia, treated the interviewer not only as a researcher or a young woman, but as a young person who can be given pieces of advice in life. The female researcher was met with paternalistic treatment wherein the researcher was treated as a daughter.

Chawla (2006) in her ethnographic research on Hindu marriages involving Indian women discusses about "shifting subjective experiences," which she experienced in the field. She states that researcher positionality is controlled by participants. Since her research participants included participants from different age groups, her experience reformed as per participants' perceptions. In her words:

> With the young group, I was accorded the role of native, thus hyper-eligible. With the middle group, I was adjusted: first as an insider, and later as an outsider and stranger. With the older group, I was 'another', a comfortable stranger, and, in more ways than one, the professional stranger of ethnographic work. Ultimately, I had to experience these eligibilities to reflect on and converse with their life-histories. . ... These eligibilities originated from my single status and not from my displaced or rooted ethnic identities. (Chawla 2006, p. 13)

In research projects where there is a relatively larger age difference between the participants, choosing appropriate questions to elicit meaningful information can be a daunting task. It is a norm that the questions used should be of equal relevance to all study participants. However, it has been proved that the researcher needs to be on the same "wavelength" as the participants especially in respect to their age. One way to minimize the impact of researcher biases is to conduct focus groups because they facilitate the undertaking of unstructured interviews, provide more freedom to the study participants, and allow the group to take control of the interview (Liamputtong 2013).

3.3 Race, Culture, and Ethnicity Issues

It has been suggested in the literature that researchers who undertake cross-cultural research should be an "insider," i.e., they should share the same social, cultural, and linguistic characteristics with the research participants (Merriam et al. 2001;

Al-Makhamreh and Lewando-Hundt 2008; Ergun and Erdemir 2010; Liamputtong 2010). This is what Ramji (2008) refers to as cultural commonality.

According to Banks (1998), there are two main types of researchers in cross-cultural research – cultural insiders and outsiders. Cultural insiders have commonality with the research participants as they share the same social background, culture, and language. Banks (1998) notes that being a cultural insider is the best approach for successful fieldwork. It is often argued that researchers sharing same cultural characteristics as their participants are in a better position to discover research ideas, arguments, and opinions. This is because they are seen as a "legitimate member of the community" (Liamputtong 2010) and provide better insights when describing the social and cultural characteristics of the participants with whom they undertake research (Tillman 2002; Liamputtong 2010). Cultural outsiders refer to the outsider researcher who enters a local area to conduct research. They hold different views, values, beliefs, and knowledge from the community where they undertake the research.

Cultural insiders may be able to conduct research "in a more sensitive and a responsible manner" than outsiders (Bishop 2008, p. 148). Due to cultural commonalities, they are better placed to gain the trust of the research participants and build relationships (Shariff 2014). This can often reduce the difficulties in building rapport with the research participants. One of the Thai-speaking respondents from Suwankhong and Liamputtong's research in Australia (2015) remarked that: "I am very happy to help..... at least I can share my story with others... I always like to support other Thai people when I can..." Further, in their research with traditional healers in Thailand, Suwankhong was seen as a cultural insider as she spent most of her life in a Thai communal environment and showed the same norms as that of the local community. They highlight that being accepted as a cultural insider is crucial in cultural research that allowed to cultivate a trusting relationship with the participants. Liamputtong (2010) contends that one of the key reasons to be seen as an "insider" by the research participants in cross-cultural research is the use of same language. In their work in Thailand and Australia with Thai women, they used Thai language and were able to avoid difficulties regarding language issues. They further highlight that having a shared ethnic identity helps in being recognized as a cultural insider and build a trusting relationship.

In our research with Indian migrants in Sydney, we noted that being a cultural insider helped facilitate the interviews due to the relationship of trust between the research participants and the interviewer. Families were able to share their personal stories on the migration process as researcher and the research participant spoke the same language.

However, cultural outsiders may also be able to get deeper understanding and explanations of a phenomenon under investigation (Al-Makhamreh and Lewando-Hundt 2008; Liamputtong 2010). This is because they may not know much about the lives of the research participants and therefore may want to get a closer and a detailed look into what is being researched. In fact, Merriam et al. (2001, p. 411) contend that "insiders have been accused of being inherently biased, and too close to the culture to be curious enough to raise provocative questions." To them, "the insider's strengths become the outsider's weaknesses and vice-versa" (p. 411).

Additionally, in regard to "insider perspective," a term called "diversity in proximity" has been used (Ganga and Scott 2006). It illustrates that as an insider, the researcher is better able to recognize not only the ties that bind him/her and participants but also the social fissures that divide them (Ganga and Scott 2006). Insider or outsider positionality can influence the researcher's objectivity, and furthermore can influence the social dynamics of interviews. Martiniello (cited in Bousetta 1997, p. 6) cautions:

> During data collection, for example, an ethnic background can be very helpful. Ethnic researchers can have privileged relations with immigrant groups, which can facilitate access to the field. Similar advantages arise from familiarity with the languages and the physical space of the researched group. On the other hand, such closeness between a researcher and his/her subject can also harm the research process.

In certain research scenarios, race of the researcher plays a very influential role in participant interviewing. For instance, Fletcher (2014) points out that in cross-cultural research, the ethnic minority participants often distrust the white researchers as they are considered as cultural outsiders. One of his participants said: "For most you're fine. Some probably won't ever speak to you. They're (the British Asians) happy to train and play alongside you, but they probably won't sit and talk to you" (p. 252). On the other hand, Pasquini and Olaniyan (2004), in their research involving Nigerian farmers, experienced that Pasquini being a white researcher enjoyed a favorable position whereas Olaniyan being a local Nigerian but of different tribe made him an outsider. A Nigerian would prefer a white person over a fellow "black" person, probably due to perception that white people are wealthier and/or even superior. In this case, Pasquini was the white person who was held in high regard and respected due to her race. Additionally, the farmers, based on their previous experience with white people, found them to be trustworthy, organized, and influential with government authorities in order to get help for them. This might be another reason due to which the farmers were forthcoming in regard to sharing their opinions and thoughts with her. Moreover, Nigerians consider white people as guests and since she was interacting with them, the famers might have felt obligated to respond. Africans have a tradition to treat well the foreign as well as African guests, as they feel something good might come of it. Olaniyan, on the other hand, was a Nigerian, but of a different tribe which did not make him an "outsider" per se but was still not considered one of them. The point to consider in this research-participant interaction is that despite of being from a different tribe, Olaniyan, due to the common language, was able to gain acceptance and trust of the farmers, and served as an interpreter between "madam" and farmers.

3.4 Social Class Issues

In addition to the issues discussed earlier, differences in social class of the researcher and participants also pose as barrier to the research process (Rashid 2007; Ramji

2008). For example, it is believed that if researcher is of a middle-class background, differences of authority and privilege within a research setting could not only negatively affect the marginalized groups but such researchers may also be unable to adequately comprehend or represent the lives of marginalized groups (Mellor et al. 2014). Mellor et al. (2014) further highlight that despite enough research evidence, there seems to remain anecdotal preferences among class researchers for class matching, particularly when research involves working-class participants. This preference of class matching is likely to be guided by the assumption that if middle-class researcher interviews the marginalized groups, the differences in the power could potentially harm the marginalized groups. Further, there has been criticism that middle-class researchers may not understand the lives of the marginalized groups. In our research with vulnerable families in Australia, we did not experience that class of the researcher influenced the research process in any way. In fact, in the focus group discussions the research participants felt quite privileged to share their personal stories with medical researchers (who they considered as health and education idols).

Mellor and her research team (2014) felt that middle-class students were more open and able to challenge their representations as privileged when interviewed by working-class researchers. One of the respondents explained the sacrifices which his parents had to make to send him to a prestigious school; the authors felt that such response would not have been received if that student was interviewed by a fellow middle-class researcher and in fact he would have highlighted his privileges rather than his parent's sacrifices. Nevertheless, at many times, it was observed that working-class interviewers shared a strong affiliation with working-class interviewees. The researchers felt like "insiders" and students openly discussed their difficult backgrounds and current circumstances without any shame or prejudice.

Hoskins (2015), in her research relating to researcher positionality involving female professors, contends that her social class, gender, age, and ethnicity all influenced the quality of interviews and the research process itself. She was able to establish strong connections with several participants from working-class backgrounds as she could relate to some of their experiences and instances of feeling as an outsider at the workplace.

In some religious communities, caste plays a major role to determine the acceptability of researcher. Ramji (2008), who herself is Hindu, conducted research within London's Hindu Gujarati community. She experienced being questioned about whether she belonged to the same or a higher caste as to the respondents, the reason primarily being that caste is a crucial factor in Gujarati Hindus. Ramji asserts that "lower caste Gujaratis were somehow thought to be lesser Hindus. This is indicative of the inter-connected relationship between caste status and religion in Indian culture: a similar caste Gujarati would share a similar perception of Hinduism" (p. 106). Also, her linguistic accent was judged to be non-Londoner while London was seen as "a natural place" for Gujarati Hindu identity, which made Ramji an outsider. However, this "outsider" status had a positive implication in her research process since some female respondents were able to openly discuss their lives without the prejudice of offending other community members.

To say the least, many researchers (see Rashid 2007) are of the opinion that social-class differences in the research process are a reality, which enable respondents to draw unrealistic assumptions towards the researchers. However, the researchers can position themselves and adopt specific interviewing methods to markedly minimize the gap between them and the respondents.

3.5 Shared Experiences

The way the participants place the researchers inside or outside the shared experiences is another important placing issue. In their research on working with their own communities, Yakushko et al. (2011) reflected on their personal experiences and highlighted that belonging to the same community and having shared similar life experiences helped them undertake high quality research in migrant communities. Similarly, Egharevba (2001) described that she shared similar skin color with the participants that had a positive impact on the research. In her research, she described that many participants told her that they would not have felt comfortable discussing their views and experiences on racism with a white researcher. Egharevba concluded:

> What they saw in me was another minority person who was living in a racist country, a commonality which made them feel less vulnerable than they would have felt with a white researcher. This shared experience transcended many of the apparent differences between us. (p. 239)

As part of recent research we conducted in Greater Western Sydney, Australia, with migrant Asian parents on child oral health (Arora et al. 2012b, 2014), we were able to get personal insights into the health behavior patterns of their children. Mothers in the study said that they felt more comfortable to talk about the diet practices of their children with us who were from Asian ethnic background and had many shared experiences regarding dietary practices with them. One mother from our study said:

> I don't like to give him (the child) tap water. You (the researcher) are Chinese, so you know what I mean. Chinese avoid tap water as we know water quality is poor in China. I really haven't changed my opinion on it in Australia.

However, in our previous research with Australian women on infant feeding practices (Arora et al. 2012a), we provide an interesting example of shared experiences between the researchers and our research participants. We anticipated that the interviewer characteristic (male researcher) might be a problematic issue during the research process. The literature suggests that male researchers conducting research on women's health may be challenging due to the fact that men's status in feminism is marginalized no matter how committed they are towards women's problems. Further, we anticipated that a researcher who did not share the same experiences (without children) with the research participants (with children) may compromise the quality of the interviews. On the contrary, our experiences on researching with women on infant feeding experiences brought authenticity and a different dimension in breastfeeding research. Interestingly, women in our recent work highlighted that they were able to share their personal stories as families had built rapport with the research team. One participant said:

> We have known you for almost a year since she (the child) was born. I don't mind talking about my experiences with you. You are a health professional and I don't think about it in a sexual way although we talk about breastfeeding.

4 Conclusion and Future Directions

In his recent book, Cresswell and Poth (2017) acknowledges that the position of researcher has an impact on the research process and research outcomes. The case examples presented in this chapter recognize the importance of insider/outsider issues in cross-cultural and sensitive research. This is particularly important in a country such as Australia, which has culturally and linguistically diverse population and may be vast differences in social class. Hence, it is most likely that health and social science researchers will work with research participants from different socio-cultural backgrounds and therefore researchers need to acknowledge the importance of positionality in terms of wide-ranging placing issues such as gender, age, ethnicity, social class, and shared experiences.

We would encourage readers to ascertain their positionality and think about their own status carefully before commencement of the research process, and continue to reflect on it throughout the data collection process. In particular, they should be aware of their position among those being researched, how they conduct research, and their understanding of the phenomenon and/or context being examined. The insider and outsider status of the researcher may shift during the course of research project, and researchers will need to clarify and maintain their position during the entire course of cross-cultural and sensitive research.

References

Al-Makhamreh SS, Lewando-Hundt G. Researching at home as an insider/outsider: gender and culture in an ethnographic study of social work practice in an Arab society. Qual Soc Work. 2008;7(1):9–23.

Al-Natour RJ. The impact of the researcher on the researched. J Media Culture. 2011;14(6). Retrieved from http://journal.media-culture.org.au/index.php/mcjournal/article/viewArticle/428

Arora A, McNab MA, Lewis MW, Hilton G, Blinkhorn AS, Schwarz E. 'I can't relate it to teeth': a qualitative approach to evaluate oral health education materials for preschool children in New South Wales, Australia. Int J Paediatr Dent. 2012a;22(4):302–9.

Arora A, Liu MN, Chan R, Schwarz E. 'English leaflets are not meant for me': a qualitative approach to explore oral health literacy in Chinese mothers in Southwestern Sydney, Australia. Community Dent Oral Epidemiol. 2012b;40(6):532–41.

Arora A, Nguyen D, Do QV, Nguyen B, Hilton G, Do LG, Bhole S. 'What do these words mean?': a qualitative approach to explore oral health literacy in Vietnamese immigrant mothers in Australia. Health Educ J. 2014;73(3):303–12.

Banks JA. The lives and values of researchers: implications for educating citizens in a multicultural society. Educ Res. 1998;27(7):4–17.

Berger R. Now I see it, now I don't: researcher's position and reflexivity in qualitative research. Qual Res. 2015;15(2):219–34.

Bishop R. Freeing ourselves from neocolonial domination in research: a Maori approach to creating knowledge. In: Denzin Y, Lincoln Y, editors. The landscape of qualitative research. 3rd ed. Thousands Oaks: Sage; 2008. p. 145–83.

Bourke B. Positionality: reflecting on the research process. Qual Rep. 2014;19(33):1–9.

Bousetta H. Personal experience and research experience: a reflexive account. MERGER – Newsletter of the Migration and Ethnic Relations Group. 1997. Retrieved from http://www.ercomer.org/merger/vol3no1/debate.html.

Brandes S. The things we carry. Men Masculinities. 2008;11(2):145–53.

Breen LJ. The researcher 'in the middle': negotiating the insider/outsider dichotomy. Aust Community Psychol. 2007;19(1):163–74.

Chawla D. Subjectivity and the "native" ethnographer: researcher eligibility in an ethnographic study of urban Indian women in Hindu arranged marriages. Int J Qual Methods. 2006;5(4):13–29.

Cohen L, Manion L, Morrison K. Research methods in education. 7th ed. Abingdon: Routledge; 2011.

Cresswell JW, Poth CN. Qualitative inquiry and research design: choosing among five approaches. 4th ed. London: Sage; 2017.

Edmonds-Cady C. A view from the bridge: insider/outsider perspective in a study of the welfare rights movement. Qual Soc Work. 2012;11(2):174–90.

Egharevba I. Researching an-'other' minority ethnic community: reflections of a black female researcher on the intersections of race, gender and other power positions on the research process. Int J Soc Res Methodol. 2001;4(3):225–41.

England KV. Getting personal: reflexivity, positionality, and feminist research. Prof Geogr. 1994;46(1):80–9.

Enguix B. Negotiating the field: rethinking ethnographic authority, experience and the frontiers of research. Qual Res. 2012;14(1):79–94.

Ergun A, Erdemir A. Negotiating insider and outsider identities in the field: "Insider" in a foreign land; "outsider" in one's own land. Field Methods. 2010;22(1):16–38.

Finch J. Its great to have someone to talk to: the ethics and politics of interviewing women. In: Hammersley M, editor. Social research, philosophy, politics and practice. London: Sage; 1993. p. 166–80.

Fletcher T. 'Does he look like a Paki?': an exploration of 'whiteness', positionality and reflexivity in inter-racial sports research. Qual Res Sport Exerc Health. 2014;6(2):244–60.

Galam RG. Gender, reflexivity, and positionality in male research in one's own community with Filipino seafarers' wives. Forum Qual Soz Forsch Forum Qual Soc Res. 2015;16(3):26–51.

Ganga D, Scott S. Cultural "insiders" and the issue of positionality in qualitative migration research: moving "across" and moving "along" researcher-participant divides. Forum Qual Soz Forsch Forum Qual Soc Res. 2006;7(3):9–20.

Gill F, Maclean C. Knowing your place: gender and reflexivity in two ethnographies. In: Atkinson P, Delamont S, editors. Sage qualitative research methods. Thousand Oaks: Sage; 2011. p. 2–18.

Hayfield N, Huxley C. Insider and outsider perspectives: reflections on researcher identities in research with lesbian and bisexual women. Qual Res Psychol. 2015;12(2):91–106.

Hoskins K. Researching female professors: the difficulties of representation, positionality and power in feminist research. Gend Educ. 2015;27(4):393–411.

Järviluoma H, Moisala P, Vilkko A. Gender and qualitative methods. London: Sage; 2004.

Kezar A. Reconstructing static images of leadership: an application of positionality theory. J Leadersh Org Stud. 2002;8(3):94–109.

Kusek WA, Smiley SL. Navigating the city: gender and positionality in cultural geography research. J Cult Geogr. 2014;31(2):152–65.

Kusow AM. Beyond indigenous authenticity: reflections on the insider/outsider debate in immigration research. Symb Interact. 2003;26(4):591–9.

Liamputtong P. Researching the vulnerable: a guide to sensitive research methods. London: Sage; 2007.

Liamputtong P. Performing qualitative cross-cultural research. Cambridge: Cambridge University Press; 2010.

Liamputtong P. Qualitative research methods. 4th ed. South Melbourne: Oxford University Press; 2013.

Manohar NN. 'Yes you're Tamil! But are you Tamil enough?' An Indian researcher interrogates 'shared social location' in feminist immigration research. Int J Mult Res Approaches. 2013; 7(2):189–203.

Maylor U. Is it because I'm black? A black female research experience. Race Ethn Educ. 2009; 12(1):53–64.

Mellor J, Ingram N, Abrahams J, Beedell P. Class matters in the interview setting? Positionality, situatedness and class. Br Educ Res J. 2014;40(1):135–49.

Merriam SB, Johnson-Bailey J, Lee M-Y, Kee Y, Ntseane G, Muhamad M. Power and positionality: negotiating insider/outsider status within and across cultures. Int J Lifelong Educ. 2001; 20(5):405–16.

Pante MBLP. Female researchers in a masculine space: managing discomforts and negotiating positionalities. Philipp Sociol Rev. 2014;62:65–88.

Pasquini MW, Olaniyan O. The researcher and the field assistant: a cross-disciplinary, cross-cultural viewing of positionality. Interdiscip Sci Rev. 2004;29(1):24–36.

Pingol A. Remaking masculinities: identity, power, and gender dynamics in families with migrant wives and househusbands. Quezon City: University Center for Women's Studies/University of Philippines; 2001.

Ramji H. Exploring commonality and difference in in-depth interviewing: a case-study of researching British Asian women. Br J Sociol. 2008;59(1):99–116.

Rashid SF. Accessing married adolescent women: the realities of ethnographic research in an urban slum environment in Dhaka, Bangladesh. Field Methods. 2007;19(4):369–83.

Riessman CK. Narrative approaches to trauma. In: Riessman CK, editor. Qualitative studies in social work research. London: Sage; 1994. p. 67–71.

Savin-Baden M, Major CH. Qualitative research: the essential guide to theory and practice. Abingdon: Routledge; 2013.

Shariff F. Establishing field relations through shared ideology: insider self-positioning as a precarious/productive foundation in multisited studies. Field Methods. 2014;26(1):3–20.

Sikes P. Methodology, procedures and ethical concerns. In: Clive O, editor. Doing educational research: a guide to first-time researchers. London: Sage; 2004. p. 15–33.

Stiedenroth KS. Female, young, unmarried: the role of positionality while conducting fieldwork in Pakistan. Orient Anthropol. 2014;14(1):81–95.

Suwankhong D, Liamputtong P. Cultural insiders and research fieldwork: case examples from cross-cultural research with Thai people. Int J Qual Methods. 2015;14(5). https://doi.org/10.1177/1609406915621404.

Takeda A. Reflexivity: unmarried Japanese male interviewing married Japanese women about international marriage. Qual Res. 2012;13(3):285–98.

Tillman LC. Culturally sensitive research approaches: an African-American perspective. Educ Res. 2002;31(9):3–12.

Underwood M, Satterthwait LD, Bartlett HP. Reflexivity and minimization of the impact of age-cohort differences between researcher and research participants. Qual Health Res. 2010;20(11):1585–95.

Wegener C. 'Would you like a cup of coffee?' Using the researcher's insider and outsider positions as a sensitising concept. Ethnogr Educ. 2014;9(2):153–66.

Yakushko O, Badiee M, Mallory A, Wang S. Insider outsider: reflections on working with one's own communities. Women Ther. 2011;34(3):279–92.

Considerations About Translation: Strategies About Frontiers

93

Lía Rodriguez de la Vega

Contents

1 Introduction	1618
2 A Brief Review of Studies on Translation	1619
2.1 Lexical Issues	1619
2.2 Syntactic and Semantic Issues	1620
2.3 Ethical Issues	1620
3 Current Situation	1621
4 Translation	1622
5 Linguistic Treatment	1623
6 Addressing the Task of Translation	1624
7 Difficulties, Strategies, and Rules in Translation	1626
8 Translation Competence	1629
9 Implicit Ethical Issues	1630
10 Conclusion and Future Directions	1632
References	1635

Abstract

The translation of any text focuses action and attention on the transference of a source language into a target language. It entails different methodological issues that range from linguistic treatment, grammatical issues of the languages considered, lexical issues and how to approach them, space and time considerations in a given textual construction, implied ethical questions, and so on. The approach to such issues varies from one translator to another. While some look for lexicographic patterns between the original text and its translation, others seek to convey the sense of the original text in the contexts of translation. Against this background, this chapter reviews some theoretical questions underpinning methodological approaches and those approaches as used in translation, along with

L. R. de la Vega (✉)
Ciudad Autónoma de Buenos Aires, University of Palermo, Buenos Aires, Argentina
e-mail: liadelavega@yahoo.com

© Springer Nature Singapore Pte Ltd. 2019
P. Liamputtong (ed.), *Handbook of Research Methods in Health Social Sciences*,
https://doi.org/10.1007/978-981-10-5251-4_36

normative, linguistic, and ethical issues, concluding with a reflection about translation in cross-cultural research. For such purpose, this chapter draws on a bibliographic review and the experiences of researchers who translate or have translated different types of texts – mainly written ones – in a nonprofessional manner.

Keywords

Translation · Lexicographic pattern · Methodological approach · Linguistic treatment · Grammatical issue · Lexical issue · Space and time consideration · Theoretical question · Internationalization of English

1 Introduction

Communication is a phenomenon inherent in human nature, and, as such, it entails a shared experience – a bond. In reviewing the polysemic character of the term, Rizo García (2012, p. 22) provides a general definition of communication as "a basic process for the construction of life in society, as a mechanism that produces senses, activates dialogue and coexistence between social subjects." Communication allows the social fabric to be conceived as a network of interactions and is interwoven with culture(s) and identity(ies).

Within this framework, it can be stated that culture – seen as a network of meanings – is closely related to identity(ies), which evidence(s) the subjective processing of cultural matrices. Both culture and identity are connected with the notion of frontier – a diffuse concept – that is characterized by its duplicity as an object/concept and as a concept/metaphor. Thus, there are on the one hand physical, territorial frontiers, and, on the other, cultural, symbolic ones (Grimson 2011) (Grimson (2011) conceives cultural frontiers as significance regimes that are distinguished and perceived by their own participants.).

Against this backdrop, translation is the task of transferring a text from the source language to the target language, which also involves the adoption of a specific attitude toward certain frontiers. In this sense, it is worth noting that fidelity to the original played a central role in translation for a long time, until it was replaced by an equivalence-oriented approach (Snell-Hornby 1990, p. 80, cited in Carbonell 1996, p. 143). Hence, the subject matter of translation studies has gradually shifted toward prioritizing the communicative aspect of translation (Carbonell 1996).

Thus, in the management of the everyday world through language, whether the target text becomes part of the canon of the target culture or emphasizes its differences depends on the choice made by the translator. This also brings to the fore other issues associated with the translation task, such as the position to be adopted with respect to textual-, cultural-, and identity-related polyphony, that is, ideological issues and issues relating to disparities between the different narratives and cultural agency.

In this chapter, I review some theoretical questions underpinning methodological approaches and those approaches as used in translation, along with normative, linguistic, and ethical issues concluding with a reflection about translation in cross-cultural research. For such purpose, this chapter draws on a bibliographic review and the experiences of researchers who translate or have translated different types of texts – mainly written ones – in a nonprofessional manner.

2 A Brief Review of Studies on Translation

2.1 Lexical Issues

Vázquez, Fernández and Martí (2000) present a classification of verb-centered mismatches between Spanish and English. Such classification is based on a proposal for lexical representation of linguistic knowledge, thus falling within the scope of lexical semantics. The authors' proposal relies on a model for lexical description that considers meaning components, event structure and diathesis alternations. The paper describes how these elements allow creating a framework for the analysis of mismatches. Along the same lines, the authors suggest how to use conceptual transfer to cope with these mismatches in a Machine Translation system.

Martínez-Melis (2008) examines the translation of the Heart Sutra, a Buddhist text of the prajñaparamita sutra literature, which has been highly influential in Chinese culture since it was first translated in the fifth century. Drawing on a translational approach (based on the categories proposed by the Marpa Term group), the author compares the versions of Kumarajiva, Xuanzang, Dharmacandra, Prajña and Liyan, Dharmasiddhi, Prajñacakra, and Danapala, all of which belong to the so-called Chinese Buddhist canon and aims to identify the methodological and technical translation choices of the translators. The author concludes that the early periods of translation of Buddhist texts in China, in the third and fourth centuries, saw the adoption by the translator monks of methodological choices consisting of the assimilation of Taoist terms in Buddhist texts. The author further wonders whether the translation rendered by Xuanzang in the seventh century (considered to be the best one) contains an overuse of loans. Finally, the author points out that the Chinese language allows another possibility, consisting of the combination of two Sanskrit-Chinese lexical elements, according to which the Sanskrit element belongs to the phonetic translation, and it is thus a combination of the Sanskrit phonetic translation with a Chinese character that creates a new Chinese term.

Ayadi (2009–2010) carries out a qualitative and quantitative study aiming to identify the reason why learners of English are unable to find the appropriate equivalents of English phrasal verbs in Arabic. The author concludes that, according to the findings of the study, the students' inability to translate English phrasal verbs into their most appropriate Arabic equivalents is based on their total ignorance of, and insufficient exposure to, phrasal verbs, while at the same time students tend to depend on the context – which does not always prove useful – and to translate literally. This renders the translation of phrasal verbs unacceptable.

2.2 Syntactic and Semantic Issues

Li (2010) addresses the problems and challenges existing in current United Nations English-Chinese document translation practice. The paper presents a detailed analysis of examples taken from the official UN document system and investigates the major grammatical and lexical problems influencing the readability of UN translations and the translation strategies adopted by UN translators. The author concludes by identifying three major problems: readability (stating that poor readability is due to translators' failure to choose the appropriate translation strategies), incomplete sentence structure (which mainly originates in the lack of a subject, which in turn is caused by passive voice or the lack of a subject in the original text), and accuracy of individual words/phrases (stating that mistranslations are unavoidable. However, translators are responsible for minimizing mistakes and ensuring that translations are accurate).

Miličević (2011) relates translation studies to the theory of language acquisition, by examining recent findings on some grammatical properties of translated texts on the one hand and the findings of acquisitional studies dealing with such properties on the other. As well as reviewing research into the well-known phenomenon of pronoun overuse, the author focuses on a less explored problem: the overuse of possessive adjectives, for which the author considers preliminary data from English to Serbian translations. Relying on a comparison with the results obtained in acquisitional studies of possessive adjectives, the author argues that the different cases of patterning between translation and language acquisition – in particular second language acquisition and first language attrition – show a similarity in the linguistic systems of translators and language acquirers and speak in favor of closer collaboration between both fields.

2.3 Ethical Issues

Mansourabadi and Karimnia (2013) reviewed the ideological differences between Hoseini's novel *A Thousand Splendid Suns* and its two Persian translations, by Ganji and Soleimani and Ghebrai. Based on Fairclough's approach – according to which ideology in discourse is encoded in the lexical, grammatical, and textual elements, and changes in them are indicative of a different ideology – the authors analyzed the lexical choices of the source book and the two translations in order to establish ideological differences between them. The results of chi-square showed that there were no ideological differences between the source text and the two translations. In addition, the authors found that the translators had chosen similar vocabularies in order to represent the ideology of the original author.

Kruger and Crots (2014) conducted a survey of 31 members of the South African Translators' Institute (SATI), in which respondents were asked about the translation strategies they were most likely to select in order to address particular ethical challenges. In addition, in order to understand the factors affecting the choice of translation strategies, the impact of the experience and age of the translators and the text type and the type of ethical problem encountered were investigated. In a second stage, the survey sought to find out the reason why respondents chose particular strategies and their views on ethical responsibility. The findings show that there is an overwhelming preference for faithful translation and also reveal an interplay between personal and professional ethics as the motivation for such preference, with certain differences across text type and type of ethical problem.

3 Current Situation

In the globalized present, English is a central language that has become the privileged means of expression of science and technology, as well as of international political and trade relations, among others. English has given rise to the existence of different types of bilingualism. These types have in turn led to the dialectal diversification of English by virtue of its geographic expansion (related to the emergence of diverse English in the world) and contact with other languages, its internationalization and its influence on other languages – (The Englishization of the world's languages) which have thus been transformed – with the consequent emergence of different sociolects (Ferro Mealha 2012).

In the lexical field, that internationalization of English has had an impact on international organizations, where working languages tend to be reduced in favor of English or where English is considered to be a *lingua franca*, with its consequent influence as a supersubstrate. This is evidenced by the importation of concepts and terms created in the English-speaking world, the significant development of neologisms (This is the case of terms such as "Third Way," "weapons of mass destruction," bypass, airbag, e-mail, leasing, link, buffer, bit, and so on (Montero Fleta 2004; Muñoz Martín y Valdivieso Blanco 2007; Gutiérrez Rodilla 2014).), and so on (Muñoz Martín and Valdivieso Blanco 2007).

Therefore, within this framework of contact of languages, we can argue that translation is indeed a special case of such contact, given that the translator him or herself is subject to the interference of one of the languages in the other (source and target language) and, in addition, because translation allows for contact to be established in a different manner.

Within the context of a same language, it is well-known that not all speakers of a language use the same variety. Such linguistic variety can be verified by observing all of the levels of a given language. For example, terms and expressions such as "che" ["Hey"], "fiaca" ["feel lazy"], "hacerme una siesta" ["I'm gonna take a nap"], and so on are indicative of the Spanish spoken in Argentina. These varieties depend on extralinguistic factors such as the origin of the speaker, his or her age, sex, socioeconomic status, among others, and give rise to social linguistic variations or sociolects (which are thus defined on the basis of the social features of the speaker).

In addition to the speaker's features, the study of linguistic variation also takes into account where the speakers are (context), who the speakers are (interlocutor), and what they talk about (topic) in the conversational exchange. All of these issues define the situational linguistic variety known as *register*.

Within the framework of language variety, we as speakers recognize a variety of the language (Spanish in the case of Argentina) which is common to all speakers and such variety is taken as a model of the language considered and is known as *written standard variety* or *written normative language*. The different oral varieties of the language considered are variants of such language and speakers tend to attribute a greater prestige to one of the oral varieties to which they are exposed, generally the one that speakers identify with the most powerful social group (in cultural, political and/or economic terms) and the one which is closest to the written standard variety. This variety is known as *oral standard variety* or *oral cultured norm*.

Moreover, it is worth noting that languages often coexist with other languages (which presupposes the existence of bilingual speakers), and in such circumstances of *linguistic contact*, the varieties of the language spoken in such places have linguistic features that may be attributed to the contact with another language. These are the so-called *contact varieties*. A situation in which more than one language is spoken is known as *diglossia*, that is, a situation where two language varieties coexist within the same population or territory. In such contexts, one of the languages is more prestigious and is usually considered as a language of culture or of official use, while the other is relegated to socially inferior situations such as oral communications, folklore, and family life (An example of this is the one between the Spanish and Guarani in Paraguay. In this regard, Rubin (1974) notes that Spanish is the language that speakers choose to issues related to education, government, high culture, and religion, while Guarani is used to "matters of privacy or primary group solidarity" (pp. 121–122). The influence of Spanish in the Paraguayan Guarani led to the variety of Guarani known as "yopará," while also observed in the Paraguayan Spanish, the influence of Guarani (Ferrero and Lasso-Von Lang 2011). Similarly, the contact of Spanish with Brazilian Portuguese originated Portuñol, which records differences in the spoken language in different places, such as the province of Misiones (Argentina) and northern Uruguay (Lipski 2011). Similarly, in northwestern Argentina, it is verified in the contact of Spanish and Quechua, showing different contributions from Quechua to the first one in the lexical field, in the phonetic loans, and in the pronunciation of the "s" (Nardi 1976–1977).). Along these lines, when three or more languages coexist in such context, the situation is one of *polyglossia* or *multiglossia* (Suriani 2008; Supisiche et al. 2010).

4 Translation

I agree with Muñoz Martín (2014) that translation and interpretation are skills that are specific to bilingual competence in the translation task. Mayoral Asensio (1999) argues that there is no homogeneous concept of translation and identifies the properties of communication in general, taking into account the maxims of Grice (1975) (Grice points out that in all those situations the readers of a translation have expectations as to the original text and its meaning, which the translator must in turn satisfy within the framework of his or her ethics. This satisfaction can be assessed by using what Grice presents as perspectives of analysis, that is, an analysis of an assertion as to its truthfulness (quality), brevity (quantity), relevance (relation), and observance of set patterns (manner). Grice thus identifies the conditions of efficiency (maxims) in the communication of the linguistic variation, namely, adjusting to the context and the situation as specified in the translation instructions (Maxim of Quality and Maxim of Relation), adjusting the communicative strategy to the translation instructions (Maxim of Quality and Maxim of Relation), using only markers with which the reader is familiar (Maxim of Relation), maintaining only those distinctions that the reader can appreciate (Maxim of Quantity) not maintaining in the translated text those distinctions made in the original text that have no communicative function (Maxim of Quantity), using the minimum amount of markers that, along with other contextualization hints, allow identifying situational features and creating the desired effect (except in cases of deliberate alliteration) (Maxim of Quantity), not introducing unjustified ambiguities in the definition of situational features (Maxim of Quality), avoiding inconsistencies (in the case of cultural parameters, such inconsistencies may be caused by mixing features that are specific to each culture) (Maxim of Manner), and maintaining consistency in the type of markers used to point to a certain feature and the set of features of a text (Maxim of Manner) (pp. 171–172).) which allow explaining a variety of conversational translation situations and of different solutions.

Within this framework, it can be stated that a translator is faced with different circumstances related to lexical semantic, grammatical, syntactic, rhetorical, pragmatic, and cultural issues. Lexical semantic issues are related to terminological alternation, neologisms, contextual synonymy and antonym issues, lexical networks, semantic contiguity, and so on. Grammatical issues comprise aspects such as tenses and pronouns. Syntactic issues may arise from syntactic parallelism, passive voice, rhetorical figures of speech such as anaphora (the repetition of a word or phrase at the beginning of a verse or sentence) and hyperbaton (inversion of the natural order of speech) reaction, and so on. Rhetorical issues are related to the identification and recreation of figures of thought (comparisons, metaphors) and diction. Pragmatic issues are concerned with differences in use (e.g., the difference between "tú," "vos" and "usted" to address a person in Spanish), idiomatic expressions, proverbs, humor, and so forth.

Finally, cultural issues concern the differences between cultural references. Coseriu (1977) examines the translation of culture in his analysis of the relationship between signification (signifier-signified relationship), designation (the relationship between sign and referent), and sense (the meaning as actualized in a text). He argues that the transposition of the three is not always possible when translating and that the translator must privilege designation and sense in the translation (Gentile 2012).

5 Linguistic Treatment

As far as linguistic treatment is concerned, it is worth pointing out that the norm constitutes the foundation that gives both cohesion and continuity to a language, which can thus maintain its usefulness. The norm can be understood in two different ways: from a prescriptive perspective, with reference to an imperative of the use/correctness that tends to be unique and does not admit variations; and from a descriptive perspective, according to which the norm is what is normal, a set of characterizing features that distinguish a given language, what the majority of speakers habitually use. Thus, while speakers can deviate from the norm, such deviation is not without limits, which are given by the need to understand the message.

The speaker can – and usually does – resort to authority as a supplement to, or substitute for, the norm. In this respect, we can distinguish between (a) power-based authority, that which has a prescriptive function in the field of language, assigned in a formal manner, or which is vested with an institutional power of another type but that has an impact on the field of language (language academies, public powers, and so on and b) credit-based authority, that which the speaker trusts because he or she assigns to it a certain measure of credit (dictionaries of language use, specialists, the internet) (Muñoz Martín and Valdivieso Blanco 2007).

As an example of the above, the commentators of Sanskrit texts during the Vedic period enjoyed an authority and preeminence that could be as important as that of the author to whom the text was attributed. When commenting the texts, those commentators identified different meanings of specific terms, provided examples, cited other texts, and so forth (see Levman 2014).

6 Addressing the Task of Translation

The task of translation can be addressed in different ways, and it depends on the stance to be adopted by the translator him or herself. Chesterman (1995, 1997, cited in Schjoldager et al. 2008) describes the evolution of translation theory by distinguishing eight interrelated stages, to each of which he assigns a metaphor to describe the prevailing view of translation in each of the stages. Chesterman points out that there is a general trend to alternate between a stage of source-text dominance and another of target-text dominance in the process of translation. The stages are presented below.

Stage 1 begins in 1000 BC, and the metaphor assigned to it by the author is "translating is rebuilding." A set of "units," namely, words in one language, are taken and rebuilt in another set of "units," that is, words in another language (Schjoldager et al. 2008).

Stage 2 starts in the fourth century, and the metaphor assigned to it by the author is "translating is copying." In this stage, translators were faced with the dilemma of how to translate without changing meaning. St. Jerome, a translator from this period, presented a three-term taxonomy that has had an influence over translation studies since then: (1) faithful word-for-word translation, (2) faithful sense-for-sense translation, and (3) unfaithful free translation.

Stage 3 begins in the fourteenth century, and the metaphor assigned to it is "translating is imitating." In this stage, many translators, instead of making a copy-like translation, sought to imitate the creative process of the original author, thus taking a freer approach (with respect to the original text) to translation.

Stage 4 begins in the nineteenth century, and the metaphor assigned to it by the author is "translating is creating." In this stage, some translators started to experiment with a translation style that deliberately created a foreignness in the target texts, thus giving rise to an overview of equivalence-based work in translation studies (Schjoldager et al. 2008).

Stage 5 starts in the twentieth century, and the metaphor assigned to it is "translating is recoding." In this stage, there was considerable experimentation with the so-called machine translation. Of particular note in this period is the Prague school, which was formed by a circle of linguists that emphasized the analysis of language as a system of functionally related units. Eugene Nida, a renowned linguist and Bible translator – whose most significant contribution is the principle of dynamic (or functional) equivalence over the principle of formal equivalence (or correspondence) – is the most outstanding scholar from this period. Another significant contribution to the linguistics-oriented approach is that of J. C. Catford, his most influential contribution probably being the notion of shifts.

Stage 6 begins in the 1970s, and the metaphor assigned to it is "translating is communicating." In this stage, translation studies became a separate discipline. Some developments in general linguistics seem to have paved the way for the independence of translation studies as an academic discipline. Language studies no longer revolved around the phrase, and the context acquired more importance (Snell-Hornby 2006). Translation studies were greatly influenced by text linguistics and pragmatics, among which Grice's theory of conversational implicature and the contributions of de Beaugrande and Dressler and, more recently, of Hatim and Mason have been highly influential. Many scholars from this period rejected the notion of equivalence and replaced it with target-text functionality as the controlling factor in translation processes. Such studies were developed by German scholars, with the *Skopos* theory by Hans J. Vermeer being the most influential (Schjoldager et al. 2008). This new approach gives more visibility to the translator, who, according to Christiane Nord, remains solely responsible for the loyalty of the translation (Vidal Claramonte 1995). While loyalty can be understood from different points of view, this new approach to translation is particularly relevant in the field of intercultural translation, given that the translator is in the privileged position of deciding how to translate what he or she translates, thus conveying one message or another in the target culture.

Stage 7 also begins in the 1970s, and the metaphor assigned to it is "translating is manipulating." During this period, concurrently with the developments in Stage 6, a group of Israeli and Belgian scholars developed a new approach to translation studies. This was a break with the trend toward prescriptivism followed by scholars from previous stages. These scholars followed Holmes' (1972) proposal for emphasis on descriptive translation studies. A particularly influential scholar in this stage was Gideon Toury, who focused on the translator as a social agent.

Stage 8 begins in the 1980s, and the metaphor assigned to it by Chesterman is "translating is thinking." This stage ushered in a growing interest in the translation process and the use by scholars of cultural studies tools for the study of translation from a cultural, sociological, and cultural perspective (Schjoldager et al. 2008).

The idea that "translating is not neutral" became increasingly prominent. Along these lines, Robinson (1997) points out that translation has always served as a mechanism of conquest and occupation. That is, during the colonization process a network of hierarchical relations is created through discourse that eventually consolidates and spreads across time and space. In addition, the concept of otherness gains impetus within the theory of postcolonial translation, due to its presence in the works of authors that live in contact with two different civilizations and cultures, who look for specific forms of expression that give voice to both languages and cultures, with consequent social and political implications. Not only are the messages conveyed important but also the manner in which they are conveyed, making it necessary to review the role of translation in the representation of other cultures (Sales Salvador 2004) (Dovidio et al. (2010) sustain that stereotypes are cognitive schema used by social percipients to process information about others, reflecting not only beliefs regarding the characteristic features of group members but also information about the diverse qualities of those "others," influencing the emotional reactions toward the members of that group. Associated to this, Pedulla (2012) points out that stereotypes may play an important role in molding attitudes, often being related to discrimination and prejudice.).

Translation studies have currently evolved in so many directions that it is difficult to determine which of them is more typical (Schjoldager et al. 2008). Within this framework, the colleagues whom we interviewed refer to their own positions and translation logics. An Indian colleague, aged 45, who translates literary texts – moving in a universe of English, Hindi, and Spanish – states that: [...] "I need to acquire the sense of the text. I always want to drown in the original." An Argentine colleague, aged 65, who also translates literary texts (essays, theory and poetry) from French, English, and exceptionally Japanese (classical poetry) into Spanish suggests that she usually reads the text in the original language first, checking if there are terms that may give rise to ambiguities and solving that using appropriate sources. After that, she begins to translate, taking into consideration not only the correct syntax but also the appropriate rhythm. Another Argentine colleague, aged 50, who does interlinguistic translation work of literary, philosophical, religious, and legal texts elaborates that as a university professor and scientific researcher, she uses translation not as an end in itself but as a didactic tool. She uses the method of grammar and translation, meaningful learning, and contrastive linguistics.

7   Difficulties, Strategies, and Rules in Translation

In analyzing the possible strategies to be adopted when translating, Venuti (1995) sets a series of parameters that mark and define the translation process. He provides a classification based on the degree of alienation of a translation, its fluidity, and the type of existing relationships between cultures. He makes reference to what he considers a good or bad translation in those terms, considering thus that while a fluid translation is that where the language used combines the literary standards of the target language with the "marginal" aspects of the source in such a way that cultural and linguistic differences show, a bad translation shapes an ethnocentric domestic attitude revolving around the foreign culture.

Different areas of study offer diverse terminological features, and the manner in which they are approached varies from one translator to another. The Indian colleague notes that:

> The most common difficulties are not grammatical (...) and I think that nothing is untranslatable. It all comes down to finding the appropriate equivalent. So I look for such an equivalent in the social experiences of the culture of the target language, in its myths and legends and in its society... I think that we are all human and that all societies have the same experiences. Then, all of us have the necessary vocabulary to convey those experiences and it is just a matter of knowing where to look and go for it.
>
> On some occasions, for example, when translating a "Chick-lit" novel from English into Hindi, I was having some lexical issues... (as) I don't believe I have the necessary vocabulary in Hindi to express feelings/intimate experiences (taking into account her own Hindi background and leaving the possibility open for more considerations from and about translators from the same perspective).

On the other hand, one of my Argentine colleagues comments that:

> The difficulties of English and French texts usually lie in idiomatic expressions, in not over-expanding where the language is concise, that is, problems related to the lexical and syntactic fields. In the case of Japanese, from which I've translated some poetry, we encounter more problems as there are no morphological coincidences: elements like number are not necessarily marked in Japanese, which also has a smaller variety of tenses and, as a result, there arises a need to choose from the different possibilities in Spanish. Besides, as this is a culture that carries a strong stamp concerning objects, habits, clothing, etc., I consider that the term should be kept in the source language and notes or comments should be included.

Similar resources are used by prestigious indologists Tola and Dragonetti (1999; Dragonetti and Tola 2004), who make use of introductions which put translated texts in context and describe the development of their study, including notes and sections with key concepts in order to clarify terms and concepts under analysis and to facilitate the access to the semantic network that sheds a light on the text and the subject itself. Another Argentine colleague suggests that:

> One of the most common difficulties when translating from a very rich and concise language is finding equivalents in other languages that lack those characteristics. For example, a complex structure, including subordinate clauses, is needed in Spanish in order to translate an idea that in Sanskrit is conveyed by a nominal phrase.

In the lexical field, a highly interesting example is that of the multidisciplinary group Marpa Term of the Departament de Traducció I d'Interpretació de la Universitat Autònoma de Barcelona, which studies Tibetan Buddhism in the West from a translation-related perspective and in this context develops a terminological database. The group points out the coexistence of diverse areas of study of

Buddhism that fostered different translation methods, which in turn result in the coexistence of different equivalent terms seeking to translate the same term in the source language. The source of its database is Tibetan, and the target languages are Spanish and Catalan. The group carries out a prescriptive task that aims to generate proposals to normalize the use of terminology in Spanish and Catalan in the context of Tibetan Buddhism. Each entry includes the Tibetan term in its original form, in the Tibetan alphabet, and its Wylie transliteration, together with its equivalent in Spanish and Catalan (also including its equivalent in English, French, and Sanskrit). It also includes a section which refers to the translation technique applied so that the user may understand how the proposed equivalent was achieved. With respect to the translation techniques considered, an order of priority is proposed as follows: equivalent, context equivalent, coinage, periphrastic translation, calque, and loan (Martínez-Melis and Orozco 2008).

In addition to any lexical issues and their equivalents, the creation of neologisms, syntactic adaptations, and the solution to agreement difficulties, there are space-time considerations involved in the text constructs under analysis. In this respect, it is worth noting that a translation, both as an activity and as a product, falls within a specific social context, determined by specific linguistic, political, cultural, and socioeconomic coordinates, which at the same time renders a text that also accounts for such coordinates. The sensitivity of this issue is related to the fact that time and space are founding variables of social relationships, paramount both for the creation of the text as well as its reception and understanding. This raises the question of how effectively establish a co-participation in space times considered by translating and if that practice contributes to a kind of intellectual genealogy of the own translator, in the deep heuristic sense of the question (Personally, I consider that the way such issues are addressed has scope to rescue the interpellation of the translated texts into the present affairs (with a corresponding impact on translation competence).).

Vibha Maurya (2008), in her translation of Don Quixote, from Spanish into modern Hindi writes:

> Besides, I consider that there is something even more meaningful. As a translator, I find myself in a very different place culturally speaking and in a fairly distant historical time. Consequently, it was necessary to make a temporal and linguistic adjustment. I think that, as a "(re)writer" of a text in its translation, it is my duty to bring the readership closer to the text as well as the text closer to the readership, so that the translation is aimed at treading a double path. I am fully aware that in this process as a whole there are several losses but also a few gains for the target language. The translation of Don Quixote into Hindi is intended to make the literary world of our language richer, I believe.
>
> Throughout my career as a translator of literary works, I have witnessed a process which is composed of two branches which are interdependent of the social system: the linguistic branch and the sociocultural branch. (pp. 546–547) (Her assertions are complemented by those of Maurya (2015, p. 4), where she stated that "faithfulness has nothing to do with literality; this practice not only ruins a translation but also distorts the relationship between the translator and the original author (. . .) The language of the author is not, after all, just a bundle of words or a fortuitous lexical collection, a random syntax. Each word, each lexical item, is carefully selected, its direct meanings and its connotations are so complex that they seem to be constantly rebelling. In fact, the rendering process is like a vortex of meanings — where a translator sinks and spins in order to obtain equivalents for words in both languages

(...) That is why faithfulness may be a sincere ideal goal of the translator of literary works but it is completely utopian (...).")

In considering an intercultural context, translators, who find themselves between two cultures, must know them both, their rules and differences, in order to be able to apply the necessary strategies in an appropriate manner, thus obtaining different results in each case (Toury 1995). Venuti (1995) argues that the translator must break certain rules shared by the members of the target culture to guarantee the rendering of the message and the source identity, but he or she must also comply with some of those rules to prevent the translation from being rejected, thereby finding the middle ground. The greater the level of acceptance of a text in the target culture, the more changes made to the target text with respect to the source. This will result in the text being closer to the readership and will make the translator invisible (as a discourse effect that depends on how the translator handles the translation and how the public reads and assesses it).

Nord (1989, 1997) meanwhile proposes a model to develop the translation process from a functional approach called "the looping model." It considers the translation process as a circle rather than as a line and makes reference to the responsibility translators have toward their partners, referred to as "loyalty," a principle first introduced by the *Skopos* (Translation theory proposed by the German translator Vermeer in 1978, who claims that the principle that defines the translation process is the *Skopos* (purpose) of the transactional action (Jabir 2006).) theory, which tries to account for the cultural features of translation concepts by setting an ethical limit to the range of possible *skopoi* for the translation of a specific source text (Nord 2006).

Hatim and Mason (1995), who start with the elements that comprise the text structure, analyze the relationship between discourse processes, and translation practice, taking into account the motivations of both the author and the translator. They analyze language as text within a cultural and situational context, giving importance to the pragmatic, semiotic, and communicative values of discourse, the context elements whereby the translator captures the intention of texts. House (1997), on the other hand, a predecessor of Hatim and Mason, may be classified within the functionalist translation theories because of her sociocultural perception and her influence in functionality concepts, both of the source text and the translation.

8 Translation Competence

In line with the above statements, besides the bicultural perspective presented by Hatim and Mason (1995), several authors acknowledge the development of a translation competence. For example, Muñoz Martín (2014) proposes a construct for translation expertise consisting of five dimensions: (1) knowledge, (2) adaptive psychophysiological traits, (3) problem-solving skills, (4) regulatory skills, and (5) the self-concept. These dimensions are regarded as scopes in a complex behavior, and they do not imply any "internal" separate mental activity (p. 17). On the other hand, in 1997, the PACTE (Process Acquisition of Translation Competence and Evaluation) group was created. The group was formed in order to analyze the acquisition of translation competence (ATC) in written translation, and in 1998, they developed a holistic model which established several basic premises.

In this regard, the group conceives translation competence as the underlying system of knowledge necessary to translate; it is, therefore, regarded as expert knowledge comprising both procedural and declarative knowledge. It is by virtue of this competence that the translator is able to perform the translation process. They also consider that translation competence comprises several sub-competences (language, extralinguistic, knowledge about translation, instrumental, psychophysiological and strategic) as well as psychophysiological components.

Considering the different sub-competences, we can say that the language sub-competence refers to the expert's ability to switch between both languages, and it comprises pragmatic, sociolinguistic, textual, grammatical, and lexical knowledge (PACTE Group 2003). The extralinguistic sub-competence consists of "[...] declarative knowledge, both implicit and explicit, about the world in general and special areas" (PACTE Group 2003, p. 58), whereas the "knowledge about translation" sub-competence implies that the translator needs to know how translation functions: "types of translation units, processes required, methods and procedures used (strategies and techniques), and types of problems, as well as knowledge related to professional translation practice: knowledge of the work market [...]," and so on (PACTE Group 2003, p. 59). The instrumental sub-competence refers to knowledge related to the use of documentation sources and information and communication technologies applied to translation – the use of dictionaries, encyclopedias, glossaries, and so on (PACTE Group 2003).

These sub-competences are complemented by certain psychophysiological components: (1) cognitive components such as memory, perception, attention, and emotion; (2) attitudinal aspects such as intellectual curiosity, perseverance, rigor, critical spirit, knowledge of and confidence in, one's own abilities, the ability to measure one's own abilities, and motivation; (3) abilities such as creativity, logical reasoning, analysis and synthesis, and so on (PACTE Group 2003, p. 59). When translation is performed, these sub-competences interrelate establishing a hierarchy that depends on the specific work, and different variants are possible depending on different factors (text, context, translation experience, and so forth). Here, the strategic sub-competence plays an essential role in that it regulates and compensates all the other sub-competences.

9 Implicit Ethical Issues

Inghilleri (2009, p. 102, cited in House 2016, p. 27) states that "once the space between the translator and the text (...) is acknowledged as irrefutably ethical, the task of the translator cannot be viewed as simply linguistic transfer, while this is understood as segregated from an ethical injunction." The idea that translation, the social context, and the translator's position/stance are closely related is not new but has been emphasized in recent years. In this respect, House (2016) considers that this new emphasis lies in conceiving culture as a space of ideological struggle and the translator as a "stimulator" resisting the influence of hegemonic structures. In line with this approach, translation may also be considered ideology. Along the same lines, there has been significant production regarding the "politics of translation" and the perception of translation as a space for political action (Baker 2013), these considerations arising within the framework of the relation between translation and conflict scenarios.

Ethical issues in translation, therefore, seem to be related to a large and complex tension between the literal rendering of the source text and a translation that considers the expectations of the target language's readership taking into account the cultural context and the text's function in the target culture. Alwazna (2014) proposes a middle ground between these approaches, paying special attention to the nature of the text, the purpose of the translation, and the readership. İçöz (2012), in turn, aims to identify those situations in which a translation becomes unethical. The author refers to a prejudiced version, misinformation affecting the translation on purpose or by mistake, among others. In any case, it is clear that translating requires the translator to perform a constant critical and reflective analysis over his or her work and over him or herself.

The colleagues whom I interviewed explain their positions with respect to this tension that raise, is resolved, and rise again. In an Indian colleague's view, "a translator should make sure the translation is an echo of the original without losing sight of the target language's reader." An Argentine colleague adds that:

> I try to be faithful to the original; I'm not keen on recreating, adapting or looking for rhymes when the original text makes use of them. I always try to make the resulting text nice to the ear and make it adapt to syntactic rules in Spanish. I have never ventured into the translation of texts which are avant-garde or experimental in the source language.

Another Argentine colleague also remarks that:

> It is essential to be as faithful as possible to the original, producing a text equivalent to the original and, for this purpose, subjective considerations should interfere as little as possible in the end result of the translation process. Translation should not be used as an ideological weapon.

Regarding the unavoidable reflection about the power that operates in translation, the respondents allude to the relationship of language with spatial semiotics, which opens another dimension to deepen: the relationship between language, space, and power (It seems relevant to remember Tally, when he says that "*To draw a map is to tell a story*" (Daricci 2015)). Thus, for an Indian colleague:

> I find great pleasure in translating from Spanish into Hindi and from Hindi into Spanish. In my opinion, translating directly from Spanish into Hindi works in two ways. Hispanic literature in India is mostly known through translations from the U.S. and Indian literature – I'm referring only to literature in Hindi – is not known in the foreign world, except for religious texts. This is why it is so important for me as a translator to work with these two languages.

An Argentine colleague also remarks that:

> Language is a defining and essential aspect of all peoples and nations. History shows the unavoidable association between power and language: for example Latin's role in the invasion and dominance of the Roman Empire, or Spanish in the discovery and colonization of the Americas. In this age, rather than through weapons and war, power is imposed through other levels: technology, commerce, and science, and in this sense, it becomes difficult to find answers or solutions. Spoken language, if not recorded, will be lost – there always were and still are many communities that don't use a written language. In my opinion, in large societies, the power of the dominant language pervades fields like commerce and science; whereas in the minority languages the arts (literature, songs) are preserved.
>
> Within the framework of this analysis, I think it is essential to address the subject of the written word: This applies to China, which despite its industrial and commercial power, uses a complex (spoken and written) language that is very much restricted and is usually replaced by English for the signature of treaties, conventions, etc. (Rodriguez (2014), links the promotion of language to the notions of "soft power" and "nation branding" and points out the successful cases of the United States and China in this task.)

Another Argentine colleague adds that:

> I think the translator's task as an agent that promotes cultural visibility is extremely important. This transference should be serious, scientifically founded and devoid of superimpositions germane to the culture of the target language. It is also very valuable that there are now more "languages of knowledge.

The experiences of my colleagues bring to the fore important aspects of the translation process, such as the projection of a concrete identity through translation and the resulting active and conscious participation in an intercultural dialogue process, the "reading" of language use and its geopolitical implications, and the space of empowerment of some languages in relation to others, as knowledge seems to be associated with certain specific languages (while others could appear as "repetition" of the former) (Ferro Mealha (2012) addresses the use of English as the language of academic and scientific discourse and points out the bias to which it can lead to not knowing the language and not publishing in English.). All of these matters – which require a thorough and focused analysis – relate to the observations made by Restrepo (2010) in the sense that identities (thinking about the identity of the author, those identities which the text realizes, the one of the translator, the ones of the readers, and so on) are not only concerned with differences but also with inequalities and domination; that is to say, identity demarcation practices are connected with the preservation or confrontation of different hierarchies (at the social, political, and other levels).

10 Conclusion and Future Directions

The all-important role that the translator has had throughout history as a mediator between languages/cultures acquires special significance today, in a context of enormous technological advances that on the one hand seem to facilitate the translator's work but on the other may also condition it.

Contact between cultures evidence the dynamics of power at play in those relations, which has an inevitable influence on the translator, the translation process, and the end product. In other words, "translating is not neutral" (and translators are not mere "technicians"). Therefore, since the translator decides what to convey and how to convey it, his or her stance becomes essential in the translation process and, consequently, in the representation of the others and the process of the cross-cultural research of which it is part. Furthermore, and in accord with the notion that the translator is influenced by an epistemic subjectivity (as he or she requires previous knowledge in order to translate, but he or she also knows the task of translation itself), I agree with Retamozo (2007, quoted in de la Vega Rodriguez 2015), who considers that the construction of such subjectivity is the result of a position that articulates volition and consciousness in an indivisible manner with the translator's ethical and political stance.

Thus, while an Argentine colleague notes that she cannot conceive intellectual work and research in any field without the aid and use of translations, my Indian colleague states that:

> In answer to the question of "what is translation?", Umberto Eco states that it is "saying almost the same thing". The key lies in that "almost", which makes us understand that translating can never imply an exact literal transposition of the meaning of one text into another, written in another language. On the contrary, it involves a process of cultural negotiation.

She also contends that:

> Translation is a bridge between two languages and cultures. It is important that the bridge we build is nice and strong, and translators play a very important role in this. The significance of this cultural change is undeniable, even in today's globalized world.

Regarding the above statement, an Argentine colleague adds: "I respect and adhere to the Buddhist principle that we should 'reach each person in the language he or she speaks'. In this sense, the translator is a true pontiff, in the etymological sense of the word, i.e. a 'bridge-builder.'"

I thus perceive a translator as a kind of bridge builder, an agent and manager of linguistic change, who recovers cultural agency and empowerment through translation, in the understanding that the link between knowing a language and applying that knowledge is not an innocent one. On the contrary, the construction of meanings around the subjects, their relationships, and their productions carries sociopolitical consequences, thus destroying, reconfiguring, and/or constructing new frontiers. In this context, the translation strategies do not appear as a certain and fixed itinerary but rather as a compass that guides the translation, while representations of otherness are always relational and relationships, although similar, are always unique.

Being an integral component of cross-cultural research, translation raises ethical, epistemological, and practical questions inherent in a research. Considering the possibility of imposing another conceptual framework on the translated subjects (and their correspondent cultural contexts, values, life worlds, languages, texts, and so on), it is evident that the need of reflexive approach to translation not only to overcome difficulties involved in cross-cultural research but also to take a stand on the implicit element of power in the situation (that concerning hegemonic and non-hegemonic cultures/languages). As translators, researchers, and/or researchers/translators, we have personal sociopolitical positions that impact not only in the product of research but also the interpersonal relations in the fieldwork. In that sense, our multilingual identities influence and impact our "locations." Researchers (and)/translators should then recognize the linguistic and cultural differences that data translation must negotiate, preserving and highlighting cultural differences rather than resembling the dominant values of the target culture by translation. Thus, the acknowledgement of the translator's roles as intercultural communicator and data interpreter in the research process shows the importance of the translator as an integral part of the knowledge production system (**see also Finding meanings: A cross-language mixed-methods research strategy and Conducting cross-language qualitative research with people from multiple language group**).

All these elements are important to consider the rigor of the research and that emphasizes the need to understand and recognize the critical importance of language in the generation of knowledge and its cultural interpretation. For that reason, the debate and constant reflection on translation/interpretation in cross-cultural research should involve "the hierarchies of languages, power, the situated epistemologies of the researcher, and issues around naming and speaking for people who may be seen as other" [...]. "speaking for others, in any language is a political issue, which involves the use of language to construct self and other" [...] "translating itself has power to reinforce or to subvert longstanding cross-cultural relationships but that power tend to rest in how translation is executed and integrated into research design and not just in the act of translation per se" (Temple and Young 2004, cited by Alzbouebi 2010, p. 7). Translation (and the given interpretation) is about understanding that language is connected to "local toponymies" (space, historical time, identity, culture, the social apprehension of reality and its narrative, and so so); all of which are crucial for any cross-cultural research.

It is then necessary to discern how we can be more faithful to the sources we translate (subjects, cultures, identities, texts) and assume that all this involves operations with different temporalities (that of the moment of translation and that related to the history of what or whom we translate and the one related to who does the translation), spaces (the one of the translator, that of what or who is translated and the space for which the translation is done), and the memories related to all that (the implicit and explicit memories in what is translated, who translates, and the memory of the translator, that activate the life trajectories involved). As Richard points out (2002), the name implies a cut and modeling of a category of intelligibility, and, as we have said, this is not dissociated from power as an element of social relations, of which we finally go to give account in the cross-cultural research processes we develop.

References

Alwazna RY. Ethical aspects of translation: striking a balance between following translation ethics and producing a TT for serving a specific purpose. Engl Linguist Res. 2014;3(1):51–7. Retrieved from http://www.sciedu.ca/journal/index.php/elr/article/view/4852/2841

Alzbouebi K. The splintering selves: a reflective journey in educational research. Int J Excell Educ. 2010;4(1). Retrieved from http://uklef.ioe.ac.uk/documents/papers/alzouebi2006.pdf

Ayadi A. Lexical translation problems: the problem of translating phrasal verbs. The case of third year LMD learners of English. Unpublished master Thesis, Master in Applied Language Studies, Department of English, Faculty of Letters and Languages, Mentouri University Constantine, People's Democratic Republic of Algeria. 2009–2010. Retrieved from http://bu.umc.edu.dz/theses/anglais/AYA1194.pdf

Baker M. Translation as an alternative space for political action. Soc Mov Stud. 2013;12(1):23–47.

Carbonell O. The exotic space of cultural translation. In: Álvarez R, Vidal Claramonte MCA, editors. Translation, power, subversion. Clevedon: Multilingual Matters; 1996. p. 79–98.

Chesterman A. The successful translator: the evolution of homo transferens. Perspect Stud Translatol. 1995;2:253–70.

Chesterman A. Memes of translation: the spread of ideas in translation theory. Amsterdam/Philadelphia: John Benjamins Publishing; 1997.

Coseriu E. El Hombre y su lenguaje. Estudios de teoría y metodología lingüística. Madrid: Gredos; 1977.

Daricci K. "To draw a map is to tell a story": interview with Dr. Robert T. Tally on Geocriticism. Revi Forma. 2015;11:27–36. Retrieved from https://www.upf.edu/forma/_pdf/vol11/05_darici.pdf

Dovidio JF, Hewstone M, Glick P, Esses VM. Prejudice, stereotyping, and discrimination: theoretical and empirical overview. In: Dovidio JF, Hewstone M, Glick P, Esses VM, editors. Handbook of prejudice, stereotyping, and discrimination. London: Sage; 2010. p. 3–28.

Dragonetti C, Tola F. Dhammapada. La esencia de la sabiduría budista. New Jersey: Primordia; 2004.

Ferrero C, Lasso-Von Lang N (Coords.). Variedades lingüísticas y lenguas en contacto en el mundo de habla hispana. Bloomington: Author House; 2011.

Ferro Mealha I. English as the language of science and academic discourse in (non-) translated medical Portuguese: the initial stages of a research project. In: Fischer B, Nisbeth Jensen M, editors. Translation and the reconfiguration of power relations: revisitng role and context of translation and interpreting. Austria: Universität Graz-Das Land Steiermark-CETRA; 2012. p. 221–36.

Gentile AM. La traducción de las referencias culturales: el caso de la obra de Philippe Delerm en español. 2012. Retrieved from http://citclot.fahce.unlp.edu.ar/viii-congreso/actas-2012/Gentile.pdf

Grice HP. Logic and conversation. In: Cole P, Morgan JL, editors. Syntax and semantics 3: speech acts. New York: Academic; 1975. p. 41–58.

Grimson A. Las culturas son más híbridas que las identificaciones. Diálogos inter-antropológicos. En Los límites de la cultura. Críticas de las teorías de la identidad. Buenos Aires: Siglo XXI. 2011. Retrieved from http://live.v1.udesa.edu.ar/files/UAHumanidades/Critica%20Cultural%202011/Culturas_hibridas.pdf

Gutiérrez Rodilla BM. El lenguaje de la medicina en español: cómo hemos llegado hasta aquí y qué futuro nos espera. Panace@. 2014;XV(39):86–94. Retrieved from http://www.medtrad.org/panacea/IndiceGeneral/n39-tribuna_GutierrezRodillaB.pdf

Hatim B, Mason I. Teoría de la traducción: una aproximación al discurso. Barcelona: Ariel S.A; 1995.

Holmes JS. The name and nature of translation studies. In: Venuti L, editor. The translation studies reader. London: Routledge; 1972. p. 172–85.

House J. Translation quality assessment: a model revisited. Tübingen: Narr; 1997.

House J. Translation as communication across languages and cultures. New York: Routledge; 2016.
Hui-Wen T, Martínez-Melis N. Siete traducciones chinas del Sutra del corazón. In: San Ginés Aguilar P, editor. CEIAP (Colección Española de Investigación sobre Asia Pacífico), 2, cap. 39. Granada: Editorial Universidad de Granada; 2008. Retrieved from http://www.ugr.es/~feiap/ceiap2v1/ceiap/capitulos/capitulo39.pdf
İçöz N. Considering ethics in translation. Electron J Vocat Coll. 2012;2:131–4. Retrieved from http://www.ejovoc.org/makaleler/aralik_2012/pdf/14.pdf
Inghilleri M. Translators in war zones: ethics under fire in Iraq. In: Bielsa E, Hughes ChW, editors. Globalization, political violence and translation. Basingstoke, Hampshire: Palgrave Macmillan; 2009. p. 207–21.
Jabir JK. Skopos therpy: basic principles and deficiencies. J Coll Arts. University of Basrah, 41. 2006. Retrieved from http://www.iasj.net/iasj?func=fulltext&aId=50013
Kruger H, Crots E. Professional and personal ethics in translation: a survey of South African translators' strategies and motivations. Stellenbosch Pap Linguis. 2014;43:147–81. Retrieved from http://www.ajol.info/index.php/splp/article/viewFile/111744/101509
Levman JG. Linguistic ambiguities: the transmissional process, and the earliest recoverable language of Buddhism. Unpublished PhD thesis, Doctorado en Filosofía, Department for the Study of Religion, University of Toronto. 2014. Retrieved from https://tspace.library.utoronto.ca/bitstream/1807/68342/1/Levman_Bryan_G_201406_PhD_thesis.pdf
Li X. An Investigation on grammatical and lexical problems in the English-Chinese document translation in the United Nations. Unpublished master thesis, Master of Arts in Translation and Interpreting, School of Languages and Comparative Cultural Studies, the University of Queensland. 2010. Retrieved from https://espace.library.uq.edu.au/view/UQ:210208/s41874398_MACTI_Thesis.pdf
Lipski JM. Contactos lingüísticos hispano-portugueses en Misiones, Argentina. Texto a formar parte de Homenaje a Emma Martell, editado por María del Mar Forment. 2011. Retrieved from http://www.personal.psu.edu/jml34/Misiones.pdf
Mansourabadi F, Karimnia A. The impact of ideology on lexical choices in literary translation: a case study of a thousand splendid suns. Procedia Soc Behav Sci. 2013;70:777–86. Retrieved from http://ac.els-cdn.com/S1877042813001249/1-s2.0-S1877042813001249-main.pdf?_tid=586cbab8-5d0f-11e6-99cb-00000aacb360&acdnat=1470623297_1fa0fb5313ab949ce31b1a78dc8bca54
Martínez-Melis N, Orozco M. Traducir la terminología budista. Del sánscrito y tibetano al castellano y catalán. Actas del III Congreso Internacional de la Asociación Ibérica de Estudios de Traducción e Interpretación (AIETI). Barcelona: Universitat Pompeu Fabra; 2008.
Maurya V. Traducción de "El Quijote": apuntes de una traductora. In: Dotras Bravo A, Lucía Megías JM, Magro García E, Montero Reguera J y, editors. Tus Obras los Rincones de la Tierra descubren, Actas del VI Congreso Internacional de la Asociación de Cervantistas. Alcalá de Henares, 13 al 16 de diciembre de 2006. Alcalá de Henares: Asociación de Cervantistas-Centro de Estudios Cervantinos; 2008. p. 545–51. Retrieved from http://cvc.cervantes.es/literatura/cervantistas/congresos/cg_VI/cg_VI_41.pdf.
Maurya V. La traducción de El Quijote al hindi: de la utopía a la realidad. Reflexiones de una traductora. Paper presented at the Academic Session: "Cultura y Sentido: las traducciones desde la India", organizado por el Grupo de Trabajo sobre India, comité de Asuntos Asiáticos, Consejo Argentino para las Relaciones Internacionales, Buenos Aires, República Argentina (13 de julio). 2015. Retrieved from https://www.youtube.com/watch?v=ouLP3FMbA4w (video).
Mayoral Asensio R. La traducción de la variación lingüística. Hermeneus: Revista de la Facultad de Traducción e Interpretación de Soria. 1999;1:1–219. Retrieved from http://www.ugr.es/~rasensio/docs/La_traduccion_variacion_linguistica.pdf
Miličević M. Subtle gramatical problems in translated texts: how can language acquisition research inform translator education? Rivista Internazionale di Tecnica della Traduzione. 2011;13:37–47. Retrieved from https://www.openstarts.units.it/dspace/bitstream/10077/9175/1/04-Subtle-%20Maja_Milicevic.pdf

Montero Fleta B. Terminología científica: Préstamos, calcos y neologismos. Paper presentado at Congreso de Segovia: " El español, puente de comunicación" (25 al 30 de julio de 2004), Segovia; 2004. Retrieved from http://cvc.cervantes.es/ensenanza/biblioteca_ele/aepe/pdf/congreso_39/congreso_39_07.pdf

Muñoz Martín R. Situating translation expertise: a review with a sketch of construct. In: Schwieter JW, Ferreira A, editors. The development of translation competence: theories and methodologies from psycholinguistics and cognitive science. Cambridge: Cambridge Scholars Publishing; 2014. p. 2–57.

Muñoz Martín J, Valdivieso Blanco M. Autoridad y cambio lingüístico en la traducción institucional. Revista Electrónica de Estudios Filológicos. 2007;13. Retrieved from https://www.um.es/tonosdigital/znum13/secciones/tritonos_C_Mu%F1oz-Valdivieso.htm

Nardi LJ. Lenguas en contacto. El sustrato quechua en el noroeste argentino. Filología, XVII, XVIII; 131–150. 1976–1977. Recuperado de http://www.adilq.com.ar/Nardi04.htm

Nord C. Loyalität statt Treue. Vorschläge zu einer funktionalen Übersetzungstypologie. Lebende Sprachen. 1989;XXXIV(3):100–5.

Nord C. Translating as a purposeful activity: functionalist approaches explained. Manchester: St. Jerome; 1997.

Nord C. Loyalty and fidelity in specialized translation. Confluências. Revista de Tradução Científica e Técnica. 2006;4:29–41. Retrieved from http://web.letras.up.pt/egalvao/TTCIP_Nord%20loyatly%20and%20fidelity.pdf

PACTE Group. Building a translation competence model. In: Alves F, editor. Triangulating translation: perspectives in process oriented research. Amsterdam: John Benjamins; 2003. p. 43–66.

Pedulla DS. The positive consequences of negative stereotypes: race, sexual orientation, and the job application process. Working Paper. Center for the Study of Social Organizaton, Department of Sociology, Princeton University. 2012. Retrieved from http://www.princeton.edu/csso/working-papers/WP7.pdf

Restrepo E. Identidad: apuntes teóricos y metodológicos. In: Castellanos Llanos G, Grueso DI, Rodriguez M, editors. Identidad, cultura y política. Perspectivas conceptuales, miradas empíricas. México: Honorable Cámara de Diputados-Universidad del Valle-Miguel Ángel Porrúa; 2010. p. 61–70.

Retamozo M. Cuaderno de Trabajo n 9. El método como postura. Apuntes sobre la conformación de la subjetividad epistémica y notas metodológicas sobre la construcción de un objeto de estudio. México: Universidad Nacional Autónoma de México; 2007.

Richard N. Saberes Académicos y Reflexión Crítica en América Latina. In: Mato D, editor. Estudios y otras prácticas intelectuales latinoamericanas en cultura y poder. Caracas: CLACSO y CEAP, FACES, Universidad Central de Venezuela; 2002. p. 363–72.

Rizo García M. Imaginarios sobre la comunicación. Algunas certezas y muchas incertidumbres en torno a los estudios de comunicación, hoy. Bellaterra: Institut de la Comunicació, Universitat Autònoma de Barcelona; 2012.

Robinson D. Translation and empire: postcolonial theories explained. Manchester: St. Jerome Publishing; 1997.

Rodriguez PL. The construction of cultural softpower and nation branding through the promotion of language: the cases of the American Binational Centers and Chinese Confucius Institutes. Paper presented at the 18th International Conference on Cultural Economics, 2014, Montreal. 2014. Retrieved from https://editorialexpress.com/cgi-bin/conference/download.cgi?db_name=ACEI2014&paper_id=257

Rubin J. Bilingüismo nacional en el Paraguay. México: Instituto Indigenista Interamericano; 1974.

Sales Salvador D. Puentes sobre el mundo. Cultura, traducción y forma literaria en las narrativas de transculturación de José María Arguedas y Vikram Chandra. Bern: Peter Lang; 2004.

Schjoldager A, Gottlieb H, Klitgard I. Understanding translation. Denmark: Academica; 2008.

Snell-Hornby M. Linguistic transcoding or cultural transfer? A critique of translation theory in Germany. In: Bassnett S, Lefevere A, editors. Translation, history and culture. London: Pinter; 1990. p. 79–86.

Snell-Hornby M. The turns of translation studies new paradigms or shifting viewpoints? Amsterdam/Philadelphia: John Benjamins Publishing; 2006.

Supisiche PM, Cacciavillani C, Renzulli S. Variedad Léxica en Intercambio lingüístico entre adolescentes, Serie de Materiales de Investigación, 3 (5). Córdoba: Universidad Blas Pascal; 2010. Retrieved from http://www.ubp.edu.ar/wp-content/uploads/2015/03/352010MI-Veriedad-L%C3%A9xica-en-Intercambio-Ling%C3%BC%C3%ADstico-entre-Adolescentes.pdf

Suriani BM. El tratamiento de la variación lingüística en intercambios sociales. In: Fundamentos en Humanidades, vol. IX(I). Argentina: Universidad Nacional de San Luis; 2008. p. 27–41. Retrieved from http://fundamentos.unsl.edu.ar/pdf/articulo-17-27.pdf.

Temple B, Young A. Qualitative research and translation dilemmas. Qual Res. 2004;4(2):161–78.

Tola F, Dragonetti C. El Sutra del Loto de la Verdadera Doctrina. Saddharmapundarikasutra. México: El Colegio de México; 1999.

Toury G. The nature and role of norms in translation. In: Descriptive translation studies and beyond. Amsterdam: John Benjamins; 1995. p. 53–69.

Vázquez G, Fernández A, Martí MA. Dealing with lexical semantic mismatches between Spanish and English. In: Proceedings of the international conference of knowledge based computer systems, Mumbai; 2000. p. 308–19. Retrieved from http://grial.uab.es/archivos/2000-4.pdf

Rodriguez de la Vega L. The role of context and culture in quality of life studies. In: Tonon G, editor. Qualitative studies in quality of life: methodolgy and practice, Social indicators research series, vol. 55. Heidelberg: Springer; 2015. p. 37–52.

Venuti L. The translator's invisibility: a history of translation. London: Routledge; 1995.

Vidal Claramonte MCA. Traducción, manipulación, desconstrucción. Salamanca: Ediciones Colegio de España; 1995.

Finding Meaning: A Cross-Language Mixed-Methods Research Strategy

94

Catrina A. MacKenzie

Contents

1 Introduction	1640
2 Finding Capable Interpreters	1641
3 Interpreter Social Position and Subjectivity	1643
4 Survey Translation Challenges	1646
5 Self-Reported Disease Load and Implications for Conservation Research	1647
6 Conclusion and Future Directions	1648
References	1650

Abstract

The literature devoted to methodological issues arising from working through an interpreter is surprisingly sparse. References that exist tend to be dated anthropological works or tend to focus on interviews in social work and medicine. The older literature tends to focus on the mechanics of translation and how to conduct an interview with an interpreter, while more recent works start to address the issues of whether the interpreter should be "invisible" or whether the changing dynamics of the interview with an interpreter present merits the rigorous treatment of the role and influence of the interpreter with respect to power and subjectivity. Interpreters are fundamental to the research process when a foreign researcher is conducting research with an indigenous culture, and when the researcher is not fluent in the local language. In this chapter, an experientially developed cross-language research strategy is discussed, including choosing and assessing the linguistic skill of interpreters, the influence of interpreter social position and subjectivity on transcript data, and the challenges encountered when

---

C. A. MacKenzie (✉)
Department of Geography, McGill University, Montreal, QC, Canada

Department of Geography, University of Vermont, Burlington, VT, USA
e-mail: catrina.mackenzie@mail.mcgill.ca

© Springer Nature Singapore Pte Ltd. 2019
P. Liamputtong (ed.), *Handbook of Research Methods in Health Social Sciences*,
https://doi.org/10.1007/978-981-10-5251-4_37

translating and conducting a household survey, including questions about self-reported illness. The chapter ends with a summary of the components needed for a successful cross-language strategy, including the need to acknowledge the limitations introduced as a result of working through an interpreter, and the need to make the role, credentials, social position, and subjectivity of the interpreter explicit in published results.

Keywords

Language interpretation · Translation · Social position · Subjectivity · Cross-language research

1 Introduction

Using qualitative, quantitative, and Geographic Information Science methodologies, my research aims to understand the spatial distribution of perceived and realized benefits and losses accrued by communities as a result of the creation of protected areas. My research is positioned within the local socioeconomic context of the protected area and investigates how incentives, disincentives, and household well-being influence conservation attitudes and behaviors of local people around Kibale National Park in Uganda (MacKenzie 2012; MacKenzie and Hartter 2013). Working in 25 villages located around the park, I and my team of Ugandan research assistants first collected general perceptions about the benefits and losses associated with living near the park during focus groups, and then conducted a household survey to collect more detailed information about perceptions and attitudes about the park and specific data to assess household well-being. Household well-being can be assessed in many ways: reported income (Montgomery et al. 2000), objective capital assets (Takasaki et al. 2000), access to education, medical care and clean water (Vyas and Kumaranayake 2006), self-reported health status (Michalos et al. 2005), or subjective happiness (Bookwalter et al. 2006). Since the residents around Kibale National Park are primarily smallholder farmers with no permanent income stream, I elected to assess well-being based on the ownership of capital assets, educational attainment, access to sanitation and clean water sources, and self-reported diseases suffered by household members in the prior 5 years.

Uganda recognizes 32 distinct languages (Mukama 2009). In school, children are taught English and Kiswahili, but only 24% of the adults in my household survey sample (596 households) had completed primary school and, therefore, only spoke local tribal languages. Around the northern border of the park, a majority of households affiliated with the Batoro tribe and spoke Ratoro, while around the southern borders people affiliated with the Bakiga tribe and spoke Rakiga (Hartter et al. 2015). Since native Ratoro and Rakiga speakers are rare outside of western Uganda, I did not have the opportunity to learn these languages prior to starting my field research. I, therefore, had to incorporate language translation into my research design. Searching the literature at the time (2008), I found few papers on the effects that language interpretation could have on qualitative and quantitative data collection, especially how the social position and subjectivity of interpreters could influence the discourses I sought to study (although see Temple 2002; Temple and Edwards 2002; Temple and Young 2004). In this chapter, I discuss my experiences creating a cross-language strategy for my research, including how I chose the interpreters for my project, learning about the influence of interpreter social position and subjectivity on focus group transcripts, and the challenges encountered when translating and conducting the household survey questions for self-reported illness.

Interpreters are fundamental to the research process when conducting research in a location where the researcher is not fluent in the local language (Liamputtong 2010; see ▶ Chap. 95, "An Approach to Conducting Cross-Language Qualitative Research with People from Multiple Language Groups"). Interpretation involves, not only finding meaning based on vocabulary and grammar but also infusing that meaning with the local context and culture of the study location (Esposito 2001). Our interpreters often do much more than just translate words. They can also act as cultural brokers, modify translated meaning to protect certain interests, provide insight into local customs, and use their position as assistants to a foreign researcher to gain status within the local community (Schumaker 2001; Liamputtong 2010; Chilisa 2012; Caretta 2015); leading some researchers to describe interpreters as "gatekeepers of meaning" (Heller et al. 2011, p. 75). Methods to improve interpretational accuracy will be discussed in this chapter; however, the influence of differences in social position, gender, age, and education between the participant, interpreter, and researcher will still influence what is communicated, translated, and understood (Temple 2002).

2 Finding Capable Interpreters

Upon my arrival in Uganda, fellow researchers and staff at the Makerere University Biological Field Station were quick to suggest potential research assistants for my project. However, most of the people working as research assistants in and around Kibale National Park did so in support of forest ecology and biology projects. Few research assistants had experience in social research methods and those who did were fully employed with long-term land-use and zoonotic disease transmission research programs. It is recommended that interpreters who have achieved proficiency and accreditation in their language skills be employed for research purposes (Kapborg and Bertero 2002; Squires 2009). However, accredited interpreters were not readily available at my remote study location. Therefore, I initially hired three potential research assistants who had been recommended to me: Peter, Mark, and Brian (Mark and Brian are pseudonyms, but Peter requested his real name be used).

There are numerous means of adding rigor to the interpretation process in cross-language research. For my research, I included triangulating between interpreters and data collection methods (Esposito 2001; Temple and Edwards 2002), and ensuring my research assistants were familiar with the technical terms to be translated and the research aims prior to interviewing participants (Irvine et al. 2007). In addition, my research assistants and I discussed interpretation challenges throughout data collection

(Larkin et al. 2007) and used forward-backward translation of the focus group question guide and the household survey (Werner and Campbell 1973; Liamputtong 2010). Forward-backward translation, where one interpreter translates the questionnaire into the local language and then another interpreter translates it back into English, highlights translation discrepancies and inappropriate use of terms, identifies ambiguous interpretations, and can be used as a means of familiarizing the interpreters with the aims and tone of the research (Edwards 1998). Forward-backward translation also gave me the means to assess the language proficiency of my research assistants.

Residents local to Kibale National Park do illegally extract resources from the park (MacKenzie et al. 2012) and often perceive the park as a source of problems rather than benefits (MacKenzie 2012). These same residents also experience many diseases but are located far from medical care which is why McGill University students and researchers working in Kibale National Park raised money to start a small medical clinic near Makerere University Biological Field Station to try and help local residents and improve community perceptions of the park (Chapman et al. 2015). As part of the medical clinic start-up, a community needs assessment survey was conducted. I offered to pay my three assistants to translate the survey into Ratoro and Rakiga, then the clinic nurse back-translated the survey into English. Performing the forward-backward translation process found that Peter and Mark were very good translators, but that Brian was not. Therefore, Peter and Mark continued employment with me as my research assistants for the remainder of that first field season and in many future field seasons.

The mechanics of the interview change when working through an interpreter. The interview will take at least twice as long to allow time for language translation (Freed 1988). Since the translator and research participant are the ones directly speaking to one another, eye contact between the researcher and the participant becomes difficult, and therefore, creating a triangular seating arrangement is recommended to try and retain the researcher/participant connection (Phelan and Parkman 1995; Edwards 1998).

It is also important to conform to local social etiquette and a locally hired interpreter can be especially helpful, guiding the researcher about acceptable decorum in a given situation (Freed 1988). For example, when my research project started, local people were apprehensive of allowing their voices to be recorded, so Peter and Mark advised me that I should not use a recording devise and that the focus group participants requested no pictures be taken. Since the interviews and focus groups could not be recorded, transcripts had to be written quickly while interacting with the research participants. Employing two interpreters minimized losing valuable participant comments. Peter was the lead interpreter during individual interviews and facilitated the focus groups. Mark also attended all interviews and focus groups. Peter would ask questions in the local languages, verbally interpreted what people said, and I wrote down Peter's interpretations. This allowed me to redirect questions, something I could not have done if the interviews had been recorded and then post-interview interpretations had been performed. Mark directly wrote down in English everything he heard the participants say in the local languages. Pooling both sets of transcripts I gained a more nuanced understanding of the discussion that had taken place, captured more participant comments, and highlighted some differences in interpretation between Peter and Mark. After interviews we would discuss the transcript notes and any differences in interpretation of what had been said. This process led to a reasonably rigorous means of capturing the meaning of what the participants had said. However, 10% of the comments interpreted by Peter and Mark during focus groups did convey different meaning and/or emotional context.

3 Interpreter Social Position and Subjectivity

As researchers, our own social position and subjectivity can influence our research because our status, gender, age, and ethnicity influence how and what our participants choose to reveal to us, and because we interpret qualitative data through the lens of our own world view (England 1994; Kobayashi 1994; Dowling 2005; see ▶ Chaps. 91, "Space, Place, Common Wounds and Boundaries: Insider/Outsider Debates in Research with Black Women and Deaf Women," and ▶ 92, "Researcher Positionality in Cross-Cultural and Sensitive Research"). Reflexivity, defined as "self-critical sympathetic introspection and the self-conscious analytical scrutiny of self as researcher" (England 1994, p. 82), can help us be aware of the differences between researcher and participant, allowing us to look for potential bias and adding rigor to our qualitative research (Baxter and Eyles 1997; Rose 1997; Valentine 2002). I knew my social position could influence knowledge production; as a white academic, I was seen as a person who might be able to help with school fees or could potentially direct nongovernmental organizations to aid local people, so people might present their situation to be worse than reality in the hope of enlisting my help. However, the social positions of Peter and Mark might also have influenced the participant responses or how my assistants interpreted interview responses. I not only needed to reflect on my social position but I also needed to understand the social positions of my research assistants. To do this, I developed a key informant interview as recommended by Edwards (1998) and Temple (2002), asking each assistant about their age, education, status in the local society, standard of living, power dynamics during interviews, and their perceptions about the research topic (see also ▶ Chap. 96, "The Role of Research Assistants in Qualitative and Cross-Cultural Social Science Research").

Comparing the differences in focus group comment interpretation with these key informant interviews did identify themes that may have resulted in interpretational differences as a result of Peter and Mark's social positions and research topic subjectivity. For example, Mark grew up in a village located right next to the park and had previously worked for researchers studying disease transmission between primates inside the park and zoonotic disease transmission between primates and local residents. This prior work experience coupled with Mark's desire to become a nurse might account for the clinical terms used when interpreting focus group comments about disease transmission; indicating Mark might have elaborated upon participants' comments.

> Peter: Tsetse flies are infecting people and animals.
> Mark: Vectors like mosquitoes and tsetse flies that bite people and make them sick are here.

Mark's friendship with Uganda Wildlife Authority rangers, and having both his and his brother's employment dependent upon following the park rules, may have led Mark to omit issues of Uganda Wildlife Authority ranger corruption from his interpretations. Mark emphasized the park rules and minimized the punishments for breaking the rules when interpreting participants' experiences with the rangers.

> Peter: If we enter the park, we are arrested by the rangers and they squeeze money from us and take our firewood too. Even if we cut a tree, they fine us. If we go to the park for medicinal plants or if the cows go into the park, the rangers catch and squeeze some money out of us.
> Mark: We are not allowed to access some resources like thatching grass, poles, hoe handles and firewood. If we do, we are always arrested. We are not allowed to pick medicinal herbs. When cows cross to the park, we are arrested.
> Peter: When digging and burning the garden, if the fire goes across to the park we are arrested. The arrest involves caning. The arrest is quite bad.
> Mark: When we clear and burn our plot and the fire crosses to the park, we are arrested. We have even sustained corporal punishment.

Peter's background also influenced how participant comments were interpreted. Peter admitted that he "used paraphrasing where possible to try and ensure understanding for both the researcher and respondents," prioritizing cultural interpretation over literal accuracy. Mark, on the other hand, kept his interpretations more anecdotal reflecting the way people do really speak in the area.

> Peter: When things are decided top down, they don't work out. Therefore, they should start from the grass roots to ensure the right things happen. We have bitterness towards the LC5 [District], LC3 [Sub-County] and LC1 [Village] chairmen. We don't want them to get the money. The money should come straight to us so we can manage it.
> Mark: Park fund management policy should start from LC1 [Village] to LC5 [District], ascending order not from LC5 [District] to LC1 [Village], descending order. The 20% that comes back is very small. It is like when you are very hungry and you come across a person with a pan full of cooked bananas, and they just give you one piece. Will you really be happy?

Peter knew I was researching the potential for local community reciprocity by improving conservation behaviors in return for park-based benefits received. Peter's interpretations tended to speak to the need to modify management processes, potentially embellishing the words of the participants to indicate intensions to reciprocate with good conservation behaviors for certain types of benefits from the park. In his subjectivity interview, Peter could not decide if he was pro-park or pro-people (a contrived dichotomy introduced into the interview to discern the potential for bias), and if people did reciprocate for park-based benefits this would put the people living next to the park in a good light and support the provision of future benefits from the Uganda Wildlife Authority as a valid mechanism for improved conservation. Mark, on the other hand, quickly stated he was pro-park because of all the benefits his family had received due to living next to the park and a wildlife authority outpost.

> Peter: We should act communally. UWA [Uganda Wildlife Authority] provide the trench and we should work together to maintain it and work together with UWA. It is our duty to protect the park, if UWA gives us the money for a trench.
> Mark: We should be good stakeholders. If the park gives money for a project, let us raise our hands and support it because the park itself cannot come and provide local labour.

Being able to link differences in interpreted transcripts to my assistants' social positions and subjectivities confirmed the importance of interviewing Peter and Mark, and reinforced the need to critically reflect on how the focus group transcripts could and should be analyzed. Since only 10% of the comments were different, this gave me confidence that most of the interpretations were representative of what participants had said. However, what was I to do with the interpretations that did differ?

The words of the participants had obviously been filtered, therefore any knowledge created from those comments were situated by the interpreter's subjectivity (Caretta 2015). To deal with this dilemma, I took a multistep approach. Firstly, I tried to be particularly vigilant, reflexive, and transparent about methodology (Bailey et al. 1999), especially acknowledging the existence and role of interpreters in publications, demonstrating that I was aware of the limitations introduced by working through an interpreter in cross-language studies (Squires 2009). Secondly, the fact that I had used more than one interpreter added rigor as interpretations could be triangulated (Baxter and Eyles 1997; Esposito 2001). I also triangulated data sources using a household survey, key informant data logs, and ecological evidence of resource extraction from the national park, to cross-check the qualitative data. Thirdly, having interviewed the interpreters, I could at least partially understand how these interpretation differences altered the discourses I sought to analyze. However, as Temple (2002, p. 851) discovered when she conducted interpreter interviews, "I was not sure if the extent to which the views I was picking up were those of the two support workers [her assistants] or of the people they interviewed." Accepting that knowledge is situated and constructed by the social relationship that occurs during data collection, by our social position, and interview style (England 1994; Kobayashi 1994; Rose 1997; Pezalla et al. 2012), I was able to reflect about how my assistants projected their own positions into focus group discussions. Although I worried about the experiences of participants being misrepresented due to interpretation differences, I also recognized that having two interpreters provided me with two additional lenses through which to produce knowledge. Since interpreters act as cultural brokers (Liamputtong 2010; Caretta 2015), our qualitative writing can be enriched by the perspectives and verbiage that multiple interpreters can bring to the qualitative data. For example, Peter's explanatory interpretations, coupled with the literal anecdotal responses as interpreted by Mark, provided deeper understanding of what participants wanted to convey while retaining the context of how they said it. Each member of the research team brings different skills and perceptions to the research process, situated through their own subjectivity and positionality (Turner 2010; see ▶ Chap. 92, "Researcher Positionality in Cross-Cultural and Sensitive Research"). Explicitly acknowledging the role, identity, and subjectivity of interpreters requires us to recognize and reflect upon the pitfalls and opportunities that finding meaning in interpreted discourse presents.

4 Survey Translation Challenges

Although I did use forward-backward translation of the survey as the most appropriate means of ensuring a "correct" translation, this approach was flawed in that I assumed there was a "correct," equivalent meaning in the local tribal languages and that the words used in the survey would not be filtered by cultural context (Larkin et al. 2007). I had been warned by a zoonotic disease transmission researcher working around Kibale National Park that translations for illnesses would be problematic (Goldberg et al. 2012). For instance, local people call a fever *Omuswija* in both Ratoro and Rakiga, but they also use exactly the same term for malaria. During the first forward-backward translation cycle of the household survey, both the question whether household members had suffered from a general fever in the last 5 years and the question asking about whether household members had suffered from malaria came back translated as identical questions. In the next translation iteration, I added some clarification to the malaria question having the term literally translated as "fever caused by mosquitoes." When we collected survey data in 2009, 98% of respondents claimed household members had experienced malaria in the last 5 years while only 89% said household members had suffered from a general fever. Although the prevalence of malaria seems very high, my study site lies in a high malaria transmission zone of Uganda where almost every household has reported malaria (World Health Organization 2015). As a lesson learned for future surveys, the disease questions will be followed with clarification questions about how the respondent knows someone in the household had the disease, and whether the patient was tested for the disease. This will at least confirm the verified cases, but given the poor access to malaria testing in rural Uganda, the verified cases significantly underestimate reality.

The other problem I encountered during forward-backward translation of diseases was the lack of familiarity of my research assistants with the Ratoro and Rakiga words for certain diseases; although this lack of familiarity did minimize the potential for Peter and Mark to interpret and bias responses based on their own prior experiences (Kapborg and Bertero 2002). Health care research is particularly prone to gaps in conceptual equivalence during translation because medical terms often have no directly equivalent translation in the participant's language (Frederickson et al. 2005). Since it is essential for interpreters to be familiar with all terms being used prior to starting data collection (Irvine et al. 2007), I asked Peter and Mark to speak with local doctors and nurses to find out how to translate the terms for yellow fever, trypanosomiasis, and tuberculosis among others. However, when we piloted the survey, the pilot respondents did not know what many of the translated disease names meant, demonstrating that the translations provided by medical personnel had no conceptual meaning for local residents. Since I still wanted to collect data on disease load for households near the park, I worked with a medical professional to create picture cards for each disease showing a person and pointing to the symptoms that one would experience in different parts of the body if one did suffer from the disease. This approach was far from medically rigorous, reducing the validity of the data to base symptoms rather than reliably recording self-reporting of specific diseases, and seriously limited what use could be made of the results. To at least partially address this issue of data validity, survey results were triangulated through cross-correlation of diseases within the survey data, and comparison of the proportion of respondents self-reporting the disease with country wide and regional statistics. For example, respondents reporting household members suffered from heart disease and those claiming to have high blood pressure in the household were strongly associated ($\varphi = 0.441$, $p < 0.001$), and the self-reported rate of cancer (4.9% of households) was similar to the cancer death rate for Uganda reported by the WHO (5%; World Health Organization 2014).

5 Self-Reported Disease Load and Implications for Conservation Research

Although the original intent of collecting information about household disease was to use the number of diseases self-reported by respondents as a proxy for household well-being, having collected disease data allowed me to check a claim made in focus groups by participants stating they suffered from more illness as a result of living near the park; a claim supported by 95% of respondents in the household survey. Of the 19 illnesses listed in the survey questionnaire (see Fig. 1), respondents reported members of their household suffering on average from 5.6 of those 19 diseases within the prior 5 years (range 0–15).

The number of self-reported diseases, or the disease load, was higher the closer a household was located to the park boundary ($r_{spearman} = -0.094$, $p = 0.021$). In particular, trypanosomiasis/sleeping sickness (Mann-Whitney $p = 0.043$), tuberculosis (Mann-Whitney $p = 0.026$), and pneumonia (Mann-Whitney $p = 0.094$) were self-reported more frequently closer to the park. Tsetse flies, the carrier of trypanosomiasis, are more prevalent in national parks in sub-Saharan Africa (Gondwe et al. 2009), so higher rates of the disease closer to the park is feasible. However, having used the symptom charts during the survey, and since both pneumonia and tuberculosis result in severe coughing, I believe the tuberculosis and pneumonia results may be more accurately interpreted as pulmonary distress, as many survey respondents commented that the women of the household were suffering from severe coughing. In rural Uganda, cooking is done by women using open fires, often within small huts

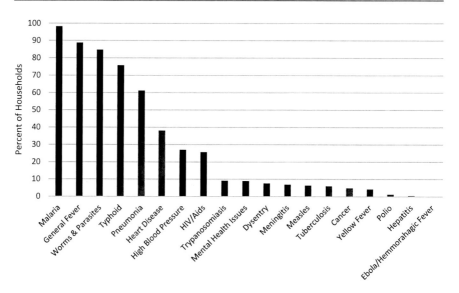

Fig. 1 Self-reported diseases by household respondents around Kibale National Park, Uganda

(Naughton-Treves et al. 2007), resulting in significant smoke inhalation (Wallmo and Jacobson 1998). Much of the rural landscape in Uganda is becoming denuded of trees as people convert land for agriculture (Hartter and Southworth 2009), so trees inside protected areas are becoming a much sought after resource (MacKenzie et al. 2012). The more frequently a household respondent admitted taking firewood from the park, the more likely the household was to self-report pneumonia (χ^2 likelihood ratio $= 19.2$, $p = 0.002$). Since households closer to the park are more likely to illegally harvest trees from the protected area (MacKenzie and Hartter 2013), this access to firewood may be increasing the cases of pulmonary distress closer to the park as more wood is used for cooking, making bricks, brewing alcohol, and creating charcoal to sell to urban centers (Naughton-Treves et al. 2007; MacKenzie and Hartter 2013). Given the self-reported nature of the disease data, these results are far from conclusive, but they do provide sufficient evidence to warrant future research into the linkage between protected areas and human illness. Providing medical facilities near protected areas may not only be good for improving relations with local communities (Chapman et al. 2015) but may also be needed to mitigate the higher disease load associated with living near a protected area.

6 Conclusion and Future Directions

Although becoming fluent in the language of one's study location is the gold standard for qualitative research, many researchers arrive in a foreign country with little or no knowledge of local languages (Liamputtong 2010). This presents a significant challenge to the research design, as a strategy for cross-language interpretation must be developed. Ideally, accredited interpreters should be used (Squires 2009; Liamputtong 2010), but often interpreters with the necessary credentials are not available at the study site or research budgets preclude being able to afford accredited personnel. The choice of interpreter may be dictated by circumstance but the researcher may have their own preferences for the interpreter to hold common traits with the interviewees to facilitate the comfort of the interviewee (Edwards 1998), while other researchers believe the best interpreter is aligned with the researcher's background and education (Freed 1988). These two criteria rarely exist in one person in a foreign research context; however, it is important to acknowledge the characteristics of the interpreter and their role during the research process (Kapborg and Bertero 2002; Squires 2009). If the researcher hires local people as interpreters, the researcher must assess the linguistic skill of the hired research assistants, and assess whether the social position or research topic subjectivity of these assistants bias the qualitative and quantitative data collected.

My cross-language research experience in Uganda leads me to recommend testing the linguistic proficiency of the interpreter by having them participate in a forward-backward translation exercise or having a trusted colleague who is fluent in the local language oversee a pilot interview translation to determine if what is being said by the participant is interpreted into the researcher's primary language with appropriate vocabulary, grammar, and cultural context. Both of these approaches have worked well for me as I worked around different protected areas in Uganda where language translation from six different tribal languages into English was required. However, even if the interpreter possesses excellent linguistic and interpretation skills, their social position within the local community, their prior lived experiences, and how they feel about the research topic can filter the discourse and data that the research project is designed to analyze.

As researchers, our positionality and subjectivity are managed by self-critical introspection and constant questioning of how our own personal bias may enter into the interview situation and the interpretation of participant comments. Similarly, this same diligent awareness of potential bias also needs to be conducted by or for the interpreters. Teaching the interpreters about the influence social position and research topic subjectivity can have on data validity needs to be part of the initial training program as the interpreter is introduced to the research project. After the interpreters are made aware of how knowledge is socially constructed and the need to be vigilant for personal biases to creep into the data, the interpreters and researchers can keep and share reflexive journals to understand how their own positions could be projected into interviews and focus group discussions (Caretta 2015). Alternatively, the interpreters can be interviewed to collect information about the interpreter's social position and research topic subjectivity (Edwards 1998; Temple 2002). These interviews can then be analyzed and assessed against the interpreted transcripts to look for potential issues in the data.

The ability to triangulate within the research design adds rigor and validity to the data collected (Baxter and Eyles 1997; Bryman 2016). This triangulation can be done through the use of different data collection methods, the use of different data sources, and in the case of language interpretation, through the use of multiple interpreters. Although hiring multiple interpreters is an added expense for the project, having the ability to compare more than one translated transcript provides confirmation of interpretation accuracy and can identify issues with paraphrasing, omissions, or elaborations of participant comments. Even if multiple interpreters are used for only a short period of the project, the insight that juxtaposing two transcripts provides can highlight the biases the researcher needs to be aware of for the rest of the project.

The limits that working through an interpreter might introduce to the research project need to be recognized and documented during the publication process. Although early ethnographers often acknowledged their assistants and the cross-language role they played, today it is rare to explicitly see the interpreter in research papers (Temple 2002). Research assistants, and interpreters are noticeably absent from most scientific writings (Sanjek 1993; Schumaker 2001), with only the voice of the manuscript authors visible in the text (Clifford 1983), as if the authors spoke directly with the research participants. A review of cross-language qualitative nursing studies found 85% of the papers did not acknowledge interpretation as a methodological limitation in their study, and only 55% of the papers mentioned that the work was conducted with a bilingual assistant but failed to mention the scope of the interpreter's role or their credentials (Squires 2009). Therefore, the cross-language strategy does not end when data collection is complete, the researcher also has the duty to acknowledge the existence, role, qualifications, and subjectivity of the interpreter in research papers as the project findings go to publication; although I have found this to be a challenge when word count limits what can be included or when editors and reviewers fail to see the value in demonstrating rigor with regard to interpretation.

Ultimately, each cross-language research project is different and the opportunities for improved cultural understanding, local networking, and access to the lived experiences of research participants are far more beneficial than the extra work required to implement a cross-language research strategy to ensure validity of the data collected. However, we as researchers need to be transparent about the challenges introduced by conducting research in a cross-language context, document the strategy used to address these challenges, and broaden critical reflection to include interpreters as we strive to find meaning in both qualitative and quantitative data collected in a foreign language.

References

Bailey C, White C, Pain R. Evaluating qualitative research: dealing with the tension between 'science' and 'creativity'. Area. 1999;31(2):169–83.

Baxter J, Eyles J. Evaluating qualitative research in social geography: establishing rigour in interview analysis. Trans Inst Br Geogr. 1997;22(4):505–25.

Bookwalter JT, Fuller BS, Dalenberg DR. Do household heads speak for the household? A research note. Soc Indic Res. 2006;79:405–19.

Bryman A. Social research methods. 5th ed. Oxford: Oxford University Press; 2016.

Caretta MA. Situated knowledge in cross-cultural, cross-language research: a collaborative reflexive analysis of researcher, assistant and participant subjectivities. Qual Res. 2015;15(4): 489–505.

Chapman CA, van Bavel B, Boodman C, Ghai RR, Gogarten JF, Hartter J, et al. Providing health care to improve community perceptions of protected areas. Oryx. 2015;49(4):636–42.

Chilisa B. Indigenous research methodologies. Thousand Oaks: SAGE; 2012.

Clifford J. On ethnographic authority. Representations. 1983;1(2):118–46.

Dowling R. Power, subjectivity, and ethics in qualitative research. In: Hay I, editor. Qualitative research methods in human geography. 2nd ed. Melbourne: Oxford University Press; 2005. p. 19–29.

Edwards R. A critical examination of the use of interpreters in the qualitative research process. J Ethn Migr Stud. 1998;24:197–208.

England KVL. Getting personal: reflexivity, positionality, and feminist research. Prof Geogr. 1994;46(1):80–9.

Esposito N. From meaning to meaning: the influence of translation techniques on non-English focus groups. Qual Health Res. 2001;11:568–79.

Frederickson K, Rivas Acuna V, Whetsell M. Cross-cultural analysis for conceptual understanding: English and Spanish perspectives. Nurs Sci Q. 2005;18(4):286–92.

Freed AO. Interviewing through an interpreter. Soc Work. 1988;33:315–9.

Goldberg TL, Paige S, Chapman CA. The Kibale EcoHealth project: exploring connections among human health, animal health, and landscape dynamics in western Uganda. In: Aguirre AA, Ostfeld RS, Daszak P, editors. New directions in conservation medicine: applied cases of ecological health. Oxford: Oxford University Press; 2012. p. 452–65.

Gondwe N, Marcotty T, Vanwambeke SO, De Pus C, Mulumba M, Van de Bossche P. Distribution and density of tsetse flies (Glossinidae: Diptera) at the game/people/livestock interface of the Nkhotakota game reserve human sleeping sickness focus in Malawi. EcoHealth. 2009;6(2): 260–5.

Hartter J, Southworth J. Dwindling resources and fragmentation of landscapes around parks: wetlands and forest patches around Kibale National Park, Uganda. Landsc Ecol. 2009;24:643–56.

Hartter J, Ryan SJ, MacKenzie CA, Goldman A, Dowhaniuk N, Palace M, et al. Now there is no land: a story of ethnic migration in a protected area landscape in western Uganda. Popul Environ. 2015;36:452–79.

Heller E, Christensen J, Long L, MacKenzie CA, Osano PM, Ricker B, et al. Dear diary: early career geographers collectively reflect on their qualitative field research experiences. J Geogr High Educ. 2011;35(1):67–83.

Irvine FE, Lloyd D, Jones PR, Allsup DM, Kakehashi C, Ogi A, et al. Lost in translation? Undertaking transcultural qualitative research. Nurs Res. 2007;14(3):46–58.

Kapborg I, Bertero C. Using an interpreter in qualitative interviews: does it threaten validity? Nurs Inq. 2002;9(1):52–6.

Kobayashi A. Coloring the field: gender, "race", and politics of fieldwork. Prof Geogr. 1994;46(1): 73–80.

Larkin PJ, Dierckx de Casterle B, Schotsmans P. Multilingual translation issues in qualitative research: reflections on a metaphorical process. Qual Health Res. 2007;17:468–76.

Liamputtong P. Performing qualitative cross-cultural research. Cambridge: Cambridge University Press; 2010.

MacKenzie CA. Accruing benefit or loss from a protected area: location matters. Ecol Econ. 2012;76:119–29.

MacKenzie CA, Hartter J. Demand and proximity: drivers of illegal forest resource extraction. Oryx. 2013;47(2):288–97.

MacKenzie CA, Chapman CA, Sengupta R. Spatial patterns of illegal resource extraction in Kibale National Park, Uganda. Environ Conserv. 2012;39(1):38–50.

Michalos AC, Thommasen HV, Read R, Anderson N, Zumbo BD. Determinants of health and the quality of life in the Bella Coola Valley. Soc Indic Res. 2005;72:1–50.

Montgomery MR, Gragnolati M, Burke KA, Paredes E. Measuring living standards with proxy variables. Demography. 2000;37(2):155–74.

Mukama R. Theory and practice in language policy: the case of Uganda. Kiswahili. 2009;72(1). online.

Naughton-Treves L, Kammen DM, Chapman C. Burning biodiversity: woody biomass use by commercial and subsistence groups in western Uganda's forests. Biol Conserv. 2007;134:232–41.

Pezalla AE, Pettigrew J, Miller-Day M. Researching the researcher-as-instrument: an exercise in interviewer self-reflexivity. Qual Res. 2012;12(2):165–85.

Phelan M, Parkman S. How To Do It: Work with an interpreter. BMJ 1995;311(7004):555–557.

Rose G. Situating knowledges: positionality, reflexivities and other tactics. Prog Hum Geogr. 1997;21(3):305–20.

Sanjek R. Anthropology's hidden colonialism: assistants and their ethnographers. Anthropol Today. 1993;9(2):13–8.

Schumaker L. Africanizing anthropology: fieldwork, networks, and the making of cultural knowledge in Central Africa. Durham/London: Duke University Press; 2001.

Squires A. Methodological challenges in cross-language qualitative research: a research review. Int J Nurs Stud. 2009;46:277–87.

Takasaki Y, Barnham BL, Coomes OT. Rapid rural appraisal in humid tropical forests: an asset possession-based approach and validation methods for wealth assessment among forest peasant households. World Dev. 2000;28:1961–77.

Temple B. Crossed wires: interpreters, translators, and bilingual workers in cross-language research. Qual Health Res. 2002;12(6):844–54.

Temple B, Edwards R. Interpreters/translators and cross-language research: reflexivity and border crossings. Int J Qual Methods. 2002;1(2):Article 1.

Temple B, Young A. Qualitative research and translation dilemmas. Qual Res. 2004;4:161–78.

Turner S. The silenced assistant: reflections of invisible interpreters and research assistants. Asia Pacific Viewpoint. 2010;51(2):206–19.

Valentine G. People like us: negotiating sameness and difference in the research process. In: Moss P, editor. Feminist geography in practice: research and methods. Oxford: Blackwell; 2002. p. 116–26.

Vyas S, Kumaranayake L. Constructing socio-economic status indices: how to use principle components analysis. Health Policy Plan. 2006;21(6):459–68.

Wallmo K, Jacobson SK. A social and environmental evaluation of fuel-efficient cook-stoves and conservation in Uganda. Environ Conserv. 1998;25:99–108.

Werner O, Campbell DT. Translating, working through interpreters, and the problem of decentering. In: Naroll R, Cohen R, editors. A handbook of method in cultural anthropology. Irvington-on-Hudson: Columbia University Press; 1973. p. 398–420.

World Health Organization. Uganda: non-communicable disease (NCD) country profiles. http://www.who.int/nmh/countries/uga_en.pdf?ua=1. (2014). Accessed 2 Aug 2016.

World Health Organization. World malaria report 2015: Uganda. http://www.who.int/malaria/publications/country-profiles/profile_uga_en.pdf?ua=1. (2015). Accessed 10 Aug 2016.

An Approach to Conducting Cross-Language Qualitative Research with People from Multiple Language Groups

95

Caroline Elizabeth Fryer

Contents

1 Introduction 1654
2 Philosophical Approach to the Research 1655
3 Research Team 1656
   3.1 Interpreters and Translators 1657
   3.2 Preparing Yourself 1658
4 Participant Recruitment 1659
   4.1 Informed Consent 1662
   4.2 Co-Participants 1663
5 Data Collection 1664
   5.1 Interview Guide 1664
   5.2 Briefing Interpreters 1665
   5.3 Language use during Interview 1667
   5.4 Second Interviews 1668
   5.5 Confidentiality 1669
   5.6 Data Analysis 1669
6 Transcription of Interview Data 1670
7 Reporting Findings 1670
8 Budget 1671
9 Conclusion and Future Directions 1672
References 1672

Abstract

Language expression and comprehension is fundamental for in-depth research interviews, representing both the data and the communication process by which data are generated. A lack of shared preferred language between researcher and participant creates complexity and additional challenges in the research process, particularly when there are participants from multiple language groups. A

C. E. Fryer (✉)
Sansom Institute for Health Research, University of South Australia, Adelaide, SA, Australia
e-mail: caroline.fryer@unisa.edu.au

© Springer Nature Singapore Pte Ltd. 2019
P. Liamputtong (ed.), *Handbook of Research Methods in Health Social Sciences*,
https://doi.org/10.1007/978-981-10-5251-4_38

common solution is to exclude participants on the basis of language preference; yet, there is a need for studies to reflect the diversity of contemporary communities. This chapter introduces a research approach and methods which have been successfully used to conduct in-depth interviews with people from multiple language groups in a constructivist grounded theory study. The approach requires the researcher to be both reflexive and adaptable in their research practice and to develop good relationships with participants and language interpreters. Key strategies are presented for conducting culturally competent and rigorous research in this unique context at modest cost. Adoption of this approach can enable health researchers to take "able to speak English" out of the inclusion criteria of studies and conduct inclusive research with culturally diverse communities.

Keywords

Language · Interpreters · Translation · Interviews · Qualitative · Grounded theory research · Constructivist paradigm

1 Introduction

Including people from multiple language groups in a study when the languages they speak are not determined before starting recruitment is a unique situation in qualitative cross-cultural research. The diversity of the participant sample restricts a researcher's ability to immerse in a single culture or to have sole reliance on either their own language skills or those of a research assistant. A common solution to this complexity is to only recruit people who are proficient in the preferred language of the researcher; yet, there is a clear need to avoid such exclusionary practice. Qualitative research is increasingly relied on to contribute to the evidence-base that informs health care practice (Sandelowski and Leeman 2012; Liamputtong 2016, 2017; see also ▶ Chap. 63, "Mind Maps in Qualitative Research"). The voices that are heard need to be carefully considered as it is often the people who do not share the identity and language of the larger cultural groups in society whose experience is missing from research findings. Yet, it can be these individuals whose voice may be the more important to hear regarding the performance of health services. For example, an important premise of my own study, which I will draw further examples from in this chapter, was to inform the practice of health care after stroke. In a contemporary Australian city, the people who present to hospital for emergency care after stroke are likely to speak one of at least 19 different languages other than English. And so I chose to recruit people with stroke irrespective of which language they preferred to speak.

A second argument to raise here in support of including people from multiple language groups in health research is one of distributive justice. All people should have the opportunity to participate in research that informs the type and quality of services provided to their community so that there is fair distribution of the benefits and burden of research (NHMRC et al. 2007). Yet I challenge that too easily the words "able to speak English" are listed in study inclusion criteria. This practice threatens the equity of the research process for our society and the credibility of research as a method of informing how contemporary communities do, and can, operate. Hence, why an approach to how it can be different for qualitative research is presented in this chapter.

This justification for inclusive cross-language research practice does not mean it is an easy approach for the researcher to take. Language expression and comprehension is fundamental for in-depth qualitative interviews, representing both the data and the communication process by which data are generated (Hennink 2008; Liamputtong 2010). When participants are from multiple language groups, there is greater complexity and additional challenges in the research process to ensure the quality and trustworthiness of interview data and its interpretation. This does not mean it should not be, or cannot be, undertaken. However, it *does* mean that researchers need to be aware of, and responsive to, related methodological issues (Mabel 2006). In this chapter, I provide guidance for how researchers can successfully navigate such issues to undertake rigorous and ethical qualitative research using interviews with people from multiple language groups when the languages are not determined prior to recruitment. The approach I present is based on my own research experience conducting cross-language research in an Australian metropolitan context and the advice of others from the literature. Each stage of a study is addressed with a pragmatic presentation of the decisions to be made and strategies that can be used. The chapter begins with a brief consideration of the philosophical orientation of studies which aim to include people from multiple language groups, and the chapter concludes with considerations for research budgeting and future directions.

2 Philosophical Approach to the Research

It is the philosophical premises of a study that influence the relationship between the researcher and the participants and the aim of the study (Wuest 2007). Identifying the methodology of the research is important as a foundation to the choice of its methods (Liamputtong 2013). For qualitative research that aims to understand the experience of people who are culturally and linguistically different to the researcher when languages are not determined prior to recruitment, there are four key requirements of the chosen philosophical approach. It needs to be sensitive to the diversity of cultural understandings and communication, provide a way to conceptually recognize the active role of the interpreter within the study, support innovation in methods to achieve a rigorous process in a complex context, and facilitate production of knowledge that is relevant to the study's objective and how others may understand and use that knowledge (Giacomini 2010).

For my study, I chose a constructivist grounded theory approach as a good match to these conceptual requirements (Wong et al. 2017). The inductive nature of grounded theory allows the researcher to focus on the experience and issues of a minority group of people without requiring comparison with a dominant group or using a theoretical framework that may not be culturally appropriate (Dilworth-Anderson and Cohen 2009). Importantly, it supports innovation and flexibility in the research process as researchers are encouraged to choose methods that will answer research questions with "ingenuity and incisiveness" (Charmaz 2006, p. 15). The approach's social constructionist philosophy conceptualizes each person's reality as constructed within a cultural context of particular historical or social conditions (Charmaz 2015), allowing a person's individual cultural understandings to be explored and recognized in the data. Its pragmatic foundation fosters reflexive questioning of both the participants' and researcher's own standpoints and assumptions about the topic (Charmaz 2009). Importantly, the voice of the participant is afforded power in the process (Temple and Young 2004) as the researcher is positioned as an interpreter of the studied phenomenon, "not as the ultimate authority defining it" (Bryant and Charmaz 2007, p. 52).

From a constructivist perspective, the language interpreter in research is not a neutral conduit of talk but a third active agent in the research interaction producing knowledge (Temple and Young 2004). In keeping with the pragmatist tradition, the interpreter is also considered to bring their past, their perspectives, and their expectations of the situation to the construction of meaning in a research interview (Wadensjö 1998). This is not unproblematic as interpreters make choices in the words they use to convey research participants' meaning. Many interpretations can be judged "correct," but it is the conceptual equivalence, or "essence of the process" (Stern 2009, p. 57), which is integral to grounded theory data and its analysis.

Constructed grounded theories are partial and conditional, situated in time, space, positions, action, and interactions for relevance to contemporary practice (Charmaz 2015). This methodology fits with a culturally sensitive approach that acknowledges culture as dynamic and views all interpretations as provisional. The inductive nature of grounded theory is particularly relevant to gaining understanding in areas where there is little previous knowledge, or when what is known from a theoretical perspective does not satisfactorily explain what is going on (Wuest 2007). Growing migrant populations provide a challenge to the monocultural and monolingual lens of much health care knowledge and so a grounded theory approach can address issues of health care practice with an explicit focus on diversity (Green et al. 2007).

Constructivist grounded theory has also been used in focus group research with people from multiple language groups (Garrett et al. 2008), but it is not the only philosophical orientation for this research context. Other published studies have used interpretive descriptive (Asanin and Wilson 2008; Clark et al. 2014), action participatory (Ellins and Glasby 2016), and narrative (Edwards et al. 2005) methodology. It is for the researcher to decide the philosophical approach which is the best fit to the research requirements as described above, the study topic, and their own values.

3 Research Team

Any research project aiming to recruit people from multiple language groups will need a team of people to provide the primary researcher with expert advice and practical support including language assistance. A mix of perspectives and experiences in the team can benefit the creativity needed to meet challenges in the research process and rigorously ensure that no assumptions go unquestioned. My study's research team included health professionals, a methodology expert, and a language expert.

The research team can support the cultural sensitivity of the research. An advisory group with members of local participant communities can advise the researcher on individual cultural contexts known to them and also more broadly on issues associated with language difference, migration experience, and research participation (Franks et al. 2007; Ellins and Glasby 2016). Although an advisory group may not cover all language groups that are recruited, the demographics of the languages spoken in the community can inform group membership and many aspects of migration experience are shared between cultural groups. Alternatively, a stakeholder group of staff who are experienced working with the population can advise on the conduct of the research process as they will be familiar with the cultural understandings of different groups and aware of communication issues (Clark et al. 2014). These groups may also assist the researcher to access potential participants for recruitment. Another strategy for cultural sensitivity, as used in my study, is to work with language interpreters as members of the research team. Interpreters can act as a type of cultural broker to help the researcher prepare for the interview and to then understand the cultural context of interview data (Dysart-Gale 2005). This role needs to be discussed with each interpreter at the beginning of their research involvement (see Sect. 5).

3.1 Interpreters and Translators

When multiple languages are spoken by participants, it is necessary to work with multiple interpreters and translators. Interpreters enable the shared understanding between researcher and participants for informed consent and data collection in qualitative research (Liamputtong 2010; see also ▶ Chap. 94, "Finding Meaning: A Cross-Language Mixed-Methods Research Strategy"). All interpretations are situated within the context of the interpreter-mediated interview and the translation memories and histories of the interpreters (Temple and Koterba 2009), so who does the interpretation does matter to the research (Temple and Young 2004). Using the one interpreter per language group aids consistency of interpretations for study rigor (Liamputtong 2010). For my study, I chose to work with professional interpreters who were accredited with the National Accreditation Authority for Translators and Interpreters (NAATI) in Australia, requiring them to abide by a Code of Ethics. Sourcing all professional interpreters from a single agency reduced my workload to negotiate individual arrangements, ensured an interpreter would be available for the majority of languages likely to be spoken by participants, and I was able to develop a relationship with the manager to request interpreters who were experienced in health care terminology and contexts which I would be discussing with participants.

Some participants may prefer to use a family member or friend to interpret for them as often occurs during health care (Alexander et al. 2004; Schenker et al. 2011; Fryer et al. 2013). It is well documented that miscommunication is more likely to occur in interpreted health care conversations when untrained interpreters are used (Flores et al. 2012). However, informal interpreters can provide qualities of personal trust, familiarity, and comfort which can be lacking in professionally interpreted interactions (Alexander et al. 2004; Gray et al. 2011). In my study, I encouraged the presence of family members during the informed consent and interviews with professional interpreters to help the participant feel comfortable and supported during their research involvement (see Sect. 4.2).

Translators facilitate written study information to be available to participants from multiple language groups. For my study, I chose to use accredited professional translators listed on the website of the Australian National Accreditation Authority for Translators and Interpreters (http://www.naati.com.au). The use of a national accredited online source widens the pool of translators available to do work in a timely fashion (within five days). Direct communication between the translator and researcher via email or telephone allows the researcher to introduce the study purpose, clarify the key concepts to be communicated by the flyer, and answer any questions. It also allows the individual translator to set their own competitive fees which can benefit the research budget.

3.2 Preparing Yourself

The findings in qualitative research are influenced by the researcher's questions, choices, and strategies in the research process and so are inseparable from the researcher's perspective (Charmaz 2015). Therefore, it is important for the researcher to prepare themselves for the culturally sensitive approach needed in this research context and to interrogate their research choices (Finlay 2002). When the languages spoken by participants are not determined before recruitment, the researcher cannot be expected to gain thorough knowledge of each participant's cultural group. Yet, the research still needs to be sensitive to each participant's cultural understandings, values, and communication for both ethical and rigorous research practice (Liamputtong 2008, 2010). Developing cultural awareness begins with a person recognizing their own cultural values and understanding that the possibility of cultural difference between researcher and participants is not isolated to comparisons of language; culture is a dynamic concept which refers to meanings built on contemporary and historical legacies, social systems, and discourses, acting as a framework for the interpretation and agency of people at different times and in different contexts (Mallinson and Popay 2007; see also ▶ Chap. 88, "Culturally Safe Research with Vulnerable Populations (Māori)"). Cultural competency training can support a researcher in their development of culturally sensitive research skills; there may be workshops available locally through health care organizations or via online modules. I attended to cultural sensitivity and reflexivity in my study by examining my own background, standpoints, and values in field notes and analytic memos, then discussing this reflexive examination with my research team, and including my position regarding the study topic in research reports. I also attended a local workshop on working with interpreters in health care.

To conduct the research approach as I have presented in this chapter, a researcher needs to become competent in working with interpreters. Conducting a research interview with an interpreter is pragmatically similar to a health professional conducting a patient assessment with an interpreter. Both contexts require specific attention to role preparation, introductions, positioning, and style of interpreting. There are many excellent resources to guide health professionals in the skill of working with an interpreter which are also useful to researchers, for example, the Interpreter information sheets available from the Australian Institute of Interpreters & Translators (http://www.ausit.org/) and Centre for Culture, Ethnicity and Health (http://www.ceh.org.au/). There may also be local training available. The key is to practice new interview skills prior to commencing data collection (see Sect. 5).

4 Participant Recruitment

The recruitment pathway for the cross-language study approach discussed in this chapter needs to accommodate for not knowing the language of participants prior to recruitment. Using a person's preferred language for recruitment is important to convey to them that the study wants to include them (Feldman et al. 2008) and to ensure that informed consent is conducted appropriately (Liamputtong 2008, 2010). Engaging recruiters who are known to potential participants can help recruitment (Liamputtong 2008, 2010). Some studies have used bilingual people working within a health care service or community organization to assist recruitment (Asanin and Wilson 2008; Clark et al. 2014). I achieved successful recruitment of older people from multiple language groups with the help of English-speaking health professionals and the strategic involvement of professional translators and interpreters. The recruitment pathway for my study is given in Fig. 1. The health professional recruiter spoke to the participant about the study and gave a promotional flyer to potential participants during a normal care interaction when the interpreter used by that person was present, either a professional interpreter or an informal interpreter such as adult family or friend.

Promotional flyers can be a useful way to advertise for study recruitment and collect contact details from potential participants. They can be translated prior to recruitment only if the languages are known for the community being sampled from and it is affordable for the study. If this is the case, the flyer can be advertised in language-specific newspapers, displayed at social meetings in the target communities, or distributed by recruiters. Alternatively, flyers can be translated on an "as needs" basis if potential participants are identified early in a window of recruitment opportunity. This second option was suitable for my study that was conducted on a limited budget and recruited from a community of more than ten language groups. When a potential participant was admitted as an inpatient to a recruitment site, the staff notified me what language they spoke. I then arranged for translation of the study's promotional flyer by professional translators (see Sect. 3). Translation of the flyer is needed only once for each language as it can then be copied and re-used for recruitment of other people who prefer to speak the same language.

For translation of the same promotional flyer to many languages, it needs to be of simple design and terms yet meets requirements of Human Research Ethics committees. An example of a flyer is shown in Fig. 2. It is likely, and to be encouraged, that a person with limited English proficiency will want to discuss their potential involvement in the study with people they know and trust, usually the same people who assist

Fig. 1 Example of recruitment pathway

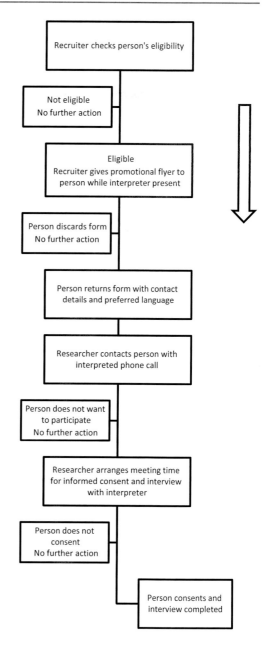

them with English language documents from official sources such as government organizations, banks, and so on. The recruiter will, therefore, need to provide both English and translated versions of the flyer to potential participants so that the literacy of the person is not assumed and the information can be shared with family or friends who may speak but not read the person's preferred language (Irvine et al. 2008).

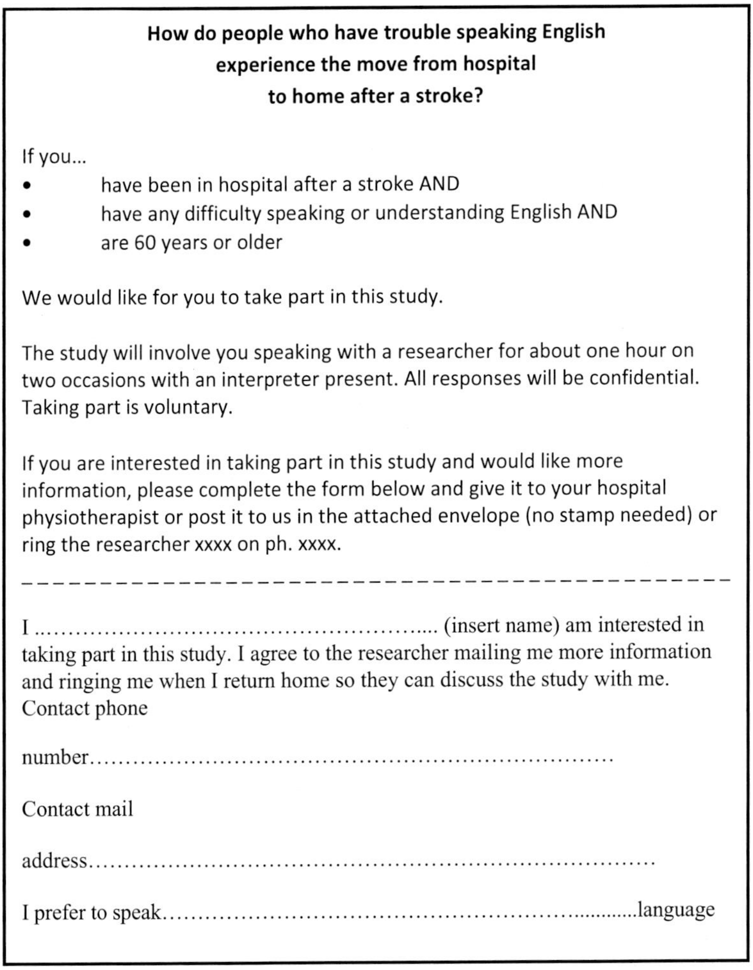

> **How do people who have trouble speaking English experience the move from hospital to home after a stroke?**
>
> If you...
> - have been in hospital after a stroke AND
> - have any difficulty speaking or understanding English AND
> - are 60 years or older
>
> We would like for you to take part in this study.
>
> The study will involve you speaking with a researcher for about one hour on two occasions with an interpreter present. All responses will be confidential. Taking part is voluntary.
>
> If you are interested in taking part in this study and would like more information, please complete the form below and give it to your hospital physiotherapist or post it to us in the attached envelope (no stamp needed) or ring the researcher xxxx on ph. xxxx.
>
> --
>
> I .... (insert name) am interested in taking part in this study. I agree to the researcher mailing me more information and ringing me when I return home so they can discuss the study with me. Contact phone
>
> number...
>
> Contact mail
>
> address...
>
> I prefer to speak...language

Fig. 2 Example of promotional flyer

When contact details are returned on the flyer, an interpreted phone call in that language can be used by the researcher to establish if the person is still interested in participating, that they meet the inclusion criteria, and to answer any questions about the research. Some people in my study preferred the initial contact to be made with an English-speaking family member on their behalf. A convenient time to meet for informed consent and the research interview can be arranged if the person chooses to be involved, and the researcher can ask if the person has any preferences regarding the professional interpreter to be present, such as dialect or gender. For the interpreted phone calls in my study, I used a commercial provider of interpreting as the conversations were brief (less than 10 minutes) so costed at a minimum rate and did not require me to prepare the interpreters for research involvement beyond introducing the study and purpose of the conversation.

4.1 Informed Consent

Informed consent in cross-cultural research has been described as a "thorny subject" (Liamputtong 2008, p. 3) and is often poorly reported in research publications (Fryer et al. 2011). If a person's preferred language is not used during the informed consent process, then it cannot be certain they are fully informed and able to make good choices about their participation (Hunt and de Voogd 2007). The potential for misunderstanding during recruitment and consent also has implications for the credibility of collected data as the words participants use and the stories they tell in interviews are influenced by their relationship with the interviewer (Mishler 1986). The Australian National Statement on Ethical Conduct in Human Research requires researchers to provide study information in the participant's first language (NHMRC et al. 2007; see also ▶ Chap. 106, "Ethics and Research with Indigenous Peoples").

Study information sheets can be lengthy and detailed documents to meet the requirements of Human Research Ethics committees and are, therefore, an expensive document to translate in written form. Such costs can be prohibitive in a study when many language translations will be required. An alternative is verbally communicating study information. A professional interpreter is already present at the arranged meeting between the person and researcher in the described approach, to assist with the interview if it proceeds. Verbally communicating study information addresses potential issues of illiteracy (Liamputtong 2008, 2010) or reading difficulty associated with a health condition, such as stroke. The researcher will also need to provide written study information in English so that it can be shared with family members or friends of the participant who do not share literacy in the person's preferred language.

The formality of consent procedures can be intimidating to people from cultural and linguistically diverse backgrounds (Liamputtong 2008). The only time when a participant was obviously hesitant during the informed consent process in my study was when an interpreter read the information sheet in a formal manner straight from the printed document rather than consecutively interpreting me, as researcher, speaking the information. It was obvious from this experience that providing study information in a conversational format between researcher and participant, with discussion and questions in their preferred language, is a more sensitive and individualized process. This discussion can incorporate a teach-back or teach-to-goal approach by the researcher (Kripalani et al. 2008; Sudore et al. 2006) to ensure the person understands what is being asked of them and why.

The consent form is a shorter document than the study information sheet, and ethics committees usually require they are in a written form in the participant's preferred language. The translation of consent forms can be arranged after the phone call to the participant, working with the same translator as for the promotional flyer.

Staggering translations in this way ensures limited research funds are not lost if a person declines to participate. The written translation of each consent form can be "checked" prior to its first use by the professional interpreter assisting with interviews in that language, see "Interpreters" for further discussion. The consent form can also be read aloud by the researcher with consecutive interpreting in case of literacy or reading issues, and several participants in my study accepted this offer.

4.2 Co-Participants

Family members often accompany people with limited language proficiency to health care appointments. Participants from all language groups in my study appeared accustomed to and comfortable with the presence of their family during the research interviews. Sometimes the family member sat quietly in the background to listen or moved in and out of the room completing household tasks. Others were more involved in the interview, mostly to assist the researcher to understand an aspect of the participant's story. For one couple, it was obviously an accustomed way for them to tell a story together (Box 1). And when the frailest participant in the study fatigued, his two sons continued to talk about their father's experience while he remained in the room listening to them. Including family members as co-participants in the qualitative research is an example of a response to a culturally sensitive situation encountered by a researcher in the cross-language context (Liamputtong 2010).

> **Box 1 Example of participant and co-participant telling a story together**
> **Interviewer (English):** Is that the wife who's a nurse?
> **Rosa (English):** Yes, yes. She's English.
> **Gino (English):** Actually, that's end nurse, was a nurse but now a boss.
> **Rosa:** A big boss now.
> **Interviewer:** But now a boss?
> **Rosa:** In a nursing home.
> **Interviewer:** In a nursing home, ok.
> **Gino:** She do just three days a week something.
> **Rosa:** Not much now, before work more but now stopped.

The term "co-participant" recognizes that the self-selected contribution by family is part of the participant's story and not a separate one. In my study, these "co-participants" were not purposively recruited so there was not a planned recruitment strategy for this group. Rather than disrupt the co-operative story telling during the interview, I waited until the interview had ended to conduct informed consent with co-participants. However, researchers interviewing people who prefer to speak another language may anticipate the preference for a family member to be present and proactively consent these co-participants.

Table 1 Interview protocol

When	Researcher actions
Prior to interview	Develop own awareness of language and cultural group of participant
	Prebrief with professional interpreter
	Complete informed consent
During interview	Encourage participant to speak language as they wish
	Follow interview guide
	Work with a professional interpreter to seek clarification if needed
	Digitally record interview
Immediately after interview	Debrief with professional interpreter
	Record field notes
Second interview	Check interpretation of meaning from first interview
	Introduce emerging concepts from analysis
Immediately after second interview	Debrief with professional interpreter
	Record field notes

5 Data Collection

The credibility of collected data is influenced by the participant's relationship with the interviewer (Mishler 1986). This can be challenging across a language gap. It takes time and effort to establish a rapport with participants to encourage them to share their experiences (Shenton 2004). A good relationship between researcher and participant can be facilitated by giving the participant the choice of interview location and time (Liamputtong 2010, 2013). In my study, all but one participant chose to be visited at home and interviews were conducted in the living room or around the kitchen table. I also spent time in conversation with participants and their family, often over a cup of tea or coffee and looking at cherished photo albums. A token of appreciation for the person's participation was always given. Following consultation with the interpreter about appropriateness of the gift, all participants were given either cholesterol-free almond bread or summer fruit.

The protocol for interviews in my study is shown in Table 1 and explained in the following sections.

5.1 Interview Guide

Semi-structured interviews which aim to explore each participant's experience have an open format with some preplanned questions, relying on the interviewer to follow up on participant's answers and to probe for more information to gain rich data for analysis (Kvale and Brinkmann 2009). It is usually expected that the participant will relate their experience to the interviewer in a narrative format. The idea of a set narrative structure may be variable across language groups and therefore, not appropriate for all participants in this cross-language research context (Temple and

Koterba 2009). Interview guides can encourage participants to choose the structure of how they relate their experience and allow the participant to identify what is important to them in their story. Cues can be used to prompt the participant where to locate its start, but this may not always be needed or appropriate. Probes can then be used by the interviewer to extend the participant's response or to draw out and understand specific parts of their experience. In Box 2 is an example of the interview opening and a probing question used in my study.

> **Box 2 Example of interview opening and probing question for participant**
> **Interview opening:**
> Now, I'm going to ask you to tell me about what happened after your stroke, right up to being here at home today. I want you to just keep talking in your own time and I will listen. I'd really like to hear your story. Later I may ask you a few questions, but for now I'll just listen to you. *(optional: So, when did you have your stroke?)*
> **Probing question:**
> Can you tell me what it is like for you when the doctor talks to your son and cannot talk to you?'

Drafting of a semi-structured interview guide can begin by the researcher looking at published interview guides from instructional texts or published papers in the topic area and then using practice and pilot interviews to develop a final interview guide. If the researcher is new to qualitative interviewing, practice interviews conducted in their own preferred language are invaluable to gain confidence in interview skills such as phrasing, intonation, sensitive listening, as well as digital recording. For the specific research context of this chapter's approach, it is essential for the researcher to also be practiced in conducting interviews with an interpreter present.

In practice interviews with an interpreter, the researcher can trial the positioning and timing of asking interview questions with interpreting and the success of probes in a cross-language context. Pilot interviews with people from the population to be studied are then important to check the appropriateness and usefulness of the interview guide for the topic; that it gains the data that the researcher is seeking. It is not possible to pilot the interview guide with all language groups and interpreters in the context of not determining languages prior to recruitment, and so pre-brief sessions with professional interpreters are important to achieving sensitivity and rigor of the interview process.

5.2 Briefing Interpreters

Interpreters need to be prepared for their role in the research (Kosny et al. 2014). A "pre-brief" meeting before their first interview facilitates a shared understanding of the study purpose, the aim of the interview, how the interview will be conducted, and the anticipated role for the interpreter (Hennink 2008). Copies of the study information sheet, consent form, and interview guide can be shared and the interview guide can be reviewed for any anticipated translation difficulty. For cultural sensitivity, the interpreter can be asked for any insights into appropriate communication and conduct during interviews with a person from their language group. In my study, pre-brief meetings were scheduled with each interpreter for one hour duration approximately a week prior to their first interview.

Real-time interpretation allows each participant to use a language they are proficient in and allows the researcher to hear the interview dialogue as it occurs and to intervene to clarify responses or to rephrase questions at the time of the interview. Data analysis can be started by the researcher directing new questions while data are being collected (Larkin et al. 2007). Consecutive interpreting works well when participants are relating experiences in interviews. In this mode, the interpreter waits for the speaker to finish talking for a reasonable length of time before interpreting it. The idea of conceptual equivalency in the interpretation rather than word equivalence is important for qualitative interpretive research (Liamputtong 2010). This can be achieved by asking the interpreter to actively select words or phrases to convey the meaning as presented by the participant to the researcher, rather than a direct translation of words used (Larkin et al. 2007). A way to explain this in a pre-brief meeting is given in Box 3.

> **Box 3 Explanation of interpretation method sought for interviews**
> **Researcher to interpreter:**
> I recognize and appreciate how important your interpreting skills are in helping me to gain useful and accurate information from the interview. As I do not speak a second language, it is important for this research that I understand the 'idea' of what the participant is telling me – what meaning they give to events that happened and how it made them feel. I realize this may sometimes be challenging to achieve but I hope we can work together to gain a good understanding. Today I will read through the planned interview questions with you and discuss their translation to the preferred language of the participant. I would also appreciate any advice you have for successful interviews with people from this cultural background.

The "pre-brief" meeting is also useful to explore each interpreter's experience and perceptions of the research topic. In my pre-brief with each interpreter, I asked them about their perspectives on the research topic; their experience with healthcare interpreting work; if they were from the same community as the participant; and how they related to or perceived the participant's community (Hennink 2008). A few interpreters closely identified with the study through personal experiences with relatives who had suffered a stroke and this allowed me to question in debriefs how the participant's experience differed from their own. Another interpreter had a strong belief that the Australian healthcare system was much better than health care in his country of birth, and we spent time at his prebrief and in debriefs discussing its relevance to the experiences being reported by participants.

A debrief after the participant interview is useful to get feedback from the interpreter on the interview conduct and management of the discussion, for example, how they perceived the ease of the participant in answering questions or any difficulties noted. The researcher can also give the interpreter feedback on their role in the discussion and discuss possible interpretations of the participant's data as explained above. Clarification can be sought for any phrases or cultural references in the dialogue that remain uncertain for the researcher. In my study, this included historical information as background to the participant's reported migration or work experiences. Record all briefing discussions in field notes to refer to during data analysis.

5.3 Language use during Interview

Not all participants will want to use their preferred language during research interviews (Liamputtong 2010). Many of the participants in my study prided themselves on their ability to converse in English and wanted to demonstrate this ability to the researcher. Even the participants who reported speaking English "not well" used occasional English words. This "code-switching" between languages is well documented for native/non-native speaker interactions (Holmes 2001). I introduced the interpreter at the start of the interview as described in Box 4 to give participants control over the language they used to tell their story.

> **Box 4 Encouraging participant to decide language use**
> **Interviewer to participant:**
> (Interpreter name) is here today to help us by interpreting our talk. I am also happy for you to speak in English at any time if you wish to. (Interviewer, English).

When participants from different language groups speak in English, their speech will often have grammatical and pronunciation differences to the researcher's speech. Professional interpreters can help to interpret accented English talk that is difficult for the researcher to understand as explained in Box 5. Another strategy a researcher can use is to repeat the sentence back to the participant for them to confirm the accuracy of understanding. If still unsuccessful, a researcher can apologize for their difficulty in understanding and respectfully ask the participant to repeat the information in their preferred language so that the interpreter can communicate their meaning to the researcher.

> **Box 5 Explanation for interpreters about participant language use**
> **Researcher to interpreter:**
> During the interview the participant can choose how much or little they wish to speak their preferred language. Sometimes it is difficult for me to understand English spoken with an accent. If I am unsure of a term or phrase, I will ask for your help to clarify it.

For participants who decline an interpreter, it can be a greater challenge for the researcher to understand their accented speech and may necessitate frequent requests to repeat or clarify talk. This is a sensitive but essential balancing act between interrupting the flow of the interview conversation and ensuring a good understanding of the participant's meaning (Marshall and While 1994). The researcher must then record extensive field notes of the conversation after each interview to refer to during transcription of the interview talk.

5.4 Second Interviews

Second interviews are an important tool for the credibility of the data analysis when researchers do not share the same preferred language as the participant. Second interviews allow the researcher to check with the participant that they have interpreted and understood the meaning of an experience as the participant intended. This is not always a straightforward task in the context of interpreter-mediated talk but was a successful strategy when used with sensitivity in my study as demonstrated in the example in Box 6.

> **Box 6 Example of use of second interview to check researcher's interpretation of data**
> **Interviewer (English):** Last time you told me you liked to have help in Greek, in case you answered wrong and you might damage yourself.
> **Andreas (Greek):** I don't know if I have said that. That is the first time I have heard this because when I have an interpreter there, the interpreter tells me, I answer properly. I never said that I made a mistake, that I said this.
> **Interviewer:** I think you meant that you were concerned that if you did not have an interpreter you may answer something wrong. Is that right?
> **Andreas:** That's it, that's what I was thinking. Then they will think, "What is he talking about? He can't go home." Just keep me in.

Second interviews also allow the researcher to ask more focused questions about the topic and gain a thicker description of aspects of a participant's experience that are important to emerging concepts in the analysis (Charmaz 2006). A question I introduced to explore how the involvement of an interpreter is decided during health care is given in Box 7.

> **Box 7 Example of focused question in second interview**
> **Interviewer to participant:**
> I'm interested in how you decide when to get help with language during your health care. So, when you are going to be talking with someone about your stroke, say a doctor or a nurse or a physio, how do you decide if you need help to understand?

5.5 Confidentiality

Confidentiality of participant data is an important part of responsible research conduct (NHMRC et al. 2007), but often causes dilemmas in practice for qualitative researchers (Lahman et al. 2015). While the researcher needs to de-identify data to minimize the potential risk of harm to the participant, enough identified data need to be kept for integrity of the research and its transferability to communities. For example, the age, gender, and language spoken by participants were reported in my study to situate the experience of the participant for readers of the research. However, I anonymized other characteristics such as their place of residence, type of stroke, and site of recruitment. I also gave participants the opportunity to choose a pseudonym. When a participant was unable to provide a pseudonym, the professional interpreter and I chose a name that we believed to be appropriate to the participant's cultural background (Lahman et al. 2015).

Another strategy to protect participant confidentiality in cross-language research is to use professional interpreters who have their own code of ethics. Professionalism is particularly important if the interpreter comes from the same small migrant community as the participant; on a few occasions in my study, the interpreter and participant recognized each other from a previous job or a shared church community. Confidentiality statements can be used to explicitly establish expectations with the interpreter; however, I suggest that the use of such statements is more to satisfy an ethics committee than what is needed when working with a professional group.

5.6 Data Analysis

A particular issue for credibility in qualitative research when participants speak multiple languages is how to analyze data (Liamputtong 2010). An ideal solution would be to employ bilingual analysts for each language group, but due to cost and availability this is often impractical. The compromise reached in my study with seven different languages spoken by participants was to analyze all data from transcriptions of the professional interpretation of the interview. Therefore, all data analysis was in English. This method of analysis has been demonstrated to produce the same major conceptual categories as when data are analyzed in the participant's spoken language despite some minor differences in text (Twinn 1997). It is also supported for a grounded theory methodology as it is the meaning rather than the grammar of the interview data which needs to be understood correctly (Stern 2009). The method does rely on successful conduct of the interview using conceptual equivalency in interpretation as described in "Data collection."

When data are analyzed in the language of the researcher rather than the participant, the credibility of the data analysis needs to be confirmed. One way to do this is by taking study findings back to the field (Adamson and Donovan 2002) as occurred with second interviews in my study. Emerging ideas from the analysis were also explored with participants in second interviews to check the "fit" with their own experience (Charmaz 2006). Regular discussions about the analysis with an advisory or supervisory group, modeled on a group process used by Van De Weyer et al. (2010), can sensitize the analysis to different nuances in the data and the possibilities for alternative interpretations. Field notes made by the researcher during data collection, such as participant behavior and feedback from the interpreter, can also assist with the interpretation of meaning during data analysis.

6 Transcription of Interview Data

Data transcription is a time-consuming task in research that cannot always be completed by the researcher alone. In my study, I chose to use a professional transcription agency for interview transcriptions. By using a single transcription agency, I was able to establish a relationship with the manager so that she gave my work to more experienced transcribers and matched familiarity with language if possible. Accented speech is often charged at a higher fee to standard interview transcription and this needs to be acknowledged in research budgeting. Transcribed data will need careful checking by the researcher due to the difficulty that transcribers experience with accented English. The researcher can use their memory of the interview and detailed field notes to assist with corrections, but there may still be words or phrases that cannot be understood. These can be followed up at second interviews with participants, or the uncertainty acknowledged on the interview transcription omitting the unclear data from the analysis.

7 Reporting Findings

Open recognition of the decisions made during the research process and the preferred languages of participants is important to the integrity of the research when multiple languages are spoken (Temple 2006) yet is often missing from research reports (Fryer et al. 2011). Strategies to recognize the language in which data were spoken and enable transferability of findings in this research context include reporting the language profile of the community the sample is drawn from; each participant's preferred language and language of interview; the language of analysis and explanation of why it was used; and the language in which words were spoken when quotes of data are presented (Box 8). The active role of interpreters and translators can be recognized by including a description of their influence and involvement, as demonstrated in Fryer et al. (2013).

> **Box 8 Examples of how to report language quoted data was originally spoken in**
> **Quote incorporated within text:**
> As Thao explained, *"It just happened out of the blue"* (Vietnamese).
> **Quote separate to text:**
> *"I do because I want to do well. I want to rehabilitate myself"* (Gino, English).
> **Quoting an excerpt of conversation:**
> Ming (Cantonese): *There's a long corridor, and the nurse go to me, 'Walk.'*
> Interviewer (English): *Why did she ask you to do that?*
> Ming: *I don't know. Exercise [pause], but I just want to.*

Research findings also need to be shared with study participants and co-participants. This can be achieved by providing a plain language summary of the findings translated to each participant's preferred language, with an English language copy of the summary to share with family who may speak but not read the person's preferred language.

8 Budget

One of the challenges of conducting qualitative research with people from multiple language groups is achieving rigor on a tight budget. The involvement of multiple interpreters and translators can add significant costs. To be financially capable of optimal practice, researchers must provide an estimate of these costs in funding proposals and funding bodies need to recognize that the costs are legitimate and necessary to enable research participation for an often excluded population (Bustillos 2009). When the languages spoken by participants are not determined prior to recruitment, estimates of required resources can be made using the language profile of the population of interest. Being familiar with how interpreting and translating services can be sourced and charged including minimum costs and cancelation fees, helps the researcher to use them efficiently. Resources to be costed for the research approach presented in this chapter include pre-brief with each interpreter; interpreted phone call to each participant; two interpreted interviews with each participant; debrief time with interpreter after each interview; and translation of promotional flyer, consent form, and summary of research findings for each language group.

9   Conclusion and Future Directions

The aim of this chapter is to communicate the importance and achievability of conducting qualitative research with people from multiple language groups, particularly in the context of participant languages not being determined prior to recruitment. The nature of multicultural societies is that health care providers work with people from many different language groups. Research that informs contemporary health care practice therefore needs to respect and reflect this cultural diversity both to be useful and for the fair distribution of its burdens and benefits across communities. The approach to cross-language research presented in this chapter relies on the researcher's ability to be culturally sensitive and reflexive, from the selection and preparation of the research team through participant recruitment, data collection, and analysis, and reporting of research findings. It is an approach that requires both flexibility and creativity in research methods to adapt to culturally sensitive situations, including accommodation to different language preferences of participants through all stages of the research process. Multiple languages spoken by participants require the researcher to develop good working relationships with multiple interpreters and translators using communication in face-to-face, email, or telephone formats. Pre-briefing and debriefing with interpreters is essential to ensuring a trustworthy shared understanding between researcher and participant. Finally, publication of completed studies is encouraged as acceptance and broader application of cross-language methodology will only be strengthened by more published examples of rigorous and cost-effective studies. It is hoped that this chapter encourages researchers to take cross-language research out of the "too hard" basket and progress an ethical and inclusive research agenda that reflects the contemporary communities we, as health researchers, seek to benefit.

References

Adamson J, Donovan JL. Research in black and white. Qual Health Res. 2002;12(6):816–25.

Alexander C, Edwards R, Temple B, Kanani U, Liu Z, Miah M, Sam A Access to services with interpreters: user views. 2004. Retrieved from York: http://www.jrf.org.uk/publications/using-interpreters-access-services-user-views

Asanin J, Wilson K. 'I spent nine years looking for a doctor': exploring access to health care among immigrants in Mississauga, Ontario, Canada. Soc Sci Med. 2008;66(6):1271–83.

Bryant A, Charmaz K. Grounded theory in historical perspective: an epistemological account. In: Bryant A, Charmaz K, editors. The Sage handbook of grounded theory. London: Sage; 2007. p. 31–57.

Bustillos D. Limited English proficiency and disparities in clinical research. J Law Med Ethics. 2009;37(1):28–37.

Charmaz K. Constructing grounded theory: a practical guide through qualitative analysis. London: Sage; 2006.

Charmaz K. Shifting the grounds. In: Morse JM, Stern PN, Corbin J, Bowers B, Charmaz K, Clarke AE, editors. Developing grounded theory: the new generation. Walnut Creek: Left Coast Press; 2009. p. 127–54.

Charmaz K. Constructing grounded theory: a practical guide through qualitative analysis. 2nd ed. London: Sage; 2015.

Clark A, Gilbert A, Rao D, Kerr L. 'Excuse me, do any of you ladies speak English?' Perspectives of refugee women living in South Australia: barriers to accessing primary health care and achieving the quality use of medicines. Aust J Prim Health. 2014;20(1):92–7.

Dilworth-Anderson P, Cohen MD. Theorizing across cultures. In: Bengston VL, Gans D, Pulney NM, Silverstein M, editors. Handbook of theories of aging, vol. 2. New York: Springer; 2009. p. 487–98.

Dysart-Gale D. Communication models, professionalization and the work of medical interpreters. Health Commun. 2005;17(1):91–103.

Edwards R, Temple B, Alexander C. Users' experiences of interpreters: the critical role of trust. Interpreting. 2005;7(1):77–95.

Ellins J, Glasby J. "You don't know what you are saying 'yes' and what you are saying 'no' to": hospital experiences of older people from minority ethnic communities. Ageing Soc. 2016;36(1):42–63.

Feldman S, Radermacher H, Browning C, Bird S, Thomas S. Challenges of recruitment and retention of older people from culturally diverse communities in research. Ageing Soc. 2008;28(4):473–93.

Finlay L. 'Outing' the researcher: the provenance, process, and practice of reflexivity. Qual Health Res. 2002;12(4):531–45.

Flores G, Abreu M, Barone CP, Bachur R, Lin H. Errors of medical interpretation and their potential clinical consequences: a comparison of professional versus ad hoc versus no interpreters. Ann Emerg Med. 2012;60(5):545–53.

Franks W, Gawn N, Bowden G. Barriers to access to mental health services for migrant workers, refugees and asylum seekers. J Public Ment Health. 2007;6(1):33–41.

Fryer C, Mackintosh S, Stanley M, Crichton J. Qualitative studies using in-depth interviews with older people from multiple language groups: methodological systematic review. J Adv Nurs. 2011;68(1):22–35.

Fryer CE, Mackintosh SF, Stanley MJ, Crichton J. 'I understand all the major things': how older people with limited English proficiency decide their need for a professional interpreter during health care after stroke. Ethn Health. 2013; 18(6):1–16.

Garrett PW, Dickson HG, Lis Y, Whelan AK, Forero R. What do non-English-speaking patients value in acute care? Cultural competency from the patients' perspective: a qualitative study. Ethn Health. 2008;13(5):479–96.

Giacomini M. Theory matters in qualitative health research. In: Bourgeault I, Dingwall R, de Vries R, editors. The Sage handbook of qualitative methods in health research. London: Sage; 2010. p. 125–57.

Gray B, Hilder J, Donaldson H. Why do we not use trained interpreters for all patients with limited English proficiency? Is there a place for using family members? Aust J Prim Health. 2011;17(3):240–9.

Green DON, Creswell JW, Shope RJ, Clark VLP. Grounded theory and racial/ethnic diversity. In: Bryant A, Charmaz K, editors. The sage handbook of grounded theory. London: Sage; 2007. p. 472–92.

Hennink MM. Language and communication in cross-cultural qualitative research. In: Liamputtong P, editor. Cross-cultural research: ethical and methodological perspectives. Dordrecht: Springer; 2008. p. 21–34.

Holmes J. An introduction to sociolinguistics. 2nd ed. Essex: Pearson; 2001.

Hunt LM, de Voogd KB. Are good intentions good enough?: informed consent without trained interpreters. J Gen Intern Med. 2007;22:598–605.

Irvine F, Roberts G, Bradbury-Jones C. The researcher as insider versus the researcher as outsider: enhancing rigour through language and cultural sensitivity. In: Liamputtong P, editor. Doing cross-cultural research: ethical and methodological perspectives. Dordrecht: Springer; 2008. p. 35–48.

Kosny A, MacEachen E, Lifshen M, Smith P. Another person in the room: using interpreters during interviews with immigrant workers. Qual Health Res. 2014;24(6):837–45.

Kripalani S, Bengtzen R, Henderson LE, Jacobson TA. Clinical research in low-literacy populations: using teach-back to assess comprehension of informed consent and privacy information. IRB Ethics Hum Res. 2008;30(2):13–9.

Kvale S, Brinkmann S. Interviews: learning the craft of qualitative research interviewing, vol. 2. Thousand Oaks: Sage; 2009.

Lahman MKE, Rodriguez KL, Moses L, Griffin KM, Mendoza BM, Yacoub W. A rose by any other name is still a rose? Problematizing pseudonyms in research. Qual Inq. 2015;21(5):445–53.

Larkin PJ, Dierckx De Casterlé B, Schotsmans P. Multilingual translation issues in qualitative research: reflections on a metaphorical process. Qual Health Res. 2007;17(4):468–76.

Liamputtong P. Doing research in a cross-cultural context: methodological and ethical challenges. In: Liamputtong P, editor. Doing cross-cultural research: ethical and methodological perspectives. Dordrecht: Springer; 2008. p. 3–20.

Liamputtong P. Performing qualitative cross-cultural research. Cambridge: Cambridge University Press; 2010.

Liamputtong P. Qualitative research methods. 4th ed. Melbourne: Oxford University Press; 2013.

Liamputtong P. Qualitative research and evidence-based practice in public health. In: Liamputtong P, editor. Public health: local and global perspectives. Melbourne: Cambridge University Press; 2016. p. 171–87.

Liamputtong P. The science of words and the science of numbers. In: Liamputtong P, editor. Research methods in health: foundations for evidence-based practice. 3rd ed. Melbourne: Oxford University Press; 2017. p. 3–28.

Mabel LSL. Methodological issues in qualitative research with minority ethnic research participants. Res Policy Plann. 2006;24(2):91–103.

Mallinson S, Popay J. Describing depression: ethnicity and the use of somatic imagery in accounts of mental distress. Sociol Health Illn. 2007;29(6):857–71.

Marshall SL, While AE. Interviewing respondents who have English as a second language: challenges encountered and suggestions for other researchers. J Adv Nurs. 1994;19(3):566–71.

Mishler EG. Research interviewing: context and narrative. Harvard: Harvard University Press; 1986.

NHMRC, ARC, AVCC. National statement on ethical conduct in human research. Canberra; 2007.

Sandelowski M, Leeman J. Writing usable qualitative health research findings. Qual Health Res. 2012;22(10):1404–13.

Schenker Y, Pérez-Stable EJ, Nickleach D, Karliner LS. Patterns of interpreter use for hospitalized patients with limited English proficiency. J Gen Intern Med. 2011;26(7):712–7.

Shenton AK. Strategies for ensuring trustworthiness in qualitative research projects. Educ Inf. 2004;22:63–75.

Stern PN. Glaserian grounded theory. In: Morse JM, editor. Developing grounded theory: the second generation. Walnut Creek: Left Coast Press; 2009. p. 55–65.

Sudore RL, Landefeld CS, Williams BA, Barnes DE, Lindquist K, Schillinger D. Use of a modified informed consent process among vulnerable patients: a descriptive study. J Gen Intern Med. 2006;21(8):867–73.

Temple B. Representation across languages: biographical sociology meets translation and interpretation studies. Qual Sociol Rev. 2006;2(1):7–21.

Temple B, Koterba K. The same but different – researching language and culture in the lives of Polish people in England. Forum Qual Soc Res. 2009;10(1).

Temple B, Young A. Qualitative research methods and translation dilemmas. Qual Res. 2004;4(2):161–78.

Twinn S. An exploratory study examining the influence of translation on the validity and reliability of qualitative data in nursing research. J Adv Nurs. 1997;6:418–23.

Van De Weyer RC, Ballinger C, Playford ED. Goal setting in neurological rehabilitation: staff perspectives. Disabil Rehabil. 2010;32(17):1419–27.

Wadensjö C. Interpreting as interaction. London: Longman; 1998.

Wong P, Liamputtong P, Rawson H. Grouded theory in health research. In: Liamputtong P, editor. Research methods in health: foundations for evidence-based practice. 3rd ed. Melbourne: Oxford University Press; 2017. p. 138–56.

Wuest J. Grounded theory: the method. In: Munhall P, editor. Nursing research: a qualitative perspective. Sudbury: Jones and Bartlett; 2007. p. 239–71.

The Role of Research Assistants in Qualitative and Cross-Cultural Social Science Research

96

Sara Stevano and Kevin Deane

Contents

1 Introduction	1676
2 Research Assistants: Necessary to the Research Process?	1677
2.1 Who is the Research Assistant?	1677
2.2 How to Assess When Research Assistants Are Needed	1678
3 Working with Research Assistants	1680
3.1 An Employment Relation: Recruiting Research Assistants	1680
3.2 Research Assistants' Tasks	1684
4 Conceptual and Ethical Issues	1686
5 Conclusions and Future Directions	1688
References	1689

Abstract

Cross-cultural research frequently involves working with research assistants to conduct data collection activities. Due to the range of different functions that research assistants end up fulfilling, from translator to guide to gatekeeper, it is clear that their participation in the research project has implications for the quality of the study design, its process and outcomes. However, their role is not always explored. Drawing on our own research as well as that of others, this chapter discusses a set of key practical decisions researchers need to make when planning their fieldwork – from assessing whether a research assistant is needed to managing a work relation. We show how these practical considerations are

S. Stevano (✉)
Department of Economics, University of the West of England (UWE) Bristol, Bristol, UK
e-mail: Sara.Stevano@uwe.ac.uk

K. Deane
Department of Economics, International Development and International Relations, The University of Northampton, Northampton, UK
e-mail: Kevin.Deane@northampton.ac.uk

© Springer Nature Singapore Pte Ltd. 2019
P. Liamputtong (ed.), *Handbook of Research Methods in Health Social Sciences*,
https://doi.org/10.1007/978-981-10-5251-4_39

intertwined with the power asymmetries rooted in the employment relation between researcher and research assistant. We also explore how the triangular power dynamics between research participants, research assistants, and researchers influence the research process and outcomes, as well as how these power dynamics reflect the broader institutional research landscape, in which questions of power, ownership, and extraction are prominent. Researchers need to reflect, discuss, and write more on this topic to fulfil a crucial gap in the literature on research methodology, to provide practical guidance for future researchers, and to identify the basis for fairer collaborations between North and South research institutions.

Keywords

Research assistants · Qualitative research · Social science · Development · Fieldwork · Africa

1 Introduction

Qualitative research in the social sciences is often reliant on the work of research assistants, even more so when it is conducted in lower income countries and cross-cultural settings. Whilst anthropological, ethnographic, and feminist literature has historically been more open to the interrogation of the role of research assistants in the research process and potential influences they may have on the quality and integrity of data collected, in general this is a topic that remains underexplored. The literature on research assistants in qualitative primary research is conspicuous in its absence from most social science qualitative research handbooks (e.g., Ritchie and Lewis 2003; Flick 2009; Silverman 2013; Merriam and Tisdell 2015), with a few exceptions such as Devereux and Hoddinott (1992) and Liamputtong (2010, 2013). Therefore, researchers have very little practical guidance on how to find, recruit, train, and work with research assistants, and the range of associated ethical, conceptual, and theoretical issues that this entails. Further, graduate methodological training frequently overlooks this issue, and as Middleton and Cons (2014), p. 282 note, "for even the most established scholars, the subject of research assistants can make for uncomfortable conversation."

Drawing on the thin but growing body of work on this issue, as well as our own experiences of conducting fieldwork in a low-income and cross-cultural setting, in this chapter we set out the practical steps researchers go through when they plan their fieldwork and work with research assistants. We show how these practical considerations are intertwined with the power asymmetries rooted in the employment relation between researcher and research assistant. We also explore how the triangular power dynamics between research participants, research assistants, and researchers influence and shape the research process and outcomes, as well as how these power dynamics reflect the broader institutional research landscape and the political economy of research more generally, in which questions of power, ownership, and extraction are prominent.

Whilst all research projects that require the employment of research assistants will face common challenges, it is also clear that the role and influence of research assistants will also be shaped by the specificities of each project. Without a concrete set of guidelines that can be mechanically followed, in this chapter we provide a number of examples designed to illustrate how these issues have been considered by different researchers. We hope that this chapter will aid researchers at all levels of experience in the fieldwork process, as well as to stimulate much-needed further discussion and exploration of the role of research assistants.

2 Research Assistants: Necessary to the Research Process?

2.1 Who is the Research Assistant?

Working as a research assistant may refer to a variety of roles and associated tasks, depending on the context in which the job is performed. Research assistants in academic and nonacademic institutions are often employed by more senior colleagues to carry out desk-based tasks such as background literature searches, annotated bibliographies, and the like (Molony and Hammett 2007). In this area of research, research assistants locate, read, and review *secondary* literature and data.

A rather different role is performed by research assistants who facilitate processes of *primary* data collection. First, the implementation of large-scale surveys relies on the work of interviewers or enumerators who are normally recruited locally, where the survey is to take place, and trained by experienced researchers. As large-scale surveys are the most widely used instrument for the generation of data on household welfare, poverty, health, employment, and so forth, there is some material on training enumerators (Grosh and Glewwe 2000; Iarossi 2006). Most manuals for the implementation of large-scale surveys contain guidance and procedures on the recruitment, training, and supervision of enumerators and instructions on how to ask questions and fill out the questionnaires (see IFC Macro 2009 for an example of this in relation to the implementation of Demographic Health Surveys). Nonetheless, the literature concerned with understanding the *power* of interviewers/enumerators in shaping the process of quantitative data collection as well as its outcomes remains very limited (Randall et al. 2013; Flores-Macias and Lawson 2008).

Second, research assistants can be interpreters. Interpreters provide translation in the course of research activities and are needed in all cases when researchers are not fluent in the language spoken by the participants. There is some literature on interpreters mostly reflecting on how to make translation work in order to collect meaningful data (Temple and Edwards 2002; Temple and Young 2004; Bujra 2006; Liamputtong 2010) and, even thinner material on the relative benefits of different interpreting techniques (Williamson et al. 2011; MacKenzie 2016). However, some argue that the role of interpreters is rarely explored in the literature as the researcher claims and maintains full ownership of the research process and outcome (Berman and Tyyskä 2011) (see also ▶ Chap. 95, "An Approach to Conducting Cross-Language Qualitative Research with People from Multiple Language Groups").

Third, research assistants facilitate and mediate data collection in qualitative research, which is the focus of this chapter. Many researchers conducting qualitative research in a variety of disciplines, ranging from anthropology to political economy, rely on research assistants to carry out their fieldwork. Whilst initially regarded as interpreters and "conduits" in the research process (Freed 1988), this view has been abandoned thus enabling the expanded role that research assistants in reality play, especially in cross-cultural settings, to be acknowledged. Research assistants tend to be familiar with the context and local language(s) where research takes place and accompany the researcher through interviews and other research activities, either leading the activities themselves or assisting the leading researcher (Deane and Stevano 2016). Research assistants may contribute to the design of research activities prior to their implementation, through informing the selection of research site(s) and participants, as well as the type, structure, and content of interviews or other activities such as focus groups or participant observation (Deane and Stevano 2016). A research assistant also takes part in the ongoing analysis of the material collected through regular discussions with the researcher, an important but often unacknowledged form of informal analysis, giving researchers the space and opportunity to think through and articulate emerging themes and narratives. This expanded role of the research assistant, therefore, requires a more detailed understanding of how they influence data collection activities and the quality of data collected.

2.2 How to Assess When Research Assistants Are Needed

When planning their fieldwork, one of the first decisions researchers are to take is whether they will need the support of one or more research assistants. The most immediate aspect that ought to be considered in this respect is the degree of knowledge, familiarity, and insiderness the researcher has with the studied context, and whether working with a research assistant can facilitate the process of data collection. We suggest a set of key guiding questions that can help researchers making this assessment.

First, the most important questions revolve around the position of the researcher vis-à-vis the research participants. If the researcher is, or is perceived to be, an outsider in the setting where research takes place, then a research assistant with a greater degree of insiderness can help bridge important gaps. These can include the researcher's lack of institutional contacts as well as her or his limited ability to verify the accuracy of the information collected, linguistic and cultural barriers, and respondents' uneasiness to speak to an outsider (Liamputtong 2010; see also ▶ Chap. 92, "Researcher Positionality in Cross-Cultural and Sensitive Research"). It is, however, important to consider that being an outsider has advantages too and researchers may find that in some contexts respondents are more comfortable giving information to an outsider (Liamputtong 2010). For example, in a study that explored women's understandings and experiences of cervical cancer screening in the UK involving many participants who were from different ethnic and religious backgrounds, it was found that by emphasizing their status as an outsider and the lack of knowledge they had about the participants' cultural and religious backgrounds, the researcher managed to elicit detailed accounts. In part, this was due to the empowerment of participants as "experts" vis-a-vis the researcher (Tinkler and Armstrong 2008). At the center of these questions lie the relations of power between researcher, researched, and research assistant, which will be discussed below. This goes to show the importance of reflecting on these threefold dynamics matters right from the beginning of any process of fieldwork planning.

Second, there are some practical questions relating to the research time frame and funding. Depending on the desired sample size and number of interviews, working with a research assistant can produce some time gains when researcher and research assistant conduct research activities separately and in parallel or when more than one research assistant can be employed. If the sample size is medium to large, given the available time, then sharing the workload with a research assistant may be an advantage. Likewise, if the researcher plans to conduct a high number of interviews in a given amount of time, then working with a research assistant can help reach the target. However, these benefits do not apply to cases where researcher and research assistant conduct research activities jointly, if translation is constantly needed for example, or where the research assistant leads the implementation of the research activity by herself/himself (Deane and Stevano 2016).

Another critical practical point has to do with funds available to remunerate the work of research assistants. As we discuss below, the relation with research assistants is first and foremost one of employment (Molony and Hammett 2007; Deane and Stevano 2016), therefore, the availability of adequate funds is necessary to employ a research assistant. The duration of employment as well as the tasks that the research assistant will be responsible for may well depend on the available funding. This is important both in the context of large research projects, where budgets need to cost research assistance appropriately, and in doctoral and other smaller research projects, where researchers are subject to tight financial constraints. Although these practical considerations are often neglected in the literature, they do shape the nature of research projects, their objectives, and the associated involvement of research assistants. To find out what is feasible with the available funds, it is advisable to investigate the ongoing hourly, daily, or monthly rates for research assistants in the context where research is to take place.

Third, there are issues that relate more specifically with the scope of each research project. A general guide on how to make a decision on whether a research assistant is needed is provided by the exercise of linking research questions to research methods. We break down the overarching research question into several sub-questions, and then we consider the best equipped method to address each research question (We both have Deborah Johnston to thank for this approach). For each research question and associated method, we reflect on the role a research assistant can play. What is the role of a research assistant if we plan to use participant observation to study intra-household decision-making among poor households? What is the role of the assistant if we plan to conduct interviews with farmers, NGO workers, or government officials? When participant observation is used in contexts where the researcher is not fluent in the participants' language, the presence of the research assistant is necessary for the entire duration of the research activities. However, there are good reasons why, in some cases, it may be appropriate that research assistants lead the research activities. This may be due to the nature of the topics at hand, or with the respondents' ease to speak to the research assistant, rather than to the researcher, or better understand their accent. For example, during a qualitative study conducted in Tanzania on the relationship between population mobility and HIV risk (Deane et al. 2016), it was decided that in-depth interviews with participants that covered sensitive topics such as sexual behavior were to be led by the research assistants alone, in part to create a more conducive and safe environment for participants, as well to address the distorting presence that an observing lead researcher (who was not fluent in the local language), would have on the interview dynamic.

While there are no uniform answers to these questions, these are important issues that need to be carefully considered when planning field research. It is clear that the role of the research assistant is nested within broader historical and political processes underlying the nature of research, its time line, and the availability of research funding as well as the power asymmetries between researcher and participants. Researchers can follow key guiding questions to make decisions on whether they should work with one or more research assistants, but, crucially, they must consider how the practicalities do not transcend from the web of power relations operating at both the individual and the institutional level, which we discuss below.

3 Working with Research Assistants

3.1 An Employment Relation: Recruiting Research Assistants

As Molony and Hammett (2007) note, the relation between researcher and research assistant is shaped by a range of wealth and power asymmetries that are rooted in a broader set of international historical relations, but it is essentially one of employment. The researcher-employer needs to recruit a research assistant, establish a work relation, and manage it.

Once it is established that one or more research assistants are needed to carry out primary data collection, then researchers need to find suitable candidates for the job. For some researchers, it may be the first time they act as employers, therefore the process of finding research assistants, even more so in contexts that may be unfamiliar, should not be taken for granted. A number of practical routes can be taken. First, some attempts can be made prior to travelling to the research site. It may be useful to use research networks and connect with other researchers who have conducted research in the same settings as they may have contacts of research assistants they directly work with or have worked with in the past.

Second, it would be highly advisable to establish an affiliation with a local research institute. These collaborations are crucial not only as channels to find research assistants (Molony and Hammett 2007) but also to get to know the local research environment, meet other researchers, and eventually disseminate findings where they are closer to the realities they describe. Further, partnerships with local research institutes are strongly encouraged by funding institutions, representing one positive, if often tokenistic, aspect of the current politics of research funding.

If collaborations with local research institutes are not possible and other attempts were unsuccessful, local university students may be willing to take up part-time jobs as research assistants. Students may be reached through teachers or announcements on the university campuses. Finally, it is also possible that the nature of research is such that the *best* research assistant is someone who is very familiar with the context, despite lacking connection with research, studies, or previous research assistance experience. This may be the situation of researchers working in remote areas, on sensitive topics, or in unsecure settings (Jenkins 2015). In these cases, building personal networks and using word of mouth may be the most effective ways to find suitable assistants.

In a formalized recruitment process, the researcher holds interviews and selects from a pool of candidates. What are the criteria to select a research assistant? The decision, as Molony and Hammett (2007), p. 295 suggest, entails considering a number of factors:

> The selection of a research assistant is a key decision, where one must balance the academic qualifications with the experience and personality of the potential assistant in relation to the physical and social environment(s) in which the research is to be conducted.

Some have suggested that matching research assistants and research participants on a number of sociodemographic characteristics (Temple and Edwards 2002; Liamputtong 2010) is a viable selection technique. This would entail considering language, age, gender, education, socioeconomic status, and residence location. For example, the research assistant needs to be fluent in the language(s) spoken by research participants. However, it becomes immediately clear that the usefulness of the matching exercise is limited, unless it is complemented with a broader assessment of the triangular power relations between researcher, research assistant, and research participants (Deane and Stevano 2016). Consider a research project focused on women: It would seem appropriate to employ a woman as a research assistant. However, in many settings it is necessary to negotiate with male gatekeepers to gain access to women, which makes it possible that male research assistants may be better placed to navigate this process (Mandel 2003; Deane and Stevano 2016). Further, insights can be gained from research assistants themselves. In a reflective article in which Turner (2010) interviewed two former female research assistants on their experiences of the research process in Vietnam and China, both research assistants noted that they found men easier to interview than women, in part because the cross-gender dynamic created space for the development of strategies to deal with this, and also because within those contexts, it was more culturally acceptable for men to talk and share information, whilst women often deferred to their husbands and so interviews with them were often difficult. As Turner (2010), p. 212 notes, this is a surprising dynamic that contradicts perceived wisdom, as it is often assumed that "female assistants will be more comfortable interviewing females; likewise male assistants interviewing men."

A related concern is the ethnicity of research assistants, and how this may influence the research process. Whilst it is common in cross-cultural settings to hire research assistants who have a similar cultural background to participants, often for linguistic reasons, this can blur the insider/outsider status of the researcher and influence the research process through the ways that research assistants have to mediate cultural norms, expectations, and structures. Ensuring that the influence that the ethnicity of the research assistant has on the research process is at least acknowledged, if not considered from the outset, is of vital importance. For example, in research conducted in the Mozambican province of Cabo Delgado (Stevano 2014), inhabited by three main ethnic groups – Macua, Maconde, and Mwani – the research assistant, a Maconde, was more of an insider with Maconde respondents and less so with Macua and Mwani interviewees.

A final dimension when trying to recruit the most suitable research assistant in cross-cultural qualitative research relates to the methodological expertise of the prospective research assistants. This can be a challenge in some settings, as due to the predominance of large-scale surveys as instruments for primary data collection, research assistants may have had prior experience of conducting structured interviews but not of more complex qualitative approaches, leading to difficulties in recruiting appropriately experienced research assistants (Molyneux et al. 2009). In this case, time must be taken to ensure that the prospective research assistant, especially if they have previously worked with questionnaire-based interviews, is informed and trained on the range of qualitative methods that will be used.The importance of recruiting research assistants who have experience and knowledge of qualitative methods is reflected in the influence that this may have on the research design itself. For example, in the Tanzanian study noted above, the research assistants' lack of experience with qualitative methods required a change in approach, with the planned life-history approach replaced with more straightforward semi-structured in-depth interviews. While this certainly influenced the research process, it highlights the ways in which researchers have to respond to the capabilities and experience of their research assistants during the fieldwork process.

Of course, it is entirely acceptable that the recruitment process is less formalized, especially when there is only one candidate, or time constraints are such that finding one suitable research assistant is preferred to selecting from a group. Nonetheless, even in informal recruitment processes, it is crucial to take into consideration power relations along the lines of gender, class, ethnicity, age, and so forth, as well as other requirements such as degree of insiderness, previous experience working as a research assistant, and commitment to the role. The absence of deterministic answers to these questions does not make them irrelevant. Reasoning on these issues requires weighing up the relative importance of each factor in the context of specific research projects, and this will lead to the best empirical choices under the given circumstances and constraints.

The labor relation highlights the responsibility of the *employer* to ensure that *employees* enjoy fair working conditions and conduct the research activities to the required academic and ethical standard. A central component of establishing a fair employment relation lies in clear contractual arrangements and adequate pay, issues that are discussed in detail by Molony and Hammett (2007). Research assistants may be paid directly by the researcher or by the research institute through which they are employed. Depending on the payment channels used, the labor relation between research and research assistant will be more or less explicit (Deane and Stevano 2016). Either way, it is crucial that the employment offer is decent and provides very clear indication of tasks, payment structure, and duration of employment. An agreement on each of these issues should be sought prior to the beginning of the job. However, this financial relationship is not always easily managed, due to the existence of an often significant wealth asymmetry between researcher and research assistant (Molony and Hammett 2007), and also because the role that research assistants fulfil are multiple, overlapping, fluid and susceptible to change, and thus, not always easy to pre-define upfront. Further, as the relationship between researcher and research assistant develops due to the time spent together and the blurred boundary between employer and friend, the integration of the researcher into the personal network of the research assistant can lead to situations in which researchers are expected to help out financially in times of need beyond paying the agreed wage. And in some situations, research assistants, aware of the wealth asymmetries, may attempt to extract as much money as possible from the researcher (Molony and Hammett 2007). There are also other smaller financial issues to deal with, such as setting and managing expectations in terms of who pays for soda/tea or lunch during extended periods in the field. Managing these financial arrangements is thus not always an easy task, involving both obligational and philanthropic urges and the need to maintain the relationship as a contractual, business arrangement (Molony and Hammett 2007).

The duration of employment should reflect the time necessary for training, before the start of the research activities and also while data collection takes place. Training is critical to make sure that the research assistant can perform her/his role to the highest scientific and ethical standards (Liamputtong 2010). As noted above, there is some literature on training enumerators who conduct large-scale surveys. However, there is a gap in the literature on training of research assistants in qualitative social science research. Molyneux et al. (2009) make a strong case for placing training at the outset of the methodology in multi-method research. A number of activities can be carried out during the training sessions. It is necessary to explain the rationale and objectives of the research project as well as the scope of each research activity. What is the expected outcome of each research activity and what is the best way to obtain it? The research assistant can take up an active role in shaping the format and content of research activities, based on their knowledge of the context. This process can carry on for the entire duration of data collection as qualitative research is not subject to the fixity of questionnaires. Constant revision and adaptation of qualitative interviews is arguably one of the strengths of qualitative research, and the research assistant can play an important role in this process.

It has been argued that the employment relation between researcher and research assistant is multifaceted because the research assistant is also a friend and a companion (Turner 2010), on the one hand, and because there are other power asymmetries that mark the labor relation, on the other hand (Molony and Hammett

2007). Here, we draw attention on the risks that research assistants may be exposed to due to their position of insiders in relation to participants and their communities (As described by Cramer et al. (2016), we recognize that researchers are also exposed to risks when conducting primary research but, due to the objectives of this chapter, we focus on the risks potentially facing research assistants.). While research assistants are employees, they also guide the researcher through the process of data collection, facilitating interactions with authorities, building trust with participants and their communities, *de facto* leading the process through which researchers gain access to informants and, eventually, data. Thus, Jenkins (2015) notes, when the trust-building process is compromised – often due to factors that are beyond the control of research assistant and researcher, such as ethnic divides and any event that creates unjustified suspicion in the researched communities – assistants are exposed to risks. Cramer et al. (2016) reflect on a challenging episode when research assistants were detained by local security officials for having held interviews with farmworkers at their homes without permission from their employers, thus highlighting the risks of operating within networks of structures of power based. Researchers may lose access to research communities but research assistants, as people who live in the research context, may face risks that carry beyond the duration of the data collection process. Thus, it is the researcher's responsibility to try to anticipate and minimize a variety of risks that may emerge during the research process.

In sum, the recruitment of research assistants requires a degree of professionality, transparency, and trust on the part of the researcher, who acts primarily as an employer. There are, however, a set of economic, political, and ethical issues we, as researchers, need to consider to establish a fair working relation with the research assistant. These include, quite obviously, adequate pay and working conditions, clear agreement on what the job entails, and high-quality training but also an open assessment of the potential risks and ways to minimize them.

3.2 Research Assistants' Tasks

Thus far, we have described the decision-making process researchers go through, from deciding if they need to work with a research assistant to employing one, but what will the research assistant eventually do? In this section, we discuss a set of tasks research assistants may be asked to perform before, during, and after fieldwork.

When funding allows it, it is advisable that the researcher arranges a scoping research trip to recruit an assistant and carry out the preliminary work to then run the research activities smoothly. Setting the scene includes establishing contacts with governmental and nongovernmental institutions and obtaining permission to carry out research, when needed. Research assistants may provide valuable contributions in the preliminary phases of fieldwork, by facilitating contacts with local institutions, authorities, and key informants. For example, in our recent fieldwork on food consumption among schoolchildren in Accra, Ghana, the research assistant managed the communication with the Accra Metro Education Office to obtain an updated list of private and public schools and then permission to conduct interviews in a sample of schools. These networks of contacts and paperwork are a crucial step to ensure access to research participants and their communities. They also represent a first move toward building trust between the research assistant and the respondents, thus possibly making the subsequent research activities more welcomed.

To then consider the variety of tasks research assistants can perform during the implementation of research activities, it is useful to go back to the exercise of matching questions, methods, and role of research assistant described in an earlier section (How to assess when research assistants are needed). Depending on a number of factors, the research assistant may only translate, assist the researcher, colead interviews or take up full leadership, in either absence of presence of the researcher. Practically, the basic decision that needs to be taken is on who leads the activities. As mentioned above, there are different criteria that can be used. First, it depends on the research method used. For example, in focus groups or other collective interviews it may be useful that the researcher and the assistant perform different tasks, with one leading the interview and the other one taking notes. Second, the nature of the topic at hand matters. Sensitive topics may be handled better by local researchers. But, at the same time, it is worth considering the possible risks arising for the assistants, especially if they have a more visible leadership role. Third, considerations on the preference of respondents to speak to an insider or to an outsider may lead the researcher to decide that the research assistant leads the interviews, and vice versa. Different configurations have implications on the methods used, the ability to go "off-script" and follow up on interesting themes that emerge in the course of the interviews, and on the research flow, with simultaneous translation making activities longer and more tiring (Deane and Stevano 2016). For example, during a study in Bangladesh that aimed to investigate women's experiences of emergency obstetric care, Pitchforth and van Teijlingen (2005) encountered challenges related to the flow of interviews when conducted with their research assistant as a translator. Not only did this make the interviews time consuming and disjointed due to the constant flow of information from researcher to research assistant to participant and back again, they also noted that the research assistant did not always interpret some questions or comments, or did not want to ask specific questions that seemed obvious to them. Reflecting on this issue, they came up with an alternative approach in which the research assistant led the interview and at key points during the interview summarized the conversation and gave the researcher the opportunity to input additional questions or lines of inquiry. This enabled the interviews to work more like a structured conversation and improved the flow but required the researcher to cede control of the interviews to the research assistant (Pitchforth and Van Teijlingen 2005). The cessation of control to research assistants was even more evident in the project conducted in Tanzania mentioned above. In this project, the researcher was not even present, and thus was completely reliant on the research assistants to ask the right questions and follow up on anything unexpected that arose (Deane and Stevano 2016). The exact configuration will have an impact of the depth of training and preparation required before research activities commence, as well as further muddying the already complex power dynamic between researcher and research assistant.

Importantly, research assistants participate in the on-going analysis of data collected. This can be done as regular informal conversations or more structured debriefs at the end of each research activity. More structured check-ups may be needed when research assistants lead the interviews (Molyneux et al. 2009; Deane and Stevano 2016). Different techniques have implications for the quantity and quality of data collected (Deane and Stevano 2016). For the purpose of planning, it is important to put time aside for the on-going data analyses. As much as it may look like a trivial observation, it is, on the contrary, meaningful in a scenario where conducting interviews tends to be given priority over reflecting on the material collected. It is time consuming to take notes on a regular basis and often what is not written will then be forgotten and lost later on.

Finally, in the post-fieldwork phase, the research assistant may still have a role to play. First, the researcher may decide to ask the assistant to read and comment on field reports, as an extension of the interim analyses described above. Second, joint publications can also be considered. This depends on the willingness of the research assistant to participate in writing up research. Research assistants who work for research institutes may have an interest in appearing as coauthors of research articles. However, researchers are also constrained by the boundaries of disciplinary practice. In some disciplines, such as medical science, nutrition sciences, and public health, researchers are expected to list as coauthors all people involved in the research process, including research assistants. Yet, other disciplines, such as economics, anthropology, development studies, normally include only the writing authors, and, additionally, researchers in these disciplines are encouraged to have solo-authored publications. It is more common in these disciplines to recognize the work of research assistants in the acknowledgements.

Research assistants' work is precious throughout field research. It is, therefore, important that researchers consider carefully how to manage the assistant's involvement, and acknowledge their contributions, not only to do justice to their work but also to reflect on the influence research assistants have on the shape of the research process and on the quality of the output.

4 Conceptual and Ethical Issues

By tracing the decision-making process researchers go through to recruit and work with assistants, it becomes evident that most practical decisions are intertwined with the web of power relations governing the interaction between researcher, research assistant, and participants. Previous work, especially by anthropologists and feminist scholars, has reflected on the positionality of researcher and researched and the need for a reflexive approach to research (Harding 1987; England 1994; Pack 2006; Ryan 2015; Suwankhong and Liamputtong 2015). Some have extended these reflections to include the research assistant as a third important actor (Temple and Edwards 2002). These approaches stress the subjective nature of these relations, thus focusing on the "values," "beliefs," "assumptions" that each actor – the researcher, the research assistant, the researched – bring in the research process, creating a "triple subjectivity"

(Temple and Edwards 2002, p. 11). As much as the inclusion of the research assistant in the picture is of paramount importance for thorough reflections on research methodology, the emphasis on subjectivities falls short of considering the materiality of these relations of power. We need to understand how the distribution of power along the lines of age, gender, nationality, class, race comes to shape the interactions between the people involved in the research process and how, in turn, research itself is shaped by them. And thus, we extend this analysis to include a consideration of the objective relations between researcher, research assistant, and participants that overlap with but are distinct from the triple subjectivities noted above (Deane and Stevano 2016).

The objective relations between research assistant and participant, and how these shape the research process and outcomes, are also important to consider. For instance, in the Tanzanian example, research assistants were frequently interviewing participants that were older, thus conferring a certain power dynamic, which was offset by other aspects, such as the fact that the research assistants were more educated. In other settings, it may be the ethnicity or the social class of research assistants vis-à-vis participants that matter. How these different dynamics play out in the research process is often difficult to disentangle, but they must be acknowledged. The relations between researcher and participant are also mediated by the research assistant, whether the researcher is present or not, and thus presents an extra layer of complexity when considering the role of research assistants. Whether these influences are addressed up front or reflected on during and after fieldwork activities, it is undeniable that they will impact the outcomes of the project.

Acknowledging that the researcher-research assistant relation is one of employment highlights fundamental material conditions of the relation, such as payment, contractual arrangements, and the responsibilities of both employer and employee. It also raises the issue of research ownership. For, as much as research assistants perform well at their job and contribute to the quality of research, the official ownership of research remains ultimately in the hands of the researcher-employer. This can have implications for how the research project is run, how and why some conflicts of interest arise, and may also help researchers understand the underlying dynamics of difficult relationships with research assistants – is your research assistant just being difficult because that is the way they are or because of these broader dynamics? As Molony and Hammett (2007) note, the relationship between researcher and research assistant brings into sharp focus the extractive nature of cross-cultural research, especially in low-income settings, and the unequal benefits that different parties derive from the process.

The question of ownership helps us see how the power dynamics we experience as researchers conducting fieldwork are nested in institutional relations, which set the boundaries within which we do research. Thus far, we have discussed the decisions, and the underlying power dynamics, we face as individual researchers. From this angle, the practice of silencing the research assistant is considered to be harmful because it does not acknowledge the invaluable work of assistants and their influence over the research process and output (Molony and Hammett 2007; Turner 2010; Caretta 2015; Deane and Stevano 2016; Jenkins 2015). As Jenkins (2015), p. 24 notes, "silencing the research assistant not only does a disservice to the extent of their influence over our research – in both its positive and negative manifestations – but it also prevents an honest, open, and fundamentally important discussion of how we can collaborate with these figures in a more ethical manner." However, the extent to which we can improve the ethical terms upon which we work with research assistants fundamentally depends on how research institutions and funding bodies govern the relations with partners in lower income countries.

Given the centrality of North-South research partnerships, especially in the fields of development and global health research (Bradley 2006; Murphy et al. 2015; Spiegel et al. 2015), it is essential to contextualize the experience of individual researchers within the broader picture. Although a discussion on the nature and implications of these partnerships is beyond the scope of this chapter, we want to underline the importance of the structural inequalities embedded in these collaborations in shaping the work relation between researcher and research assistant. As Bradley (2006), p. 15 notes, "this asymmetry [between northern and southern partners] manifests itself in the form of inequitable access to information, training, funding, conferences, publishing opportunities, and disproportionate influence of Northern partners in decision-making on the research agenda, project administration and budget management." A more equitable engagement of research assistants is critically constrained by these asymmetries, in that ownership of research remains in the hands of the researcher, acting on behalf of their institution. Thus, there are ways, as described above, in which researchers can make the employment relation with their assistant fairer in terms of pay and working conditions. But, reflecting and writing about the role of research assistants does not resolve all ethical issues, many of which should rather be dealt with at the institutional level.

5 Conclusions and Future Directions

Researchers conducting qualitative social science research do not have much guidance on how to make decisions to find, recruit and work with research assistants. This chapter is important because it is an initial attempt to fulfil some of these gaps. Drawing on our own research as well as that of others, we have discussed a set of key practical considerations researchers need to address when planning their fieldwork – from assessing whether a research assistant is needed to managing a work relation with assistants. The intention of the exercise was not to provide uniform answers, but rather to identify key issues and reasoning to address them. While their impact on the overall research process will not always be known, an awareness of how research assistants can influence the quality of data collected, as well as the final analysis, will improve the data collection process.

It is evident that the practicalities of working with research assistants are entrenched with the power relations that shape interactions between researchers, research assistants, and participants. The central relation we discussed is that between the researcher and the research assistant, which leads to considering a set of key responsibilities and obligations the researcher has as the employer. Thus managing the relation with the assistant requires thinking about a series of practical issues, such as adequate training, adequate pay, fair working conditions, and clear contractual arrangements. But, crucially, it is based on understanding the essential character of the relationship as one of employment. Researchers need to be more explicit in their acknowledgement of the work of research assistant and also need to reflect more on the influence they have on the research process and outcomes. However, there are important ethical concerns, stemming from the structurally unequal relations between partner institutions, that will not be resolved by individual researchers, or in the context of specific research projects.

Future contributions on this topic are much needed to fulfil a crucial gap in the literature on research methodology. Practical guidance for researchers would be significantly enriched if researchers doing cross-cultural research in the area of social science reflected, discussed, and wrote more on this topic. In particular, it would be useful to know more about different ways in which research assistants are employed by researchers and the type of assistance they provide in different contexts. At the same time, contributions exploring the institutional determinants of the habitual silence on the role of research assistant would be critical to help us see this issue within the bigger picture and reflect on what is needed to make the research landscape more conducive to fair collaborations between North and South institutions.

References

Berman RC, Tyyskä V. A critical reflection on the use of translators/interpreters in a qualitative cross-language research project. Int J Qual Methods. 2011;10(2):178–90.

Bradley M. North-South research partnerships: literature review and annotated bibliography: Special Initiatives Division, International Development Research Centre; 2006.

Bujra J. Lost in translation? The use of interpreters in fieldwork. In: Desai V, Potter RB, editors. Doing development research. London: SAGE; 2006. p. 172–9.

Caretta MA. Situated knowledge in cross-cultural, cross-language research: a collaborative reflexive analysis of researcher, assistant and participant subjectivities. Qual Res. 2015;15(4):489–505.

Cramer C, Johnston D, Oya C, Sender J. Research note: mistakes, crises, and research independence: the perils of fieldwork as a form of evidence. Afr Aff. 2016;115(458):145–60.

Deane K, Stevano S. Towards a political economy of the use of research assistants: reflections from fieldwork in Tanzania and Mozambique. Qual Res. 2016;16(2):213–28.

Deane K, Samwell P, Ngalya L, Boniface GB, Urassa M. Exploring the relationship between population mobility and HIV risk: evidence from Tanzania. Glob Public Health. 2016. Published online May 27th 2016. https://doi.org/10.1080/17441692.2016.1178318.

Devereux S, Hoddinott J. Fieldwork in developing countries. Hemel Hempstead: Harvester Wheatsheaf; 1992.

England KVL. Getting personal: reflexivity, positionality, and feminist research. Prof Geogr. 1994;46(1):80–9.

Flick U. An introduction to qualitative research. London: SAGE; 2009.

Flores-Macias F, Lawson C. Effects of interviewer gender on survey responses: findings from a household survey in Mexico. Int J Public Opin Res. 2008;20(1):100–10.

Freed AO. Interviewing through an interpretor, Social work. 1988 July–August, p. 315–19.

Grosh M, Glewwe P. Designing household survey questionnaires for developing countries: lessons from 15 years of the living standards measurement study, vol. 1, 2 & 3. Washington, DC: The World Bank; 2000.

Harding S. Feminism and methodology. Bloomington: Indiana University Press; 1987.

Iarossi G. The power of survey design: a user's guide for managing surveys, interpreting results, and influencing respondents. Washington, DC: The World Bank; 2006.

ICF Macro. Training field staff for DHS surveys. Calverton: ICF Macro; 2009.

Jenkins SA. Assistants, guides, collaborators, friends: the concealed figures of conflict research. J Contemp Ethnogr. 2015. Published online December 18th 2015. https://doi.org/10.1177/0891241615619993.

Liamputtong P. Performing qualitative cross-cultural research. Cambridge: Cambridge University Press; 2010.

Liamputtong P. Qualitative research methods. 4th ed. Melbourne: Oxford University Press; 2013.

MacKenzie CA. Filtered meaning: appreciating linguistic skill, social position and subjectivity of interpreters in cross-language research. Qual Res. 2016;16(2):167–82.

Mandel JL. Negotiating Expectations in the Field: Gatekeepers, Research Fatigue and Cultural Biases. Singapore Journal of Tropical Geography. 2003;24(2):198–210.

Merriam SB, Tisdell EJ. Qualitative research: a guide to design and implementation. San Francisco: Wiley; 2015.

Middleton T, Cons J. Coming to terms: reinserting research assistants into ethnography's past and present. Ethnography. 2014;15(3):279–90.

Molony T, Hammett H. The friendly financier: talking money with the silenced assistant. Hum Organ. 2007;66(3):292–300.

Molyneux C, Goudge J, Russell S, Chuma J, Gumede T, Gilson L. Conducting health-related social science research in low income settings: ethical dilemmas faced in Kenya and South Africa. J Int Dev. 2009;21(2):309–26.

Murphy J, Hatfield J, Afsana K, Neufeld V. Making a commitment to ethics in global health research partnerships: a practical tool to support ethical practice. J Bioeth Inq. 2015;12(1):137–46.

Pack S. How they see me vs. how I see them: the ethnographic self and the personal self. Anthropol Q. 2006;79(1):105–22.

Pitchforth E, Van Teijlingen E. International public health research involving interpreters: a case study from Bangladesh. BMC Public Health. 2005;5(71):1. https://doi.org/10.1186/1471-2458-5-71.

Randall S, Coast E, Compaore N, Antoine P. The power of the interviewer: a qualitative perspective on African survey data collection. Demogr Res. 2013;28(27):763–92.

Ritchie J, Lewis J. Qualitative research practice: a guide for social science students and researchers. London: SAGE; 2003.

Ryan L. "Inside" and "outside" of what or where? Researching migration through multi-positionalities. Forum: Qual Soc Res. 2015;16(2):Art. 17. http://nbn-resolving.de/urn:nbn:de:0114-fqs1502175.

Silverman D. Doing qualitative research: a practical handbook. London: SAGE; 2013.

Spiegel JM, Breilh J, Yassi A. Why language matters: insights and challenges in applying a social determination of health approach in a north-south collaborative research program. Glob Health. 2015;11(9):1–17. https://doi.org/10.1186/s12992-015-0091-2.

Stevano S. Women's work, food and household dynamics: a case study of northern Mozambique. Unpublished PhD thesis, Department of Economics, SOAS University of London. 2014.

Suwankhong D, Liamputtong P. Cultural insiders and research fieldwork: case examples from cross-cultural research with Thai people. Int J Qual Methods. 2015;14(5):1–7. https://doi.org/10.1177/1609406915621404.

Temple B, Edwards R. Interpreters/translators and cross-language research: reflexivity and border crossings. Int J Qual Methods. 2002;1(2):1–12.

Temple B, Young A. Qualitative research and translation dilemmas. Qual Res. 2004;4(2):161–78.

Tinkler C, Armstrong N. From the outside looking in: how an awareness of difference can benefit the qualitative research process. Qual Rep. 2008;13(1):53–60.

Turner S. Research note: the silenced assistant. Reflections of invisible interpreters and research assistants. Asia Pac Viewpoint. 2010;51(2):206–19.

Williamson DL, Choi J, Charchuk M, Rempel GR, Pitre N, Breitkreuz R, Kushner KE. Interpreter-facilitated cross-language interviews: a research note. Qual Res. 2011;11(4):381–94.

Indigenous Statistics

97

Tahu Kukutai and Maggie Walter

Contents

1 Introduction	1692
2 Methodology	1693
2.1 What Is a Methodology	1693
2.2 Why Methodology Matters	1694
3 Exposing the Orthodoxy of Indigenous Statistics	1695
4 Indigenous Data Sovereignty	1696
5 Getting to Understand Indigenous Methodologies	1697
6 Indigenous Quantitative Methodology	1698
7 Indigenous Quantitative Methodology in Practice	1699
7.1 Case Study 1: Aotearoa NZ: Māori Concepts of Family	1699
7.2 Case Study 2: Australia: How Do Indigenous Children Grow Up Strong in Education	1702
8 Conclusion and Future Directions	1703
References	1704

Abstract

Statistics about Indigenous peoples are a common feature of Anglo-colonizing nation states such as Canada, Australia, Aotearoa New Zealand, and the United States (CANZUS). The impetus for the production of most Indigenous statistics is the shared position of Indigenous disadvantage in health and socioeconomic status. In this chapter, we contrast statistics *about* Indigenous peoples with statistics *for* Indigenous people and statistics *by* Indigenous people. There are very significant differences between these categories of Indigenous statistics. At

---

T. Kukutai (✉)
University of Waikato, Hamilton, New Zealand
e-mail: tahuk@waikato.ac.nz

M. Walter
University of Tasmania, Hobart, Tasmania, Australia
e-mail: Margaret.Walter@utas.edu.au

© Springer Nature Singapore Pte Ltd. 2019
P. Liamputtong (ed.), *Handbook of Research Methods in Health Social Sciences*,
https://doi.org/10.1007/978-981-10-5251-4_40

the heart of these differences is the methodology that informs the research processes and practices. Statistics *about* Indigenous peoples often reflect the dominant social norms, values, and racial hierarchy of the society in which they are created. In the CANZUS states, these statistics are deficit focused and, at times, victim blaming. Also missing from these statistical portrayals is the culture, interests, perspectives, and alternative narratives of the Indigenous peoples that they purport to represent. We contrast these statistics with those from statistical research using processes and practices that are shaped by Indigenous methodologies. Indigenous methodologies are distinguished by their prioritization of Indigenous methods, protocols, values, and epistemologies. We conclude with two examples of what Indigenous quantitative methodologies look like in practice from Aotearoa NZ and Australia.

Keywords

Indigenous · Statistics · New Zealand · Australia · Colonization · Methodology

1 Introduction

The estimated number of Indigenous peoples ranges between 300 and 370 million, and comprises thousands of distinct polities covering all of the world's continents (Gracey and King 2009; Hall and Patrinos 2012). Statistics about Indigenous peoples (Given the diversity of Indigenous peoples, the United Nations does not have an official definition of "Indigenous" but rather invokes the following criteria: (1) Self- identification as indigenous peoples at the individual level and accepted by the community as their member; (2) Historical continuity with pre-colonial and/or pre-settler societies; (3) Strong link to territories and surrounding natural resources; (4) Distinct social, economic or political systems; (5) Distinct language, culture and beliefs; (6) Form non-dominant groups of society; (7) Resolve to maintain and reproduce their ancestral environments and systems as distinctive peoples and communities.) are a common feature of Anglo-colonizing nation states such as Canada, Australia, Aotearoa New Zealand, and the United States (the so-called CANZUS group; Meyer 2012). The impetus for the production of most of these Indigenous statistics is the shared position of socioeconomic and health disadvantage. In all of the four CANZUS nations, Indigenous peoples are far more likely to die younger, to experience much poorer health, to be unemployed, to be homeless, to be incarcerated, and to not have the same level of educational achievement as non-Indigenous citizens (Anderson et al. 2006; Cooke et al. 2007; Gracey and King 2009; Anderson et al. 2016; see also ▶ Chaps. 87, "Kaupapa Māori Health Research," and ▶ 88, "Culturally Safe Research with Vulnerable Populations (Māori)").

In this chapter, we not only discuss statistics *about* Indigenous peoples but also statistics *for* Indigenous people and statistics *by* Indigenous people. There are very significant differences between these categories of Indigenous statistics. At the heart of these differences is the methodology that informs the research processes and practices. Methodology matters and we demonstrate how the methodology informing the standard trope of statistics about Indigenous people in the CANZUS states are deficit focused and, at times, victim blaming. We contrast these statistics with those from statistical research using processes and practices shaped by Indigenous methodologies.

2 Methodology

2.1 What Is a Methodology

The terms "method" and "methodology" tend to be used interchangeably within the health and social science research literature. However, they mean quite different things. While both are related to the practice of doing research, they differ conceptually. Researchers need to have both a methodology and a method for the conduct of good research. All kinds of research, not just research related to Indigenous peoples, have a methodology (see also ▶ Chaps. 6, "Ontology and Epistemology," ▶ 87, Kaupapa Māori Health Research," and ▶ 90, "Engaging Aboriginal People in Research: Taking a Decolonizing Gaze").

The term "method" has a straightforward meaning. It refers to the method of collecting and/or analyzing data. For a qualitative research project exploring how Aboriginal women experience breast cancer treatments, the method might be in-depth interviews. For a research project exploring heart disease rates among urban Māori, the method would be statistical analysis. Despite what is written in some texts, methodology is not primarily related to whether the research has a qualitative or a quantitative base.

So what is a methodology? Basically, it is the framework that guides how the researcher approaches the research. This framework is not always consciously understood by researchers, especially those from dominant social, cultural, and racial groups. As has been argued elsewhere (Walter 2010; Walter and Andersen 2013), the basis of this guiding framework is the social positioning of the researcher. Social positioning relates to the race, class, gender, and social and cultural space that the researcher/s occupy. With different social positioning attributes comes different sets of values and belief systems (axiological elements) that can, for example, help determine what research questions the researcher thinks are important. The social position of the researcher will also be important in shaping what data or knowledge sets will be gathered (epistemological elements) and, if there is more than one set of knowledge, which knowledge set is prioritized. Social positioning also influences how researchers see the world, their place in it, and the place of others who are not like themselves (ontological elements).

Methodology can also affect the choice of method. This is because it is important for a method to be able to gather the sort of data that the researcher needs to address the research questions. Sometimes this means developing new research methods. For example, Yarning is an Aboriginal research method built around Aboriginal ways of communication (see also ▶ Chap. 90, "Engaging Aboriginal People in Research: Taking a Decolonizing Gaze"). But Yarning is more than just Aboriginal people talking. As argued by Bessarab and Ng'andu (2010), Yarning as a research method is also a process of meaning making and communicating in culturally appropriate ways. It is, therefore, likely to be much better method fit for researchers working with Aboriginal people than the traditional (Western) method of in-depth interviewing.

2.2 Why Methodology Matters

Methodology matters to the way research is done and to the findings that result. Quality research should always make obvious the methodology that informed the research process. When the research relates to Indigenous populations or cultural minorities, a clear articulation of the research methodology is even more crucial. This is because while Indigenous peoples are the frequent objects of health and social science research, they are far less likely to be the commissioners, research designers, or data interpreters of that research. In all CANZUS countries, the vast majority of Indigenous-related research is still undertaken by non-Indigenous researchers and commissioned by non-Indigenous policy makers (Taylor 2008; Kukutai and Walter 2015).

The imbalance is important because the social position of the subject "knower" (e.g., policy analyst within government) and the social position of the object of statistical study (Aboriginal and Torres Strait Islander and Māori people) are not even remotely the same. If these differences are reflected in the values that inform the research, the prioritization of knowledge, the analysis, and the interpretation of results, then the outcome is likely to focus on Indigenous "deficits" (Valencia 2012). Deficit research focuses on Indigenous problems and locates the source of those problems within Indigenous populations and culture. The validity of this approach and its methodological underpinnings has long been challenged by Indigenous scholars (see Tuhiwai Smith 1999) but still remains a dominant trope in Indigenous data (see next section).

To demonstrate how methodology operates in practice, let's relook at our research examples. In the qualitative example of exploring the experience of breast cancer treatment of Aboriginal women, there are some key methodological questions. More critically, differing answers to these questions will produce very different research projects and different findings. For example, do the researchers decide that the data are only going to be from Aboriginal women? Or will treatment personnel be included? If so, what will happen if the report from the treatment personnel and Aboriginal women differ? Whose perspective will be deemed more accurate? Will and how will the research process and practice be adjusted from mainstream models to capture the specific experiences of Aboriginal women? And which women? Are we talking about urban or remote experiences or both? Aggregation of "the" Aboriginal population into one category is a common practice in Australian research. This practice, however, ignores the reality of over 500 Aboriginal nations in Australia, all whom have different cultural, historical, and contemporary realities.

For our second example, cardiovascular disease (CVD), the statistical data already exists. However, the researcher's methodology will still shape the research outcomes. In Aotearoa NZ, ischemic heart disease accounts for over half of all cardiovascular disease mortality and the age-standardized ischemic heart disease mortality rate among Māori (35+ years) is more than twice as high as that among non-Māori (RR 2.14, CI 2.02–2.27) (Age standardized rates can be accessed at: http://www.health.govt.nz/our-work/populations/Māori-health/tatau-kahukura-Māori-health-statistics/nga-mana-hauora-tutohu-health-status-indicators/cardiovascular-disease.). What factors might explain why Māori have higher rates of heart disease at younger ages? Researchers without a strong understanding of Māori culture, values, and life circumstances might line up the usual suspects of heart disease: smoking, diet, and exercise. Their choice of such variables may be influenced by negative stereotypes of Māori people that circulate in the public discourse and which define the "problem" of Māori CVD as primarily one of poor individual choices and health behaviors. Subsequent policy interventions may also be focused on promoting individual lifestyle changes. This is despite the substantial evidence that such an approach has a limited effect in disadvantaged populations because of the failure to address the issues that gave rise to the behaviors.

By contrast, a Māori researcher who is embedded in both their discipline and their culture will likely include elements of the social determinants of health, which are the underlying economic and social conditions that drive racial health inequities (Commission on Social Determinants of Health 2008). These factors are inclusive of the heavy socioeconomic disadvantage experienced by Māori related to dispossession, colonialism, and ongoing marginalization including institutional racism and unmet needs in access to high quality and culturally appropriate healthcare services (Ajwani et al. 2003; Kerr et al. 2010; Axelsson et al. 2016). Focusing on these distal determinants of health, and how they shape the distribution of more immediate risk factors such as poor diet and ultimately CVD, engenders a different understanding of health inequities and approaches to reducing them. Policy responses might include engaging Māori in the design and delivery of culturally grounded health services, addressing the institutional barriers to timely diagnosis and treatment pathways, and taking a broader whānau (family) approach to health promotion rather than a narrow individualistic focus (Durie 2003; Kerr et al. 2010; see also ▶ Chaps. 87, "Kaupapa Māori Health Research," and ▶ 88, "Culturally Safe Research with Vulnerable Populations (Māori)").

The question then arises, if methodology is so important to the research process, why is it so frequently not articulated within research? The answer seems to be that researchers whose social positioning places them in the dominant racial or cultural group have not been trained to recognize that their social positioning directly affects how they "do" research. In cross-cultural research, such a lack of researcher reflexivity is a recipe for at best, poor quality research, and at worst, research that does harm to the group it is professing to research.

3 Exposing the Orthodoxy of Indigenous Statistics

In cross-cultural health and social science research, the traditional way of doing Indigenous research flows from the dominant model of what Indigenous statistics looks like within Aotearoa New Zealand, Australia and other first-world colonized nations. The privileging of mainstream "mental models" to frame and explain Indigenous peoples has real-life consequences for Indigenous peoples.

Indigenous researchers and communities have made numerous criticisms of how statistical agencies collect, disseminate, and analyze Indigenous data. The criticisms include a tendency to focus on Indigenous "problems" rather than strengths; a failure to recognize Indigenous culture, values, and practices in the measures and processes used to gather and analyze data; a failure to prioritize Indigenous needs in data system development; ineffective measures to address longstanding data quality issues such as Indigenous undercounting; and a tendency to use token consultation rather than meaningful Indigenous engagement and partnership (Taylor 2009; Robson and Reid 2001; Prout 2012). In response to these problems, Kukutai and Walter (2015) proposed five development principles aimed at enhancing the functionality of official statistics for both Indigenous peoples and national statistics agencies.

These concerns are not limited to domestic policy making. Global forums, such as the United Nations Permanent Forum on Indigenous Issues and the Special Rapporteur on the Rights of Indigenous Peoples, have stressed the importance of high quality and meaningful data for enabling Indigenous development. However, the extent to which governments recognize the existence of Indigenous peoples in official statistics varies widely. Preliminary findings from the *Ethnicity Counts?* project show that, of the 150 countries and territories that encompass Indigenous peoples, only 45% identify Indigenous peoples in the population census (Taylor and Kukutai 2015). In some countries, there are multiple questions relating to Indigenous identity. In Aotearoa NZ, for example, Māori can be identified by ethnicity, ancestry, tribal affiliation, and language. However, in the majority of countries, Indigenous peoples are statistically invisible. Ironically, some of these countries, such as Sweden and Norway, have some of the most well-developed official statistics systems in the world.

The census is the flagship of official statistics in many countries. It provides the population-level denominator for many indicators of well-being within countries, as well as for many of the UN's Sustainable Development Goal indicators (United Nations General Assembly 2015). The extent of Indigenous invisibility in the census has far-reaching implications for the ability to monitor Indigenous development on a global scale. The 2015 State of the World's Indigenous Peoples report noted that it is still often difficult to obtain a global assessment of Indigenous peoples' health status because of the lack of data (United Nations Department of Economic and Social Affairs 2015).

4 Indigenous Data Sovereignty

One of the questions raised by Indigenous quantitative methodologies is who has the power to control Indigenous data. In the CANZUS states, there has been a growing call for greater control over the collection, dissemination, analysis, and storage of Indigenous data. This call for "Indigenous data sovereignty" (Kukutai and Taylor 2016) is founded on Indigenous rights to self-determination which emanate from their inalienable relationships to lands, waters, and the natural world, and which are encapsulated in Articles 3 and 4 on the United Nations Declaration on the Rights of

Indigenous Peoples (The full text of the UNDRIP can accessed at: http://www.un.org/esa/socdev/unpfii/documents/DRIPS_en.pdf.). The idea of data sovereignty is a recent development of the digital age referring to the management of information in a way that is consistent with laws, practices, and customs of the nation-states where data are located. Indigenous data sovereignty sees Indigenous data as subject to the laws of the nation from which it is collected and requires a relocation of authority over relevant information from nation states back to Indigenous peoples (Snipp 2016). Indigenous data is broadly understood as data about Indigenous peoples, their territories, conditions (including health conditions), and ways of life. Such data includes genetic samples, linked "mega" datasets, digitized health records, and data on land and other natural resources. In the context of cross-cultural research, the implications of Indigenous data sovereignty are far reaching because it has the potential to transform power relationships in terms of who owns, governs, and controls access to and management of Indigenous data.

In the CANZUS states, Indigenous peoples are giving practical expression to various forms of Indigenous data sovereignty. In Canada, there are the First Nations' principles and practices of ownership, control, access, and possession over First Nations data known as OCAP® (First Nations Information Governance Centre 2014). OCAP® was created by the First Nations Information Governance Centre to help guide the development of the First Nations Regional Health Survey (FNRHS), the only First Nations-governed, national health survey in Canada that collects information about First Nation on-reserve and northern communities. The development of OCAP® was motivated by negative experiences with research projects conducted by non-First Nations people that did not benefit First Nations people or communities. OCAP® ensures that First Nations own their information and respects the fact that they are stewards of their information, much in the same way that they are stewards over their own lands. It also reflects First Nation commitments to use and share information in a way that maximizes the benefit to a community, while minimizing harm. First Nation communities have passed their own privacy laws, established research review committees, entered data-sharing agreements, and set standards to ensure OCAP® compliance. Other Indigenous data sovereignty initiatives are being driven by Te Mana Raraunga, the Māori Data Sovereignty network in Aotearoa NZ, the US Indigenous Data Sovereignty Network, and the Yawuru Native Title holders of Broome in Western Australia (Yap and Yu 2016). Collectively, these networks and organizations, and others like them, are developing new ways of "doing" Indigenous data that are challenging conventional methods and methodologies.

5 Getting to Understand Indigenous Methodologies

The deficiencies of traditional Western research methodologies for Indigenous peoples have led Indigenous scholars, globally, to develop Indigenous methodologies. Indigenous methodologies are a paradigm rather than a category of methodologies. Each, however, shares a philosophical base. This base is concisely summed up by Sami scholar Porsanger (2004) when she states that Indigenous methodologies all reflect Indigenous ways of knowing, doing, and being. In doing so, they make visible what is meaningful and logical for Indigenous people and Indigenous understandings of the world.

The field of Indigenous methodology scholarship was led by the ground-breaking work of Linda Tuhiwai Smith (1999). Smith's book, *Decolonizing Methodology: Research and Indigenous Peoples*, details the tenets of Kaupapa Māori, a methodology intricately connected to Māori philosophy and principles, the validity and legitimacy of Māori, Māori language (Te Reo Māori) and culture, and Māori autonomy over their own cultural well-being. Moewaka Barnes (2000) emphasizes three defining principles of this approach:

- It is by Māori for Māori.
- Māori worldviews are the normative frame
- Research is for the benefit of Māori.

In a similar vein, Native Hawaiian scholar Ku Kukahalau, the first person to earn a PhD in Indigenous education, highlights the importance of Hawaiian cultural protocols in her integration of existing heuristic methodology and Indigenous epistemology (2004). In Australia, Aboriginal scholar Karen Martin (2003, 2008), aligns the philosophical underpinnings of Indigenous methodology into theoretical principles. These require a recognition of Aboriginal worldviews, knowledge, and realities; the honoring of Aboriginal social mores; the social, historical, and political contexts which shape Aboriginal experience, lives, positions, and futures; and the privileging of the voices, experiences, and lives of Aboriginal people and Aboriginal lands. Native American scholar Margaret Kovach (2009) focuses on qualitative research practices in her theorizing of Indigenous methodologies. She argues that Indigenous methodologies are distinctive from Western and other methodological frames, and are distinguished by their prioritization of Indigenous methods, protocols, meaning making, and epistemologies in how to undertake research processes and research practice.

Indigenous methodology scholarship is also emerging from non-Anglo colonized nation states. Botswanan scholar Bagele Chilisa (2011) for, example, uses a postcolonial frame to demonstrate how methodologies are not restricted to academic knowledge systems. Her Indigenous methodological stance focuses on how the paradigms and practices of research can support Indigenous epistemologies and honor integrative knowledge systems (see also ▶ Chap. 15, "Indigenist and Decolonizing Research Methodology").

6 Indigenous Quantitative Methodology

It is fair to say that, within the diverse spectrum of Indigenous methodologies, there is a strong preference toward qualitative methods and a widely held view that statistical research sits in tension with "Indigenous ways of knowing" (Kovach 2009). This is largely due to the perception that quantitative research methodologies are rooted in a Western *positivist* tradition that relies on "external evidence, testing and universal laws of generalizability...contradict[s] a more integrated, holistic and contextualized Indigenous approach to knowledge" (Kovach 2009, p. 78). The question then arises – what does a quantitative methodology built on Indigenous ways of knowing look like?

In their book *Indigenous Statistics*, Tasmanian Aboriginal scholar Maggie Walter and Metis scholar Chris Andersen (2013) tackle this question directly, proposing a way to move the understanding of Indigenous methodologies into the field of quantitative research. Dominant ways of doing Indigenous statistics, they argue, shortchange Indigenous peoples and communities through their narrow portrayal of who Indigenous peoples are, and their circumscription of how Indigenous people can be understood. Mainstream narratives of Aboriginal and other Indigenous populations in Anglo-colonizing nation states are based on data about Indigenous peoples that the nation state, rather than Indigenous peoples, deem to be important. The result is a depressing familiar role call that Walter (2016) calls 5D data: data about Indigenous people that focuses on disparity, deprivation, disadvantage, dysfunction, and difference.

The central problem of Indigenous statistics is that population or racial group statistics are not neutral data. Rather, they reflect the dominant social norms, values, and racial hierarchy of the society in which they are created. In Australia and Aotearoa NZ, these dominant social norms and values typically reflect those of Anglo/European settler descendants. Norms can be thought of as the shared expectations for social behavior around what is culturally desirable or acceptable. Norms are evident in everyday interactions, in institutions such as schools and healthcare services, and in policy approaches. The power of these norms comes from their "taken for granted nature" – very rarely are they made explicit or visible like formal rules. Statistics, and especially official statistics, embody norms but hold an aura of objectivity and tend to be presented and understood as "facts." The trouble is that these "facts" only tell a very small, and specifically framed, part of the reality of Indigenous peoples. What is not present in these statistical portrayals is the culture, interests, perspectives, and alternative narratives of the Indigenous peoples that they purport to represent.

Indigenous quantitative methodologies, in contrast, can support the development of statistical portrayals that go beyond the narrow, frequently pejorative, reflections that dominate official statistics of Indigenous peoples. Moreover, Indigenous statistics developed from an Indigenous methodological frame can, as argued by Walter and Andersen (2013, p. 73) "speak back" to the state in a way that both incorporates Indigenous knowledge and is ontologically translatable to state actors. We illustrate this by way of our two case studies below.

7 Indigenous Quantitative Methodology in Practice

7.1 Case Study 1: Aotearoa NZ: Māori Concepts of Family

In this section, we discuss two examples of what Indigenous quantitative methodologies look like in practice from Aotearoa NZ and Australia. The first case study is from a project exploring Māori expressions of whānau or family (Kukutai et al.

2016). Families are a fundamental social unit in all societies but vary greatly in terms of their form, function, and meaning. Families are also an important focus for research, public policy, and service delivery, from the immunization of children, to state-funded assistance for single parents, and elder care. In Aotearoa NZ, statistical studies of Māori families have tended to focus on household structure and circumstances and, more recently, on vulnerable children and family violence (Vulnerable Children Act 2014). These portrayals are often deficit focused and viewed through the lens of Western theoretical models. Missing from these statistical narratives are Māori perceptions of who their whānau are, how their whānau are doing, and what whānau well-being entails (Cunningham et al. 2005; Tibble and Ussher 2012).

The whānau concept and well-being project is a collaboration between all-Māori research team and government policy agencies (We thank our colleagues at the Social Policy Evaluation and Research Unit (Superu), Te Puni Kōkiri, the Ministry of Māori Development, and the Superu Whānau Reference Group.). Much of the analysis is drawn from "Te Kupenga," a nationally representative postcensal survey of well-being among Māori adults, which was conducted for the first time in 2013 (Kukutai et al. 2016). Unlike other official surveys such as the Census and General Social Survey, Te Kupenga was specifically designed with Māori values and priorities in mind and had substantial input from Māori researchers, communities, and policy makers (Statistics New Zealand 2009). The initial stage of the project focused on two key questions:

1. How do Māori define who belongs to their whānau?
2. How are expressions of whānau related to factors such as cultural identity, household living arrangements, and social context?

The word whānau literally means to "to be born" or to "give life." While there is no univocal definition of whānau, there is a broad consensus that genealogical relationships form the basis of whānau, and that these relationships are intergenerational, shaped by context, and given meaning through roles and responsibilities (Lawson-Te Aho 2010). From a Māori standpoint, to be part of a whānau is to share common "whakapapa." In a traditional sense, whakapapa is understood as descent-based relationships which extend from the physical world to the spiritual world (Kruger et al. 2004). Whakapapa also refers to the layers of relationships that connect individuals to ancestors, to the living, and to the natural environment (Te Rito 2007). Whakapapa relationships are not just ways of situating individuals within a kin group but are connected to roles, responsibilities, and obligations including mutual acts of giving and receiving, and the intergenerational transmission of knowledge.

The literature also refers to the concept of kaupapa whānau which is based on a common purpose or shared interests (Lawson-Te Aho 2010). In kaupapa whānau, "family-like" relationships of support and reciprocity are established as individuals purposefully engage to achieve a common goal. An oft-cited example is that of Māori language revitalization and preschool Māori language nests called kōhanga reo (Smith 1995). This expansive understanding of family is far removed from Euro normative concepts of family, especially those emphasizing the household as the economic unit of production. But, how do these culturally grounded understandings of whānau play out in the context of a representative national survey?

In defining whānau, the approach taken in Te Kupenga was to acknowledge kinship and interest-based whānau and leave it to the individual to define their own whānau within four broad relationship categories (Tibble and Ussher 2012). The question and response categories are shown in Fig. 1. Respondents could select as many categories as they needed. For the statistical analysis they were grouped into one of four mutually exclusive categories describing the broadest concept of whānau category reported, ranging from nuclear family to friends and others. The distribution can be seen in Table 1.

Just over 40% of respondents in Te Kupenga reported that their whānau *only* comprised immediate relatives, that is, parents, partner/spouse, brothers, sisters, brother, sister, parent in-laws, and children. A further 15% reported that their whānau included grandparents and grandchildren, and about one-third included extended whānau such as aunts, uncles, and cousins. Interestingly, nearly 13% of Māori counted close friends and others as part of their whānau.

Regression analyses showed that household-based living arrangement – the conventional way of measuring family in Aotearoa NZ – is a very poor predictor of how Māori see their whānau. More important are demographic factors (age, region) and cultural factors including connectedness to customary communities, access to cultural support, and having a high regard for Māori culture. Māori with

Describe whānau (qWHAWhanauDescribe)

Which group or groups include those you were thinking about as your whānau? You can select as many as you need.

A parents, partner/spouse, brothers and sisters, brothers/sisters/parents in-law, children	C aunts and uncles, cousins, nephews and nieces, other in-laws
B My grandparents, my grandchildren	D close friends, others

Fig. 1 Whānau question from Te Kupenga 2013

Table 1 Broadest concept of whānau (family) reported by Māori respondents in Te Kupenga 2013

	Per cent
A. Parents, partner/spouse, brothers and sisters, brothers/sisters/ parents-in-law, children	40.2
B. Grandparents/grandchildren	15.2
C. Aunts, uncles, cousins, nephews, nieces, other in-laws	31.9
D. Close friends/others	12.5

strong cultural connections tend to have a broader concept of whānau. The analysis has also helped to clarify the contexts within which nongenealogical relationships are perceived as being "whānau-like." Interestingly, those who have participated in Māori language education and lived in homes where Māori is spoken are more likely to include friends and others as part of their whānau. Similarly, Māori who provide support to people living in other households, and those in challenging economic circumstances, are also more likely to count nonrelatives as part of their whānau. The project has important implications for research and policy focused on families. It suggests that, for Māori, household-based measures of family are a very poor proxy for the more complex set of whānau relationships that exist and that policy responses based on these narrow Western concepts may have limited relevance.

7.2 Case Study 2: Australia: How Do Indigenous Children Grow Up Strong in Education

The following case study demonstrates practically that it is not the method, in this case statistical analysis, but the methodological frame that shapes research. A key element is that the focus is not describing or investigating "the problem" of lower educational achievement for Aboriginal and Torres Strait Islander children as is the traditional research approach. Rather, the focus is on identifying the causes and the best ways to achieve good educational outcomes.

The all Aboriginal research team are researching Aboriginal and Torres Strait children's (0–18 years) lived experience of schooling and education. The study's objectives are to:

1. Identify the critical intersections of events that impact on Indigenous children's educational chances across the childhood life course across locations.
2. Identify the pathways, protective factors, and resilience dimensions that support educational achievement for Indigenous children irrespective of disadvantage.

The project uses data from the Longitudinal Study of Indigenous Children (LSIC) a national longitudinal panel study conducting annual waves of data collection, with Wave 1 (2008) surveying families of 1,670 Indigenous children from 11 sites across Australia. Face-to-face interviews are conducted between the study child's primary parent and locally employed Indigenous research administration officers. Use of the LSIC data is a key is part of the research project's methodological frame. The study is guided by an Indigenous-led Steering Committee and its question topics, question design and conduct and are overtly shaped by Aboriginal and Torres Strait Islander perspectives and values.

The starting premise of this *Strong in Education* research project is that while a lot is known about Indigenous children, it is a certain sort of knowledge from a particular perspective. Government statistics tell us that Aboriginal and Torres Strait Islander children are much more likely to live in poor households and do far less well in the education system than non-Indigenous children. They also consistently record that Aboriginal and Torres Strait Island children are more likely to miss school, be suspended, and less likely to go on to higher education. What such existing statistics do not do is tell us what factors support good education and resilience for Aboriginal and Torres Strait Islander children. Identifying those factors is the key aim of this research. Within this, epistemologically, the analysis centers Indigenous people's knowledge, concepts, and worldviews.

The research's key concepts also reflect an Aboriginal and Torres Strait Islander methodological frame. The term "Strong" is conceptualized as the deployment of resilience to achieve good education despite adverse life circumstances. "Good education" refers to academic achievement to non-Indigenous median norms but also to cultural and community education (Malin and Maidment 2003; Andersen and Walter 2010). "Resilience" refers to the ability to cope with stress and adversity and do well in life despite difficulties (Gunnestad 2006). Its conceptualization within the research recognizes the interface of Indigenous social and cultural resilience with individual/family resilience and that social, cultural, and identity practices that support positive adaptation are integrally connected to resilience (Lalonde 2006).

Recent results from examinations of LSIC data in relation to educational outcomes find that parental and child social and emotional well-being are strong predictive factors for children's reading scores (Anderson et al. in press) and that how well the primary parent thought their child's school understands the needs of Indigenous families was a consistent predictor in how involved parents were with their child's schooling (Trudgett et al. in press).

8 Conclusion and Future Directions

In the CANZUS states, governments continue to invest substantial time and resources in monitoring the well-being outcomes of Indigenous peoples. In recent decades, governments in these countries have amassed a wealth of statistical data on Indigenous populations, all of whom are a significant focus of population research and policy in their respective countries. However, the categories and contexts employed in statistics about Indigenous peoples typically reflect dominant group norms, and their social and economic institutions. Because statistics about Indigenous peoples rarely encompass Indigenous methodologies, key aspects of Indigenous life are either missing or misrepresented. These epistemological and methodological shortcomings have stimulated calls for approaches for statistics that are *by* and *for* Indigenous peoples, rather than simply *about* them. In this chapter, we have identified key differences between these statistical approaches and the crucial importance of methodology for determining which questions are asked, and which processes and practices are employed.

In terms of future directions, major data transformations will raise new challenges along with potential opportunities. Technological innovations in the private sector involving big data are changing how data are used, most evident in the area of health. In the United States, genomics and big data science are being exploited in new ways to provide targeted, predictive, and personalized care in a "precision health" approach. Official data practices are also being transformed as governments seek alternatives to traditional data collection practices. Aotearoa NZ is at the forefront of these changes with several major initiatives that will fundamentally alter the national data ecosystem. These include legislative reform to enable greater data sharing across agencies (NZ Government 2014), a greater emphasis on extracting social and economic value from data, and the use of linked data on individuals and families to inform the government's social investment spending through targeted inventions (The Treasury 2016). Given that Māori are disproportionately the subject of government interventions, these shifts raise a number of key issues about Māori data governance, ownership, and access. The rise of linked mega datasets and broader data sharing is likely to become a standard feature of official statistics in all of the CANZSUS states in the near future. The capacity of Indigenous peoples to benefit from the "data revolution" will likely depend on the extent to which they are able to exert meaningful influence and oversight of the practices, processes, and principles that emerge over the next decade.

References

Ajwani S, Blakely T, Robson B, Tobias M, Bonne M. Decades of disparity: ethnic mortality trends in New Zealand 1980–1999, Public Health Intelligence Occasional Bulletin Number 16. Wellington: Ministry of Health and University of Otago; 2003.

Andersen C, Walter M. Indigenous perspectives and cultural identity. In: Hyde M, Carpenter L, Conway R, editors. Diversity and inclusion in Australian schools. South Melbourne: Oxford University Press; 2010. p. 63–87.

Anderson I, Crengle S, Kamaka ML, Chen T, Palafox N, Jackson-Pulver L. Indigenous health in Australia, New Zealand, and the Pacific. Lancet. 2006;367(9524):1775–85.

Anderson I, Robson B, Connolly M, Al-Yaman F, Bjertness E, King A, . . ., Yap L. Indigenous and tribal peoples' health (The Lancet–Lowitja Institute Global Collaboration): a population study. The Lancet. 2016;388(10040):131–57.

Anderson I, Lyons JG, Luke JN, Reich HS. Health determinants and educational outcomes for indigenous children. In: Walter M, Martin KL, Bodkin-Andrews G, editors. Growing up strong children: indigenous perspectives on the longitudinal study of indigenous children. London: Palgrave MacMillan; in press.

Axelsson P, Kukutai T, Kippen R. The field of indigenous health and the role of colonisation and history. J Popul Res. 2016;33(1):1–7.

Bessarab D, Ng'andu B. Yarning about yarning as a legitimate method in indigenous research. Int J Crit Indig Stud. 2010;3(1):37–50.

Chilisa B. Indigenous research methodologies. London: SAGE; 2011.

Commission on Social Determinants of Health. Closing the gap in a generation: health equity through action on the social determinants of health, Final report of the Commission on Social Determinants of Health. Geneva: World Health Organization; 2008.

Cooke M, Mitrou F, Lawrence D, Guimond E, Beavon D. Indigenous well-being in four countries: an application of the UNDP's human development index to indigenous peoples in Australia, Canada, New Zealand, and the United States. BMC Int Health Human Rights. 2007;7(9):1–11.

Cunningham C, Stevenson B, Tassell N. Analysis of the characteristics of whānau in Aotearoa. Report for the Ministry of Education. Palmerston North: Massey University; 2005.

Durie MH. The health of indigenous peoples: depends on genetics, politics, and socioeconomic factors. Br Med J. 2003;326(7388):510–1.

First Nations Information Governance Centre. Ownership, control, access and possession (OCAP™): The path to First Nations information governance. Ottawa: The First Nations

Information Governance Centre; 2014. Retrieved from http://fnigc.ca/sites/default/files/docs/ocap_path_to_fn_information_governance_en_final.pdf.
Gracey M, King M. Indigenous health part 1: determinants and disease patterns. Lancet. 2009;374(9683):65–75.
Gunnestad A. Resilience in a cross-cultural perspective. How resilience is generated in different cultures. J Intercult Commun. 2006;11. Retrieved from http://www.immi.se/intercultural/nr11/gunnestad.htm.
Hall GH, Patrinos HA. Indigenous peoples, poverty, and development. Cambridge, UK: Cambridge University Press; 2012.
Kahakalau K. Indigenous heuristic action research: bridging western and indigenous research methodologies. Hulili: Multidiscip Res Hawaii Well-Being. 2004;1(1):19–33.
Kerr S, Penney L, Moewaka Barnes H, McCreanor T. Kaupapa Māori action research to improve heart disease services in Aotearoa, New Zealand. Ethn Health. 2010;15(1):15–31.
Kovach M. Indigenous methodologies: characteristics, conversations, and contexts. Toronto: University of Toronto Press; 2009.
Kruger T, Pitman M, Grennell D, McDonald T, Mariu D, Pōmare A, Mita T, Maihi M, Lawson-Te Aho K. Transforming whānau violence – a conceptual framework. Wellington: Second Māori Taskforce on Whānau Violence; 2004.
Kukutai T, Taylor J, editors. Indigenous data sovereignty: towards an agenda, CAEPR Research Monograph, 2016/34. Canberra: ANU Press; 2016.
Kukutai T, Walter M. Recognition and indigenizing official statistics: reflections from Aotearoa New Zealand and Australia. Stat J IAOS. 2015;31(2):317–26.
Kukutai T, Sporle A, Roskruge M. Expressions of whānau. In: Social policy evaluation and research unit, families and Whānau status report 2016. Wellington: Superu; 2016. p. 52–77.
Lalonde CE. Identity formation and cultural resilience in aboriginal communities. In: Flynn RJ, Duding P, Barber J, editors. Promoting resilience in child welfare. Ottawa: University of Ottawa Press; 2006. p. 52–67.
Lawson-Te Aho K. Definitions of whānau: a review of selected literature. Wellington: Families Commission; 2010.
Malin M, Maidment D. Education, indigenous survival and well-being: emerging ideas and programs. Aust J Indigenous Educ. 2003;32:85–100.
Martin K. Ways of knowing, ways of being and qays of soing: a theoretical framework and methods for indigenous re-search and indigenist research. J Aust Stud. 2003;27(76):203–14.
Martin K. Please knock before you enter: aboriginal regulation of outsiders and the implications for researchers. Teneriffe: Post Press; 2008.
Meyer WH. Indigenous rights, global governance, and state sovereignty. Hum Rights Rev. 2012;13(3):327–47.
Moewaka Barnes HM. Kaupapa Māori: explaining the ordinary. Pac Health Dialog. 2000;7(1):13–6.
New Zealand Government. Government ICT strategy and action plan to 2017. ICT Action plan 2014. Retrieved from: https://www.ict.govt.nz/assets/Uploads/Government-ICT-Strategy-and-Action-Plan-to-2017.pdf (2014).
Porsanger J. An essay about indigenous methodology. Retrieved from http://munin.uit.no/bitstream/handle/10037/906/article.pdf..?sequence=1 (2004).
Prout S. Indigenous wellbeing frameworks in Australia and the quest for quantification. Soc Indic Res. 2012;109(2):317–36.
Robson B, Reid P. Ethnicity matters: Māori perspectives. Wellington: Statistics New Zealand; 2001.
Smith G. Whakaoho whānau: new formations of whānau and an innovative intervention into Māori cultural and economic crises. He Pukenga Korero. 1995;1:18–36.
Snipp M. What does data sovereignty imply – what does it look like? In: Kukutai T, Taylor J, editors. Indigenous data aovereignty: towards an agenda, CAEPR Research Monograph, 2016/34. Canberra: ANU Press; 2016.
Statistics New Zealand. He kohinga whakaaro/Māori social survey discussion document. Wellington: Statistics New Zealand; 2009.
Taylor J. Indigenous peoples and indicators of well-being: Australian perspectives on United Nations global frameworks. Soc Indic Res. 2008;81(1):111–26.

Taylor J. Indigenous demography and public policy in Australia: population or peoples? J Popul Res. 2009;26(2):115–30.

Taylor J, Kukutai T. Indigenous data sovereignly and indicators: reflections from Australia and Aotearoa New Zealand. In: Paper presented at the UNPFII Expert Group Meeting on "The Way Forward: Indigenous Peoples and the 2039 Agenda for Sustainable Development", United Nations, HQ, New York, 22–23 Oct. 2015.

Te Rito JS. Whakapapa and whenua: an insider's view. MAI Rev. 2007;3:1–8.

The Treasury. Social investment. Retrieved from: http://www.treasury.govt.nz/statesector/socialinvestment (2016).

Tibble A, Ussher S. Kei te pēwhea tō whānau? Exploring whānau using the Māori social survey. Wellington: Statistics New Zealand; 2012.

Trudgett M, Page S, Bodkin-Andrews G, Franklin C, Whittaker A. Another brick in the wall? Parent perceptions of school educational experiences of indigenous Australian children. In: Walter M, Martin KL, Bodkin-Andrews G, editors. Growing up strong children: indigenous perspective on the longitudinal study of indigenous children. London: Palgrave MacMillan; in press.

Tuhiwai Smith L. Decolonizing methodologies: research and indigenous peoples. London/New York: Zed Books; 1999.

United Nations Department of Economic and Social Affairs. State of the world's indigenous peoples, 2nd vol.: indigenous people's access to health services. New York: UNDESA; 2015.

United Nations General Assembly. Resolution adopted by the General Assembly on 25 September 2015. 70/1. Transforming our world: the 2030 Agenda for Sustainable Development. Retrieved from: http://www.un.org/ga/search/view_doc.asp?symbol=A/RES/70/1&Lang=E (2015).

Valencia RR, editor. The evolution of deficit thinking: educational thought and practice. London: Routledge; 2012.

Vulnerable Children Act. Retrieved from http://www.legislation.govt.nz/act/public/2014/0040/latest/DLM5501618.html (2014).

Walter M. The politics of the data: how the Australian statistical indigene is constructed. Int J Crit Indigenous Stud. 2010;3(2):45–54.

Walter M. Data politics and indigenous representation in Australian statistics. In: Kukutai T, Taylor J, editors. Indigenous data sovereignty: towards an agenda, CAEPR Research Monograph, 2016/34. Canberra: ANU Press; 2016.

Walter M, Andersen C. Indigenous statistics: a quantitative methodology. Walnut Creek: Left Coast Press; 2013.

Yap M, Yu E. Data sovereignty for the Yawuru in Western Australia. In: Kukutai T, Taylor J, editors. Indigenous data sovereignty: towards an agenda, CAEPR Research Monograph, 2016/34. Canberra: ANU Press; 2016.

A Culturally Competent Approach to Suicide Research with Aboriginal and Torres Strait Islander Peoples

98

Monika Ferguson, Amy Baker, and Nicholas Procter

Contents

1	Introduction	1708
2	Understanding Suicide in Aboriginal Communities	1709
3	Awareness of the Historical Research Context	1711
4	Understanding and Applying Ethical Research Principles	1712
5	Cultural Competency Throughout the Stages of the Research Journey	1713
	5.1 Developing Relationships	1713
	5.2 Deciding on a Research Topic	1716
	5.3 Applying an Appropriate Research Methodology	1716
	5.4 Seeking Ethical Approval	1717
	5.5 Engaging with Research Participants	1718
	5.6 Reporting and Disseminating Findings	1718
6	Conclusion and Future Directions	1719
References		1719

Abstract

Despite the strength and resilience of Aboriginal peoples, suicide has profound and ongoing impacts for individuals, families, and communities, and has been identified as an area requiring further research. This chapter outlines a culturally competent approach for conducting social and emotional well-being research, from the perspective of non-Aboriginal researchers. The chapter begins by outlining the topic of suicide in the context of Aboriginal peoples and history, as a base from which to understand approaches to researching this complex topic.

M. Ferguson (✉) · N. Procter
School of Nursing and Midwifery, University of South Australia, Adelaide, SA, Australia
e-mail: Monika.Ferguson@unisa.edu.au; Nicholas.Procter@unisa.edu.au

A. Baker
School of Health Sciences, University of South Australia, Adelaide, SA, Australia
e-mail: Amy.Baker@unisa.edu.au

© Springer Nature Singapore Pte Ltd. 2019
P. Liamputtong (ed.), *Handbook of Research Methods in Health Social Sciences*,
https://doi.org/10.1007/978-981-10-5251-4_41

Sensitivities associated with conducting research as non-Aboriginal researchers are outlined, stressing the importance of developing a consciousness toward the historical relationship between Aboriginal and non-Aboriginal Australians. The chapter then introduces important ethical principles, which can be used to guide culturally competent practice throughout the research journey. Specific methodological approaches are outlined, with an emphasis on those that are participatory in nature. Although the topic of suicide is utilized as a backdrop in this chapter, the approaches discussed here are transferrable to research exploring a range of social and emotional well-being concerns experienced by Aboriginal communities.

Keywords
Social and emotional well-being · Collaboration · Participatory research · Strengths-based research · Suicide · Aboriginal people

1 Introduction

Despite their resilience, strength, and ongoing connection to culture, Aboriginal peoples experience profound inequities across a spectrum of social and emotional well-being concerns. In particular, suicide, which is recognized as a global public health priority (WHO 2014), has been identified as an area in critical need of research among Aboriginal communities (King and Brown 2015). In light of a history of poor research practices and experiences, research with Aboriginal people needs to be carried out in a culturally competent manner. Cultural competency has been argued to be a vital strategy for reducing inequalities in health experienced by Aboriginal people (Bainbridge et al. 2015; see also ▶ Chaps. 88, "Culturally Safe Research with Vulnerable Populations (Māori)," and ▶ 89, "Using an Indigenist Framework for Decolonizing Health Promotion Research").

Although the importance of culturally competent practice is widely recognized in the context of health care, its application to health research appears to have received less attention. This is surprising, given that cultural competency is essential to all stages of the research process, including building and maintaining relationships, designing and implementing methodological and data collection approaches, analyzing and interpreting results, and disseminating findings. Papadopoulous and Lees (2002) have proposed a four-concept model for cultural competency, adapted from their work in health care. This includes: cultural awareness – whereby researchers reflect on, understand, and challenge their own values, perceptions, behavior, and presence, and how these relate to the research process; cultural knowledge – which involves engaging with diverse cultural groups and disciplines to understand similarities, differences, and inequities in health, and how these might be socially determined; cultural sensitivity – through partnering with research participants, and building trust, respect, and empathy throughout the research; and cultural competency itself – which is demonstrated through synthesis and application of awareness, knowledge, and sensitivity, and involves a commitment to engaging participants, organizations, and communities in all stages of the research. Working toward cultural competency is an ongoing process and is considered vital as it can "lead to high quality, valid research irrespective of research design which can be used to inform the delivery of relevant health-care to all members of society" (Papadopoulous and Lees 2002, p. 263).

This chapter draws on our commitment and ongoing learning toward cultural competency as non-Aboriginal researchers, with experience and interest in mental health and suicide prevention, particularly among diverse, vulnerable, or marginalized communities. Our combined experiences in the specific area of suicide research with Aboriginal peoples include: literature reviews exploring the issue of suicide among Aboriginal young people and adults; as well as being part of a team of researchers and clinicians who engaged with the Aboriginal health sector in a project which involved the development, delivery, and evaluation of a community-wide suicide prevention training program throughout regional South Australia. The learning presented in this chapter has been greatly assisted by the time, wisdom, sharing, and expertise of our Aboriginal colleagues, to whom we would like to express our sincere thanks (see also ▶ Chap. 90, "Engaging Aboriginal People in Research: Taking a Decolonizing Gaze").

We would also like to acknowledge the diversity of the Aboriginal people of Australia and recognize that it is preferable to refer to specific language and/or cultural group names where possible. Given the broad nature of this chapter, the term "Aboriginal" will be used throughout to collectively refer to Aboriginal and Torres Strait Islander peoples, in line with recommendations suggested by Australia's National Aboriginal Community Controlled Health Organization (NACCHO 2016).

2 Understanding Suicide in Aboriginal Communities

Developing an ongoing awareness of suicide and its impacts for Aboriginal communities is an important precursor to undertaking research in this area. Despite being considered a rare occurrence prior to the 1970s, rates of suicide have "increased dramatically over the last three decades from levels that were previously much lower than the wider Australian population" (Hunter 2007, p. 89). The most recent statistics show that suicide rates were twice as high compared to rates for non-Aboriginal Australians in 2015 (ABS 2016). Between 2011 and 2015, suicide was the leading cause of death for Aboriginal people aged 15–34 years, and the second leading cause of death for those aged 35–44 years (ABS 2016). Overall suicide rates are particularly concerning among certain age groups. For example, Aboriginal children (5–17 years of age) died by suicide at a rate of 9.3 deaths per 100,000 persons, compared to 1.8 per 100,00 for non-Aboriginal people, and accounted for 27% of Australian children who died by suicide between 2011 and 2015. Adding concern to these already high figures is the widely held belief that suicide rates are an underestimate of the true values. For Aboriginal suicide deaths specifically, this is likely attributed to "issues relating to identification of race and cause of death" (Hunter 2007, p. 90). Tatz (2001) suggests that actual suicide rates might be at least two to three times greater than the official recordings. Further, Hunter and Milroy (2006) caution that rates of suicide have changed over time and also differ between communities, indicating the need for an ongoing effort to understand this issue for suicide prevention to be effective.

Although there is a paucity of research exploring the heightened risk of suicide among Aboriginal communities (De Leo 2012), or Aboriginal experiences of suicide (Elliott-Farrelly 2004), many explanations and contributing factors have been proposed. While some of the factors, such as substance use and unemployment (Hanssens 2007; Silburn et al. 2010), are also associated with suicide deaths of non-Aboriginal people, numerous authors have described the unique contributors to suicide for Aboriginal people. In his book, *Aboriginal Suicide is Different*, Colin Tatz, a leading author in this field, urges: "To understand Aboriginal suicide, one has to understand Aboriginal history" (Tatz 2001, p. 8). This understanding encompasses a wide range of issues, including: persistent loss and grief resulting from a history of colonization and intergenerational trauma (Tatz 2001); disintegrated cultural identity (Tatz 2001); loss of connection to culture and land (The Elders' Report 2014); an ongoing cycle of grief associated with suicide and other deaths (Tatz 2001; Silburn et al. 2010); scarcity of role models and mentors, particularly for young people (Tatz 2001); and racism (Silburn et al. 2010; The Elders' Report 2014). Hunter and Milroy (2006, p. 150) summarize the combined impact of these factors:

> Considering life as a narrative or story, the desire to end one's personal story abruptly, prematurely and deliberately can be seen to stem from the complex interplay of historical, political, social, circumstantial, psychological and biological factors that have already disrupted sacred and cultural continuity; disconnecting the individual from the earth, the universe and the spiritual realm – disconnecting the individual from the life affirming stories that are central to cultural resilience and continuity.

This highlights the important need for Aboriginal social and emotionally well-being research to consider the interconnectedness between cultural, spiritual, social, and physical influences (King and Brown 2015), in contrast to the more biomedically focused approach adopted in Western understandings. For example, for Aboriginal people, the term mental health comes more from an illness or clinical perspective, implying a greater focus on the individual (SHRG 2004). Instead, the term "social and emotional well-being" is preferred, which is considered to differ in important ways to non-Indigenous concepts of "mental health" (SHRG 2004). Social and emotional well-being recognizes the importance of connection to land, culture, spirituality, ancestry, family, and community (SHRG 2004), with these factors serving as sources of strength and recovery when Aboriginal peoples experience stress and adversity (Kelly et al. 2009).

The impact of suicide on Aboriginal communities has received little attention, but is being increasingly recognized as an important area for research. With profound, ongoing implications for individuals, families, and communities (Department of Health and Ageing 2013), it is not surprising that suicide results in persistent experiences of bereavement (Tatz 2001), with grief and mourning often spreading between communities, particularly where families and communities are interconnected, and where cultural obligations exist, such as funeral attendance and observance of "sorry business" (Silburn et al. 2010). For many, the frequency of deaths – through suicide and other causes – often means that the grieving process is cut short, resulting in complex trauma experiences, with families and communities experiencing an ongoing state of mourning, grief, and bereavement (Silburn et al. 2010). In addition, suicide "clusters" are not uncommon, particularly among young people (Hanssens 2007), whereby one suicide in a community can spark more, often with the same method and by people of the same gender and similar age of the deceased (Elliott-Farrelly 2004). This creates a perpetual cycle of grief.

3 Awareness of the Historical Research Context

A culturally competent approach requires awareness of and sensitivity toward the historical context of research with Aboriginal people and communities more broadly. As non-Aboriginal researchers, we came into this space with a consciousness toward the historical relationships between Aboriginal and non-Aboriginal Australians. We have continued to remain aware of, and feel concern for, this historical context, so as not to perpetuate power imbalances or other negative outcomes during our interactions with Aboriginal peoples.

Our awareness of the historical relationships between Aboriginal and non-Aboriginal Australians also extends to the research process itself, as explained by Laycock et al. (2011, p. 5):

> This history of research for Indigenous peoples is tied to the history of colonisation. In the eighteenth and nineteenth centuries, Europeans explored and 'discovered' other worlds, expanded trade and established colonies. Western scientific thought developed. As Indigenous peoples were systematically colonised, their societies and cultures began to be studied from the point of view of groups with more power and privilege, and with different systems of knowledge.

This historical approach to research has been viewed as unhelpful for addressing Aboriginal experiences and concerns (Humphrey 2001). It is not surprising, then, that research can be viewed as serving academic interests rather than benefiting Aboriginal health (King and Brown 2015), and that Aboriginal people have been critical of the research undertaken on their communities and cultures (Humphrey 2001), skeptical about its usefulness (Greenhill and Dix 2008), and hesitant to participate (Ralph et al. 2006).

Although there has been a gradual shift away from this approach where Aboriginal people are seen as "subjects" of research and a movement toward a more collaborative focus (Ralph et al. 2006; Laycock et al. 2011), having this awareness is important for understanding the potential perspectives of the people and communities we engage with during the research journey. Concerns about conducting work against a backdrop of political and social exploitation are not uncommon for researchers working in the cross-cultural space (Liamputtong 2008, 2010). Greenhill and Dix (2008) highlight how having a consciousness toward this history of research can challenge researchers to continuously work toward building relationships, engage in respectful interactions, and harbor a willingness to learn. This emphasizes the need for collaboration, consultation and relationship building (see also ▶ Chaps. 15, "Indigenist and Decolonizing Research Methodology," ▶ 89, "Using an Indigenist Framework for Decolonizing Health Promotion Research," and ▶ 90, "Engaging Aboriginal People in Research: Taking a Decolonizing Gaze").

4 Understanding and Applying Ethical Research Principles

Against a history of poor research practices, including exploitation, with Aboriginal communities, the 1970s saw the emergence of Indigenous activism in research. This was evidenced, for example, by the drafting of research guidelines, which called for "Aboriginal control of, and participation in, research, the adoption of non-invasive and culturally sensitive methodologies, the pursuit of research of need and benefit to communities, and full Aboriginal control over the dissemination of findings" (Humphrey 2001, p. 198). By the 1980s, this led to the formalization of ethical research guidelines (Humphrey 2001).

In Australia, there are several guidelines which specifically address ethical research practices when working with Aboriginal communities, and researchers should endeavor to understand and apply these from the outset. These guidelines include the National Health and Medical Research Council's (NHMRC 2003), *Values and ethics: Guidelines for ethical conduct in Aboriginal and Torres Strait Islander health research (Values and Ethics)* and *Keeping Research on Track*, which was developed by the NHMRC in 2005, translating *Values and Ethics* into a community guide for participating in research.

Culturally competent research requires an understanding of the ethical principles underpinning research with Aboriginal peoples, irrespective of the topic of interest. There are six core values considered important to all Aboriginal and people – spirit and integrity, reciprocity, respect, equality, survival and protection, and responsibility (as defined in *Keeping Research on Track*). While specific values and protocols vary between societies, these core six are considered common to all, and "each community or organization has the right to express how these core values, and any unique values, will be addressed in research" (NHMRC 2005, p. 8). The Australian Institute of Aboriginal and Torres Strait Islander Studies (AIATSIS) and The Lowitja Institute (2013, p. 9) summarize these six core values:

- **Spirit and integrity**: A connection between the past, present, and future, and the respectful and honorable behavior that holds Aboriginal values together.
- **Reciprocity**: Shared responsibilities and obligations to family and the land based on kinship networks, also includes sharing of benefits.
- **Respect**: For each other's dignity and individual ways of living. This is the basis of how Aboriginal peoples live.
- **Equality**: Recognizing the equal value of all individuals. Fairness and justice, the right to be different.

- **Survival and protection**: Of Aboriginal cultures, languages, and identity. Acknowledging shared values is a significant strength.
- **Responsibility**: Is the recognition of important responsibilities, which involve country, kinship, caring for others, and maintenance of cultural and spiritual awareness. The main responsibility is to do no harm to any person or any place. Responsibilities can be shared so others can be held accountable.

Both the NHMRC's *Values and Ethics* and *Keeping Research on Track* are currently being reviewed. In an examination of ethical guidelines for research with Aboriginal communities, *Researching the Right Way*, AIATSIS and The Lowitja Institute (2013) note that both NHMRC documents have been developed with a view to *prevent* unethical practices in health research involving Aboriginal peoples. In contrast, Jamieson et al. (2012) developed a set of guidelines which aim to *promote* ethical practices in research among Aboriginal populations, based on best practice from their own experience and the literature (AIATSIS and The Lowitja Institute 2013). Table 1 presents the five essential principles relevant to health research and suggestions for how these could be applied to a culturally competent approach to suicide research with Aboriginal peoples. These ethical principles, and those outlined by the NHMRC (2003), should be used as a guide throughout the research journey, from project conception to dissemination of results (Laycock et al. 2011).

5 Cultural Competency Throughout the Stages of the Research Journey

The Lowitja Institute, Australian's National Institute for Aboriginal and Torres Strait Islander Health Research, provides a range of valuable resources for conducting research in this area. This section draws attention to some of the key stages in the research journey, and readers are encouraged to consult The Lowitja Institute resources for additional information.

5.1 Developing Relationships

Trustworthy and respectful relationships are integral to research that is both meaningful and sustainable. This involves relationships with collaborating organizations and researchers, as well as research participants themselves. Preferably, research priorities will be identified by, or through collaboration with, Aboriginal communities (Laycock et al. 2011). A commitment to partnering with Aboriginal communities helps to ensure that research is developed from the ground up and contributes to building capacity to develop and implement programs within the community of interest. Elliott-Farrelly (2004) highlights the various benefits that can result from suicide prevention programs that have been both developed and implemented by the Aboriginal communities they are targeting, including: heightened personal and

Table 1 Key principles of research and application to practice for ethical, culturally competent approaches to suicide research

Essential principle (Jamieson et al. 2012)	Application to practice in culturally competent approaches to suicide research with Aboriginal peoples
Addressing a priority health issue as determined by the community	Addressing issues which have been determined by the community ensures that research focuses on issues that are of most importance – and which are likely to lead to the most benefits – to address the health and well-being of Aboriginal peoples. Suicide has been identified as a priority by numerous leaders within Aboriginal communities (e.g., The Elders' Report 2014), as well as peak bodies representing Aboriginal people, e.g., Aboriginal Health Council of Australia (King and Brown 2015). Community identification of priorities for research, such as in the area of suicide, also helps to foster community ownership and build relationships
Conducting research within a mutually respectful partnership framework	Respectful relationships characterized by openness, trust, and transparency underpin research with Aboriginal communities that is successful and mutually beneficial (AIATSIS and The Lowitja Institute 2013). Building such relationships can take considerable time and involves genuine commitment on the part of researchers. The burdens and benefits of research should be distributed equitably. At the outset of a project, researchers need to work closely with communities to make joint decisions on various arrangements. In the sensitive area of suicide research, arrangements to ensure Aboriginal peoples are provided with sufficient support to participate and are protected when necessary, throughout the research process, are critical
Capacity building is a key focus of the research partnership, with sufficient budget to support this	Just as the capacity building of communities to take action in response to suicide is important (Department of Health and Ageing 2013), building the capacity of Aboriginal people as researchers and consultants in this area is vital. Working in partnership with Aboriginal people and communities provides opportunities for empowerment and to build their own research capacity (King and Brown 2015). Importantly, Aboriginal peoples bring unique knowledge and understanding, distinctive to their communities and history, which contribute significantly to the evidence base (Department of Health and Ageing 2013). As community engagement is a key aspect in any such

(*continued*)

Table 1 (continued)

Essential principle (Jamieson et al. 2012)	Application to practice in culturally competent approaches to suicide research with Aboriginal peoples
	research (Gwynn et al. 2015), particularly in an under-researched area such as suicide, it is important for researchers to factor in additional time and community engagement activities when budgeting for projects in this area
Flexibility in study implementation while maintaining scientific rigor	In the sensitive area of suicide research, it is critical that researchers remain flexible throughout the research process and are prepared to: "tread carefully if the research concerns sensitive issues or is likely to occur at crucial time of life" (King and Brown 2015, p. 9). For example, being sensitive toward rituals related to loss and grief, e.g., "sorry business," means that research may need to be postponed. The choice to not participate or withdraw participation should be respected at all times
Respecting communities' past and present experience of research	It is important that researchers are mindful of the context in which suicide research occurs – in particular the history of research that has taken place with Aboriginal peoples. Understanding wider sociopolitical, historical factors are also vital to research in this area (Tatz 2001)

community awareness, increased self-respect and dignity, higher levels of commitment to achieve desired outcomes, decreased dependency, and empowerment.

In some situations, relationships will already exist between non-Aboriginal researchers and Aboriginal communities. However, this may not be the case and developing relationships may be a new experience. In these instances, relationship building may initially seem an elusive task, particularly in situations where the researchers have a few, or no, existing relationships with Aboriginal peoples or the Aboriginal health sector, or in the area (geographic or subject) where they are conducting the research. For any researchers, Aboriginal Community Controlled Health Organizations (ACCHOs) can be an important starting point for building relationships.

Central to relationship building should be respect and understanding for what each person or community brings to the research – whether that be knowledge, perspectives, skills, experiences, or ways of working (Laycock et al. 2009). This can be achieved through openly discussing what each person, organization, or community would like to contribute, and being open to this changing as the project progresses. Consideration should also be given to when and where meetings take place. In some instances, external pressures, such as funding, limit the time that can be allocated to this process. However, efforts should be made to factor this in to the early stages of the research project. This might involve, for example, informal meetings (e.g., over coffee) for different parties to get to know one another and begin to build rapport, before specific research agendas are discussed. We see this as an ongoing process, and something that continues over time, between individual projects.

5.2   Deciding on a Research Topic

Until more recently, research topics exploring Aboriginal suicide have largely been developed and investigated from a Western, positivist perspective. Reflective of trends in suicide research more broadly, these studies have typically sought to understand rates and prevalence (e.g., Clayer and Czechowicz 1991; Cantor and Slater 1997; Parker and Ben-Tovim 2002; Pridmore and Fujiyama 2009; De Leo et al. 2011; Luke et al. 2013; Soole et al. 2014), methods (e.g., De Leo et al. 2011; Soole et al. 2014), and place of suicide (Soole et al. 2014). Similarly, risk factors have received attention (Clough et al. 2006; Silburn et al. 2010; Calabria et al. 2010; Jamieson et al. 2011; Priest et al. 2011; Zubrick et al. 2011; Luke et al. 2013; Soole et al. 2014). Less emphasis has been placed on understanding suicide from the perspective of Aboriginal peoples (Lindeman et al. 2014), how suicide might be prevented (Capp et al. 2007; Lopes et al. 2012), or how effective prevention programs are (Harlow et al. 2014; Ridani et al. 2015).

It should now be clear that research priorities should be set by the people to whom they relate (King and Brown 2015). These priorities can be identified through consultation and collaboration with individuals, communities, and the ACCHO sector, as well as through an awareness of the policy context. For example, the National Aboriginal and Torres Strait Islander Suicide Prevention Strategy (Department of Health and Ageing 2013) reflects calls for unique approaches to understanding and addressing this issue (e.g., Tatz 2001; Elliott-Farrelly 2004; Procter 2005), and was developed through extensive community consultation, Australia-wide. At its heart, the policy "has a holistic and early intervention focus that works to build strong communities through more community-focused and integrated approaches to suicide prevention," with an emphasis on Aboriginal people developing "local, culturally appropriate strategies to identify and respond to those most at risk" (Department of Health 2014, p. 2). Two of the six key action areas are to build on strengths and capacity in communities, and to build strengths and resilience in individuals and families. The policy's explicit focus on strengths-based approaches was confirmed more recently during the first National Aboriginal and Torres Strait Islander Suicide Prevention Conference, held in 2016, where there was an emphasis on the need for strengths-based, community driven solutions to suicide (Finlay 2016).

5.3   Applying an Appropriate Research Methodology

The focus on positivist research to date has favored quantitative methodologies in suicide research, such as retrospective analyses of death records. This emphasis means that Aboriginal understandings of suicide have predominantly been excluded. Further, these approaches can be viewed as ignoring the diversity of Aboriginal people and communities (King and Brown 2015). As such, qualitative approaches are seen to be more favorable, and Laycock et al. (2009, p. 6) highlights how:

> some research approaches are better suited to Indigenous health research than others. Collaborative, participatory and multidisciplinary research approaches are often used in community settings because they provide more opportunities for communities to set priorities and guide research processes, to build Indigenous ways of doing things into the project and to 'privilege' the voice of Indigenous participants.

While peer-reviewed qualitative research on this topic is sparse, examples can be seen in the gray literature. For example, The Elders Report (2014) incorporates voices of 31 elders from over 17 communities in Western Australian, Queensland, and the Northern Territory, who have experienced the impacts of suicide in their communities, and who offer solutions for how to best address it. Despite the diversity of experiences of each individual, some common themes were identified, including the important healing and protective role of culture.

Tsey and colleagues (2007) describe the development of several empowerment-based research methodologies to improve the social and emotional well-being of Aboriginal people, including participatory action research (PAR). Research approaches such as PAR seek to shift unequal power relations between participants and researchers as participants become researchers in their own right to address issues of concern that are a priority to them (Tsey et al. 2007; Liamputtong 2013; Higginbottom and Liamputtong 2015). This approach can offer a range of benefits for participants, including increased personal empowerment and control over their lives and situations, stronger and longer-term research partnerships which are based on mutual respect and trust, and more sustainable outcomes which are driven by the priorities of communities (Tsey et al. 2007). In some situations, research projects may involve participants from a range of backgrounds and sectors, and it may not be possible to develop a methodology that best suits all participants. In these instances, a mixed-methods approach could be considered. For example, while our recent research included a survey component, it also involved an opportunity for participants to engage in qualitative interviews.

5.4 Seeking Ethical Approval

Ethics approval is an important process in any research (Ramcharan 2017; see also ▶ Chap. 106, "Ethics and Research with Indigenous Peoples"). For projects involving Aboriginal people and communities, approval should be sought from an Aboriginal Health Research Ethics Committee (AHREC). Engaging with AHRECs should not be viewed as simply a process in the research journey, but also as an opportunity to ensure the safety of the proposed research and to strengthen the research plan. In our experience, this process has been invaluable for building new connections with the ACCHO sector, ensuring the cultural appropriateness of the content of our suicide prevention training program, and promoting the research to interested participants. These processes have been essential for ensuring steps were taken to maximize opportunities for participation.

5.5 Engaging with Research Participants

Researchers should be aware that although there has been a long history of research "on" Aboriginal people, many Aboriginal people may not have had experiences being involved in research and may not have a clear understanding of the process (e.g., their rights regarding participation). As with any research, researchers have a responsibility to facilitate this understanding. Examples of how this can be achieved include: recruiting participants in collaboration with the aforementioned ACCHOs; providing participants with access to *Keeping Research on Track*; developing research materials (e.g., information sheets and consent forms) using appropriate language (Liamputtong 2008, 2010); and being available and approachable to discuss the research. In our experience of recruiting participants for a community-based suicide prevention education project across rural South Australia, valuable reciprocal conversations were had when individuals phoned to express an interest in participating, rather than doing so by email. Participants were able to ask questions about the research project, and we were able to learn more about their community prior to meeting in person during the intervention and data collection.

Steps can also be taken to demonstrate cultural sensitivity toward participants, communities, and culture throughout the implementation of research. For example, during our suicide prevention training program, our team's Aboriginal and Torres Strait Islander Project Coordinator made efforts to organize a Welcome to Country by an Elder or other recognized community leader in each town, where possible. Further, the team agreed that cultural considerations for health professionals working with Aboriginal people should be discussed early in the training days, rather than added towards the end of the day.

5.6 Reporting and Disseminating Findings

Researchers need to give careful thought to how the findings from research with Aboriginal people will be reported and shared. In particular, the types of reports or strategies used to disseminate research needs to be considered, including whether a combination of approaches is more appropriate. Researchers should engage with people or organizations connected to the communities, as early on as possible, to share decisions on aspects such as: processes for involving community members in the design of reports and other mediums for presenting findings; whether consent needs to be sought to present aspects of the findings, e.g., use of photos; and who needs to be acknowledged in reports and other approaches to dissemination (NHMRC 2005). Research findings should be delivered in way that is understandable and accessible, with the NHMRC (2005) recommending that a Plain English Community Report should be made available.

6 Conclusion and Future Directions

It is clear that despite enormous strengths and resilience among Aboriginal peoples, suicide is an ongoing, complex concern and can have profound consequences. It is critical that researchers in this space remain conscious of the ongoing impacts of colonization and intergenerational trauma for Aboriginal peoples, and be open to a holistic understanding of social and emotional well-being. Similar to research related to other aspects of social and emotional well-being, to date, research into suicide among Aboriginal communities has been limited by being primarily undertaken from a positivist, Western perspective. This chapter has outlined the need for a culturally competent approach to research with Aboriginal people and communities, through all stages of the research process. Although discussed in the context of suicide, the principles described in this chapter apply to a range of sensitive topics, including other social and emotional well-being and health concerns. Similarly, these are not specific to researchers working in Australia, but could be considered when working with Indigenous peoples in other parts of the world (e.g., Canada and New Zealand). Across this range of sensitive topics and participant groups, researchers are encouraged to strive for ongoing cultural competency, with an emphasis on projects that have been developed by, or in collaboration with, the communities in which they hold relevance, and through adopting a participatory, strengths-based approach. Working together, Aboriginal and non-Aboriginal researchers have important roles to play in progressing the research agenda on the sensitive and complex topic of suicide. Conducted in the spirit of respect, shared goals, and shared responsibilities, ongoing partnerships can lead to achieving mutual understanding, joint capacity building, and research processes which not only are sustainable but also have positive impacts for Aboriginal people, families, and communities.

References

ABS. Causes of death, Australia, 2015, cat. no. 3303.0. Canberra: ABS; 2016.

Australian Institute of Aboriginal and Torres Strait Islander Studies (AIATSIS) & the Lowitja Institute. Researching right way – Aboriginal and Torres Strait Islander health research ethics: a domestic and international review. https://www.nhmrc.gov.au/health-ethics/ethical-issues-and-further-resources/ethical-guidelines-research-involving-aboriginal-. (2013). Accessed 20 June 2016.

Bainbridge R, McCalman J, Clifford A, Tsey K. Cultural competency in the delivery of health services for Aboriginal people, Issues paper no. 13. Produced for the closing the gap clearinghouse. Canberra: Australian Institute of Health and Welfare & Melbourne, Australia: Australian Institute of Family Studies; 2015.

Calabria B, Doran CM, Vos T, Shakeshaft AP, Hall W. Epidemiology of alcohol-related burden of disease among indigenous Australians. Aust N Z J Public Health. 2010;34:S47–51.

Cantor CH, Slater PJ. A regional profile of suicide in Queensland. Aust N Z J Public Health. 1997;21(2):181–6.

Capp K, Deane FP, Lambert G. Suicide prevention in Aboriginal communities: application of community gatekeeper training. Aust N Z J Public Health. 2007;25(4):315–21.

Clayer JR, Czechowicz AS. Suicide by Aboriginal people in South Australia – comparison with suicide deaths in the total urban and rural populations. Med J Aust. 1991;154(10):683–5.

Clough AR, Lee KSK, Cairney S, Maruff P, O'Reilly B, d'Abbs P, et al. Changes in cannabis use and its consequences over 3 years in a remote indigenous population in northern Australia. Addiction. 2006;101(5):696–705.

De Leo D. Mental disorders and communication of intent to die in indigenous suicide cases, Queensland, Australia. Suicide Life Threat Behav. 2012;42(2):136–46.

De Leo D, Sveticic J, Milner A. Suicide in indigenous people in Queensland, Australia: trends and methods, 1994–2007. Aust N Z J Psychiatry. 2011;45(7):532–8.

Department of Health and Ageing. National Aboriginal and Torres Strait Islander suicide prevention strategy, May 2013. Canberra: Commonwealth of Australia; 2013.

Elliott-Farrelly T. Australian Aboriginal suicide: the need for an Aboriginal suicidology? Aust e-J Adv Ment Health. 2004;3(3):138–45.

Finlay SM. Conference highlights report. http://www.atsispep.sis.uwa.edu.au/natsispc2016. (2016). Accessed 20 June 2016.

Greenhill J, Dix K. Respecting culture: research with rural aboriginal community. In: Liamputtong P, editor. Doing cross-cultural research: ethical and methodological perspectives. Dordrecht: Springer; 2008. p. 49–60.

Gwynn J, Lock M, Turner N, Dennison R, Coleman C, Kelly B, et al. Aboriginal and Torres Strait Islander community governance of health research: turning principles into practice. Aust J Rural Health. 2015;23(4):235–42.

Hanssens L. Indigenous dreaming: how suicide in the context of substance abuse has impacted on and shattered the dreams and reality of indigenous communities in Northern Territory, Australia. Aborig Islander Health Worker J. 2007;31(6):26–34.

Harlow AF, Bohanna I, Clough A. A systematic review of evaluated suicide prevention programs targeting indigenous youth. Crisis. 2014;35(5):310–21.

Higginbottom G, Liamputtong P, editors. Participatory qualitative research methodologies in health. London: SAGE; 2015.

Humphrey K. Dirty questions: indigenous health and 'Western research'. Aust N Z J Public Health. 2001;25:197–202.

Hunter E. Disadvantage and discontent: a review of issues relevant to the mental health of rural and remote Indigenous Australians. Aust J Rural Health. 2007;15(2):88–93.

Hunter E, Milroy H. Aboriginal and Torres Strait Islander suicide in context. Arch Suicide Res. 2006;10(2):141–57.

Jamieson LM, Paradies YC, Gunthorpe W, Cairney SJ, Sayers SM. Oral health and social and emotional well-being in a birth cohort of Aboriginal Australian young adults. BMC Public Health. 2011;11:656.

Jamieson LM, Paradies YC, Eades S, Chong A, Maple-Brown L, Morris P, et al. Ten principles relevant to health research among Indigenous Australian populations. Med J Aust. 2012;197(1):16–8.

Kelly K, Dudgeon P, Gee G, Glaskin B. Living on the edge: social and emotional wellbeing and risk and protective factors for serious psychological distress among Aboriginal and Torres Strait Islander People, Discussion Paper No. 10. Darwin: Cooperative Research Centre for Aboriginal Health; 2009.

King R, Brown A. Next steps for Aboriginal health research: exploring how research can improve the health and wellbeing of Aboriginal people in South Australia. Adelaide: Aboriginal Health Council of South Australia; 2015.

Laycock A. with Walker D, Harrison N, Brands J. Supporting researchers: a practical guide for supervisors. Melbourne: The Lowitja Institute; 2009.

Laycock A. with Walker D, Harrison N, Brands J. Researching indigenous health: a practical guide for researchers. Melbourne: The Lowitja Institute; 2011.

Liamputtong P. Doing research in a cross-cultural context: methodological and ethical challenges. In: Liamputtong P, editor. Doing cross-cultural research: ethical and methodological perspectives. London: Springer; 2008. p. 1–20.

Liamputtong P. Performing qualitative cross-cultural research. Cambridge: Cambridge University Press; 2010.

Liamputtong P. Qualitative research methods, 4th end. Melbourne: Oxford University Press; 2013.

Lindeman MA, Kuipers P, Grant L. Front-line worker perspectives on Indigenous youth suicide in Central Australia: contributors and prevention strategies. Int J Emerg Ment Health Human Resilience. 2014;17(1):191–6.

Lopes J, Lindeman M, Taylor K, Grant L. Cross cultural education in suicide prevention: development of a training resource for use in Central Australian Indigenous communities. Adv Mental Health. 2012;10(3):224–34.

Luke JN, Anderson IP, Gee GJ, Thorpe R, Rowley KG, Reilly RE, et al. Suicide ideation and attempt in a community cohort of urban Aboriginal youth: a cross-sectional study. Crisis. 2013;24(4):251–61.

National Aboriginal Community Controlled Health Organisation (NACCHO). Definitions. http://www.naccho.org.au/aboriginal-health/definitions/. (2016). Accessed 20 June 2016.

National Health and Medical Research Council (NHMRC). Values and ethics: guidelines for ethical conduct in Aboriginal and Torres Strait Islander health research. Canberra: Commonwealth of Australia; 2003.

NHMRC. Keeping research on track: a guide for Aboriginal and Torres Strait Islander peoples about health research ethics. Canberra: Australian Government; 2005.

Papadopoulous I, Lees S. Developing culturally competent researchers. J Adv Nurs. 2002;37(3):258–64.

Parker R, Ben-Tovim DI. A study of factors affecting suicide in Aboriginal and 'othpopulations in the Top End of the Northern Territory through an audit of coronial records. Aust N Z J Psychiatry. 2002;36(3):404–10.

Pridmore S, Fujiyama H. Suicide in the Northern Territory, 2001–2006. Aust N Z J Psychiatry. 2009;43(12):1126–30.

Priest NC, Paradies YC, Gunthorpe W, Cairney SJ, Sayers SM. Racism as a determinant of social and emotional wellbeing for Aboriginal Australian youth. Med J Aust. 2011;194(10):546–50.

Procter NG. Parasuicide, self-harm and suicide in Aboriginal people in rural Australia: a review of the literature with implications for mental health nursing practice. Int J Nurs Pract. 2005;11(5):237–41.

Ralph N, Hamaguchi K, Cox M. Transgenerational trauma, suicide and healing from sexual abuse in the Kimberly region, Australia. Pimatisiwin. 2006;4(2):117–36.

Ramcharan P. What is ethical research? In: Liamputtong P, editor. Research methods in health: foundations for evidence-based practice. 3rd ed. Melbourne: Oxford University Press; 2017. p. 49–63.

Ridani R, Shand FL, Christensen H, McKay K, Tighe J, Burns J, et al. Suicide prevention in Australian Aboriginal communities: a review of past and present programs. Suicide Life Threat Behav. 2015;45(1):111–40.

Silburn S, Glaskin B, Henry D, Drew N. Preventing suicide among indigenous Australians. In: Purdie N, Dudgeon P, Walker R, editors. Working together: Aboriginal and Torres Strait Islander mental health and wellbeing principles and practice. Canberra: Commonwealth of Australia; 2010. p. 91–104.

Social Health Reference Group (SHRG). National Strategic Framework for Aboriginal and Torres Strait Islander peoples' mental health and social and emotional wellbeing 2004–2009. Canberra: Australian Government; 2004.

Soole R, Kolves K, De Leo D. Factors related to childhood suicides: analysis of the Queensland Child Death Register. Crisis. 2014;35(5):292–300.

Tatz C. Aboriginal suicide is different: a portrait of life and self-destruction. Canberra: Aboriginal Studies Press; 2001.

The Elders' Report. The elders' report into preventing indigenous self-harm and youth suicide. Melbourne; 2014. http://www.cultureislife.org/resources/#elders.

Tsey K, Wilson A, Haswell-Elkins M, Whiteside M, McCalman J, Cadet-James Y, et al. Empowerment-based research methods: a 10-year approach to enhancing indigenous social and emotional wellbeing. Australas Psychiatry. 2007;15:S34–8.

World Health Organization (WHO). Preventing suicide: a global imperative. Geneva: WHO; 2014.

Zubrick SR, Mitrou F, Lawrence D, Silburn SR. Maternal death and the onward psychosocial circumstances of Australian Aboriginal children and young people. Psychol Med. 2011;41(9):1971–80.

Visual Methods in Research with Migrant and Refugee Children and Young People

99

Marta Moskal

Contents

1 Introduction	1724
2 Children as Social Agents	1724
3 Participatory Methods with Children and Young People	1725
4 The Use of Visual Methods as Participatory for Children and Young People	1726
5 Collecting and Analyzing: Migrant and Refugee Children's Images in Research	1728
5.1 Maps and Drawings	1728
5.2 Photographs and Videos	1732
6 Conclusion and Future Directions	1734
References	1735

Abstract

This chapter examines how visual methods have been used in understanding and interpreting children's worlds. Focusing on social sciences engagement with diverse visual methods (like drawing, maps, photographs, and videos), the chapter contributes to the discussion about their value and limitations. The chapter broadly reviews the body of knowledge on the use of children's images in research. It provides some references to the research in health social sciences, however, focuses particularly on migrants and refugees. Although visual images play a meaningful role in the lives of young people, social sciences still privileges approaches based on words and numbers. Children's visual methods, however, are gaining increasing interest as many social scientists search for methods that align with the current conceptualization of children as social agents and cultural producers. It has been argued that visual methods can secure participant

---

The chapter has been developed based on the paper (Moskal 2010).

M. Moskal (✉)
Durham Univeristy, Durham, UK
e-mail: marta.z.moskal@durham.ac.uk

© Springer Nature Singapore Pte Ltd. 2019
P. Liamputtong (ed.), *Handbook of Research Methods in Health Social Sciences*,
https://doi.org/10.1007/978-981-10-5251-4_42

engagement and reflexivity among a group who may not be comfortable with a traditional survey, interview, or focus group methods. Exploring specific examples of the studies with migrant and refugee children and young people, the chapter demonstrates how visual methods can be evaluated as a research strategy.

> **Keywords**
> Visual research · Participatory research · Migrant and refugee children · Research with young people · Children's perceptions of health and illness

1 Introduction

This chapter focuses on visual research methods that could be effective in research with children and young people who have experienced transitions involved in forced or voluntary migration and resettlement. Researching migrants and refugees in global times requires an approach able to reach into people's transient, volatile lives, understanding the complexity of multiple places, and diverse cultural and intellectual heritages (Alasuutari 2004). Particularly in the migration and refugee context, the research has focused on health indicators and schooling rather than on children's everyday lives, and has used traditional research tools such as structured questionnaires and focus group interviews. As a result, there are only small number of studies that document their lives as told in their own voices (Oh 2012, p. 282). There has been an increased interest in research processes that are child-oriented, and which model participatory research designs that place children at the center of the process as active participants. This has particularly been the case of qualitative research methods that are designed to examine children's experiences and perspectives, especially about children in vulnerable situations such as from refugee or migrant backgrounds (see Crivello et al. 2009; Due et al. 2014). In a process of a wider re-evaluation of research with children and young people (Lewis et al. 2004; Christensen and James 2008), visual, participatory methods have become increasingly used. Researchers and practitioners across a range of disciplines employ them as a means of exploring the children and young people's experiences, relationships, and lifestyles (Hart 1992; Barker and Weller 2003; Thompson 2008).

2 Children as Social Agents

This chapter sees children as competent beings whose views, actions, and choices are of values (Alanen and Mayall 2001; see also ▶ Chap. 115, "Researching with Children"). This refers to the recent scholarship of human geography, social anthropology, and sociology of children and childhood that treats children as active participants in their own socialization (Zelizer 1985; James et al. 1998; Johnson et al. 1998; Orellana 1999; Punch 2002; White 2002). I explain how this approach extends to visual research methods.

Over the past 20 years, many social scientists begin to view minors, as not merely the reproducers of culture, but as "cultural agents and social actors in their right" (Mitchell 2006, p. 60). For children to be able to participate in research, it might be necessary to develop different nonadult centered methods (Mitchell 2006). Boyden (2003) suggests a need for "age-appropriate" methods that "empower children" and lead to "valid child-led data." Mitchell (2006) argues that visual methods are said to be "child-centered" in the sense that they may be familiar, even enjoyable to the child. When activity is familiar and pleasant to the child, it can be particularly "useful in bringing out the complexities of their experience" (Nieuwenhuys 1996). Although they have been promoted as a corrective to the disempowering positivism, the child-centered methods, including visual methods, have been criticized for limiting the research to the level of individual child perception and being unable to account adequately for the ways in which political and social forces work to shape children' lives (Ansell 2009; Marshall 2013). However, following Ansell, I argue that visual, participatory methods with children and young people have a potential take into account wider social and political contexts (Ansell 2009; Mitchell and Elwood 2012; White et al. 2012).

In my research with migrant children and young people, drawing, mapping, and storytelling were used to understand the ideas and practices of home and belonging among transnational families in Scotland (Moskal 2010, 2015). These child-centered methods helped to reveal how children and young people experience mobility and construct a sense of home while resettling. The stories help us to look at and simultaneously interpret the local and the global experiences of participants. Children's involvement in migration upsets the notion of children as innately local beings and clearly illustrates how children's everyday lives are shaped by structures and constraints originating beyond the local scale. By viewing children's agency as a process, Ansell (2009, p. 194) argues that "too often local, concrete and agency are conflated into an acceptable focus for research, in opposition to a global, abstract or structuralist perspective that is viewed with suspicion as too 'distant' from real children." I demonstrate how children's stories seem to destabilize the dualism of the global and local (using Massey's 2005 idea) as the transnational social spaces in which they live continue to shape their social relations, cultural practices, and identifications (Moskal 2015).

3 Participatory Methods with Children and Young People

The participatory practices start from the beliefs that young people have the capacity to express themselves and the right to do so and that expressing themselves can include visual means (Thompson 2008; see also ▶ Chaps. 100, "Participatory and Visual Research with Roma Youth," and ▶ 117, "Participant-Generated Visual Timelines and Street-Involved Youth Who Have Experienced Violent Victimization"). There has been increasing emphasis, within the social sciences, on working participatory with children so that they might define research agendas and participate in fair ways (Thompson 2008; Lomax 2012). The participatory research is increasingly associated with the rapid growth in the application of an immense range of creative and visual methods including photography and video walking tours, mapping, and art-based approaches such as drawing and collage. This makes a substantial and genuine attempt to include children in the production of knowledge where previously their experiences have been marginalized or absent (Lomax 2012).

Lodge (2009) insists that images used in research require the participation of young people to provide contextual information to make meaning. She further argues that this is one of the most emancipatory aspects of using images in research because to understand or read the images we need the participation of the young people who produced them. Importantly, child-centered methodologies are not focused solely on what may be meaningful to a child participant, but what is meaningful to larger contexts of children's lives.

4 The Use of Visual Methods as Participatory for Children and Young People

Visual methods present several advantages as participatory methods for engaging children in research. Therefore, they are being used creatively in diverse social and cultural contexts. For example, to explain children's perspectives on health and illness (Geissler 1998b; Ross et al. 2009; Fernandez et al. 2015), poverty (Sime 2008), tourism (Gamradt 1995), identity (Cowan 1999), identity and consumption (Croghan et al. 2008), time (Christensen and James 2000), and place and belonging (Orellana 1999; den Besten 2010). There are many ways of collecting visual data, and the choice of method depends on the aims and theoretical perspective of the researcher (Bagnoli 2009). This may include drawings (e.g., Geissler 1998a, b; Guillemin 2004; Van Blerk and Ansell 2006; Fernandez et al. 2015; Liamputtong and Fernandez 2015), maps and diagrams (e.g., Bromley and Mackie 2009; Moskal 2010; Copeland and Agosto 2012), and photographs and videos (e.g., Radley and Taylor 2003; Tinkler 2008; Due et al. 2014; Marshall 2013).

Visual methods can be used alone, in combination with verbal data or as one of some multisensory methods. For example, Bagnoli (2009) describes the use of multimethod biographies to holistically explore young people's identities. Methods included oral interviews, written diaries, and visual methods such as self-portraits, video diaries, relational maps, and diagrams. The aim is not only to use visual methods as a tool to assist with interviews but as an important method of eliciting and understanding experience in its right (Bagnoli 2009; see also ▶ Chap. 71, "Self-portraits and Maps as a Window on Participants' Worlds"). Visual research offers an accessible way for young people to become active in the research process themselves, to reverse the typical role of having research done to them, and to allow them to participate more in this process (Thomson 2008). Lodge (2009) claims that the creation of images (drawings, video, or photography) can offer opportunities for the usually silenced and marginalized to participate, and perhaps to alter prevailing power relationships. She gives an example of the project by Kaplan and Howes

(2004) who describe the creation of a contact zone (a website) where differently empowered people – teachers and students – could interact. As their project's title suggests "Seeing through Different Eyes" invites participants to consider alternative views and meanings through dialogue about images produced by young people. Sometimes, using images created by young people will challenge accepted wisdom and assumptions underpinning practice in schools (Lodge 2009).

Likewise, Geissler et al.'s (1997) study with Luo schoolchildren in western Kenya showed that the children confronted in school with entirely different traditions of health knowledge without relation to their daily life and the experience of illness and healing. Geissler (1998b, p. 133) suggested that the children moving at the interface of indigenous knowledge conveyed informally in the family and biomedical ideas, mainly transmitted in school, are creatively contributing to the integration of old and new and actively shaping their ideas about health and the body. In Geissler's study, drawing was used together with the written narrative ("composition"). The children were asked to draw images of the body and its inside, partly on paper and partly with sticks in the sand in the schoolyard. The same children were asked to write compositions in Dholuoh (local language) on the topic of "worms." Both compositions and body maps proved very useful in that the children enjoyed their tasks and felt more confident than in the immediate face-to-face interview. Body maps, demanded the creation of some order in space rather than in time, giving different insight (than written and oral expressions) into how the children thought about worms (Geissler 1998b).

Listening to children' ideas about worms and their role in the body, Geissler (1998b) showed that the children active contribution to the creation of medical culture. The children's drawings and written narratives demonstrate that "Luo medical culture" is dynamic, open to innovation, and draws on all available sources of knowledge that can render experience coherent and guide action. Thus, the Luo children are agents of cultural production at the interface of different kinds of medical knowledge and not mere recipients of prefabricated health messages.

The value of visual methods as a catalyst for more conventional interviewing techniques was reflected, for example by Bromley and Mackie (2009) who found the mapping task was useful for prompting discussion with street working children in Peru. Bromley and Mackie (2009) employed a mapping task (along with a card selection game and interview) to explore the experiences of these children. The mapping work required no literacy or verbal skills and therefore was the most suitable for working with street children as it enabled all of them to express themselves fully. The second phase of the task did demand some oral skills in giving reasons for the patterns, so explanations varied in detail. The mapping work firstly sought to determine the specific locations where the children liked and disliked working. It also aimed to investigate why children have such preferences. A basic map of the city center was drawn, and a copy was given to the children. Each child was given two colored pencils, and they were asked to color in red the locations where they like to trade and in blue the locations where they do not like to trade. Once the child was happy with their map, they were asked to explain why they had used specific colors in particular locations and these explanations were recorded on the map (Mackie 2011). The authors reflect on the process saying that the mapping task not only prompted for discussion but, in contrast to an interview, gave the children time to think, rather than give a spontaneous response. Thus, the finished product was a more accurate representation of the child's opinions.

To summarize, these methodological considerations, researchers in diverse disciplines have sought research methods that may be particularly well suited to working with, rather than "on" children. Enabling children's perspectives through image can make their knowledge and concerns visible to adults and can be the basis for involving children in identifying and solving issues that affect them. In my research among migrant children in Scotland, drawing was a strategy to collect the research material itself and to facilitate the narrative interviews, especially with the younger children. In the next session, I detail and evaluate these drawing activities.

5 Collecting and Analyzing: Migrant and Refugee Children's Images in Research

5.1 Maps and Drawings

Several studies (e.g., Dockett and Perry 2005; Einarsdottir et al. 2009) have used drawings to capture children's knowledge and experience. However this approach has been relatively scarce in migration research (the exceptions include Mitchell 2006; Van Blerk and Ansell 2006; White et al. 2012; Fernandez et al. 2015; Liamputtong and Fernandez 2015). Mental maps have been used in research with migrant children, for example by den Besten (2010), to describe their local, urban experience in order to map out the fears and dislikes in Berlin and Paris.

In my research on the experiences of families and children of Polish migrants in Scotland (Moskal 2010, 2015; Moskal and Tyrell 2016), children and young people (between the ages of 5 and 17) participated in individual and small group interviews as well as drawing and mental map making. Children drew maps from memory that helped outline their spatial awareness, the locations of their activities, as well as a sense of belonging to the particular place. Participants, sometimes, had difficulty constructing their maps, and this was based on a lack of spatial concepts among the children. Perspective, symbolization, and other standard map qualities were very rarely observed. However, some of the children demonstrated the use of national symbols, namely the Polish and Scottish flags. I did not ask participants to focus on particular localities. The mental maps produced a wide diversity of images regarding the number of elements included and the perspective taken to show the varied childhood experiences as well as the organization and meaning of migrant childhood (Anning and Ring 2004).

Lynch (1960) claims that most often our perceptions of a locality are not sustained, but are fragmentary and mixed with other concerns. His assertion resonates with the migrant children's images of the localities they were in. For example, Kate, aged nine, drew a subjective map (Fig. 1) representing one street, along which she placed her house and school in Edinburgh next to her home, garden, and playground in the Polish town that she came from. Kate's everyday routines and practices marked a way for her to bridge the gap between past and present, and between

Fig. 1 Kate's mental map

here and there. Kate's map illustrated this particular sense of connection and a sense of temporary belonging. Kate's map evidences that children are able to imagine translocal life, as opposed to a transnational life. The similarities she perceives between her Polish and Scottish places of residence constitute a powerful translocal tie and do much to make her feel at home when she is abroad. This type of evidence emphasizes the importance of different sites of belonging, connected with the various spheres of life that children encounter. Children's lives are not necessarily statically attached to their physical experience of space because they can imagine distant places and the process of moving between places (Van Blerk and Ansell 2006).

The awareness that people and places "back home" were changing while they were not there was unsettling for some children (see Moskal and Tyrell 2016). The family is crucial in the children's and young people's constructions of home and belonging, regardless of whether they refer to the family back in their country of origin, the family in their current place of residence, or in both places. In a second technique, children were asked to draw a tree with roots and then to draw or write beside the roots the things that they were attached to.

This activity also resulted with the drawings demonstrating a sense of connection with the home country that exists in reality (Internet conversations, phone calls, more or less frequent visits to and from the country of origin) but also in the imaginary realm. "I drew the phone to call my family in Poland and a computer to talk to them," said 10-year-old Vicky to describe her drawing (see Fig. 2). Then, she explained further:

> I have got four cousins and grandma and granddad and three aunts and three uncles and many friends in Poland. We call often grandparents and I talk with my friends on Skype and there is one friend from Poland who went to Ireland and I contact her by Skype too.

The material objects, phone and computer, present on Vicky picture, were not unique as many participants drew the same objects under their trees of attachment or inside of their houses on the mental maps. For example, in his house Mathew,

Fig. 2 Vicky's tree of attachment

8 years old, drew himself next to the computer with a person and phone on the screen (see Fig. 3). These are "material and imaginative" aspects of these representations, which show that communication, social relations always stretched beyond their localized presence. The very coexistence of closeness and remoteness is what makes the position of migrant children and their families problematic at all times. Spatial proximity is involved in terms of the immediacy of family members and friends, but at the same time home includes spatial distance. Children defined homes by the inclusion of various elements (people) but also by what or rather who was absent.

Letting the children create more than one drawing was a way to allow them to express multiple ideas about themselves; however, some children chose to make only one drawing or were tired after the first drawing. Younger children sometimes did not understand the concept of the thematic drawings or deviated somewhat from the instructions (e.g., drawing a tree with some significant things around it), or they

Fig. 3 Matthew's inside of his house

asked for another sheet of paper to draw something entirely different – whatever they felt inspired to draw.

Strict visual analyzes without the ability to engage with the child seem to be difficult when we try correctly identify the images on the drawings and to determine the most important features. This is an important methodological issue, as visual data collection strategies can become so "child-centered" that the researcher has difficulty with interpretation. Therefore, in this study, pictures were used to prompt more detailed oral information, keeping the images as the central reference point. I also ask the capable children (excluding those over 8 years old) to sign the drawn objects. To connect a drawing to the social life, intent and interests of its producer, the analysis of drawings should move reflexively between "the image and verbalization" (Harrison 2002, p. 864). In working with children, embedding the analysis of an image within its producer's account of that image is especially relevant since it is often assumed that children need someone to speak for them. However, as with adults, children vary in their ability and inclination to talk about their visual productions. My general experience was that children were willing to describe the elements of their drawings verbally, but rich narrative accountings were uncommon. There were few examples of the children who did not want to talk. This observation highlights more than children's differing communicative competence or the ethnographic fact that adults rarely ask for children's opinions in this community. Drawings are not a substitute for children's voices and the absence or muting, or fragmentation of children's speak about their images means researchers need to be particularly cautious about overinterpreting their images. Therefore, I placed greater importance on using the maps and the drawings as "catalysts for further oral discussion" to properly interpret the images. However, children are not used to interviews, so the structure of the interviews depended on the particular child, and its ageless structured interviews were conducted with younger children. Therefore, for individual interviews with young children, even children as young as 5 and 6 years olds, the drawings were essential to provide a point of reference and to enable communication.

5.2 Photographs and Videos

Photo-elicitation methodologies commonly introduce photographs into a research interview to obtain information (Harper 2002). In photo-elicitation, pictures are used to drive the conversation, to evoke more and different hint of information than in the interviews using words alone. Croghan et al. (2008) suggest that the photo interview offers young people an opportunity to show rather than "tell" aspects of their identity that might have otherwise remained hidden. It may therefore be a useful tool for researching contentious or problematic identity positions. Croghan et al. (2008) have also argued that the features of visual representation influence the versions of young people's identity that are presented. Photo-elicitation differs from photovoice approach, which is the community-based participatory research method also known as "participatory photography" (see also ▶ Chap. 65, "Understanding Health Through a Different Lens: Photovoice Method"). Photovoice was developed by Wang and Burris (1994) who created "photo novella," now known as photovoice, as a way to empower rural women in China to influence the policies and programs affecting them. Applying photovoice to public health promotion, the authors describe the photovoice methodology and analyze its value for participatory needs assessment (Wang and Burris 1997; Wang et al. 2000). Matthews and Singh (2009) undertook a photovoice approach, which sought to promote understanding of the needs of African refugee young people while at the same time stimulating the development resources and pedagogies with a group who were struggling with the English language in the local high schools in Australia. In their Narrating Our World (NOW) project, visual methods enabled refugee young people to demonstrate their resilience, their enjoyment of their new lives, and their capacity to appropriate youth culture to their ends (Matthews and Singh 2009). The (NOW) Project trialed three forms of visual communication (Ramirez and Matthews 2008): digital photography, drawing and painting, and sand tray (that involved placing miniature figurines of people, animals, houses, water, bridges, fences, and so on, into a tray of sand to create stories and served as an interesting point of comparison with the other media). Photography was found to be the most productive in grasping the experience of refugee young people. Matthews and Singh also highlight that many young refugee people did not want to represent themselves in ethnic, cultural, or national terms. Neither were they interested in generating accounts of themselves as victims of inadequate educational and resettlement regimes. Thus, visual methods enabled refugee young people to demonstrate their resilience, their enjoyment of their new lives, and their capacity to appropriate youth culture to their ends.

In the similar vein, Oh (2012) found photovoice used in photo-friend program with Burman refugee children in Thailand as an alternative to research that pictures children who have experienced war within the narrow framework of "victims." Oh found a photovoice technique particularly useful in research with refugee children living in camps. She used photographs and photo-elicitation to glean information about the material circumstances of their everyday lives, as represented by them. The interpretation and analysis of the research did not center on the composition of the images. Instead, the key data came from the interviews and conversations with the children. The children's narratives provided context and meaning, giving us a window into their social worlds and their interpretations of everyday life and their surroundings. Moreover, the technique induces an indirect way of gaining access to refugee children's experiences of conflict, displacement, poverty, and food insecurity, thus reducing the possibility of causing them distress.

Photovoice works within photo-friends program, which engaged postsecondary students who receive training on research and interview skills and practice using digital cameras. These students were introduced to the boarding house children to pair students (aged between 18 and 23 years) with children who are either orphaned or separated from their parents for mentoring sessions so they could teach them how to use the cameras. The children are asked to take photographs of whatever strikes their fancy. They are then asked to choose and talk about the photographs, which are significant to them. An unstructured approach to the interview was used, where the children were encouraged to talk about their chosen pictures. The rest of the session included questions designed to elicit information about the children's physical security, food, relationship with adults and peers, health, school, play, movement, work, and everyday activities. The program has run for four to six sessions with each child over a 2-month period (Oh 2012).

Oh evaluates photovoice in photo-friend program as being a subtle and sensitive tool that allows generating a rich, valid, and meaningful visual and textual representation in an indirect and unobtrusive way, thus reducing the likelihood of causing harm to children during the research. As she further argues, research about refugee children tends to focus on indicators of their nutrition, health, and schooling, rather than on their lived experience as a whole. For many of the children, photovoice was the first time that their thoughts and opinions were solicited by adults and incorporated into a project that would have an impact on their well-being. Thus, photovoice as used in photo-friends as an alternative to research that conceptualizes children who have experienced war within the narrow framework of "victims" (Oh 2012, p. 287). As such, this method has often been used to promote change in the lives of oppressed and disenfranchised groups, including refugees (Green and Kloos 2009; Oh 2011).

In the ultimate example, Marshall (2013) reports on the project, which conducted a variety of qualitative, visual research with small (about six children each) groups. Activities included guided tours of the camp, photo-diaries, participatory video projects, mental mapping, drawing, and focus group interviews. Marshall works with Palestinian refugee children aged 10–13 in the schools and community centers of Balata Refugee Camp. The project examined the ways in which Palestinian children variously perform and transform the discourse of trauma and the aesthetic of suffering that have come to dominate representations of Palestinian childhood and the Palestinian struggle in general. The data gathered demonstrates everyday beauty in the lives of Palestinian refugee children, as found in mundane spaces and enacted through interpersonal relationships, constitutes an aesthetic disruption to the dominant representation of trauma as put forward by international humanitarian-aid organizations and development agencies (Marshall 2013, p. 57).

This funding resonates with Matthews and Singh (2009) who expected photographic images, artwork, and narrative accounts to deliver negative reports of traumatic presettlement experience, as well as the trials of postsettlement of young African refugees in Australia. Instead, they received images and accounts of full, busy, and happy postsettlement lives. Through the participatory activities, they also realized that participants were less concerned with communicating about the trials and tribulations of school, than in the opportunity to meet people with similar experiences and overcome their social isolation. Visual methods enable us to understand better the resilience born of the cultural and political conditions of refugee experience.

6 Conclusion and Future Directions

Two issues seem to be of particular importance when discussing migrant and refugee children participation in visual research: their representation and recognition. As Marshall (2013) points out, the notion of ethics as the distribution of what can be seen and heard and aesthetics as the disruption of the dominant distribution of the senses not only presents a challenge to conceptions of the political, but also challenges social science research itself, in particular research with children.

While attempting to balance adult-centered and child-centered methods, many research presented in this chapter, including my research with migrant children in Scotland, did not use these research method in isolation but for example combine visual child-centered methodologies with adult-centered research examining the production of children's discourses (Marshall 2013). In my research, migrant children drawings and maps clearly brought to my attention the children's ideas and concerns about their transnational experiences and practices (Moskal 2015). Visual methods were not used exclusively but were employed as supplementary methods to the narrative interview and observation in the family or/and school context to maximize opportunities for researchers to understand the children's experiences. The combination of methods also allowed researchers to be more conscious about the ways children expressed themselves with the visual methods (Liamputtong and Fernandez 2015; Kurban and Liamputtong 2017). The value of eliciting and analyzing visual methods is now well established and widely used in ethnographic research among adults (Prosser 1998; Pink 2001), although surprising little of it examines drawing. As the following overview makes evident, a growing number of researchers are taking seriously Wagner's (1999, p. 4) suggestion that "placing images in the foreground of our talk with children can increase opportunities for getting a clearer sense of what kids think." I found that asking children to draw a picture or a map related to the topic and then to tell a story to go with this is a good strategy to facilitate an interview. Particularly, a standard, lengthy series of questions and answers may not work as well for children as for adults. My research in the

Scotland provides evidence of the value of using drawing and mental maps accompanied by interviews as a research strategy among migrant children.

Visual research methods have frequently been discussed as not only suitable for a child-oriented research process, but even more so for research with children and young people with limited language skills or who may have complex experiences of trauma (see Young and Barrett 2001; Crivello et al. 2009; Due et al. 2014; Liamputtong and Fernandez 2015; Kurban and Liamputtong 2017). In working with refugee and asylum children and young people, commonly used approaches include a photovoice (Matthews and Singh 2009), photo-friend (Oh 2012), photo-diaries, and participatory video projects (Marshall 2013), or mapping exercises (Gifford and Sampson 2010; Kurban and Liamputtong 2017), to mention a few examples. Over the years, some concern has been raised if these approaches are to be more widely adopted. The requirement for anonymity usually prohibits publication or public display – for example, on the project websites – of images (photographs/photo-collages/videos) produced by and depicting participants. Thus, the research participants (refugee young people) might not receive the recognition they initially desired (Matthews and Singh 2009, pp. 64–65). It is an argument against reaction that assumes that anonymity is always desirable. Anonymity can act to silence as well as to protect. Smudging or pixelation of images is the visual equivalent of anonymizing text (Lodge 2009). The anonymizing of images reinforces, in fact, the invisibility of research participants and vanish their efforts to locate themselves as creative agents and active research subjects. While research project could give young people access to technologies and other resources, which enabled them to represent and subvert negative images and narratives of themselves and their families, the strictures of research ethics concerning anonymity meant that the project limited participant's capacity to challenge dominant victim-orientated narratives of refugees (Matthews and Singh 2009).

Some scholars have pointed out that the discourse on and conceptualization of refugee children (Hart and Tyrer 2006) is often framed around their vulnerability, helplessness, and role as victims. Conventional wisdom on research methods with children also highlights their vulnerability to persuasion, adverse influence, and harm in research, as in the rest of life. Many research, however, demonstrate the visual participatory method can elucidate children's understanding of their everyday experiences and allow young people to express themselves.

References

Alanen L, Mayall B. (eds) Conceptualizing child - adult relations. London: RoutladgeFalmer; 2001.
Alasuutari P. The globalization of qualitative research. In: Seale C, Gobo G, Gubrium JF, Silverman D, editors. Qualitative research practice. London: SAGE; 2004. p. 595–608.
Anning A, Ring K. Making sense of children's drawings. Maidenhead: Open University Press; 2004.
Ansell N. Childhood and the politics of scale: descaling children's geographies? Prog Hum Geogr. 2009;33(2):190–209.

Bagnoli A. Beyond the standard interview: the use of graphic elicitation and arts-based methods. Qual Res. 2009;9(5):547–70.

Barker J, Weller S. 'Never work with children?' The geography of methodological issues in research with children. Qual Res. 2003;3(2):207–27.

Boyden J. Children under fire: challenging assumptions about children's resilience. Children, Youth and Environments. 2003;13(1). Spring. Retrieved 15 Mar 2017 from http://colorado.edu/journals/cye.

Bromley RDF, Mackie PK. Child experiences as street traders in Peru: contributing to a reappraisal for working children. Child Geogr. 2009;7(2):141–58.

Christensen P, James A, editors. Research with children: perspectives and practices. New York: Falmer Press; 2000.

Christiansen P, James A. Research with Children: Perspectives and Practices, Second edition, London: RoutledgeFalmer; 2008.

Copeland AJ, Agosto DE. Diagrams and relational maps: the use of graphic elicitation techniques with interviewing for data collection, analysis, and display. Int J Qual Methods. 2012;11(5):513–33.

Cowan P. Drawn into the community: re-considering the artwork of Latino adolescents. Vis Sociol. 1999;14:91–107.

Crivello G, Camfield L, Woodhead M. How can children tell us about their wellbeing? Exploring the potential of participatory research approaches within young lives. Soc Indic Res. 2009;90:51–72.

Croghan R, Griffin C, Hunter J, Phoenix A. Young people's constructions of self: notes on the use and analysis of the photo-elicitation methods. Int J Soc Res Methodol. 2008;11(4):345–56.

den Besten O. Local belonging and 'geographies of emotions': immigrant children's experience of their neighbourhoods in Paris and Berlin. Childhood. 2010;17(2):181–95.

Dockett S, Perry B. Children's drawings: experiences and expectations of school. Int J Equity Innov Early Child. 2005;3(2):77–89.

Due C, Riggs DW, Augoustinos M. Research with children of migrant and refugee backgrounds: a review of child-centered research methods. Child Indic Res. 2014;7(1):209–27.

Einarsdottir J, Dockett S, Perry B. Making meaning: children's perspectives expressed through drawings. Early Child Dev Care. 2009;179(2):217–32.

Fernandez S, Liamputtong P, Wallersheim D. What makes people sick: Burmese refugee children's understanding of health and illness. Health Promot Int. 2015;30(1):151–61.

Gamradt J. Jamaican children's representations of tourism. Ann Tour Res. 1995;22(4):735–62.

Geissler PW. 'Worms are our life', part I: understandings of worms and the body among the Luo of western Kenya. Anthropol Med. 1998a;5:63–79.

Geissler PW. Worms are our life, part II: Luo children's thoughts about worms and illness. Anthropol Med. 1998b;5(2):133–44.

Geissler PW, Mwaniki D, Thiongo F, Friis H. Geophagy among primary school children in western Kenya. Tropical Med Int Health. 1997;2:624–30.

Gifford S, Sampson R. Place-making, settlement and well-being: the therapeutic landscapes of recently arrived youth with refugee backgrounds. Health Place. 2010;16(1):116–31.

Green E, Kloos B. Facilitating youth participation in a context of forced migration: a photovoice project in Northern Uganda. J Refug Stud. 2009;22:460–82.

Guillemin M. Understanding illness: using drawings as a research method. Qual Health Res. 2004;14(2):272–89.

Harper D. Talking about pictures: a case for photo elicitation. Vis Stud. 2002;17(1):13–26.

Harrison B. Seeing health and illness worlds – using visual methodologies in a sociology of health and illness: a methodological review. Sociol Health Illn. 2002;24(6):856–72.

Hart R. Children's participation: from tokenism to citizenship. Innocenti essays no. 4. Florence: UNICEF International Child Development Centre; 1992.

Hart J, Tyrer B. Research with children living in situations of armed conflict: concepts, ethics & methods. RSC Working Paper No. 30. Oxford: Refugee Studies Centre; 2006. https://www.rsc.

ox.ac.uk/files/publications/working-paper-series/wp30-children-living-situations-armed-conflict-2006.pdf.

James A, Jenks C, Prout A. Theorizing childhood. New York: Teachers College Press; 1998.

Johnson V, Ivan-Smith E, Gordon G, Pridmore P, Patta S, editors. Stepping forward: children and young people's participation in the development process. Southampton Row: Intermediate Technology Publications; 1998.

Kaplan I, Howes A. 'Seeing through different eyes': exploring the value of participative research using images in schools. Camb J Educ. 2004;34(2):143–55.

Kurban H, Liamputtong P. Perceived social attitudes and connections with new social environment: The lived experience of young Middle-Eastern refugees in Melbourne. Youth Voice J. 2017. http://www.youthvoicejournal.com.

Lewisa A, Porter G. Interviewing Children and Young People with Learning Disabilities. British Journal of Learning Disabilities. 2004;32(4):191–97.

Liamputtong P, Fernandez S. The drawing method and Burmese refugee children's perceptions of health and illness. Australas J Early Childhood. 2015;40(1):23–32.

Lodge C. About face: visual research involving children. Int J Prim Elem Early Years Educ. 2009;37(4):361–70.

Lomax H. Contested voices? Methodological tensions in creative visual research with children. Int J Soc Res Methodol. 2012;15(2):105–17.

Lynch, K. (1960) The Image of the City, Cambridge MA: MIT Press.

Mackie P. Using visual methods with vulnerable children. Cardiff Case Studies Geography. 2011. Online at http://www.cardiff.ac.uk/cplan/sites/default/files/CCS-VisualResearchVulnerableChildren.pdf.

Marshall DJ. 'All the beautiful things': trauma, aesthetics and the politics of Palestinian childhood. Space Polity. 2013;17(1):53–73.

Massey D. For space. London: SAGE; 2005.

Matthews J, Singh P. Visual methods in the social sciences: refugee background young people. Int J Interdiscip Soc Sci. 2009;4(10):59–70.

Mitchell LM. Child-,? Thinking critically about children's drawings as a visual research method. Vis Anthropol Rev. 2006;22(1):60–73.

Mitchell K, Elwood S. Mapping Children's politics: the promise of articulation and the limits of nonrepresentational theory, Envinronement and Planning D: Sociaty and Space 2012;30(5):788–804.

Moskal M. Visual methods in researching migrant children's experiences of belonging. Migr Lett. 2010;7(1):17–32.

Moskal M. 'When I think home I think family here and there': translocal and social ideas of home in narratives of migrant children and young people. Geoforum. 2015;58:143–52.

Moskal M, Tyrell N. Family migration decision-making, step-migration and separation: children's experiences in European migrant worker family. Child Geogr. 2016;14(4):453–67.

Nieuwenhuys O. Action research with street children: a role for street educators. Theme Issue, Children's Participation, PLA Notes 25 1996.

Oh S-A. Rice, slippers, bananas and caneball: children's narratives of internal displacement and forced migration from Burma. Glob Stud Child. 2011;1:104–19.

Oh S-A. Photofriend: creating visual ethnography with refugee children. Area. 2012;44(3):382–288.

Orellana MF. Space and place in an urban landscape: learning from children's views of their social worlds. Vis Sociol. 1999;14:73–89.

Pink S. Doing visual ethnography: images, media and representation in research. London: SAGE; 2001.

Prosser J, editor. Image-based research: a sourcebook for qualitative researchers. New York: Routledge; 1998.

Punch S. Research with children: the same or different from research with adults? Childhood. 2002;9(3):321–41.

Radley A, Taylor D. Remembering one's stay in hospital: a study in photography, recovery and forgetting. Health Interdisc J Soc Stud Health Illn Med. 2003;7(2):129–59.

Ramirez M, Matthews J. Living in the NOW: young people from refugee backgrounds pursuing respect, risk and fun. J Youth Stud. 2008;11(1):83–92.

Ross NJ, Renold E, Holland S, Hillman A. Moving stories: using mobile methods to explore the everyday lives of young people in public care. Qual Res. 2009;9(5):605–23.

Sime D. Ethical and methodological issues in engaging young people livinging poverty with participatory research methods. Child Geogr. 2008;6(1):63–78.

Thompson P. Children and young people: voices in visual research. In: Thomson P, editor. Doing visual research with children and young people. London: Routledge; 2008. p. 1–20.

Tinkler P. A fragmented picture: reflections on the photographic practices of young people. Vis Stud. 2008;23(3):255–66.

Van Blerk L, Ansell N. Imagining migration: placing children's understanding of 'moving house' in Malawi and Lesotho. Geoforum. 2006;37(2):256–72.

Wagner J. Visual sociology and seeing kid's worlds. Vis Sociol. 1999;14:3–6.

Wang C, Burris MA. Empowerment through photo novella: portraits of participation. Health Educ Q. 1994;21:181–6.

Wang C, Burris MA. Photovoice: concept, methodology, and use for participatory needs assessment. Health Educ Behav. 1997;24:369–87.

Wang C, Cash JL, Powers LS. Who knows the streets as well as the homeless? Promoting personal and community action through photovoice. Health Promot Pract. 2000;1:81–9.

White S. Being, becoming and relationship: conceptual challenges of a child rights approach in development. J Int Dev. 2002;14:1095–104.

White A, Bushin N, Carpena-Mendez F, Ni Laoire C. Using visual methodologies to explore contemporary Irish childhoods. In: Hughes J, editor. Sage visual methods, vol. 4. London: SAGE; 2012.

Young L, Barrett H. Adapting visual methods: action research with Kampala street children. Area. 2001;33:141–52.

Zelizer V. Pricing the priceless child: the changing social value of children. New York: Basic Books; 1985.

Participatory and Visual Research with Roma Youth

100

Oana Marcu

Contents

1 Introduction: Ethical Research with Migrant and Ethnic Minority Groups	1740
2 Participatory Action Research with Youth	1741
3 Visual Research with Marginalized Youth	1742
4 Methods and Instruments in PAR with Youth: Two Examples	1744
4.1 Roma Youth and Drug Use: How to Design Visual and Participatory Research	1744
4.2 Peer Research on Migration, Gender Scripts and Cultural Change	1748
5 Conclusion and Future Directions	1752
References	1753

Abstract

Drawing from two examples of research carried out with Roma youth, this chapter discusses applications of participatory research, backed up with visual methods and creative group techniques. It describes methods, tools, and strategies which can be used in peer-research with young people belonging to minorities, from migrant backgrounds or marginalized ethnic groups. In the context of ongoing political debate regarding the minority status and migration rights of the Roma in all European countries, knowledge construction processes are particularly sensitive to issues of age, class, gender, and ethnicity, intersecting in transnational processes. The lack of self-representation of the Roma in the public sphere, as a group discriminated against on an ethnic basis, leaves a need for participation in knowledge-making processes, and research can contribute in this direction. While participation addresses some ethical issues in cross-cultural research, by leaving space for participants' perspectives, it also means having to address competence, class, and power distances that may exist between the researcher and the participants. This goal invites the researchers to use new

O. Marcu (✉)
Università Cattolica del Sacro Cuore, Milan, Italy
e-mail: Oanamarcu@gmail.com

tools in order to engage young people in creative and reflective explorations. Research strategies such as the participatory design process and choosing specific levels of participation in all stages are discussed. Visual and participatory methods are illustrated here with examples from two research projects: the first one on the representation of drugs and the second one on the migratory experience from a gendered perspective.

> **Keywords**
>
> Participatory action research · Visual research · Peer research · Migrant and ethnic minority groups · Youth

1 Introduction: Ethical Research with Migrant and Ethnic Minority Groups

When conducting research with migrant and ethnic minority (MEM) groups, researchers need to consider the ethics of ethnic data collection and the management of power gaps, to work in the group's best interest without increasing stigma. When social change is also a goal, effective action and reflection cycles should be backed up by the necessary authority and power to improve the condition of the group we are interested in. These are typical issues that have been put at the center of the Participatory Action Research (PAR) approach, which I will briefly present below, focusing on its educational and empowerment potential in working with young people.

I refer in this chapter to the Roma as a migrant and ethnic minority group, present mostly throughout Europe, but also on other continents (Piasere 2004). The situation of the Roma can reveal exclusion patterns and identity construction processes shared with other MEM groups, and can be taken, contextual factors ascertained, as a case study for the condition of many minority groups.

Different Roma groups are present in most European countries, with different legal statuses, cultural, linguistic, and historical backgrounds (Piasere 2004). In Western European countries, they have been traditionally associated to nomadism (Piasere 2004). Within Europe, some groups are recognized as national minorities, while others are citizens with no official recognition. Others still are migrants, EU citizens, or third-country citizens. Local groups tend to identify themselves as culturally different from one another (Marcu 2014). Piasere (2004) points out that "Roma" is a politethical category, meaning that its members have some similar traits, but without a common core of traits that could define this belonging.

This is one of the reasons why talking about "Roma health" can be a misnomer, as the different ways in which we interpret ethnicity and define this group lead to quite different lines of research and intervention (Matras 2016).

Researchers need to pay special attention to those determinants of health that particularly work against stereotypical interpretations of difference, pointing out how inequalities weigh on marginalized groups, often as a result of institutional and structural discrimination. Such determinants might include legal status, differential rights, and barriers in access to health, inadequate welfare support, poor living conditions, homelessness and substandard housing, informal, precarious and unstable work, stress connected to poverty, and fear of discrimination and of xenophobic violence.

The risk of essentializing culture and using it to explain behaviors can have negative consequences for the group, at the level of public opinion and policy-making. Accepting and reinforcing reified definitions of "culture" and "ethnicity" gives them facticity outside of the human and social processes that create them (Berger and Luckmann 1966). If we do not examine their genealogy and their processuality (Baumann 1999) and, thus, assume that the typical characteristics of the group uniformly apply to all members, we can give rise to even stronger, and apparently "evidence-based," racisms. This risk can increase when findings are interpreted as ethnic differences, without accounting for the processes that construct them.

Qualitative research methods and, even more, participatory methods, allow us to connect health behaviors and challenges to the meaning attributed to them and to the more general social context, from the point of view of those who can best describe and reflect on their own experience (Higginbottom and Liamputtong 2015; see also ► Chaps. 63, "Mind Maps in Qualitative Research," and ► 17, "Community-Based Participatory Action Research"). Health behaviors, just as cultural consumption or gender norms, express a dynamic process of relating to the "other" and to one's own stigmatized position and can thus reveal good indications for shared, responsible, and respectful social intervention.

2 Participatory Action Research with Youth

Participatory action research is a process in which research, education, and action are intertwined, as participants take an active role in addressing issues affecting themselves, their families, and their communities, with the goal of producing social change and empowerment (Brydon-Miller 2001; Higginbottom and Liamputtong 2015). It has very often involved groups that have been exploited or oppressed (Brydon-Miller 2001).

Participatory action research is connected to two main ideals: emancipation and empowerment (Boog 2003; Higginbottom and Liamputtong 2015). Emancipation refers to contesting the distribution of power in the society and to transforming it in order to reach a more balanced state, and is inspired by critical theory and Marxist approaches. Empowerment was initially connected to developing self-consciousness, self-actualization, and self-advocacy, at a more individual level. It is connected to the development of the capacity to solve problems while generating critical knowledge of the system in which the problems exist (Boog 2003). Both researcher and participants are more actively involved in the process than in traditional social research, as they share responsibility for a process whose impact does not rely on the insufficient feedback loops that exist between mere academic knowledge production and social change (Higginbottom and Liamputtong 2015).

PAR projects may range from institutionally funded social research and intervention (as is the case with both of our examples presented below), to classroom educational methods (Kennedy 1989; May 1993), and to academic activism (Chatterton et al. 2007). Critiques of participatory methods argue that they are not always capable of generating equality and empowerment (Cooke and Kothari (2001); using participatory techniques in otherwise researcher-driven processes can be misleading, and surely does not achieve full democratic participation or social change (Pain and Francis 2003). Instead, various authors have discussed a participation continuum, ranging from co-option to collective action (De Negri et al. 1999; Arnstein 2004; Kindon et al. 2007; Higginbottom and Liamputtong 2015; see also ▶ Chap. 17, "Community-Based Participatory Action Research").

More specifically, PAR with youth, which I will refer to also with the term peer-research, works on the inequality axis of age, aiming to give young people more freedom of expression and involvement in the relevant decisions for their lives (Nairn and Smith 2003). Just like in other forms of participatory research, involvement in all research stages, from the design to the dissemination of results, is crucial in order to produce empowerment or social change (Nairn and Smith 2003; Conrad et al. 2015). Kennedy (1989), in an early attempt to use peer-research, underlines the educational value of this approach, but remains skeptical regarding the value and validity of the knowledge produced by young people carrying out research.

Price and Hawkings (2002) use peer ethnography in order to conduct research on sexual and reproductive health, starting from the assumptions that it is easier to establish the trust relationships needed to approach intimate topics between peers. From this point of view, peer research uses the insider role of young people in their own networks to allow for a stronger involvement of the community in the research process. However, the authors realize that peer relationships are not always characterized by consensus, but also by conflict and mistrust. Thus, their insider status is not always of help in data collection (Price and Hawkings 2002).

Peer research is connected to peer education, which is based on a natural way of learning and education, established between young people sharing the same life context and networks, bringing educational intervention closer to the communities (Youth Peer Education Network 2005).

3 Visual Research with Marginalized Youth

Although not directly connected to the history of PAR, visual studies initiated their own research experiments aiming for social change. Various assumptions are at the core of these research methods. These include: the power of self-representation in challenging established discriminatory practices and worldviews, the potential of visual messages to reach larger, targeted, and more engaged audiences, the possibility of producing change by decolonizing communication processes (Frisina 2013).

These assumptions find fertile ground in social research, in particular with marginalized groups, in order to reach various goals. First, images are used to enrich social research, by exploiting their specific qualities, as they are polysemic, vivid, metaphorical (Gauntlett 2007; Grady 2008). Paying attention to the visual material culture also means the recognition of popular culture as legitimate for academic interest, an idea pioneered by the cultural studies school (Frisina 2013).

Second, visual methods facilitate the engagement of young people by enriching the possible means of expression, allowing them to spend more time with the questions and ideas stemming from research topics (Gauntlett 2007). This contributes to in-depth and sensitive social research. As a consequence of engaging in the process and gaining more expressive competence, the potential of visual methods to bring about transformative effects, typical of PAR, grows. This is the case for the **photovoice** method, considered to be a PAR method, which consists of asking participants to shoot photos on an important topic for their everyday life, to comment and discuss them and then to present them to a relevant audience (Wang and Burris 1997; see also ▶ Chaps. 65, "Understanding Health Through a Different Lens: Photovoice Method," and ▶ 99, "Visual Methods in Research with Migrant and Refugee Children and Young People").

Third, images are useful to disseminate better and to talk to wider audiences, supporting the diffusion of counter-narratives and enhancing the impact of research and action.

Due to their metaphorical and artistic potential, their role in generating reflective thought is also not to be ignored. This introduces a new, aesthetic dimension in the knowledge production process that is also related to the effectiveness of the dissemination and, thus, to the impact of visual research. An aesthetically and communicatively efficient product might need the involvement of experts or artists that can support the creation of vivid representations and transmit the competencies needed to use techniques and media. The balance between the input of experts or artists and the input of peer participants is negotiated in each step, but financer's standards, time, and resources have a large impact (Bugli, personal communication, July 13, 2016).

In research-oriented visual production, the subjectivity of the image reflects the relationship established between researcher and participants. When we involve artists and experts, we introduce "forms of polyphonic authorship" (Frisina 2013, p. 11) in the process and imagine a wider audience for the products.

The **participatory video** is a process where the complexity of the medium does require different professionals to be involved, starting with the producer/director, who, in this case, should put him or herself in the position to tell other people's story, not to create his original story, as is common in artistic production (Seidl 2003). With this method, researchers involve grassroots stakeholders in producing video messages as tools for self-definition, empowerment, education, community building, and activism (White 2003).

Visual methods can, therefore, strongly contribute to PAR by supporting wider participation of those excluded from the academic and public writing processes, and reaching wider audiences for research products.

4 Methods and Instruments in PAR with Youth: Two Examples

In any qualitative research design, most of the planning is creative, but in PAR it should be even more flexible because decisions are not taken by an expert researcher alone. Similarly, methods and instruments to be used have to be easily managed by people who do not have extensive training in social sciences.

The flexibility and interdisciplinarity of PAR link it to a more artisan style of planning, and PAR facilitators, often professionals in the educational field, have included pretty much everything that could have a transformative and engaging potential in the process: arts and theater, music, video, live performance, and so on. These methods can actually be combined in many ways, when the researchers allow themselves to be guided by the resources and interests of the people involved. I will show and discuss, with two examples, how my team and I organized the PAR processes and the decisions we took in various steps, in order to make the most of each situation.

The first example focuses on three key elements, from a methodological point of view: involving stakeholders in designing the research, using the visual instrument of the collage, and applying metaphorical analysis. With the second example, I will present various research and action methods used with Roma young people (theater, photovoice, interviews, participant observation, focus group) along with some instruments (concept maps, collage, decision-making techniques) in order to creatively facilitate the work with youth with only basic literacy skills.

4.1 Roma Youth and Drug Use: How to Design Visual and Participatory Research

"Addiction Prevention Within Roma and Sinti" (SRAP), was a 3-year project, financed by the Executive Agency for Health and Consumers of the European Union, comprising research, training, and piloting of addiction prevention interventions with Roma youth, in six countries. The consortium, made up of institutions responsible for providing social and health services, research bodies, and NGOs working in the field of education and health, carried out the activities in a coordinated manner, transnationally. The research was conducted in the first project year and managed by a central team composed of myself and two colleagues from Codici Research and Intervention, a private firm active in research and consultancy based in Milan (I will use the first-person plural to refer to the collective work done by the scientific coordination and the local teams forming the consortium.) (Marcu 2015).

Little was known about the topic, as few surveys have explored indicators of drug use among the Roma. Results were quite alarming because of the significant differences in the incidence and level of use, abuse, and addiction, when compared with the general population. Previous research identified tendencies such as early ages for the onset of tobacco and alcohol consumption, diffuse problematic alcohol consumption, wide use of self-medication with tranquilizers, sleeping pills and relaxants, higher lifetime prevalence for all types of drugs, and stigma and concealment of consumption (EMCDDA 2002, 2008; Fundación Secretariado Gitano 2009). Most scholars attributed the wide health disparities between the Roma and the general population to the conditions of poverty in which most members of this ethnic group live (Földes and Covaci 2011) and on the barriers that this group meets in access to appropriate health care (Ivanov 2004).

Drug use and addiction was perceived as a problem also among the professionals that were involved in the project. In order to test this hypothesis with young people themselves, and to get better guidance regarding the priorities, from their point of view, we needed to involve them from the research design phase.

The **participatory design process** (Bergold and Thomas 2012) is a strategy that aims to overcome a common limitation: that research design is usually decided by experts (most of the time in the phase of the financing application) and remains the stage which participants control the least, although it is the stage with the biggest impact on the research (Nairn and Smith 2003).

Building the research design together with the target groups helped us understand how to approach the drug "issue" in a nonstigmatizing manner, working with parents' and young people's existing interest and motivation to tackle this problem in their communities.

For this planning stage, Roma youth (74 participants), decision-makers in institutions or NGOs (41 participants), and professionals in the field of addiction and social work with Roma (62 participants) gave their contribution in 23 group meetings, with the specific goal of offering indications for the research design regarding targets, contexts, recruitment, and ethical issues. The procedure of these meetings consisted of a brainstorming task on the topic "Roma youth and drugs," followed by the instigation of a concept map.

Concept maps are graphical tools for organizing knowledge, for meaningful learning, and can be used as group facilitation tools. As Novak and Cañas (2006, p. 1) explain, "they include concepts, usually enclosed in circles or boxes of some type, and relationships between concepts indicated by a connecting line linking two concepts." The main categories of concepts were then ranked by participants on the basis of their importance and of their accessibility for research and prevention actions. They then prioritized the problems, ordered the resources, identified the critical issues, and warned about potential errors to be avoided in the research process.

The results of these meetings contributed to the development of a conceptual framework that guided the construction of tools for data collection and analysis. Results were summarized transnationally in a group process with professionals that had facilitated the local meetings.

The data collection and the analysis of the results were then carried out by the same professionals (outreach, educational, and social workers) with the help of Roma mediators. They were trained in research methodology and in the use of the specific tools designed for this study by the scientific coordination team. As a part of this transnational coordination process, my team and I conducted shadowing and onsite training of field researchers, facilitated constant exchange between them, periodically summarized results and observations, reviewed the instruments after the pilot study, and gave feedback on the results.

Surely, in this research, the voices of social workers and mediators, who had a more constant role in all stages, were stronger than those of the young people and other stakeholders involved in the design.

Following this preparatory stage, the local teams conducted 23 visual focus groups and 58 life-story interviews in all countries, with 199 young Roma, out of which 98 were aged 11–13, while 101 were 14–16 years old.

The **visual focus groups** used the technique of the **collage**, a less common method, recommended by some positive experiences (Awan 2007; Reavey 2011). It involved participants in the process of pasting various materials on a white sheet of paper, such as newspaper clippings or parts of photographs, around the topic of their relationship with alcohol, tobacco, and other drugs. Researchers interacted with participants during and after the execution of this task, eliciting meanings involved in the choice and placement of the images. Following the production and presentation of the collages, the researcher guided a group discussion about the shared themes, in order to reach an understanding of the issues with input from all participants, interacting with one another (Fig. 1).

The **elicitation**, usually conducted with photos, here with collages, helps researchers gain more insight into participants' points of view, by relying on the verbalization of intended meanings. This process, as Harper (2000, p. 725) suggests, can be seen as an insight into cultural explanations, a "cultural Rorschach test."

It is common to use print media (such as magazines and newspapers) for the collage task, but it is important to acknowledge that it can introduce bias in the direction of socially accepted or commercially promoted images and discourses. As concerns minority groups, media communication has often been found to misrepresent and to stereotype negatively; or the lack of depiction of these groups in the media constructs by omission their social identity as marginal. There have not been many studies regarding the representation of the Roma in the media, although a few examples pointed out negative, stereotypical discourse. Also notable is the lack of representation and absence of the voices of Roma and Sinti: they are pictured by the voices of others (Waringo 2005; Sigona 2006).

Visual focus groups were backed-up with individual life-story interviews with conducted with the older youths (17–24 years old), who were purposely selected as consumers of different drugs, 37 male and 21 female. Given the focus of this chapter on visual methods, I will not detail the interviews, a more traditional way of gathering data, which followed the nondirective approach proposed by Bichi (2002).

The interpretation of the data put participants' motivations, interests, and perspectives, subjectively raised during the interviews, at the center of the

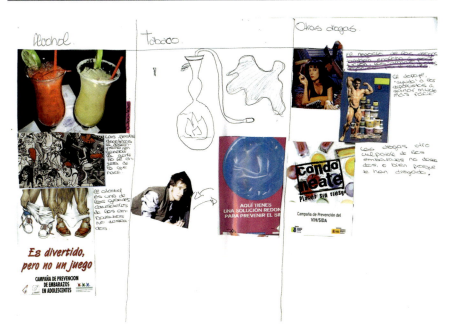

Fig. 1 Collage produced by participant in Madrid, Spain, divided in three sections corresponding to alcohol, tobacco, and other drugs

explanations. With visual materials, we used **metaphorical analysis** (Schmitt 2005), drawing on the seminal work of Lakoff and Johnson (1980), who argue that social thought, language, and action processes can be understood when studying individual metaphorical models and collective ones. We used the steps proposed by Gauntlett (2007): listed the signifier and the signified for each metaphor, identified recurrences, similarities, contrasts or other relationships between metaphors, then mapped source and target domains in order to access broader areas of meaning. In later stages, it was possible to link metaphors by means of interpretation, to categorical knowledge and to the initial conceptual areas. Some participants wrote texts in the collage, ranging from just a few words to page-long stories, next to images which were also used as an input for analysis.

Other sources of interpretation were the image composition, disposition, closeness, relative dimensions of the components, and types of elements (persons, objects, and their characteristics), following the argument made by Bohnsack (2008) regarding the analysis of the formal structure of images. In this manner, it is possible to reconstruct preiconographic knowledge that is difficult for participants to conceptualize or verbalize.

One of the most interesting findings of this research was describing meanings that linked health behaviors to the construction of ethnic, class, and gender identities. The use of particular types of drugs expressed identification with a nonethnicized peer group and culture, and were linked to social change within the younger generation of Roma, while other drugs are seen as "drugs of the poor," their use was explained as a reaction to continuous stress related to poverty and to the marginal status. Girls' consumption had different patterns than boys' and was more stigmatized. The rich meanings produced using these research methods produced evidence that drug consumption is connected to the negotiations between keeping tradition and being part of a globalized youth culture.

With this research example, I intended to show how the participation of various stakeholders, especially of youth, can be central in the initial phases of the research, even if, given the transnational dimension and the resources constraints, it could not be implemented in all phases. Although this does not ensure the creation of a profoundly democratic knowledge-making process, it sets up a good level of collaboration with young people, strengthening the representation of their own perspectives, as compared to traditional social research, where they are involved only as respondents. Other stakeholders (specifically, professionals in social work) had the central role in training, in data collection and data analysis, while my team of expert researchers kept control of the entire process, consulting and collaborating with the partners at each step.

4.2 Peer Research on Migration, Gender Scripts and Cultural Change

The second example I will discuss involved only Romanian Roma in Italy, and focused on their experiences regarding their migrant condition. It was part of a series of projects dealing with irregular Romanian Roma's migration, especially that of young people involved in street work (begging, playing music, or doing petty crime). The projects followed an action research approach, relying mainly on street and institutional ethnography, but also on biographical interviews, peer-research, and group work with social services and NGO professionals (Marcu 2014). It was conducted by a team from Codici Research and Intervention including myself, and two colleagues, and coordinated by a senior researcher.

Romanian Roma's migrant condition is characterized by extreme precariousness: undocumented work, precarious housing conditions ranging from abandoned buildings, tents, or shacks to authorized encampments, and (very seldom) social housing or rented apartments. Young Roma coming to Italy are involved, especially initially, in street work, with all the other members of the family. When supported by volunteering organizations or social services, however, they tend to reach higher school levels than their parents and successfully follow job integration programs (Fondazione Casa della Carità 2012; Marcu and Bacigalupo 2013). Migration, therefore, brings many changes in young people's lives, and with their participation in nonsegregated street, school, and work environments, intergenerational gaps grow wider.

Within this context, the peer-research was intended to create better knowledge regarding the meaning that young Romanian Roma attribute to their migratory experience, while developing empowerment and giving them a public voice, using expressive means familiar to them, put in the context of systematic, participatory social research. Seven young people aged 15–18 participated in the process, which was framed as a research scholarship, accompanied by three facilitators with different expertise (social research, educational work, photography) and two undergraduate students working as interns. I will use the first person plural to refer to the more expert research group, formed by the three facilitators and interns. The peer-research involved various other experts, such as a DJ, an actor, and an ethno-psychologist, who supervised the entire process.

The participants were chosen using the snowball sampling method, with the mediation of outreach workers from local organizations that worked on the transition from being vulnerably housed to more stable housing and from undocumented work to official employment. This was done in order to find volunteers with more stable life conditions than those we met in street ethnography, as the request of sustained participation excluded those living in the harshest conditions. We also asked that participants have at least basic literacy skills. All the volunteers ensured us of their interest, motivated by the topic, which was relevant to their life experience, but also by the financial reward. In the end, we did not operate any selection, but covered the available places as people applied and accepted our offer.

The first phase of a **theater laboratory** involved a professional actor who trained the group on self-expression, self-presentation, group cohesion, and trust (Fig. 2). Divided in two smaller groups, participants prepared two stories, freely chosen, interpreted only using gestures, facial expressions, objects, space, and movement. The first group represented a love and migration narrative, in which a young girl wins a scholarship to study abroad where she falls in love and lives her story freely. The second group represented a story that was quite close to their own migratory experience, comprising migration forced by poverty, ending up begging and squatting.

Fig. 2 Peer researchers and facilitators during the theater laboratory (Photography: Luca Meola)

The next instrument we used in order to explore participants' stories was the **collage**. Each participant represented his/her past, present, and future identity. I described this instrument in the previous example, and here we followed a similar procedure. The initial stories and the collages produced a wealth of material on young Roma's migratory experiences which we grouped and analyzed by building together a **concept map** (as described above) around the central topic of "Young Romanian Roma in Migration." In order to choose just one topic on which to focus in-depth in the following stages, we used an adaptation of a well-known decision-making technique, the **Six Thinking Hats** (De Bono 2000). We asked participants to rate the main themes of the concept map, taking into consideration how much they liked the topic, how emotionally connected they felt to it, the capacity it had to represent young Roma life, and the interest it presented to the general public. The topic of romantic relationships was chosen, confirming the importance of the affective sphere in young people's life.

With this choice, the process we initiated was adopted in an intergenerational and cultural protest/negotiation, thus becoming more than just a knowledge-making process: a ground for young people to express their freedom of choice as well as their belonging.

With this topic, facilitators proposed that an in-depth exploration of the perspective of Romanian Roma could be conducted.

The **research design** was unfortunately proposed as an abstract and academic scheme. What seemed like a natural task to expert researchers proved to be boring and frustrating for young participants, leading to the goals of this phase not being achieved.

This generated a discussion on the choice of the groups to be interviewed in research. In a research-oriented approach, bearing in mind the ideal of triangulation, facilitators suggested that various stakeholders were interviewed, including the parents. Participants strongly resisted, as they wanted to voice their own experiences, and would not accept hearing others, more powerful than themselves, on such a sensitive topic. We finally agreed with peer-researchers to interview only young people, single or couples, and to include both endogamous and mixed couples.

The following step was exploring their everyday life, by means of a **photovoice** process (applied as described above). Participants took photographs of their life settings, focusing on the general context rather than specifically on romantic relationships, which we considered a sensitive topic to explore with this medium, for privacy reasons, as some participants were minors. They represented living conditions, relationships between peers, and friendship, chose five photos each to be presented to the group, and were interviewed by the group on the content and intended meanings (Fig. 3).

Then we passed on to **interviews** with other peers, specifically on the topic of romantic relationships in youth and adolescence. The group built an interview guide eliciting stories of various stages of "love," also asking about relationships with parents and their opinion and reaction to the love story being recounted. The peer-researchers were **trained with simulations** for in-depth interviewing as they tested the guidelines within the group of peer-researchers, and received feedback on their questioning and listening skills. They then conducted 25 interviews, in our office or in the places where they lived, with young people they knew more or less: friends, neighbors, schoolmates.

Fig. 3 Photo produced by during the photovoice process by one of the peer-researchers around the topic of the living conditions of migrant Roma in Italy

The interviewed persons were chosen by convenience sampling and snowball sampling, all were Romanian Roma aged 14–25, 4 girls and 21 boys.

The context where the interviews were carried out greatly influenced the content and constituted a topic for further reflection. We backed up the interviews with participant observation, but instead of having peer-researchers write fieldnotes, we discussed them in the group meetings, which were documented by the two interns. When interviews were carried out in camps or squats, there was no privacy, and they became informal conversations mostly among young men, comprising socially accepted versions of gender and sexuality scripts. In more private spaces, such as in family shacks or in parks, it was possible to gather more personal experiences.

Another issue concerned the relationship of trust between interviewers and interviewees. We assumed that it would have been easier to talk about intimate relationships between peers, but the thick network of power relationships in young people's living contexts leads to differential levels of trust, even among those who share the same age. The small number of girls that agreed to be interviewed is an indicator of this, as the consequences of revealing the intimate details of their romantic relationships can be devastating for their reputation. A solution to this drawback is to use indirect, third-person questions or projective questioning techniques (Price and Hawkings 2002; Bichi 2007).

Aware of the role that Romanian Roma music genre *manele* plays in young people's lives, we invited participants to summarize the results of this research by telling the story with a **music compilation**. They selected the songs and prepared a mix, with the support of a professional DJ.

For a more formal **analysis of the results**, we used a **focus group** format, relying on interactional dynamics, generated by the stimuli that the team of facilitators: transcribed interviews, selected excerpts, photos, collages, the concept map, and the music mix. The intense discussion allowed us to record multiple positions and perspectives, which were presented to social workers, project partners, and the commissioning foundation. Facilitators also prepared an online blog, gathering all the materials produced above (except for interviews, which were summarized into anonymous stories).

Results pointed out that gender and sexuality norms for migrant groups, especially for youth, express a complex negotiation between tradition and change and have a crucial role for ethnic identities. Many migrant groups affirm positive cultural identity through the reinforcement of traditional rules regarding gender and sexuality (Das Gupta 1997; Espiritu 2001). At the same time, young people challenge these norms, grow feelings of belonging to more than one culture and enact different ways of growing up, parallel to the traditional model. We assisted these complex negotiations in the areas of consumption, body display, and dress code, in cultural and goods consumption (Conte et al. 2009) and also with drugs, as shown above. In the area of gender and sexuality, young people talked about everyday battles. For example, they claimed the freedom to choose their spouse, criticizing arranged marriages and sometimes using elopements to force parents' acceptance, or the freedom to use birth control methods, in spite of the elders' family planning projects.

With this example, I intended to present a wider variety of methods and tools that can be used in order to engage young people in knowledge-making processes and, at the same time, to give an example of a peer-research where young people were involved in all research stages. The empowerment process was generated by numerous opportunities for self-reflection and expression designed to work on personal and group experiences, and polyphonic authorship was present throughout the process. Even if expert researchers maintained control and designed the entire research, at moments peer-researchers took control and claimed some decisions as their own.

5 Conclusion and Future Directions

Participatory action research with young people can be a fruitful option for knowledge production and a tool for education, empowerment, and activism, working on multiple axes of difference: age, class, gender, and ethnicity. Both examples I analyzed show a capacity to shed light on complex identity and behavior processes, starting from personal stories and working intensely with group interaction and reflexive tools.

The wealth of methods and instruments that can be creatively combined in PAR should take into account the capacities and resources of the people involved, the constant attention to authorship and direction, and the potential for engagement and reflexivity. Visual methods have a privileged position from this point of view, given the decolonizing potential of visual counter-narratives.

The two examples present different levels of participation of the target groups in the research process, and are both ascribable to PAR. The first one is closer to consultation, as young people gave precious advice regarding the research design, while the second one, closer to co-learning, with young people engaging in knowledge sharing and generating communicative action with our facilitation (De Negri et al. 1999). If we want research to be scientifically recognized while completely sharing decisions and tasks with participants, we find ourselves teaching some competencies that are far from young people's daily experiences and difficult to recycle in other areas of their activity (research design, data analysis). It, therefore, makes sense to diversify roles and tasks while reducing the level of their involvement in more abstract tasks, or in tasks that are not interesting to them. It is valuable though to clearly identify the levels of participation, authorship, and direction in the various phases of PAR research.

While more horizontal and engaged participation of young people in research and social change should remain one of the main goals of scholars working with marginalized youth, problem-specific engagement and consultation practices can still improve the collaboration between researchers, practitioners, and young people.

PAR with marginalized youth is highly engaging for the expert researchers too. Overlapping roles as scholars, facilitators, employers, and friends can create confusion and might require the high emotional involvement typical of long-term, ethnographic work. Documenting this process with personal fieldnotes can be useful for greater reflexivity and interaction with youth.

Visual communication and the bottom-up creation of visual culture is ever more present today, thanks to the new media, and young people have the possibility and capacity to express their minds and fight their struggles using these tools. Still, the access to the public sphere remains uneven, especially if we keep in mind members of the migrant and ethnic minority groups. PAR can give its contribution to filling up this gap by bringing in support, expertise, resources, as well as drawing from its ability to question the taken for granted and deconstruct hegemonic narratives that are detrimental to migrant and ethnic minority groups.

References

Arnstein SR. A ladder of citizen participation. J Am Inst Plann. 2004;35(4):216–24.

Awan F. Young people, identity and the media: A study of conceptions of self identity among youth in Southern England [Unpublished PhD Thesis]. Bournmouth University; 2007.

Baumann G. The multicultural riddle: rethinking national, ethnic and religious identities. New York: Routledge; 1999.

Berger PL, Luckmann T. The social construction of reality: a treatise in the sociology of knowledge. Harmondsworth, Middlesex: Penguin Books; 1966.

Bergold J, Thomas S. Participatory research methods: a methodological approach in motion. Forum Qual. Soc. Res. 2012;13(1):30.
Bichi R. L'intervista biografica. Una proposta metodologica. Milano: Vita e Pensiero; 2002.
Bichi R. La conduzione delle interviste nella ricerca sociale. Roma: Carocci; 2007.
Bohnsack R. The interpretation of pictures and the documentary method [64 paragraphs]. Forum Qual. Soc. Res. 2008;9(3):26. http://www.qualitative-research.net/index.php/fqs/article/view/1171.
Boog B. The emancipatory character of action research, its history and the present state of the art. J Community Appl Soc Psychol. 2003;13:246–38.
Brydon-Miller M. Education, research and action: theory and methods of participatory action research. In: Tolman D, Brydon-Miller M, editors. From subjects to subjectivities: a handbook of interpretive and participatory methods. New York: New York University Press; 2001. p. 76–89.
Chatterton P, Fuller D, Routledge P. Relating action to activism. In: Kindon S, Pain R, Kesby M, editors. Participatory action research approaches and methods: connecting people, participation and place. London: Routledge; 2007. p. 216–22.
Conrad D, Hogeveen B, Minaker J, Masimira M, Crosby D. Involving children and youth in participatory research. In: Higginbottom G, Liamputtong P, editors. Participatory qualitative research methodologies in health. London: Sage; 2015. p. 109–35.
Conte M, Marcu O, Rampini A. Giovani rom: Consumi e strategie di affermazione sociale. In: Visconti LM, Napolitano ME, editors. Cross generation marketing. Milan: Egea; 2009. p. 283–302.
Cooke B, Kothari U. Participation: the new tyranny? London: Zed Books; 2001.
Das Gupta M. "what is Indian about you": a gendered, transnational approach to ethnicity. Gend Soc. 1997;11(5):572–96.
De Bono E. Six thinking hats. Westminster: Penguin; 2000.
De Negri B, Thomas E, Ilinigumugabo A, Muvandi I, Lewis G. Empowering communities. Washington, DC: The Academy for Educational Development; 1999.
EMCDDA. Update and complete the analysis of drug use, consequences and correlates amongst minorities. Lisbon: European Monitoring Centre for Drugs and Drug Addiction; 2002.
EMCDDA. Drugs and vulnerable groups of young people. Luxembourg: Office for Official Publications of the European Communities; 2008.
Espiritu YL. "we don't sleep around like white girls do": family, culture, and gender in Filipina American lives. Signs. 2001;26(2):415–40.
Földes ME, Covaci A. Research on Roma health and access to healthcare: state of the art and future challenges. Int J Public Health. 2011;57(1):37–9.
Fondazione Casa della carità "Angelo Abriani". EU inclusive. Rapporto nazionale sull'inclusione lavorativa e sociale dei rom in Italia. 2012. http://www.casadellacarita.org/eu-inclusive/rapporto.html.
Frisina A. Metodi visuali e trasformazioni socio-culturali. Torino: Utet; 2013.
Fundacion Secretariado Gitano. Health of the Roma population, analysis of the situation in Europe. In: Bulgaria, Czech Republic, Greece, Portugal, Romania, Slovakia, Spain. Madrid: Fundacion Secretariado Gitano; 2009.
Gauntlett D. Creative explorations: new approaches to identities and audiences. Oxford: Routledge; 2007.
Grady J. Visual research at the crossroads [74 paragraphs]. Forum Qual. Soc. Res. 2008;9(3):38. http://www.qualitative-research.net/index.php/fqs/article/view/1173/2619.
Harper D. Reimagining visual methods: Galileo to Neuromancer. In: Denzin NK, Lincoln YS, editors. The sage handbook of qualitative research. 2nd ed. Thousand Oaks: Sage; 2000. p. 717–33.
Higginbottom G, Liamputtong P, editors. Participatory qualitative research methodologies in health. London: Sage; 2015.
Ivanov I. Reflections on the access of Roma to health care. 2004. Retrieved 24 Nov 2010, from Sito di ERRC – European Roma Rights Centre. http://www.errc.org/cikk.php?cikk=2067.

Kennedy G. Involving students in participatory research on fatherhood: a case study. Fam Relat. 1989;38:363–70.

Kindon S, Pain R, Kesby M. Participatory action research: origins, approaches and methods. In: Kindon S, Pain R, Kesby M, editors. Participatory action research approaches and methods: connecting people, participation and place. Abingdon: Routledge; 2007. p. 9–18.

Lakoff G, Johnson M. Metaphors we live by. Chicago: University of Chicago Press; 1980.

Marcu O. Malizie di strada. Una ricerca azione con giovani rom romeni migranti. Milano: Franco Angeli; 2014.

Marcu O. Using participatory, visual and biographical methods with Roma youth. Forum Qual. Soc. Res. 2015;17(1):5.

Marcu O, Bacigalupo A. Money needs to move: financial literacy and educational opportunities for Roma and Sinti in Bologna and Piacenza. 2013. Retrieved 29 Mar 2016, from www.project-finally.eu. http://finally.splet.arnes.si/files/2014/03/Finally-National-Report-IT-italian.pdf.

Matras Y. Roma and health. Paper presented at Roma populations & Health Inequalities workshop. Liverpool: University of Liverpool, Institute of Psychology, Health and Society; 2016.

May WT. Teachers as researchers or action research: what is it, and what good is it for art education. Stud Art Educ. 1993;34(2):114–26.

Nairn K, Smith A. Young people as researchers in schools: the possibilities of peer-research. Paper presented at the Annual Meeting of the American Educational Research Association. Chicago; 2003.

Novak JD, Cañas AJ. The theory underlying concept maps and how to construct them. Technical Report IHMC CmapTools 2006–01. Florida Institute for Human and Machine Cognition. 2006. Available at: http://cmap.ihmc.us/Publications/ResearchPapers/TheoryUnderlyingConceptMaps.pd.

Pain R, Francis P. Reflections on participatory research. Area. 2003;35(1):46–54.

Piasere L. I rom d'Europa. Una storia moderna. Bari-Roma: Laterza; 2004.

Price N, Hawkings K. Researching sexual and reproductive behaviour: the peer ethnographic approach. Soc Sci Med. 2002;55:1325–36.

Reavey P. Visual methods in psychology: using and interpreting images in qualitative research. Hove: Psychology Press; 2011.

Schmitt R. Systematic metaphor analysis as a method of qualitative research. Qual Rep. 2005;10(2):358–94.

Seidl B. Candid thoughts on the not-so-candid camera: how video documentation radically alters development projects. In: White SA, editor. Participatory video: images that transform and empower. New Delhi: Sage; 2003. p. 157–49.

Sigona N. Political participation and media representation of Roma and Sinti in Italy. 2006. Retrieved 23 Sept 2014, from Osservazione. Centro di ricerca azione contro la discriminazione di rom e sinti. http://www.osservazione.org/documenti/OSCE_ITALYv1.pdf.

Wang C, Burris MA. Photovoice: concept, methodology, and use for participatory needs assessment. Health Educ Behav. 1997;24(3):357–68.

Waringo K. Gypsies, tramps and thieves: the portrayal of Romani people in the media. 2005. Retrieved 23 Sept 2014, from Europäisches Zentrum für Antiziganismusforschung. http://www.ezaf.org/down/IIIAZK23.pdf.

White SA. Participatory video: a process that transforms self and others. In: White SA, editor. Participatory video: images that transform and empower. New Delhi: Sage; 2003. p. 63–101.

Youth Peer Education Network. Standards for peer education programs. New York: Family Health International; 2005.

Drawing Method and Infant Feeding Practices Among Refugee Women

101

June Joseph, Pranee Liamputtong, and Wendy Brodribb

Contents

1	Introduction	1758
2	Situating Ourselves	1758
3	The First Author's Personal Encounter with the Drawing Methodology	1759
4	The Postmodern Turn: The Use of Drawing Among Vulnerable Populations	1761
5	The Drawing Method	1761
6	Drawing in the Context of Participants from Refugee Backgrounds	1762
7	The Real-World Application of Drawing in Our Research	1763
	7.1 Participant: *"I Am Sorry, I Cannot Draw"*	1763
	7.2 Research Team: "That Is Okay...It Must Be Hard for You... Let's Move On.... Now This an Interesting and Brilliant Concept.... Would You Like to Sketch It for Me."	1765
	7.3 Participant: *"This really unleashes my past... I have never seen myself capable"*	1766
8	The Refugee Perspectives of Infant Feeding	1768
9	What Women Use Drawing For	1768
	9.1 To Illustrate a Location	1768
	9.2 To Illustrate the Concepts of Motherhood	1769
	9.3 To Illustrate Vulnerability	1770
	9.4 To Illustrate How Mothers Counter the State of Vulnerability	1771
	9.5 To Illustrate the Freedom in Australia	1771
	9.6 To Illustrate the Goodness of Breast Milk	1772
10	Conclusion and Future Directions	1772
References		1773

J. Joseph (✉) · W. Brodribb
Primary Care Clinical Unit, Faculty of Medicine, The University of Queensland, Herston, QLD, Australia
e-mail: j.joseph@uq.edu.au; w.brodribb@uq.edu.au

P. Liamputtong
School of Science and Health, Western Sydney University, Penrith, NSW, Australia
e-mail: p.liamputtong@westernsydney.edu.au

© Springer Nature Singapore Pte Ltd. 2019
P. Liamputtong (ed.), *Handbook of Research Methods in Health Social Sciences*,
https://doi.org/10.1007/978-981-10-5251-4_44

Abstract

The pressures exerted by political agendas in third world nations continue to displace many individuals daily, with mothers being greatly impacted due to their dual childbearing and child-rearing roles. Mothers arriving in a new Western country are confronted with a need to adapt into a new societal norm and healthcare system. This "shift" frequently impedes their ability to breastfeed optimally due to the clashing belief systems. Often, mothers are judged and discriminated for the way they choose to "mother" their infants. Their cultural beliefs and perspectives of infant feeding, compounded by the stressful trail of resettlement, are unknown to authorities in Western nations due to their silent unassertive nature. This chapter uses the postmodern methodological framework to unravel the multiple truths that drive the perceptions and perspectives of infant feeding among Myanmarese and Vietnamese mothers from refugee backgrounds in Brisbane. Since the research trend of gaining visual access to the lives of mothers from refugee backgrounds is new, we also outline some tips and tricks that steered our initially rocky data collection journey. The chapter continues with ways in which women from refugee backgrounds conceptualize motherhood and infant feeding. Finally, we delineate the implications for practice and the usefulness of using drawing as research method for practitioners who work around this scope.

Keywords

Drawing methods · Refugee · Mothers · Breastfeeding · Infant feeding · Displacement

1 Introduction

Arts-based research has been gaining popularity in qualitative social and health research (see ▶ Chap. 64, "Creative Insight Method Through Arts-Based Research"). Bergum and Godkin (2008) contend that the application of art in research processes could cover a wide spectrum of areas namely where it can be used as an inspiration, method, intervention, data, or as a mode of dissemination too. This chapter focusses on the use of arts-based research with particular attention to drawing methodology as a means of gathering data to grasp the thematic essence of experiences (Van Manen 1997). An increasing number of social researchers (Broadbent et al. 2004; Cross et al. 2006; Fernandes et al. 2014; Gill and Liamputtong 2014; Guillemin 2004; Victora and Knauth 2001) have used drawing as a method of eliciting data and have found it to yield rich and meaningful information.

2 Situating Ourselves

This chapter is based on the work of June Joseph, the first author who is in her final year of a PhD that focuses on the experiences of infant feeding among refugee women. The study was supervised by Pranee Liamputtong and Wendy Brodribb.

With the suggestion and encouragement of Pranee Liamputtong, whom June asked to be her external supervisor, the exploration of drawings as research method/tool began three years ago. June's PhD research focusses on understanding the lived experiences of infant feeding and its connection with motherhood among women from Asian refugee backgrounds in Australia. The research incorporated mothers from two countries of origin namely Vietnam and Myanmar. These women have endured *journeys* of oppression, persecution, scarcity of resources, and disconnection from familiar support networks. June is not a refugee, neither is she a mother. Borrowing the concept of sociological imagination which echoes the connection between personal trouble and public concerns (Mills 2000), her interest in this research surfaced from an inner desire to understand how her late mother whom she conceptualize lived a "refugee" lifestyle managed her infant feeding days. In attempt to gain an extensive understanding of how women from refugee backgrounds theorize their infant feeding experiences, beyond the common notions available in published journals and medical textbooks, her research challenge was to dig deeper into other ways of "knowing" and engaging women of refugee backgrounds. As her late mother relayed her experiences in verbal and written forms, June desired an emergence of new concepts that visual art methodologies could entail – since the richness of subjective experiences that could emanate from drawing is currently a trend of discussion in the qualitative research paradigm (Knowles and Cole 2008). Her PhD research aims to engage mothers and to have them share their perceptions and experiences of motherhood and infant feeding visually through drawings as they narrate their stories. However, inasmuch as infant feeding is ingrained within generational belief systems, mothers communicated their entire childbirth experience – which explains why the perspectives of childbirth will also be covered in later sections.

Despite being inspired and enthusiastic about the richness of data that could be unraveled from the use of drawing as research methodology, almost all initial attempts we struggled to obtain artwork from the participants. After several "trial and error" attempts, the process of gaining insightful visual data became easier with time. Since the trend of forced migration is escalating, we believe that it would be priceless to publish the journey, tips and tricks of working with women from refugee backgrounds. We will commence the chapter with June's personal encounter with art.

3 The First Author's Personal Encounter with the Drawing Methodology

It is not uncommon for anyone, when subjected with a request to produce an artwork, to question their ability to draw and later deny it– with exception to children who willingly uptake the challenge. Somewhere along the growth curve, June had become conscious of judgments and the sense of unworthiness. Despite having been heavily immersed in various published reading materials, steered by the notion of *sociological imagination*, June knew she had to equip herself with the experience of being subjected to drawing in order to successfully implement it in a research endeavor. Prior to embarking on the fieldwork, she was privileged to attend a workshop conducted by the mother of the drawing methodology herself, Marilys Guillemin in April 2015.

As enthusiastic as the delegates were, interestingly, it was dramatically short-lived. They were filled with shivers as they took their seats as they stumbled upon a sheet of paper and some colored felt pens "generously" provided on each table. June vividly recalls whispering into the ears of the participant seated beside her: *"Why did we land ourselves here? She is going to ask us to dra*w!" Yes, how true!! At the beginning of the workshop, they were each asked to introduce themselves through an image. It was an undeniable struggle. Everyone initially tried to have a peep at each other's sheet. Ironically, everyone received a mutual peep in return, followed by a burst of laughter, *celebrating* the empty sheets.

The process of image construction and production took time. Marilys was experientially and professionally gentle – quietly performing "participant observation" from behind – acting "disengaged" but indeed very "engaged." Pondering deeper, June took up an orange felt pen and drew a calendar, highlighting the sixth month. Beneath the sheet, she drew the image of some flowers in red (Image 1).

Fascinatingly, what a "story"! Never in life had June deciphered "herself" in this manner. The sixth month reflects *her name* "June," flowers representing the month of June which in *her understanding* is associated with springtime. In terms of colors, orange, to her signifies realms of *nature*.....The very nature of her *conceptualization of self* and her "name" at baseline! The red flowers signified her fiery anticipation for her PhD confirmation viva then approaching. Wow, how rich a data asset! In a real sense the process of art production makes the *knowledge of self* and its *levels of development* "visible." This exercise illuminated the importance and richness of this humble methodology and how June should strive to best incorporate it into her research despite some prior disagreements that drawing would never fit the social-medical research arena.

Image 1 June's personal encounter with the art of "self"

4 The Postmodern Turn: The Use of Drawing Among Vulnerable Populations

Phenomenological research is frequently a way in which art is used in an attempt to grasp the essence of experience in a thematic way (Van Manen 1997). However, in deciding on the methodological framework, we were propelled to also incorporate postmodernism, along with phenomenology (to capture essences of lived experiences), in hope that it would enable us to dredge deeper into subjective realms that have yet to be explored. Postmodernism argues that realities are constructed within specific social and cultural contexts (Myanmarese and Vietnamese in our research), which enable the meaning of realities to be understood within a particular context (Liamputtong 2013). Within the envelope of postmodernism, all insights are treated as legitimate without preference over the other (Grbich 2004; Liamputtong 2006). Interestingly, postmodernism permits the emergence of multiple identities on the basis that historical, social, and cultural knowledge shaped within the confines of race, class, gender, and religion (Angrosino 2007; Liamputtong 2006; Borer and Fontana 2012). This explains the similarities and dissimilarities in the expressions of drawings among mothers from both countries and even within Myanmar's subethnic groups.

Access to knowledge is understood as the ability of one to provide "warranted assertions" – warranted with regard to the truth, while assertion being language (Stein 2000; Eisner 2008). Various ethnographic studies in the scope of infant feeding among immigrants have been conducted in the past to "listen to" the narratives of migrant and refugee women who have resettled in Australia (Liamputtong and Naksook 2001; Maharaj and Bandyopadhyay 2013; Gallegos et al. 2014). However, in the mid-twentieth century, it became increasingly evident that knowledge and understanding is always not reducible to language alone (Eisner 2008). This awakening persuaded the academic environment to be more responsive to new methods of investigation (Arnheim 1966). We contend that in this scope of study, the postmodern methodological framework has the capacity to steer the research in a new direction with the advent of arts-based inquiry (Knowles et al. 2008; Liamputtong and Rumbold 2008). Furthermore, an increasing number of arts-based research has successfully used postmodernism as its backbone (Fernandes et al. 2014; Liamputtong and Fernandes 2015; Liamputtong and Suwankhong 2015; Suwankhong and Liamputtong 2016; Benza and Liamputtong 2017).

5 The Drawing Method

This study was driven by the suggestion that gathering visual data from participants of vulnerable backgrounds could enable us to access the subtle, hard-to-put-into-words aspect of knowledge that might otherwise remain obscure or overlooked (Weber 2008). Additionally, artistic productions also give rise to a new symbolic visual twist to plain old texts, shattering our commonplace perceptions and beckoning us to think outside the theoretical box. Thus, the power of art helps to project the research findings across a wider audience in a stronger manner because "seeing is believing," both literally and figuratively. Drawings also powerfully help ratify someone else's gaze and viewpoint and allow us to be absorbed into their experience for a moment. In short, artistic images (drawings) creatively help us generate new insights, ways of understanding, and also promotes ethical awareness (Bergum et al. 2008). The process of drawing is not solely a rational or emotional response but simultaneously involves both the heart (through artistic expression) and mind (through theoretical and analytical considerations) to work together. This was evident in all the images obtained from this study as mothers made meaning (rationalized) their experience as they drew. The rationalization was not solely confined to responding to the drawing prompts but it initially helped mothers unleash stories of their life.

Despite being a powerful tool to capture valuable subjective experiences, drawing is best used in conjunction with other research methods such as interviews or focus group discussions (Guillemin 2004; Liamputtong et al. 2008, 2015). For a number of decades, researchers have successfully engaged children in studies involving "draw and talk" or "draw and write" techniques. In agreement that children struggle to articulate themselves through words, these techniques have contributed to rich data pertaining their perceptions, views, reflections, and phenomena (Angell et al. 2011; Liamputtong and Fernandes 2015). However, of late, scholars have successfully incorporated drawing as a research method among women and adults in the area of health and illness (see Broadbent et al. 2004; Guillemin 2004; Liamputtong and Suwankhing 2015; Suwankhong and Liamputtong 2016; Benza et al. 2017). Despite adults naturally having the greater ability to convey their narratives compared to children, the depths of data gained from adult participants through drawing of images has been surprising (Guillemin 2004; Guillemin and Westall 2008; Liamputtong and Suwankhong 2015; Suwankhong and Liamputtong 2016; Benza et al. 2017), thus, suggesting that images produced by adults could be a hallmark in gaining deeper access to the sub- and unconscious experiences that construct our worldview.

6 Drawing in the Context of Participants from Refugee Backgrounds

It is not uncommon that research involving the "stories of life" and experiences of individuals from refugee backgrounds are being expressed in sheer words (Liamputtong 2002; Niner et al. 2014). Here, we do not disagree that the aforementioned means of data collection is untrue, but instead contend that the data obtained by those forms could be limited due to their nature and that trauma, displacement, and oppression render participants unassertive. As suggested by Lenette and Boddy (2013), we too agree that prompting participants to produce artworks could seek nuanced perspectives of sensitive themes, allowing the emergence of richer sets of data rather than focusing on speech alone.

Before going further, the use of arts-based research methods, with attention to data in forms of drawings, is not new in the "refugee" research sphere. A study was conducted to conceptualize the understanding of health and illness among refugee children from Burmese ethnic backgrounds (Fernandes et al. 2014; Liamputtong et al. 2015), while Benza et al. (2017) used this method of data collection to understand the experiences of motherhood among Zimbabwean women in Melbourne. Driven by their success, we contend that the usage of this method to understand how women of refugee backgrounds construct their patterns of beliefs surrounding motherhood and infant feeding pre- and post-resettlement would lend a significant strength to the literature.

7 The Real-World Application of Drawing in Our Research

While the abovementioned researchers gained meaningful artwork from their participants, this journey of using drawing methodology among women of refugee backgrounds from Myanmar and Vietnam in our study was a struggle. Our initial months of attempting to obtain visual data from mothers was exceedingly bleak. Despite applying the techniques gathered from the workshop, having Pranee's prominent qualitative expertise and the privilege of working alongside two experienced bicultural workers, we were repeatedly faced with rejection when it came to artwork. We knew every individual had the capacity to draw but it became increasingly evident the main problem was June's ethnicity during immersion in the field work (She is Malaysian Indian by ethnicity). June's different ethnic identity made mothers feel reluctant to share their visual stories with her. At the beginning, June attempted to distance herself from the mothers during the phase of art production, either by walking out of the research "area" during the art exercise or suggesting that mothers draw during the interval between the first and second interview visits. However, Pranee suggested that this method would hamper June's ability to observe mothers through the process of art production, which is vital for analysis of drawings (Gill and Liamputtong 2014). Pranee, Wendy, both bicultural workers, and June, at several times, gathered to discuss a solution to this issue.

After about six months of fieldwork, we eventually learnt the "art" of requesting images in a culturally appropriate manner. This consolidates that increased engagement and exposure in fieldwork enhances trust between researcher, bicultural worker, and participants (Liamputtong 2006). Below we share some tips, tricks, challenges, and solutions gathered through the journey of fieldwork which could inspire researchers who share a similar interest in obtaining visual data from women of refugee backgrounds.

7.1 Participant: *"I Am Sorry, I Cannot Draw"*

The study initially focused on recruiting 30 mothers from Myanmar (Karen, Karenni and Chin ethnic groups). However, after the first nine interviews, despite follow-up interviews and consent to participate in artwork, we did not succeed in getting any visual data, and mothers were also unwilling to share their lived experiences verbally (transcribed verbatim were closed ended answers). This led us to take a step back to review/reframe the research approach should we desire to proceed on this research path. We began with an awareness that only a handful of researchers have thus far worked alongside women of refugee backgrounds from Myanmar (Niner et al. 2013). Additionally, we were also informed that access to these participants itself has been difficult, and it would be more so with more attempts to request visual data from them. We contend that women from refugee backgrounds have battled oppression and struggle with trust and confidence because of their contact with persecution. Also, limited access to resources pre-resettlement has led to the feelings of powerlessness, insufficiency, and incapacity. This cascade of experiences has contributed to an enormous propensity to suppress feelings and stories.

The initial plan was to conduct the drawing exercise at the end of the interview. It was only then that the art materials were handed to mothers. The reason to this approach was to encourage comfort with the research process and team first, and using drawing downstream as means of validating verbal data. Since the plan was unsuccessful, a few "trial-and-error" adjustments were made, such as: (1) Adding another cultural group (Vietnamese) just in case the research fails to reach its purpose. A balanced number of Myanmarese and Vietnamese mothers were recruited. Obtaining artwork from Vietnamese mothers was easier due to their artistic nature; however they too rejected drawing initially. (2) Adding another subethnic group from Myanmar (Kachin) who is known to be more outspoken and expressive. The incorporation of the Kachin ethnic group gave us the confidence to derive proper techniques when researching mothers from Myanmar. (3) Drawing materials would be visibly introduced as soon as the mother consented to artwork, in hope the participant will understand what the process entails from the start. In this manner, the invitation to draw would not shock the mother. (4) The drawing of perceptions and experiences will be encouraged from the start and mothers were encouraged to sketch when they felt like it. (5) Participants will also be prompted and encouraged to draw at the juncture when the team discerns one's struggle to express the idea verbally. The team became sensitive to cue words such as "I don't know how to explain," "It is hard to describe."

Here is an image and some verbatim quotations reflecting the pathways from which the research team formulated our conversation among ethnic Kachin mothers. The ethnic Kachin mothers were the first group of women from Myanmar who got us all inspired that gaining images from Myanmarese women was not a "mission impossible" (Image 2)

Image 2

June: Can for tell me an experience to having your baby at your breast for the first time?
Participant: It's beautiful.... I just don't know how to describe it?
June: Of wow... That must be incredible... Would you like to draw something about the feeling? It could be anything at all... There is no right or wrong ... You experience is so special to my research....
Participant: Hmmm..... It's just like this.. I am smiling as soon as a saw my baby and feeling her latch on my breast. I'm carrying her and she is attaching to me.... This is our first time being together.... I used the brown color because our skin was touching and I am Kachin and my skin is brown...

Image 2 Experiences of seeing and breastfeeding their babies for the first time after delivery. This is an experience that mothers struggled to narrate. Prompting an explanation of the indescribable engendered powerful images, concepts of this image are founded in their religiosity

7.2 Research Team: "That Is Okay...It Must Be Hard for You... Let's Move On.... Now This an Interesting and Brilliant Concept.... Would You Like to Sketch It for Me."

After obtaining rich, visual data from the Kachin mothers, the now blooming research technique was carried to mothers from other ethnic groups. Along the process of data collection, it was understood that empathizing with the mother's rejection to draw was important as it established participant-researcher rapport, trust, and comfort. It was observed that the quality and richness of data increased as the research team gently affirmed and encouraged the mothers. Their past experiences of constant discouragement led to their struggle with confidence and self-esteem. Thus, *empathy* and *encouragement* was the *key* to unleashing their artistic ability. This approach finally yielded visual data from Karen and Karenni mothers. They could share their visual perception of experiences, while waiting patiently for their construction of thoughts. Imaged 4 and 5 were produced by a Karen and Karenni mother. Through observation, the sketches from Karen and Karenni mothers are simpler compared to those of the Kachin mothers whose images were more colorful. Karen and Karenni mothers choose not to use the provided felt pens and crayons but instead asked for the pen that June

Image 3 This image illustrates the bonding during breastfeeding. It was initially difficult to get the mothers from both subcultures to share their stories. However, this rich piece was attained through continued engagement

was using while jotting her fieldnotes. This portrayed deep humility and simplicity. The sketches by Karen and Karenni mothers were delicate (Image 3).

Image 3
Participant: Breastfeeding makes my baby close to me...   **June:** Thank you. Would you like to tell me more?   **Participant :** No   **June:** Would you like to draw that sense of closeness for me?   **Participant:** I cannot.... But I will try...   **Participant:** (Here this mother seems unsure how to hold the pen but I kept affirming her that she is doing well as she drew this beautiful stick-human image)... During breastfeeding I <u>always hold my baby</u>.

7.3 Participant: *"This really unleashes my past... I have never seen myself capable"*

Here, we highlight drawing as a means of empowerment especially when used with women from vulnerable backgrounds. The iterative process of emphasizing, encouraging, and affirming gave mothers the power to convert what they deemed as an incapacity into a form of power (through their art work). During fieldwork, June observed empowerment among ethnic Chin and Vietnamese women. The Chin and Vietnamese refugees arrived in Australia without their extended family members, following a deeply traumatic resettlement journey. In countering this, the research team persistently reinforced with mothers that their stories were very important and could be a means of encouraging mothers in the near future.

Here, we argue that the process of visual image production unraveled some traumatic aspects which they had never discussed in the past. The process of artwork production, hence, had a therapeutic effect in their lives. While the struggle in artwork production among ethnic Chin mothers was nested around oppression, poverty, and discouragement, the Vietnamese relayed their struggle with artwork as a status quo issue. In contemporary Vietnam, art was only encouraged in primary school as it was seen as a task of children or the "immature." Culturally, it is also perceived that the job as an artist is undertaken by the outcast and least privileged in society. The conversations in the charts below are aspects of communications that took place between the research team and participants during the process of engaging and building trustful relationships. The verbatim reflects questions pertaining motherhood and the meaning of infant feeding in their cultural perspectives.

Conversation A (Chin)	Conversation B (Vietnamese)
Participant: "In our Chin culture…the child is a mirror image of the mother." **June:** Would you like to draw this thought on motherhood? **Participant:** Yes I could since I already started talking on it. It makes it easier…. Well, I rejected all your initial drawing prompts.. I am so sorry.. I struggle to draw… When I was little I loved drawing … but my mother told me I'm wasting paper and wasting money by drawing… So I stopped… This is the next time (after 22 years) that I am drawing again" "Looking back .. I still can draw… This really unleashes my past… I have never seen myself more capable"	**Participant:** I needed to undergo some radiation treatment… it will harm the baby. I felt like I only had one choice which was to opt for the formula milk. Just like the fish… it only has one choice, it can only be in the water to continue living…" " Now that I have drawn this picture I wonder.. I have never expressed this feeling to anyone before….Thank you for asking me to draw… I feel a big burden released"

In short, women from refugee backgrounds struggled to produce visual data initially. However, after some interventions, we succeeded in gaining valuable insights from their life stories. First and foremost, it was fundamental, that we were sensitive to their resettlement issues as subjectively as possible – by "going into" their situation. This enabled the participant to establish a sense of security and trust with the research team. This subtheme while highlighting the multiple challenges that refugee women undergo, also demonstrates ways in which we can enhance maternal sense of agency over the stresses of daily living.

8 The Refugee Perspectives of Infant Feeding

The visual data (in the form of drawings) described the participants' lived experiences in multiple dimensions. Mothering and infant feeding in their cultural understanding begins with pregnancy where the mother is watchful of her food and bodily conducts. This extends through parturition in order to counter the perceived thermodynamic disequilibrium due to the loss of blood and energy from the childbirth process. Traditionally, the new mother is surrounded and supported by female family members who provide her with moral, verbal, and physical support in terms of house-keeping and preparation of culturally prescribed meals. However, mothers experience a "disconnect" in Australia where they are not only separated from their primary support networks but also faced with the need to navigate the Western healthcare system and new societal culture. While breastfeeding was the norm of infant feeding among Myanmarese mothers, the cultural "shift" caused by resettlement led some mothers to opt for formula feeding. Knowledge on the goodness of breast milk and rooted spiritual beliefs encouraged breastfeeding in Australia. Conversely, formula feeding is the current trend of infant feeding in contemporary Vietnam. Despite being from the less privileged population groups, most Vietnamese mothers in this study portrayed some degree of mimicking the perceived modern-world trends, while those who breastfed their babies were driven by determination and sacrificial love that was rooted in their conceptualization of God.

From our observation, there was no vast difference between the type of images produced by mothers from Vietnam and Myanmar, despite mothers from Vietnam being more corporeal about their drawings expressions – focusing on bodily notions. Chin mothers struggled to engage with drawings, while the Kachin mothers drew confidently when prompted. The Karen and Karenni mothers tended to draw fine-stick sketches, gaining confidence through successive interviews.

9 What Women Use Drawing For

Through the journey of data collection, a vast spectrum of artwork that dictates many stories and notions was obtained. Visual data unravels a deeper understanding of subjective experiences. Participants in this study used drawing to express several themes:

9.1 To Illustrate a Location

Mothers from the Karen culture traditionally bury the placenta of their newborn immediately after delivery. Mothers in this study did not have the opportunity to continue this cultural practice due to lack of cultural information within the Australian healthcare system. This artwork helped mothers unleash the importance of cultural beliefs for their child's well-being. While mothers were able to list the locations for burial, they were unaware of the meanings these burial sites held. The process of meaning-making that takes place during the process of art production

Image 4 Location (*dotted*) is where the placenta is buried. The color (*pink*) used in this sketch aided in meaning-making of the burial site: for the child will be loved due to the frequency of people passing by the busy junction

enabled mothers to point out the accurate site of the burial and later explain the meaning behind the cultural location (Image 4).

9.2 To Illustrate the Concepts of Motherhood

Mothers, including those in this study, conceptualized themselves as carriers of multiple identities. They juggle roles as nurturers, protectors, and role models. They too feel that by nature they contribute to the well-being of their children and battle struggles and sacrifices. The use of drawing as a research method unleashed some notions of motherhood that may have not been possible via conventional research methods.

9.2.1 Shelter

Mothers from refugee backgrounds used the image concepts of hands and trees to illustrate their worldview as nurturers. In terms of shelter to their children, mothers also used the concept of trees to illustrate this theme. In the Myanmarese culture, trees hold a significance of strength. An ethnic Karen mother mentioned that only a "strong tree could shelter people from the sun." She attributed this to the torrid summer heat in Brisbane, a season that her children associated with picnics. She adds: "Everytime we go to the park, the first thing we do is to find a tree that could give us shade." Here, she conceptualizes that a good mother is one who could be a good shelter and refuge for her children (Image 5).

9.2.2 Mirror

An ethnic Chin mother conceptualized motherhood as a mirror (Image 6). She believes that her characteristic traits will be mirrored in her child. She adds that this cultural belief encourages her to watch her conduct vigilantly

Image 5 Motherhood as a good tree that could shelter from heat

Image 6 The weakness after childbirth, like a snake that is shedding its skin

9.2.3 Shadow

Mothers also perceive breastfeeding as a hidden act. While discussing this, a Vietnamese mother explained as she drew: "Mother's role is so important in the child's development, however it's…often undervalued." This feeling reminded her of the traditional Vietnamese wedding family photograph where their mothers look mundane and stand at a distance.

9.3 To Illustrate Vulnerability

The initial postpartum period in traditional cultures is perceived to be a phase of vulnerability for the mother. While this vulnerability is frequently described in literature (Liamputtong 2000; Chu 2005), it was interesting to understand how women pictured

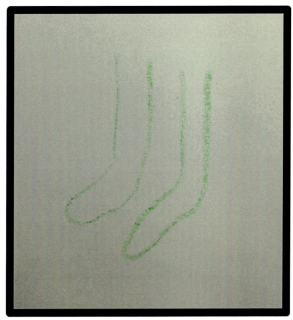

Image 7 Socks to keep the new mother's body warm after childbirth

this state. Vietnamese women in Liamputtong (2000) call it a period similar to that of "crab shedding its shell." Likewise, a Vietnamese mother delved into her memory of seeing snakes changing their skin during her childhood years. She exclaimed: "The period after childbirth is that is like a snake changing its skin We are so weak... We need to hide in a safe and dark place to be protected from danger" (Image 6).

9.4 To Illustrate How Mothers Counter the State of Vulnerability

During the period of vulnerability, mothers abide by some culturally prescribed rituals that perpetuate the preservation of bodily heat to counter the state of vulnerability. While Vietnamese mother perform a ritual known as "mother-roasting" to regain their health for the motherhood tasks ahead (Rossiter 1992; Liamputtong 2000; Groleau et al. 2006), Myanmarese mothers keep their bodies warm by sitting by a fire. A Karen mother drew an image of a pair of socks to illustrate how she kept her body warm in Australia. She wore thick socks instead as homes in Australia are structurally different and not suited for fire-based rituals, which is vital during the confinement period (Image 7).

9.5 To Illustrate the Freedom in Australia

Mothers from Myanmar and Vietnam underwent moments of persecution and oppression prior to arrival in Australia. Upon arrival, they expressed the freedom of being a

Image 8 Experience of freedom, peace and tranquility in Australia

big country of valleys and terrains as birds flying in the sky. They expressed that they could hardly bask in such beauty during the distressful journey of fleeing their homelands. These hardships helped mothers to build resilience and endurance which was vital for adaptation into a new dimension of motherhood in Australia. Image 8 portrays an expression of their freedom and joy of living in Australia (Image 8).

9.6 To Illustrate the Goodness of Breast Milk

Breast milk has been long known to be the best food for infants. Mothers from refugee backgrounds too agree with this. While breastfeeding is recommended for the numerous health benefits it bestows upon mother and baby, it also provides the baby with a unique species-specific nutritional and immunological treasures – the mother's bodily goodness and antibodies. An ethnic Karen mother drew this image when illustrating her perception of "breast milk is best." She highlighted that the nutritional and immunological properties from her body helps her baby, which she describes as a flower "blooming" in radiance, health, and glory. She believes that breast milk is the continuance of nurturing that begins at pregnancy. In her perspective, this image is a representation of her Christian perspective of "bloom where you are planted," which expresses that a child is a gift to the family in the Christian perspective (Image 9).

10 Conclusion and Future Directions

Through the valuable images gathered, we could conclude that visual data in the form of drawings from mothers of refugee backgrounds, driven by postmodernism, will be a powerful means of driving healthcare practice and policy development to

Image 9 The radiantly blooming flower, a representation of "breast milk is best"

the next level as it unravels perspectives beyond cultural understandings. Despite this research method facing rejection as a "mission impossible" initially, its success here is a proof that women from refugee backgrounds could engage in visual arts-based research. It is hoped that the tips and tricks listed above could be a means of encouraging more researchers to invest in studies that involve women from refugee backgrounds. The messages projected from visual data are powerful as it helped participants to contemplate and delve deeper into level of consciousness that are difficult to express by words. The use of colors related to their feelings also unmasked perspectives that words could not describe. We found the postmodern research approach appropriate as it seeks multiple truths. It was interesting to see various principles emerge as participants were given the freedom to express their worldview in whichever manner suits them best.

The rise in political agendas worldwide continues to displace a many vulnerable people from third world nations daily. This has a significant impact on women in particular, as they not only battle losses, but also shoulder the need to shelter their innocent charges. While from the public health perspective, we are unable to control their displacement, we argue through these findings that drawing could be a powerful means of gaining access to their subjective and hidden experiences of trauma for intervention purposes. Additionally, drawing could also be a means of soothing disquieting feelings of their journey, as art could be a meaningful therapy for people who have never had a chance to be heard.

References

Angell C, Alexander J, Hunt JA. How are babies fed? A pilot study exploring primary school children's perceptions of infant feeding. Birth. 2011;38(4):346–53.

Angrosino, M. (2007). Doing ethnographic and observational research. Retrieved from http://methods.sagepub.com/book/doing-ethnographic-and-observational-research

Arnheim R. Toward a psychology of art: Collected essays. Berkeley: University of California Press; 1966.
Benza S, Liamputtong P. Becoming an 'amai': Meanings and experiences of motherhood amongst Zimbabwean women living in Melbourne, Australia. Midwifery. 2017;45:72–8.
Bergum V, Godkin D. Nursing research and the transformative value of art. In: Knowles GJ, Cole AL, editors. Handbook of the arts in qualitative research: Perspectives, methodologies, examples and issues. Thousand Oaks: Sage; 2008. p. 603–12.
Borer MI, Fontana A. Postmodern trends: Expanding the horizons of interviewing practices and epistemologies. 2nd ed. Thousand Oaks: Sage; 2012.
Broadbent E, Petrie KJ, Ellis CJ, Ying J, Gamble G. A picture of health – myocardial infarction patients' drawings of their hearts and subsequent disability: A longitudinal study. Journal of Psychosomatic Research. 2004;57(6):583–7.
Chu CM. Postnatal experience and health needs of Chinese migrant women in Brisbane, Australia. Ethnicity & health. 2005;10(1):33–56.
Cross K, Kabel A, Lysack C. Images of self and spinal cord injury: Exploring drawing as a visual method in disability research. Visual Studies. 2006;21(2):183–93.
Eisner E. Art and knowledge. In: Knowles JG, Cole AL, editors. Handbook of the arts in qualitative research. Thousand Oaks: Sage; 2008. p. 3–12.
Fernandes S, Liamputtong P, Wollersheim D. What makes people sick? Burmese refugee children's perceptions of health and illness. Health Promotion International. 2014;30(1):151–61.
Gallegos D, Vicca N, Streiner S. Breastfeeding beliefs and practices of African women living in Brisbane and Perth, Australia. Maternal and Child Nutrition. 2014;11(4):727–36.
Gill J, Liamputtong P. The drawing method: Researching young people with a disability. Sage Cases in Methodology (SCiM) (an online resource for students and researchers). London: Sage; 2014.
Grbich C. New approaches in social research. London: Sage; 2004.
Groleau D, Soulière M, Kirmayer LJ. Breastfeeding and the cultural configuration of social space among Vietnamese immigrant woman. Health & Place. 2006;12(4):516–26.
Guillemin M. Embodying heart disease through drawings. Health. 2004;8(2):223–39.
Guillemin M, Westall C. Gaining insight into women's knowing of postnatal depression using drawings. In: Liamputtong P, Rumbold J, editors. Knowing differently: Arts-based and collaborative research methods. New York: Nova Science Publishers; 2008. p. 121–39.
Knowles JG, Cole AL. Handbook of the arts in qualitative research: Perspectives, methodologies, examples, and issues. Thousand Oaks: Sage; 2008.
Lenette C, Boddy J. Visual ethnography and refugee women: Nuanced understandings of lived experiences. Qualitative Research Journal. 2013;13(1):72–89.
Liamputtong P. Rooming-in and cultural practices: Choice or constraint? Journal of reproductive and infant psychology. 2000;18(1):21–32.
Liamputtong P. Infant feeding practices: The case of Hmong women in Australia. Health Care for Women International. 2002;23(1):33–48.
Liamputtong P. Researching the vulnerable: A guide to sensitive research methods. London: Sage; 2006.
Liamputtong P. Qualitative research methods. 4th ed. Melbourne: Oxford University Press; 2013.
Liamputtong P, Fernandes S. What makes people sick?: The drawing method and children's conceptualisation of health and illness. Australasian Journal of Early Childhood. 2015;40(1):23–32.
Liamputtong P, Naksook C. Breast-feeding practices among Thai women in Australia. Midwifery. 2001;17(1):11–23.
Liamputtong P, Rumbold J, editors. Knowing differently: Arts-based and collaborative research methods. New York: Nova Science Publishers; 2008.
Liamputtong P, Suwankhong D. Therapeutic landscapes and living with breast cancer: The lived experiences of Thai women. Social Science & Medicine. 2015;128:263–71.
Maharaj N, Bandyopadhyay M. Breastfeeding practices of ethnic Indian immigrant women in Melbourne, Australia. International Breastfeeding Journal. 2013;8(1):1.

Mills CW. The sociological imagination. Oxford: Oxford University Press; 2000.

Niner S, Kokanovic R, Cuthbert D. Displaced mothers: Birth and resettlement, gratitude and complaint. Medical Anthropology. 2013;32(6):535–51.

Niner S, Kokanovic R, Cuthbert D, Cho V. "Here nobody holds your heart": Metaphoric and embodied emotions of birth and displacement among Karen women in Australia. Medical Anthropology Quarterly. 2014;28(3):362–80.

Rossiter JC. Maternal-infant health beliefs and infant feeding practices: the perception and experience of immigrant Vietnamese women in Sydney. Contemporary Nurse. 1992;1(2):75–82.

Stein E. Knowledge and faith: The collected works of Edith Stein, vol. 8. Washington, DC: ICS Publications; 2000.

Suwankhong D, Liamputtong P. Breast cancer treatment: Experiences of changes and social stigma among Thai women in southern Thailand. Cancer Nursing. 2016;39(3):213–20.

Van Manen M. Researching lived experience: Human science for an action sensitive pedagogy. London, ON: Althouse Press; 1997.

Victora CG, Knauth DR. Images of the body and the reproductive system among men and women living in shantytowns in Porto Alegre, Brazil. Reproductive Health Matters. 2001;9(18):22–33.

Weber S. Visual images in Research. In: Knowles JG, Cole AL, editors. Handbook of the arts in qualitative research. Thousand Oaks: Sage; 2008. p. 41–55.

Understanding Refugee Children's Perceptions of Their Well-Being in Australia Using Computer-Assisted Interviews

102

Jeanette A. Lawrence, Ida Kaplan, and Agnes E. Dodds

Contents

1 Introduction	1778
2 Theoretical Grounding	1779
2.1 Grounded in Respect for Refugee Children	1780
2.2 Grounded in Refugee Children's Expressions of Their Well-Being	1781
3 Features of CAIs	1781
3.1 CAIs Affirm Children's Agency and Control	1781
3.2 CAIs Present Accessible and Attractive Interfaces	1782
3.3 CAIs Yield Data Suitable for Quantitative and Qualitative Analyses	1782
4 Two CAIs for Researching Aspects of Refugee Children's Well-Being	1783
4.1 People and Places in Your Life	1783
4.2 Appropriateness and Relevance of the People CAI	1784
4.3 Living in Australia	1787
4.4 Appropriateness and Relevance of *Living* CAI	1787
5 Conclusion and Future Directions	1789
References	1792

Abstract

Children from refugee backgrounds have the right and the ability to contribute to research knowledge. But they need researchers to develop methods that enact

---

J. A. Lawrence (✉)
Melbourne School of Psychological Science, The University of Melbourne, Melbourne, VIC, Australia
e-mail: lawrence@unimelb.edu.au

I. Kaplan
The Victorian Foundation for Survivors of Torture Inc, Brunswick, VIC, Australia
e-mail: kaplani@foundationhouse.org.au

A. E. Dodds
Melbourne Medical School, The University of Melbourne, Melbourne, VIC, Australia
e-mail: agnesed@unimelb.edu.au

© Springer Nature Singapore Pte Ltd. 2019
P. Liamputtong (ed.), *Handbook of Research Methods in Health Social Sciences*,
https://doi.org/10.1007/978-981-10-5251-4_45

respect and are theoretically appropriate. This chapter describes a methodological approach to understanding the well-being of children from refugee backgrounds from their own perspective. Two computer-assisted interviews (CAIs) were developed as research tools that enact respect for refugee children by facilitating young refugee participants' agency and engagement, using accessible interfaces with child-friendly, age and culture appropriate tasks, and instructions that enable children to express their views with confidence and comfort. *People and places in your life* and *Living in Australia* invite children to work on and evaluate quantitative and qualitative tasks and questions. Two illustrative studies grounded in respect show how the data constructed in these CAIs are suitable for analyzes of themes and trends, standout features, and personal meanings as the basis of group comparisons and textual analysis of individual profiles. The usefulness of the methodology is discussed in relation to the need to understand the perspectives of refugee children and other children about their well-being.

Keywords

Refugee children · Respect · Theoretical appropriateness · Computer-assisted interviews · Thick description · Textual analysis · Individualized profiles

1 Introduction

Children constitute over half the 60 million people who are currently leaving their homes to seek refuge from war, violence, or persecution. Many children resettle in other countries, either with their families or guardians, or as unaccompanied minors. Their plight causes great concern among governments, practitioners, and researchers about how well these refugee children are able to adjust and thrive. International, national, and local authorities and agencies need positive and negative indicators as the evidence base for fostering the well-being of refugee children (Lippman et al. 2009; Measham et al. 2014). These bodies are not well informed, however, about children's views of what supports and inhibits children's well-being in resettlement. Consequently, Lippman et al. (2009), in reporting to UNICEF, argued that any measures designed to uncover disadvantaged children's well-being and "well-becoming" needs information from the children themselves if it is to "identify the factors that make them happy, motivated, and successful as children" (p. 626) (see also ▶ Chaps. 115, "Researching with Children," ▶ 99, "Visual Methods in Research with Migrant and Refugee Children and Young People," ▶ 100, "Participatory and Visual Research with Roma Youth," and ▶ 107, "Conducting Ethical Research with People from Asylum Seeker and Refugee Backgrounds").

This chapter discusses the development of a methodological approach to understanding the well-being of children from refugee backgrounds that appropriately centers on children's perspectives. The method is the computer-assisted interview grounded in terms of ethical appropriateness and theoretical relevance.

Computer-assisted interviews (CAIs) have been used by other researchers with vulnerable children when research*er* sensitivity and research*ee* confidence and enjoyment are important, for example, for children with autism spectrum disorders (Barrow and Hannah 2012), for children anxious or concerned about clinical interviews (Dolezal et al. 2012; Bokström et al. 2015), and for children reporting their sexual behaviors (Connolly 2005). The accessibility of digital programs gives young research participants the kind of control and confidence not available in paper and pencil or face-to-face methods. They give the young participant autonomy in responding to attractive, reasonable, and safe means of expressing their views (e.g., de Leeuw et al. 2003; Barrow and Hannah 2012).

The chapter is organized in five sections. The theoretical grounding of computer-assisted interviews (CAIs) is followed by a description of how CAIs translate respect into action and how they enable refugee children to express their sense of well-being. A fourth section presents a pair of CAIs with illustrative studies of their usefulness. The final section discusses the suitability of using this CAI approach with refugee and other vulnerable children and the implications for cross-cultural research.

2 Theoretical Grounding

Methodological development properly begins with theoretical questions that frame the appropriateness and usefulness of the measures used. Toomela (2011) makes a critical analysis of how contemporary psychologists employ quantitative and qualitative methods in ways that violate the principle of "questions come before methods" (p. 47). In the physical sciences, Toomela argues, methods are typically used to address theoretically generated questions – by being selected from available methods on their suitability for those questions or being specifically created to address new questions. In psychology (and other social sciences), it is not unusual for the choice of methods to precede the formulation of research questions, and for measures then to dictate analyses, and in the process, to severely limit contributions to knowledge. A prime example can be found in evaluations of social programs. When method selection dominates, questions about the value of a program may be sacrificed when peripheral details are reported instead of the effects of program delivery. Researchers may fall back on reporting, for example, the number of people who attended a program and how often, because the methodology cannot address more theoretically telling issues, such as the immediate or lasting effects of the intervention (Patton 2011). Cairns and Dawes (1996) make a similar observation about the tendency to count instances rather than effects of violence: "We can count exposure to violence items but this does not help us gauge their differential impact" (p. 136).

The tendency to method-driven research can be particularly dangerous in cross-cultural and culturally oriented research. Method-by-group-mismatches occur when methods are "parachuted" from one culture (usually Western) to another, despite the inappropriateness of asking some cultural groups to respond to essentially Western tasks and questions with little meaning for them (Goodnow 2014). Callaghan et al. (2011), for instance, criticize "researchers who parachute their procedures from Western labs into cultures where even asking a question one knows the answer to is considered odd" (p. 112). Some methods for asking seemingly universal questions do not transpose well from one culture to another. Community workers, for example, report that asking young people from Middle Eastern Cultures about their self-esteem is to ask them about something that has no reference point and no linguistic meaning for them. Ignoring cultural differences has promulgated deficit models of indigenous cultural groups, despite successful integrations of their traditional approaches to healing with biomedical care (Hirch 2011). In cross-cultural research, methods need to be appropriately grounded in respect and the application of theory to particular groups. Accordingly, the CAI methodology is grounded in two concerns: the ethical concern to demonstrate respect for refugee children whose perspectives are under-represented in cross-cultural research and the theoretical concern to enable these children to express their views on their well-being.

2.1 Grounded in Respect for Refugee Children

Respect is an ethical principle that defines people's rights, including the rights of young research participants in the UN Convention on the Rights of the Child (United Nations 1989; Beazley et al. 2009). It is a guiding rule for applying rights directly to specific human interactions. Rights do not become practical realities automatically because they are rights. Rights must be enacted, and in research, enactment comes from the decisions and choices that researchers make specifically under the direction of the respect principle. Researchers have a particular obligation, for instance, to *enable* the self-expression that is the right of refugee children (United Nations 1989; Lawrence et al. 2015).

Coupled with a strong focus on person-oriented data, respect directs CAI research to treat every research participant as an individual with unique as well as common experiences to contribute to knowledge (von Eye 2010; Bergman and Wangby 2014). This avoids reducing individuals' contributions to quantitative analyses of group trends where the individuals' responses are treated as replaceable slot-fillers (von Eye and Bogat 2006). It also avoids handling qualitative data as entries in banks of quotes from which researchers may selectively "cherry-pick" material supporting their own positions (Morse 2010; Toomela 2011; Mazzei and Jackson 2012).

The patterns and themes in people's data may only be revealed with careful analysis (Steinberg 1995) that is dependent on appropriate representation of what people said or did in context so that analyses yield thick, contextualized descriptions (Ponterotto 2006). Participants' meanings emerge in the themes they reiterate, links they make between questions and sections of data, and standout comments by which they express both typical and atypical ideas as individuals or as members of a certain group (von Eye 2010).

The CAI method is well suited to investigating children's views of well-being, especially in light of Ben-Arieh's (2005) complaint that children are not usually consulted about their evaluative responses to their lives.

2.2 Grounded in Refugee Children's Expressions of Their Well-Being

In general, well-being is used as an umbrella concept that reflects children's thoughts and feelings about different aspects of their lives and covers the numerous everyday experiences that add to their more and less good impressions of how life is for them (Ben-Arieh et al. 2014). Variously defined and measured, the well-being concept has elastic boundaries and referents of present and future contentment, and it is always contextualized and associated with routines and expectations (Weisner 2014). Children's senses of well-being and belonging can be affected positively or negatively by their contextual habitats and circumstances as well as by events and losses, intrusive thoughts and personal characteristics (Ben-Arieh et al. 2014).

Two custom-built CAIs, *People and places in your life* and *Living in Australia*, were developed to enact respect for refugee children and their views about a range of positive and negative aspects of their well-being in Australia. Procedures were designed to give due recognition to their vulnerability to adult power (Lawrence et al. 2015). Tasks and questions were based on indicators of their well-being that emerged in practice at Victorian Foundation for the Survivors of Torture (Kaplan 2013), in consultations with professionals from multiple disciplines working with refugee communities through the Victorian Foundation (VFST) and from interviews of VFST clients and counselors (McFarlane et al. 2011).

3 Features of CAIs

3.1 CAIs Affirm Children's Agency and Control

Organizing research environments to affirm children's agency and control is a direct application of respect for vulnerable refugee children and a practical recognition of their rights. At recruitment, each child is asked explicitly for his or her informed consent to participate, irrespective of prior parental consent. The accessibility of digital programs gives young research participants the kind of control and confidence not available in paper and pencil questionnaires or face-to-face interviews (e.g., de Leeuw et al. 2003; Barrow and Hannah 2012). Young participants are invited to interact with the computer program as the primary focus of the interview session. The researcher is presented as someone who sits beside the child to assist, but does not take charge. Most children prefer to work on the keyboard themselves and to be helped with words and spelling. This change in the usual center of control goes some way towards dispelling the reluctance to disclose information that other researchers have found with young refugee samples, particularly among unaccompanied minors (Kohli 2006; Chase 2013; Goodnow 2014). Confidence is critical, but only when it is genuinely related to agency.

In the data construction phase, the CAIs provide the children with a facility to personalize their expressions; create and manipulate personally generated diagrams of the people who help them and review and if they wish, revise responses (e.g., changing entries between "helps me feel better in myself" to "doesn't help me feel better in myself" lists). An invitation to evaluate their research experiences is an integral part of agency grounded in respect. This innovation is easily presented in the digital environment.

Digitization makes it possible for the exchange to follow a participant's line of reasoning using branching sequences and asking for explanations, with prompts. Pathways and specifics emerge out of a participant's patterns of response in real time. For example, in *Living in Australia*, participants rate their level of worrying by choosing a color-coded and labeled button on an ascending scale. The next screen then returns their chosen button and asks for an explanation as a sentence completion: "I worry a little bit/more than a little bit because...."

Standard off-the-shelf measures do not readily encourage the emergence of novel insights that young participants can generate when typing in a sequence of exchanges. One 8-year-old Iraqi girl, for example, made a surprising choice of computers as what helped her the most. She explained this uncharacteristic choice. Her brother whom she adored was still in Iraq. "Because I have facebook. I can write to my brother, say nice things to him, chat."

3.2 CAIs Present Accessible and Attractive Interfaces

All story agents are illustrated as universal human images, colored blue to avoid racial discrimination or stereotyping. Instructions are brief and presented in uncluttered working spaces supported by pop-up menus and simple navigation buttons and arrows. Font size is large and clear. The language is simple English, following the advice of cross-cultural counselors. For a new intake of refugees from Syria and Iraq, however, a new CAI, *My life in Australia* offers the choice of a simple Arabic or simple English version. Practice tasks and animated demonstrations introduce children to making rating and choice responses. Scenarios are illustrated and animated, with colored headings specifying whether a current question is about the story agent's family or the participant's. The scenarios develop in ways that are meant to convey to children that thoughts and feelings about refugee experiences are recognized as legitimate, giving serious attention, for example, to the story agent's worries about family members.

3.3 CAIs Yield Data Suitable for Quantitative and Qualitative Analyses

Tasks and questions yield a mix of quantitative and qualitative, standard and innovative activities giving children multiple ways of expressing their thoughts and feelings about specific aspects of their well-being. All data can be downloaded as complete sequences of tasks and participants' responses and revisions, for contextualized representation as individualized profiles. Examples of individualized profiles are shown in Lawrence et al. (2013).

4 Two CAIs for Researching Aspects of Refugee Children's Well-Being

Two CAIs ask children to comment on different aspects of their well-being in Australia in a research environment constructed with appropriate attention to respect and to children's expressed understanding of particular aspects of their well-being.

4.1 People and Places in Your Life

The *People* CAI uses accessible and well-explained choices, sorts, ratings, and open-ended comments to ask young participants to express their thoughts and feelings about: (i) significant places in their lives, (ii) being nurtured, (iii) what helps them feel better, and (iv) the research experience.

(i) Children express their positive feelings about places by sorting named tags for home, school, and suburb into pairs of labeled baskets to describe their feelings about each place. The baskets are presented as nine randomly ordered pairs, each labeled with a positive feeling or its negation (e.g., loved or not loved, scared or not scared) building up a description of their positive feelings. They then sort the same places into "important" or "not important" baskets. Positive feelings scores for each place are constructed by reversing four negative items and range from 0 to 9. There was reasonable consistency for a sample of 49 children from refugee backgrounds, with alpha levels around 0.70 (Dodds et al. 2016).

At the end of this sorting activity, children are reshown their "important" and "not important" choices for the three places and how they can change their choices if they wish. They then are asked to type in what makes a place important for them. They are reshown the feelings they specified for home, and how they can click on any feeling to change it to its opposite (e.g., changing "lonely" to "not lonely"). These revised choices are used in analyzes.

(ii) Experiences of being nurtured at home are expressed as children respond to the stem question, "At home how much do you think you are..." for each of nine randomly presented experiences with thumbnail illustrations: "guided, loved, looked after, listened to, protected, misunderstood, treated unfairly, hurt, upset." Children are first introduced to rating these experiences with an animated demonstration of four glasses being filled to different levels with associated progressive labels (0, "not at all," 1, "a little bit," 2, "more than a little bit," 3, "a lot"). Ratings, with four reversals, form a "being nurtured" scale, with typically good internal consistency for nine ratings ($\alpha = 0.80$ for 49 children) (Dodds et al. 2016).

(iii) Discriminations are made among things that do and do not help by clicking on icons for each of 13 "helps" to send them to one of two illustrated lists: "helps me feel better in myself" or "doesn't help me feel better in myself." The

13 helps include social interactions (e.g., spending time with family/friends), personal achievements (e.g., people saying you did well, being proud, feeling understood), and activities (e.g., playing sport, listening to music, time alone). Children are given opportunity to change on reflection their entries to the illustrated lists and then to choose which of the 13 things help most and explain why with an open-ended comment.

(iv) Children evaluate the CAI experience by rating it with 0–5 stars on its easiness, fun, understandability, and suitability for their age and whether or not they would recommend to a friend and why.

4.2 Appropriateness and Relevance of the People CAI

The CAI is useful as a research tool to the extent that it appropriately enacts respect for refugee children by enabling them to express themselves with confidence and ease, and to the extent that it provides evidence of their personal perspectives on their well-being.

Evaluations of *People* made by 49 children from refugee backgrounds illustrate their comfort with saying what they thought about it with their personal reasons (Dodds et al. 2016). These positive ratings were consistent with those made by 90 refugee, immigrant, and local children (Lawrence et al. 2013).

The 49 children generally gave *People* mean ratings above 4 (4.31–4.63, with standard deviations lower that 0.8), assigning it 0–5 stars on each of four bases: ease of using it, fun, understandability, and suitability for people their age. The mean rating on a "happy with the program" scale was 3.65 (SD, 0.56) on 0–4 scale. Comments about whether they would recommend to a friend or not revealed the children's engagement and their concerns. Table 1 shows the patterns of recommendations with coded categories of comments and examples typed-in by 32 children (65%) who said "Yes, I would tell a friend to do this program"; 13 (27%) who were "Not sure"; and the 3 only (6%) who said "No, I would not tell a friend to do this program."

The comments reveal children's spontaneity and insight into their own concerns and those of their friends, and a level of sophistication in relating to the *People* activities, not only as "fun," but for some, as something related to their own psychology, for example, "it takes away the bad stuff," "it's like someone was interviewing you." Several commented that participating friends would "get more smart" or "more ideas for life" and that researchers could "get more to participate." Undecided and declining children also displayed some sophisticated reasons: "it's hard to explain," "they will be angry." In summary, these children seriously took up the opportunity to exercise their agency and to judge their CAI experience, in the process providing evidence of their abilities as well as the facility the CAI offered them for self-expression.

Evidence of different aspects of refugee children's well-being emerged in the multiple forms of self-expression and levels of engagement that *People* invited from children. Analyzes within the person-oriented approach (von Eye and Bogat 2006;

Table 1 Patterns of Forty-nine Children's Evaluations of '*People*' CAI with Illustrative Type-in Responses to Question: "*Would you tell your friend to do this program or not?*"

Response (Coded)	No.	Example – because … (age in years group gender)
Yes (Helpful)	11	It could help your friend with what he's feeling (11yo O-C B) They now what they can do when they are worried (11yo I-C B) I'd tell them how awesome it is like someone was interviewing you getting to know someone (11yo O-C G)
Yes (Fun)	7	It's fun answering the question (10yo I-C G)
Yes (Fun and Helpful)	1	It is fun and it takes away bad stuff that I am worried about. I always get sad when I talk about people and my life, but this made me think how they help me and give me strategies (12yo I-C B)
Yes (Easy)	4	It is an easy program, and it will keep all your personal details safe (11yo O-C G)
Yes (Helps researchers)	4	You get more people to participate (11 yo O-C B) It will help the researchers understand what children do (11yo O-C B) It may be helpful to them like if they have any problems you might know about it (10yo O-C G)
Yes (Informs)	3	To get more ideas for life (10yo I-C B) So that they can get more smart (9yo I-C B)
Yes (Idiosyncratic)	2	It's not something that helps you get rid of bad thoughts (12yo O-C G) Some of them go to schools that teach them maths so if they look at this program, they won't need to go there anymore (9yo I-C G)
Not sure	13	It's hard to explain the program (9yo I-C G) It could feel bad for the person having a bad time (10yo I-C B) Some of it I didn't understand and I might not be able to explain what it is about (11yo O-C B) I'm not sure he likes it (11yo I-C B)
No	3	They wouldn't want to do it (10yo I-C B) They will be angry (9yo I-C B) It's a secret (9yo I-C G)
No answer	1	I do know. All my friends have done it already (9yo I-C B)

von Eye 2010) revealed, for example, a range of perspectives in what 90 children indicated helped them to feel better (Lawrence et al. 2013). Six clusters specified distinctive patterns of "helps me"/"doesn't help me" lists, and these clusters could not be identified as having refugee, migrant, or local Australian-European backgrounds, nor in relation to their age or gender, but only in their personally generated and checked lists. Clusters differed not in the possible helps that most children endorsed (such as spending time with family/friends, doing your best, someone saying nice things about you), but more finely, in the possible helps that they differentially rejected as not helpful.

While one happy cluster said all 13 things helped them, another cluster rejected sport and another rejected both having time alone and listening to music. Compared with the "everything helps" cluster, the "not sport" cluster also expressed less positive feelings about home and school and about being nurtured at home; and the "not being alone or listening to music" cluster less positive feelings about home but not about school. This level of discrimination goes beyond simple group differences by supporting children to make considered, reflective choices and then by taking their choices seriously in identifying discriminating subgroups.

The Dodds et al.'s (2016) study also illustrates the different patterns of data across multiple forms and tasks from two groups of children from similar refugee backgrounds. These groups were living in different types of state houses in different Melbourne suburbs: 29 in high rise inner-city state housing flats and 20 in outer-suburb individual houses. Correlations between different aspects of the groups' well-being are shown in Fig. 1a, b.

Figure 1a shows how for the inner-city flat dwellers, their quite high-level positive feelings about home were not correlated with positive feelings about school or suburb or about being nurtured at home. For outer-city house dwellers, however, the comparable correlations were strong (Fig. 1b). These children were not simply expressing socially desirable feelings. They were making discerning choices when sorting their feelings about places and discriminating ratings, and these choices were not always similar. For example, one 10-year-old outer-city girl recorded no positive feelings about school. Rather than treating her extreme responses as outlier data, inspection of the patterns across all her responses suggested the probable meaning behind her unusual expression of discomfort. In typing in her reason for recommending the CAI to a friend, she said; "Because it helps people who get bullied to tell people what they feel." In her response to a final optional question about how the CAI could be improved, she was explicit, "These programs could ask how you get bullied and how you feel when you do."

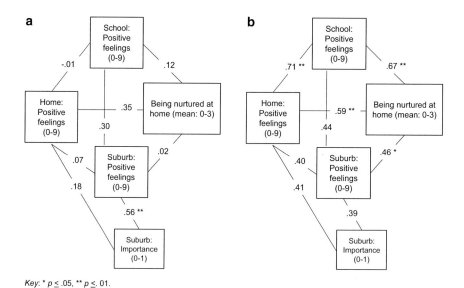

Key: * $p \leq .05$, ** $p \leq .01$.

Fig. 1 Correlations for Three Places and Experience of Being Nurtured, for: (a) Twenty-nine Inner City Children, and (b) Twenty Outer City Children

Patterns of similarities and differences – of both central and outlying tendencies in the quantitative and qualitative data – pointed to the children's considered expressions of their thoughts and feelings. The quantitative/qualitative mix of data is useful for uncovering different aspects of the children's views about well-being, while fulfilling the criterion of respect for their individual as well as common contributions to knowledge. Children as young as 8 years were able to express thoughts showing their engagement with the tasks and their freedom to express their individuality.

4.3 Living in Australia

In the *Living* CAI, participants are introduced to Aya (girl) or Ali (boy), as "about your age." In young child and teenage versions, the CAI uses illustrations and animations to present Aya/Ali's story in a series of situations with prompts asking participants to comment on Aya/Ali's experiences and any related experiences of their own (the teenage form has more requests for explanations). Young participants are told that Aya/Ali came to Australia after some bad things happened back in the home country. One situation describes how Aya/Ali is worried about family left in other countries, and another that she/he is worried about people at home at present who are sad, then angry. The CAI asks participants whether or not they have experienced situations similar to those experienced by Aya/Ali and what they were like. It then tells participants that Aya/Ali finds it difficult to engage in each of five adolescent activities because of worries (hanging about with friends, making friends, sleeping at night, getting out of bed, talking with adults). Participants then are asked to construct personal casts of the people in their lives by clicking on figures (adult, adolescent, and child males and females) and giving each person a role (e.g., father, friend). Later participants use this cast to build another diagram of who among those people help them. Then they are asked to identify what they would like to achieve or change in their lives in the next year and who could help them achieve that and how. Participants use four simple forms of response: forced choice (no, not sure, yes), Likert scale ratings with numbers and words for clarity (0, none, 1, a little bit, 2, more than a little bit, 3, a lot), constructed diagrams, and open-ended comments. The CAI records all quantitative and qualitative responses for downloading as each participant's profile.

4.4 Appropriateness and Relevance of *Living* CAI

Understandably given their vulnerability, refugee and asylum seeker children who travel without their parents are often reluctant to talk about themselves or their families to people they may see as official in any way (Kohli 2006). The specific enactment of respect for unaccompanied minors involved constructing a research environment and a set of exchanges that gave them a secure way of communicating their individualized views, or to decline to answer questions in comfort.

Theoretically, the *Living* CAI focused on worries and hopes that are prominent features of life for refugee young people (McFarlane et al. 2011). Worries have the potential to disrupt a person's well-being and to impact on their hopes for the future. Young people who are isolated from their families are mostly hoping for some form of family reunion, either under the provisions of invitational humanitarian programs or for asylum seekers, by gaining economic stability so they can independently provide for family members. Their hopes as well as their worries then are personal. Respect for these young people extends to how their views are represented in analyses (Mazzei and Jackson 2012). Lawrence et al. (2016), accordingly, recently used a close, exegetical textual analysis of the complete CAI interviews of four teenagers who arrived in Australia as unaccompanied minors.

The contextualized analysis pays due respect to the young people's constructed knowledge by noting their personalized expressions and language and by tracing their themes across their complete data transcript that is downloaded from the CAI immediately and intact. A special feature of the analysis is reproduction in table form of each participant's complete exchange with the relevant tasks, questions, and probes. The tables present the primary data that are the bases of researcher inferences. Textual analysis of the tabulated data is similar to keyed exegetical analyses used in Lawrence and Valsiner (2003) to trace the sequenced processing of a young shoplifter's crime by two police officers and by Welsh (2009–2010) to represent the multiple indicators of beginning teachers' classroom performances. This approach to data management of qualitative produces descriptions of research participants' expressions with their contextualized meanings that in Ponterotto's (2006) terms make them thick rather than thin, and anchored and contextualized rather than decontextualized, liberally interpreted descriptions of people's ideas.

The table-anchored textual analysis gives young participants the appropriate respect of working with their considered and developed thinking as the themes they repeat across time and tasks and as the specific concerns they identify. For example, all four unaccompanied minors typed comments about their worries about their families, but referred to their isolation in personal ways. One young woman made continual references to her preoccupation with family reunion, turning questions away from her foster family in Australia and onto her preoccupation with her own family overseas, commenting; "I worry a lot because I live with other family while my family is alive." A young man who had been an asylum seeker had no prospect of family reunion under the humanitarian program. He instead concentrated on his studies and his own loneliness, but commented on his need to, "concentrate on my studies and stop worrying about my family."

The fully tabulated data gave particular meaning to the young woman's ambivalence about her well-being as "a bit OK and a bit not OK," and her despair that anyone could help her. The data also made sense of the lonely young man's feeling that things were "mostly not OK," and his lack of response on who does and could help him. The tables also make the researchers' inferences and interpretations open to the inspection and scrutiny that, according to Feyerabend (1975), is a criterion for empirical research. While the full tables shown in Lawrence et al. (2016) may be confined to small numbers, they are amenable to identifying cluster subgroups and reproduction of group patterns, consistent with the person-oriented approach (e.g., Martinez-Torteya et al. 2009).

5 Conclusion and Future Directions

The CAI methodology was developed to promote better understanding of refugee children's well-being. Two CAIs were designed to enable these children to report their thoughts and feelings about multiple aspects of their well-being and to provide representations of their meanings and emphases that allow researchers to identify the themes and stand-out features in children's expressed perspectives.

Refugee children have the right and the ability to be active and respected contributors to research knowledge. They have been through situations far beyond the normal range of family and other experiences of their mainstream peers. Many are dealing with isolation from family members along with their own dislocation, and others are trying to settle in situations that are less than optimal for children's development. Their strengths and resilience can be easily overlooked in the face of the magnitude of their traumas and vulnerability (Kaplan 2013). Whether they are reporting positive or negative aspects of their well-being, these children have significant contributions to make to a better understanding of well-being.

The input of refugee children is valuable for those seeking to promote their positive development in resettlement, and in general, valuable for cross-cultural research involving children. If service delivery and program development are to alleviate disadvantage and support development, the voices of refugee children need to be heard, as Lippman et al. (2009) reported to UNICEF. Nevertheless, the involvement of these and other disadvantaged children in research is not simply a matter of open invitations or easy applications of standard measures. Criteria apply, particularly the criteria of grounding methods in respect for the children and in the theoretical relevance of the perspectives they are asked to express.

Well-being has many facets (Ben-Arieh et al. 2014), and its meanings and manifestations in an individual child's life cannot be most productively captured with a single, umbrella concept. Nor can well-being be assumed to have the same indicators in the lives of refugee children whose background and settlement circumstances appear to be similar. The two CAIs provide evidence of some consistent concerns for younger and older refugee children and evidence of some diverse concerns that cannot be explained by age, gender, or other demographic characteristics.

Although the importance of family, for instance, cannot be overestimated for dislocated and separated children (McFarlane et al. 2011; Measham et al. 2014), family-related worries take on different proportions in how children interpret events and relationships, as the illustrative examples clearly demonstrate. Even family reunion can be framed differently in relation to unaccompanied minors' circumstances and immigration status. Places carry special significances in terms of relationships and personal identity (Christensen 2003), but again, the CAIs revealed that significant places are not uniformly interpreted. Along with the suburban neighborhood, children's homes and schools elicit different configurations of expressed feelings, when measures are sensitive enough to capture children's personal interpretations of events and relationships. In summary, refugee children cannot be grouped together as a single, disadvantaged class that is expected to express uniform views of similar experiences.

Standard analyses of group means using general purpose, off-the-shelf measures are not likely to reveal the different views of individual group members (von Eye and Bogat 2006), or how individuals infuse the same or similar events with distinctive meanings (Lawrence and Valsiner 2003). Personal themes and preoccupations call for full, contextualized, and thick representations and multilayered analyses (Ponterotto 2006; Mazzei and Jackson 2012; Morse 2010). Unusual and novel perspectives may only come to the surface when a young person's full complement of choices, ratings, and explanations can be inspected as a whole set. Preoccupations and worrying concerns may only be highlighted when young participants are offered a variety of tasks and opportunities to reflect upon with options to revise their initial choices. The CAI method specializes in translating respect into digitized constructions and representations that give children such options.

Digitization in itself, however, is not sufficient to achieve appropriate levels of reflection and authenticity. There is a contemporary trend for data-greedy researchers to use online and computer-presented barrages of standard questionnaires. Admittedly, these are economical of researcher costs and effort, but they may offer attractive environments and participant incentives without showing due respect. Computerized or online delivery does not guarantee the appropriate address of theoretical questions. Parachuted measures can be delivered online without theoretical grounding. Theoretically generated questions also must be viewed with suspicion whenever they are presented with rough regard for cultural sensitivity or for the personal worth of a vulnerable sample (Spyrou 2011).

Attending to respect and theoretical relevance involves tailoring materials and procedures to children's interests and cultural backgrounds. Participant agency and engagement, interface accessibility, suitable tasks, and instructions all address the construction of respect-driven and theoretically relevant research using CAIs. The quality and relevance of the quantitative and qualitative data constructed using the present CAIs encourages further use in refugee studies and other areas in need of children's contributions to knowledge.

What future directions could be added to this program of methodological development? Presenting the interface in the children's own language is one possibility. This courtesy, however, requires more than translation and back-translation. It demands deep attention to the cultural nuances imbedded in language. For example, in English a "grown-up" is an adult, but in Arabic it may also be any relative older than oneself, so another child. Superficial translation may lead to participant misinterpretation of the question or researcher misinterpretation of responses. Choice of language also requires understanding that the educational language of some groups may not be the language of their heritage culture, instead, it may be the language of oppression, as for example, for Assyrian-Chaldean and Coptic groups from Syria and Iraq who had to do all their schooling in Arabic. Similarly, the relationship of minority and newly settled groups to mainstream language is by no means easily discerned (Goodnow and Lawrence 2015). Readily acquired Street English does not serve children well for school learning.

Another direction, officially mandated in the UK, required that children be involved in research decisions and policy-making. That seemingly empowering policy, however, has not been uniformly successful in giving children confidence that they or their input are respected (Tisdall 2008).

Naturally, well-being does not exhaust the area of refugee studies. Other aspects of personal development could be investigated together with aspects of cognitive and neurological development. While the general idea of using CAIs may be suitable, partly because of the groundwork already achieved, it would be disrespectful in the extreme to succumb to using quick adaptations. Toomela's (2011) point outweighs convenience – theoretical questions must be addressed in advance of methodological attractions.

This point pertains also to adaptations of CAI methodology to the wider area of cross-cultural studies. CAIs are suitable for bringing to the fore different aspects of children's contributions to knowledge. The tasks presented to young contributors in the present CAIs were grounded in issues of theoretical and practical relevance for refugee populations. The different aspects of well-being that are particularly challenging for refugee children in an Australian urban setting may be similar to those that challenge children in other settings, but these CAIs were developed for their specific settings and concerns, and not with an eye to universalization. While worries about family dangers and family reunion are widespread across refugee and migrant groups (Goodnow 2014), the applicability of the indicators of well-being for these refugee children needs testing for their relevance to other groups. Just as no one method is a one-to-one match to theoretical concepts, neither is it automatically applicable to the concerns and needs of all disempowered and dislocated groups. Other children with well-being issues are likely to belong to indigenous and marginalized groups whose prospects are likely to be compromised by either poverty, disempowerment, or prejudice. It is advisable to consider specific aspects related to their well-being, for example, issues related to land and identity for indigenous populations. Any methodological applications also would depend on appropriate translations of respect for a specific group. In the case of asylum seekers in detention, for instance, it may not be ethically appropriate to ask for their research participation at all, when it is not possible to offer what they may see as appropriate advocates and champions with the immigration system.

Respect applies some further restrictions on cross-cultural comparisons. Theoretical questions and empirical measures are not equally suitable for all cultural groups. Their unevenness is especially pertinent when one group in a comparative study is a minority or disadvantaged group seen through a deficit lens, when compared with a mainstream group. As Callaghan et al. (2011) pointed out, measures may be a good fit for a Western group but a bad fit for an non-Western group.

In conclusion, it is not only the right of refugee and other children to communicate their perspectives on their well-being (UNHCR 1983), it is the responsibility of governments and agencies to understand their communications about what helps or hinders their health and well-being. Researchers bring rights and responsibilities closer together to the extent that they appropriately enable children's expressions of their thoughts and feelings, represent and analyze those expressions with attention to themes and trends, standout features, personal meanings, and above all, are guided by respect for refugee children.

Acknowledgments We gratefully acknowledge programming, artwork, animation and data collection by Abi Brooker, Hugh Campbell, Amy Collard, Al MacInnes, and Cat MacInnes, and funding by The Victorian Foundation for the Survivors of Torture and The University of Melbourne.

References

Barrow W, Hannah EF. Using computer-assisted interviewing to consult with children with autism spectrum disorders: an exploratory study. Sch Psychol Int. 2012;33(4):193–8.

Beazley H, Bessell S, Ennew J, Watson R. The right to be properly researched: research with children in a messy, real world. Child Geogr. 2009;7(4):365–78.

Ben-Arieh A. Where are the children? Children's role in measuring and monitoring their well-being. Soc Indic Res. 2005;74:573–96.

Ben-Arieh A, Casas F, Frønes I, Korbin JE. Multifaceted concept of wellbeing. In: Ben-Arieh A, Casas F, Frønes I, Korbin JE, editors. Handbook of child well-being. Dordrecht: Springer; 2014. p. 1–27.

Bergman LR, Wangby M. The person-oriented approach: a short theoretical and practical guide. Estonian J Educ. 2014;2(1):29–49. Retrieved from https://doi.org/10.12697/eha.2014.2.1.02b. 30 Aug 2016

Bokström P, Fängström K, Calam R, Lucas S, Sarkadi A. 'I felt a little bubbly in my tummy': eliciting pre-schoolers' accounts of their health visit using a computer-assisted interview method. Child Care Health Dev. 2015;42(1):87–97.

Cairns E, Dawes A. Children: ethnic and political violence – a commentary. Child Dev. 1996;67(1):129–39.

Callaghan T, Moll H, Rakoczy H, Warneken F, Liszkowski U, Behne T, Tomasello T. Early social cognition in three cultural contexts. Monographs of the Society for Research in Child Development. 2011;76:1–142.

Chase E. Security and subjective wellbeing: the experiences of unaccompanied young people seeking asylum in the UK. Sociol Health Illn. 2013;35(6):858–72.

Christensen P. Place, space and knowledge: children in the village and the city. In: Christensen P, O'Brien M, editors. Children in the city: home, neighbourhood and community. London: RoutledgeFalmer; 2003. p. 13–28.

Connolly P. Children, assessment and computer-assisted interviewing. Child Abuse Rev. 2005;14:407–14.

de Leeuw E, Hox J, Kef S. Computer-assisted self-interviewing tailored for special populations and topics. Field Methods. 2003;15(3):223–51.

Dodds AE, Brooker A, Collard A, Lawrence JA. Children's sense of place: perceptions of children from refugee families living in inner-city flats and outer-city suburban houses. In: Paper presented at 24th Biennial meeting of the international society for the study of behavioural development. July 14. Vilnius; 2016.

Dolezal C, Marhefka SL, Santamaria EK, Leu C-S, Brackis-Cott E, Mellins CA. A Comparison of audio computer-assisted self-interviews to face-to-face interviews of sexual behavor among perinatally HIV-exposed youth. Archives of Sexual Behavior. 2012;41:401–410.

Feyerabend P. Against method. London: Verso; 1975.

Goodnow JJ. Refugees, asylum seekers, displaced persons: children in precarious positions. In: Melton G, Cashmore J, Goodman G, Ben-Arieh B, editors. The Sage handbook of child research. New York: SAGE; 2014. p. 339–60.

Goodnow JJ, Lawrence JA. Children and cultural context. In: Bornstein M, Leventhal T, editors. Handbook of child psychology and developmental science, volume 4: ecological settings and processes in developmental systems. Editor-in-Chief: R. M. Lerner. Hoboken: Wiley; 2015. p. 746–86.

Hirch M. Self-determination in indigenous health: a comprehensive perspective. Fourth World J. 2011;10(2):1–30.

Kaplan I. Trauma, development and the refugee experience: the value of an integrated approach to practice and research. In: de Gioia K, Whiteman P, editors. Children and childhoods 3: immigrant and refugee families. Newcastle upon Tyne: Cambridge Scholars Publishing; 2013. p. 1–22.

Kohli R. The sound of silence: listening to what unaccompanied asylum-seeking children say and do not say. Br J Soc Work. 2006;36:707–21.

Lawrence JA, Valsiner J. Making personal sense: an account of basic internalization and externalization processes. Theory Psychol. 2003;13(6):723–52.

Lawrence JA, Collard A, Kaplan I. What helps children from refugee, immigrant and local backgrounds feel better. In: de Gioia K, Whiteman P, editors. Children and childhoods 3: immigrant and refugee families. Newcastle upon Tyne: Cambridge Scholars Publishing; 2013. p. 111–34.

Lawrence JA, Kaplan I, Dodds AE. The rights of refugee children to self-expression and to contribute to knowledge in research: respect and methods. J Hum Rights Pract. 2015;7(3):411–29.

Lawrence JA, Kaplan I, Collard A. Understanding the perspectives of refugee unaccompanied minors using a computer-assisted interview. Forum: Qualitative Social Research. 2016;17(2). http://nbn-resolving.de/urn:nbn:de:0114-fqs160268.

Lippman LH, Anderson Moore K, McIntosh H. Positive indicators of child well-being: a conceptual framework, measures, and methodological issues. Innocenti working paper, UNICEF. 2009. Retrieved from https://www.unicef-irc.org/publications/pdf/iwp_2009_21.pdf. 30 Aug 2016.

Martinez-Torteya G, Bogat A, von Eye A, Levendosky AA. Resilience among children exposed to domestic violence: the role of risk and protective factors. Child Dev. 2009;80(2):562–77.

Mazzei L, Jackson AY. Complicating voice in a refusal to 'Let participants speak for themselves'. Q Inq. 2012;18(90):745–51.

McFarlane CA, Kaplan I, Lawrence JA. Psychosocial indicators of wellbeing for resettled refugee children and youth: conceptual and developmental directions. Child Indic Res. 2011;4(2):1–31.

Measham T, Guzder J, Rousseau C, Pacione L, Blais-McPherson M, Nadeau L. Refugee children and their families: supporting psychological well-being and positive adaptation following migration. Curr Probl Pediatr Adolesc Health Care. 2014;44:208–15.

Morse JM. "Cherry-picking writing" from thin data. Editorial. Qual Health Res. 2010;20(1):3.

Patton MQ. Developmental evaluation: applying complexity concepts to enhance innovation and use. New York: The Guildford Press; 2011.

Ponterotto JG. Brief note on the origins, evolution, and meaning of the qualitative research concept "thick description". Qual Rep. 2006;11(3):538–49. Retrieved from http://nsuworks.nova.edu/tqr/vol11/iss3/6. 2 Feb 2016.

Spyrou S. The limits of children's voices: from authenticity to critical, reflexive representation. Childhood. 2011;18(2):151–65.

Steinberg L. Commentary: developmental pathways and social contexts in adolescence. In: Crockett LJ, Crouter AC, editors. Pathways through adolescence: individual development in relation to social contexts. Mahwah: Lawrence Erlbaum Associates; 1995. p. 245–53.

Tisdall EKM. Is the honeymoon over? Children and young people's participation in public decision-making. Int J Chil Rights. 2008;16:419–29.

Toomela A. Travel into fairyland: a critique of modern qualitative and mixed methods psychologies. Integr Psychol Behav Sci. 2011;45:21–47.

United Nations. Convention on the rights of the child (CRC). Geneva: United Nations; 1989.

United Nations High Commission for Refugees (UNHCR). UNHCR guidelines on reunification of refugee families. 1983. Retrieved from http://www.unhcr.org/3bd0378f4.pdf. 10 Feb 2016.

von Eye A. Commentary: developing the person-oriented approach: theory and methods of analysis. Dev Psychopathol. 2010;22:277–85.

von Eye A, Bogat A. Person-oriented and variable-oriented research: concepts, results, and development. Merrill-Palmer Q. 2006;52(3):390–420.

Weisner TS. Culture, context, and child well-being. In: Ben-Arieh A, Casas F, Frønes I, Korbi JE, editors. Handbook of child well-being: theories, methods and policies in global perspective. New York: Springer; 2014. p. 87–103.

Welsh BH. Text-based analytic philosophical analysis. Curric Teach Dialogue. 2009–2010;12(1 & 2): 27–39.

Conducting Focus Groups in Terms of an Appreciation of Indigenous Ways of Knowing

103

Norma R. A. Romm

Contents

1 Introduction	1796
2 Defining Some Contours of Indigenous Epistemologies	1797
3 Some Literature on Focus Group Research and Lacunae Identified	1800
4 An Illustrative Example: A Focus Group Session with Teachers in KwaZulu-Natal	1802
5 Conclusion and Future Directions	1806
References	1807

Abstract

This chapter offers deliberations around the facilitation of focus groups in a manner that takes into account Indigenous ways of knowing. Indigenous knowing (within various Indigenous cultural heritages) can be defined as linked to processes of people collectively constructing their understandings by experiencing their social being in relation to others. This chapter explores how the conduct of focus groups can be geared toward taking into account as well as strengthening knowing as a relational activity defined in this way. I suggest that once facilitators of focus groups appreciate this epistemology, they can set up a climate in which people feel part of a research process of relational discussion around issues raised. This requires an effort on the part of facilitators to make explicit the type of orientation to research that is being encouraged via the focus group session to participants. In this chapter, I offer an illustrative example of an attempt to practice such an approach to facilitation in a rural setting in South Africa.

N. R. A. Romm (✉)
Department of Adult Education and Youth Development, University of South Africa, Pretoria, South Africa
e-mail: norma.romm@gmail.com

© Springer Nature Singapore Pte Ltd. 2019
P. Liamputtong (ed.), *Handbook of Research Methods in Health Social Sciences*,
https://doi.org/10.1007/978-981-10-5251-4_46

Keywords

Indigenous ways of knowing · Collective exploration as relational · Facilitator orientation · Focus group research · Participant feedback

1 Introduction

In this chapter, I consider the conduct of focus groups underpinned by an outlook that takes account of Indigenous ways of knowing. The chapter is set in the context of Horsthemke's (2008) statement that "Indigenous knowledges, as a discursive framework in the academy, are relatively new and have only gained currency in the last twenty years" (as cited in Wane 2013, p. 100).

When considering Western-oriented epistemologies since the beginning of the so-called Enlightenment in the seventeenth century, Ladson-Billings (2003, p. 398) states her concerns that these (still) dominate the way in which knowing is conceptualized in research literature, to the detriment of "ethnic epistemologies." She sees René Descartes' rationalist approach to knowing which he propounded in his *Discours de la méthode* (1637) as offering an epitome of how knowledge-making is defined as resting on individual thought patterns, where "the individual mind is the source of knowledge" (p. 398). She contrasts this with

> the African saying "Ubuntu," translated "I am because we are," [which] asserts that the individual's existence (and knowledge) is contingent upon relationships with others. (Ladson-Billings 2003, p. 398)

Ladson-Billings suggests that these divergent epistemological perspectives are not merely matters of "preferences." She argues that the "preference" for the former kind of epistemology serves to reproduce a "dominant worldview and [attendant] knowledge production and acquisition processes" (p. 399). In the light hereof, it is crucial to revitalize alternative traditions of knowledge-construction and ways of appreciating collective processes of knowing, as offered within ethnic/Indigenous worldviews.

Like the word *ethnic,* the term *Indigenous* can be used here to denote "Indigenous peoples and culture" in different contexts (Kovach 2009, p. 20; see also ▶ Chap. 15, "Indigenist and Decolonizing Research Methodology"). Notably, however, authors such as Ladson-Billings and Kovach do not consider "culture" as harboring monolithic or static meanings but rather as harboring symbolic expressions which can form a basis for continuing conversation around the symbols, and a revitalization of their meanings in use (see also ▶ Chaps. 88, "Culturally Safe Research with Vulnerable Populations (Māori)," and ▶ 108, "Ethical Issues in Cultural Research on Human Development").

The point is that, as Rajagopalan (2016, p. 235) contends, it needs to be appreciated that the languages/cultural heritages of many Indigenous people "facilitate meta-rational understandings of their (human) conditions" and a "participative approach" to the cosmos (including our connectedness to all human and non-human life) but "Western science has made them ... into the enemies of rational analysis" and in this process denigrated them. Chilisa (2012) argues that in such a context, a postcolonial (Indigenous) research paradigm offers an orientation aimed at strengthening Indigenous worldviews (which are premised on an experience of connectedness), epistemologies (premised on the idea of knowing as a communal endeavor), and axiologies (premised on relational accountability) (see also ▶ Chap. 87, "Kaupapa Māori Health Research").

With reference to research undertaken in post-apartheid South Africa, Ndimande (2012, p. 215) suggests that this would imply, *inter alia*, being attentive, especially when organizing research work that involves Indigenous participants, not to enforce still dominant understandings of what it means to conduct research – and to be attuned to the spirit of "decolonizing research." He argues that decolonizing research presents a challenge to traditional ways of conducting qualitative research, as well as to ways of doing quantitative research. He summarizes the movement toward decolonizing research as follows:

> Decolonizing methodologies can help researchers interrogate the very notion of "knowing" as well as what it is that we know and who benefits from that knowledge (Rogers and Swadener 1999). This study [referred to in his article, where Ndimande conducted focus groups with Indigenous parents of school children] involved marginalized communities whose cultural epistemologies rarely find due recognition or even acknowledgment in academe and/or in educational research institutions. Interrogating the constitution of knowledge and its function is what can translate to decolonizing research. (Ndimande 2012, p. 223)

Speaking about the premises of decolonizing research across the globe, and more specifically with reference to examples of working with Aboriginal communities, Nicholls (2009, p. 120) explains how "Indigenous epistemologies and axiologies can inform the undertaking of participatory and collaborative research." This is by taking into account that "the individual person is constituted through his or her communicative and interactive relations with others" (p. 121). Chilisa (2012) for her part notes the similarity between Maori arguments concerning the essential connectivity of people, who exist in relation, and the African concept of Ubuntu (see also ▶ Chaps. 15, "Indigenist and Decolonizing Research Methodology," ▶ 87, "Kaupapa Māori Health Research," ▶ 88, "Culturally Safe Research with Vulnerable Populations (Māori)," ▶ 89, "Using an Indigenist Framework for Decolonizing Health Promotion Research," ▶ 90, "Engaging Aboriginal People in Research: Taking a Decolonizing Gaze," and ▶ 97, "Indigenous Statistics").

In this chapter, I offer some suggestions for how, in the conduct of focus groups, one can consciously inform the research in terms of an appreciation of Indigenous ways of knowing, where knowing is understood as a relational exercise. That is, I offer detail on how focus group facilitators can give due recognition to epistemological questions in the process of conducting research, while being alert to Indigenous ways of knowing.

2   Defining Some Contours of Indigenous Epistemologies

Considering Indigeneity in the context of Africa, Ossai (2010, p. 5) suggests that one of the prime qualities of Indigenous ways of knowing is that they are relevant to solving problems as identified in communities and they involve the community's participation in the knowledge-construction process. In order to spell out what "African Indigenous Knowledge Systems" (AIKS) can put forward as a viable alternative (or as an important complement) to "Western Scientific Knowledge," Ossai points to several features of AIKS, which he contrasts with more Western-oriented styles of knowing as follows:

- Indigenous styles of knowing encapsulate intuition in the sense of intuiting connections, and therefore are more holistic than (Western-oriented) analytic and reductionist approaches.
- Oral story telling is paramount in this style of knowing, based on people exploring their stories (experiences) together.
- The way of creating/generating data is "slow/inconclusive" – in that it is stressed that experiences/data can be subjected to reformulation in the light of further experience and discussion around it. This is in some contrast to Western scientific approaches which he sees as fast and selective (for example, selecting specific factors which are isolated for attention).
- Indigenous knowing is more capable of taking into account long-term cycles of feedback, rather than being focused on short-term analyses. It is also less linear in its thinking – and does not try to link "effects" with particular causes in linear fashion.
- And lastly, it caters for the inexplicable (which can also be linked to the spiritual) – rather than trying to orient truth-seeking primarily in terms of what can supposedly be "scientifically" understood (Ossai 2010, p. 10)

Goduka (2012) compares these understandings of the processes of AIKS knowledge construction with Western-oriented epistemologies by focusing attention on the question of how truth-seeking as a collective enterprise is regarded in AIKS. She argues, along with other Indigenous-oriented authors, that what is specific about Indigenous modes of knowing is that they are *intentionally communally oriented*. As she puts it:

> Communal knowledge ensures that knowledge is not collected and stored for personal power and ownership by individual specialists, but is rather developed, retained and shared within Indigenous groups for the benefit of the whole group. (Goduka, 2012, p. 5)

This conception of knowing as an expression of communal togetherness is also emphasized by Harris and Wasilewski (2004), in their setting out of what they see as an alternative worldview and attendant epistemology as offered by Indigeneity across the globe. They maintain that part of the strength of Indigenous cultural symbols (as they have described them) is that they can serve to create the groundwork for a "dynamically inclusive dialogic space." This space, they note, "includes you, me, all of our relationships, taking place in our various personal, social, political, cultural, physical and spiritual contexts" (p. 494). They explain too that one of the characteristics of such a dialogue is that people appreciate that our strength is increased by sharing [in the process of developing communal wisdom]. We can affirm our view, expand our view, or sometimes alter or even give up our current view when we encounter a new one. We can also allow others to have contrastive views as long as they do not impose their views on us and vice versa (p. 498).

Harris and Wasilewski's (2004) reference to not imposing views is set in a global context where they observe that thus far Western outlooks have become imposed as ways of knowing and living, to the detriment of more relational ways of knowing and problem-solving.

It is worth highlighting here that I am not claiming that this approach to knowing can be "found" in all Indigenous cultural heritages independently of the hermeneutic process of *interpreting the heritages*. As Hallen (2002, p. 66) points out in the context of Africa, the process of "identifying and reexamining Africa's Indigenous 'traditions'" is itself not free of an interpretation based on some values that are brought to bear. Serequeberhan (2000, p. 67) too argues that this involves actively "sifting through [various] legacies, retaining that which is alive, ... [and] casting off that which is lethargic." This hermeneutic process is at the same time a call for regenerating what is considered to be valuable and alive – and worthy of nurturing – "in" the traditions (as located). This fits in with a constructivist position which recognizes that the interpretation of history and traditions is indeed a value-laden process, guided by an intention, in this case the intention to revitalize aspects of the heritages (see also Romm 1998, 2010, 2016; Magnat 2012; Lincoln and Guba 2013; Quan-Baffour and Romm 2015, for more detailed expositions of how the hermeneutic process of interpretation can be admitted.)

A question remains as to whether people who are not brought up within cultural heritages, where it is understood that the individual is inseparable from the community, can fully appreciate the flavor of epistemologies based on communal knowledge construction. This is the question of the translatability of paradigms, including epistemological ones. My position in this regard is that stances are not completely untranslatable and that points for comparison between them can be made, albeit that the terms used in talking "across" paradigms may still have different meanings/ understandings associated with them, as people engage in conversation to enrich their appreciation of their possible meanings (cf. Flood and Romm 1997; Romm 1998, 2001, 2007, 2010, 2015; Hallen 2002; McIntyre-Mills 2000, 2014; Pollack 2006; Osei-Hwedie 2007; Denzin and Lincoln 2008; Midgley 2011; Cisneros and Hisijara 2013; Mertens et al. 2016). Chilisa (2012, p. 25) supports this stance when she suggests that her locating of an "Indigenous Research Paradigm" need and should not imply adopting "an either-or approach, where, in the discussion of "Euro-Western paradigms and postcolonial Indigenous paradigms, these paradigms become essentialized, compelling thought along binary opposites." Likewise, Nicholls (2009) mentions that she does not agree with a position which essentializes the difference between cultures and reinforces dichotomies between cultural groups.

What Nicholls (2009) proposes is that researchers involved in cross-cultural research should be alert to spaces of potential connections between themselves and participants (as played out in the research process). I suggest that this was attempted in the case of the facilitation of the focus group discussed below, where I as White cofacilitator was involved (with other facilitators, White and Black) in facilitating focus group discussion with Black school teachers. In considering these racialized categorizations, I take the view, along with a myriad of other authors/actors that the categories Black and White are social constructions, which are nevertheless socially "real" in their consequences (see, for example, Kiguwa 2006, p. 113; Ansell 2007, p. 329; Romm 2010, pp. 10–13; Naidoo 2011, p. 628).

In the case discussed (as an illustrative example), I offer an account of an attempt to give credence to Indigenous epistemological orientations during the focus group session, by the facilitators expressing explicitly (albeit not in jargonized language) that the purpose of the discussion is to generate ideas collectively, as a process of being and learning together. As some authors advise (Gregory and Romm 2004; Dickson-Swift et al. 2006; Mapotse 2012; Romm et al. 2013; Austin 2015; Rajagopalan 2016), we also tried to blur the boundary between ourselves as "professional researchers" and the participants, by seeing ourselves as part of the community of teachers concerned with "making schools better" and by seeing the participants (lay researchers in this case) as part of the research process in which we all were involved in coexploring the issues. This intention resonates with Higginbottom and Liamputtong's (2015, p. 4) delineation of a participatory approach, which is, *inter alia*, "multidirectional" and "may create new insights for the professional researcher[s] and the communities involved in the research."

In offering their reflections around decolonized methodologies in cross-cultural research, Vannini and Gladue (2008, p. 141) consider that a sharing approach as adopted by researchers can express a heartfelt caring and is a way of admitting that the construction of realities (as understood) is relational. Through the example which I outline briefly in this chapter, I offer an instance of how we as facilitators became involved in the discussion to forward its development. But before I turn to this, I provide some background to my considerations around sensitivity to the notion of knowledge-making as a communal process, by engaging with some of the literature on focus group research as a methodology.

3 Some Literature on Focus Group Research and Lacunae Identified

It is specified in much of the literature on focus group research that the "data" that are developed via focus group discussion should be seen as being a product of the group interactions. For example, Hollander (2004, p. 362) expresses this understanding of focus group discussion as follows: "Participants in a focus group are not independent of each other, and the data collected from one participant cannot be considered separate from the social context in which it was collected." Gray et al. (2007, p. 362) suggest that this should be considered as one of the strengths of focus group research. They explain: "Perhaps the most important benefit of focus groups is that the give-and-take among participants fosters reflection on other people's ideas." Clavering and McLaughlin (2007, p. 400), writing in the context of considering qualitative health research, likewise suggest that "focus groups are an important element of qualitative health research, valued for the forms of knowledge and understanding that emerge from interactions among participants." And Mkandawire-Valhmu and Stevens (2010, p. 684), in their account of what they call the "critical value of focus group discussions" (in their case in research with women living with HIV in Malawi), write up the research by choosing to "identify collective insights revealed in focus group interactions whereby participants reacted to and built on the experiences, interpretations, and evaluations of other participants" (p. 688).

Nevertheless, although many authors have pointed to the important function that focus groups can serve in aiding collective generation of insights (see Liamputtong 2011), most authors do not indicate how discussions *can be set up* so that participants can appreciate that collective researching of the topic(s) is what is being encouraged, as a process of people thinking together about the issues being raised (if this is the case). Nor it is common for authors to render explicit the orientation to "knowing/understanding" that facilitators may be adopting when orienting the discussion process. Hence, Farnsworth and Boon remark (2010, p. 605) that "plenty of attention has been paid to the development of the focus group as a research tool but, oddly, very little attention has been paid to the relational dynamics that are intrinsic to its use."

There is thus arguably a dearth of literature on how the processes of interaction and relational dynamics in focus group discussions lead to the data that become generated via the discussions. And (linked to this), there is scant reference to how facilitators might pay attention to *epistemological issues concerning styles of knowing* that can be activated during the discussion and, more importantly for this chapter, on how *cultural sensitivity to knowing practices of participants* can be displayed. In this chapter, I suggest that to express sensitivity to Indigenous ways of knowing, facilitators need to find a way of alerting participants that the focus group session ideally involves a *process of being-in-relation as well as knowing-in-relation*. In other words, one can alert participants that the facilitators value cultural styles where coconstruction of views is a result of people feeling "in connection." This would then fit in with the advocacy of decolonizing research as explicated by, for example, Chilisa (2012), Ndimande (2012), and Nicholls (2009), as discussed earlier.

Rodriguez et al. (2011) specifically raise the issue of cultural sensitivity as a concern that needs to be made more central in focus group research settings. With acknowledgment to the work of Morgan (2002), they make the point that although there is a large body of literature on focus groups, the goal when undertaking methodological reflections should also be to develop the focus group method in this direction. In trying to develop the method, they suggest that this can be done along the lines of "illuminating the importance of using culturally responsive research practices ... to guide qualitative methodology and, in particular, for focus group development" (Rodriguez et al. 2011, p. 401). Liamputtong (2008, 2010, 2011) similarly states, it is crucial for researchers facilitating (qualitative) research in cross-cultural contexts to exhibit culturally appropriate communication. These arguments are applicable to the focus group session discussed below, where this was an issue that was tied to the possibility of defining "knowing" with reference to knowing practices which resonate with Indigenous traditions for communal coconstruction. This is also affirmed in Ndimande's (2012) reflections on "decolonizing research." Considering in particular the use of focus groups, Ndimande indicates that what he sees as important about the potential of focus group discussions, used in decolonizing fashion, is that "the emphasis on collective or group participation versus individual participation has stronger impact in the discussions" (2012, p. 216). He sees focus groups as able to encourage this emphasis.

4 An Illustrative Example: A Focus Group Session with Teachers in KwaZulu-Natal

In this section, I turn briefly to one of the focus groups conducted with teachers at a rural school in Estcourt, KwaZulu-Natal as part of a national project in South Africa (which took place from 2013 to 2015) entitled "Making Schools Better." This project targeted 500 schools across five (out of nine) provinces in South Africa, with questionnaires and focus group sessions being employed as part of the research process. Forty-six researchers from the University of South Africa (Unisa) were involved in facilitating focus group discussions with subsampled schools (http://www.unisa.ac.za/cedu/news/index.php/2015/12/500-schools-project-making-schools-better-closing-seminar/).

In the exemplar discussed in this chapter, which involved focus group discussion with six teachers at the school, the facilitators used both mother tongue (isiZulu) and English (with different facilitators using different languages at various points in the discussion). We hoped to show respect for Indigenous cultural expressions through introducing the session in isiZulu. Also we believed that by introducing the session in mother tongue, the participants would recognize that they could feel free to speak the language with which they were most comfortable. They would presumably be aware that we were trying to give cognizance to their cultural styles of expressing themselves. In this respect, we concur with Colucci (2008, p. 244) that comoderators/facilitators "are essential in focus groups where the moderator has limited understanding of the language, habits, shared knowledge and beliefs of the group … as it is likely to be in a great part of cross-cultural research."

But besides this, it was important to make explicit the epistemological orientation that we were bringing to the session. Therefore, one of the isiZulu speakers on our facilitation team opened the discussions with reference to a piece in the field guide, part of which I had written (the part cited below). This was written to be used as an introduction to all the focus groups conducted during this project. The facilitator thus translated the relevant section of the guide, which gave guidance as follows:

> **Focus Group Interviews with Teachers**
> Introduction for facilitators to share with participants:
> The purpose of the focus group sessions [in this research project] is to gather some information about the teaching and learning of various school subjects, which can become the basis for "making schools better." The idea is that together we can explore issues connected with teaching and learning more fully. This will supplement some of the data that we have obtained from questionnaires that have been filled in by a large sample of teachers across five different provinces. This project is supported by the Department of Education [in South Africa]. Any information or ideas that you share with us will, however, remain confidential in that no one will know who has said what in any of our reports.
> Please note that if you are feeling that you do not want to answer some of the questions asked when you hear them, you can mention this to us – we do not want you to feel under any pressure here. But your answers will be helpful to us to gain a better understanding of what we are asking you about. You also may learn by hearing our questions and thinking about your answers to them. And hopefully you will learn from one another too.

Following this introduction (which included all the facilitators introducing ourselves), we proceeded to ask the participants to introduce themselves in terms of what grades they were teaching; and thereafter we proceeded with asking questions, which included asking for suggestions for "making schools better." The main questions that we asked had been pre-prepared for the purposes of comparing answers from the different focus groups across the whole research project; nevertheless, the style of conversation followed an informal style of discussion, as suggested also by Ndimande (2012) as appropriate for research with Indigenous participants. As he states:

> The focus group, as part of research design [in the case of the research upon which he reports], was appropriate with Indigenous parents because it is a technique for interviewing that straddles the line between formal and informal interviewing ... Thus, a focus group interview allows for an informal environment (p. 216).

As the discussion proceeded, we intermingled English with isiZulu (with the isiZulu-speaking facilitators being fluently bilingual), albeit that the talk was mainly in English and even when facilitators spoke in isiZulu, the teachers answered for the most part in English. This could be because the teachers were teaching English as a first additional language (FAL) at the school as one of the subjects, and they might have wished to show us that they were proficient in the language.

Space in this chapter does not permit any detail on how the discussion proceeded, but by way of example, I refer to a set of statements around the issue that was raised (by a Grade 4 Mathematics teacher) regarding the language transition that the learners had to go through when they reached Grade 4. In this grade,

English (instead of home language isiZulu in this case) became the medium of instruction for all subjects. He expressed concern that the transition was difficult for the learners. One of the other teachers then mentioned that: "English is allocated less time in Grade 3 and the Department [of Education] always sticks on the mother tongue in this phase, so we just visit English." Another teacher added that the government's Annual National Assessment (ANA) process does not concentrate on assessing English in this phase, which is one of the reasons why it becomes sidelined. As she put it: "It is because we are afraid of ANA and we don't like our school to be named as belonging to the ones that have done very badly so this is why we always do isiZulu and Mathematics most of the time and just visit English and life skills because they [the learners] are examined on that which is isiZulu and Mathematics."

I then asked – in relation to the Maths teacher's original statement of concern – "And do you think the learners are being prepared for making the change over to the language of instruction which is English later on?" One of the teachers stated: "Yes we try our best." One of the facilitators (isiZulu mother tongue, but speaking English here) followed this up: "But mam [some of] you just said you just visit English, do you think when the learners reach Grade 4 … when are they supposed to do everything in English, are they prepared?" The teacher replied that "They are not really." And another teacher reiterated that it became problematic because "they are expected [to do] all the subjects in English." At this juncture, I asked: "And what could be a solution to that?"

The conversation went on by participants expressing that the problem should not be addressed by merely looking at the challenge of English on the part of learners; the problem was more widely that learners were not exposed to an adequate amount of reading matter in either isiZulu or English. There was no library for them and problems arose because learners were not doing much reading. Some suggested that a solution to this could be that newspapers and other material could be used to supplement the dearth of reading matter. Others used the opportunity to indicate to us that the lack of reading material is a matter that needed to be relayed to the government as they needed support in this regard. This concern (along with others) was again expressed by this group when we organized our "member checking" visit to the school a few months later. Their expressions of concern on this matter, which were echoed by many teachers and learners across the various provinces, and which we discussed with government officials in the Research division of the Department of Basic Education (DBE), could have been a contributing factor which led to a government program of providing readers in the 11 official languages to supplement the government provision of workbooks. As noted by Minkler (2010, p. 85, citing Guthrie et al. 2006), "most policy work involves multiple players 'hitting' numerous leverage points. In this complex system, it is difficult to sort out the distinct effect of any individual player or any single activity."

Meanwhile, returning to the focus group itself, as had been specified in the focus group field guide for all facilitators, toward the end of the session we requested feedback from participants regarding their experience of the session. I stated this request as follows: "We are interested to know how you experienced the discussion today. How did you feel about talking with us? And would you want to raise any other questions?"

A few of the participants indicated that they were happy that we had been there as they enjoyed the refreshments that we had brought along (biscuits, nuts, and fruit juice). Others laughed and said that these had been good. Although this may seem an unimportant feedback, it lends substance to Liamputtong's (2011, p. 139) statement that "the sharing of food ... is an essential part of conducting focus groups in cross-cultural settings." She gives examples of research with Latina women, Emirati women, immigrant participants in Canada, Aboriginal communities, Mexican migrant farmworkers, and Pacific Northwest Indian communities. The food given in our case (which for us was a sign of respect for their time and a way of offering something immediately concrete in return) was clearly also significant to these participants. We later presented them all with Unisa-logo'd pens, which were also well received.

Having heard their feedback regarding the value of the refreshments that we had brought, I then asked: "Do you think you learned something from hearing each other talk?" To this, certain participants said "yes" and I asked if they could offer some examples. One teacher stated that she had learned about group teaching as being a strategy that the DBE advises (dividing up the class into groups and spending some time with one before moving to the next and hoping that the first groups will keep themselves busy). But, she said that in a class of 48 people, it is chaos. Another participant added that even when they mention to the District officials that this advice is impractical, they are told to "keep trying." Various participants echoed this frustration with the way the advice is given, while they also found somewhat amusing – and perhaps cathartic – the manner in which their colleagues were expressing to us about their being told repeatedly to "keep trying." In this way, they were collectively coconstructing the experience of frustration at the attitude that they were seeing in the injunction for them to continue trying against all odds. This was one of the other issues that we mentioned in our report that we subsequently shared with government officials in the Research division of the DBE, as part of our effort to engender further research dialogues.

One of the other facilitators in the focus group feedback session then asked: "Do you sometimes sit together and discuss problems that you have at school?" They answered in chorus: "Yes." He asked: "What makes this session different from discussing amongst yourselves?" One of the teachers answered: "It is nearly the same." I summarized: "So it is not very different from your talking here today with us being here?" And another replied: "It is like our normal way of talking together." Another stated: "But the answers that we give when we discuss together are different." When I asked how this was so, she said that when we were there the answers involved more what could be done within the system to support them, for instance, in terms of workshops for teachers that could be set up – including how to introduce inclusive educational teaching strategies so that all children could be catered for, especially in large classes.

What is important in relation to the argument in this chapter is that the teachers did not feel that the process of discussion in the focus group session – that is, of sharing ideas and thinking together – was different in terms of process from their "normal," that is, culturally familiar, style of interacting. They remarked merely that the content in terms of thinking about how they might harness support from "the system" differed somewhat from their usual meetings together. It can thus be said that they experienced the discussion with our being there as involving a similar style of conversation that they normally practice, which involves seeking together "answers" to experienced problems. This implies that the form of inquiry as used in the focus group resonated with what Ndimande (2012, p. 223) calls "cultural epistemologies." That is, the participants did not feel that the facilitators brought in a culturally unfamiliar mode of developing understanding as a collective enterprise. Whether our use of their Indigenous language to open the discussion helped in this regard is difficult to say, especially that the participants themselves chose to switch to English most of the time. But what can be said is that the beginning introduction at least set the tone for what Ndimande calls an informal conversation (p. 216), where people participated in being-in-relation with one another toward knowledge-construction/collective inquiry.

What is also worth highlighting is that participants often used the collective "we" in their way of speaking, as, for example, in the set of statements referred to above – where one of the participants indicated that "we don't like our school to be named as belonging to the ones that have done very badly," and another stated in relation to learner preparedness in English that "we try our best." Also, in their recounting their experience of frustration with subject advisors who address them as if they are not trying hard enough to manage the group work in classes, the collective "we" was used by participants.

Mkandawire-Valhmu and Stevens (2010, p. 688) likewise favorably report that in the focus group sessions that they organized with women living with HIV in Malawi, "the collective *we* was frequently used in their explanations and responses to each other." They see this as a sign that the focus group discussion was a forum for collective generation of insight as well as a forum for activating feelings of togetherness.

5 Conclusion and Future Directions

In this chapter, I have explored possibilities for conducting focus groups underpinned by an Indigenous-oriented epistemological position which does not consider individuals (individual selves) as being the route to knowledge production, but sees knowledge-construction as a relational process of developing insights (which are linked to practical ways of living). I pointed to lacunae generally in the literature regarding efforts of facilitators to pay attention to epistemological issues and more particularly regarding efforts to orient focus group discussion to take into account Indigenous styles of knowing. Although the example I used in this chapter is situated within education, many of the issues I have discussed can be applicable to any other health and social issues within the health social sciences.

In view hereof, I would suggest that especially when focus group research is being conducted with Indigenous participants, an endeavor should be made to introduce the sessions in a way that indicates that relational styles of knowing are being encouraged. Without having to use jargon such as "epistemology," "relational thinking," and so on, facilitators can still find ways of indicating that they are not gearing the research to replicating "dominant" (Western-oriented) styles of knowing. I would suggest that culturally attentive researchers can experiment with types of "introductions" that signal this; they can in turn seek feedback from participants later in order to see how their storylines about, and attempts to encourage, relational ways of knowing which are familiar to participants have been received.

In this way, more case material on possibilities for disrupting the (sole) legitimacy of dominant styles of knowing can be developed. With reference to such case material, researchers can make comparisons between different experiences of attempting to foreground epistemological questions as part of the research process, also considering participant feedback (and exact expressions of participants) in relation to this. I have left partly in abeyance the issue of language use and how this might be handled when organizing cross-cultural research. But, Austin (2015, p. 25) reminds us that when the nurturing of "relationship" (as in a relational axiology) is given primacy by researchers, it needs to be borne in mind that "there may be differences in comfort level with the language used for the research, which needs to be recognized and addressed." Apart from this issue, I have suggested that it is important to find ways of signaling to participants that relational styles of knowing are being encouraged.

Acknowledgement This chapter is based on my article that appeared in January 2015 in *Forum: Qualitative Social Research (FQS)*, 16, 1, Article 2. The article was entitled: Conducting Focus Groups in Terms of an Appreciation of Indigenous Ways of Knowing: Some Examples from South Africa. Permission was granted by Katja Mruck – editor of FQS – on 31 May 2016. She stated (by email) that "all FQS texts are published under a Creative Commons Attribution 4.0 International License," which means that I have the copyright for use of the material.

References

Ansell AE. Two nations of discourse: mapping racial ideologies in post-apartheid South Africa. In: Coates RD, Dennis RM, editors. The new black: alternative paradigms and strategies for the 21st century. Amsterdam: Elsevier; 2007. p. 307–36.

Austin W. Addressing ethical issues in participatory research: the primacy of relationship. In: Higginbottom G, Liamputtong P, editors. Participatory qualitative research methodologies in health. London: Sage; 2015. p. 22–39.

Chilisa B. Indigenous research methodologies. London: Sage; 2012.

Cisneros RT, Hisijara BA. A social systems approach to global problems. Cincinnati: Institute for 21st Century Agoras; 2013.

Clavering EK, McLaughlin J. Crossing multidisciplinary divides: exploring professional hierarchies and boundaries in focus groups. Qual Health Res. 2007;17(3):400–10.

Colucci E. On the use of focus groups in cross-cultural research. In: Liamputtong P, editor. Doing cross-cultural research: ethical and methodological perspectives. Dordrecht: Springer; 2008. p. 233–52.

Denzin NK, Lincoln YS. Introduction: critical methodologies and indigenous inquiry. In: Denzin NK, Lincoln YS, Smith LT, editors. Handbook of critical and indigenous methodologies. Thousand Oaks: Sage; 2008. p. 1–20.

Dickson-Swift V, James EL, Kippen S, Liamputtong P. Blurring boundaries in qualitative health research on sensitive topics. Qual Health Res. 2006;16(6):853–71.

Farnsworth J, Boon B. Analyzing group dynamics within the focus group. Qual Res. 2010;10(5):605–24.

Flood RL, Romm NRA. From metatheory to multimethodology. In: Mingers J, Gill A, editors. Multimethodology. Chichester: Wiley; 1997. p. 291–322.

Goduka, N. (2012). Re-discovering indigenous knowledge – ulwaziLwemveli for strengthening sustainable livelihood opportunities within rural contexts in the Eastern Cape Province. Indilinga Afr J Indig Knowl Syst, 11(1),1-19.

Gray PS, Williamson JB, Karp DA, Dalphin JR. The research imagination: an introduction to qualitative and quantitative methods. Cambridge: Cambridge University Press; 2007.

Gregory WJ, Romm NRA. Facilitation as fair intervention. In: Midgley G, Ochoa-Arias A, editors. Community operational research: OR and systems thinking for community development. New York: Kluwer Academic/Plenum Publishers; 2004. p. 157–74.

Hallen B. A short history of African philosophy. Bloomington: Indiana University Press; 2002.

Harris L-D, Wasilewski J. Indigeneity, an alternative worldview: four R's (relationship, responsibility, reciprocity, redistribution) vs. two P's (power and profit): sharing the journey towards conscious evolution. Syst Res Behav Sci. 2004;21:489–503.

Higginbottom G, Liamputtong P. What is participatory research? Why do it? In: Higginbottom G, Liamputtong P, editors. Participatory qualitative research methodologies in health. London: Sage; 2015. p. 1–21.

Hollander JA. The social contexts of focus groups. J Contemp Ethnogr. 2004;33(5):602–37.

Kiguwa P. Social constructionist accounts of intergroup relations and identity. In: Ratele K, editor. Inter-group relations: South African perspectives. Cape Town: Juta; 2006. p. 111–36.

Kovach M. Indigenous methodologies: Characteristics, conversations, and contexts. Toronto: University of Toronto Press; 2009.

Ladson-Billings G. Racialized discourses and ethnic epistemologies. In: Denzin NK, Lincoln YS, editors. The landscape of qualitative research: theories and issues. 2nd ed. London: Sage; 2003. p. 398–432.

Liamputtong P. Doing research in a cross-cultural context: methodological and ethical challenges. In: Liamputtong P, editor. Doing cross-cultural research: ethical and methodological perspectives. Dordrecht: Springer; 2008. p. 3–20.

Liamputtong P. Performing qualitative cross-cultural research. Cambridge: Cambridge University Press; 2010.

Liamputtong P. Focus group methodology: principles and practice. London: Sage; 2011.

Lincoln YS, Guba EG. The constructivist credo. Walnut Creek: Left Coast Press; 2013.

Magnat V. Performative approaches to interdisciplinary and cross cultural research. In: Smith L-H, Narayan A, editors. Research beyond borders. Plymouth: Lexington Books; 2012. p. 157–77.

Mapotse TA. The teaching practice of senior phase technology education teachers in selected schools of Limpopo province: an action research study. Unpublished Doctor of Education thesis. Pretoria: University of South Africa; 2012.

McIntyre-Mills J. Global citizenship and social movements: creating transcultural webs of meaning for the new millennium. Amsterdam: Harwood; 2000.

McIntyre-Mills J. From Wall Street to wellbeing. New York: Springer; 2014.

Mertens DM, Bazeley P, Bowleg L, Fielding N, Maxwell J, Molina-Azorin JF, Niglas K. Thinking through a kaleidoscopic look into the future. J Mixed Method Res. 2016;10(3):221–7.

Midgley G. Theoretical pluralism in systemic action research. Syst Pract Action Res. 2011;24:1–15.

Minkler M. Linking science and policy through community-based participatory research to study and address health disparities. Am J Public Health. 2010;100(S1):81–7.

Mkandawire-Valhmu L, Stevens PE. The critical value of focus group discussions in research with women living with HIV in Malawi. Qual Health Res. 2010;20(5):684–96.

Morgan DL. Focus group interviewing. In: Gubrium JF, Holstein JA, editors. Handbook of interview research. Thousand Oaks: Sage; 2002. p. 141–59.

Naidoo K. Poverty and socio-political transition: perceptions in four racially demarcated residential sites in Gauteng. Dev South Afr. 2011;28(5):627–39.

Ndimande BS. Decolonizing research in post-apartheid South Africa. Qual Inq. 2012;18(3):215–26.

Nicholls R. Research and indigenous participation: critical reflexive methods. Int J Soc Res Methodol. 2009;12(2):117–26.

Osei-Hwedie K. Afro-centrism: the challenge of social development. Soc Work. 2007;43(2):106–16.

Ossai NB. African indigenous knowledge systems (AIKS). Simbiosis. 2010;7(2):1–13.

Pollack J. Pyramids or silos: alternative representations of the systems thinking paradigms. Syst Pract Action Res. 2006;19:383–98.

Quan-Baffour KP, Romm NRA. Ubuntu-inspired training of adult literacy teachers as a route to generating "community" enterprise. J Lit Res. 2015;46(4):455–74.

Rajagopalan R. Immersive systemic knowing: rational analysis and beyond. Unpublished PhD thesis, University of Hull, Kingston; 2016.

Rodriguez KJ, Schwartz JL, Lahman MKE, Geist MR. Culturally responsive focus groups: reframing the research experience to focus on participants. Int J Qual Methods. 2011;10(4):400–17.

Romm NRA. Caricaturing and categorizing in processes of argument. Sociol Res Online. 1998;3. http://www.socresonline.org.uk/socresonline/3/2/10.html. Accessed 6 Jan 1999.

Romm NRA. Accountability in social research: issues and debates. New York: Springer; 2001.

Romm NRA. Issues of accountability in survey, ethnographic, and action research. In: Rwomire A, Nyamnjoh FB, editors. Challenges and responsibilities of social research in Africa: ethical issues. Addis Ababa: The Organization for Social Science Research in Eastern and Southern Africa (OSSREA); 2007. p. 51–76.

Romm NRA. New racism: revisiting researcher accountabilities. New York: Springer; 2010.

Romm NRA. Reviewing the transformative paradigm: a critical systemic and relational (Indigenous) lens. Syst Pract Action Res. 2015;28:411–27.

Romm NRA. Researching indigenous ways of knowing-and-being: revitalizing relational quality of living. In: Ngulube P, editor. Handbook of research on theoretical perspectives on indigenous knowledge systems in developing countries. Pennsylvania: IGI Global Publications; 2016. p. 22–48.

Romm NRA, Nel NM, Tlale LDN. Active facilitation of focus groups: exploring the implementation of inclusive education with research participants. South Afr J Educ. 2013;33(4):1–14. Article #811

Serequeberhan T. Our heritage. New York: Rowman and Littlefield; 2000.

Vannini A, Gladue C. Decolonized methodologies in cross-cultural research. In: Liamputtong P, editor. Doing cross-cultural research: ethical and methodological perspectives. Dordrecht: Springer; 2008. p. 137–59.

Wane N. [Re]Claiming my indigenous knowledge: challenges, resistance, and opportunities. Decoloniz Indig Educ Soc. 2013;2(1):93–107.

Visual Depictions of Refugee Camps: (De)constructing Notions of Refugee-ness?

104

Caroline Lenette

Contents

1 Introduction	1812
2 Why Discuss the "Visual"?	1814
3 Visual Depictions of Refugee Camps	1816
4 Humanitarian Sentimentalism	1816
5 Visual Analysis Process	1818
5.1 Tropes Emphasized	1818
6 Adding to Kurasawa's Framework: Feminization, Childhood, and Criminalization	1822
6.1 Feminization	1822
6.2 Childhood	1823
6.3 Criminalization	1824
7 Conclusion and Future Directions	1826
References	1827

Abstract

Visual representations of asylum seekers and refugees in precarious situations can have a significant impact on how such individuals are imagined in politically stable contexts, particularly in Western nations. Visual methodologies are relatively underused to examine how notions of refugee-ness are constructed and perpetuated to shape public opinion about asylum seekers and refugees. This topic is of particular relevance considering the intense media coverage of the recent and continuing Syrian refugee crisis in Europe during 2015–2016, and the abundance of images about this humanitarian catastrophe. Using refugee camps as example, I apply Kurasawa's framework of "humanitarian sentimentalism" where he describes four typifications (or tropes) associated with images of

---

C. Lenette (✉)
Forced Migration Research Network, School of Social Sciences, University of New South Wales, Kensington, NSW, Australia
e-mail: c.lenette@unsw.edu.au

© Springer Nature Singapore Pte Ltd. 2019
P. Liamputtong (ed.), *Handbook of Research Methods in Health Social Sciences*,
https://doi.org/10.1007/978-981-10-5251-4_47

humanitarian crises (Personification, Massification, Care, and Rescue) to visual depictions of refugee camps. This process of categorization of themes highlights what is emphasized for viewers in such imagery. As part of a broader *reflective* approach, I also discuss the themes of Feminization, Childhood, and Criminalization as key conventions in visual representations of asylum seekers and refugees. While there are ongoing tensions in relation to visual representations of people in precarious situations, visual methodologies can provide a rich dimension to critical discussions on complex and multifaceted issues.

Keywords
Visual representations · Asylum seekers · Refugees · Visual analysis · Refugee camps

1   Introduction

Images can shape and convey specific understandings of global events in the imagination of viewers, particular among Western audiences. Amidst the largely chaotic circumstances that characterize different stages of refugee-ness, photographs are often used to convey the precariousness, misery, and urgency that asylum seekers and refugees can experience. The United Nations High Commissioner for Refugees [UNHCR] (2017) estimates that there were approximately 22 million refugees at the end of 2016. The most *visible* humanitarian crisis occurring in the world in 2015 and 2016 was the global movement of asylum seekers and refugees from Syria and northern Africa as they attempted to reach European countries. This situation received intense media coverage and a number of images have been used to convey the urgency and precariousness of Syrian refugees (see Lenette and Cleland 2016). Indeed, visibility is crucial to denouncing injustices and chaos (Szörényi 2006), and visual means are often used to trigger pity or sympathy among Western audiences (Kurasawa 2013). Some images have been instrumental in shaping the world's understanding of crisis situations, and indeed in *changing* it. For example, the notorious photograph of Phan Thi Kim Phuc, the 9-year-old Vietnamese girl running from her village after being severely burnt in a napalm attack, conveyed the horrors of the Vietnam War in a powerful way, adding to pressures on political leaders to end the conflict. Closer to our times, photographs of 3-year-old Syrian Aylan Kurdi, who drowned while crossing the Mediterranean Sea with his family in September 2015, and whose body washed up on a Turkish shore, was influential in urging nations to act in the face of the major refugee crisis unfolding (Lenette 2016).

Even when situations of humanitarian crises do not make front-page news, key actors such as international nongovernment organizations, humanitarian agencies, and Western media outlets have continued to use images as a strategy "deployed to break through numbing conditions of denial" (Haaken and O'Neill 2014, p. 82). For instance, dominant visual depictions of refugee children usually show emaciated and sad-eyed African infants requiring urgent medical care, to mobilize donations from audiences in Western countries (Malkki 2015; Thompson and Weaver 2014).

Photographs can hence contribute to how viewers imagine refugees with implicit and explicit aims of triggering compassion or empathy (Haaken and O'Neill 2014). Visual depictions of asylum seeking and refugee situations are assumed to convey a certain reality, but by doing so, tend to shape viewers' ability to "imagine" refugees as distant and threatening (Johnson 2011). The outcome tends to be that the distance between viewers (usually from Western nations) and those depicted in photographs is reinforced. Such photographs tend to "produce spectacle rather than empathy" (Szörényi 2006, p. 24) and contribute to induce fear by further dehumanizing refugees, rather than challenging conservative political rhetoric and sociopolitical disparities at the source of forced migration (Kurasawa 2013).

Negative notions linked to asylum seekers and refugees through imagery remain a dominant theme in the literature. Bleiker et al.'s (2014) analysis of how asylum seekers were visually portrayed in front-page news in two prominent Australian newspapers highlight the potency of visual imagery in producing problematic constructs about refugees and asylum seekers in the media, which, when combined with deep-seated ideas of state sovereignty, border security, and national identity, create a culture of *inhospitality*. As a result, a sense of moral panic (Martin 2015) can arise in the nation's psyche, where "[f]ear of strangers outweighs the moral obligation to help them" (Bleiker et al. 2014, p. 192). Furthermore, photographic collections like the UNHCR online repository that document various refugee situations emphasize a particular "image" of refugees as victimized, depoliticized, and female (Johnson 2011; see also Malkki 1996; Haaken and O'Neill 2014). Therefore, it is useful to refer to what Christmann (2008) describes as a photographic *reality* conveyed in a two-dimensional format as one among many ways of considering a person, place, or event. In this chapter, my concern is precisely to explore the photographic realities and discourses in four images used by the UNHCR to convey the circumstances of refugees living in camps to the rest of the world.

While the focus of visual representations tends to be on negative aspects linked to refugee-ness, scant discussions on "positive" imagery linked to refugees can contribute (albeit modestly) counternarratives to dominant visual discourses (Gilligan and Marley 2010). Yet, it is fairly unusual to come across visual depictions of refugee camps, in particular, that convey more positive aspects linked to resilience, livelihood, and community (see one example in Lenette 2016). Furthermore, in the United Kingdom, for instance, positive portrayals of refugees in the media tend to focus on their artistic contributions; apart from that, meaningful contributions in socioeconomic terms, for instance, are largely ignored or overlooked (Gilligan and Marley 2010). This trend perhaps signifies an inability to recognize the vast contributions that refugees make to a nation's socioeconomic welfare. Conversely, promoting positive images solely can also "sugar-coat" the complexities of refugees' realities and project a narrow understanding of the myriad of issues they may face in conflict situations, exile, as well as in resettlement. Indeed, Gilligan and Marley (2010) caution against focusing on talented and exceptional individuals solely and instead, argue that visual representations should convey the full intricacies of refugees' lived realities.

To this end, I use Kurasawa's (2013) framework of "humanitarian sentimentalism" and his four typifications (or tropes), namely Personification, Massification, Care, and Rescue to determine which dominant tropes appear in four selected photographs. I use this approach to look beyond the surface and highlight what themes can be emphasized in such visual representations. My concern is to show how viewers can internalize specific ideas about asylum seekers and refugees (consciously and unconsciously) through visuals, and the potential for these set ideas to perpetuate public perceptions – in Western nations but also more broadly – of asylum seekers and refugees, and of forced migration situations in general. It is not my intent here to simplify complex and intricate depictions by using categories to analyze them. Rather, I would like to highlight how notions of refugee-ness can be constructed and reinforced through wide dissemination of visual patterns.

This analysis builds on my earlier work using Collier's (2004) "open viewing" to consider media images depicting asylum seekers in 2015 (Lenette and Cleland 2016), as well as discussions on an iconographic-iconologic approach to analyzing refugee photography (Lenette 2016). The "open viewing" approach is most useful at the initial stages of visual analysis where viewers let any impression emerge without rushing to find meaning or categorize. Following on from that, the iconographic-iconologic framework offers a *technical* or systematic way of "reading" photos by following a series of questions. Kurasawa's framework of using four tropes on the other hand offers a *conceptual* way of "categorizing" photos, and as such, Kurasawa's approach enriches my application of the iconographic-iconologic framework to visual representations of asylum seekers and refugees. I argue here that a broader **reflective** approach, starting with a technical description through to conceptualizing key themes represented, is most useful in the visual analysis process. A reflective approach effectively combines technical and conceptual lenses to enable the emergence of richer themes.

2　Why Discuss the "Visual"?

My interest in analyzing visual representations of asylum seekers and refugees stems from a decade-long research program using visual ethnography. Having used photo-elicitation, photovoice, and digital storytelling in collaborative research with refugee women (Lenette and Boddy 2013; Lenette et al. 2013), I began paying more attention to how the Australian media in particular portrayed stories about asylum seekers and refugees using visual means in a distinctively polemic context. The abundance of refugee photographs globally and nationally means that audiences are constantly bombarded with visual narratives on the topic, most reinforcing or justifying draconian policy measures under the banner of "security" and "sovereignty." In recent times, I have developed a particular interest in understanding how visual-based research methods can be used to influence policy directions. I have begun teaching undergraduate students how we imagine asylum seekers and refugees based on visual narratives that dominate our everyday lives, and how these constructs shape public opinion and government policy. At first, students struggle to understand the links between visual representations and policy, but once they do, wonderful insights emerge in class.

The intense media attention to the Syrian refugee crisis in Europe which peaked in 2015 has triggered more robust discussions about how images of asylum seekers and refugees are disseminated and used to influence policy and decision-making. The photographs of Aylan Kurdi's lifeless body, broadcast around the world, shocked audiences with the human tragedy of an issue long presented to the public with little humanitarian consideration. As the images became symbolic of the European refugee crisis, a number of policy measures emerged to address this precarious situation; Germany in particular showed leadership by urging nations to allow asylum seekers in dire circumstances to cross borders as they sought safety. Concurrently, the Ethical Journalism Network's *International Review of How Media Cover Migration* (2015) was released, highlighting the narrow focus of media coverage of refugee and asylum seeker issues, and recommending a broader lens for coverage of these issues, considering the media's strong influence on shaping understandings of and responses to refugee issues. Ethical Journalism Network (2015, p. 7) stated: "The inescapable conclusion is that there has never been a greater need for useful and reliable intelligence on the complexities of migration and for media coverage to be informed, accurate and laced with humanity."

In that same year, I set out to write critical commentaries on different 'forms' of visual representations of asylum seekers and refugees to add to emerging discussions on this topic. Some examples include: analyzing extant photographs using an iconographic-iconologic approach (Lenette 2016), the media's use of asylum seeker photography in times of crisis (Lenette and Cleland 2016), visual depictions of refugee deaths at border crossings (Lenette and Miskovic 2016), media representations of asylum seekers and refugees in Australian regional press (Cooper et al. 2016), international media coverage of key events linked to asylum seekers in the Australian context (Laney et al. 2016), or themes in drawings created by children living in Australian detention centers (Lenette et al. 2017). Like McDonald (forthcoming), a disabilities scholar examining visual representations of disability in art, I must admit that I was a passive viewer myself for many years and "I simply gazed and moved on." But since I turned my attention to the sociopolitical constructs conveyed in visual depictions of asylum seekers and refugees, I have become passionate about highlighting the need for a more critical stance on how such visual representations are used to influence public opinion and reinforce political rhetoric. Just like Malkki (2015), I reiterate Foucault's (1988, p. 154) definition of criticism as "pointing out on what kinds of assumptions, what kinds of familiar, unchallenged, unconsidered modes of thought the practices that we accept rest," and this is what this chapter attempts to achieve.

3 Visual Depictions of Refugee Camps

The analytical focus of refugee photography currently lies in the field of media and journalism, and explores how photographs can be used to manipulate public perceptions of asylum seekers and refugees. For instance, a systematic review of visual patterns associated with refugee photography in two major Australian national newspapers when refugee issues were at the peak of political debate revealed that images were used to reinforce detrimental and fear-mongering public discourses, ultimately depersonalizing and dehumanizing refugees (Bleiker et al. 2013; see also Esses et al. 2013). Gilligan and Marley (2010) concur that the image of the "overcrowded boat" is common and probably the most negative trope used in media representations of asylum seeking as threat-to-the-west. But the conceptualization of asylum seekers and refugees as "threatening" begins *well before* they (attempt to) reach a Western nation's borders. More precisely, visual constructs linked to refugee camps also play a key role in shaping the nation's perception of asylum seekers and refugees.

Refugee camps are said to be "the only place for people who don't fit in with our image of the world" (Møller 2015, p. 134) and are increasingly becoming the norm (Bulley 2014). Such camps have been theorized as "non-places" where "refugees are reduced to a bare form of life that merely maintains and manages their raw biological existence" (Bulley 2014, p. 65). There is an abundance of literature detailing poor living conditions in refugee camps (Malkki 1995, 2015; Agier 2011; Ramadan 2013; Bulley 2014), ones that generally fail to meet an adequate standard to provide "spaces of security for individuals and communities at their most vulnerable" (Bulley 2014, p. 63). Yet, when it comes to visual representations of refugee camps, such imagery does not necessarily speak to the vulnerability of refugees. Instead, the inherent assumption that refugees are "unlawful" (Bleiker et al. 2013, 2014) rather than vulnerable is more likely to shape viewer responses to photos of refugee camps. For instance, images of refugees living in inhospitable conditions can convey strong connotations of poverty and low socioeconomic status, which can also be associated with risk factors leading to deviancy and criminal behavior (Gabbidon and Greene 2005). This assumption would then create a different notion of the "imagined refugee" in the minds of Western viewers in particular, highlighting the importance of looking at images used by agencies like the UNHCR to portray refugee camps.

4 Humanitarian Sentimentalism

Kurasawa (2013, p. 202) argues that there is a sense of "humanitarian sentimentalism" created through the use of pictures to convey suffering and precarious situations linked to forced migration and displacement, which refers to:

> a set of narratively and visually based mechanisms aiming to trigger feelings of sympathy, repugnance, pity, and nobleness amongst Northern audiences toward subjects represented as

victims whose suffering is directly attributable to events constituted as humanitarian catastrophes in the global South.

The framework of humanitarian sentimentalism as outlined in Kurasawa's description of four "tropes" or iconographic conventions is based on major visual typifications of crises used systematically mainly by international nongovernment organizations and Western media. Each trope or convention of Personification, Massification, Care, and Rescue aims to elicit the respective moral sentiments or qualities of "pity (sorrow toward the state of victims); repugnance (revulsion or guilt toward the conditions of victims); nobleness (greatness of character directed at saving victims); and sympathy (compassion toward the suffering of victims)" (p. 206):

i. **Personification (evokes pity)**: This convention involves representing a single victim (or a very small group of people) in a state of distress to emphasize their vulnerability and precariousness; the typically close-up nature of such depictions conveys a sense of isolation that magnifies the person's suffering, and by doing so, usually provides a decontextualized narrative of the issue.
ii. **Massification (evokes repugnance)**: In contrast with Personification, people in situations of upheaval or displacement are represented in large groups or "en masse" to emphasize the magnitude of the crisis, resulting in a dehumanizing and depersonalizing effect; as such, subjective narratives are not privileged.
iii. **Rescue (evokes nobleness)**: This convention portrays an unequal rescuer-"victim" dynamic, with the "rescuer (almost invariably a white Westerner) being the only one in the frame granted the agency to intervene to transcend the deadly or unjust circumstances under which dwell those needing to be rescued (almost invariably poor non-Westerners of color)" (p. 208). This contributes to a passive or deficit-focused image of individuals at the center of crises, and the viewer is reassured that humanitarian agencies are fulfilling their roles.
iv. **Care (evokes sympathy)**: Unlike the relatively unequal and "heroic" dynamic inherent in the Rescue convention, the interpersonal and intersubjective nature of relationships between humanitarian actors and "victims" is more apparent; humanitarian involvement is based on an ethic of concern and care and a moral obligation to privilege the well-being and recovery of individuals.

These conventions have dominated visual imagery due to the need to convey the circumstances of those affected by disasters, humanitarian crises, and forced migration to mostly Western audiences, by translating "their suffering in ways that are comprehensible to such viewers because they correspond to familiar or typical interpretive patterns" (p. 207). Kurasawa acknowledges that, despite these conventions' capacity to elicit emotive responses from Western viewers, sociopolitical contexts and dominant policy approaches at the cores of such situations remain largely unchallenged. Importantly, similar themes as described in Kurasawa's work are explored in recent literature on visual representations of refugees and asylum seekers (see Szörényi 2006; Johnson 2011; Bleiker et al. 2013; Lenette 2016; Lenette and Cleland 2016), making Kurasawa's framework particularly relevant to cross-cultural analyses of constructions of refugee-ness.

5 Visual Analysis Process

World Refugee Day is celebrated each year on 20 June. In 2015, the UNHCR website featured a story entitled *Ending the second exile* to highlight the precarious situations of refugee groups around the globe. The story featured a series of photographs taken for the UNHCR and illustrating refugee situations in different settings (Figs. 1, 2, 3, and 4). I used Kurasawa's framework of humanitarian sentimentalism to identify which trope(s) described above was/were represented predominantly in these figures.

5.1 Tropes Emphasized

5.1.1 Massification

Massification: In Fig. 1, the panoramic shot of a large number of refugee tents firmly lined up into the distance conveys a sense of endlessness and permanency to the issue. The caption contributes to the notion that this is a long-term and ongoing situation, referring to the setup of Dadaab refugee camp in 1991, the growth from an initial 90,000 to 325,000 refugees in 2014, and its status as the world's largest refugee camp. The barren nature of the surrounds adds to the "distance" of the

Fig. 1 Dadaab refugee camp (Source: UNHCR 2015)

situation as a context that would be largely foreign to Western viewers. While there are individuals in the foreground (who look like children), no connection is made possible with the viewer as their features are indistinguishable and their stories remain unknown. The cloudy background adds to the harshness and gloom of the setting.

Fig. 2 Ban Mai Nai Soi refugee camp (Source: UNHCR 2015)

Fig. 3 Chad refugee camp in Iriba (Source: UNHCR 2015)

Fig. 4 Dollo Ado camp (Source: UNHCR 2015)

Rescue: The prominence of the UNHCR name and logo on the side of most tents (the unmistakable blue letters on a white tarpaulin background) conveys the organization's mandate, or more specifically, its mission to "save" vulnerable individuals by providing them with shelter and basic needs while they live in exile. The framing of the photograph ensures that the organization's presence and role in the camp is acknowledged, so that Western viewers can witness the intervention from the comfort of their homelands.

5.1.2 Personification

Personification: In Fig. 2, the close-up depiction of two women and a child in a small dwelling with basic amenities in a Thai refugee camp is presented with a narrative of intergenerational refugee-ness. Through the captions, the state of uncertainty linked to refugee camps is juxtaposed with the story of the older woman, Baw Meh, a Burmese refugee for 18 years, and the mother and grandmother of the other persons captured in the photograph. The framing emphasizes the modest and cramped living conditions: the women are squatting and sitting on the floor, and the cooking takes place over a naked flame. The women's traits are identifiable, and we learn Baw Meh's name, making them more relatable. The woman at the center of the photograph is depicted performing the traditionally female role of cooking. There is no adult male figure in the photograph, which may suggest that the women are alone to care for the child. Importantly, there is no indication of the sociopolitical factors in Myanmar that resulted in this family living in a refugee camp in Thailand.

Care: While less evident here (in relation to Kurasawa's definition), the trio is intentionally depicted as a family in the same space and engaged in a mundane activity. This may convey the impression that they are cared for and safe in this refugee camp, and that despite the conditions in which they live, they are together. The humble camp living conditions though dominate the photograph.

5.1.3 Massification

Massification: As in Fig. 2, the focus in Fig. 3 is on a small group of refugee women. However, none of the women are named and we do not know anything about their relationships, except that they are all Sudanese. The women are not identifiable and the effect is rather depersonalizing. They are pictured sitting on the ground in bare surroundings with clothes hanging in the background, with no real sense of what they are doing. The caption suggests boredom because of lack of opportunity to undertake paid work. One woman bows down her head, and all of the women wear scarves over their heads and around their faces, adding to the space between viewers and those depicted. The overall impression is deficit-oriented and suggests that they are passive actors in this setting. The brown and bare landscape extends into the distance, suggesting a lack of infrastructure in the immediate vicinity. Although the women are wearing colorful clothes, the overall impression of the photograph is that it is quite gloomy.

5.1.4 Personification

Personification: In Fig. 4, the smiling faces of the woman and child in the foreground, as well as the solar lamp, are the focal points of the photograph. The woman is named as Kadija; the caption establishes that the children are hers and that they are Somali refugees. Their traits are distinguishable, and Kadija looks straight at the camera (and by extension, at the viewer). Her posture is different to that of the women portrayed in Fig. 3, and suggests a sense of confidence. The setting is identified as *her* house, which also points to agency. In contrast with the gloomy undertones in Figs. 1, 2, and 3, the light enhances Kadija's bright and colorful clothing.

Rescue: While in the background, the blue UNHCR name and logo on a white background is still discernable. The framing of the photograph ensures that there is a constant reminder of the organization's presence and role in supporting refugees in camps. Kadija is depicted alone with her four children; perhaps she is a sole parent or the lack of a presence of an adult male figure is intentional. Either way, the framing of the photograph suggests her vulnerability as a sole carer of four, with the UNHCR (literally) watching over her.

Care: The caption refers to Ikea, a large multinational retail company popular in Western countries for its designs and ready-to-assemble furniture, and a philanthropic partner of UNICEF. The reference suggests that a highly profitable company on the other side of the world shows concern for vulnerable families in precarious situations. Solar lamps are particularly vital for women and young children who are more at risk of assaults in camps, particularly at nightfall; ownership of a lamp in this setting can mean the difference between safety and risk.

6 Adding to Kurasawa's Framework: Feminization, Childhood, and Criminalization

My ongoing work on the topic of visual representations of asylum seekers and refugees and their influence on public opinion and policy highlights three other important dimensions or key themes that can complement Kurasawa's framework, namely Feminization (represented in Figs. 2, 3, and 4), Childhood (represented in Figs. 1, 2, and 4), and Criminalization (represented in Fig. 1).

6.1 Feminization

Over the past few decades, the UNHCR and other international humanitarian organizations' visual rhetoric has adopted an approach of victimization, feminization, and racialization whereby "the image of a third world mother and child are emblematic of the refugee" (Johnson 2011, p. 1032). Images of women, strategically utilized by aid agencies in particular to trigger empathy and mobilize donations from Western audiences, tend to convey a lack of political agency and a sense of powerlessness. Bleiker et al. (2014, p. 194) explain:

> Women are typecast as emotionally fragile and vulnerable, as helpless and needy. They are passive and powerless, overwhelmed by the circumstances surrounding them. Women and their social roles become frozen into a stereotypical pattern of familial duties combined with a sense of passivity and dependence upon others for rescue and survival.

This is particularly apparent in Fig. 3, but also evident in Figs. 2 and 4. In relation to this theme, Wright (2002) suggests that dominant depictions of refugees such as those serve to reinforce stereotypes based on Christian iconography, with the aim of stimulating empathy or responses among Western viewers using "familiar" imagery (see also Malkki 2015). Besides the theme of "exodus" replicated in depictions of mass migration and forced displacement, the recurring "Madonna and child" pose is increasingly used to convey vulnerability and precariousness through images of starving mothers with young children. In fact, Johnson (2011, p. 1032) explains that "the frequently used phrase 'refugee women and children' collapses the two groups into one undifferentiated whole. The cliché *womenandchildren* (...) serves to identify men as the norm, to reiterate the notion that women are family members." The abundance of images depicting refugee women and "womenandchildren" is problematic (A Google Images search using the terms "refugee woman" will often result in photographs of women holding children predominantly.), because it reinforces the ingrained "colonial attitudes toward the developing world" (Bleiker et al. 2014, p. 194) that perpetuate stereotypical gender roles, and produces unagentic ideas about refugee women.

What these points also suggest is that the theme of Feminization can easily be overshadowed by notions of Massification and Rescue, despite the widespread use of images of women (alone or more commonly with children) to represent notions of

Fig. 5 Feminization. Refugee women in a refugee camp in Sudan (Source: Australian Broadcasting Corporation 2012)

asylum seeking and refugee-ness. I argue that Feminization, due to its increasing prevalence, should in fact be considered as a stand-alone convention or trope. Considering the growing number of women in forced migration statistics worldwide (where numbers are often lumped together as "women and children," see UNHCR 2013), a gendered lens should be integral to visual analysis of depictions of asylum seekers and refugees. Not in a simplistic and problematic manner as critiqued by Bleiker et al. (2014), but using a critical and intersectional approach (Lenette et al. 2013) that identifies how – and which part(s) of – women's identities are purposely utilized, and when they are conveniently *ignored* (Fig. 5).

6.2 Childhood

I draw here on the important work of Malkki (2015) on figurations of the child and the infantilization of humanitarian work, and how this appeals to Western audiences. Images of children as representations of humanity as a whole have long been used to raise concern and compassion among donor countries (Wright 2002; Johnson 2011; Thompson and Weaver 2014), and to shape conceptualizations of migration in public discourse (Anderson 2012). Malkki (2015, p. 79) explains that the "child is often made to appear as the exemplary human, and as politically harmless and neutral – the most neutral of neutrals, *hors combat*" and as such, in the Western imagination at least, is considered as "a potent ritual and political actor in war zones and genocides." The author argues that visual representations of children are most effective when they are used "(1) as embodiments of a basic human goodness and innocence,

Fig. 6 Childhood. A Syrian refugee child in a camp in Erbil, Iraq (Source: Itv plc (UK) 2015)

(2) as sufferers, (3) as seers of truth, (4) as ambassadors of peace (and symbols of world harmony), and (5) as embodiments of the future" (Malkki 2015, p. 80).

In Fig. 1, children are standing in the foreground but their traits are unidentifiable; however, we still perceive a notion of suffering because of the bleak surrounds. In Fig. 2, most of the child's face has been intentionally left out of the photograph, perhaps to respect his privacy. However, his "absence" is even more potent as a representation of goodness and innocence; his identity as the grandchild not only emphasizes the intergenerational nature of the refugee status, but also symbolizes looking towards the future, with perhaps hopes for a different situation for him. In Fig. 4, the children are smiling and embody all the conventions outlined by Malkki above. As such, the inclusion of children as representatives of humanitarian concern and intervention is likely to remain a prominent form of dissemination using visuals particularly among Western audiences (Fig. 6).

6.3 Criminalization

Criminalization of asylum seekers and refugees (see Morrison 2001; Fàbos and Kibreab 2007; Møller 2015) has emerged as a theme in the literature due to popular skepticism as to the motives of individuals seeking refuge, paired with nations' fear over potential risks or burdens that meeting international obligations would entail. Concern regarding the "influx" of refugees and threats to national security through terrorism (Esses et al. 2013; Milton et al. 2013) has grown to such degree across the globe that it has given rise to a moral panic over the issue. One of the reasons for this trend is that conditions in refugee camps are deemed as similar to those of a prison environment, characterized by the presence of fences, surveillance towers, and ominous boundaries to convey a notion of refugees as illegal, threatening, and unlawful (Bleiker et al. 2013; Milton et al. 2013). Discussions of refugees as "enemies at the gates" of Western nations, as Esses et al. (2013, p. 519) suggest, further emphasize the separation established between the viewer and the "victim." Moreover, limited information on immigration detention centers like in Australia, for instance, also convey living conditions, layouts, and structures akin to prisons, with "extensive security and monitoring measures, and the omnipresent surveillance features, including high wire and razor wire fences, [and] surveillance cameras" (Coffey et al. 2010, p. 2073; see also Zion et al. 2010).

This consideration is further reinforced in observations likening the structured layout of refugee tents in a camp to a prison dorm, conjuring ideas of criminality, threat, and risk. Often, boundaries and observation buildings similar to the exterior façade of a prison are prominent in camp images; refugees are presented as being justifiably confined in settings similar to low security prisons. The layout and concentration of refugee camp accommodation alludes to high intensity accommodation in a small space, giving rise to an environment likely to foster conflict (Gaes 1985). Furthermore, the erroneous belief that ethnic minorities are commonly perceived as more susceptible to criminality (Gabbidon and Greene 2005) only reinforces increased perceptions of asylum seekers and refugees as criminals. Figure. 1 conveys underlying notions of Criminality; the impression of endlessness with numerous refugee tents lined up into the distance shows both the extent of the issue of living in exile in a rather permanent structure, and emphasizes the unregulated status of individuals and families waiting to be recognized as refugees and possibly resettled (Fig. 7).

Fig. 7 Criminalization. Asylum seekers at the Manus Island detention centre (Source: Australian Broadcasting Corporation 2016)

7 Conclusion and Future Directions

One question that commonly comes up when my students discuss visual representations of asylum seekers and refugees is whether there is in fact a "right" way to portray issues linked to forced migration, detention, exile, humanitarian crises, or resettlement. How do we avoid reinforcing negative ideas or stereotypical framings and capture positive images without "sugar-coating" refugees' realities as Gilligan and Marley (2010) caution against? How does one strike a balance between showing what happens without being voyeuristic and adding to "poverty porn"? This is arguably one of the key challenges for photographers in this context. From a research perspective, these questions will continue to guide visual analyses and are by no means easy to answer. The growing body of literature on the topic signifies a shift in the nature of the conversation on visual representations of asylum seekers and refugees.

Kurasawa's framework offers a useful starting point to explore dominant themes in visual representations of refugees and asylum seekers. Through this process, it can be identified what key aspects can contribute to how refugees are imagined in the minds of Western audiences, and how these notions influence strong (and mostly negative) public opinion and (mostly harsh) government policy. The important dimensions of Feminization, Childhood, and Criminalization have become increasingly evident in contemporary visual depictions and as such, deserve more attention; they add rich dimensions to Kurasawa's tropes and make the focus of visual analysis in this field more specific. A broader reflective approach to visual analysis can therefore illuminate some of the more intricate aspects of photography and images that convey the experiences of asylum seekers and refugees, by drawing on technical as well as more conceptual approaches. This way, cross-cultural research in this field can achieve the aim of challenging what Foucault (1988, p. 154) refers to as "unconsidered modes of thought," to provide broader perspectives on the topic.

Future research in this field could focus on different sources across timeframes. In terms of sources, visual representations of asylum seekers and refugees are abundant. Common sources include: international NGO and government websites and their publications, public and private accounts on social media, photographic collections and online repositories, cartoons and photographs in the media (print and online), artwork and drawings (by asylum seekers and refugees, or by other artists), videos, posters and advertisements, or graffiti (like Banksy) to name but a few. Comparisons between sets of images across these sources using a reflective approach would add rich dimensions to this discussion by showing commonalities and differences in how similar issues are represented across sources.

In relation to timeframes, visual representations of asylum seekers and refugees clearly shift over time, and tend to change drastically around catastrophic events that receive worldwide attention, like the drowning of Aylan Kurdi or the terrorist attacks in the United States in 2001, in France in 2015 and 2016, and in Belgium and Germany in 2016. It would be enriching to compare whether contemporary representations differ from previous eras such as the coverage of the Gulf War in the 1990s (see Lenette and Miskovic 2016). Comparing when and how these shifts occur could provide important insights on the motivations behind emphasizing one aspect of refugee-ness over another, and how major events like terrorist attacks in Western nations affect the tropes used in visual representations. Visual analysis can provide an engaging and textured way of understanding complex issues and there are many opportunities to develop this area of research in cross-cultural research.

References

Agier M. Managing the undesirables: refugee camps and humanitarian government. Cambridge: Polity; 2011.

Anderson Z. Borders, babies, and "good refugees": Australian representations of "illegal" immigration, 1979. J Aust Stud. 2012;36(4):499–514.

Australian Broadcasting Corporation [ABC]. Doro refugee camp, 3 Apr 2012. 2016 Accessed 5 Aug 2016. http://www.abc.net.au/news/2012-04-03/women-sit-in-a-queue-in-doro-refugee-camp/3928974.

Australian Broadcasting Corporation. The Manus asylum seeker case awaits our next government. 2016 27 Jun 2016. Accessed 5 Aug 2016. http://www.abc.net.au/news/2016-06-27/barns-manus-asylum-seeker-case-awaits-our-next-government/7546464.

Bleiker R, Campbell D, Hutchison E, Nicholson X. The visual dehumanisation of refugees. Aust J Polit Sci. 2013;48(4):398–416.

Bleiker R, Campbell D, Hutchison E. Visual cultures of inhospitality. Peace Rev J Soc Justice. 2014;26(2):192–200.

Bulley D. Inside the tent: community and government in refugee camps. Secur Dialogue. 2014;45(1):63–80.

Christmann GB. The power of photographs of buildings in the Dresden urban discourse: Towards a visual discourse analysis. Forum: Qualitative Social Research. 2008;9(3):Art. 11.

Coffey GJ, Kaplan I, Sampson RC, Montagna Tucci M. The meaning and mental health consequences of long-term immigration detention for people seeking asylum. Soc Sci Med. 2010;70:2070–9.

Collier M. Approaches to analysis in visual anthropology. In: Van Leeuwen T, Jewitt C, editors. The handbook of visual analysis. London: SAGE; 2004. p. 35–61.

Cooper S, Olejniczak E, Lenette C, Smedley C. Media coverage of refugees and asylum seekers in regional Australian press: a critical discourse analysis. Media International Australia. 2016; 162(1): 78-89.

Esses VM, Medianu S, Lawson AS. Uncertainty, threat, and the role of the media in promoting the dehumanization of immigrants and refugees. J Soc Issues. 2013;69(3):518–36.

Ethical Journalism Network. International review of how media cover migration. 2015. Accessed 13 Sept 2015. http://ethicaljournalismnetwork.org/en/contents/moving-stories-international-review-of-how-media-cover-migration.

Fàbos A, Kibreab G. Urban refugees: introduction. Refug Can J Refug. 2007;24(3):3–10.

Foucault M. Practicing criticism. In: Kritzman LD, editor. Politics, philosophy, culture: interviews and other writings 1977–1984. New york: Routledge/Chapman & Hall; 1988. p. 152–8.

Gabbidon SL, Greene HT. Race, crime, and justice: a reader. Abingdon: Routledge; 2005.

Gaes G. The effects of overcrowding in prison. Crime Justice. 1985;95(6):95–146.

Gilligan C, Marley C. Migration and divisions: thoughts on (anti-) narrativity in visual representations of mobile people. Forum Qual Soc Res. 2010;11(2):32.

Haaken JK, O'Neill M. Moving images: psychoanalytically informed visual methods in documenting the lives of women migrants and asylum seekers. J Health Psychol. 2014;19(1):79–89.

Itv plc. Dumfries & Galloway prepares to welcome Syrian refugees. 2015. 16 Nov 2015. Accessed 5 Aug 2016. http://www.itv.com/news/border/2015-11-16/dumfries-galloway-prepare-to-welcome-syrian-refugees/.

Johnson HL. Click to donate: visual images, constructing victims and imagining the female figure. Third World Q. 2011;32(6):1015–37.

Kurasawa F. The sentimentalist paradox: on the normative and visual foundations of humanitarianism. J Glob Ethics. 2013;9(2):201–14.

Laney H, Lenette C, Kellett A, Karan P, Smedley C. "The most brutal immigration regime in the developed world": International Media Responses to Australia's Asylum-Seeker Policy. Refuge. 2016;32(3):135–149.

Lenette C. Writing with light: an iconographic-iconologic approach to refugee photography. Forum Qual Soc Res. 2016;17(2):Art. 8.

Lenette C, Boddy J. Visual ethnography: promoting the mental health of refugee women. Qual Res J. 2013;13(1):72–89.

Lenette C, Cleland S. Changing faces: visual representations of asylum seekers in times of crisis. Creat Appr Res. 2016;9(1):68–83.

Lenette C, Miskovic N. "Some viewers may find the following images disturbing": visual representations of refugee deaths at border crossings. Crime Media Cult Int J. 2016. https://doi.org/10.1177/1741659016672716.

Lenette C, Brough M, Cox L. Everyday resilience: narratives of single refugee women with children. Qual Soc Work. 2013;12(5):637–53.

Lenette C, Karan P, Chrysostomou D, Athanasopoulos A. What is it like living in detention? Insights from asylum seeker children's drawings. Australian Journal of Human Rights. 2017. Published online first.

Malkki LH. Purity and exile: Violence, memory, and national cosmology among Hutu refugees in Tanzania. Chicago, IL: University of Chicago Press; 1995.

Malkki LH. Speechless emissaries: refugees, humanitarianism, and dehistoricization. Cult Anthropol. 1996;11(3):377–404.

Malkki LH. The need to help: the domestic arts of international humanitarianism. Durham: Duke University Press; 2015.

Martin G. Stop the boats! Moral panic in Australia over asylum seekers. Continuum. 2015;29(3): 304–22.

McDonald D. Visual narratives: contemplating the storied images of disability and disablement. In: Wexler A, Derby J, editors. Contemporary art and culture in disability studies. Syracuse/New York: Syracuse University Press; forthcoming.

Milton D, Spencer M, Findley M. Radicalism of the hopeless: refugee flows and transnational terrorism. Int Interact Empir Theor Res Int Relat. 2013;39(5):621–45.

Møller B. Refugees, prisoners and camps: a functional analysis of the phenomenon of encampment. London: Palgrave Macmillan; 2015.

Morrison J. The dark-side of globalisation: the criminalisation of refugees. Race Class. 2001;43(1): 71–4.

Ramadan A. Spatialising the refugee camp. Trans Inst Br Geogr. 2013;38(1):65–77.

Szörényi A. The images speak for themselves? Reading refugee coffee-table books. Vis Stud. 2006;21(1):24–41.

Thompson B, Weaver CK. The challenges of visually representing poverty for international non-government organisation communication managers in New Zealand. Publ Relat Inq. 2014;3 (3):377–93.

United Nations High Commissioner for Refugees. Stories from Syrian refugees. 2013. Accessed 17 July 2016. http://data.unhcr.org/syrianrefugees/syria.php.

United Nations High Commissioner for Refugees. Ending the second exile. 2015. Accessed 1 July 2016. http://www.unhcr.org/news/stories/2015/6/56ec1e9b15/ending-the-second-exile.html.

United Nations High Commissioner for Refugees. Figures at a glance. 2016. Accessed 25 August 2017. http://www.unhcr.org/figures-at-a-glance.html.

Wright T. Moving images: the media representation of refugees. Vis Stud. 2002;17(1):53–66.

Zion D, Briskman L, Loff B. Returning to history: the ethics of researching asylum seekers health in Australia. Am J Bioeth. 2010;10(2):48–56.

Autoethnography as a Phenomenological Tool: Connecting the Personal to the Cultural

105

Jayne Pitard

Contents

1	Introduction	1830
2	The Purpose of a Phenomenological Study	1831
3	Autoethnography as a Phenomenological Research Tool	1832
4	The Influence of Culture in Building Relationships	1834
5	Structured Vignette Analysis	1836
6	Explanation of Research Terms	1840
7	An Example of a Structured Vignette Analysis	1840
	7.1 Vignette: Leaving the Past behind	1841
8	Conclusion and Future Directions	1842
References		1843

Abstract

Autoethnography retrospectively and selectively writes about experiences that have their basis in, or are made possible by, being part of a culture and/or owning a specific cultural identity. Telling about the experience though must be accompanied by a critical reflection of the lived experience in order to conform to social science publishing conventions (Ellis et al., Forum Qualitat Social Res 12, 2001). In researching my role as the teacher of a group of vocational education professionals from Timor-Leste, I conducted a phenomenological study using autoethnography to portray the existential shifts in my cultural understanding. I used vignettes to firstly place me within the social context, and then to explore my positionality as a researcher, carefully monitoring the impact of my biases, beliefs, and personal experiences on the teacher–student relationship. Initially, I lacked structure in my vignettes, and found it difficult to maintain a format which would guide the reader through my developing cultural awareness.

J. Pitard (✉)
College of Arts and Education, Victoria University, Melbourne, VIC, Australia
e-mail: jayne@pitard.com.au

© Springer Nature Singapore Pte Ltd. 2019
P. Liamputtong (ed.), *Handbook of Research Methods in Health Social Sciences*,
https://doi.org/10.1007/978-981-10-5251-4_48

In searching for analytical and representational strategies that would enable me to increase self-reflexivity and honor my commitment to the actual, I used vignettes to describe (show) moments of cultural existential crises, and then explore my experiences by reflecting on the reactions I had, and the actions I subsequently took, in dealing with these crises (telling). My structured vignette analysis framework helped me to reveal layers of awareness that might otherwise remain experienced but concealed, and to take the reader on a collaborative journey of cultural discovery. In this chapter, I present to you my framework as used in a cross-cultural setting.

Keywords

Autoethnography · Vignettes · Phenomenology · Cultural understanding · Structured vignette analysis

1 Introduction

An autoethnographic anecdote:

> As they arise from their seats the two women's eyes meet nervously. They approach the front of the classroom and commence their PowerPoint presentation. The first student addresses the slides in a quiet, hesitant voice struggling with pronunciation. The second student stands aside, silent, a look of consternation clouding her expression. At the conclusion of the slide presentation, the student announces that her colleague will lead the learning activity. She steps forward as a slide appears on the screen with what looks like words of a poem or song written in Tetun, the indigenous language of Timor-Leste. She starts singing in a melodious voice and encourages her co-students to join in. It appears that the song is well known to all the students as their voices rise in unison. The student leading the song becomes animated, smiling, clapping her hands and moving her body in rhythm. She looks happy for the first time since I met her one week ago. In an explosion of understanding, I realize this student is not speaking English. Memory flashes of other students speaking to her in whispers, and her puzzled facial expressions during my explanations of our activities, astound me. I hear myself congratulating the pair on their presentation and call for a ten minute break.

Autoethnography is a valuable tool for researchers when undertaking a study of the self, interacting with people within a culture. It is a research tool for writing from the heart about the researcher's experience of being part of or studying a cultural experience. Culture refers beyond ethnicity to include personal differences in the way one orients to the world, such as through privilege (being born into privilege or the absence of privilege), physical ability, emotional trauma, and religious, sexual, and political difference. Culture can also refer to a professional culture such as an experienced health worker interacting with someone who has no previous experience of interacting with the health industry.

My field of work and research is education, but the framework I developed to assist my methodology in my autoethnography adapts very well to the health and social sciences, and I wish to present it to you in this chapter. In researching my role as a white Australian teacher of a group of vocational education professionals from

Timor-Leste, I searched for a representation of shifts in my cultural understanding which would allow me to identify the very personal progression of my cultural emergence. I conducted a phenomenological study using autoethnography to highlight the existential shifts in myself, as I developed my relationship with my culturally different students. I used vignettes to place myself within the social context, to explore my positionality as a researcher and to carefully self-monitor the impact of my biases, beliefs, and personal experiences on the teacher–student relationship (Pitard 2016). To assist my analysis, I developed a structured method for analyzing each vignette to reveal layers of awareness that might otherwise remain experienced but concealed. This method of structured vignette analysis is well suited to research within the health and social sciences fields, where the researcher wishes to explore interactions (the lived experience) with those whose cultural background might give them a different perspective. In this chapter, I explain the purpose of a phenomenological study (the study of the lived experience), how autoethnography is useful in revealing the self in a cultural context, the influence of culture in building relationships, and how my structured vignette analysis builds a story without the need for an overarching narrative. I provide an example to highlight the usefulness of this method.

2 The Purpose of a Phenomenological Study

Phenomenology is essentially the study of lived experience or an event as experienced in the life world (Van Manen 1990). It considers the world as lived by a person, and not as an experience that is separate from a person (Laverty 2003). Phenomenology is a methodology that is helpful for us to understand the nature and meaning of everyday experience. With this methodology, we can investigate the meaning of participants' experiences of a phenomenon, in which the researcher is either an observer or a participant (Van Manen 1990). In essence, "phenomenology describes how one orients to lived experience" (Van Manen 1990, p. 4). Husserl (1970 [1936]) asserts that although all knowledge begins with experience, not every experience produces knowledge. How we interpret the lived experience determines whether developed knowledge will result. Generally, we view the experience as something that happens to us (beyond our control) and which we react to in the moment, and then we can see the experience as something we become conscious of and begin to interpret (Pitard 2016).

Take for example the death of a patient. The prereflective stage is when the experience happens to us, before we consciously start thinking about it. We suspend our judgment and set aside our assumptions, to instead analyze the phenomenon itself, in its purity. We are no longer holding the hand of a patient with a beating heart. We have experienced the passing of life. The sensation is real and yet it is not. Husserl describes it as "an epoche – we call it the 'transcendental reduction' ... an accomplishment of a reduction of 'the' world to the transcendental phenomenon 'world'" (p. 58). This stage is referred to variously in the literature as the *epoche*, transcendental reduction, phenomenological reduction, or bracketing (Husserl 1970

[1936]; Laverty 2003; Friesen et al. 2012; Van Manen 2014). We place ourselves at the moment of impact and describe our sensations before our mind has time to process and analyze these sensations (Pitard 2016). According to Husserl, the *epoche* dictates that any phenomenological description must be written in the first person to ensure it is described as it is experienced. The writing of this epoche is very personal to us, to the physiological sensations this death produces in us, and the feelings it invokes in us. We might then become aware of our surroundings, of family who must be consoled, and of procedures which must be followed.

3 Autoethnography as a Phenomenological Research Tool

Autoethnography is a contentious qualitative research methodology which speaks from the heart about existential crises or transformational experiences (Anderson 2006; Denzin 2006; Ellis et al. 2011). It allows researchers to focus on "ways of producing meaningful, accessible and evocative research grounded in personal experience" (Ellis et al. 2011, p. 2) by encouraging the researcher to write from the heart. Autoethnography is self-focused and context-conscious (Reed-Danahay 1997; Ngunjiri et al. 2010). It is a constructive method for researching the health practitioner–patient relationship where the practitioner and patient or patients' relatives are from diverse cultures and economic backgrounds, as it allows an in-depth exploration of the researcher as a health practitioner interacting within cultural difference. Writing about the experience of a phenomenon within an autoethnography can be a very powerful research methodology (see also ► Chap. 30, "Autoethnography").

The use of autoethnography within a phenomenological study produces real-world knowledge as experienced by the person at the center of the experience. However, in merely relating the experience as a story, it is possible the insight gained from the experience may elude the reader. As I have written previously (Pitard 2016), within a phenomenological framework, the use of autoethnography as a research tool places the self at the center of a cultural interaction, as it explores the impact of an experience on the writer. Autoethnography is an approach to research and writing to "describe and systematically analyse (*graphy*) personal experience (*auto*) to understand cultural experience (*ethno*)" (Ellis et al. 2011, p. 1). Autoethnography retrospectively and selectively indicates experiences based on, or made possible by, being part of a culture or owning a specific cultural identity. It draws the reader into a very personal experience which encourages empathy between the reader and the author, and promotes a deeper cultural understanding (Ellis and Bochner 2000). However, in order to conform to social science publishing conventions, the experience must be accompanied by a critical reflection, an analysis, of the lived experience which invites the reader to carefully consider the cultural experience being described. Autoethnographers must "use personal experience to illustrate facets of cultural experience" (Ellis et al. 2011, p. 9). The use of vignettes to examine and analyze lived experiences can provide a window through which the reader can gain an understanding of the insight which comes from placing a person with one cultural identity in a setting of different cultural norms (Pitard 2016).

Chang (2007) warns against self-indulgent introspection in autoethnography which tends to distance the reader from the cultural interaction taking place. She argues that "autoethnography should be ethnographical in its methodological orientation, cultural in its interpretive orientation, and autobiographical in its content orientation" (p. 207). Autoethnography should emphasize "cultural analysis and interpretations of the researcher's behaviours, thoughts, and experiences in relation to others in society" (p. 207). I contend that autoethnography should also draw the reader into "the inner workings of the social context studied, thereby enhancing the reader's understanding and knowledge of the culture studied. This could be explained as a collaborative journey between the reader and ...the author" (Pitard 2016, p. 5).

Reed-Danahay (1997) proposes that autoethnography connects the social to the cultural by exploring the self within the cultural context, to extend knowledge and understanding of the sociological setting within a culture. Chang (2007) "emphasizes the cultural (ethnographic) nature of autoethnography stating that this characteristic of autoethnography distinguishes it from other forms of narrative writing by connecting the personal to the cultural, the self to the social, where the self refers to the ethnographer self" (Pitard 2016, p. 6). Alexander (2005, p. 423) states that autoethnography "engages ethnographical analysis of personally lived experience." We can argue then that the prolific and significant writers on autoethnography agree that the cultural interaction which takes place is equally as important as the experience itself. Remember what was stated earlier in this chapter – culture does not rely on ethnicity. It refers to different ways of being.

Difference can produce misunderstanding which can prevent us from fully comprehending the essence of our experience with another culture. Ellis and Bochner (2000) suggest that the development of autoethnography has been driven by a desire to produce significant, accessible, and evocative accounts of personal experience to intensify our ability to empathize with people who are different from us. These accounts may provide health practitioners with an opportunity to develop a depth of understanding they may not already have developed through their own experience or through contact with different "others." In other words, this empathy can be learned through the telling and critical reflection as experienced by others. Therein lies the collaborative journey between reader and author.

As well as the benefit to readers, autoethnography permits the researcher a wider research lens with which to study their lived experience because it acknowledges the researcher's influence. Autoethnography accommodates and even embraces subjectivity. The process of autoethnography involves writing about and analyzing selected epiphanies that stem from interactions involving being part of a culture (Ellis et al. 2011). Even within the selection of which epiphanies to consider, the researcher acknowledges their influence on the research. Aligned with researcher influence though, Ellis et al. propose that it is a duty of autoethnographers in analyzing their personal experience to consider ways others may experience similar epiphanies. They further contend that it is in the analyzing of personal experience in such epiphanies that the characteristics of a culture become familiar to those inside and outside the culture. In the autobiographical style of writing used in autoethnography, it is important to show through personal descriptive writing how the epiphany was invoked through thoughts, emotions, and actions. Emphasizing ethnographic performance, Alexander (2005, p. 423) states that showing is "less about reflecting on the self ... as an act of critically reflecting culture, an act of seeing the self through and as the other." This showing can make writing emotionally rich; but to enable the reader to consider the events in a more abstract way, it needs to be balanced with some telling (Pitard 2016). Telling is a style used by an author to state what happened from a less emotional, involved standpoint (Ellis et al. 2011).

I contend that the challenges of Chang (to focus on the cultural interaction to distinguish it from other forms of narrative; 2007), Ellis et al. (to consider the ways others may have experienced our epiphanies; 2011), and Alexander (to balance our writing with showing and telling; 2005) urge writers to search for analytical and representational strategies to enable increased self-reflexivity (Pitard 2016). This honors the commitment to the actual while providing the reader with an opportunity to think about the events in a more abstract way. To conform to this version of autoethnography, I used anecdotes to describe (show) moments of cultural existential crises and explored my experiences, by reflecting on my reactions and subsequent actions, in dealing with these crises (telling) (Pitard 2016). Adopting this method of self-conscious reflexivity (Ellis and Bochner 1996) specifies the researcher's exact relation to self and to culturally different others (Alexander 2005). In striving for this process of showing and telling through the use of narrative vignettes and memory recall based on notes recorded as the event happened, an analysis of the researcher's reflexivity to these crises will reveal how the researcher responds at a more academic level. This reflexivity is a demonstration of how the professional reacts to these very personal lived experiences, which helps to emphasize the cultural impact in the analyses of the narrative vignettes. They are not a simple story of the researcher's life. They are a story of the researcher's interaction with another culture. Understanding our interactions with another culture first requires us to understand what culture is, and how it impacts our relationships.

4 The Influence of Culture in Building Relationships

Among researchers and authors, there appears to be a common premise that culture is a learned set of shared interpretations about "beliefs, norms and social practices" (Lustig and Koester 2006, p. 142), of a "historically shared system of symbolic resources through which we make our world meaningful" (Hall 1959, p. 4), a "collective programming of the mind" (Hofstede 2011, p. 3) that differentiates groups of people. Culture then, rather than stemming from genetic factors, is derived from the social environment. It is learned, not inherited (Spencer-Oatley 2008). Hofstede (1991, p. 8) contends that although certain aspects of culture are physically visible, their meaning is invisible; their cultural meaning "lies precisely and only in the way these practices are interpreted by the insiders." Insiders refer to those of a similar cultural background. Hofstede's model of the three levels of uniqueness in human mental programming – human nature, culture, and personality (Hofstede 1994, p. 6) – places human nature as common to all humans. It represents our ability to feel fear, anger, love, joy, sadness; the need to associate with others; to play and exercise; plus the ability to observe the environment and talk about it with others (Spencer-Oatey 2008). How we communicate these feelings is dictated by our culture, and the dichotomy exists that what is perfectly acceptable to one culture can be absolutely repugnant to another. Spencer-Oatey (2008) contends that culture is as much an individual, psychological construct as it is a social construct. Personality is individual as it represents a person's characteristics which are partly inherited and partly dictated by culture. Social and cognitive processing may vary in individuals from the same culture, in accordance with their life experience, and forming conclusions about a culture based on stereotypes should be avoided. Lastly, culture has both universal (etic) and distinctive (emic) elements. An example of an etic element is that common to all cultures, people feel closer to their family and relatives and those they view more similar to themselves than they do to those they view as different. However, this may be adapted in diverse cultures to include broader or smaller groups, thus making it emic. It is the emic elements of a culture which are of greatest interest to researchers. Avruch (2004, p. 20) states culture is "rooted deeply in on-going or past social practice and is to some extent situational, flexible and responsive to the exigencies of the worlds that individuals confront" (see also ▶ Chap. 88, "Culturally Safe Research with Vulnerable Populations (Māori)").

In dealing with unknown cultures, the inability to predict outcomes of intercultural communication often results in anxiety, stemming from a lack of understanding of the implied rules by which the interaction will occur (Gudykunst and Nishida 2001). Gudykunst (2005) regards uncertainty in intercultural communication as a "cognitive phenomenon"; an inability to predict attitudes, feelings, and behavior outcomes as a result of not being able to read both verbal and nonverbal cues. In explaining his theory of uncertainty and anxiety in intercultural communication, Gudykunst (1988) emphasizes the necessity of recognizing that at least one of the participants is a stranger. He argues that although the stranger is situated within the group at the time of the intercultural communication, he/she is outside the group in terms of cultural alignment. When a health practitioner is communicating with the relatives of a patient from a different culture, she is the stranger within the group. However, once she steps outside that group into the wider health network, the patient's relatives can become the strangers (displaced) within the Australian healthcare system. In this situation, Gudykunst's theory assumes the strangers' initial experiences with a different group are experienced as a series of crises where the stranger is not cognitively sure how to behave. The stranger is basing their reactions to the group on known or previously experienced interactions within their own culture, often referred to as implicit theory or habitual reactions. This can be disorientating and disconcerting as the stranger realizes the known ground rules or reactions of habit do not apply with those from another culture, creating heightened awareness of situation-behavior sequences (Gudykunst 2005). Matsumoto et al. (2005) assert self-concepts and individual values affect communication styles across cultures and that differences exist in the use of apologies, self-disclosure, compliments, and interpersonal criticism. The pervasive impact of culture on all aspects of the communication process underpins the difficulty people from diverse cultures will have in anticipating the meaning of verbal communication based on nonverbal cues. A critical examination of intercultural experiences will reveal the process of building relationships between cultures. Autoethnography is a tool for critical examination of intercultural experiences, and vignettes, describing a sequence of such experiences, can build a profile of intercultural communication which can expose a growth in cultural adaptation. In the next section, I will describe my structured vignette analysis.

5 Structured Vignette Analysis

My practice of structured vignette analysis within my autoethnography allowed me to examine my own contribution to the development of my relationship with my culturally different students. I use anecdotes within my vignettes to describe my experience *as it happened*. Van Manen (2014) uses the term *anecdote* to describe writing about the reduction moment or *epoche*. He states that "what makes anecdotes so effective is that they seem to tell something noteworthy or important about life" (p. 250). He describes the use of the anecdote in autoethnographic writing to give voice to the unconscious, deep, and pathic sensations experienced in the reduction moment. He contends that phenomenological writing should try to find "expressive means to penetrate and stir up the pre-reflective substrates of experience as we live them ... to discover what lies at the ontological core of our being" (p. 240). We should use the manifestation of our anecdotes, the words we use to express our prereflective experience, to awaken memories of the event that remained previously concealed. In this regard, researchers should allow themselves the liberty of writing freely while expressing the lived experience within their anecdotes, and only allow themselves reflection and editing when writing analyzes of the moments described in their vignettes (Pitard 2016). This almost hypnotic, trance-like state of expressing the existential moments captured in anecdotes allows the unconscious to divulge the depth of experience. Humphreys (2005) uses embedded autoethnographic vignettes with the intention of creating stories to stimulate an emotional response and provoke understanding from his readers. He connects with his innermost feelings during periods of career stress and describes experiences that his audience can connect with also. The reader is transported to the moment of truth for Humphreys.

In an effort to make sense of the cultural impact of outstanding (impactful, transformational) interactions with the culturally different, when writing vignettes the researcher should mentally transport themselves to the prereflective moment using journal entries and photographs as prompts. While it is acknowledged the retelling of these stories has already altered the prereflective experience simply through putting the experience into words (Van Manen 2014), anecdotes (the recall of the *epoche* within a vignette) are the closest a researcher will be able to transport

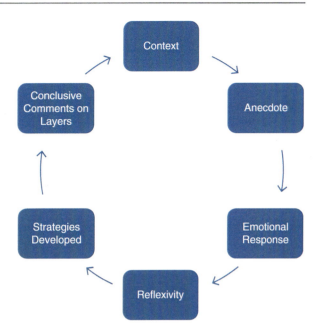

Fig. 1 Structured vignette analysis (Pitard 2016)

themselves to the prereflective moment of happening. These vignettes should record how the researcher makes sense of what happened, as a practitioner. It is essential to use reflexivity as a means to achieve an expansion of understanding in cultural interactions.

The use of a structured vignette analysis permits individual vignettes to describe distinct experiences while connecting these stories through recounting the context and revealing new strategies developed to cater for the researcher's growing cultural awareness (Pitard 2016). This is achieved through the development of a six-step framework (see Fig. 1). The framework also has the advantage of revealing several layers of awareness, described by Ronai (1995) as the layered account. The different voices of the researcher add to the richness of the analysis as the personal leads into the academic reflexive voice. Each of these layers adds a different perspective to a vignette. The six steps in my structured vignette analysis are listed in Fig. 1 and explained individually in the following section.

Context
Reality is known through socially constructed meaning (Guba and Lincoln 1994; Ritchie and Lewis 2003; Trede and Higgs 2009). Truth is negotiated through dialogue and the context of that dialogue is vital to the shaping of the data. It is through the dialectical process (the tolerance for holding apparently contradictory beliefs (Peng and Nisbett 1999) that members of a community with different cultural backgrounds come to an understanding of their social world. The researcher and the participants are "changed" by the experience and the new knowledge is a result of

this interaction, bound by the timing of the interaction and the context in which the interaction took place. According to Dervin (2003, p. 130), "context is something you swim in like a fish. You are in it. It is you." Dervin proposes that most writers about context postulate its meaning as a focus on process. She states it is "attention to process, to change over time, to emergent and fluid patterns" (p. 116). Context becomes known when the researcher turns the research lens back on the researcher. Interpretivism and constructivism manifest understanding of the meanings behind the actions of individuals to understand "the entire context, at both the macro and micro environmental level" (Pickard 2007, p. 13). Within the context relevant to each individual vignette, the reader discovers the progress of the researcher and the "other" over time (Pitard 2016).

Anecdote

In narrative anecdotes, the prereflective impact is recalled to return the researcher to the conditions before reflection or the written word impacts on the recall, to restore contact with the lived experience (Van Manen 2014). Van Manen prescribes "a certain succinctness" in the style of the anecdote (p. 252). He suggests "a set of guidelines for gathering powerful narrative material or for editing appropriate lived experience descriptions into exemplary anecdotes." These guidelines are presented below:

1. "An anecdote is a very short and simple story.
2. An anecdote usually describes a single incident.
3. An anecdote begins close to the central moment of the experience.
4. An anecdote includes important concrete details.
5. An anecdote often contains several quotes (what was said, done and so on).
6. An anecdote closes quickly after the climax or when the incident has passed.
7. An anecdote often has an effective or 'punchy' last line: it creates punctum" (p. 252).

Emotional Response

This relates to the immediate physiological and emotional responses experienced as the existential crisis unfolds. An emotional response is involuntary and unconscious. Hockenbury and Hockenbury (2011) describe three components to an emotional response: a subjective experience, a physiological response and a behavioral response.

Reflexivity

Berger's definition of reflexivity takes into account the positionality of the researcher:

> Reflexivity is commonly viewed as the process of a continual internal dialogue and critical self-evaluation of a researcher's positionality as well as active acknowledgement and explicit recognition that this position may affect the research process and outcome. (Berger 2013, p. 220)

The emphasis on reflexivity as a process, rather than an attitude or a single action, aligns with the philosophy that reality is constantly in flux, and the moment we observe reality, it is already changing. The automatic internal dialogue commences as soon as we experience something, and being aware of that internal dialogue and taking control of it is the essence of reflexivity. To take control of this internal dialogue, Berger argues that:

> researchers need to increasingly focus on self-knowledge and sensitivity; better understand the role of the self in the creation of knowledge; carefully self-monitor the impact of their biases, beliefs, and personal experiences on their research; and maintain the balance between the personal and the universal (p. 220).

Counter transference is a term used in psychodynamic language (Berger 2013) to explain the impact of clinical practitioners' own history and issues on their understanding of and reactions to the client. According to Hughes and Kerr (2000), Freud believed the practitioner's formative dynamics (assumptions based on personal life experience and important, impactful early relationships with adults such as parents) could transfer to the patient, and vice versa. He coined the term counter transference to describe this process.

Transference involves the projection of assumptions based on a previous experience to a present experience (Hughes and Kerr 2000). Reflexivity in research is the researchers' acknowledgment of and response to the impact of their own history and life issues on their interactions with their research participants. Reflexivity acknowledges that counter transference occurs in research involving participants, from the participants to the researcher, and from the researcher to the participants.

Reflexivity has been increasingly recognized as an essential strategy in the process of producing knowledge through qualitative research. Watt (2007, p. 82) explains that "since the researcher is the primary 'instrument' of data collection and analysis, reflexivity is deemed essential." According to Russell and Kelly (2002), reflection allows researchers to become aware of not only what enhances their ability to see, but also what may inhibit their seeing. Questioning one's assumptions can expose taken-for-granted attitudes that might not apply to the cultural group with whom you are interacting (Pitard 2016).

Strategies Developed
The use of reflexivity to expand understanding of an interaction between oneself and the other offers an opportunity to develop strategies to transform future interactions into more positive experiences. Exposing the strategies developed, as a result of the experience, to deal differently with similar experiences in the future can help the researcher and the reader to understand the lessons learned.

Conclusive Comments on Layers
Bringing together the different layers of account provides a concise summary of the effects of the experience and how it developed the researcher's understanding of a different culture.

6 Explanation of Research Terms

In this chapter, I have referred to the terms *epoche, phenomenological reduction*, and *bracketing,* used variously to describe the transcendental reduction (Husserl 1970 [1936]) of the lived experience. While I reference these terms to assist in explaining the transcendental reduction or prereflective stage of studying the lived experience, I do not provide a serious analysis and critique of these concepts. These terms are offered by way of assisting you to understand the phenomenological nature of autoethnography and are not intended to provide an in-depth analysis of the terms themselves (Pitard 2016).

In my structured vignette analysis, I use the term vignette in reference to my framework of six-stage analysis. My anecdote sits within the six stages of the vignette. The use of the terms *anecdote* and *vignette* are often interchanged in qualitative research methodology and my search for an explanation of their interchangeability resulted in my preference to use the term vignette to describe my six-step structured framework. I use the term anecdote to distinguish the narrative of my lived experience, which is Step 2 in my framework and I have written my anecdote, my lived experience, in the present tense. It is presented in italics to distinguish it within my vignette, as the anecdote is central to the vignette. The Sage Encyclopedia of Qualitative Research Methods does not contain an entry for anecdote and describes vignettes as "stimuli that selectively portray elements of reality to which research participants are invited to respond" (Hughes 2008, p. 918). No entry describes vignettes as autobiographical stories written within an autoethnography. Although Ellis et al. (2011), Humphreys (2005), and Ronai (1995) all use the term vignette to describe the writing of their personal experiences within autoethnography, they do not identify the structure of their vignettes into steps which include the writing of an anecdote within their vignettes. It is noted that earlier writers on phenomenological research use the term anecdote to describe the telling of their lived experience and Van Manen (2014) provides a set of guidelines for writing powerful narrative material in an anecdote (Pitard 2016).

7 An Example of a Structured Vignette Analysis

This example of my structured vignette analysis was developed during my PhD research and is based on my teaching practice with a group of students from Timor-Leste who came to Melbourne, Australia, to undertake study. Timor-Leste is a least developed country with a history of invasion, massacre, and destruction. I visited Timor initially to meet with strategic personnel to discuss the outcomes anticipated for these students for their study in Australia. Over the course of my research, I kept journal entries of my personal experience in understanding a different culture, developing intercultural relationships, notes on teaching strategies commenting on what worked and what did not, notes on conversations I had with students, photographs of students' mind maps (which became their favorite method of learning), and photographs of notes from the whiteboard in the classroom. These methods of recording my interactions with the students provided rich data to inform my vignettes. To capture as closely as possible the prereflective experience in my vignettes, I used my journal entries as a springboard to create a trance-like state in which I propelled myself back in time to the actual experience, the point of contact (Pitard 2016). I allowed the vision to flow, capturing the essence of body language, facial expressions, and physical sensations, such as a racing heart and altered spatial awareness. I offer a sample vignette below which is situated in my first visit to Timor, where I was accompanied by colleagues from my university in Melbourne, Australia, who had significant experience there.

7.1 Vignette: Leaving the Past behind

7.1.1 Context

My colleagues and I attended an alumni dinner, organized and hosted by my Australian university. At this point in our journey, I was finding it difficult to be present at the dinner because of ongoing chronic fatigue. Over a period of 18 years, I had learned to manage this condition through withdrawal for rest at strategic intervals, and I sensed this might be one of those occasions; however, I was strongly encouraged to attend the dinner, and I did so reluctantly.

We took a taxi to a Chinese restaurant in a part of Dili I had not visited. The restaurant was on the ground floor of a building in a street lined with two-story buildings of basic construction. The restaurant was decorated as part Chinese and part Timorese, so it gave the feeling of being both familiar and fragile. Two university colleagues, with whom I had worked in Melbourne, were in Dili with a group of Bachelor of Education students from Melbourne undertaking their school internships in Dili. These two colleagues also attended the dinner. It was hot and for the first hour everyone stood around chatting and drinking.

7.1.2 Anecdote

I am finding being on my feet very difficult. Fatigue is gripping my chest and my legs feel weak. I scan the room for a kind looking face and engage a Timorese alumna in conversation about her study and what this means to her in light of the declaration of independence in Timor in 2002. As she smiles broadly throughout our conversation, I feel the need to know how she can be happy when her country has been so devastated by invading forces. I ask if she lost family members during the invasion and massacre. Her smile disappears. I feel the blood rushing to my face as I recognise my intrusion into her private world. She describes those members of her family whose lives have been sacrificed. Almost too quickly, I express admiration for her positive attitude and, as her smile returns, she explains there is no point in looking back, that she and her people must build a new, strong nation and to do this, they need the assistance of the Indonesians. Confusion overwhelms me as I recognize my own anger towards the invaders. Gripped by fatigue, tears well in my eyes. I do not understand. I move swiftly towards the exit for a breath of fresh air.

7.1.3 Emotional Response

At first glance, this entry in my notes may not seem to be the intense sort of experience which might provoke an epiphany in general, but it had a very striking impact on me, which remains with me still. I felt confused by the forgiveness inherent in the Timorese attitude to the loss of family, culture, and wealth. Based on my cultural experience, I anticipated anger would be an undercurrent; but it was not. I was humbled by the happiness and excitement I was feeling all around me that night, and I had anticipated that it stemmed from the release from invasion and violence. However, the violence and loss they had experienced was in the past and, although they mourned for lost loved ones, they seemed not to place any blame as they looked to the future. Stunned by this, and fatigued beyond endurance, I was overwhelmed by guilt for my own feelings of anger.

7.1.4 Reflexivity

It was not until a few months later when I was reading in *The Age* online newspaper an article (Green 2013) based on an interview with Ms. Rosa Storelli, the displaced principal of Methodist Ladies College in Kew, Victoria, that I understood this attitude at a deeper level. The author stated his admiration that Ms. Storelli held no ill feeling to those who had moved against her to displace her from her position as a successful and much loved principal of a girls' school. When asked how she could manage this, Ms. Storelli remarked that if she harbored ill feeling, it would affect her ability to move forward in her life. She stated that she was not going to allow anybody else to dictate her future, so she had to shed any feeling she had about people who had influenced her past. I was deeply moved by this statement. In thinking about this, I recalled my conversation with the university alumna in the restaurant in Dili and experienced an epiphany of understanding so profound I realized that the Timorese were conceivably more deeply spiritual and wiser than me. In terms of coping with cultural adversity, this was very humbling.

7.1.5 Strategies Developed and Concluding Comments on Layers

I learned from this experience that if I was to serve my students' forward development, I had to adopt their attitude. What had happened in the past must stay in the past. Our task would be to work together toward the future development of the TVET system in TL. My position as the facilitator of their learning compelled me to address current and future issues. The past was theirs to deal with.

8 Conclusion and Future Directions

Autoethnography connects the personal to the cultural. It provides a qualitative research method which allows us to explore the emotions, concerns, misunderstandings, blunders, and frustration which can emerge in developing relationships with people within a cultural setting. Parks (1997) argues for a scholarly representation of the autoethnographic research experience, stating that to be evocative is insufficient to make a work scholarly. He calls for deeper levels of reflection and analysis, where the researcher develops the relationship between the personal experience and broader theoretical concepts. Others such as Garrett and Hodkinson (1999), Bochner (2000), Richardson (2000), Sparkes (2000), Holt (2003), Duncan (2004), and Wall (2006) discuss the role of criteria in judging the validity of autoethnography, and conclude that traditional criteria do not apply to autoethnography. There is an acknowledgment that different epistemological and ontological assumptions inform autoethnography simply because it involves an individual account. It is interpretivist, postmodernist, and does not rely on neutrality and objectivity. It is not disembodied, but rather relies on the tenet that we cannot separate ourselves from how we experience our lives. Richardson (2000, p. 11) argues that beyond criteria for judging validity, narratives (including autoethnography) should "seek to meet literary criteria of coherence, verisimilitude, and interest." A scholarly presentation of autoethnography can be achieved through my structured vignette analysis. This framework of analysis exposes my sense-making of teaching students from Timor-Leste for the first time. It is my unique story which captures my shock and my reflexivity. My structured vignette analysis provided me with the framework to structure my responses in an organized (coherent), predictable method (verisimilitude/reliability) to guide the reader through my experience with my culturally different students. I wrote 15 vignettes, without an overarching narrative, which take readers on a collaborative journey (interest) into my growing cultural awareness, and how I adapted my teaching to cater for the difference I perceived (connecting theoretical concepts). To balance my research (for the students' voices to be heard), I also conducted a case study of their experience. The result is a story of developing relationships amidst an emerging understanding of the cultural differences and human sameness of people of diverse backgrounds, flung together to achieve a common objective. Developing your own understanding of your experience of cultural difference could enhance your professional practice. Using autoethnography to write from the heart in cross-cultural experiences can be painful, shocking, and revealing. Moving beyond these revelations to the analytical and reflexive phases has the positive impact of adding to the body of knowledge in cross-cultural, professional interactions. This, in turn, develops the possibility that those whose experience in such matters has not afforded them the learning which can be gained from the moment of impact, may read your structured vignette analysis and experience the moment of truth for themselves. Be courageous in your research. Seek to inspire.

References

Alexander BK. Performance ethnography: the re-enacting and inciting of culture. In: Denzin NK, Lincoln YS, editors. The sage handbook of qualitative research. 3rd ed. Thousand Oaks: Sage; 2005. p. 411–41.
Anderson L. Analytic autoethnography. J Contemp Ethnogr. 2006;35(4):373–95.
Avruch K. Culture and conflict resolution. Washington, DC: United States Institute of Peace; 2004.
Berger R. Now I see it, now I don't: Researcher's position and reflexivity in qualitative research. Qual Res. 2013;15(2):219–34.

Bochner AP. Criteria against ourselves. Qual Inq. 2000;6(2):266–72.
Chang H. Autoethnography: raising cultural awareness of self and others. In: Walford G, editor. Methodological developments in ethnography 1. Oxford: Elsevier; 2007. p. 207–21.
Denzin NK. Analytic autoethnography, or déjà vu all over again. J Contemp Ethnogr. 2006; 35(4):419–28.
Dervin B. Given a context by any other name: methodological tools for taming the unruly beast. In: Dervin B, Foreman-Wernet L (with E. Lauterbach), editor. Sense-making methodology reader: selected writings of Brenda Dervin. Cresskill: Hampton Press; 2003. p. 111–32.
Duncan M. Autoethnography: critical appreciation of an emerging art. Int J Qual Methods. 2004; 3(4):Article 3.
Ellis C, Bochner AP. Composing ethnography: alternative forms of qualitative writing. Walnut Creek: AltaMira Press; 1996.
Ellis C, Bochner AP. Autoethnography, personal narrative, reflexivity: researcher as subject. In: Denzin NK, Lincoln YS, editors. Handbook of qualitative research. 2nd ed. Thousand Oaks: Sage; 2000. p. 733–68.
Ellis C, Adams TE, Bochner AP. Autoethnography: an overview. Forum Qualitat Social Res. 2011;12(1):Art. 10.
Friesen N, Henriksson C, Saevi T. Hermeneutic phenomenology in education. Method and practice. Rotterdam: Sense Publishers; 2012.
Garrett D, Hodkinson P. Can there be criteria for selecting research criteria? A hermeneutical analysis of an inescapable dilemma. Qual Inq. 1999;4:515–39.
Green S. Lessons learned from the college of life; 2013. The Age. Retrieved from http://www.theage.com.au/national/education/lessons-learnt-from-the-college-of-life-20131127-2y9wk.html
Guba EG, Lincoln YS. Competing paradigms in qualitative research. In: Denzin NK, Lincoln YS, editors. Handbook of qualitative research. Thousand Oaks: Sage; 1994. p. 105–17.
Gudykunst WB, editor. Language and ethnic identity. Clevedon: Multilingual Matters; 1988.
Gudykunst WB, Nishida T. Anxiety, uncertainty, and perceived effectiveness of communication across relationships and cultures. Int J Intercult Relat. 2001;25(1):55–71.
Gudykunst WB. Theorizing about intercultural communication. Thousand Oaks: Sage. 2005.
Hall ET. The silent language. New York: Doubleday; 1959.
Hamilton ML, Smith L, Worthington K. Fitting the methodology with the research: an exploration of narrative, self-study and autoethnography. Stud Teach Educ. 2008;4(1):17–28.
Hockenbury DH, Hockenbury SE. Discovering psychology. New York: Worth Publishers; 2011.
Hofstede G. Dimensionalizing cultures: the hofstede model in context. Online Read Psychol Cult. 2011;2(1).
Holt N. Representation, legitimation, and autoethnography: an autoethnographic writing story. Int J Qual Methods. 2003;2(1):Article 2.
Hughes P, Kerr I. Transference and countertransference between in communication between doctor and patient. Adv Psychiatr Treat. 2000;6(1):57–64.
Hughes R. Vignettes. In: Given L, editor. The sage encyclopedia of qualitative research methods. Thousand Oaks: Sage; 2008. p. 919–21.
Humphreys M. Getting personal: reflexivity and autoethnographic vignettes. Qual Inq. 2005;11:840–60.
Husserl E. The crisis of European sciences and transcendental phenomenology (trans: Carr D.). Evanston: Northwestern University Press; 1970 [1936].
Laverty SM. Hermeneutic phenomenology and phenomenology: a comparison of historical and methodological considerations. Int J Qual Methods. 2003;2(3):Art. 3.
Lustig MW, Koester J. Intercultural competence: Interpersonal communication across cultures. 5th ed. Boston: Pearson Allyn and Bacon; 2006.
Matsumoto D, Leroux JA, Yoo SH. Emotion and intercultural communication. Kwansei Gakuan University J. 2005;99:15–38.
Ngunjiri FW, Hernandez K-AC, Chang H. Living autoethnography: connecting life and research [editorial]. J Res Pract. 2010;6(1):Art E1.

Parks M. Where does scholarship begin? Address presented at the annual conference of the National Communication Association, Chicago; 1997 .
Peng K, Nisbett RE. Culture, dialectics, and reasoning about contradiction. Am Psychol. 1999; 54(9):741–54.
Pickard AJ. Research methods in information. London: Facet; 2007.
Pitard J. Using vignettes within autoethnography to explore layers of cross-cultural awareness as a teacher [40 paragraphs]. Forum Qualitat Social Res. 2016;17(1):Art. 11.
Reed-Danahay DE. Leaving home: schooling stories and the ethnography of autoethnography in rural France. In: Reed-Danahay, editor. Auto/ethnography: rewriting the self and the social. Oxford: Berg; 1997. p. 123–44.
Richardson L. New writing practices in qualitative research. Sociol Sport J. 2000;17:5–20.
Ritchie J, Lewis J. Qualitative research practice: a guide for social science students and researchers. London: Sage; 2003.
Ronai CR. Multiple reflections of child sex abuse: an argument for a layered account. J Contemp Ethnogr. 1995;23(4):395–426.
Russell GM, Kelly NH. Research as interacting dialogic processes: implications for reflexivity. Forum Qualitat Social Res. 2002;3(3):Art. 18.
Sparkes AC. Autoethnography and narratives of self: reflections on criteria in action. Sociol Sport J. 2000;17:21–43.
Spencer-Oatey H. Culturally speaking: culture, communication and politeness theory. 2nd ed. London: Continuum; 2008.
Trede F, Higgs J. Framing research questions and writing philosophically in writing qualitative research on practice. In: Higgs J, Horsfall D, Grace S, editors. Writing qualitative research on practice. Rotterdam: Sense Publishers; 2009. p. 13–26.
van Manen M. Researching lived experience: human science for an action sensitive pedagogy. Albany: State University of New York Press; 1990.
van Manen M. Phenomenology of practice: meaning giving methods in phenomenological research and writing. Walnut Creek: Left Coast Press; 2014.
Wall S. An autoethnography on learning about autoethnography. Int J Qual Methods. 2006;5(2): Article 9.
Watt D. On becoming a qualitative researcher: the value of reflexivity. Qual Rep. 2007; 12(1):82–101.

Ethics and Research with Indigenous Peoples

106

Noreen D. Willows

Contents

1 Introduction	1848
2 Indigenous Peoples and Right to Self-Determination in Health Research	1850
3 Barriers to Addressing Health Inequities and Health Disparities Experienced by Indigenous Peoples	1851
4 The Right of Indigenous Peoples to Participate in Decolonizing Research that Improves Health and Well-Being	1852
5 Research Frameworks for Research with Indigenous Peoples: Examples from Canada	1854
6 Institutional Research Guidelines: Canada and New Zealand	1856
6.1 Canada: Tri-Council Policy Statement: Ethical Conduct for Research Involving Humans: TCPS 2	1858
6.2 Ethical Guidelines for Health Research with Māori in New Zealand	1859
7 Guidelines for Health Research Developed by Indigenous Communities or Agencies	1860
8 Building Capacity to Do Ethical Research with Indigenous Peoples	1861
8.1 Institute on the Ethics of Research with Indigenous Peoples	1862
8.2 Centre for Excellence in Indigenous Health	1863
8.3 Community-Based Research and Evaluation Certificate Program	1863
8.4 Master of Public Health: Native Hawaiian and Indigenous Health Specialization	1864
8.5 Community Mobilization for Healthy Lifestyles and Diabetes Prevention Training Program	1864
8.6 Summer Research Training Institute for American Indian and Alaska Native Health Professionals	1865
9 Conclusion and Future Directions	1865
References	1867

N. D. Willows (✉)
Faculty of Agricultural, Life and Environmental Sciences, University of Alberta, Edmonton, AB, Canada
e-mail: noreen.willows@ualberta.ca

> **Abstract**
>
> Many Indigenous peoples have poorer health compared with the settler populations that colonized their territories. States and academic institutions have an obligation to support ethical research with Indigenous peoples that results in the elimination of health disparities. Decolonizing research is required that serves to restore health in conformity with enduring Indigenous values that affirm life. Indigenous peoples may have concerns that health research under the control of outsiders will come to conclusions about Indigenous health disparities that stereotype, pathologize, and/or marginalize Indigenous peoples; be instrumental in rationalizing colonialist perceptions of Indigenous incapacity and the need for paternalistic control of Indigenous interests, or deduce that Indigenous peoples are sick and incapable of self-care. Health research that respects Indigenous self-determination, and is safe, ethical, and useful for participants, requires increased capacity among Indigenous and non-Indigenous peoples alike. Indigenous peoples have the right to control research that generates knowledge affecting their well-being. Community members need workshops and training sessions that will inform them how to negotiate with health researchers, let them know their rights as research participants, and build their skills to conduct their own research. Non-Indigenous researchers require appropriate ethical guidelines to follow and training opportunities that offer guidance on Indigenous ways of knowing, the social determinants of health, strength-based research approaches, community-based participatory research, and how to engage in culturally appropriate ways with Indigenous peoples. Researchers wanting to pursue a specialization in Indigenous health research need support from academic leadership and funding agencies to be successful in their endeavor.

> **Keywords**
>
> Community-based participatory research · Decolonizing research · Health disparities · Health inequities · Indigenous peoples · Research ethics

1 Introduction

The United Nations Declaration on the Rights of Indigenous Peoples was adopted by the General Assembly in 2007 (United Nations 2008). Article 24 of the Declaration indicates that States will take the necessary steps to ensure that Indigenous individuals have an equal right to the enjoyment of the highest attainable standard of physical and mental health. Given this proclamation, this chapter argues that States can help to ensure that a high standard of health is achieved by Indigenous peoples by supporting ethically conducted, innovative research programs based on scientific excellence and Indigenous community collaboration. State-supported research of this nature is occurring in some countries. For instance, the Institute of Aboriginal Peoples' Health (IAPH) is one institute of the Canadian Institutes of Health Research (CIHR), Canada's federal funding agency for health research. IAPH supports health research that is conducted using the highest ethical and moral standards and that respects Aboriginal cultures, while generating new knowledge to improve the health and well-being of Aboriginal peoples (Canadian Institutes of Health Research 2011).

This chapter discusses how States, academic institutions, and Indigenous groups can support ethical research with Indigenous peoples, communities, and nations. It builds on an editorial that I wrote about the requirement for ethical principles of health research involving Indigenous peoples in Canada (Willows 2013). The editorial followed revelations by a historian in 2013 that a series of egregious nutrition studies had been conducted in Canada in the 1940s and 1950s by federal government scientists, bureaucrats, and university researchers that used malnourished First Nations children and adults as experimental material and their communities as laboratories for scientific experimentation (Mosby 2013). In the studies, researchers had exclusive control over the research process and the use of results, and they did not return meaningful results to the communities. There was no evidence that these studies resolved the malnutrition, hunger, or suffering of the children or community members included in them; rather, the beneficiaries were those who lead the research, as carrying out these studies furthered their own professional and political interests (Mosby 2013). While it is unlikely today that health research that violates the inalienable rights of research participants such as the provision of informed consent would be sanctioned within Canada, it is possible that Indigenous peoples' unique rights might be violated, or that unique considerations for conducting research with Indigenous peoples would not be followed. For example, researchers may not discern that their research creates an imbalance of power between them and Indigenous research participants; that it devalues traditional Indigenous knowledge in favor of Western scientific knowledge; that it violates community norms; or, that the findings misrepresent or stigmatize community members (Canadian Institutes of Health Research, Natural Sciences and Engineering Research Council of Canada, and Social Sciences and Humanities Research Council of Canada 2014, p. 109).

In this chapter, I discuss why scientifically rigorous health research that takes into account Indigenous rights is urgently needed as well as some of the considerations for conducting ethical research with Indigenous peoples. I will refer mostly to the Canadian literature on these topics due to my familiarity with the material. Although the chapter has broad applicability to research with Indigenous populations in many countries, it will likely have the greatest application to Indigenous groups in Australia, Canada, New Zealand, and the United States for several reasons. The Indigenous peoples in Canada have geographical and/or cultural contiguity with many of the Indigenous peoples in the United States. All four countries are Western, liberal democracies, originally settled by colonizers who were predominantly of European-ancestry. Indigenous peoples in these countries are united by similarities in their colonial treatment by settler populations whereby colonization resulted in settlers usurping the land and resources of the local Indigenous peoples and expropriating and/or suppressing Indigenous peoples' lives and identities. Indigenous peoples in all four countries have poorer socioeconomic and health outcomes in comparison to the settler majority (see also ▶ Chaps. 87, "Kaupapa Māori Health Research," ▶ 88, "Culturally Safe Research with Vulnerable Populations (Māori)").

Given the geopolitical focus of this chapter, the term Indigenous used herein has the greatest applicability to Métis, First Nations, and Inuit peoples in Canada (collectively called Aboriginal peoples in Section 35 of Canada's Constitution Act of 1982); Aboriginal peoples and Torres Strait Islanders in Australia; Māori in New Zealand; and, American Indians, Alaska Natives (i.e., Inuit, Yupik, and Aleut peoples), and Native Hawaiians in the United States. It is important to bear in mind that settler governments conjured these terms for Indigenous peoples for political and jurisdictional purposes. Indigenous populations have much greater cultural diversity than ascribed by these labels, and may choose to use their own group's cultural name when referring to themselves, as a decolonizing action, as a means to achieve self-determination, and for increased accuracy of identity (Allan and Smylie 2015).

The perspective of ethical research that I present partially reflects my professional experiences as a non-Indigenous health researcher working in an academic environment. My career has focused predominantly on nutrition research with First Nations communities in Canada, including the development, implementation, and evaluation of nutrition interventions (e.g., Pigford and Willows 2010; Triador et al. 2015). I aim to adopt a community-based participatory research (CBPR) approach to the work that I do, whereby I partner with First Nations community members to find culturally appropriate solutions to their nutrition concerns or nutrition-related health conditions. The viewpoints that I express in this chapter stems from my interest in articulating the practice and outcomes of CPBR with First Nations peoples (Pigford et al. 2013; Willows et al. 2016; Gokiert et al. under review); what constitutes ethical research with Indigenous peoples (Willows 2013); how to copartner with First Nations peoples (Genuis et al. 2014, 2015; Willows et al. 2016); and developing culturally appropriate frameworks to conceptualize Indigenous peoples' health issues (Willows et al. 2012).

2 Indigenous Peoples and Right to Self-Determination in Health Research

There is no singularly authoritative definition of Indigenous peoples under international law and policy although criteria to help identify Indigenous peoples have been established by the United Nations (United Nations 2013; United Nations, n.d.). These criteria include peoples who have historical continuity with precolonial and/or presettler societies that developed on their territories; consider themselves to have distinct social, economic, or political systems, language, culture, and beliefs distinct from other sectors of the societies now prevailing on their territories, or parts of them; have strong linkages to territories and surrounding natural resources; and form at present nondominant groups of society that resolve to maintain and reproduce their ancestral environments and systems as distinctive peoples, in accordance with their own cultural patterns, social institutions, and legal systems. Based on these criteria, in 2016 there were more than 390 million Indigenous peoples worldwide (Food and Agriculture Organization of the United Nations 2016).

The label of "indigeneity" serves a pragmatic purpose, as it allows a diversity of groups, societies, and nations to resist domination by settler populations and to demand their entitlement to their rights as Indigenous peoples. Indigenous rights is about "unfolding *in practice* such notions as equality, procedural justice and a universal right of self-determination that the idea of human rights has always promised" (Guenther et al. 2006, p. 28, italicized in original). The application of the right to self-determination for Indigenous groups, societies, and nations requires recognition of collective rights, self-governance, and autonomy and control of lands and resources (Guenther et al. 2006). Fundamental to the exercise of self-determination, Indigenous peoples have the right to control research that generates knowledge affecting their cultural heritage (including their traditional knowledge, traditional cultural expressions, and intellectual property), identity and well-being, and to construct knowledge in accordance with self-determined definitions of what is real and what is valuable (Castellano 2004; Australian Institute of Aboriginal and Torres Strait Islander Studies 2012; see also "Indigenist and Decolonizing Research Methodology").

3 Barriers to Addressing Health Inequities and Health Disparities Experienced by Indigenous Peoples

Improved health and well-being for Indigenous peoples through research is desired by Indigenous and non-Indigenous peoples alike. However, there are many barriers that prevent the undertaking of high-quality, ethical health research. Many Indigenous communities do not have the financial or internal human resources to address the health disparities that they face. For example, due to policies and practices that emerged from colonial ideologies, Indigenous peoples in Canada have low high school completion rates and often do not achieve a postsecondary education (Allan and Smylie 2015). Consequently, Canadian universities report an underrepresentation of Indigenous scholars in their professoriate (Ramos 2012) meaning that there are few academically trained health researchers who are Indigenous. Non-Indigenous health researchers may be reluctant to engage in research with Indigenous peoples, based on perceived barriers to conducting research with Indigenous peoples, some which have been articulated elsewhere (Castleden et al. 2015).

Fears that have been expressed to me by non-Indigenous colleagues about engaging in health research in Indigenous communities are that the requirements for undertaking ethical and collaborative research with Indigenous communities will prevent research from being scientifically rigorous, will lengthen the time to do research, and will limit the number of publications produced from the research. These concerns are generally unfounded *as long as academics are supported by their institutions and funding agencies to undertake ethical and collaborative research.* While research with Indigenous peoples can require an extensive commitment of time, academic scholars have established credible scientific careers based on research partnerships with Indigenous communities. I have been fortunate to receive salary awards from my Provincial Government through Alberta Innovates Health Solutions, and before that, the Alberta Heritage Foundation for Medical Research that reduced my teaching and administrative load at the University of Alberta, giving me the time to fully engage with First Nations partners. Thus, it is my contention that many of the barriers that academic researchers perceive limit the production of high-quality research that addresses Indigenous health issues could be overcome if academic scholars were better supported by academic departments and faculties to engage with Indigenous communities in a respectful way.

Indigenous peoples have the right to know about their health status; the causes, nature, and treatment of their ill-health; and the resources available to improve their health (UN Office of the High Commissioner for Human Rights 2008). The task of undertaking health research in Indigenous communities to address these topics often falls to non-Indigenous peoples, many who have been trained exclusively in Western ways of knowing and conducting science, and who may have little knowledge of Indigenous peoples. Indigenous peoples may be reluctant to engage with these health researchers due to concerns that research under the control of outsiders to Indigenous communities will come to conclusions about Indigenous health disparities that stereotype, pathologize, and/or marginalize Indigenous peoples; be instrumental in rationalizing colonialist perceptions of Indigenous incapacity and the need for paternalistic control of Indigenous interests; deduce that Indigenous peoples are sick and incapable of self-care; and appropriate or not value traditional knowledge or cultural practices (Castellano 2004; Gracey and King 2009; Reading and Wien 2009; Smylie and Adomako 2009; Willows 2013; Adam and Smylie 2015). The dilemma for Indigenous communities with pressing health concerns is that they may feel compelled to engage with researchers who lack experience with research approaches and methodologies that are appropriate for an Indigenous research context such as being respectful of traditional knowledge, cultural practices and beliefs. There may be concerns that the approaches taken will not be inclusive of Indigenous perspectives, processes, and ways of learning/knowing; will not recognize colonization and exclusionary social policies as Indigenous health determinants; will not take a strength-based approach to research but rather one based on deficits and victim-blaming; and will not use approaches that recognize the potential trajectories of the social determinants of health across the life course (Bartlett et al. 2007; Reading and Wien 2009; Pigford et al. 2013).

4 The Right of Indigenous Peoples to Participate in Decolonizing Research that Improves Health and Well-Being

Decolonizing research is required that serves to restore order to daily living in conformity with ancient and enduring Indigenous values that affirm life (Castellano 2004; see also "Indigenist and Decolonizing Research Methodology and Using an Indigenist Framework for Decolonizing Health Promotion Research"). Table 1

Table 1 Aspects of life-affirming decolonizing research with Indigenous populations

Aspect of decolonizing research	Sample reference
Aims to create knowledge for social benefit	Castellano (2004), Gray and Oprescu (2015), Robertson (2016)
Follows a code of ethics based on rules of conduct which distinguish between acceptable and unacceptable research practices, and expresses and reinforces important indigenous social and cultural values	Castellano (2004), First Nations Centre (2007)
Addresses the hierarchical relation of power that privileges academic over local, indigenous knowledge by incorporating or honoring research methods and theories rooted in indigenous knowledge	Bartlett et al. (2007), Castellano (2004), Zavala (2013)
Includes indigenous elders or other keeps of cultural knowledge in the design and execution of research, and the interpretation of findings in the context of cultural norms and traditional knowledge	Castellano (2004), Canadian Institutes of Health Research, Natural Sciences and Engineering Research Council of Canada, and Social Sciences and Humanities Research Council of Canada (2014)
Explicitly recognizes the right of indigenous communities and nations to be self-determined and self-governed in matters relating to their internal and local affairs	Castellano (2004)
Controlled by indigenous people, which overcomes the ineffectiveness of externally imposed and expert-oriented forms of research and helps to prevent the production of knowledge of little value to indigenous communities	Bartlett et al. (2007), Castellano (2004), Zavala (2013)
Use strength-based rather than deficit-based research approaches	Pigford et al. (2013)
When research involves communities, use a community-based participatory research approach whereby community members are equal partners in the research process with researchers from outside of the community	Bartlett et al. (2007), Castellano (2004), Castleden et al. (2012), LaVeaux and Christopher (2009), Pigford et al. (2013), Zavala (2013)
When working with communities, consider including a research partnership agreement "that represents a formal summary of rights, responsibilities and good faith between the parties entering into a partnership to jointly conduct research"	First Nations Centre (2007, p. 11)

outlines some aspects of such life-affirming decolonizing research. Indigenous peoples, according the United Nations Declaration on the Rights of Indigenous Peoples, have the right to maintain, control, protect, and develop their intellectual property over traditional knowledge. In Canada, Ownership, Control, Access, and Possession (OCAP® is a registered trademark of the First Nations Information

Governance Centre (FNIGC)) (www.FNIGC.ca/OCAP) research principles offer a First Nations approach to research, and data and information management, which help to ensure First Nations aspirations towards self-determination and self-governance. These principles affirm that a First Nations community owns research information collectively (**O**wnership); is within its rights to seek control over all aspects of research and information management processes which impact it (**C**ontrol); has access to the information and data about their community (**A**ccess); and can have physical control of research data (**P**ossession) (First Nations Centre 2007).

Below, I expand on the practices and considerations for decolonizing and ethical research with Indigenous communities, using examples from Indigenous research frameworks, ethic guidelines developed for academic researchers working with Indigenous peoples in Canada and New Zealand, and research guidelines developed by Indigenous agencies or communities to ensure ethical research. The guidelines and frameworks are meant to ensure that research is ethical, culturally appropriate, collaborative, meaningful, and beneficial to Indigenous communities. Concerns about the implementation of some of these practices and considerations are outside the scope of this chapter to discuss in detail. For instance, there may be inconsistencies in ethical requirements between institutional review boards/academic research ethics boards and Indigenous review boards, lack of clarity about whether community self-determination is more important than individual autonomy in decisions about research participation, concerns that the rigidity of institutional review boards/academic research ethics boards requirements to "protect" Indigenous communities ironically undermines community self-determination, contested issues around academic freedom and research findings as intellectual property, and the expense and challenges of conducting community-based participatory research (Smith-Morris 2007; Ritchie et al. 2013; Angal et al. 2016; Brunger and Wall 2016).

5 Research Frameworks for Research with Indigenous Peoples: Examples from Canada

In Canada, several Aboriginal-specific frameworks for decolonizing research and health promotion activities exist that de-emphasize the focus on individual-level risk factors for disease and instead highlight the contributions of social and environmental conditions to the divergence in health status between Indigenous and non-Indigenous peoples. A process framework has been developed for Aboriginal-guided decolonizing research that privileges Indigenous ways of learning/knowing. It employs iterative, culturally based, and process-oriented methods (Fig. 1). It was reported that implementing this framework in research involving Métis and First Nations peoples with diabetes increased the efficiency and effectiveness of the research process (Bartlett et al. 2007).

Some Canadian frameworks use a social-ecological approach that recognizes that individuals are embedded within social, economic, and political systems that shape health behaviors and access to resources necessary to maintain health. The *Integrated Life Course and Social Determinants Model of Aboriginal Health* is a

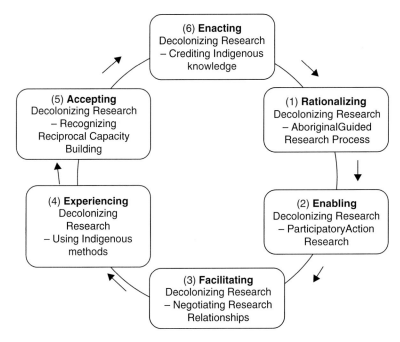

Fig. 1 Process framework for decolonizing research (Bartlett et al. 2007) (Reprinted from Social Science & Medicine, 65(11), Bartlett, J. G., Iwasaki, Y., Gottlieb, B., Hall, D., & Mannell, R. Framework for Aboriginal-guided decolonizing research involving Métis and First Nations persons with diabetes, pages 2371-2382, 2007, with permission from Elsevier)

population-focused conceptual framework for understanding the relationship between various health dimensions and the social determinants, categorized as proximal, intermediate, and distal. It also examines potential trajectories of health across the life course (Reading and Wien 2009). Not taking social-ecological factors into account can result in failed health interventions. This was the conclusion of a review of seven healthy weight interventions specifically aimed at First Nations or American Indian children and youth (Towns et al. 2014). Only two of the interventions included environmental or policy components that supported behavior change. The authors of the review concluded that the ineffectiveness of the interventions to reduce obesity or overweight was because structural factors in the social, economic, and physical environments where the Indigenous children and youth lived prevented them from making the behavioral changes required to have a healthy body weight.

The *Integrated Life Course and Social Determinants Model of Aboriginal Health* provides an analytical guide to explore the relationships between the social determinants of health (SDoH) and health outcomes among First Nations people living off-reserve as captured in the 2012 national Aboriginal Peoples Survey (Rotenberg 2016). SDoH were examined using the categories of the model as follows: proximal (health behaviors, physical and social environment); intermediate (community infrastructure, resources, systems, and capacities); and distal (historic, political, social, and economic). This approach permitted the researcher to determine which SDoH assessed at the three levels had the greatest impact on the likelihood of a First Nations person having a chronic condition, poor or fair self-rated general health, or poor or fair self-rated mental health. For example, when proximal determinants were examined, those related to negative health outcomes were smoking, obesity, living in a dwelling where major repairs were needed, having less than a high school education or being unemployed, living in a low-income household, or experiencing household food insecurity. The compounding effects of having multiple points of social disadvantage on health were explored by examining the exacerbating effects of intersecting SDoH. The results showed that the likelihood of reporting any of the three negative health outcomes increased as the number of social determinants of poor health increased. It was evident from study findings that health behaviors as well as environmental and social conditions impact the health outcomes of First Nations people living off-reserve, thus individual, family, and community interventions are all required to improve health (Reading and Wien 2009).

Willows et al. (2012) developed a *Socioecological framework to understand weight-related issues in Aboriginal children* that highlighted the need to understand childhood obesity within the context of inequities in the social determinants of health (Fig. 2). The framework focuses on the many environments at different times in childhood that influence an Aboriginal child's weight status, including prenatal, sociocultural, family, community, and policy environments. The framework highlights historical and ongoing factors related to children having an unhealthy body weight, including colonization by Europeans, dispossession of Aboriginal peoples from their traditional lands, and assimilation policies which influence all other socioecological levels (i.e., individual; intrapersonal; community, home, sociocultural; built environment; and society).

6 Institutional Research Guidelines: Canada and New Zealand

Health research ethics guidelines for Indigenous peoples should deal minimally with issues such as the nature of the relationship between the researcher and the research participant; ownership of and access to data; conflict of interest; consent to research; privacy and confidentiality; and measures to preserve human dignity (First Nations Centre 2007). Various guidelines for research with Indigenous peoples exist is Australia, Canada, New Zealand, and the United States. For example, in Australia there are ethical guidelines for research involving Aboriginal and Torres Strait Islander People, which were last revised in 2004, that indicate that research must be conducted in an ethical and culturally safe and appropriate manner as to protect the health, safety, and well-being of Aboriginal and Torres Strait Islander peoples and their communities (National Health and Medical Research Council 2016). The Australian Institute of Aboriginal and Torres Strait Islander Studies (2012) also created *Guidelines for Ethical Research in Australian Indigenous Studies* to ensure that research with and about Indigenous peoples follows a process

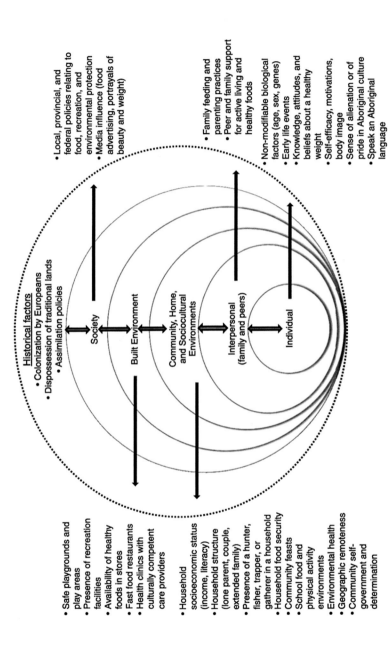

Fig. 2 Ecological model for understanding obesity in children, which illustrates the reciprocity among levels that influence active living, the consumption of healthy foods, and weight status, and which recognizes that historical factors encompass and influence all ecological levels (Willows et al. 2012)

of meaningful engagement and reciprocity between the researcher and the individuals and/or communities involved in the research. In the United States, there are various codes of ethics and guidelines for researchers and scholars working on projects related to Native Americans, Alaska Natives, or Native Hawaiian peoples. However, as recently as 2009, concerns were expressed that issues such as tribal sovereignty were still not adequately protected by current legislation (Sahota 2009). The website of the National Congress of American Indians' Policy Research Center (http://www.ncai.org/policy-research-center/initiatives/research-regulation) provides resources to support tribes and American Indian and Alaska Native communities working to develop research oversight processes and policies. For the sake of brevity, only health research ethics guidelines from Canada and New Zealand will be described in this chapter.

6.1 Canada: Tri-Council Policy Statement: Ethical Conduct for Research Involving Humans: TCPS 2

Institutions eligible to receive funding from Canada's three federal research agencies – the Canadian Institutes of Health Research, the Natural Sciences and Engineering Research Council of Canada, and the Social Sciences and Humanities Research Council of Canada – must agree to adhere to the Tri-Council Policy Statement: Ethical conduct for research involving humans (known as the TCPS 2) as a condition of funding (Canadian Institutes of Health Research, Natural Sciences and Engineering Research Council of Canada, & Social Sciences and Humanities Research Council of Canada 2014). Research Ethics Boards at Canadian Universities must ensure the application of the TCPS 2 to human research. The TCPS 2 was last revised in 2014. Chapter 9 is titled "Research Involving the First Nations, Inuit and Métis Peoples of Canada." It is designed to serve as a framework for the ethical conduct of research involving the Indigenous peoples of Canada, but is not intended to overrule or replace ethical guidance offered by Aboriginal peoples themselves. An online tutorial is available to help researchers to understand Chapter 9 (Government of Canada n.d.).

It is intended that research that follows Chapter 9 guidelines will reflect Aboriginal worldviews, and will benefit Aboriginal peoples or communities. The chapter offers guidance on how to respect a community's cultural traditions, customs, and codes of practice. For example, Article 9.15 relates to the recognition of the role of Elders and other knowledge holders in the design and execution of research, and the interpretation of findings in the context of cultural norms and traditional knowledge. Aboriginal Elders are often the most knowledgeable persons about Indigenous cultural rules and traditions, perceptions of physical and spiritual reality, the teaching and practices of ceremony, and the nuances of meaning in Indigenous languages.

Chapter 9 interprets in nine articles how the value of respect for human dignity and the core principles of Respect for Persons, Concern for Welfare, and Justice discussed in other chapters of the TCPS 2 apply specifically to research involving Aboriginal peoples in Canada. For example, Respect for Persons in an Aboriginal context goes beyond securing free, informed and ongoing consent of participants. It also includes obligations to maintain, and pass on to future generations, knowledge received from ancestors. The principle of Concern for Welfare goes beyond considerations for individual well-being to considerations for the welfare of the Aboriginal community to which participants belong. Justice may be compromised when an imbalance of power prevails between researchers and participants. With research involving Aboriginal peoples, the social, cultural, or linguistic distance between participants and researchers may be significant, thus, engagement between the community involved in research and the researchers, initiated prior to recruiting participants and maintained over the course of the research, is recommended as an integral aspect of ethical research.

The ethics review process at the University of Alberta (UAlberta) where I work in Canada will be used here as an example of how Chapter 9 of the TCPS 2 is ensured. UAlberta requires that all research involving humans conducted by staff or students affiliated with UAlberta, or involving UAlberta resources, must be reviewed and approved by one of UAlberta's Research Ethics Boards before the research starts. The Boards ensure that research projects involving human participants, identifiable data, and/or human biological material meet the requirements of the TCPS 2 and UAlberta policy, as well as any applicable provincial, federal, and other legislation and regulations. Research with Aboriginal peoples conducted by UAlberta staff or students must include community engagement to ensure that Aboriginal peoples have a role in the research that affects them. As of 2016, applicants seeking ethics approval for research involving Aboriginal peoples are required to answer questions relating to topics such as obtaining consent from Elders, leaders, or other community representatives; details about whether property or private information belonging to the group as a whole is studied or used; details about whether the research is designed to analyze or describe characteristics of the group; details about whether individuals are selected to speak on behalf of or otherwise represent the group; information regarding consent; information about the access, ownership, and sharing of research data with communities; information about how final results of the study will be shared with the participating community; and, the nature of research agreements.

Appropriate protocol must be followed when researchers seek the advice of Elders and when they acknowledge the contributions of Elders to their research. Researchers may be unfamiliar with these protocols. The Council of Aboriginal Initiatives at UAlberta has therefore published a document related to Elder protocol and guidelines to help researchers meet the guidelines of Chapter 9 of the TCPS 2 (Council of Aboriginal Initiatives, University of Alberta 2012).

6.2 Ethical Guidelines for Health Research with Māori in New Zealand

The document *Guidelines for Researchers on Health Research Involving Māori* (version 2) was created in 2010 by the Māori Health Committee (MHC) of the

Health Research Council of New Zealand (HRC) (Health Research Council of New Zealand 2010). Its aim is to assist researchers who have received Health Research Council funding to undertake biomedical, public health, or clinical research involving Māori participants or research on issues relevant to Māori health. The intent of the Guidelines is to help develop research that contributes to Māori health development whenever possible, and partnerships between health researchers and Māori communities or groups on issues important to Māori health. The Guidelines state that the principles of partnership and sharing implicit in the Treaty of Waitangi, New Zealand's founding document which describes principles for a partnership between the government and Māori, should be respected by researchers and, where applicable, should be incorporated into all health research proposals. Three principles are particularly relevant in the proceedings and processes of ethics committees relating to research with Māori. There is the principle of Partnership, which means working together with iwi (tribes or nations that form the structure of Māori society), hapu (clans or descent groups within each iwi), whanau (extended families or family groups), and Māori communities to ensure Māori individual and collective rights are respected and protected. There is the principle of Participation, meaning that Māori will be involved in the design, governance, management, implementation, and analysis of research. There is the principle of Protection, meaning that the research process will actively protect Māori individual and collective rights, Māori data, and Māori culture, cultural concepts, values, norms, practices, and language.

The Guidelines require that researchers conducting research on a Māori health issue and/or involving Māori as participants need to start initial consultation and conversations with Māori before putting the research proposal together. Ongoing consultation throughout the research is urged. Researchers must acknowledge Māori ways of knowing and conducting research. When a project involves Māori within a given geographical area as participants, researchers must contact local Māori representative organizations, advise them of the nature of the intended study, and invite their comments and/or involvement. Research must not desecrate or contribute to the erosion of Māori cultural values (see also ▶ Chaps. 87, "Kaupapa Māori Health Research," and ▶ 88, "Culturally Safe Research with Vulnerable Populations (Māori)").

7 Guidelines for Health Research Developed by Indigenous Communities or Agencies

Many Indigenous communities and organizations in Australia, Canada, New Zealand, and the United States have taken up the challenge of conducting and monitoring research to ensure local involvement. Some Indigenous communities have developed their own codes of research conduct (e.g., Kahnawake Schools Diabetes Prevention Project 2007; Pigford et al. 2013), and Indigenous organizations have formed their own ethic review panels and boards, which function separately from University Institutional Review Boards/Research Ethics Boards (e.g., Sahota 2009; Harding et al. 2012; Angal et al. 2016). The existence of community codes of research conduct as well as independent Indigenous ethics panels and boards means that a single research project may require multiple authorizations to proceed depending on the nature, scope, and location of the research activity.

Space does not permit a review of the numerous research guidelines developed independently by Indigenous agencies and communities that outline local principles of research conduct. Typically, these guidelines provide ethical considerations regarding local perspectives and values, provide for optimal community oversight of the research, and help ensure that research addresses community concerns and expectations. They may focus on ensuring that research adheres to values that University Institutional Review Boards/Research Ethics Boards might overlook such as the interdependent relationship between humans and natural elements, recognizing community as a unit of identity, awareness that any particular event or phenomenon functions as part of a larger whole, or safeguarding cultural and intellectual heritage (Harding et al. 2012; Brunger and Wall 2016). For example, all research occurring in the predominantly Inuit-occupied territory of Nunavut in northern Canada is licensed by the Nunavut Research Institute in accordance with the Nunavut's Scientists Act, which helps ensure that research is collaborative and addresses Nunavut's needs and priorities (Nunavut Research Institute 2015).

8 Building Capacity to Do Ethical Research with Indigenous Peoples

Indigenous researchers trained as Western health professionals but who are grounded in their Indigenous identities are an important, but small, cadre of researchers. Australia, Canada, New Zealand, and the United States all have too few Indigenous scholars available to conduct all of the health research required with Indigenous peoples. This under-representation among the professoriate means that training to undertake ethically conducted Indigenous health research is required for non-Indigenous scholars. Researchers who are not Indigenous need to develop a capacity to bracket Western research paradigms and assumptions in order to become knowledgeable about Indigenous paradigms (Bartlett et al. 2007). Even Indigenous scholars conducting research can benefit from formal training in decolonizing research if they have become alienated from their culture, if they do not hold traditional Indigenous world views, or if they are not familiar with how to conduct such research.

Health research with Indigenous peoples that respects Indigenous self-determination, and is safe, ethical, and useful for participants, requires increased capacity among Indigenous and non-Indigenous peoples alike. Researchers regardless of their indigeneity require, in addition to conventional health research training that emphasizes a biomedical model of illness and disease, training that focuses on understanding the SDoH in relation to disease occurrence and patterns. Conventional health research often dictates the use of positivism as a scientific paradigm, whereby researchers, in their search for an ultimate reality, are to be value free and objective in their approach to research, and detached from research participants to ensure that researchers and their research "subjects" do not influence each other (see ▶ "Indigenous Statistics"). While there is value in this paradigm, research with Indigenous peoples may require a more interpretivist approach which is receptive to capturing meanings in human interaction and making sense of what is perceived as reality (Carson et al. 2001).

As stated already in this chapter, many of the health inequities experienced by Indigenous peoples are the result of historic and current national and local policies designed to eliminate and/or assimilate Indigenous people. Research that is designed to address and eliminate the health and socioeconomic inequities faced by Indigenous peoples requires researchers to have knowledge of history, policy, health determinants, ethics, and Indigenous rights issues. Researchers must value the lived experience of participants by focusing on the meaning and interpretation that individuals place on events (Liamputtong 2013, 2017). Researchers must have an attitude of mutuality and openness, self-awareness, and self-reflexiveness; facilitative skills in interpersonal and group settings; and, a willingness to produce knowledge to empower a group of people, and to work authentically in collaboration with people to improve their lives (Liamputtong 2010, 2013). The research should be designed to coproduce culturally respectful, relevant, and empowering knowledge (Castleden et al. 2012). Healthcare training programs designed to produce researchers that can work in an Indigenous context should have content related to how structures of power rooted in colonialism continue to create health inequities and how an individual's own experiences of privilege and oppression affect their practice (Beavis et al. 2015). The acceptance by the professoriate of research that uses Indigenous methodologies to produce transformative change may require a "decolonizing" of the academy (Robertson 2016; see ▶ Chap. 15, "Indigenist and Decolonizing Research Methodology").

There are numerous examples of programs and workshops designed to build capacity among Indigenous peoples to conduct research or to partner with outsiders to conduct research. For the sake of brevity, below I provide some examples of Canadian and American programs that seek to improve the ability of health researchers to work with Indigenous peoples or in Indigenous communities, or that aim to educate Indigenous peoples about research topics. Some of the programs offer travel scholarships or tuition waivers to Indigenous applicants.

8.1 Institute on the Ethics of Research with Indigenous Peoples

Carleton University in Canada launched the Institute on the Ethics of Research with Indigenous Peoples (CUIERIP) in June of 2015 (https://carleton.ca/aboriginal/). CUIERIP is a week-long summer institute to equip Indigenous and non-Indigenous researchers with tools to implement ethical practices when working with First Nations, Inuit and Métis communities, or conducting research on traditional Indigenous territory. Its scheduling coincides with the spring term session of courses for Carleton's Masters degree concentration in Indigenous Policy and Administration. CUIERIP is intended for a diverse audience, including academic researchers, research ethics board members, graduate students, First Nations, Inuit and Métis community members, and researchers and representatives from governmental and nongovernmental organizations. Participants learn in a collaborative environment and are led by Carleton faculty, research ethics professionals, and Indigenous and non-Indigenous community-based researchers with expertise in research ethics, community engagement, and research design and review. Participants work together in small groups using case studies to work through the ethical issues involved in community engagement plans and research agreements.

8.2 Centre for Excellence in Indigenous Health

The Centre for Excellence in Indigenous Health in the Faculty of Medicine at the University of British Columbia in Canada exists to support and develop Aboriginal health programs, curriculum, research, and advocacy with Aboriginal communities and partners on local, national, and international levels (http://health.aboriginal.ubc.ca/education/). It offers an undergraduate course called Topics in Indigenous Health: A Community-Based Experience (http://health.aboriginal.ubc.ca/education/ihhs-408/). This 4-week practice-based Indigenous health elective has health sciences students live and work with students from other health disciplines within an Indigenous community in the Province of British Columbia. The course objectives seek to foster increased awareness of the core principles required to do community-based participatory research with Indigenous communities such as gaining an understanding of and respect for Indigenous perspectives on health and wellbeing; understanding, acknowledging, and exploring the implications of specific processes of colonization and related social policies for the health of Indigenous peoples; examining and identifying patterns of health and illness from multiple perspectives: epidemiology, interdisciplinary health, community, and Indigenous knowledge; and, demonstrating respectful communication with Indigenous peoples.

8.3 Community-Based Research and Evaluation Certificate Program

The UAlberta in Canada offers the embedded graduate Community-Based Research and Evaluation (CBRE) Certificate Program (https://www.extension.ualberta.ca/study/community-engagement-studies/cbre/). It is designed for graduate students who seek to develop their capacity to participate in and lead community-based research and evaluation. Though not specifically designed to train students to do research and evaluation in Indigenous communities, students can choose to partner with an Indigenous community and Indigenous community mentor if they are interested in CBRE in Indigenous contexts.

All students in the certificate program must take a graduate level course called An Introduction to CBPR. This course has historically included some guest lectures by researchers working in Indigenous communities or Indigenous community members participating in CBPR with academic researchers. In consultation with a CBRE

Advisor and their graduate supervisor, students are additionally required to take graduate level-courses in program planning and evaluation; quantitative research methods; and qualitative research methods. Through coursework, students develop an understanding of CBRE concepts, program planning and evaluation, and a variety of quantitative, qualitative, and/or mixed methods. The program has a mandatory experiential component in a community setting to apply the concepts and methods learned through their course work including relationship building and maintenance; political sensitivity; development of a partnership agreement; participation in day-to-day, project-management duties; participation in partnership decision-making; participation in the development of a specific project within a partnership; and, development and/or implementation of a process evaluation of the partnership. A Community Mentor and the CBRE program Experiential Learning Coordinator jointly supervise the CBRE community experience which is a minimum of 156 h.

8.4 Master of Public Health: Native Hawaiian and Indigenous Health Specialization

In an effort to address the disparities faced by Native Hawaiians and other Indigenous Peoples, the University of Hawai'i, Office of Public Health Studies, has a specialization in Native Hawaiian and Indigenous Health within the Master of Public Health degree. Both of the professors in the Native Hawaiian and Indigenous Health specialization are Indigenous scholars. This specialization is designed to prepare students with the public heath skills and training necessary to serve Indigenous People globally and assist in addressing their health and wellness needs by contextualizing health determinants within historical and political frameworks. It provides extensive training in culturally sensitive research ethics which is critical for safely and effectively implementing public health research and programs aimed to address and eliminate the inequities faced by Indigenous People. Students enrolled in the specialization are required to take advanced level training in Indigenous health policy, ethics, and research design. The curriculum integrates Indigenous Public Health Competencies with traditional competencies to help build a stronger, more effective public health workforce in Native Hawaiian and Indigenous communities (Taualii et al. 2013). Students participate in ongoing research and practice programs with Indigenous communities through a practicum assignment (http://manoa.hawaii.edu/publichealth/specializations/native-hawaiian-and-indigenous-health).

8.5 Community Mobilization for Healthy Lifestyles and Diabetes Prevention Training Program

The Kahnawake Schools Diabetes Prevention Project (KSDPP) is one of the longest-running community-based health research projects in Canada. It has been in operation since 1994. It occurs in the Kanien'kehàka (Mohawk First Nations) community of Kahnawake. The community offers a KSDPP Training Program in Diabetes

Prevention to enable and empower participants to begin or enhance a diabetes prevention or wellness program in their community based on the experiences of the KSDPP (http://www.ksdpp.org/elder/training_program.php). The Community Mobilization for Healthy Lifestyles & Diabetes Prevention Training Program shares the successful experiences of the health promotion model of KSDPP with Indigenous community organizations and individuals working with Indigenous communities (http://www.ksdpp.org/media/brochure.pdf). These multiday workshops provide information, facilitate discussion, and engage participants to plan community actions for healthy lifestyles and diabetes prevention. The program provides participants with skills that will help them to do research in their own communities or negotiate CBPR by providing them with an understanding of the theoretical background for successful healthy lifestyles programming; identifying community values in relation to healthy lifestyles and diabetes prevention; the importance of teamwork; how to build a Community Advisory Board (CAB), identify potential CAB members, and the nature and activities of CAB volunteers; conducting an environmental scan to identify key goals, objectives, and strategies for the planning and successful implementation of healthy lifestyles and diabetes prevention planning; different types of community intervention activities, the steps in planning activities, the development of community activity calendars, and activity evaluation; and, developing a dissemination program to promote healthy lifestyles and diabetes prevention.

8.6 Summer Research Training Institute for American Indian and Alaska Native Health Professionals

The 2016 Summer Research Training Institute for American Indian and Alaska Native Health Professionals in the United States was hosted by the Northwest Portland Area Indian Health Board and the Center for Healthy Communities at Oregon Health & Science University (http://www.npaihb.org/images/training_docs/NARCH/2016/2016_SI_Brochure_Final.pdf). The curriculum of this 3-week course was designed to meet the needs of professionals who work in diverse areas of American Indian and Alaska Native health – from administrators to community health workers, physicians, nurses, researchers, and program managers. It emphasizes research skills, program design, and implementation.

9 Conclusion and Future Directions

This chapter discussed how States, academic institutions, and Indigenous groups can support ethical research with Indigenous peoples, communities, and nations located in Australia, Canada, New Zealand, and the United States. It also discusses why they should do so. Ideally, ethically conducted research would be decolonizing research that leads to self-determination. It would empower Indigenous people through the process of constructing and using their own knowledge.

Unfortunately, there is reluctance on the part of some academic researchers and Indigenous communities to coparticipate in health research. Indigenous peoples may believe that they will not benefit from the research, or worse, that they will come to harm by participating in health research. Both Indigenous community members and health researchers perceive barriers to doing ethical research. This hesitancy to engage in research means that the research that is needed in Indigenous communities for improved health and well-being is not always being done, or is not being done well. To help ensure that beneficial research occurs, community members need to be provided with workshops and training sessions that will teach them how to negotiate with health researchers from outside of their communities, let them know their rights as research participants, and build their skills to conduct their own research or to engage in CBPR with outsiders. Research undertaken by non-Indigenous researchers can lead to improvements in the health status of Indigenous people provided that researchers have appropriate ethical guidelines to follow and training opportunities that offer guidance on Indigenous ways of knowing, the SDoH, strength-based research approaches, CBPR, and how to engage in culturally appropriate ways with Indigenous peoples. Researchers wanting to pursue a specialization in Indigenous health research need support from academic leadership and funding agencies to be successful in their endeavor.

Although conducting ethical research in Indigenous communities offers challenges to both academic and community partners, there are many community, academic, and personal rewards and benefits to adhering to the additional ethical standards and research procedures required to do research well in an Indigenous context. I suggest more understanding of the following research areas and topics to broaden the support for engaged scholarship between researchers and community members undertaking ethical research related to Indigenous health.

- Richly detailed case studies that demonstrate how and why ethical research with Indigenous peoples increases both individual and community-level self-determination, and consequently, increases Indigenous peoples' perceived control over health.
- Examples of how to best support Indigenous health research that champions democratic empowerment, whereby communities and community members assess their own interests and make decisions on how to see these interests put into action.
- Research to discover if the acquisition of new skills and knowledge by Indigenous community members in relation to conducting health research, or participating in health research, results in the ability of members to positively influence community change.
- The development of research models that incorporate an understanding of how Indigenous communities can combine the diverse types of research knowledge (e.g., biomedical) emanating from community-based participatory research with the practical and cultural knowledge of community members to create positive health outcomes, by influencing the social determinants of health, the social policy process, and important policy issues (Bryant et al. 2007).

- Examples of how academics can be supported both financially and institutionally to conduct ethical health research with Indigenous peoples and communities that emphasizes social change and positive health outcomes as an endpoint.
- Descriptions from health researchers of how they have benefited professionally and personally from undertaking ethical research in Indigenous communities.
- Descriptions from Indigenous community members of how they have benefited from undertaking ethical research with health researchers.

References

Allan B, Smylie J. First peoples, second class treatment: the role of racism in the health and well-being of indigenous peoples in Canada. Toronto: The Wellesley Institute; 2015. http://www.wellesleyinstitute.com/wp-content/uploads/2015/02/Report-First-Peoples-Second-Class-Treatment-Final.pdf. Accessed 17 Oct 2016.

Angal J, Petersen JM, Tobacco D, Elliott AJ. Ethics review for a multi-site project involving tribal nations in the Northern Plains. J Empir Res Hum Res Ethics. 2016;11(2):91–6.

Australian Institute of Aboriginal and Torres Strait Islander Studies. AIATSIS guidelines for ethical research in Australian Indigenous studies, 2nd edn. 2012. http://aiatsis.gov.au/sites/default/files/docs/research-and-guides/ethics/gerais.pdf. Accessed 17 Oct 2016.

Bartlett JG, Iwasaki Y, Gottlieb B, Hall D, Mannell R. Framework for aboriginal-guided decolonizing research involving Métis and first nations persons with diabetes. Soc Sci Med. 2007;65(11):2371–82.

Beavis AS, Hojjati A, Kassam A, Choudhury D, Fraser M, Masching R, Nixon SA. What all students in healthcare training programs should learn to increase health equity: perspectives on postcolonialism and the health of Aboriginal Peoples in Canada. BMC Med Educ. 2015;15(1):155. https://bmcmededuc.biomedcentral.com/articles/10.1186/s12909-015-0442-y. Accessed 17 Oct 2016.

Brunger F, Wall D. "What do they really mean by partnerships?" questioning the unquestionable good in ethics guidelines promoting community engagement in Indigenous health research. Qual Health Res. 2016;26(13):1862–77.

Bryant T, Raphael D, Travers R. Identifying and strengthening the structural roots of urban health in Canada: participatory policy research and the urban health agenda. Promot Educ. 2007;14(1):6–11.

Canadian Institutes of Health Research. Aboriginal people's health. 2011. http://www.cihr-irsc.gc.ca/e/8172.html. Accessed 17 Oct 2016.

Canadian Institutes of Health Research, Natural Sciences and Engineering Research Council of Canada, and Social Sciences and Humanities Research Council of Canada. Tri-Council policy statement: ethical conduct for research involving humans. 2014. http://www.pre.ethics.gc.ca/pdf/eng/tcps2-2014/TCPS_2_FINAL_Web.pdf. Accessed 17 Oct 2016.

Carson, D., Gilmore, A., Perry, C., and Gronhaug, K. (2001). Qualitative Marketing Research. London: Sage.

Castellano MB. Ethics of Aboriginal research. J Aborig Health. 2004;1(1):98–114.

Castleden H, Morgan VS, Lamb C. "I spent the first year drinking tea": exploring Canadian university researchers' perspectives on community-based participatory research involving Indigenous peoples. Can Geogr/Géogr Can. 2012;56(2):160–79.

Castleden H, Sylvestre P, Martin D, McNally M. "I don't think that any peer review committee … Would ever 'get' what I currently do": How institutional metrics for success and merit risk perpetuating the (re)production of colonial relationships in community-based participatory research involving Indigenous peoples in Canada. Int Indigenous Policy J. 2015;6(4.) http://ir.lib.uwo.ca/iipj/vol6/iss4/2. Accessed 17 Oct 2016

Council of Aboriginal Initiatives, University of Alberta. Elder protocols and guidelines. 2012. http://www.provost.ualberta.ca/en/~/media/provost/Documents/CAI/Elders.pdf. Accessed 17 Oct 2016.

First Nations Centre. OCAP: ownership, control, access and possession. Sanctioned by the first nations information governance committee, assembly of first nations. Ottawa: National Aboriginal Health Organization; 2007. http://cahr.uvic.ca/nearbc/documents/2009/FNC-OCAP.pdf. Accessed 17 Oct 2016

Food and Agriculture Organization of the United Nations. Indigenous peoples. 2016. http://www.fao.org/indigenous-peoples/en/. Accessed 17 Oct 2016.

Genius SK, Willows N, Nation AF, Jardine C. Through the lens of our cameras: children's lived experience with food security in a Canadian indigenous community. Child: Care Health Dev. 2014;41:600–10.

Genuis SK, Willows N, Nation AF, Jardine CG. Partnering with Indigenous student co-researchers: improving research processes and outcomes. Int J Circumpolar Health. 2015;74:27–38.

Gokiert RJ, Willows N, Georgis R, Stringer H, Alexander Research Committee. Wahkohtowin: The governance of good community-academic research relationships to improve the health and well-being of children in Alexander First Nation. Int Indigenous Policy J. under review.

Government of Canada. Panel on Research Ethics. Module 9: Research involving First Nations, Inuit and Metis Peoples of Canada. n.d. http://www.pre.ethics.gc.ca/education/Module9_en.pdf. Accessed 17 Oct 2016.

Gracey M, King M. Indigenous health part 1: determinants and disease patterns. Lancet. 2009;374(9683):65–75.

Gray MA, Oprescu FI. Role of non-indigenous researchers in indigenous health research in Australia: a review of the literature. Aust Health Rev. 2015;40(4):459–65.

Guenther M, Kenrick J, Kuper A, Plaice E, Thuen T, Wolfe P, Zips W, Barnard A. The concept of indigeneity. Soc Anthropol. 2006;14(1):17–32.

Harding A, Harper B, Stone D, O'Neill C, Berger P, Harris S, Donatuto J. Conducting research with tribal communities: sovereignty, ethics, and data-sharing issues. Environ Health Perspect. 2012;120(1):6–10.

Health Research Council of New Zealand. Guidelines for researchers on health research involving Māori. Version 2. 2010. http://www.hrc.govt.nz/sites/default/files/Guidelines%20for%20Researchers%20on%20Health%20Research%20Involving%20M%C4%81ori%202010.pdf. Accessed 17 Oct 2016.

Kahnawake Schools Diabetes Prevention Project. Code of research ethics. 2007. http://www.ksdpp.org/media/ksdpp_code_of_research_ethics2007.pdf. Accessed 17 Oct 2016.

LaVeaux D, Christopher S. Contextualizing CBPR: key principles of CBPR meet the indigenous research context. Pimatisiwin. 2009;7(1):1–16. https://www.ncbi.nlm.nih.gov/pmc/articles/PMC2818123/

Liamputtong P. Performing qualitative cross-cultural research. Cambridge: Cambridge University Press; 2010.

Liamputtong P. Qualitative research methods. 4th ed. Melbourne: Oxford University Press; 2013.

Liamputtong P. The science of words and the science of numbers. In: Liamputtong P, editor. Research methods in health: foundations for evidence-based practice (chapter 1). Melbourne: Oxford University Press; 2017.

Mosby I. Administering colonial science: nutrition research and human biomedical experimentation in aboriginal communities and residential schools, 1942–1952. Hist Soc/Soc Hist. 2013;46(1):145–72.

National Health and Medical Research Council. Ethical guidelines for research involving Aboriginal and Torres Strait Islander Peoples. 2016. https://www.nhmrc.gov.au/health-ethics/ethical-issues-and-further-resources/ethical-guidelines-research-involving-aboriginal-. Accessed 17 Oct 2016.

Nunavut Research Institute. About us. 2015. http://www.nri.nu.ca/about-us. Accessed 17 Oct 2016.

Pigford AE, Willows ND. Promoting optimal weights in aboriginal children in Canada through ecological research. In: O'Dea JA, Eriksen M, editors. Childhood obesity prevention: international research, controversies, and interventions. Toronto: Oxford University Press; 2010. p. 309–20.

Pigford AE, Dyck Feherau D, Ball GDC, Holt NL, Plotnikoff RC, Veugelers PJ, Arcand E, Nation AF, Willows ND. Community-based participatory research to address childhood obesity: experiences from Alexander first nation in Canada. Pimatisiwin: J Indigenous Aborig Commun Health. 2013;11(2):171–85. http://www.pimatisiwin.com/online/wp-content/uploads/2013/10/02PigfordFehderau.pdf. Accessed 17 Oct 2016

Ramos H. Does how you measure representation matter?: assessing the persistence of Canadian universities' gendered and colour coded vertical mosaic. Can Ethn Stud. 2012;44(2):13–37.

Reading CL, Wien F. Health inequalities and social determinants of aboriginal peoples' health. Prince George: National Collaborating Centre for Aboriginal Health; 2009. http://www.sac-conference.ca/wp-content/uploads/2016/04/Fred-Wien-Social-Determinants-of-Health-Among-Aboriginal-Populations-in-Canada-Handout-21.pdf. Accessed 17 Oct 2016

Ritchie SD, Wabano MJ, Beardy J, Curran J, Orkin A, VanderBurgh D, Young NL. Community-based participatory research with indigenous communities: the proximity paradox. Health Place. 2013;24:183–9.

Robertson DL. Decolonizing the academy with subversive acts of Indigenous research. A review of Yakama Rising and Bad Indians. Soc Race Ethn. 2016;2(2):248–52.

Rotenberg, C. Social determinants of health for the off-reserve First Nations population, 15 years of age and older, 2012. 2016. http://www.statcan.gc.ca/pub/89-653-x/89-653-x2016010-eng.htm. Accessed 17 Oct 2016.

Sahota, P. C.. Research regulation in American Indian/Alaska Native communities: Policy and practice considerations. 2009. NCAI Policy Research Center. http://www.ncaiprc.org/files/Research%20Regulation%20in%20AI%20AN%20Communities%20-%20Policy%20and%20Practice.pdf. Accessed 17 Oct 2016.

Smith-Morris C. Autonomous individuals or self-determined communities? The changing ethics of research among native Americans. Hum Organ. 2007;66(3):327–36.

Smylie, J., & Adomako, P.. Indigenous children's health report: Health assessment in action. 2009. http://caid.ca/IndChiHeaRep2009.pdf. Accessed 17 Oct 2016.

Taualii M, Delormier T, Maddock J. A new and innovative public health specialization founded on traditional knowledge and social justice: native Hawaiian and indigenous health. Hawai'i J Med Publ Health: J Asia Pac Med Publ Health. 2013;72(4):143–5.

Towns C, Cooke M, Rysdale L, Wilk P. Healthy weights interventions in aboriginal children and youth: a review of the literature. Can J Diet Pract Res. 2014;75(3):125–31.

Triador L, Farmer A, Maximova K, Willows N, Kootenay J. A school gardening and healthy snack program increased aboriginal first nations children's preferences toward vegetables and fruit. J Nutr Educ Behav. 2015;47(2):176–80.

UN Office of the High Commissioner for Human Rights (OHCHR), Fact sheet No. 31, The right to health. 2008. http://www.refworld.org/docid/48625a742.html. Accessed 29 Nov 2016.

United Nation. United Nations declaration on the rights of Indigenous peoples. 2008. http://www.un.org/esa/socdev/unpfii/documents/DRIPS_en.pdf. Accessed 17 Oct 2016.

United Nations. Indigenous peoples and the United Nations human rights system. Fact sheet No. 9/Rev. 2. New York and Geneva, United Nations. 2013. http://www.ohchr.org/Documents/Publications/fs9Rev.2.pdf. Accessed 17 Oct 2016.

United Nations. United Nations Permanent Forum on Indigenous Issues. Indigenous peoples, Indigenous voices fact sheet. n.d.. http://www.un.org/esa/socdev/unpfii/documents/5session_factsheet1.pdf. Accessed 17 Oct 2016.

Willows N. Ethical principles of health research involving indigenous peoples. Appl Physiol Nutr Metab. 2013;38(11):iii–v. http://www.nrcresearchpress.com/doi/pdf/10.1139/apnm-2013-0381

Willows ND, Hanley AJG, Delormier T. A socioecological framework to understand weight-related issues in aboriginal children in Canada. Appl Physiol Nutr Metab. 2012;37(1):1–13. http://www.nrcresearchpress.com/doi/pdfplus/10.1139/h11-128

Willows N, Dyck Fehderau D, Raine KD. Analysis grid for environments linked to obesity (ANGELO): framework to develop community-driven health programmes in an Indigenous community in Canada. Health Soc Care Community. 2016;24(5):567–75.

Zavala M. What do we mean by decolonizing research strategies? Lessons from decolonizing, Indigenous research projects in New Zealand and Latin America. Decolonization: Indigeneity Educ Soc. 2013;2(1):55–71.

Conducting Ethical Research with People from Asylum Seeker and Refugee Backgrounds

107

Anna Ziersch, Clemence Due, Kathy Arthurson, and Nicole Loehr

Contents

1	Introduction	1872
2	A Note on Terminology	1873
3	Research Design and Methodologies	1874
	3.1 Participatory Research Methodologies	1875
	3.2 Quantitative Research Methods	1876
	3.3 Qualitative Research Methods	1877
4	Research with Children and Young People	1879
5	Data Collection	1880
	5.1 Building Rapport and Cross-Cultural Competency	1881
	5.2 Sampling Issues	1881
	5.3 Interpreting and Translation	1882
6	Ethical Considerations	1882
7	Advocacy	1885
8	Conclusion and Future Directions	1885
References		1886

Abstract

This chapter outlines issues to be considered when working on health and other research with people from asylum seeker or refugee backgrounds in countries of resettlement. The chapter highlights the utility of a Social Determinants of Health framework and outlines the importance of ethical research, which balances the considerations of formal ethics committees by ensuring that the voices of the most

---

A. Ziersch (✉) · C. Due · K. Arthurson
Southgate Institute for Health, Society and Equity, Flinders University, Adelaide, SA, Australia
e-mail: anna.ziersch@flinders.edu.au; clemence.due@flinders.edu.au; kathy.arthurson@flinders.edu.au

N. Loehr
School of Social and Policy Studies, Flinders University, Adelaide, SA, Australia
e-mail: nicolemarialoehr@gmail.com

© Springer Nature Singapore Pte Ltd. 2019
P. Liamputtong (ed.), *Handbook of Research Methods in Health Social Sciences*,
https://doi.org/10.1007/978-981-10-5251-4_50

vulnerable people within this population are able to be heard. In addition, the chapter highlights the need to facilitate the full participation of people with asylum seeker and refugee backgrounds, including in the governance structures of the research project and initial research design, in order to ensure that the outcomes of the research are relevant and address community needs and concerns. The chapter also outlines appropriate methodologies, including emerging and innovative research methods such as visual scales, photovoice, photo-language, and digital storytelling, as well as discussing the ways in which these data collection methods contribute to high quality quantitative and qualitative data. Finally, the chapter also covers the challenges of working cross-culturally such as the use of standardized scales and interpreting and translation, and the need to ensure that research is culturally appropriate, consultative, and meaningful.

Keywords

Refugee · Asylum seeker · Social determinants of health · Participatory approaches · Ethics

1   Introduction

The number of refugees and asylum seekers worldwide currently stands at the highest level since World War II, with estimates of over 60 million people displaced or seeking refuge (UNHCR 2015). This group of people face a disproportionate burden of mental and physical ill health, due to issues such as a lack of infrastructure or health systems in countries of origin, frequently unequal access to healthcare in resettlement countries, and the ongoing impact of war and trauma (Schweitzer and Steel 2008). In addition, people with asylum seeker or refugee backgrounds face a range of complex and interplaying challenges when arriving in a resettlement country. These include access to education, difficulty finding employment, problems finding appropriate housing, disrupted social and family networks, and barriers to health service access. As such, and given the increasing numbers of people who are displaced or seeking refuge, it is critically important that there is empirical research into the health and well-being of this vulnerable group of people (Ellis et al. 2007; Hugman et al. 2011).

The World Health Organization (WHO) defines (1948, p. 100) health as "a state of complete physical, mental and social well-being and not merely the absence of disease or infirmity." WHO (2014) also defines mental health as "a state of well-being in which every individual realizes his or her own potential, can cope with the normal stresses of life, can work productively and fruitfully, and is able to make a contribution to her or his community." These broader notions of health and well-being are particularly important in planning and conducting research with refugees and asylum seekers as they allow for an inclusion of different cultural understandings of what constitutes health and/or well-being and extends the focus away from narrow disease or illness notions of health (Hugman et al. 2011).

This chapter draws on a Social Determinants of Health (SDoH) framework to consider best practice in health research with refugees and asylum seekers. The social determinants of health are defined by the World Health Organization as "the conditions in which people are born, grow, work, live, and age, and the wider set of forces and systems shaping the conditions of daily life. These forces and systems include economic policies and systems, development agendas, social norms, social policies and political systems" (WHO 2016). Drawing on a SDoH approach focuses the research "lens" onto the range of factors impacting on refugee and asylum seeker health and well-being ranging from the structural determinants of socioeconomic and political context and social position through to the material circumstances that people live in, psychosocial factors, behaviors, and biological factors, as well as the health system itself (Solar and Irwin 2010; Liamputtong et al. 2012). Improving refugee and asylum seeker health requires action at all these levels and research has an important role to play across this endeavor (McMichael 2016).

While research into refugee and asylum seeker health is vital, many previous researchers working in this area have documented a range of challenges to research with refugee communities, including the ethical challenges of working with vulnerable populations, and methodological challenges relating to cross-cultural research, valid research tools, and equivalency of meaning when considering data (Ellis et al. 2007; Due et al. 2014). As argued by Ellis et al. (2007, p. 460), research with refugees and asylum seekers "demands that researchers think beyond standard recommendations." While there are existing research guidelines, such those pertaining to human research ethics, that outline the importance of rigorous and ethical research, there is a need to critically examine these documents when working across cultural and ethnic divides, and when the research concerns a new or emerging social or community issue (Ellis et al. 2007). In this chapter, we highlight a range of considerations for undertaking health-related research with refugees and asylum seekers including research design, data collection, and ethical considerations and the role of advocacy.

2 A Note on Terminology

The 1951 United Nations Refugee Convention defines a refugee as "a person who is outside his/her country of nationality or habitual residence; has a well-founded fear of persecution because of his/her race, religion, nationality, membership in a particular social group or political opinion; and is unable or unwilling to avail himself/herself of the protection of that country, or to return there, for fear of persecution" (UNHCR 1967, p.14). Asylum seekers can be defined as people who are displaced and likely seeking refugee status, but who have either not have their claims reviewed or have not yet have them approved. While both refugees and asylum seekers face a range of challenges, it is important to note that asylum seekers are likely to experience added difficulties in resettlement countries, relating to issues such as experiences of detention, visa restrictions, and restrictions on employment and education (McMaster 2006). Despite this extra vulnerability, most research in this area focuses on refugee health rather than asylum seeker health. In this chapter, we similarly refer to "refugees and asylum seekers" for ease of expression, but acknowledge the heterogeneity in this population, the risks of assuming that all asylum seekers or refugees experience similar challenges, and also that some people from refugee backgrounds may not identify themselves as refugees once they have resettled. We deal with this issue throughout the chapter, where appropriate.

3 Research Design and Methodologies

Numerous researchers (e.g., Ellis et al. 2007; Block et al. 2012; Hanza et al. 2016) have highlighted that research with refugees and asylum seekers can be challenging methodologically. Not least of the issues that may be presented is that of community engagement and involvement in the research process. Moreover, Hanza et al. (2016) argue that this is particularly the case in health research, where refugees or asylum seekers are frequently either completely overlooked, or at least under-represented, in clinical trials or other forms of research concerning health and well-being. As a result, many of the conclusions drawn from "mainstream" health research may not be applicable to refugee or asylum seeker populations, leading to a dangerous gap in knowledge. This gap is compounded by the fact that the results of mainstream research are unlikely to be generalizable to refugee and asylum seeker communities given the diversity of experiences and compounded risk factors these groups face. As such, it is important that researchers take up the challenges to include refugees and asylum seekers in their research protocols, and design their research accordingly. Furthermore, and as outlined elsewhere (Pittaway and Bartolomei 2003; Ellis et al. 2007; Hugman et al. 2011) and above, it is important that research designs take into account community understandings of the constructs under consideration in the research, including physical or mental health. Correspondingly, a central aspect of research design with refugee or asylum seeker communities is that of close community collaboration to ensure that the research results will be valid.

In relation to theory, previous research with refugees and asylum seekers has spanned a broad and diverse range of theoretical frameworks, clearly depending on the research aims and questions of specific research. Here, we argue that whatever theoretical orientation is used in research, there are two important considerations. First, theories should not constrain research such that the meanings of concepts or constructs are limited and do not take into account cultural understandings. This is particularly important with regard to health research, as outlined throughout this chapter. Secondly, and relatedly, research should be designed with close collaboration with community members. Below, we discuss participatory research methodologies as one example of how research can centrally involve community members.

3.1 Participatory Research Methodologies

Participatory research methodologies (see Pittaway and Bartolomei 2003; McMichael et al. 2014; Higginbottom and Liamputtong 2015 for an overview; see also ▶ Chaps. 17, "Community-Based Participatory Action Research," and ▶ 100, "Participatory and Visual Research with Roma Youth") are well suited to research designs in the area of refugee health and well-being (Hanza et al. 2016). Participatory research designs foreground community knowledges and promote close collaboration between researchers and communities by placing community leaders at the center of the research structure such that the research becomes collaborative rather than "top down" (Hugman et al. 2011). Participatory research methodologies aim to work side by side with community leaders and members in all elements of research design, including defining research questions, the choice of research methods, the conduct of the research, and the application of research findings and outcomes (Pittaway and Bartolomei 2003; Block et al. 2012; Due et al. 2014; Higginbottom and Liamputtong 2015; Hanza et al. 2016). As such, participatory research methodologies work to bridge the gap between science and practice through community engagement and social action to decrease health disparities. Such designs employ strategies to reduce power imbalances, encourage knowledge translation, and incorporate community perspectives and theories into the research (Wallerstein and Duran 2006). This close community collaboration at each research stage goes some way to overcoming some of the challenges noted above, particularly in relation to ensuring cultural relevancy and correctly interpreting results.

Importantly, a range of methods and theoretical standpoints can be used within an overarching participatory research framework, although grounded theory or deductive frameworks are arguably best suited given the focus on community knowledges (see Charmaz 2014; Corbin and Strauss 2015; see also for an overview of grounded theory). Nevertheless, participatory research may adopt an empirical, theory-driven methodology, with previous health research with refugees and asylum seekers focusing heavily on trauma theory (Schweitzer and Steel 2008). However, regardless of the epistemological stance taken, participatory research methodologies with refugee and asylum seeker populations should be flexible in their approach, in that the sociopolitical situation facing many refugees is also flexible and likely to change during the course of the research itself. As argued by Schmidt (2007), such highly charged political and social contexts mean that environmental situations often lead to methodological constraints, and research cannot proceed without taking into account these contexts. For example, the health focus of the research may be heavily influenced by sociopolitical change during the research (e.g., the start or end of conflict), and thus, a narrow focus on only individual factors may overlook important predicators of health and well-being.

This engagement of refugees and asylum seekers is vital in research concerning health and well-being, given the culturally and socially contingent definitions of these terms. Correspondingly, research in this area should be collaborative and take into account the role of SDoH in health outcomes. This is especially pertinent for refugees and asylum seekers, since issues such as employment, education, housing, language, and community networks and the broader sociopolitical elements that influence them have all been previously shown to be closely related to health outcomes for this population.

3.2 Quantitative Research Methods

Quantitative research tools (such as standardized or validated physical or mental health scales) provide important information about the health of refugee and asylum seeker communities, particularly since the use of such tools allows comparison to other community groups or previous research outcomes (Schweitzer and Steel 2008; see also the ▶ Chap. 63, "Mind Maps in Qualitative Research"). However, standardized instruments measuring mental or physical health should be used with caution in research with refugee and asylum seeker populations. This is particularly the case since instruments may not be validated for the populations under consideration, leading to misunderstandings or inaccurate conclusions on the basis of the results (Ellis et al. 2007). Indeed, there is a dearth of appropriately validated instruments for measuring health and well-being in refugee populations, including children (Ehntholt and Yule 2006). In addition, the use of standardized tools such as mental or physical health measures may obscure cultural understandings of health and illness, such that individual scores on their own may provide little insight into the actual experiences of research participants themselves (Schwietzer and Steel 2008). This issue is compounded by the fact that a few health or mental health measures take a SDoH approach. Therefore, the sole use of such measures may obscure more culturally relevant understandings of aspects of health or the impact of particular factors on health outcomes (such as the impact of housing, employment, education, or family relationships). Correspondingly, quantitative methods may be best used side-by-side with qualitative methods, in order to ensure that the data gained from quantitative scales may be informed by qualitative work that provides insight into cultural understandings and knowledge through which to interpret the results (Gifford et al. 2007; Block et al. 2012).

Where quantitative methods such as surveys or standardized scales are used, it is important that these are translated and back-translated into first languages (or languages in which research participants are fluent) in order to ensure equivalency of meaning of questions, particularly if the research involves collecting data from diverse participants. Nevertheless, even such translation should be used with caution, particularly in the case of research with cultures with oral-based methods of communication, where direct translation of English text may obscure nuanced meanings and understandings (Ellis et al. 2007; Liamputtong 2010). If appropriate, a growing body of research includes the use of visual scales in quantitative measures – such as "smiley faces" ranging from very unhappy through to very happy – for questions that ask about happiness or satisfaction (Due et al. 2014). However, again, consideration should be paid to the impact that the use of such scales has on the original validity of the scale in question (e.g., in relation to the number of points on the original Likert scale and the associated scoring procedures). The use of such scales should also take into account cultural norms around portraying the human face.

One final issue of importance in quantitative research lies with ensuring that the sample used in the research is sufficiently representative of the population under consideration to meet standards for generalizability typically held in quantitative research (Jacobsen and Landau 2003). For example, in many countries of resettlement there is limited data concerning new arrivals, and thus statistics concerning the population may be difficult to obtain (Spring et al. 2003; Ellis et al. 2007), meaning that it may be hard to recruit a sufficiently stratified sample (Vigneswaran and Quirk 2012). Attention should also be paid to the specific cultural, ethnic, and linguistic background of participants when recruiting from key community leaders who may only have access to particular groups of people (Ellis et al. 2007). Nevertheless, the inclusion of refugee participants is central to research designs in this area, and clear recruitment strategies should be in place to ensure that this occurs. This issue is discussed further later in the chapter.

3.3 Qualitative Research Methods

Given the limitations noted above in relation to quantitative research with refugee and asylum seeker populations, qualitative research methods may be used either as standalone methods, or in conjunction with quantitative designs, in order to provide a nuanced understanding of the research focus. Here, we outline several innovative qualitative research methods used in research with populations who may be considered "vulnerable," including photovoice, photolanguage, and digital storytelling. Importantly, many of these research methods are based on visual information rather than text or solely spoken communication, meaning that they are a more likely to be empowering and appropriate for use with people for whom English (or the language in which the research is being conducted) is an additional language (see Crivello et al. 2009; Correa-Velez et al. 2010; Due et al. 2016 for some examples of previous uses of visual research methodologies). Below, we consider photovoice and photolanguage, digital storytelling, and the use of visual prompts in interviews.

Photovoice typically involves participants being provided with a camera and asked to take photos according to a particular theme related to the research aims or questions. Participants are then invited to discuss their images in either a focus group or interview setting (Drew et al. 2010; Liamputtong 2010; Haque and Eng 2011; see also ▶ Chap. 65, "Understanding Health Through a Different Lens: Photovoice Method"). Correspondingly, the photovoice method allows research participants to "lead" the direction of their responses in relation to the photographs they took, rather than following a predetermined interview schedule. In addition, the photovoice method is suitable for research with participants who may have limited English language skills (or language skills in the language of the country in which they are living). Photovoice may also be particularly relevant where the subject of the research is sensitive, as is often the case in health and well-being related research.

For example, photovoice allows participants to discuss topics of sensitivity on their own terms, which may lead to more nuanced and relevant data than direct questions (Drew et al. 2010; Haque and Eng 2011). Photovoice has been used previously in health-related research, such as research exploring understandings of well-being and cultural interpretations of health (Crivello et al. 2009).

In analyzing photovoice images, it is important to analyze the "talk" around the images, the broader context around the images, and also what is *not* photographed or discussed. For example, in a study of socioeconomically contrasting neighborhoods in South Australia, Browne-Yung et al. (2016) used the example of two photos selected by participants depicting graffiti as something they did not like about their neighborhoods. The participant from the wealthier area discussed the graffiti on a fence as unsightly and likely put there by people living *outside* the area. In the more disadvantaged area, the graffiti depicted in the photo was the "tag" of a local gang living *in* the area known to terrorize people living in the area and the image was chosen to reflect the fear that this caused for residents. Thus, the meaning ascribed to graffiti and the implications for feelings of safety and cohesion within neighborhoods was quite different in each context. A cursory analysis of the photos would have indicated that graffiti was an issue in both areas but not provided these more nuanced understandings. Likewise, a photovoice exercise with refugee children in school where they were asked to photograph places they felt safe revealed a lack of images of shared outside play spaces, suggesting that many children who were newly arrived in the school were generally relegated to the periphery of school spaces (Due and Riggs 2011).

Photolanguage involves asking participants to choose an image or images from a selected array that best represents the issue of interest (e.g., what does "home" mean to you) and to explain why they chose that image or images. It is related to photovoice methods but varies in that the images are predetermined prior to interview. This is similar to the photo elicitation method. The technique is valued by researchers and therapists alike for its "ability to challenge the viewer to thoughtful reflection" (Cooney and Burton 1986, p. 2). In addition, Fullana et al. (2014) have shown how photo elicitation methods may facilitate more inclusive participation in qualitative research. Although initially developed in 1965 as a therapeutic tool in counseling and group therapy, its merits have since been recognized in community development, pedagogical, and research contexts (Freire 2005).

Like other qualitative data-collection methods, photolanguage enables the collection of data in the form of opinions, descriptions, memories, and anecdotes. However, the use of images as a symbolic medium sets photolanguage (and photovoice) apart from some other qualitative tools. It is thought that when individuals choose an image in response to a question or probe, they are drawn to the image because it resonates with their most essential perceptions in response to that probe. This is considered to be an effective way of reducing more descriptive responses because individuals are not limited to expressing themselves through words; they can use the image to gain insight and clarity into their reaction to the question. There are some photolanguage card kits that can be used (e.g., Seamer 2007). While a predetermined set of images may constrain choices, the more abstract array can access different experiences and responses than more concrete images generally taken in photovoice exercises. An example of photolanguage methods used in research with refugees involved asking participants about the meaning of home through presenting a selection of Seamer's (2007) images (Loehr 2016). The selection by participants of images such as a sunrise to signify the new hope associated with home ownership and a key in a lock representing the importance of security of tenure, prompted discussion of deeper issues such as the need for a fresh start and previous experiences of home not being a sanctuary from violence and fear (Loehr 2016).

Digital storytelling encompasses a short form of video media in which participants are able to narrate or provide information about various aspects of their lives using still images (photographs), video, and music (Meadows 2003; see also ▶ Chap. 74, "Digital Storytelling Method"). Digital storytelling has been used in a range of settings such as education and community development and more recently in a research capacity, including – in small number of cases – research with refugee populations (see, for example Lenette and Boddy 2013; Lenette et al. 2015). However, this work is in its infancy and there remain potential barriers for those who are not fluent in the dominant language of the resettlement country, with little previous research exploring digital storytelling as a viable research method when language barriers are present. Nevertheless, and as with the photovoice method, digital storytelling may also provide research participants with a powerful medium in which to share information about their lives. In one project, "Residents Voices," digital story telling was utilized to create opportunities for social housing tenants to develop and express their own knowledge and understanding of the links between place and disadvantage. Tenants not only created a digital story through the workshops, but more importantly, they acquired the skills and knowledge to create and teach others how to create additional digital stories in the future (Rogers et al. in press).

4 Research with Children and Young People

Much of the existing research concerning the health and well-being of refugee or asylum seeker children has involved adults close to children (such as parents, teachers, or service providers) as research participants rather than children themselves (Due et al. 2014). However, a growing body of research has argued that children, including refugee and asylum seeker children, are able to provide important information about their own lives. As such, instead of adopting a "cognitive deficit" approach that assumes that children do not have the capacity to respond to research, such research designs use methodologies that ensure that children can participate in research on their own terms. Such research methodologies generally include a 'toolkit' of approaches (see Crivello et al. 2009; Gifford et al. 2007; Due et al. 2014; McMichael et al. 2014) which allows researchers to give children a range of options in terms of research activities, thereby increasing the likelihood that some of the activities will suit the child's preferred communication channels, as well as aiding in breaking down power relations. "Toolkit" approaches to research methodology also allows researchers to examine a number of different forms of data, enabling cross-checking of results and comparison of data, and reducing the issue of missing data

(McMichael et al. 2014). In addition, using a toolkit of approaches facilitates one of the central tenets of participatory research with children: for the researcher to enter the world of the child and, in doing so, modify research agendas to ensure that the experiences of the participants involved are reflected in the research process (O'Kane 2000). As with adult research participants, this process assists in ensuring that the research is relevant to children, particularly in relation to highly context-dependent constructs such as well-being (see also ▶ Chap. 115, "Researching with Children").

In terms of the specific research methods within such a 'toolkit' methodology, previous research has identified the utility of photovoice designs in particular with children with refugee or migrant backgrounds (Due et al. 2014). Specifically, photovoice has been identified as a child-focused, flexible approach to research that allows children's views to be communicated (Darbyshire et al. 2005; Newman et al. 2006; Due and Riggs 2011). In particular, photovoice has the potential to allow children to capture aspects of their lives which adults may otherwise not have access to (Young and Barrett 2001). As with adults, photovoice also allows children to highlight the issues of most importance to them, and may facilitate communication of sensitive issues relating to health and well-being, such as exposure to traumatic events.

Other visual research methods can be used as appropriate for the research aims. Examples include getting children to draw images of their happy and sad experiences (Liamputtong and Fernandes 2015), using images such as a circle and asking students to draw themselves in terms of where they felt they belong in that community (that is, towards the center of the circle if they felt they belong, and outside if they did not), asking children to draw pictures of themselves, and then to discuss aspects of the pictures that they drew that were important to them (Gifford et al. 2007; McMichael et al. 2014), or drawing 'social network maps' in order to consider social inclusion and belonging (Gifford et al. 2009; Block et al. 2012; Kurban and Liamputtong 2017). As noted above with adult research participants, visual or smiley-face scales can also aid the collection of quantitative data, and other novelty scales such as lolly jars may also be used (Due et al. 2014). It is also important to note that the issues noted above concerning the lack of appropriately validated health and well-being measures for refugee populations are magnified with children, with very few culturally appropriate tools for measuring health and well-being (See also ▶ "Visual Methods in Research with Migrant and Refugee Children and Young People," and ▶ 100, "Participatory and Visual Research with Roma Youth").

5 Data Collection

People with refugee or asylum seeker backgrounds are typically seen as a hard-to-reach research population and are frequently in very vulnerable positions (Liamputtong 2007, 2010). As such, ethical and careful data collection strategies are central to research with this population. As noted above, participatory research methods may go some way to assisting with issues of data collection, since they enable trust to be developed between the researcher and the potential participants (Pittaway and Bartolomei 2003; Kabranian-Melkonian 2015). In this section, we consider three elements of data collection: building rapport and cross-cultural competency, sampling issues, and interpreting and translation.

5.1 Building Rapport and Cross-Cultural Competency

Building rapport and trust with the communities that the research is concerned with is a central aspect of research methodologies involving people with refugee and asylum seeker backgrounds (Pittaway and Bartolomei 2003). Such relationships enable not only access to participants, but also crucial information about cultural or ethnic norms that need to be considered in the research process. An advisory group, bilingual researcher, or research assistant from the country of origin may assist with this process, although again attention should be paid to the nuances of relationships between different ethnic or language groups in the areas relevant to the study (Lee et al. 2014; Kabranian-Melkonian 2015).

Rapport building is also of central importance to research designs, in particular where either children or unaccompanied minors are included (Vervliet et al. 2015). This process may involve (where appropriate) spending time with children prior to primary data collection, in order to ensure that the children participating in the research feel comfortable to share their experiences (Due et al. 2014; Vervleit et al. 2015). This is particularly important given that children are typically keen to please adults and thus may be likely provide socially desirable or confirmatory responses if they do not feel comfortable in a research setting (Zeinstra et al. 2009). Building rapport also assists with ensuring appropriate strategies are taken to gain ongoing assent from children, in that rapport with an individual child places a researcher in a better position from which to determine when children may be distressed, hesitant, or unsure, and therefore when to move forward in the research or stop altogether.

However, building rapport and trust can lead to challenges to researchers' perceptions of their role in participants' lives (Vervliet et al. 2015). As noted by Vervliet, researchers can become heavily involved in participants' lives (especially in the case of longitudinal research), and sometimes even build relationships with family members or support workers. In this sense, researchers may struggle with their role, leading to challenges at both a personal level and within the research itself (Due et al. 2014; Verviet et al. 2015). In this sense, research concerning health and well-being should be implemented with clear governance structures and referral pathways for research participants, such that there are clear protocols to follow where individual research participants become distressed. Some of these broader ethical issues are discussed further below.

5.2 Sampling Issues

Rapport and trust is also important so that researchers should ensure that they are able to recruit as diverse a sample as possible. As argued by Kabranian-Melkonian (2015) and Bloch (2007), much of the research conducted with refugee populations uses as its recruitment strategy one or two community or service-provider organizations through which to access people. Recruiting using this method necessarily leads to a relatively homogenous sample (e.g., those accessing specific services or those in specific communities), and correspondingly constrains results, particularly in relation to quantitative research where the sample may not be sufficiently robust to meet standards for generalizability. Bloch (2007) notes that researchers should attempt to employ cluster sampling methods, with as many starting points as possible, in order to ensure a diverse sample which represents the broader population of interest. In addition, research that engages refugee and asylum seeker participants – such as through participatory models – may also benefit from wide snowball sampling, whereby the research participants themselves become advocates for the research and pass information on to their own networks (Hanza et al. 2016).

5.3 Interpreting and Translation

It is well established that scales, measures, and interview questions should be administered with either translation or interpreting where participants do not speak the language in which the research is being conducted (Kabranian-Melkonian 2015). Survey tools or scales should also be back-translated into their original language to ensure accurate translation, particularly for nuanced concepts or health constructs (Liamputtong 2010; Kabranian-Melkonian 2015). Nevertheless, attention should also be paid to the impact that translation may have on scale validation, and at times comparison to broader research should be done with caution.

As noted above, ensuring that interpreters' own ethnicity or culture match those of participants is also important in order to not to cross boundaries with countries and cause discomfort or distress to research participants (Liamputtong 2010; Kabranian-Melkonian 2015). This is particularly important in situations where there is conflict within a country, or where particular ethnicities are the subject of ongoing persecution. However, there can also be some confidentiality concerns about the use of interpreters from the same ethnic or cultural group particularly in small communities and around some health issues such as sexual health or domestic violence (Gartley and Due 2016). In these cases, participants may have preferred interpreters or request someone from outside their direct community but who speaks a shared language. Researchers need to be mindful of these nuances when considering approaches to interpreting (see ▶ Chaps. 93, "Considerations About Translation: Strategies About Frontiers," ▶ 95, An Approach to Conducting Cross-Language Qualitative Research with People from Multiple Language Groups," and ▶ 94, "Finding Meaning: A Cross-Language Mixed-Methods Research Strategy").

6 Ethical Considerations

As highlighted above, there are diverse considerations in relation to ethics when working with people from refugee or asylum seeker backgrounds (Birman 2006; Block et al. 2012). One of the foremost of these relates to the issue of gaining informed consent and vulnerability (Liamputtong 2010). By virtue of their previous experiences, many people with refugee backgrounds can be considered "vulnerable" in relation to their involvement in research (Ellis et al. 2007; Liamputtong 2010; Hugman et al. 2011), although the term should be used with caution (Levine 2004). This status requires consideration in research designs, particularly in relation to the ethical issue of power relations when obtaining consent to research participation. In health or mental health research, research questions can be sensitive, and researchers should be aware of issues such as the similarity of interviews or surveys to the interviews or surveys conducted as part of refugee processing, and the fact that some people may therefore not understand that participation in the research is voluntary (Ellis et al. 2007). Issues with informed consent also arrive when working with collectivist cultures who may be reluctant to decline to participate in order to benefit the broader community of which they are part (Ellis et al. 2007), and from cultural norms concerning who gives consent for others (including gender relations and the status of the elderly or children).

Many people with refugee or asylum seeker backgrounds may be hesitant to sign consent forms due to previous negative experiences with signing documents or cultural norms concerning placing a signature (Liamputtong 2010; Kabranian-Melkonian 2015). As such, it may be more appropriate to gain recorded verbal consent prior to the start of an interview, or to state that consent is assumed if participants agree to complete a survey. Critically, it is important to ensure that all project information is translated into a language in which potential participants can understand, or that there is scope in a project to have a neutral person explain the project to participants, if appropriate. Moreover, the process of informed consent should be seen as an iterative, rather than once off, process, such that participants who are involved in multiple stages of data collection or long data collection process are provided with an opportunity to provide ongoing assent to the project, and withdraw if they wish (Mackenzie et al. 2007). In this sense, then, the process of gaining consent should not be seen as a once-off, single event, but rather an ongoing situation of negotiation and shared understanding about the research process (Mackenzie et al. 2007). Recording of interviews can also be challenging for participants who may be concerned about such recordings being used against them, for example, in determining the outcome of their claim for asylum. Again, for the researcher, this is a delicate balance of being able to report on people's experiences in their own words and the need to be responsive to people's apprehension about recorded interviews.

Relatedly regarding assurances of confidentiality and/or anonymity attention should be paid to the possibility of identifying participants in new and emerging communities. This is particularly important in qualitative research which might involve the publication of individual extracts or participant information which is not aggregated. In this situation, particular combinations of demographic details (age, gender, country of origin, ethnicity, family size, religion, number of years in a country) may be combined in such a way that individuals from new or emerging communities can be identified. In health research specifically, such identification may lead to negative consequences for research participants. Here, then, researchers should balance the need to provide sufficient details to ensure the research can be assessed as reliable or replicable, with the imperative to protect research participants from harm (Liamputtong 2007, 2010).

Particular ethical issues arise in relation to working with children with refugee backgrounds, particularly in relation to obtaining informed consent (Due et al. 2014). Clearly, standard practice for research is the requirement to gain informed consent from parents or carers for child participants. However, the fact remains that obtaining consent in this manner does not take into account the child's own willingness to work with the researchers or participate in the study, and this is particularly the case given cultural differences and considerations in relation to obtaining consent from adults or assent from children (for example, determinations of power based on premigration experiences, see Morrow 2008). Correspondingly, where possible, ongoing "assent" should be gained from child participants, such that the project is explained to them and they have a choice regarding their participation that is on their own terms to the extent that this is possible. For example, children should not be pressed to answer questions or participate in activities that they do not wish to complete. While verbal assent should always be gained from children, it should also be noted that this process may not always reflect an autonomous agreement to participate in an activity, particularly where research is conducted in an "adult" environment such as a school (Punch 2002). As such, a similar iterative process to that noted above should be used, whereby children provide assent at every stage of the research (Block et al. 2012).

Another issue pertaining to ethics lies in relation to considering the well-being of research participants themselves (Jacobsen and Landau 2003). Researchers should consider the potential impact of the research on the possibility of retraumatization, and there should be a comprehensive plan to refer participants to services if required. Correspondingly, research with refugee participants should pay particular attention to the requirement to "do no harm" (Hugman et al. 2011). On the other hand, research has also found that participation in research can be empowering for participants (Newman and Kaloupek 2004; Ellis et al. 2007) and that research which involves interventions may be of direct benefit to communities (Ellis et al. 2007). As such, researchers should weigh the risks and benefits of their research, and ensure that the benefits outweigh the risks before they proceed (Ellis et al. 2007).

It is also important to note that some researchers have highlighted that erring too far on the side of caution in relation to "protecting" refugees from research can be paternalistic and may lead to inequalities relating to the lack of refugee voices in research (Kilpatrick 2004). This situation is in itself unethical and may lead to negative outcomes for refugees and asylum seekers, particularly if there is a lack of health and well-being research from which to develop appropriate interventions or assistance (Birman 2006). Correspondingly, researchers should ensure that where possible refugee and asylum seeker voices *are* heard in research, rather than reliance on others (e.g., service providers, healthcare professionals). In this sense, then, ethical research with refugees and asylum seekers can be seen to involve a fine balance between rigor, advocacy, benefit, and inclusion in the research process (see also ▶ Chap. 106, "Ethics and Research with Indigenous Peoples").

7 Advocacy

While most researchers agree that close community collaboration is a central aspect of research design with refugees and asylum seekers, it is also worth noting that one of the main tensions in research with refugees on any research topic is that of the role of advocacy in research. Some researchers (e.g., Jacobsen and Landau 2003) have suggested that much research with refugees is politically charged and does not meet standards for rigorous research that can be seen as reliable or valid. This lack of rigor is noted primarily due to a priori assumptions on the part of the researchers about what the research outcomes will be, as a result of a desire to enact social or political changes on the basis of the results. On the other hand, other researchers (e.g., Mackenzie et al. 2007) have suggested that ethical research with refugees *should* lead to better outcomes for those affected by the research, and that given the frequently desperate situation of many refugees, academics should also stand in solidarity with research participants through advocacy-type roles. Mackenzie et al. (2007, p. 316) state:

> When a human being is in need and the researcher is in a position to respond to that need, non-intervention in the name of "objective" research is unethical. Further, it could be argued that if researchers are in a position to assist refugees to advocate on their own behalf... that it is morally incumbent on them to do so.

Indeed, in many contexts around the world, academics have taken an advocacy stance as a result of their research, resulting in letters to politicians and the media, and petitions for issues such as asylum seeker policy and refugee education or access to healthcare (Hartley et al. 2013). Correspondingly, we argue in this chapter that research with refugees should be balanced between meeting academic standards of rigor and objectivity, while also leaving room for advocacy (Block et al. 2012). This is particularly important for research concerning refugee mental and physical health, and the social determinants of health, where the findings of a research project could be directly relevant to policy and practice. In such situations it is arguably unethical for researchers to *not* use the outcomes of their research to advocate for change.

8 Conclusion and Future Directions

As noted above, it is critical that any health research that seeks to document the lives of people with refugee and asylum seeker backgrounds also includes their voices, and takes into account specific cultural knowledges and understandings concerning health and well-being. In this regard, researchers need to balance several key considerations in design research protocols concerning refugee health. This includes particularly the need to consider the use of culturally appropriate research methods while also ensuring that research processes and tools are valid and reliable. In addition, it is important to think broadly about health and wellbeing– such as through a SDoH approach – to allow for a nuanced examination of the sociopolitical circumstances in which refugees live or have lived.

Clearly, refugee and asylum seeker health is a crucial area, and one in need of more research that can inform policy and practice as the numbers of refugees worldwide continues to increase. However, in addition to applied research concerning health and well-being, there is also a need for researchers to develop methodologies and research tools to ensure that research in this area can be rigorous and can include appropriately validated tools through which to draw conclusions. In particular, there is a critical need for physical and mental health scales that are validated cross-culturally, specific to the refugee population, and validated for diverse language groups. This is particularly the case in areas such as child trauma (Ehntholt and Yule 2006). Furthermore, there is a need for an evidence-base to consider methodologies, and the ways in which rigor can be balanced with diverse methods that include the voices of refugees, and that incorporate community understandings into research practices.

References

Bessell AG, Deese WB, Medina AL. Photolanguage. Am J Eval. 2016;28 (4):558–569

Birman D. Ethical issues in research with immigrants and refugees. In: Trimble JE, Fisher CB, editors. The handbook of ethical research with ethnocultural populations and communities. Thousand Oaks: Sage; 2006. p. 155–78.

Bloch A. Methodological challenges for national and multi-sited comparative survey research. J Refug Stud. 2007;20(2):230–47.

Block K, Warr D, Gibbs L, Riggs E. Addressing ethical and methodological challenges in research with refugee-background young people: reflections from the field. J Refug Stud. 2012;26(1):69–87.

Browne-Yung K, Ziersch A, Baum F. Neighbourhood, disorder, safety and reputation and the built environment: perceptions of low income individuals and relevance for health. Urban Policy Res. 2016;34(1):17–38.

Charmaz K. Constructing grounded theory: a practical guide through qualitative analysis. 2nd ed. London: Sage; 2014.

Cooney J, Burton K. Photolanguage Australia: human values. Sydney: Catholic Education Office; 1986.

Corbin J, Strauss A. Basics of qualitative research: techniques and procedures for developing grounded theory. 4th ed. Thousand Oaks: Sage; 2015.

Correa-Velez I, Gifford SM, Barnett AG. Longing to belong: social inclusion and wellbeing among youth with refugee backgrounds in the first three years in Melbourne, Australia. Soc Sci Med. 2010;71:1399–408.

Crivello G, Camfield L, Woodhead M. How can children tell us about their wellbeing? Exploring the potential of participatory research approaches within young lives. Soc Indic Res. 2009;90:51–72.

Darbyshire P, MacDougall C, Schiller W. Multiple methods in qualitative research with children, more research or just more? Qual Res. 2005;5:417–36.

Drew S, Duncan R, Sawyer S. Visual storytelling: a beneficial but challenging method for health research with young people. Qual Health Res. 2010;20(12):1677–88.

Due C, Riggs DW. Freedom to roam? Space use in primary schools with new arrivals programs. Online J Int Res Early Child Educ. 2011;2:1–16.

Due C, Riggs DW, Augoustinos M. Research with children of migrant and refugee background: a review of child-centered research. Child Indic Res. 2014;7:209–27.

Due C, Riggs DW, Augoustinos M. Experiences of school belonging for young children with refugee backgrounds. Educ Dev Psychol. 2016;33:33.

Ehntholt K, Yule W. Practitioner review: assessment and treatment of refugee children and adolescents who have experienced war-related trauma. J Child Psychol Psychiatry. 2006; 47(12):1197–210.

Ellis BH, Kia-Keating M, Yusuf SA, Lincoln A, Nur A. Ethical research in refugee communities and the use of community participatory methods. Transcult Psychiatry. 2007;44(3):459–81.

Freire P. Pedagogy of the oppressed. New York: Continuum International Publishing Group Inc; 2005.

Fullana J, Pallisera M, Vilà M. Advancing towards inclusive social research: visual methods as opportunities for people with severe mental illness to participate in research. Int J Soc Res Methodol. 2014;16(6):723–38.

Gartley T, Due C. 'The interpreter is not an invisible being': a thematic analysis of the impact of interpreters in mental health service provision with refugee clients. Aust Psychol. 2016;52:31.

Gifford S, Bakopanos C, Kaplan I, Correa-Velez I. Meaning or measurement? Researching the social contexts of health and settlement among newly-arrived refugee youth in Melbourne, Australia. J Refug Stud. 2007;20(3):414–40.

Gifford S, Correa-Velez I, Sampson R. Good starts for recently arrived youth with refugee backgrounds: promoting wellbeing in the first three years of settlement in Melbourne, Australia. Melbourne: Refugee Health Research Centre & Victorian Foundation for Survivors of Torture; 2009.

Hanza MM, Goodson M, Osman A, Porazz Capetillo MD, Hared A, Nigon SJ, Meiers JA, Weis ML, Wieland, Sia IG. Lessons learned from community-led recruitment of immigrants and refugee participants for a randomized, community-based participatory research study. J Immigr Minor Health. 2016;18(5):1241–5.

Haque N, Eng B. Tackling inequity through a photovoice project on the social determinants of health: translating photovoice evidence to community action. Glob Health Promot. 2011; 18(1):16–9.

Hartley L, Pedersen A, Fleay C, Hoffman S. The situation is hopeless; we must take the next step: reflecting on social action by academics in the asylum seeker policy debate. Aust Community Psychol. 2013;25(2):22–37.

Higginbottom G, Liamputtong P, editors. Participatory qualitative research methodologies in health. London: Sage; 2015.

Hugman R, Pittaway E, Bartolomei L. When 'do no harm' is not enough: the ethics of research with refugees and other vulnerable groups. Br J Soc Work. 2011;41:1271–87.

Jacobsen K, Landau LB. The dual imperative in refugee research: some methodological and ethical considerations in social science research on forced migration. Disasters. 2003;27(3):185–206.

Kabranian-Melkonian S. Ethical concerns with refugee research. J Hum Behav Soc Environ. 2015;25(7):714–22.

Kilpatrick DG. The ethics of disaster research: A special section. J Trauma Stress. 2004;17(5): 361–362

Kurban H, Liamputtong P. Health, social integration and social support: the lived experiences of young Middle-Eastern refugees living in Melbourne, Australia. Unpublished paper submitted for publication. 2017.

Lee SK, Sulaiman-Hill CR, Thompson SC. Overcoming language barriers in community-based research with refugee and migrant populations: options for using bilingual workers. BMC Int Health Hum Rights. 2014;14:11. http://www.biomedcentral.com/1472-698X/14/11

Lenette C, Boddy J. Visual ethnography: promoting the mental health of refugee women. Qual Res J. 2013;13(1):72–89.

Lenette C, Cox L, Brough M. Digital storytelling as a social work tool: learning from ethnographic research with women from refugee backgrounds. Br J Soc Work. 2015;45(3):988–1005.

Levine C. The concept of vulnerability in disaster research. J Trauma Stress. 2004;17(5):395–402.

Liamputtong P. Researching the vulnerable: a guide to sensitive research methods. London: Sage; 2007.

Liamputtong P. Performing qualitative cross-cultural research. Cambridge: Cambridge University Press; 2010.

Liamputtong P, Fernandes S. What makes people sick? The drawing method and children's conceptualisation of health and illness. Aust J Early Child. 2015;40(1):23–32.

Liamputtong P, Fanany R, Verrinder G, editors. Health, illness and well-being: perspectives and social determinants. Melbourne: Oxford University Press; 2012.

Loehr N. At home in the market: risk, acculturation and sector integration in the private rental tenancies of humanitarian migrants. Unpublished thesis, Flinders University, Adelaide. 2016.

Mackenzie C, McDowell C, Pittaway E. Beyond "do no harm": the challenge of constructing ethical relationships in refugee research. J Refug Stud. 2007;20(2):299–319.

McMaster D. Temporary protection visas: obstructing refugee livelihoods. Refugee Survey Quarterly. 2006;25(2):135–145

McMichael C. The health of migrants and refugees. In: Laimputtong P, editor. Public health: local and global perspectives. Melbourne: Cambridge University Press; 2016. p. 330–49.

McMichael C, Nunn C, Gifford S, Correa-Velez I. Studying refugee settlement through longitudinal research: methodological and ethical insights from the good starts study. J Refug Stud. 2014; 28(2):238–57.

Meadows D. Digital storytelling: research-based practice in new media. Vis Commun. 2003; 2(2):189–93.

Morrow V. Ethical dilemmas in research with children and young people about their social environments. Child Geogr. 2008;6(1):49–61.

Newman E, Kaloupek DG. The risks and benefits of participating in trauma-focused research studies. J Trauma Stress. 2004;17(5):383–394

Newman M, Woodcock A, Dunham P. 'Playtime in the borderlands': children's representations of school, gender and bullying through photographs and interviews. Child Geogr. 2006; 4(3):289–302.

O'Kane C. The development of participatory techniques: facilitating children's views about decisions which affect them. In: Christensen P, James A, editors. Research with children: perspectives and practices. London: Falmer Press; 2000. p. 136–59.

Pittaway E, Bartolomei L. Women at risk – field research report: Thailand 2003. Sydney: Centre for Refugee Research, University of New South Wales; 2003.

Punch S. Research with children: the same or different from research with adults? Childhood. 2002;9(3):321–41.

Rogers D, Darcy M, Arthurson K Researching territorial stigma with social housing tenants: tenant-led digital media production about people and place. In: Kirkness P, Tijé-Dra A, editors. Negative neighbourhood reputation and place attachment: conceptual approaches, policy responses and resistance to territorial stigmatisation. Ashgate: Surrey; in press.

Schmidt A. 'I know what you're doing', reflexivity and methods in Refugee Studies. Refugee Surv Q. 2007;26(3):82–99

Schweitzer R, Steel Z. Researching refugees: methodological and ethical considerations. In: Liamputtong P, editor. Doing cross-cultural research: ethical and methodological perspectives. Dordrecht: Springer; 2008. p. 87–101.

Seamer B. Picture this: 75 colour photographs for conversation and reflection. Bendigo: St Luke's Innovative Resources; 2007.

Solar O, Irwin A. A conceptual framework for action on the social determinants of health, Social Determinants of Health Discussion Paper 2 (Policy and Practice). Geneva: World Health Organization; 2010.

Spring M, Westermeyer J, Halcon LL, Savik K, Robertson C, Johnson DR, Butcher JN, Jaranson JM. Sampling in difficult to access refugee and immigrant communities. J Nervous Mental Dis. 2003;191(12): 813–819.

UNHCR. Convention and protocol relating to the status of refugees. Geneva: Office of the United Nations High Commissioner for Refugees; 1967.

UNHCR. Worldwide displacement hits all time high as was and persecution increase. 2015. Retrieved from http://www.unhcr.org/558193896.html

Vervliet M, Rousseau C, Broekaert E, Derluyn I. Multilayered ethics in research involving unaccompanied refugee minors. J Refug Stud. 2015;28(4):468–85.

Vigneswaran D, Quirk J. Representing 'hidden' populations: a symposium on sampling technique. J Refug Stud. 2012;26(1):110–6.

Wallerstein NB, Duran B. Using community-based participatory research to address health disparities. Health Promot Pract. 2006;7:312–23.

World Health Organization. Preamble to the constitution of the World Health Organization as adopted by the International Health Conference, New York, 19–22 June, 1946; signed on 22 July 1946 by the representatives of 61 States (Official Records of the World Health Organization, no. 2, p. 100) and entered into force on 7 April 1948. 1948. http://www.who.int/about/definition/en/print.html. Accessed 16 June 2016.

World Health Organization. Mental health: a state of wellbeing. 2014. http://www.who.int/features/factfiles/mental_health/en/. Accessed 16 June 2016.

World Health Organization. Social determinants of health. 2016. http://www.who.int/topics/social_determinants/en/. Accessed 16 June 2016.

Young L, Barrett H. Adapting visual methods: action research with Kampala street children. Area. 2001;33(2):141–52.

Zeinstra GC, Koelen MA, Colindres D, Kok FJ, de Graaf C. Facial expressions in school-aged children are a good indicator of 'dislikes', but not of 'likes'. Food Qual Prefer. 2009;20:620–624.

Ethical Issues in Cultural Research on Human Development

108

Namrata Goyal, Matthew Wice, and Joan G. Miller

Contents

1 Introduction	1892
2 Cultural Broadening of Operational Definitions	1893
3 Cultural Adequacy of Procedures and Modes of Data Collection	1895
3.1 Informed Consent	1895
3.2 Privacy	1897
3.3 Minimizing Harm	1898
4 Ethical Considerations in the Interpretation and Reporting of Data	1900
5 Conclusion and Future Directions	1902
References	1902

Abstract

This chapter addresses ethical issues in cultural research on human development. We argue for the importance of attending to culture in all phases of the research process and highlight ways that promoting the ethical sensitivity of cultural research enhances its validity and explanatory force. The first portion of the chapter focuses on early phases of the research process. We underscore the need to operationalize constructs in culturally valid ways and identify challenges that arise when objectively comparable procedures involve culturally variable meanings. The next section focuses on ethical issues in sampling, including the importance of tapping understudied populations and respecting local cultural norms in securing informed consent. We next address ethical aspects of study design and data collection, pointing out ways that harm, coercion and invasion of

---

N. Goyal (✉) · M. Wice · J. G. Miller
Department of Psychology, New School for Social Research, New York, NY, USA
e-mail: goyal@newschool.edu; wicem585@newschool.edu; millerj@newschool.edu

© Springer Nature Singapore Pte Ltd. 2019
P. Liamputtong (ed.), *Handbook of Research Methods in Health Social Sciences*,
https://doi.org/10.1007/978-981-10-5251-4_51

privacy that may result from inadequate attention to cultural meanings and practices. Lastly, we discuss the impact of drawing unsound or stereotypical conclusions about culture and human development, while discussing the insights cross-cultural research has to offer in terms of broadening psychological constructs, contributing to basic psychological theory, and making the discipline less culturally parochial. We conclude by outlining ways in which culturally sensitive research can enhance both ethics and research quality.

Keywords

Ethics · Confidentiality · Informed consent · Privacy · Harm · Attachment · Parenting · Motivation · Culture

1 Introduction

Research ethics in the field of psychology mandate that psychologists conduct their research in accord with ethical principles. As in other health and social science research, psychologists must be concerned not only with protecting the human rights of research participants, including their rights to privacy and to protection from harm, but also with insuring the welfare of research participants and of the larger community affected by psychological research findings. As we will argue, to meet these goals, psychologists must adopt practices that take into account the cultural beliefs, values, and practices of the populations involved in and affected by their work (see also ▶ Chap. 106, "Ethics and Research with Indigenous Peoples").

As an illustration of the ethical challenges that arise in taking cultural considerations into account in research on human development, we begin with an example of an ethical dilemma that we encountered while planning a study among fourth-grade elementary school children in the USA. We had designed the study to assess the relationship between young children's empathy and their interactions with pets. Our methods included: (a) basic demographic questions; (b) a short questionnaire that assessed reactions to short stories about hypothetical children; and (c) emotion recognition probes. We submitted a research proposal describing our research materials to the Institutional Review Board (IRB) at our university and received the following feedback:

> I would ask that the (demographic) question 'Are you a boy or a girl?' be changed to 'Do you see yourself as a boy, a girl, neither of these, or something else?' Likewise in the questionnaire the proper names of the (hypothetical) children in the story can replace he/she. These changes are easy to accomplish yet showcase the ways in which our university is inclusive and does not promote the binary depictions of gender and the stereotypes associated with them.

Upon first glance, this feedback may seem reasonable, as the IRB was urging us to be more inclusive of sexual and gender minority populations. However, the feedback was ethnocentric in reflecting the value system of our liberal university's culture but not the more conservative value system of the suburban community in which we would be carrying out the project. Although eliminating the pronouns in our vignettes would be nonproblematic, the rephrasing of the demographic question advocated by the IRB would be experienced as foreign and possibly intrusive by the research participants in that it did not reflect the local cultural context, but rather the culture of the IRB members with their progressive agenda. By rephrasing the demographic question in the way suggested, we would have introduced an unfamiliar and potentially uncomfortable way of conceptualizing gender to the children without having obtained prior consent from their parents to do this – consent that it is likely the parents would not have given. Moreover, we would be undermining the validity of our study, as children might have been confused about the meaning of this modified gender probe.

The response that we made to the IRB was improvised in that ethical guidelines of IRB committees do not typically address how to make accommodations for this type of case. We were faced with a situation in which two ethical challenges (inclusion vs. harm) were in direct conflict. However, our IRB had failed even to recognize this conflict much less to adjudicate which ethical issue should take precedence over the other. Rather than complying with what we judged to be an ethically problematic IRB ruling, we counter-proposed that our demographic question be modified to an open-ended probe about gender that would be completed by the child's teacher, rather than either by the child or their parents. This allowed us to be inclusive, while at the same time avoiding harming the children and their parents, and retaining the integrity of the investigation.

In this chapter, we present an overview of ethical issues that arise in cultural research on human development in all phases of the research process. Considering early phases of the research process, we begin by underscoring the need to operationalize constructs in culturally valid ways and to identify challenges that arise when objectively comparable procedures reflect culturally parochial meanings. We next address ethical issues in data collection, including the importance of sampling understudied populations and of respecting local cultural norms in securing informed consent, as well ethical aspects of study design, including ways to avoid harm, coercion and invasion of privacy that may result from inadequate attention to cultural meanings and practices. Lastly, we discuss the impact of drawing unsound or stereotypical conclusions about culture and human development, while discussing insights that cultural research has to offer in terms of broadening psychological constructs, contributing to basic psychological theory, and making the discipline less culturally parochial.

2 Cultural Broadening of Operational Definitions

To achieve fairness in formulating initial research questions, it is important to attend to cultural variation in the meaning of the constructs under consideration and in their operational definitions. Research may fail to capture the perspective of certain cultural groups as the constructs under consideration may embody culturally bound meanings and assumptions. This type of concern may be illustrated in research on parental control. Developmental psychologists make a distinction between authoritative and authoritarian parenting styles (Baumrind 1966, 1996; Darling and Steinberg 1993). Whereas authoritative parenting is based on behavioral control, involving active guidance and direction of the child's behavior, authoritarian parenting is based on psychological control, involving the use of strategies such as manipulation, guilt induction, and coercion (Baumrind 1966, 1996; Conger et al. 1992; Coplan et al. 2002). It is assumed that positive affective experiences and beneficial adaptive outcomes arise from authoritative parenting styles whereas affectively harsh affective experiences and maladaptive outcomes arise from authoritarian parenting styles (Baumrind 1966, 1996; Conger et al. 1992; Coplan et al. 2002; Baumrind et al. 2010).

Among European American samples, researchers have observed that authoritative parenting styles are associated with closeness in parent-child relationships and higher academic achievement, whereas authoritarian parenting styles are associated with affectively distant parent-child relationships and lower academic achievement (Barber et al. 1994; Chao 2001; Jackon-Newsom et al. 2008). Additionally, research conducted among a large sample drawn from the USA found that adults who remembered the parenting style of their parents as being authoritative reported less depression and greater well-being than those who remembered it as being authoritarian (Rothrauff et al. 2009).

On the widely used standardized scale of parenting, the Child Report of Parenting Behavior Inventory (CRPBI, Schaefer 1965) the following item is included to tap authoritarian parenting: "[my mother or primary care giver] says if I really cared for her, I would not do things that cause her to worry." Whereas most youth associate feelings of being controlled by their parents with this scale item, African American children associate this scale item and similar authoritarian scale items with feelings of being cared for and loved (Mason et al. 2004). Thus, when African American populations are classified as authoritarian on the basis of their CRPBI responses, it is unclear whether this classification is actually reflective of their maintaining an authoritarian parenting orientation, given the positive affective meanings African-Americans tend to associate with parental control. Likewise, research conducted among Asian-American populations indicates that controlling forms of parenting are interpreted as reflecting parental warmth rather than parental harshness (Chao 1994, 1995, 2001) and that authoritarian parenting is associated with positive academic performance (Dornbusch et al. 1987). In sum, the affective significance and behavioral correlates of authoritarian parenting as measured by indices such as the CRPBI are culturally variable and thus it cannot be assumed that the constructs measured by a scale such as the CRPBI have the same meaning in different cultural groups. In this type of case in which the constructs under consideration on a standardized scale have culturally variable meanings, a researcher should not simply adopt the scale in its original form with cultural populations on which it has not been validated. Rather, researchers must adapt the scale measure in ways that make it more culturally appropriate for the particular cultural community under consideration or design their own measures that have greater cultural validity.

Cultural sensitivity in the operationalization of psychological constructs also allows researchers to avoid potential harm caused by inappropriately applying research findings in real world settings. For example, a parental education program that is encouraging parents to adopt an authoritative parenting style, and that is discouraging their reliance on more controlling styles of parenting, may have the effect of leading children to feel rejected or unloved if the affective meanings of the parenting styles in their cultural community are different from that assumed in the concept of authoritarian parenting. Efforts made to integrate culture-specific concerns into parenting interventions for Latino communities in the USA have been well received by participating parents (Parra-Cardona et al. 2016) and highlight the ethical importance of operationalizing constructs in a culturally sensitive manner.

Additionally, parents may feel that their personal parenting practices are being disparaged if psychologists portray the parenting style that is normative in their community as flawed because it does not match that of the middle class European American cultural communities on which the constructs of authoritative versus authoritarian parenting styles are based. Taking cultural considerations into account during the early phase of construct operationalization is integral then not only to ensure the validity of research but also to protect research participants and consumers of psychological research findings.

3 Cultural Adequacy of Procedures and Modes of Data Collection

Researchers must be culturally aware not only in the formation of their research ideas but also during the process of data collection. Cultural challenges that arise in the process of data collection include being attentive to the cultural context while obtaining informed consent, taking cultural considerations into account in ensuring the privacy of participants, as well as protecting participants from culturally variable sources of harm (see also ▶ Chap. 106, "Ethics and Research with Indigenous Peoples").

3.1 Informed Consent

Informed consent is a process for obtaining permission from prospective research participants to be part of a research study while providing them with information about the nature of the study and the potential benefits or harm associated with participating in it. As children lack the legal right to refuse to participate in research and in many cases the cognitive maturity to fully understand research procedures, efforts must be made to fully explain the nature of any benefits and risks associated with research to parents and guardians of the children. Moreover, in cases in which children are able to provide verbal and/or written consent, researchers must make efforts to obtain written consent from minors through assent procedures that are written in simplified child-centered ways.

Ensuring a child's consent and full comprehension of the study procedures is a challenging but important step in conducting research among children from diverse cultural groups. For example, in a series of studies conducted in Canada, elementary school children were able to accurately describe the nature and purpose of the study but significantly underestimated the potential risks associated with participating it, such as feelings of being embarrassed or upset by their performance, or being bored by the length of a particular questionnaire or study (Abramovitch et al. 1995). Children may also overestimate the potential benefits of participation or expect benefits which may be difficult to predict based on the nature of the research being conducted (Miller and Feudtner 2016). These types of concerns may be even more pronounced in the case of populations that have limited exposure to the norms of Western experimentation. In such cases, individuals may not anticipate that psychological research is commonly impersonal and hence they may react negatively to research contexts on the grounds of its impersonality. For example, AIDS research conducted among African Americans showed that participants viewed many standardized questions as disrespectful of their feelings and experiences (Stevenson et al. 1993).

Cultural variations in conceptions of authority also need to be taken into account when obtaining informed consent. For example, among European Americans it is common practice to obtain consent from the parent or legal guardian of the child, who is bestowed the right to accept or decline their child's participation in a research study. However, in the case of American Indian or Alaskan Native tribes, it may be considered essential to consult with and obtain prior approval from the tribal leaders about whether, how, and when investigators should approach children and their families within the tribe (Beauvais and Trimble 1992; Norton and Manson 1996). Likewise, in the case of certain Asian cultural groups, consent may be viewed as a family right, and thus, it is expected to be secured from parents, guardians, and grandparents even in cases in which an adolescent child is legally permitted to give consent (Tai and Lin 2001).

In enhancing the sensitivity of their research in diverse cultural contexts, researchers need to give additional attention to ensuring that consent is voluntary. Although consent forms include a stipulation that participants are free to withdraw at any time without penalty, in research with children, especially across different cultural contexts, this freedom might not be salient. For example, Abramovitch et al. (1995) have demonstrated that children are less likely to withdraw from a study if the experimenter does not voluntarily reiterate that he/she will not be upset with the child for stopping participation. This tendency not to treat research participation as voluntary may be further exacerbated in cultural communities in which individuals are prone to defer to authority, as members of such cultural communities may be even less familiar with psychological research norms than their European American counterparts. When concerns do arise during the process of obtaining consent, not all participants may be equally willing to ask questions of the researcher. Specifically, participants of lower socioeconomic status, who tend to ask fewer questions than those of higher SES during the informed consent process, may miss an opportunity to obtain all of the information required to make a truly informed decision (Rajaramn et al. 2011).

Finally, researchers also need to take special care to ensure that compensation provided for research participation is fair, noncoercive, and afforded the same meaning in different cultural communities. For example, if participation by the family is given with the sole aim of contributing to scientific research (i.e., an altruistic act), offering compensation could be construed as offensive or infringing on family values. Moreover, if families accept compensation for their participation, efforts must be made to ensure the equality of compensation in different cultural settings, such that one cultural group does not get over-compensated or under-compensated due to different spending power associated with their currency. Additionally, the compensation offered should not be so great as to be experienced by individuals as compelling their participation but not so low that participants feel exploited.

3.2 Privacy

Privacy in the research context entails: (a) physical privacy, i.e., providing spatial seclusion; (b) informational privacy, i.e., ensuring confidentiality and protection of data; and (c) decisional privacy, i.e., allowing participants to make decisions for themselves, especially decisions concerning sex, religion, and/or reproduction (Allen 1999). As explained below, in all of these instances, special concern must be taken into account to achieve cultural sensitivity.

Contrasting cultural norms concerning physical space and openness may affect the meaning and desirability of physical and informational privacy in the conduct of research. For example, a recent public health interview study undertaken in Sri Lanka, which compared the outlooks of Buddhist, Western, and Ayurvedic healers, demonstrated that respondents were "uncomfortable" with participating in interviews in hidden spaces, such as closed offices – a behavior that was normative among Western participants. When these same participants were moved to public patio spaces, they participated in research with less hesitation and unease. Furthermore, in the public health interviews, the research participants were also found to be more comfortable coming to the interviews accompanied by their significant others, parents, or close friends, and felt safer disclosing information in the presence of these significant others (Monshi and Zieglmayer 2004). Thus, in seeking to protect the privacy of information collected, researchers should be mindful about different cultural outlooks concerning the meaning and practice of privacy.

Variation in cultural values underlying decisional privacy must also be taken into account when determining who has the right to obtain information and make decisions about the participant, especially in the case of children. One central issue that arises is the weight to be given to the rights of adolescents to control their own behavior. Since older adolescents have the same cognitive capacity as their adult counterparts as well as engage in adult-like behaviors, such as driving and drinking alcohol, arguments that hold true for younger children no longer constitute clear grounds for the parents' rights to make decision on behalf of their adolescent offspring. In resolving such conflicts, however, attention should be given to cultural norms about parental involvement.

For example, in research conducted with US samples, researchers have concluded that in the case of major health decisions when a parents and adolescent disagree on the right of the youth to control his/ her behavior, priority should be given to the rights of the adolescent (Brooks-Gunn and Rotheram-Borus 1994). Researchers commonly argue that a youth should have autonomy over his/her own behavior and that the involvement of parents may preclude the adolescent from gaining access to the health care she needs. However, in many non-Western cultural communities (or even in the case of ethnic subgroups within the USA), families play a more central role in planning the adolescent's life than among European Americans. In such cultural communities, adolescents may themselves afford greater legitimacy to parental involvement than in middle class European American cultural contexts and expect their parents to be more involved in significant everyday life decisions. Thus, in such instances, it could potentially be experienced as disrespectful or intrusive to the parent-child relationship to bypass the parent by withholding information from them about their adolescent's research participation, especially in cultural communities in which the adolescents themselves value such parental input in the research process (Casas and Thompson 1991; Fisher 2002).

3.3   Minimizing Harm

Minimizing harm in the research context entails balancing the risks that human subjects may encounter by participating in scientific research with the benefits to scientific insight. It is not surprising then that the subjective experience of risk is culturally variable. Yet, historically, most experimental procedures have been created taking into account the cultural norms and expectations of European American families. Although most experimental procedures induce some level of stress, the stress is generally judged to be temporary and to be no greater than general stress involved in daily life. However, when research procedures violate everyday socialization practices of the cultural communities of research participants, they may entail greater potential for harm or discomfort than everyday stress.

For example, the Strange Situation procedure designed to assess attachment behavior created by Ainsworth (1963) was intended to induce mid-level stress in children and their primary caregivers by subjecting them to periods of separation and communion. In this procedure the infant, for a brief period lasting several minutes, interacts with an unfamiliar adult in the absence of his/her primary caregiver and is thus at times left alone. Although this procedure invariably induces some distress in the child, this distress is deemed to be temporary and used to gain insight into the child's attachment style. This procedure is routinely approved by the Human Subjects Committees (IRB) as Ainsworth has demonstrated that the procedure is congruent with commonly experienced American practices and thus entails an acceptable level of risk. However, Takahashi (1982) notes that this type of behavior rarely occurs in Japan, as Japanese mothers rarely leave their children alone, even in the presence of other family members such as grandparents, aunts, and uncles. When the Strange Situation was administered to a sample of Japanese infants (Takahashi 1986; Takahashi and Hatano 2009), Japanese infants were interpreted as being predominantly anxiously attached to their caregivers. However, it is unclear whether the anxiety displayed by the infants was in response to the stress of the procedure or reflective of their attachment style.

To give another example, the Still Face paradigm (Tronick et al. 1978) is another experimental procedure employed with infants that is discordant with parenting practices and beliefs in cultural communities outside the dominant, white middle class model that is taken to be the default in developmental science research. In this procedure, mothers are instructed to interact in face to face play with their infants for 2 min, after which they are asked to maintain a still face for 2 min followed by another interactive face to face play for another 2 min. These interactions are later coded for the baby's emotional self-regulation as well as the mother's capacity to reengage the infant (Tronick and Cohn 1989).

Although most infants are expected to be distressed by the still face period involving the extreme nonresponsiveness, marked cultural differences occur in the process through which self-regulation is achieved (Meléndez 2005). Whereas !Kung San hunter-gatherers of Botswana tend to respond to babies' cries within 10 s, Western mothers tend to refrain from responding to their infants' cries as much as 40% of the time (Barr 1999) and may not even identify a cry as requiring a response unless it has persisted for 10 min (Small 1998). Moreover, in many cultures of Africa and Southeast Asia, infants tend to self-regulate before making a full blown cry (Papousek 2000). Given the contrasting expectations of parental responsiveness to infant distress, it is probable that the Still Face Paradigm entails more stress and thus more harm and discomfort in certain cultural communities than in others. Thus, in the case of the Strange Situation, Still Face Paradigm, and other experimental procedures that involve culturally variable levels of stress, researchers need to make significant modifications to the consent agreements, if not also to the procedures themselves, in order to adapt them for use in different cultural contexts. Although it may be reasonable to describe such procedures as involving a minimal level of harm to European American families, this type of assurance may not apply in cultural communities in which parenting practices differ markedly. It is also important to modify the procedures themselves to minimize the harm caused by their use in diverse cultural communities, as well as to substitute procedures that have greater cross-cultural validity.

4 Ethical Considerations in the Interpretation and Reporting of Data

Ethical considerations related to culture extend beyond study design and data collection to the final stages of interpreting and communicating the study results. Drawing conclusions that avoid unwarranted generalizations and provide accurate representations of diverse populations is not just a matter of producing valid scientific research but also an ethical imperative. This ethical imperative includes avoiding drawing overly broad conclusions about populations based on limited samples as well as avoiding biased and stereotypical conceptualizations of the groups being studied. Both of these can be seen as matters of ethics in that they involve fairness, in terms of the inclusivity of diverse populations and their unbiased portrayal.

Psychological research draws a disproportionate amount of data from a small segment of the world's population. This heavily sampled demography, consisting of predominantly educated Western individuals living in wealthy, democratic, industrialized countries forms the basis for many universalistic claims about human behavior (Henrich et al. 2010). An analysis conducted by Arnett of six major APA journals published between 2003 and 2007 found that only 3% of the samples were from Asia, 1% from Latin America, and 1% from either Africa or the Middle East; the remaining 95% of the samples were drawn from the USA, other English-speaking countries, or Europe (Arnett 2008). Formulating generalized claims based upon such a limited sample is made even more problematic by the fact that the individuals comprising this narrow slice of humanity are in many respects a poor representation of the rest of the world, as shown by comparative studies investigating cognition and behavior ranging from visual perception to theories of folkbiology (Henrich et al. 2010).

In research on human development, relying upon a narrow, biased sample base reflects a lack of appreciation for the role of the sociocultural environment in shaping human development. Different contexts afford access to experiences that may have significant effects on development. Studying children in varying cultural contexts provides a necessary check on assumptions that psychological development among children raised in Western, educated, wealthy, white families is the universal default. One example of this can be seen in children's folk theories of biology (Henrich et al. 2010). Much research conducted on children's folk-biological reasoning has taken place in urban settings in which children have limited experience interacting with the natural world (Henrich et al. 2010). Thus, among these urban samples, children's biological reasoning before age 7 tends to be anthropocentric, relying strongly on humans in making inferences about other living things (Carey 1985). This would be expected given the human-centered world in which urban children grow up as well as the limited exposure the children have to various living things present in the natural world. Children raised in rural environments, who have more frequent and rich interactions with the natural world, however, did not display this tendency to employ anthropocentric folkbiological reasoning (Ross et al. 2003).

In addition to giving limited attention to cultural diversity in child development, research on early child development is lacking in socioeconomic diversity, with research underrepresenting children from disadvantaged families (Fernald 2010). Developmental research typically relies disproportionately on narrow samples comprised of parents with the financial means and motivation to go out of their way to participate in studies. This is especially problematic because of the variability in access to certain types of cognitive stimulation for children of differing levels of SES, which may affect aspects of cognitive development (Fernald 2010).

It could be argued that the sampling bias in psychology is in itself an ethical violation of fairness in its failure to create a science of mind that is representative of the full spectrum of human thought and behavior. Collecting data across diverse cultural and socioeconomic contexts is an important first step in addressing this problem and making research on human development more inclusive. However, in doing so, researchers must pay adequate attention to the manner in which they represent the beliefs and attitudes of the communities involved. Specifically, great care must be taken to avoid an overly homogenized and stereotypical portrayal of the population of interest. Just as conducting research among diverse cultural communities requires careful consideration of potential variability in the interpretation and impact of study procedures, researchers choosing to undertake the important task of conducting cross-cultural research must also take seriously the challenge of adequately representing the sampled population.

Cross-cultural research conducted among East Asian populations is an example of an attempt at cultural inclusivity that often falls prey to such an overly homogenized approach to culture. It does this through making unjustifiably global claims about "Asian culture" or even "Eastern culture" based on samples composed of individuals coming from a diverse set of countries (e.g., Japan, China, Korea) with distinct sociohistorical backgrounds and strikingly different cultural traditions (e.g., Nisbett 2003). Conceptualizing culture in this way overlooks not only nationality but also variability in other factors such as religion, which lead to greater within-group diversity in attitudes and behavior than is implied by global cultural categories (Fisher 2002).

Conceptualizing cultural communities in overly homogenized terms is problematic in its failure to accurately represent the population of interest, and it is also prone to stereotypical characterizations of such groups. Data may be interpreted or summarized based on preconceived, stereotypical notions of large, heterogeneous groups. This may be exacerbated by an over-reliance in the field on attempts to measure large-scale cultural orientations, such as individualism-collectivism (Hofstede 1980) and independent-interdependent self-construal (Markus and Kityama 1991), accompanied by attempts to plot entire nations along these dimensions (e.g., Hofstede 1984). Categorizing cultural groups in this manner sacrifices a more nuanced, sensitive approach to culture by applying broad labels based solely on endorsement of decontextualized individual difference scale items.

5 Conclusion and Future Directions

In order to conduct ethically sound research on human development, culture is an issue that should be seriously addressed at every stage of the research process. Taking culture into account goes beyond considering the adequacy of procedures across cultural contexts, though this is surely of great importance. Culture should also factor into the initial theoretical constructs, which frame the study design, and carry through to the interpretation of the results, in addition to guiding all of the steps in-between. A failure to consider culture during any of these stages raises the possibility of an ethical breach and has the potential to diminish the quality of the work.

While lack of attention to culture presents a serious problem with ethical implications, psychologists who choose to conduct cross-cultural research must also give serious thought to negotiating the complexities inherent in applying ethical guidelines across diverse contexts. The necessity of being sensitive to how codes of ethics might be interpreted and experienced by local communities makes taking a cultural perspective on research ethics a challenging endeavor, but it is also one that is indispensable to making valid, meaningful contributions to the study of human development.

With globalization comes an increase in opportunities for cross-cultural collaborative research, which has the potential to paint a more representative picture of human development. Such collaborations also allow for meaningful and productive cooperation between researchers from different cultural backgrounds with regard to how best to implement ethical guidelines in diverse settings. This ease with which these research collaborations can span the globe thus provides an ideal atmosphere for not only conducting high-quality cross-cultural research on human development but for considering multiple cultural perspectives on the most ethical way to do so.

References

Abramovitch R, Freedman JL, Henry K, Van Brunschot M. Children's capacityto agree to psychological research: knowledge of risks and benefits and voluntariness. Ethics Behav. 1995;5:25–48.

Ainsworth MD. The development of infant-mother interaction among the Ganda. In: Foss BM, editor. The determinants of infant behaviour II. London: Methuen; 1963. p. 67–112.

Allen A. Coercing privacy. In: Goldman J, Choy A, editors. Privacy and confidentiality in health research, vol. 2. Bethesda: National Bioethics Advisory Commission; 1999. p. C1–C34.

Arnett JJ. The neglected 95%: why American psychology needs to become less American. Am Psychol. 2008;63:602–14.

Barber BK, Olsen JE, Shagle SC. Associations between parental psychological and behavioral control and youth internalized and externalized behaviors. Child Dev. 1994;65:1120–36.

Barr RG. Infant crying behavior and colic: an interpretation in evolutionary perspective. In: Trevarthen WR, Smith EO, McKenna JJ, editors. Evolutionay medicine. New York: Oxford University Press; 1999. p. 27–51.

Baumrind D. Effects of authoritative parental control on child behavior. Child Dev. 1966;37:887–907.

Baumrind D. The discipline controversy revisited. Fam Relat J Appl Fam Child Stud. 1996;45(4): 1405–14.

Baumrind D, Larzelere RE, Owens E. Effects of preschool parents' power assertive patterns and practices on adolescent development. Parent Sci Pract. 2010;10:156–201.

Beauvais F, Trimble JE. The role of the researcher in evaluating American Indian alcohol and other drug abuse prevention programs. In: Orlandi MA, editor. Cultural competence for evaluators: a guide for alcohol and other drug abuse prevention practitioners working with ethnic/racial communities. Rockville: US Department of Health & Human Services; 1992. p. 173–201.

Brooks-Gunn J, Rotheram-Borus MJ. Rights to privacy in research: adolescents versus parents. Ethics Behav. 1994;42:109–21.

Carey S. Are children fundamentally different thinkers and learners from adults? In: Chipman SF, Segal JW, Glaser R, editors. Thinking and learning skills, vol. 2. Hillsdale: Erlbaum; 1985. p. 485–517. Reprinted by Open University Press: Open University Readings in Cognitive Development.

Casas JM, Thompson CE. Ethical principles and standards: a racial-ethnic minority research perspective. Couns Values. 1991;35:186–95.

Chao RK. Beyond parental control and authoritarian parenting style: understanding Chinese parenting through the cultural notion of training. Child Dev. 1994;65:1111–9.

Chao RK. Chinese and European American cultural models of the self reflected in mothers-childrearing beliefs. Ethos. 1995;23:328–54.

Chao RK. Extending research on the consequences of parenting style for Chinese Americans and European Americans. Child Dev. 2001;72:1832–43.

Conger RD, Conger RD, Elder GH, Lorenz FO, Simons RL, Whitbeck LB. A family process model of economic hardship and adjustment of early adolescent boys. Child Dev. 1992;63:526–41.

Coplan RJ, Hastings DP, Lagace-Seguin DG, Moulton CE. Authoritative and authoritarian mothers' parenting goals, attributions, and emotions across different childrearing contexts. Parent Sci Pract. 2002;2:1–26.

Darling N, Steinberg L. Parenting style as a context: an integrative model. Psychol Bull. 1993;113:487–96.

Dornbusch SM, Ritter PL, Leiderman PH, Roberts DF, Fraleigh MH. The relation of parenting style to adolescent school performance. Child Dev. 1987;58:1244–57.

Fernald A. Getting beyond the "convenience sample" in research on early cognitive development. Behav Brain Sci. 2010;33:91–2.

Fisher CB. Participant consultation: ethical insights into parental permission and confidentiality procedures for policy relevant research with youth. In: Lerner RM, Jacobs F, Wertlieb D, editors. Handook of applied developmental science, vol. 4. Thousand Oaks: SAGE; 2002. p. 371–96.

Henrich J, Heine SJ, Norenzayan A. The weirdest people in the world? Behav Brain Sci. 2010;33:61–83.

Hofstede G. Motivation, leadership and organization: do American theories apply abroad. Organ Dyn. 1980;9:42–63.

Hofstede G. Cultural dimensions in management and planning. Asia Pac J Manag. 1984;1:81–99.

Jackon-Newsom J, Buchanan CM, McDonald RM. Parenting and perceived maternal warmth in European American and African American Adolescents. J Marriage Fam. 2008;70:62–75.

Markus HR, Kityama S. Culture and the self: implications for cognition, emotion, and motivation. Psychol Rev. 1991;98:224–53.

Mason CA, Walker-Barnes CJ, Tu S, Simons J, Martinez-Arrue R. Ethnic differences in the affective meaning of parental control behaviors. J Prim Prev. 2004;25:59–79.

Meléndez L. Parental beliefs and practices around early self-regulation: the impact of culture and immigration. Infants Young Child. 2005;18:136–46.

Miller VA, Feudtner C. Parent and child perceptions of the benefits of research participation. IRB Ethics Hum Res. 2016;38(4):1–7.

Monshi B, Zieglmayer V. The problem of privacy in transcultural research: reflections on an ethnographic study in Sri Lanka. Ethics Behav. 2004;14:305–12.

Nisbett RE. The geography of thought: how Asians and westerners think differently – and why. New York: Free Press; 2003.

Norton IM, Manson SM. Research in American Indian and Alaska Native communities: navigating the cultural universe of values and process. J Consult Clin Psychol. 1996;64:856–60.

Papousek M. Persistent crying, parenting, and infant mental health. In: Osofsky JD, Fitzgerald HE, editors. WAIMH handbook of infant mental health, vol. 4. New York: Guilford Press; 2000. p. 326–38.

Parra-Cardona JR, López-Zerón G, Villa M, Zamudio E, Escobar-Chew AR, Rodríguez MM. Enhancing parenting practices with Latino/a immigrants: integrating evidence-based knowledge and culture according to the voices of Latino/a parents. Clin Soc Work J. 2016. https://doi.org/10.1007/s10615-016-0589-y.

Rajaramn D, Jesuraj N, Geiter L, Bennett S, Grewal H, Vaz M. How participatory is parental consent in low literacy rural settings in low income countries? Lessons learned from a community based study of infants in South India. BMC Med Ethics. 2011;12:3.

Ross N, Medin DL, Coley JD, Atran S. Cultural and experiential differences in the development of folkbiological induction. Cogn Dev. 2003;18:25–47.

Rothrauff TC, Cooney TM, An JS. Remembered parenting styles and adjustment in middle and late adulthood. J Gerontol Ser B Psychol Sci Soc Sci. 2009;64B(1):137–46.

Schaefer E. Children's reports of parental behavior: an inventory. Child Dev. 1965;36:413–21.

Small MF. Our babies, ourselves: how babies and culture shape the way we parent. New York: Anchor Books; 1998.

Stevenson HC, DeMoya D, Boruch RF. Ethical issues and approaches in AIDS research. In: Ostrow DG, Kessler RC, editors. Methodological issues in AIDS behavioral research. New York: Plenum Press; 1993. p. 19–51.

Tai MC, Lin CS. Developing a culturally relevant bioethics for Asian people. J Med Ethics. 2001;27:51–4.

Takahashi K. Attachment behaviors to a female stranger among Japanese two-year-olds. J Genet Psychol. 1982;140:299–307.

Takahashi K. Examining the strange-situation procedure with Japanese mothers and 12 month old infants. Dev Psychol. 1986;22:265–70.

Takahashi K, Hatano G. Toward a valid application of the adult attachment interview to the Japanese culture and language: the need for ethnographic adaptations. Sacred Heart Univ J. 2009;112:42–63.

Tronick EZ, Cohn JF. Infant-mother face-to-face interaction: age and gender differences in coordination and the occurrence of miscoordination. Child Dev. 1989;60:85–92.

Tronick EZ, Als H, Adamson L, Wise S, Brazelton TB. The infant's response to entrapment between contradictory messages in face-to-face interaction. J Am Acad Child Psychiatry. 1978;17:1–13. https://doi.org/10.1016/S0002-7138(09)62273-1.

Section IV

Sensitive Research Methodology and Approach: Researching with Particular Groups in Health Social Sciences

Sensitive Research Methodology and Approach: An Introduction

109

Pranee Liamputtong

Contents

1 Introduction	1907
2 About the Section	1910
References	1915

Keywords

Sensitive research · Vulnerable people · Marginalized people · Hard-to-reach people · Hidden populations · Sensitive topics · Elites · Experts

1 Introduction

The goal of research is "that of discerning and uncovering the actual facts of [people]' lives and experience, facts that have been hidden, inaccessible, suppressed, distorted, misunderstood, ignored" (Bergen 1993; Bergen 1996, p. 200).

This section of the Handbook covers sensitive research methods and approaches when researching with particular groups of individuals and groups. The section includes chapters which are related to those who are referred to as vulnerable, marginalized, hard-to-reach people, hidden populations, as well as the elites and experts.

Within the present climate of our fractured world, it is inevitable that health and social science researchers will engage with the vulnerable, disadvantaged, and marginalized groups as it is likely that these population groups will be confronted with more problems with their health and well-being. Despite this, only a few books document and provide advice about how to go about performing research with

P. Liamputtong (✉)
School of Science and Health, Western Sydney University, Penrith, NSW, Australia
e-mail: p.liamputtong@westernsydney.edu.au

© Springer Nature Singapore Pte Ltd. 2019
P. Liamputtong (ed.), *Handbook of Research Methods in Health Social Sciences*,
https://doi.org/10.1007/978-981-10-5251-4_122

vulnerable people (see Renzetti and Lee 1993; Liamputtong 2007; Pitts and Smith 2007; Aldridge 2015; van Liempt and Bilger 2009). This lack of discussions, as Melrose (2002, p. 338) points out, may leave researchers "feeling methodologically vulnerable, verging on the distressingly incapable, because of emotional and anxiety challenges, and thus ill equipped to deal with some of the issues that may arise in this context." In this section of the Handbook, I bring together a number of chapters that discuss some important issues for the conduct of research within the vulnerable groups of people in health and social sciences.

The "vulnerability" is a socially constructed concept (Liamputtong 2007; ten Have 2016). Thus, vulnerable individuals have been referred to in multiple ways (Liamputtong 2007; Bracken-Roche et al. 2017). The word "vulnerable people" implies "the disadvantaged sub-segment of the community requiring utmost care, specific ancillary considerations and augmented protections in research" (Shivayogi 2013, p. 53). The vulnerable, according to Quest and Marco (2001, p. 1297), are people with "social vulnerability." They suggest that some population groups, including children, unemployed, homeless, drug-addicted people, sex workers, and ethnic and religious minority groups, face particular social vulnerability. When involving them in their research, these groups of people need special care from the researchers. Stone (2003, p. 149) refers to the "vulnerable" as individuals who are "likely to be susceptible to coercive or undue influence." To Stone (2003), the "vulnerable" includes children, pregnant women, mentally disabled persons, or those who are "economically or educationally disadvantaged." Punch (2002, p. 323) suggests that children are marginalized in an adult-dominated society, and as such they "experience unequal power relations with adults and much in their lives." In this sense, children are particularly vulnerable in society, particularly when it involves abusive behavior on the part of adults in their lives (see also Melrose 2002). Some groups may be vulnerable due to their so-called legal status. Some immigrants in the United States, for example, are undocumented immigrants (see Birman 2005). Due to their illegal status, they are denied access to health and social services. Most of these groups live in poverty, and most are employed in seasonal cropping industries which are prone to poor health and bad living situations.

Vulnerable individuals have also been referred to as "marginalized" people. Marginalized people, according to O'Donnell et al. (2016, p. 198), are those "populations outside of 'mainstream society.'" They are "highly vulnerable populations" who are "systemically excluded from national or international policy making forums." Marginalized individuals and groups include the homeless, drug users, sex workers, refugees, and ethnic minorities such as Roma and Irish Travellers. Often, they "experience severe health inequities" and have "poorer health status than the general population" (see also ▶ Chap. 110, ""With Us and About Us": Participatory Methods in Research with "Vulnerable" or Marginalized Groups").

Whatever definition we may use to best represent the "vulnerable" or the "marginalized," it is clear that extreme sensitivity is needed in the conduct of research with these groups (ten Have 2016). Johnson and Clarke (2003, p. 422) contend that in conducting sensitive research, "the process of gathering such information necessarily involves direct contact with vulnerable people, with whom sensitive and difficult topics are often raised and sometimes raised within difficult contexts." Therefore, undertaking research with vulnerable groups can present numerous serious difficulties for the researcher as well as the researched (see Melrose 2002; Liamputtong 2007, 2013; Dickson-Swift et al. 2008; Menih 2013; Shivayogi 2013; Aldridge 2012, 2015; Couch et al. 2014; McCauley 2015; Quinn 2015; Bracken-Roche et al. 2016, 2017; Wrigley and Dawson 2016; Medeiros 2017). It also raises many ethical concerns (see Menih 2013; Solomon 2013; Schrems 2014; ten Have 2015, 2016). Despite potential difficulties, the task of undertaking research with the vulnerable and marginalized participants can also present researchers with unique opportunities. Many chapters in this section illustrate these.

Closely linked with vulnerable and marginalized people is the concept of "sensitive research" (Liamputtong 2007). Social researchers increasingly undertake research on topics which are "sensitive" as they are concerned with behavior that is "intimate, discreditable, or incriminating" (Renzetti and Lee 1993, p. ix; see also McCosker et al. 2001; Lee and Lee 2012). Sensitive research, Dickson-Swift (2005, p. 11) suggests, "has the potential to impact on all of the people who are involved in it." Similarly, according to Lee (1993, p. 4), sensitive research poses several threats to the people: intrusive threat, a threat of sanction, and political threat. Research that intrudes into private lives of the research participants will create stressful experiences as well as pose intrusive threats to them. Barnard (2005, p. 2) refers sensitive research to those projects dealing with the "socially-charged and contentious areas of human behaviour" such as the impact of parental drug problems on the well-being of their children, or underage sex work (see ▶ Chap. 121, "Researching Underage Sex work: Dynamic Risk, Responding Sensitively, and Protecting Participants and Researchers"). Sensitive topics would be "often the 'difficult' topics – trauma, abuse, death, illness, health problems, violence, crime – that spawn reflection on the role of emotions in research" (Campbell 2002, p. 33).

Renzetti and Lee (1993, p. 6) point to some areas that will make research sensitive and pose more threats and create vulnerability of the researched. These include:

- Studies which are concerned with deviance and social control
- Inquiries which exercises coercion or domination
- Research that intrudes into the private lives or deeply personal experiences of the research participants
- Research that deals with sacred things

Nevertheless, there are individuals who are not referred to as vulnerable or marginalized that some researchers in the health and social sciences have also engaged with in their research. They are referred to as the elites or the experts (see Zuckerman 1996; Odendahl and Shaw 2002; Payne and Payne 2004; Stephens 2007; Harvey 2011; Taylor 2011; Mikecz 2012; Williams 2012; Abbink and Salverda 2013; Aguiar and Schneider 2013). They too have particular issues and challenges that health and social science researchers need to consider in conducting research with them. Elite research is about "studying up." It is research that involves individuals who have more power than the researcher (Williams 2012; Straubhaar 2015). As Straubhaar (2015) suggests, "elite" is defined as "one who, due to a potentially variable combination of social privileges (on the basis of social class, educational opportunity, etc.), has access to desirable and powerful social networks, through which he or she regularly has the ability to exercise power."

Neil Stephens and Rebecca Dimond say this clearly in their chapter (▶ Chap. 126, Researching Among Elites"): "Where there is power, there are elites; and there is power everywhere." Studying up is important as these powerful people are not only "the cause and cure of social problems," or "may be inefficient or abuse their power," but they also can "create positive change and social progress" (Williams 2012, p. 1). Chapters written by Neil Stephens and Rebecca Dimond (see ▶ Chap. 126, "Researching Among Elites"), Jyoti Belur (see ▶ Chap. 125, "Police Research and Public Health"), and Robert Campbell (see ▶ Chap. 127, "Eliciting Expert Practitioner Knowledge Through Pedagogy and Infographics") will illustrate many things that health and social science researchers can learn from.

Sensitive research, Lee (1993, p. 3) argues, stretches beyond the consequences of carrying out the research, but the methodological issues are also inherently essential in doing such research. Lee advocates the need to examine this issue from both the researchers and the researched. As readers will see, Lee's standpoints have been embraced by several authors in the section. As sensitive researchers, we must make our judgments on the impact of our research on not only the participants but also ourselves as researchers. As such, we must think carefully about the methodology used in collecting their data and the procedures that must be observed as sensitive to research participants, whoever our research participants might be. These are things that are included in chapters in this section.

2 About the Section

This section of the book comprises 18 chapters. In "'With us and about us': Participatory methods in research with 'vulnerable' or marginalized groups," Jo Aldridge writes about participatory methods in research with "vulnerable" or marginalized people (▶ Chap. 110, ""With Us and About Us": Participatory Methods in Research with "Vulnerable" or Marginalized Groups"). She argues that most health and social science research that include "vulnerable," "hard-to-reach," or marginalized people do not enhance participant engagement and "voice." For those who claim to use participatory research, the validity of the claims is unclear. Often, "the nature and extent of participant involvement in such studies are not always defined and the value and efficacy, as well as the challenges, of using participatory methods are often misunderstood." In this chapter, Aldridge explores these issues drawing on her own extensive research with marginalized groups using participatory models which "promote and enhance participant engagement and emancipation in research processes." She contends that "such approaches see 'vulnerable,' marginalized or socially excluded research participants in transformative roles in research, including as co-researchers, co-analysts and as designers and producers of their own research agendas and projects."

The chapter on inclusive research (▶ Chap. 111, "Inclusive Disability Research") for disability research is written by Jennifer Smith-Merry. More researchers are becoming involved in inclusive research: research that "people with a lived experience of the field of research under study are included as part of the research team." In this chapter, Smith-Merry provides an overview of inclusive research with people with disability. Examples from existing projects are illustrated to give a sense of the field and its limitations as well as its possibilities.

Tinashe Dune and Elias Mpofu write about understanding sexuality and disability using interpretive hermeneutic phenomenological approaches in their chapter (▶ Chap. 112, "Understanding Sexuality and Disability: Using Interpretive Hermeneutic Phenomenological Approaches"). In particular, the chapter "discusses the rational and processes for applying IHPA to engage participants in these sensitive and complex discussions on their lived experiences of understandings of sexuality." Procedural guidelines for applying IHPA to studying sexuality with CP as well as the strengths and limitations of this approach are also provided. The authors contend that "IHPA provides a unique advantage to studying heath issues with hidden populations or socially sensitive topics with the general population."

Ethics and practice of research with people who use drugs are presented by Julaine Allan (▶ Chap. 113, "Ethics and Practice of Research with People Who Use Drugs"). Effective global harm-reduction strategies aim to prevent or reduce the severity of problems associated with nonmedical use of dependence-causing drugs including alcohol need to fit the personal, social, and environmental context of people using drugs. The best way to develop these strategies is "to research and understand drug use practices including how, why and when drugs are used." In this chapter, Allan discusses a number of ethical and practical factors to consider when planning and conducting research with people who use drugs. Recruitment of a marginalized and hidden population, gaining consent, ensuring anonymity, and responding to harm and distress are discussed using examples from the author's research on alcohol and other drug use in rural Australian settings including farming and fishing workplaces, on illicit fentanyl use, and with people in treatment.

Jane McKeown writes about researching with people with dementia (▶ Chap. 114, "Researching with People with Dementia"). She contends that doing research with people with dementia has historically been seen as problematic, particularly where there are concerns over the capacity to make decisions and give informed consent among these individuals. She argues that "by not involving people with dementia across the trajectory of the condition, we are failing to develop important understandings from the perspectives of people living with the condition across a range of research topics." Moreover, by not involving, people with dementia, them in the research process, "we may not be exploring the most relevant research topics or not considering the most relevant methods and approaches to capture their experiences." In this chapter, McKeown draws on the evidence base as well as personal experience in the United Kingdom when discussing approaches to consent and approaches and methods that "seek to include rather than exclude people living with dementia throughout the research process."

Researching with children is presented by Graciela Tonon, Lia Rodriguez de la Vega, and Denise Benatuil (▶ Chap. 115, "Researching with Children"). Research with children presents many challenges to researchers, for example, finding methods which are appropriate and interesting to children and which "recognize the importance of children's experience and agency." They suggest that these methods "should promote a respectful approach based on ethics." The chapter discusses "the possibility of using quantitative, qualitative and mixed methods and the emerging of new proposals such as the inclusion of technologies and arts-based methods."

Optimizing interviews with children and youth with disability is presented by Gail Teachman (▶ Chap. 116, "Optimizing Interviews with Children and Youth with Disability"). This chapter provides readers with innovative techniques, strategies, and methods that would engage disabled children and youth in qualitative interviews. In this chapter, a child-interview methodological approach is discussed with "an emphasis on three key elements: assembling a range of customizable interview methods; partnering with parents; and considering of the power differential inherent in child-researcher interactions." Examples which are drawn from her own research are used to illustrate the methods as well as discuss how they were modified as the research unfolded.

Kat Kolar and Farah Ahmad write about participant-generated visual timelines in their research with street-involved youth who have experienced violent victimization (▶ Chap. 117, "Participant-Generated Visual Timelines and Street-Involved Youth Who Have Experienced Violent Victimization"). In this chapter, drawing from a study that explores resilience among street-involved youth, they examined "how participant-created visual timelines inform verbal semi-structured interviewing with persons who have experienced personal victimization in the form of violence, as well as structural marginalization." They discuss the process of timeline implementation in depth. In their discussion section, they discuss "the potential of visual timelines to supplement and situate semi-structured interviewing" and "illustrate how the framing of research is central to whether that research facilitates increased participant authority in the research process, enhances trust, and ensures meaningful, accountable engagement."

A feminist application of Bakhtin to examine eating disorders and child sexual abuse is written by Lisa Hodge (▶ Chap. 118, "Capturing the Research Journey: A Feminist Application of Bakhtin to Examine Eating Disorders and Child Sexual Abuse"). Drawing from her research which explored the nature of the relationship between women's experiences of child sexual abuse and eating disorders, Hodge demonstrates how Bakhtin's theoretical constructs can "expose the hidden mechanisms of control found in these gender-based oppressive practices." She also shows "how drawing and poetry, when used in qualitative research methodologies, can create space for interactional discovery and give voice to the unspeakable."

Lizzie Seal discusses feminist dilemmas in researching women's violence (▶ Chap. 119, "Feminist Dilemmas in Researching Women's Violence: Issues of Allegiance, Representation, Ambivalence, and Compromise"). She contends that the use of violence among women is a sensitive topic for feminist researchers. This is "because feminists have sought to delineate the role of male violence in continuing women's subordination." Highlighting women's violence can detract from this position. She argues that "researching women's violence using feminist methodologies, which place value on creating knowledge from women's experiences, hearing marginalized voices and democratizing the research process, raises dilemmas." In this chapter, Seal considers these dilemmas across three areas – questions of allegiance, questions of representation, and questions of ambivalence and compromise. A number of examples are highlighted in order to illustrate these issues.

Andi Spark presents a very interesting discussion regarding researching in the animated visual arts and mental welfare fields (▶ Chap. 120, "Animating Like Crazy: Researching in the Animated Visual Arts and Mental Welfare Fields"). She argues that her research "straddles the divide between creative arts practice and social science methodologies, with a focus on outlining a practical approach to developing short-form mixed-media format animated projects that address serious issues such as postnatal depression." This chapter discusses "how animation can be utilized for both communicative, informative and entertaining purposes." In order to deliver authentic and authoritative projects for a targeted audience, Spark explains "how animation works in emphasizing symbols and metaphors to elicit empathetic responses" using a number of examples from other independent creative practitioners.

Researching underage sex work is written by Natalie Thorburn (▶ Chap. 121, "Researching Underage Sex Work: Dynamic Risk, Responding Sensitively, and Protecting Participants and Researchers"). This chapter tells us the challenges inherent in her own experience of undertaking research with adolescent sex workers who have experienced complex trauma histories. The chapter also discussed the "dynamic nature of risk as it relates to research with vulnerable populations, particularly in regard to physical safety, emotional and psychological safety, consent, confidentiality, and interpersonal power within the research relationship." Strategies of identifying and managing these risks are provided in the chapter.

Lauren Rosewarne writes about the Internet and research methods in the study of sex (▶ Chap. 122, "The Internet and Research Methods in the Study of Sex Research: Investigating the Good, the Bad, and the (Un)ethical"). She contends that "the Internet has thoroughly revolutionized sex." The technology has become a key source for individuals to explore sexuality and participate in the erotic activity. For researchers, the Internet has provided great "access to academic databases and archives, to social media sites and public diaries, and notably to a world of possible research participants, in turn dramatically altering the ways sex gets studied." In this chapter, drawing on a wide range of literature on research ethics as well as her own background as a sex researcher, an author of the Internet research, a supervisor of research students on new media, and a long-time member of her university's human ethics committee, Rosewarne outlines, analyzes, and problematizes the use of the Internet in sex research that we can learn from.

Emotions and sensitive research are presented by Virginia Dickson-Swift (▶ Chap. 123, "Emotion and Sensitive Research"). Researching sensitive topics is often an emotional journey, not only for the participants but also for others that may be involved in the research. The emotional challenges that researchers face when doing fieldwork cannot be ignored. Drawing on her earlier empirical work with researchers in Australia and published accounts, Dickson-Swift provides an overview of the emotional challenges inherent in this type of research. She also provides suggestions for researchers, research supervisors, and others involved in the research team. She suggests that these suggestions can be adopted by academic or research institutions to ensure that researchers have the necessary support to conduct sensitive research.

Doing reflectively engaged, face-to-face research in prisons is written by James E. Sutton (▶ Chap. 124, "Doing Reflectively Engaged, Face-To-Face Research in Prisons: Contexts and Sensitivities"). Sutton provides "fundamental features of prisons and prisoners' lives" which make them "sensitive settings and populations for researchers to study." He then presents ethical issues that researchers need to be mindful of when conducting research in prisons. Ultimately, the chapter "endorses being reflectively engaged with the setting, the research process, and oneself when doing face-to-face research in prisons, regardless of the substantive goals of one's study or the particular research methods one employs." Sutton contends that the issues raised in this chapter will be valuable to health and social science researchers who enter prisons to study prisoners.

Jyoti Belur writes about police research and public health (▶ Chap. 125, "Police Research and Public Health"). The chapter discusses the challenges in conducting face-to-face research with police officers regarding their work in sensitive areas or with hard-to-reach groups. She suggests that the "changing nature of policing" and "a rise in societal demands for security" have resulted in the "overlap between law enforcement and public health." In this chapter, based on her personal experience and drawing upon the research experience of others working in this area, Belur provides some basic guidelines to assist researchers who work in this space. She identifies fundamental features of policing culture which make it "a sensitive organization to access and research." She then discusses "the difficulties in approaching gatekeepers and negotiating access to police data and individual officers." Ethical and practical considerations in carrying out police research are also discussed in this chapter.

Researching among elites is presented by Neil Stephens and Rebecca Dimond (▶ Chap. 126, "Researching Among Elites"). Often, health and social science researchers conduct research with those who have less power than them. But the authors argue that researchers should also engage with those who are seen as the "health elites" as they are "powerful actors in the medical domain." However, there is a specific set of methodological challenges that involve in doing this kind of research. In this chapter, drawing on their own research practice, the authors discuss several main issues and challenges that researches need to consider before undertaking research with health elites. They argue for the continued need for qualitative social science to engage with health elites and also for researchers "to be informed by methodological awareness of the challenges and rewards of doing so."

In the last chapter of this section, Robert H. Campbell writes about eliciting expert practitioner knowledge through pedagogy and infographics (► Chap. 127, "Eliciting Expert Practitioner Knowledge Through Pedagogy and Infographics"). Campbell suggests that qualitative research routinely obtains knowledge of expert professionals. However, eliciting tacit or implicit knowledge effectively can be problematic. In this chapter, Campbell discusses a method in which pedagogy and infographics were combined to elicit the knowledge of expert practitioners. The focus of the chapter is on the underpinning principles and deployment of the method. He suggests that this approach can be easily transferred into a range of qualitative research domains.

References

Abbink J, Salverda T. The anthropology of elites: power, culture, and the complexities of distinction. New York: Palgrave Macmillan; 2013.

Aguiar LLM, Schneider CJ. Researching amongst elites: challenges and opportunities in studying up. Hoboken: Taylor and Francis; 2013.

Aldridge J. Working with vulnerable groups in social research: dilemmas by default and design. Qual Res. 2012;14(1):112–30.

Aldridge J. Participatory research: working with vulnerable groups in research and practice. Bristol: Policy Press; 2015.

Barnard M. Discomforting research: colliding moralities and looking for 'truth' in a study of parental drug problems. Sociol Health Illn. 2005;27(1):1–19.

Bergen RK. Interviewing survivors of marital rape: Doing feminist research on sensi- tive topics. In C. M. Renzetti & R. M. Lee (Eds.), *Researching sensitive topics* (pp. 197–211). Newbury Park, CA: Sage; 1993.

Bergen RK. Wife rape: understanding the responses of survivors and service provides. Thousand Oaks: Sage; 1996.

Birman D. Ethical issues in research with immigrants and refugees. In: Trimble JE, Fisher CB, editors. Handbook of ethical research with ethnocultural populations and communities. Thousand Oaks: Sage; 2005. p. 155–77.

Bracken-Roche D, Bell E, Racine E. The "vulnerability" of psychiatric research participants: why this research ethics concept needs to be revisited. Can J Psychiatr. 2016;61:335–59.

Bracken-Roche D, Bell E, McDonald ME, Racine E. The concept of 'vulnerability' in research ethics: an in-depth analysis of policies and guidelines. Health Res Policy Syst. 2017;5:8. https://doi.org/10.1186/s12961-016-0164-6.

Campbell R. Emotionally involved: the impact of researching rape. New York: Routledge; 2002.

Couch J, Durant B, Hill J. Uncovering marginalised knowledges: undertaking research with hard-to-reach young people. Int J Mult Res Approaches. 2014;8(1):15–23.

Dickson-Swift V. Undertaking sensitive health research: The experiences of researchers. Unpublished PhD thesis, Department of Public Health, School of Health and Environment, La Trobe University, Bendigo, Australia; 2005.

Dickson-Swift V, James E, Liamputtong P. Undertaking sensitive research in the health and social sciences: managing boundaries, emotions and risks. Cambridge, MA: Cambridge University Press; 2008.

Harvey WS. Strategies for conducting elite interviews. Qual Res. 2011;11(4):431–41.

Johnson B, Clarke JM. Collecting sensitive data: the impact on researchers. Qual Health Res. 2003;13(3):421–34.

Lee RM. Doing research on sensitive topics. London: Sage; 1993.

Lee R, Lee Y. Methodological research on "sensitive" topics: a decade review. Bull Sociol Methodol. 2012;114(1):35–49.

Liamputtong P. Researching the vulnerable: a guide to sensitive research methods. London: Sage; 2007.

Liamputtong P. Qualitative research methods. 4th ed. Melbourne: Oxford University Press; 2013.

McCauley K. Vulnerable populations in research. In: Cautin RL, Lilienfeld SO, editors. The encyclopedia of clinical psychology. Hoboken: Wiley; 2015. p. 1–4. https://doi.org/10.1002/9781118625392.wbecp214.

McCosker H, Barnard A, Gerber R. Undertaking sensitive research: issues and strategies for meeting the safety needs of all participants. Qual Soc Res. 2001;2(1):327–53.

Medeiros P. A guide for graduate students: barriers to conducting qualitative research among vulnerable groups. Conting Horiz: York Univ Stud J Anthropol. 2017; Retrieved on 4 July 2017 from https://contingenthorizons.com/2017/02/25/a-guide-for-graduate-students-barriers-to-conducting-qualitative-research-among-vulnerable-groups/.

Melrose M. Labour pains: some considerations on the difficulties of researching juvenile prostitution. Int J Soc Res Methodol. 2002;5(4):333–51.

Menih H. Applying ethical principles in researching a vulnerable population: homeless women in Brisbane. Current issues in. Crim Justice. 2013;25(1):527–39.

Mikecz R. Interviewing elites: addressing methodological issues. Qual Inq. 2012;18(6):482–93.

O'Donnell P, Tierney E, O'Carroll A, Nurse D, MacFarlane A. Exploring levers and barriers to accessing primary care for marginalised groups and identifying their priorities for primary care provision: a participatory learning and action research study. Int J Equity Health. 2016;15:197. https://doi.org/10.1186/s12939-016-0487-5.

Odendahl T, Shaw AM. Interviewing elites. In: Gubrium JF, Holstein JA, editors. Handbook of interview research: context & method. Thousand Oaks: Sage; 2002. p. 299–316.

Payne G, Payne J. Key concepts in social research. Sage key concepts. London: Sage; 2004.

Pitts MK, Smith AMA, editors. Researching the margins: strategies for ethical and rigorous research with marginalised communities. London: Palgrave Macmillan; 2007.

Punch S. Research with children: the same or different from research with adults. Childhood. 2002;9(3):321–41.

Quest T, Marco CA. Ethics seminars: vulnerable populations in emergency medicine research. Acad Emerg Med. 2001;10(11):1294–8.

Quinn CR. General considerations for research with vulnerable populations: ten lessons for success. Health Justice. 2015;3:1. https://doi.org/10.1186/s40352-014-0013-z.

Renzetti CM, Lee RM, editors. Researching sensitive topics. Newbury Park: Sage; 1993.

Schrems BM. Informed consent, vulnerability and the risks of group-specific attribution. Nurs Ethics. 2014;21:829–43.

Shivayogi P. Vulnerable population and methods for their safeguard. Perspect Clin Res. 2013;4(1):53–7. https://doi.org/10.4103/2229-3485.106389.

Solomon SR. Protecting and respecting the vulnerable: existing regulations or further protections? Theor Med Bioeth. 2013;34:17–28.

Stephens N. Collecting data from elites and ultra elites: telephone and face-to-face interviews with macroeconomists. Qual Res. 2007;7(2):203–16.

Stone TH. The invisible vulnerable: the economically and educationally disadvantaged subjects of clinical research. J Law Med Ethics. 2003;31(1):149–53.

Straubhaar R. The methodological benefits of social media: "studying up" in Brazil in the Facebook age. Int J Qual Stud Educ. 2015;28(9):1081–96.

Taylor P. Talking to terrorists: a personal journey from the IRA to Al Qaeda. London: Harper Press; 2011.

ten Have H. Respect for human vulnerability: the emergence of a new principle in bioethics. J Bioeth Inq. 2015;12:395–408.

ten Have H. Vulnerability: challenging bioethics. Abingdon: Routledge; 2016.

van Liempt I, Bilger V. The ethics of migration research methodology: dealing with vulnerable migrants. Brighton: Sussex Academic Press; 2009.

Williams C. Researching power, elites and leadership. London: Sage; 2012.

Wrigley A, Dawson A. Vulnerability and marginalized populations. In: Barrett DH, Ortmann LW, Dawson A, Saenz C, Reis A, Bolan G, editors. Public health ethics: cases spanning the globe. Dordrecht: Springer Open; 2016. p. 203–40. https://doi.org/10.1007/978-3-319-23847-0_7.

Zuckerman H. Scientific elite. Transaction: New Brunswick; 1996.

"With Us and About Us": Participatory Methods in Research with "Vulnerable" or Marginalized Groups

110

Jo Aldridge

Contents

1 What is Participatory Research?	1920
2 "Vulnerability," Marginalization and Enhancing "Voice" in PR	1922
3 Towards a Participatory Model	1926
4 Principles of Participation	1930
5 Conclusion and Future Directions	1932
References	1933

Abstract

In much health and social scientific research that includes "vulnerable," "hard-to-reach" or marginalized groups, claims are often made about participatory methods and techniques that enhance participant engagement and "voice." In many cases, however, the validity of these claims remains unclear – the nature and extent of participant involvement in such studies is not always defined and the value and efficacy, as well as the challenges, of using participatory methods are often misunderstood. In many respects, these oversights can be explained by the lack of cognate *and* applicable participatory models or frameworks that can help researchers work more effectively *with* marginalized participants. This chapter explores these issues drawing on the author's own extensive research with marginalized groups and participatory models of working that both promote and enhance participant engagement and emancipation in research processes. Such approaches see "vulnerable," marginalized, or socially excluded research participants in transformative roles in research, including as co-researchers, co-analysts, and designers and producers of their own research agendas and projects.

J. Aldridge (✉)
Department of Social Sciences, Loughborough University, Leicestershire, UK
e-mail: J.Aldridge@lboro.ac.uk

© Springer Nature Singapore Pte Ltd. 2019
P. Liamputtong (ed.), *Handbook of Research Methods in Health Social Sciences*,
https://doi.org/10.1007/978-981-10-5251-4_126

Keywords

Participatory research · Vulnerable groups · Participant voice · Social exclusion

1 What is Participatory Research?

In their 1995 review of participatory research, Cornwall and Jewkes asked a key question: "If all research involves participation, what makes research participatory?" (p. 1668). This question is perhaps even more pertinent today given the increasing interest in, and use of, participatory research (PR) methods in social (and health) sciences since its emergence in the 1970s and its relevance for working more effectively with "vulnerable" or marginalized groups. Nevertheless, despite advances made in PR over the past four decades, and the claims made for it in terms of working more successfully with marginalized participants (using methods that enhance participation), the call for greater theoretical and methodological rigor in PR remains pertinent (see Chevalier and Buckles 2013). In part, this is due to the lack of formal participatory models or frameworks that help lend clarity and validity to PR, as well as serve to demonstrate more precisely the link between PR and working effectively with marginalized individuals or groups. Therefore, questions such as *what makes research participatory?* and *how can PR be used effectively to enhance participation among marginalized participants?* are perhaps even more pertinent today than they were four decades ago.

Since the 1970s, new approaches to research, particularly in the qualitative field, facilitated more creative methods of investigation (see chapters in the "Innovative Research Methods in Health Social Sciences" Section). The result was that more diverse and inventive empirical approaches began to flourish across a range of different disciplines that also enabled researchers to address issues such as the meaning of, and relationship between, vulnerability, inclusion, and participation in research – although the connection and congruence between these concepts were not always made explicit in research design and discourses; to some extent this is still the case today. One of the reasons for this is that the terms "participatory research" and "participatory action research" (PAR) are often not fully understood, or exaggerated claims are made for studies that (may or may not) use these kinds of methods. PR is a broad umbrella term under which a number of participatory, collaborative, or inclusive research methods and approaches are located. PAR, on the other hand, emerged in the 1940s following the pioneering work of Kurt Lewin and the Tavistock Institute (see Chevalier and Buckles 2013) and focuses specifically on *social change* outcomes for participant groups, organizations, or communities (as opposed, specifically, to individuals) through *action* research. Ongoing developments in PAR have seen a diversity of approaches and techniques emerge in the qualitative field, including, for example, participatory rural appraisal (PRA) and participatory learning and action (PLA) and demonstrating, "a well documented tradition of active-risk taking and experimentation in social reflectivity backed up by evidential reasoning and learning through experience and real action"

(Chevalier and Buckles 2013, p. 4) (see also ▶ Chap. 17, "Community-Based Participatory Action Research").

Chataway's (1997) study with indigenous communities of North America is a good example of the PAR approach, which embraced principles of collaboration and mutuality. Focusing specifically on issues of identity, security, and self-government, Chataway adopted a PAR approach through the use of focus groups and intergroup discussions, and enabling community participants to draw on "non-native" research strategies to identify reasons for community divisions and the barriers to change. Given the level of political oppression, the indigenous population had experienced under "Euro-American dominance," and the subsequent distrust of "outsiders" within communities, the researchers were required to be flexible and adaptable with respect to PAR approaches and methods. They understood that the study needed to be both "internally directed" as well as in the best interests of the participants if it was to be successful. Chataway (1997, p. 748, emphasis added) concluded that without a PAR approach, "we would not have been able to overcome the barriers to research in this context. With PAR, we were able to complete four successive pieces of *collaborative* research." For a full discussion about different kinds of PAR approaches, see Chevalier and Buckles (2013).

The intention of PR more broadly – both conceptually and philosophically, for example – is to promote greater inclusion and collaboration in research and to recognize and give credence to the voices of *individuals* both within and outside communities (see Fals Borda 1988; Whyte 1989; McTaggart 1997; Goodley and Moore 2000; O'Neill et al. 2002; Aldridge 2007, 2012a, b; Higginbottom and Liamputtong 2015). Thus, PR draws on philosophical principles and objectives that relate to mutuality and understanding in research practices and which are "designed to promote active involvement" in the research process by participants who, in studies that use more conventional methods, may be treated simply as the objects of research (Chataway 1997, p. 747; see also Fals Borda 1988; Whyte 1989; Bourdieu 1996; McTaggart 1997). Rapoport's participatory research in the 1970s (1970, p. 499) demonstrated both the distinction and advantages of PR compared to other qualitative methods in the ways in which it facilitated the active involvement of researchers and participants in research processes, with a specific focus on "joint collaboration with a mutually acceptable ethical framework."

Although the different types of research methods used in PR (and PAR) studies can be diverse and often also adopt and adapt techniques from other disciplines and practices (see Higginbottom and Liamputtong 2015), the fundamental difference between PR and other research methodologies lies in "the location of power in the various stages of the research process" (Cornwall and Jewkes 1995, p. 1667). Further, the stories and "voices" of participants are placed center stage, both in the design and objectives of PR studies. Walmsley and Johnson (2003, p. 10) recognize the methodological diversity in PR studies that have been described variously as "participatory, action or emancipatory," but all of which have a common objective and intention to engage participants in more empathic and democratizing research relationships. Walmsley and Johnson describe this type of approach, as well as their own PR with people with learning difficulties, as "inclusive" research, where participants "are active participants, not only as subjects but also as initiators, doers, writers and disseminators of research" (2003, p. 9; see also Atkinson 1986; Flynn 1989).

O'Neill et al. (2002, p. 69) have pointed to the ways in which PR methods have helped to "transgress conventional or traditional ways of analyzing and representing research data" at a time when ideas about "hard-to-reach" or marginalized groups in research, as well as in health and social care discourses, were also changing. Since the development of PR four decades ago, and certainly more recently, greater emphasis has been placed on the needs of "socially excluded," "vulnerable," or "marginalized" people in research, many of who would have been overlooked or excluded from research in the past because they were deemed too "difficult" to reach and include (see Liamputtong 2007; see also ▶ Chaps. 99, "Visual Methods in Research with Migrant and Refugee Children and Young People," and ▶ 100, "Participatory and Visual Research with Roma Youth").

However, despite these methodological advances and the opportunities offered by PR, critical questions still remain, not least about the nature and extent of participation in PR studies, and especially in those that do not make reference to formal participatory frameworks or models, or to the (participation) needs of vulnerable or marginalized populations. It is important to note, however, that not all research with vulnerable or marginalized groups is participatory in design or intent, nor necessarily aligns itself with PR or PAR objectives; neither are participatory methods only used among "vulnerable" respondents. Additionally, not all research that lays claim to an inclusive, participatory agenda promotes the principles commonly associated with participatory approaches, such as understanding, mutuality, emancipation, collaboration, and giving "voice." In which case, it is necessary to explore the relevance of, and alliance between PR and working with vulnerable or marginalized individuals or groups in research much more closely, and specifically regarding the ways in which individual and collective "voices" are facilitated and heard in PR methods and research praxis.

2 "Vulnerability," Marginalization and Enhancing "Voice" in PR

When considering the role and efficacy of PR in research that involves vulnerable or marginalized participants, it is necessary to clarify what is meant by "vulnerability" both definitionally and conceptually. In many respects, "vulnerability" remains a mutable, even contestable, concept and especially when considering the various definitions and classifications adopted in research governance and ethical frameworks, in health and social care discourses, and with respect to the perceptions of those people defined as "vulnerable." It is also clear that in some studies, even though claims are made about working inclusively with vulnerable people using PR techniques, vulnerability as a concept is not always defined nor explained in sufficient detail to demonstrate clearly the connection between participatory research objectives and research praxis.

In her work on vulnerable groups in health and social care, Mary Larkin (2009) notes the extensive use and relevance of the term "vulnerable" both conceptually and in practice, but also recognizes that a precise definition remains "elusive," acknowledging that "its meaning also varies according to the context in which it is used" (p. 1). This is evident both in the various and sometimes inconsistent definitions used in health and social care discourses – where it is most often used to denote susceptibility to harm or risk, for example, or as an indicator of enhanced need – and in research governance frameworks where, for example, lack of capacity for self-care determines a person's vulnerable status (see, Department of Health 2000, Sect. 2.3) and where individuals are often grouped together under more general categories of vulnerability, for example, people with mental health problems, people with disabilities, and people in prison. Twenty years ago, Rogers (1997) was more "inclusive" in his definition of vulnerability, which included the "very young" and the "very old," those susceptible to illness, black and minority ethnic (BME) groups, people on low income or who were unemployed, as well as women. Since that time, the term "vulnerable" has been assigned to groups of people who, for example, lack the capacity for self-protection, for developing resilience or effective coping strategies (see Parrott et al. 2008), or who "lack the ability to make personal life choices, to make personal decisions, to maintain independence and to self-determine" (Moore and Miller 1999, p. 1034; see also Liamputtong 2007).

In terms of research governance and ethical frameworks, in the UK, the Department of Health's (2005) *Research governance framework for health and social care* emphasizes participants' capacity and willingness to provide informed consent as well as describes vulnerable or potentially vulnerable participants as children and adults with mental health problems or learning difficulties. Economic and Social Research Council (ESRC 2010) descriptors, on the other hand, focus on research that puts participants at "more than minimal risk" and "potentially vulnerable groups" as children and young people, those with learning difficulties or cognitive impairments, those who lack the mental capacity to give consent, and "individuals in a dependent or unequal relationship" (p. 8).

Helpfully, in her health and social care research, Mary Larkin makes the distinction between someone who is individually, uniquely, or innately vulnerable (through chronic illness or disability, for example) and those who are vulnerable because of their circumstances, environment or as a result of structural, systemic factors. It is worth noting, further, that subjective, self-perceptions of need may not always accord with objective or external identifiers or classifications of vulnerability. Steel (2001, p. 1), for example, acknowledges that some people, "would not describe themselves as vulnerable or marginalised at all," and that self-perceptions of vulnerability are both socially constructed and again, contextual – "[it depends on] where you are standing at the time, and in relation to who, or what."

From a research governance and ethical perspective, however, individual self-perceptions of vulnerability or marginalization, and philosophical debates about such concepts, are unimportant; what is important is that research participants themselves are not put at further risk of harm or their vulnerability exacerbated by research processes and that researchers and institutions are equally protected.

Thus, research governance frameworks and ethical guidelines provide extensive advice and guidance about the type and extent of ethical clearance required in order for research studies that include vulnerable or marginalized groups to proceed. For some researchers, this is welcome; for others, there are concerns that strict ethical regulation may serve to further exclude vulnerable or marginalized individuals and groups from having their voices heard, or from being included in research studies in the first place (Boddy and Oliver 2010; Hurdley 2010).

Despite these arguments, it is important that researchers are conversant with and understand issues and debates relating to vulnerability and marginalization when planning research studies with participants who may have particular and distinct needs and thus who may also require different kinds and levels of participation. It is equally important that researchers consider carefully the link between vulnerability (or marginalization) and research participation with respect to developing a clear understanding of equality and power in research relationships; what is central to the relevance of, and relationship between, the participatory project and working effectively with vulnerable or marginalized people – both in terms of research processes and outcomes – is the intention to address inequality and powerlessness by giving "voice" to individuals who may otherwise be overlooked or excluded from research. Individuals such as those with mental health problems or learning difficulties, for example, are often denied full participation in public or political life and can be overlooked in research studies that adopt more conventional, non-PR approaches simply because they are deemed hard to access or to recruit on to research projects (see Aldridge 2012b).

This is why the notion of "voice" should both inform and underpin PR approaches in social scientific and health research (as well as in other fields and disciplines), as a way of ensuring the experiences of participants are located center stage in research agendas and processes by enabling participants to speak or "tell" their individual or collective stories in their own ways – that is, in ways that are deemed most appropriate to each individual or group. Notably, "voice" in this context can be understood and interpreted in different ways – theoretically, culturally and/or politically, for example. Equally, a number of different methodological approaches are available to the participatory researcher keen to enhance participant "voice"; for example, participatory visual or narrative methods, life history research, diary methods, and so on. A good example of the first of these is Thomson's (2008, p. 3) visual research with (vulnerable) children and young people in which visual techniques (photography and video, for example) are used in order to "find ways to bring previously unheard voices into scholarly and associated professional conversations." Working with Britzman's (1989) multiconceptual understanding of "voice" (literal, metaphorical, and political), Thomson recognizes the importance of conferring competency and agency on (vulnerable) children and young people – of "giving voice to the voiceless" (Britzman 1989) – by engaging with them in more direct and inclusive ways using less conventional methods. For Thomson and her colleagues, the use of participatory visual research methods presents valuable opportunities for involving children as the co-producers of research – by giving them cameras, video equipment, art materials, and so on in order to collect their own evidence (see also

▶ Chaps. 99, "Visual Methods in Research with Migrant and Refugee Children and Young People," and ▶ 100, "Participatory and Visual Research with Roma Youth").

Other participatory methods which have borrowed techniques from visual sociology, for example, in order to enhance participants' visual "voices" and to work more inclusively with marginalized individuals and groups, combine the visual with first-person oral or written accounts and introduce a range of techniques, including photographic diaries, photovoice techniques, and photographic elicitation methods (Aldridge 2012a, 2015; see also Sempik et al. 2005; Aldridge and Sharpe 2007; Joanou 2009; Catalani and Minkler 2010; see also ▶ Chap. 65, "Understanding Health Through a Different Lens: Photovoice Method"). The intention here is not just to uncover new insights into lived experience and needs that other more conventional methods might miss, but also to foster understanding and more empathic *responses* – from the audience/reader as well as the academic researcher – to the visual and narrative "voices" of participants who might otherwise be overlooked or excluded entirely from research.

A good example of this kind of empathic testimonial approach is Rapport's (2008, p. 1) in-depth "research conversation" with survivors of the Holocaust. Rapport used a combination of poetic (textual) and photographic narrative methods in order to take both the reader and the researcher on a textual-visual journey. Her objective – congruent with Bourdieuean concepts of understanding and mutuality in research – was to develop greater in-depth understanding of the researcher–participant relationship as well as research processes, including in the final output and communication phase. She described these revelatory aspects of the study as a process of "coming to know" the data and argues that this approach makes visible "what is often invisible in more traditional approaches" (2008, p. 1). Without a doubt, the combination of the visual and the prose or poetic-style of the personal survivor narratives are compelling and engender an emotional as well as empathic response in the reader/viewer that may be missed by other more conventional research methods. The photograph of the barracks and execution wall at Auschwitz, for example, with its austere and imposing red brick façade and enclosed dirt yard, taken by Rapport herself, coupled with the personal narratives of the survivors – written *by them* – are both compelling and intuitive testimonials of the survivor experience.

My own area of research has focused on working exclusively with vulnerable or marginalized research participants, including individuals with little or no connection to known communities of support (either formal or informal). In many cases, visual methods work well in terms of facilitating participants' engagement in research as data collectors and analysts – in short, as co researchers. Two of my participatory visual studies included participants who were either unable (due to profound learning difficulties) or unwilling (children who did not want to speak about their experiences) to engage with conventional qualitative methods such as interviews or focus groups and took part in photographic diary and elicitation methods that invited them to act as data collectors and analysts (see Sempik et al. 2005; Aldridge and Sharpe 2007; Aldridge 2014). The photographic diary method used in both studies engaged participants as photographers to enable them to demonstrate their experiences visually and to highlight needs that other studies might have missed.

In the first, 2005, study (Sempik et al. 2005), 19 people with profound learning difficulties who attended gardening projects across the UK were invited to take photographs (using disposable cameras) that were meaningful to them on site at the projects over a 2-week period. The photographs were developed and given to the participants to keep but were also used in visual "conversations," using the images as visual prompts in order to "tell" or show a story about project participation. The participatory visual method was critical in this study as a way of enhancing participant inclusion and action – participants took on the roles of data collectors and (to some extent) analysts – in the research process (for further discussion, see Aldridge 2012a, b).

The second, 2007 (Aldridge and Sharpe 2007), study used the same participatory photographic methods (again, over a 2-week period) with 16 children and young people who lived with and cared for parents with serious mental health problems. The children were invited to take photographs of aspects of their lives that were meaningful to them. Again, the photographs that were produced were used in the elicitation phase of the study where the children and young people engaged in selecting particular images and telling their own visual story (the story they wanted to "tell" and show) about what life was like living with and caring for a parent with a serious mental health problem. Both studies used participatory visual methods as part of a multimethod approach and generated visual data that gave important new insights into the lives of marginalized participants. The photographic method itself was also highly effective in engaging participants who might otherwise be left out of research because conventional methods would not suit or meet their needs. Following completion of both studies, photographic methods were adopted in professional practice (at both gardening and young carer projects in the UK) as a way of engaging with new service users/clients in more meaningful ways.

3   Towards a Participatory Model

Despite the efficacy and relevance of these kinds of "bespoke" participatory visual methods in PR, it is critical that new and future PR studies make clear reference to participatory frameworks or models of working that help lend validity and credence to PR methods and that also enable both researchers and participants to locate their approaches within a clear methodological frame of reference. Without this, research studies that use PR methods are in danger of remaining firmly on the margins of empirical investigation and thus may make only limited contributions to knowledge. However, only a limited number of PR models/frameworks have been available to the participatory researcher in the four decades since its development in the 1970s. Some early pioneers of PAR and PR methods proposed typologies or frameworks that enabled researchers to reflect to some extent on the efficacy of their own participatory approaches. For example, in the late 1980s, drawing on his research in agriculture, Biggs (1989) proposed four modes of PR that included contractual, consultative, collaborative, and collegiate phases in the participatory research process. Six years later, Cornwall and Jewkes (1995, p. 1669) offered some refinements to these modes to include "shallow" and "deep" participation, where researchers in the former mode controlled the research process but relinquished some of that control to participants in the "deep" phase. Hart's "Ladder of Participation" model (1992) for working more effectively with children and young people in research included seven distinct stages, with "manipulation" at the bottom of the Ladder – where participants in research: "do or say what adults suggest they do. They have no real understanding of the issues, although they may be asked for their views. They do not know what influence their views will have on any decisions that are made" (p. 8). At the top of Hart's Ladder, PR is much more emancipatory and is delineated by both control and ownership, where participants take on the role of researcher, initiating, designing, and conducting research themselves.

In terms of working more effectively with vulnerable or marginalized groups, any PR model or framework should enable researchers (whether they are academic, advocate, or participant researchers) to "locate" their own position, and role, in the research process as well as those of research participants. The Participatory Model (PM) in Fig. 1 is constructed from a participant-oriented standpoint – and is presented very much as a continuum, one that always works away from treating the research participant as object – and is intended as an aid to researchers (both within and outside the academy) who are planning or reflecting on the use of participatory methods with different participant populations, including vulnerable,

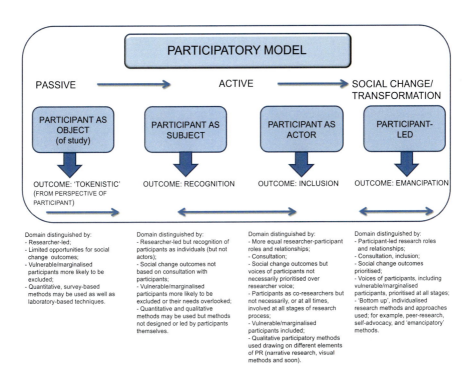

Fig. 1 Participatory model

marginalized, or socially excluded people. The PM is also intended to help promote participant emancipation and "voice" in qualitative research by encouraging researchers to reflect more carefully on research relationships and the ways in which they may work *with* participants as *actors* in research (thus, in research that, from a participant standpoint, is both "*with* us and about us").

Rather than working within a strict set of research *rules*, the PM also requires researchers to recognize and adhere to a number of *principles* (see below) and objectives when conducting PR. As has been stated, the principles of participation in PR should ensure greater equity in researcher–participant relationships (including and particularly when participants are vulnerable, marginalized or socially excluded), and engagement in dialogue with participants – about research design, ethics outputs and so on – should be part of an ongoing process throughout the duration of any PR study. Such studies should always be designed from the outset with the needs of participants in mind. As has been discussed, under the somewhat broad qualitative research umbrella, a number of different PR methods can be effective, including participatory narrative or life story methods and those produced in visual form, including visual-textual and autobiographical methods, for example (see Atkinson 1997; Aldridge 2015). These kinds of methodological techniques that reconstruct and articulate subjective experience as spoken, written, or visualized narratives produced by participants themselves can, as Goodson argues (2013, p. 30), serve as "a starting point for developing further understandings of the social construction of each person's subjectivity." These types of participatory story telling/showing (*auto*biographical) methods, while sometimes difficult for some people to initiate themselves or see through to publication or output phases on their own, may, as Walmsley and Johnson (2003, p. 149) argue, nevertheless also "[hold] the greatest potential for full and equal partnership." This is because such methods and approaches emphasize and promote the participant-as-narrator as the expert or "ultimate insider" (Walmsley and Johnson 2003, p. 149).

In some cases, of course, truly "emancipatory" research may be out of reach, inappropriate, or even irrelevant for some participants. Simply talking about their experiences, in interviews or focus groups, for example, may be sufficient and even welcomed by some research participants – it is important not to assume, for example, that even when participants may be considered "vulnerable" or marginalized, they may wish to design and lead their own research projects about their lives and/or the lives of others. What is important to remember is that "bespoke" methods in PR should be designed from the bottom up, that is, with the needs of individuals (or groups of individuals) in mind and that consultation with participants about their needs and their roles and relationships in research processes should form part of an ongoing process in any PR study – these are some of the key principles and components of the PR approach.

As has already been stated, not all research that adopts participatory methods lays claim to or aligns itself with a strict or clearly defined participatory approach. On the other hand, some PR studies make participatory claims without explaining or being clear about the nature, extent, and limitations of participant involvement. This is clearly unhelpful in terms of promoting PR as a way of building better and more democratic research relationships. With respect to advancing PR theoretically and also in lending methodological credence and validity to the PR approach, it is equally unhelpful when researchers have to search – sometimes in vain – for participatory indicators in descriptions of research studies that claim to use and promote inclusive or emancipatory ways of working but do not make either the *principles* of PR, or the research *process*, clear. The kinds of participatory methods used in PR, as well as the design and context of the research itself, and the relevant theoretical underpinnings, should be clear from the outset in any PR project; as should the ways in which participatory and, where relevant, emancipatory principles and objectives will be achieved through working collaboratively and inclusively with participants.

Part of this inclusive process must also ensure avoidance of research that is tokenistic or simply pays lip service to participatory or emancipatory principles or objectives. Thus, credible and reliable PR must include careful consideration of the best and most appropriate ways in which experiential (life story) evidence can be garnered, analyzed, and represented (as opposed to re-presented; see Aldridge 2015, 2012b). Lewis and Porter (2004, p. 196) argue that where researchers are committed to "giving voice" in research, then the process also demands "careful planning, preparation and the apportioning of appropriate time," as well as the need to "keep asking ourselves what trust we can place in our methods and check we have not overly predetermined the views that we have encouraged to be heard."

This also means that academic researchers need to demonstrate commitment not only to research participants (as well as in most cases to the academy and to research funders; see Aldridge 2012b), but also to developing and advancing PR methods in order to enhance the credibility and rigor of these kinds of qualitative approaches. It is acknowledged that studies that use less conventional, more creative and inventive qualitative techniques are not so readily accepted for their empirical or "scientific" value in all policy contexts and in some disciplinary settings. Walker et al. (2008, p. 164) argue, for example, that "the world of policy and practice tends to be more cautious in its response" to this type of evidence.

Thus, there is even greater need for PR to achieve the kind of rigor and cohesion that has been called for in other related disciplines and fields. In the late 1990s, Lieblich et al. (1998, p. 1) proposed that narrative research methods, for example, required "a deliberate investment of effort in the elucidation of working rules for such studies." I would argue that this same kind of investment is required in order to advance PR both methodologically and theoretically, but with an equal emphasis on the *principles* and *models* of working rather than on strict "rules" or formulas.

Without such models of working, there is a real danger that unrealistic or even half-hearted attempts will be made to work more collaboratively and inclusively with specific participant groups (including vulnerable or marginalized participants), thus further undermining the credibility of PR methods. Although in recent years important advances have been made in PR and in working more empathically and inclusively with people who have traditionally been left out of research studies, there is currently even greater need for care and attention to advancing both the principles and mechanisms of PR; arguably, during times of such serious fiscal retrenchment, it is even more likely that research studies that are not considered sufficiently "scientific" or "credible" may be at even greater risk of being overlooked in a competitive and increasingly restrictive funding environment.

It is in these contexts that the needs of vulnerable or marginalized people need to be considered carefully and sensitively by PR investigators (from both within and outside the academy) in order to ensure that they are included in "bespoke" PR studies as well as, where appropriate, in more "mainstream" studies that can accommodate or allow for a degree of methodological flexibility. Incorporating qualitative methods and, more specifically, clearly defined and designed PR techniques as part of a multimethod approach is one way of achieving this. This would ensure that those who have traditionally been left out of, or have had very little input into, research or public policy decision-making processes could be included in ways that, as Hill et al. (2004, p. 78) argue, "meet their wishes and felt needs" through "multidimensional participation."

The PM above (Fig. 1) is relevant and applicative in the context of both stand alone, "bespoke" PR studies and for those that are part of larger, more conventional, or "mainstream" multimethod projects. The underlying principle in PR, however, is that in each case research should always move *away from* tokenistic methods that treat participants as (passive) objects (the "participant as object" domain in the PM). Thus, the PAO domain is included only in order to serve as a point *from* which PR should always advance. At the same time, the PM can accommodate different levels of PR approaches, recognizing that it is not always possible for research to be solely "emancipatory" and participant-led (PL; Fig. 1) – some research, for example, may be designed with little or no consultation with participants, but may then involve further new or adapted elements of participation at a later stage, either through greater collaboration with participants during fieldwork phases or during the outputs phase. Thus, some research may contain participatory elements, where participants are treated as individual subjects ("participant as subject" on the PM), as well as other participatory elements (across the domains) that facilitate closer collaboration in research processes and relationships (e.g., the "participant as actor" domain). The PL domain prioritizes social change outcomes as well as methods and approaches that facilitate and promote participant voice, self-advocacy, and emancipation. PR studies that are located and operate within this domain should, wherever possible, be designed and led by participants themselves.

The PM has also been designed as a guide for researchers who are planning future PR projects, but it can also be used to evaluate past and current studies. When considering the two participatory visual projects described in this chapter with reference to the PM, we can see how both the young carer/mental health and the gardening studies are more clearly located in the PAA domain, but also with some elements borrowed from the PL domain – the purpose of the PM is to allow for such crossovers, or fluidity of movement within and across the model, *as long as the points of designation and intersection are made clear.*

4 Principles of Participation

The PM provides researchers with a point of reference and an opportunity to consider more carefully both the principles (see below) and processes involved in their own participatory approaches – locating, designating, and identifying points of intersection within and across other domains is, for example, an important part of the PR process, as discussed above. The PM also gives researchers the chance to consider more carefully the participatory claims that are made in other research studies that purport to be participatory in nature and intent. While such claims should be explained and verified through careful explication of the nature, extent, and limitations of the kind of participation involved (the PR process), it is clear that this is not always the case. Reference to a clearly designed participatory framework, typology, or model, such as the PM described and discussed in this chapter, would undoubtedly help in this respect. For those researchers who are thinking about or planning PR, and particularly with vulnerable or marginalized participants, moving away from research that is tokenistic and which treats participants simply as objects, and even as subjects, in research (the PAO and PAS domains in the PM) would help advance PR methods and approaches that are delineated clearly by their emphasis on inclusion, collaboration and emancipation (wherever the latter is possible). With these issues in mind, the following principles are proposed as a further guide or reference point for researchers (both within and outside the academy) when thinking about and planning participatory projects. In these contexts, PR should:

- Be designed with the needs of participants in mind – it should take account of the needs of participants, their conditions, and circumstances (including the nature and extent of their vulnerability or marginalization, as well as social exclusion factors, where relevant).
- Involve a process of ongoing dialogue and consultation; this should include discussion of research design issues, the needs, and rights of participants, and how "voice" is facilitated and can lead to transformative outcomes, as well as ethical issues and requirements.
- Ensure research relationships are based on mutuality, understanding, and trust and, depending on the nature and extent of the participatory principles and objectives involved (what is achievable and realistic in research terms), that the voices of participants are prioritized over those of academic researchers.
- Be clear about the opportunities for participation, as well as the extent and limitations of the participatory approach, so that research projects do not raise unrealistic expectations for participants or make false participatory claims. It is essential that PR projects are clearly defined with respect to participatory typologies or participatory models/frameworks.
- Ensure participants are given opportunities to reflect on their engagement in research projects, as well as the level/extent or limitations of the participatory methods and approach. In this way, the views of participants must "inform the link between social inclusion and participation" (Hill et al. 2004, p. 80). Academic researchers may also want to be reflexive about research processes and relationships in order to make useful contributions to contemporary methods debates and discourses.
- Be sufficiently flexible so that participatory techniques may be included in larger, mainstream, multidimensional studies. PR methods can be effective in studies that adopt multimethod approaches, and this is especially the case when less

conventional PR strategies are used and when working with vulnerable or marginalized people.
- Recognize that vulnerability (or marginalization) is both a mutable and contestable concept and that for most people, this is not a fixed identity or condition. PR should address vulnerability in this way and attempt to redress the impact of marginalization and/or social exclusion, for example, through transformative objectives.
- Recognize that transformative outcomes in PR can be personal, social, political, and so on and may occur immediately, indirectly, and/or over time.
- Recognize that the data collated and/or produced by participants in PR can be subject to different kinds of analyses, interpretation, and reflexive processes; these should serve to enhance participant "voice" and ensure that participants are not just treated as the objects of research, but are also considered as co-researchers, collaborators, disseminators, "doers," and self-advocates within what is realistically achievable in PR terms.

5 Conclusion and Future Directions

When considering the ways in which greater clarity and methodological rigor can be brought to PR, what is missing from many studies that make participatory claims, as discussed, is recognition of the nature, extent, and limitations of participation within individual projects and for specific individuals or groups of participants. Without such rigor, clarity, and focus in PR studies, the voices of "vulnerable" or marginalized individuals or groups may continue to go unheard. While various participatory typologies and models of working have been described and proposed in PAR specifically (see Biggs 1989; Hart 1992; Chevalier and Buckles 2013), few attempts have been made to evaluate their efficacy, use, and relevance in the field, particularly by participants themselves, or to bring these together under a broader PR banner. Neither do many studies that make participatory claims always align themselves with specific participatory models or frames of reference.

The PM discussed in this chapter has been developed with the intention of addressing some, if not all, of these issues and oversights by providing a participatory framework for researchers (from both within and outside the academy) to locate their own participatory projects with reference to the various domains within (and across) the model. Drawing extensively on my own experiences of conducting PR or participatory-type projects, as well as evidence from a wide range of research studies in my own and other fields, the PM has been designed for the purpose of helping researchers more clearly align or define their PR projects with reference to, or across, a particular domain or domains within the model itself. An additional intention of the PM is to advance PR as a credible and constructive approach to working more effectively with participants who may not readily or obviously fit within the boundaries of conventional research participation.

References

Aldridge J. Picture this: the use of participatory photographic research methods with people with learning disabilities. Disabil Soc. 2007;22(1):1–17.

Aldridge J. The participation of vulnerable children in photographic research. Vis Stud. 2012a;27(1):48–58.

Aldridge J. Working with vulnerable groups in social research: dilemmas by default and design. Qual Res. 2012b;14(1):112–30.

Aldridge J. Participatory research: working with vulnerable groups in social research. Sage Research Methods Cases. 2014. https://doi.org/10.4135/978144627305014540252.

Aldridge J. Participatory research: working with vulnerable groups in research and practice. Bristol: The Policy Press; 2015.

Aldridge J, Sharpe D. Pictures of young caring. Loughborough: Young Carers Research Group, Loughborough University; 2007.

Atkinson D. Engaging competent others: a study of the support networks of people with a mental handicap. Br J Soc Work. 1986;16:83–101.

Atkinson D. An auto/biographical approach to learning disability research. Aldershot: Ashgate; 1997.

Biggs S. Resource-poor farmer participation in research: a synthesis of experiences from nine national agricultural research systems. OFCOR (On-Farm Client-Oriented Research) Comparative Study Paper 3. The Hague: International Service for National Agricultural Research; 1989.

Boddy J, Oliver C. Research governance in children's services: the scope for new advice. London: Thomas Coram Research Unit, Department for Education; 2010.

Bourdieu P. Understanding. Theory Cult Soc. 1996;13(2):17–37.

Britzman D. Who has the floor? Curriculum teaching and the English student teacher's struggle for voice. Curric Inq. 1989;19(2):143–62.

Catalani C, Minkler M. Photovoice: a review of the literature. London: Centre for Narrative Research, University of East London; 2010. www.uel.ac.uk/cnr.

Chataway C. An examination of the constraints on mutual inquiry in a participatory action research project. J Soc Issues. 1997;53(4):747–65.

Chevalier JM, Buckles DJ. Participatory action research: theory and methods for engaged enquiry. Abingdon: Routledge; 2013.

Cornwall A, Jewkes R. What is participatory research? Soc Sci Med. 1995;41(12):1667–76.

Department of Health (DH). No secrets: guidance on developing and implementing multi-agency policies and procedures to protect vulnerable adults from abuse. London: DH; 2000.

Department of Health. Research governance framework for health and social care. London: DH; 2005. www.dh.gov.uk/prod_consum_dh/groups/dh_digitalassets/@dh/@en/documents/digitalasset/dh_4122427.pdf.

Economic and Social Research Council (ESRC). Framework for research ethics [updated September 2012]. 2010. www.esrc.ac.uk/_images/framework-for-research-ethics-09-12_tcm8-4586.pdf.

Fals Borda O. Knowledge and people's power: lessons with peasants in Nicaragua, Mexico and Colombia. New York: New Horizons Press; 1988.

Flynn M. Independent living for adults with mental handicap: a place of my own. London: Cassell; 1989.

Goodley D, Moore M. Doing disability research: activist lives and the academy. Disabil Soc. 2000;15(6):861–82.

Goodson IF. Developing narrative theory: life histories and personal representation. Abingdon: Routledge; 2013.

Hart RA. Children's participation from tokenism to citizenship. Florence: UNICEF Innocenti Research Centre; 1992. www.freechild.org/ladder.htm.

Higginbottom G, Liamputtong P, editors. Participatory qualitative research methodologies in health. London: SAGE; 2015.

Hill M, Davis J, Prout A, Tisdall K. Moving the participation agenda forward. Child Soc. 2004;18(2):77–96.
Hurdley R. In the picture or off the wall? Ethical regulation, research habitus and unpeopled ethnography. Qual Inq. 2010;16(6):517–28.
Joanou JP. The bad and the ugly: ethical concerns in participatory photographic methods with children living and working on the streets of Lima, Peru. Vis Stud. 2009;2(3):214–23.
Larkin M. Vulnerable groups in health and social care. London: SAGE; 2009.
Lewis A, Porter J. Interviewing children and young people with learning disabilities: guidelines for researchers and multi-professional practice. Br J Learn Disabil. 2004;32(4):191–7.
Liamputtong P. Researching the vulnerable: a guide to sensitive research methods. London: SAGE; 2007.
Lieblich A, Tuval-Mashiach R, Zilber T, editors. Narrative research: reading, analysis, and interpretation. London: SAGE; 1998.
McTaggart R, editor. Participatory action research: international contexts and consequences. Albany: State University of New York Press; 1997.
Moore LW, Miller M. Initiating research with doubly vulnerable populations. J Adv Nurs. 1999;30(5):1034–40.
O'Neill M, Giddens S, Breatnach P, Bagley C, Bourne D, Judge T. Renewed methodologies for social research: ethno-mimesis as performative praxis. Sociol Rev. 2002;50(1):69–88.
Parrott L, Jacobs G, Roberts D. Stress and resilience factors in parents with mental health problems and their children, Social Care Institute for Excellence Research Briefing, vol. 23. London: Social Care Institute for Excellence; 2008.
Rapoport RN. Three dilemmas in action research. Hum Relat. 1970;26(3):499–513.
Rapport F. The poetry of holocaust survivor testimony: towards a new performative social science. Forum Qual Soc Res. 2008;9(2):1–16. article 28.
Rogers AC. Vulnerability, health and health care. J Adv Nurs. 1997;27:65–72.
Sempik J, Aldridge J, Becker S. Health, well-being and social inclusion: therapeutic horticulture in the UK. Bristol: Policy Press; 2005.
Steel R. Involving marginalised and vulnerable groups in research: a consultation document. Involve. London: NHS; 2001.
Thomson P, editor. Doing visual research with children and young people. London: Routledge; 2008.
Walker R, Schratz B, Egg P. Seeing beyond violence: visual research applied to policy and practice. In: Thomson P, editor. Doing visual research with children and young people. Abingdon: Routledge; 2008. p. 164–74.
Walmsley J, Johnson K. Inclusive research with people with learning disabilities: past, present and futures. London: Jessica Kingsley Publishers; 2003.
Whyte WF. Advancing scientific knowledge through participatory action research. Sociol Forum. 1989;4(3):367–85.

Inclusive Disability Research

111

Jennifer Smith-Merry

Contents

1 Introduction	1936
2 How Inclusive Is Inclusive Research?	1936
3 Researching Disability: Overview of the Field	1939
4 Benefits to Research of Co-researcher Involvement	1941
5 Co-researcher Experiences of Inclusive Research	1942
6 Academic Researcher Experiences of Inclusive Research	1944
7 Methodological Issues to Consider When Carrying Out Inclusive Research	1945
7.1 Rigor	1945
7.2 Power and Co-researcher Partnerships	1946
7.3 Training	1946
7.4 Access to Co-researchers	1947
7.5 Co-writing Co-research	1947
8 Conclusion and Future Directions	1949
References	1949

Abstract

Increasingly academic researchers are becoming involved in inclusive research, in which people with a lived experience of the field of research under study are included as part of the research team. This chapter provides an overview of inclusive research with people with disability. Discussions focus on the dimensions of current inclusive research, including the ethical basis for engagement with co-researchers, and potential barriers and facilitators to successful research. The impact of co-research on both the co-researchers with a disability and academic researchers involved is discussed. Examples of existing projects are provided to give a sense of the field and its limitations and its possibilities.

J. Smith-Merry (✉)
Faculty of Health Sciences, The University of Sydney, Sydney, Australia
e-mail: jennifer.smith-merry@sydney.edu.au

© Springer Nature Singapore Pte Ltd. 2019
P. Liamputtong (ed.), *Handbook of Research Methods in Health Social Sciences*,
https://doi.org/10.1007/978-981-10-5251-4_129

Keywords

Inclusive · Disability · Methods · Rigor · Co-research · Ethics · Consumers

1 Introduction

This chapter focuses on the practice of inclusive research in disability. This approach to research involves the development of research partnerships with the people that would usually be the subjects of research. It is a diverse field of practice and a creative one, with new research approaches being developed to suit the contexts in which the research is taking place. The aim of this chapter is to introduce inclusive disability research and provide an overview of the field by referring to recent inclusive research work. This chapter will describe the dimensions of the field, the issues involved in doing inclusive research, and the benefits of the research both to the project and to *all* the researchers involved.

Why should the "subjects" of research be included as researchers? This question can be answered from a number of different angles, both practical and philosophical. Ethically, or philosophically, the often quoted and important maxim "nothing about us without us" provides a good starting point to think about the ethics of inclusive research (Woelders et al. 2015, p. 529). Many researchers have a strong belief that all research with people with a disability should be emancipatory and include co-researchers as part of this emancipatory project (Björnsdóttir and Svensdóttir 2008; Boland et al. 2008; Woelders et al. 2015).

Practically, "co-researchers" (here I refer to researchers with a disability working with other researchers) are often included because of the different insights that someone with a lived experience of the field of research can bring. They might also be included because it makes it easier to conduct research in a field where some on the research team are known and respected – they thus become gatekeepers to the field. Increasingly, funding bodies are also demanding that researchers include people with a disability in the research process in order for researchers to fulfil their contractual obligations (Case et al. 2014). This means that some researchers may feel forced to include co-researchers, which is not such a great basis upon which to develop a collaborative project. The practical and ethical dimensions of inclusive research will be discussed further later in the chapter

Throughout this chapter, the term "co-researchers" is used to signify those people with a lived experience of disability engaging with research. The term "consumers" is also used to signify people with a lived experience more generally. It is acknowledged that these terms are contested and that in other contexts the terms "service users," "users," or "clients" might be used. The term "consumers" is used because that is the most commonly used term in the context in which I teach and research.

2 How Inclusive Is Inclusive Research?

Setting out to define inclusive research is not a straightforward project. This is because the phrase is used to cover a wide range of approaches to research. Inclusive research may variously refer to research that uses co-researchers in an advisory capacity only, through to research which is led by consumers and all gradations in between. It should also be noted that conversely a wide range of terms is used to cover what is termed here as inclusive research. Frequently used terms include emancipatory research, collaborative research, participatory action research, consumer-led or user-led research, co-research and co-created, and co-developed or co-produced research. These terms are often also used to discuss noninclusive research (e.g., research which merely includes people with a lived experience as participants in the research rather than as researchers), which means that there can be significant difficulties in definition. This should be kept in mind when researching or writing about inclusive research.

Inclusivity is usually characterized according to what extent co-researchers are actually involved in the research. Rose (cited in Horsfall et al. 2007, p. 1202) states that there are three types of co-researcher involvement in research: "consultative, collaborative, and user-led." "User led" research is that where co-researchers are in control of the whole research project and is positioned in comparison with "collaborative research" which involves a more equal collaboration between co-researchers and other members of the research team (Arthur et al. 2008). Consultative approaches merely use consumers to bounce ideas off. Abma et al. (2009) describe the roles that co-researchers take in research on a hierarchy from "object" of research to "research principal" leading the research. Following this approach, a spectrum of inclusive research can be imagined where at one end sits consultative research (e.g., Lewis et al. 2008) and at the other end sits consumer-led research where co-researchers conceive of and develop the research, directing or employing academic researchers as needed to carry it out.

Along this spectrum are positioned other types of research which provide varying degrees of inclusivity. For example, some research may be co-led or be led by academics who then employ co-researchers for a role or number of roles in the project, usually data collection or analysis. Co-researchers are sometimes included in all stages of the research except the project development stage or may be included at just one or two points. Some research includes co-researchers as co-authors in publications, but others do not.

For much research, inclusivity is determined on an ad hoc basis according to project necessities. This can include issues such as funding, co-researcher capabilities and interest, research location, and time limits. An ad hoc approach is viewed as preferable to highly structured models and several authors have reflected on the demands of an ideologically based model and what this meant for both researchers and co-researchers. Nind and Vinha (2014, p. 108) comment that "as a community of researchers, we continue to juggle balancing principles and pragmatics." Likewise, Woelders et al. (2015, p. 538) feel that inclusive research "can be a rigid concept, guided by the ideal of social justice, expecting the same things of the academic researcher and the person with intellectual disabilities without a critical look at the added value of including people with intellectual disabilities." For the researchers in this team, difficulties in implementation of the model necessitated the development of a less structured approach.

In contrast to an ad hoc approach, several structured models of inclusion have been developed and promoted. The most frequently used of these are "inclusive research" (here, confusingly, described as a model rather than a general descriptor) (Bigby and Frawley 2010), participatory action research (Caldwell et al. 2009), collaborative group research (Bigby et al. 2014), and emancipatory research (Björnsdóttir and Svensdóttir 2008). Each of these models provides a set of principles or structures around the practice of inclusion, which may be useful to some people considering inclusive research. For example, a participatory action research approach sets up principles around the social purpose of research and the extent of inclusion of co-researchers (see also ▶ Chaps. 110, ""With Us and About Us": Participatory Methods in Research with "Vulnerable" or Marginalized Groups," and ▶ 17, "Community-Based Participatory Action Research"). In relation to disability research, this can be seen in the four principles developed by Selener and Balcazar et al. (cited in Buettgen et al. 2012, pp. 607, 609, 611, 613):

Principle 1. Disabled individuals articulate the problem and participate directly in the process of defining, analyzing and solving it. . ..
Principle 2. Direct involvement of disabled people in the research process facilitates a more accurate and authentic analysis of their social reality. . ..
Principle 3. The process of participatory research can increase awareness among disabled people about their own resources and strengths. . ..
Principle 4. The ultimate goal of the research endeavor is to improve the quality of life for disabled people.

Likewise, the distinct model of inclusive research developed by Walmsley and Johnson (paraphrased in Bigby and Frawley 2010, p. 53) includes the following components:

1. having ownership of research questions;
2. being collaborators; that is, involved in the doing of the work;
3. exercising some control over process and outcomes;
4. being able to access questions, reports and outcomes; and finally
5. that outcomes will further the interests of people with intellectual disability.

As these principles show, many inclusive research projects take an ideological stance to the inclusion of co-researchers. This is most clear in that research described by authors as "emancipatory." Emancipatory research seeks to actively challenge dominant hierarchies of knowledge, is shaped by a "social model of disability," and evolves from the disability social movement in the UK (Björnsdóttir and Svensdóttir 2008). It is underpinned by an explicit social justice framework and is described as a form of "political action" (Woelders et al. 2015, p. 529). The aim is not just to include or even partner with co-researchers, but to emancipate or free them from a social context which has subjugated them and invalidated their experience in the past. Boland et al. (2008) imply that the term is used too widely and make a distinction between research merely described as emancipatory (research which occurs in *partnership*) and "true emancipatory research" (research *led* by those with a disability). Some researchers have even gone so far as to state that only those people who have a disability should actually carry out research on disability (Björnsdóttir and Svensdóttir 2008). In projects led by co-researchers, a preexisting group of co-researchers, usually aligned with a service or previous research project, usually devise the research goals and set in motion the project (see Box 2 for example). They then direct the process and are usually involved in each step of the work. However, in a project described by Davidson et al. (2010), the co-researchers engaged academics to carry out the research while they oversaw the research program.

3 Researching Disability: Overview of the Field

The field of inclusive research is mainly restricted to co-research with people with a lived experience of intellectual disability (see Boxes 1 and 2, below for examples) and mental ill-health (see Box 3 for an example).

The vast majority of existing inclusive research focuses on co-research in the fields of health or social sciences and focuses either on service development, consumer experiences of health or social care, and identity (Arthur et al. 2008; Hreinsdóttir and Stefánsdóttir 2010; Buettgen et al. 2012; Hutchinson and Lovell 2013). Most papers also use a qualitative methodology. The most frequently used data collection methods were interviews and focus groups. A small number have used surveys including several with quantitative components (Conder et al. 2011; Kramer et al. 2013; Nicolaidis et al. 2015). Other less-used structured methodologies include randomized controlled trials (Hassouneh et al. 2011) and user-acceptability testing of a computer program (Oschwald et al. 2014). Non-traditional data collection methods are used frequently in inclusive research. This includes visual data collection methods where art is produced by participants (Brookes et al. 2012). Theatre has also been used to collect data via approaches such as "Theatre of the Oppressed" or "Forum Theatre" methods (Daniel et al. 2014). In other projects, existing methods such as focus groups have been refined for use with co-researchers with particular needs. For example, in research conducted by Garcia-Iriarte and colleagues (2009), focus groups were adapted in order to be accessible as a method for collecting data by people with an intellectual disability who found the moderation required in standard focus groups challenging.

These examples demonstrate the creativity of many inclusive research projects which really seek to understand the ways in which co-researchers can best participate in a project to provide meaningful data which will make sense in the organizational and community context in which the results are to be presented. A very significant proportion of papers that have been published in relation to inclusive research are self-reflexive accounts of inclusive research which do not include a structured methodology but rather provide an account of process and experience.

While an adaptive methodological approach makes sense in relation to the accessibility needs of co-researchers, it can also lead to issues with rigor when compared to other qualitative research projects and existing scales of research quality such as that produced by the National Health and Medical Research Council in Australia or McMaster University in Canada (Letts et al. 2007; NHMRC 2009). A systematic review of inclusive research by Anderson et al. (2015) has emphasized that inclusive research is generally of low quality and that all studies that they included in their review showed research bias. However, one review that has provided a structured comparison of the use of co-researchers in research versus research without co-researchers has found no difference in research outcomes (Nilsen et al. 2013). The quality of inclusive research is discussed further below in relation to rigor.

> **Box 1 Example of inclusive research including people with a lived experience of intellectual disability**
> This research has been written up in two papers: Garcia Iriarte et al. (2014) (focuses on research findings) and O'Brien et al. (2014) (focuses on process).
>
> *Research aims*: To conduct a national study of the experiences of people with an intellectual disability with respect to quality of life.
> *Co-researchers*: People with intellectual disability. The research was directed by a team which included five co-researchers, four researchers, and three assistants. Two of the assistants assisted the work of the co-researchers. Fifteen further co-researchers were engaged in the data collection and analysis.
> *Methods used*: 23 focus groups across Ireland with people with intellectual disability were conducted. Project conducted over a 4-year period. Initial data analysis was conducted by academic researchers and then co-researchers assisted. This happened in a facilitated group event. Separate coding conducted by independent academic researchers to test for validity.
> *Inclusion*: Research topic was developed before co-researchers joined. There was an inclusive advisory group, which provided advice on research in progress. People with intellectual disability on the advisory group needed some time to feel able to speak comfortably. People with a lived experience co-led the research and co-chaired the advisory group. Co-researchers involved in methodology selection, made changes to data collection questions and process (e.g., inclusion of visual prompts), led data collection with support from academic researchers, involved in data analysis (detailed above), helped to develop dissemination strategy, and involved in dissemination and writing of project report.

(continued)

Box 1 (continued)

Impact of co-research on the project: Participants opened up to co-researchers more readily than other researchers. Co-researchers made decisions about data collection which allowed more co-researchers to be involved so that research could take place across the country and be more representative. They were also instrumental in ensuring that the data collection questions were appropriate.

Comments on process: A significant amount of support from local organization was needed for co-researchers to be able to participate, but this was not considered at the start of the project. The quality of this support differed between organizations. Services needed to be better educated about inclusive research to better support co-researchers. Greater clarity was needed of the place of the research assistants in their support of the co-researchers. Meetings were not long enough and needed to be available for telecommuting in order to enable voices to be heard. Training was needed for all researchers. For co-researchers, this involved training in data collection. For other researchers, this involved learning communication strategies for use with co-researchers. Involvement of co-researchers needed to be nurtured. Flexibility of process was important.

4 Benefits to Research of Co-researcher Involvement

In addition to the ethical and emancipatory benefits of inclusive research, there are two main practical benefits of including co-researchers. These were the ability for co-researchers to bring a different type of knowledge into the research process and to act as gatekeepers to a field where academic researchers were outsiders.

Academic researchers often come to a field from either a purely theoretical perspective or with practical experience as practitioners working in the field. What they, therefore, lack is a personal understanding of what it means to receive services or live as someone with a disability. Researchers engaging with co-researchers place value on the tacit knowledge of lived experience and want to enhance the place of that knowledge in their work. While including participants with a lived experience is the traditional way of doing qualitative research, the problem with this traditional approach is that the knowledge of lived experience must be filtered through the academic researcher's lens, which may not focus on what is important or interesting for the participants themselves. Including people with a lived experience as co-researchers helps to ameliorate the subjective bias of the researcher (Abma et al. 2009; Cook and Inglis 2012). Existing literature reveals several ways that co-researcher involvement in a project can help to orient the project towards what is important for people with disability:

1. Participation in the development of project goals so that it is pertinent to the interests of people with a disability and will meet their needs.
2. Involvement in the development of data collection questions, prompts, or stimuli to make sure that they include relevant areas and are phrased in a way that makes sense to the context participants are experiencing.
3. Having a role in data analysis helps to ensure that analysis aligns more with the lived experience of the participants rather than the researcher's detached expectations of that experience.
4. Participation in dissemination and choices about dissemination mean that dissemination is more likely to be done in a way where those whose lives are the subject of the research are able to engage with it.

The other practical reason for including co-researchers is to enable better access to participants (e.g., Abma et al. 2009; Strnadová et al. 2015). Because of a history of poor practices by researchers who have not respectfully engaged with people with a disability when conducting research, there can exist a lack of trust among potential participants. This can make recruitment more difficult and participants reticent to open up to researchers about their experiences. When co-researchers are genuinely included in a project, the inclusion of co-researchers can allow trust to be built in the field because participants can see the respect for consumer knowledge inherent in collaboration (Abma et al. 2009).

5   Co-researcher Experiences of Inclusive Research

> One of the biggest challenges with having a mental illness is being invisible. When people look at you, they do not see you. They see your diagnosis. You are constantly over-looked because people think you have nothing to offer your community and society. Sometimes you even overlook yourself as you start to believe what others think of you. Taking part in this project has been a reminder to ourselves and others that we (persons diagnosed with a psychiatric illness) have something important to offer (Case et al. 2014, p. 404)

> I am confident and proud of myself. I am a real researcher and this is my story (White and Morgan 2012, p. 102)

Many inclusive researchers include a discussion of how the process of the research impacts on the co-researchers involved (de Wolff 2009; Tuffrey-Wijne and Butler 2010; Cook and Inglis 2012; Williams et al. 2015). This reflexive exercise is important when writing up research as it allows a consideration of co-researcher experiences in relation to their input into the research and its impact on the project outcomes. This needs to be carefully weighed as a project may produce good research, but would be problematic if the process had been overwhelmingly negative for co-researchers. Conversely research may be a good experience for co-researchers, but the quality of research produced is so low that it has little impact. In such a case, the process could be classified as a therapeutic one rather than research. This is discussed further below in relation to rigor.

Co-researchers generally describe their experiences in very positive terms, but are also open about the difficulties of co-research. The positive impact of co-research comes from learning (Abma et al. 2009; Carey et al. 2014), validation and respect (Bigby et al. 2014), connection to others (Bell and Mortimer 2013), work experience (Grayson et al. 2013), remuneration (Case et al. 2014), contributing to the community (Brookes et al. 2012), and the development of personal relationships (Strnadova et al. 2014). For co-researchers, these impacts may result in a changed sense of self, as evidenced in the quotations above, empower them, and have a therapeutic impact (Gillard et al. 2010; Martin 2015; Rome et al. 2015; Tilly 2015).

It is also important to reflect on what might be difficult for co-researchers. While personal accounts of co-research do not tend to speak about the research being overall a negative experience, they do highlight areas of difficulty in their accounts. Co-researchers speak about feeling anxious about the research particularly in the early stages (de Wolff 2009; Case et al. 2014). Others do not understand some of the research or feel comfortable with some parts, such as analysis, and not others (Conder et al. 2011; Strnadová et al. 2015). They also report being upset about aspects of the research including content that is disturbing or makes them recall upsetting experiences (Tuffrey-Wijne and Butler 2010; Hutchinson and Lovell 2013). Some co-researchers also feel uncomfortable about disclosing that they are co-researchers because of community stigma, for example, around mental ill-health (Lincoln et al. 2015).

Accounts of inclusive research show that structured processes core to project progress within universities can also disempower co-researchers. For example, several researchers speak about the problems with ethics processes which question the co-researcher's knowledge and abilities (see Box 3, below and e.g., Flood et al. 2013; Morgan et al. 2014; Walmsley and Central England People First History Project Team 2014). These bureaucratic processes need to be carefully managed so that they do not undermine the confidence of team members.

Box 2 Example of inclusive research led by co-researchers

This research was written up in the paper by Walmsley and Central England People First History Project Team (2014). This article was co-written with co-researchers.

Research aims: The project aimed to create an organizational history of the Central England People First.

Co-researchers: The co-researchers were all people with a lived experience of intellectual disability. They had all been part of the organization which was the focus of the study.

Methods used: Oral history interviews, "talking event" where different perspectives were collected from a group and document and photograph collection. These were used in order to develop a history of the organization.

(continued)

> **Box 2** (continued)
>
> *Inclusion*: The project was consumer-led in that all decisions were made by the co-researchers. The inclusion was holistic as described here: "Members of CEPF actively chose to do the project, were in charge of key decisions and carried out much of the work. It was a team approach. Non-disabled people played a hugely significant role, as supporters, advisors and expert consultants – it could not have happened without them – but decisions remained in the hands of the CEPF project team" (Walmsley and Central England People First History Project Team 2014, p. 40). The team employed an academic researcher (Walmsley) to create the history. They also employed a project worker to assist with the project management.
>
> *Impact of co-research on the project*: The co-researchers enjoyed the project and felt that they produced work of great value. However, they also expressed regret about some of the difficulties encountered. For example, they felt uncomfortable about speaking about or writing up research data about difficult relationships and bullying within the organization's history so made the decision not to include that information. The project successfully delivered its aims to create a history of the organization. They also produced guides for how to create inclusive oral history research.
>
> *Comments on process*: Process was hindered by instability in the organization due to staff turnover. Some potential respondents who had left the organization were unwilling to be interviewed by them. They, therefore, got an external person to do the interviews, but this was seen as "not really satisfactory" to the co-researchers who wanted to do all the data collection (p. 38). One of the Universities that they approached were unwilling to include people who are unable to read in research projects so they had to not pursue that research relationship. The project took more time than a project which was not consumer-led. Co-research worked well within this organizational context because they were already co-led by consumers, so this research made sense.

6 Academic Researcher Experiences of Inclusive Research

Several academic researchers have reflected on the process of inclusive research leading to an altered sense of themselves or themselves in relation to academia (Chappell et al. 2014; Kidd and Edwards 2016). This is as a result of researchers questioning their own knowledge or the process of inclusive research engendering a reconceptualization of their values, for example, towards a human rights perspective or as an activist social researcher (following Healy in Stevenson 2010). Gillard and colleagues (2012) write that co-research directly challenges the basis of academic knowledge because it is community driven, elevates nonacademic knowledge, and generally utilizes a nonstandard methodological approach. This can bring those academics who practice co-research into conflict with their own organizations and put them in a place where they have to justify their "nonscientific" approaches (Kidd and Edwards 2016).

The process of co-research itself can also be practically challenging (Woelders et al. 2015). In Woelders et al. (2015, p. 533), one researcher stated that they felt they "could not live up to the ideals of inclusive research" and that they faced multiple difficulties and challenges in the process. These difficulties can arise from communication problems amongst team members (McClimens 2008) and tension from different timeframes operating in academic versus co-researcher worlds (Dorozenko et al. 2016). This can lead to researchers feeling torn between the requirements of their institutions and their obligations to their research partners. Careful planning in the initial stages of a project and extra time built into the research timeline will help to ameliorate these potential issues.

7 Methodological Issues to Consider When Carrying Out Inclusive Research

7.1 Rigor

The discussion here shows a careful line being drawn between co-research which is rigorous and valid within a wider academic context and that which has important benefits for the co-researchers and their organizations but might lack rigor. As discussed earlier, the research quality of inclusive research papers is generally low. The research methodology and the exact nature of participation of co-researchers are poorly described in many inclusive research studies. For example, a significant proportion of papers state that co-researchers were involved, but do not say how they were involved, what this involvement meant for the research or how this involvement impacted on the co-researchers (Karban et al. 2013; Sherwood-Johnson et al. 2013; Linz et al. 2016). This is a serious limitation of inclusive research studies which has also been highlighted elsewhere (Bigby and Frawley 2010; Jivraj et al. 2014).

The most common methodological relates to uncontrolled and unacknowledged bias. This occurred frequently in papers where co-researchers in the study were also involved as participants and were involved in the project design or data analysis (Ollerton and Horsfall 2013; Azzopardi-Lane and Callus 2015). Critiques of co-research also state that co-research means that the researcher loses objectivity (Kiernan, cited in Tuffrey-Wijne and Butler 2010). Hancock et al. (2012, p. 219) also state that those opposed to consumer involvement cite a "perceived lack of objectivity and capacity and lack of research knowledge/skills." Some bias may be unavoidable because of the nature of the research, but it is nevertheless important that these issues are acknowledged and spoken about when writing up the research. Failure to do so or clearly think through research before it is embarked upon may mean that the research which co-researchers have put so much time into is not able to be published or change practice because of inherent faults.

7.2 Power and Co-researcher Partnerships

Imbalances of power between the academic researcher and co-researchers can be a significant problem if not explicitly considered in the interactions of a collaborative research project. This results from a history of research practices where the person with a disability has been "studied" rather than genuinely involved. Chappell et al. (2014, p. 386), drawing on the work of Corker and Shakespeare, comment that "the medical model of disability constructs persons with disabilities in terms of 'deviance, lack and tragedy' as victims of impairment ... and as objects to be studied rather than as subjects and agents of research." Inclusive research challenges this. However, if researchers feel that they are "giving voice" to people with a disability by including them in research, this can also reinforce the problematic medical model – inclusive research has to be genuinely collaborative to meet emancipatory goals. This is a warning against an empty "emancipation" that makes co-researchers feel good but does not move knowledge forward. In order to address these potential problems, academic researchers need to continually employ a self-reflexive approach where they consider their own actions and motivations.

Inclusive research often employs assistants who will help the co-researchers complete data collection or analysis (see for example the research in Box 2). If employing assistants in this way, it should be ensured that the co-researcher "supporters" do not take over or influence the co-researchers (Bigby and Frawley 2010). The danger here is that co-researchers will be co-opted into expressing someone else's interests rather than their own. The assistants should therefore be trained in research and a set of principles developed for support in order to avoid this outcome (see also ▶ Chap. 96, "The Role of Research Assistants in Qualitative and Cross-Cultural Social Science Research").

7.3 Training

Inclusive research projects a wide variety of approaches to training, from no training to on-the-job training conducted as needed, and further to highly structured training involving multiple training sessions over an extended period (Northway et al. 2013 and see also Box 4; Strnadova et al. 2014). Training is generally put in place in order to ensure a greater rigor to the data collection project and to help the co-researchers to be confident in their work (Grayson et al. 2013). However, training is not always accepted by co-researchers with Bigby and Frawley (2010) describing one project where co-researchers rejected training and would not engage. On reflection, the academic researchers were positive about this rejection as it made them reconsider the impact of training, which may have been to force co-researchers to fit into an academic mold. For these researchers and others who choose not to train or train more informally, the choice was justified because training might distort the co-researcher's unique knowledge (Nind et al. 2016).

7.4 Access to Co-researchers

The organizational context in which the research takes place is an important factor in facilitating research progress. In two of the exemplars provided (Boxes 2 and 3), the inclusive research approach worked because the organizational context was already oriented towards collaborative practice. Consumer knowledge was already validated within the organizations, and therefore, the co-researchers were taken seriously and could conduct their elements of the research without having to counter any stigmatized perceptions of their capacity. In other organizational contexts, however, this could be a significant confounder to the research progress (Baart and Abma 2011). Organizational values oriented towards participation make organizational support more likely, assists with co-researcher involvement, and therefore, makes research easier.

In existing inclusive research, co-researchers have generally been physically present to conduct data collection and be otherwise involved in the research. However, this approach significantly limits the participation of co-researchers whose disability may prevent them from co-research in person, including people with certain forms of physical disability or mental ill-health (e.g., social anxiety disorder). Recognizing this limitation, one study including co-researchers with autism developed processes for collaborating online (Nicolaidis et al. 2011). The accessibility of co-researchers should be considered in all inclusive research so that co-research is not just limited to people with the capacity to be physically present to complete research.

7.5 Co-writing Co-research

A significant percentage of peer-reviewed journal articles which write up inclusive research are co-written in some way by people with a lived experience of disability. Co-writing with co-researchers should be considered particularly when they have had a significant part in the design, data collection, or analysis. Inclusion of co-researchers will allow new perspectives and new modes of expression to enter academic discourse. Co-written papers mainly include the co-researchers generally as part of the writing team or by allocating them part of the paper to write from their own perspective (Nicolaidis et al. 2011; Makdisi et al. 2013; Rome et al. 2015). In the consumer-led research reported in Box 2, the co-researchers co-wrote the paper in a style where they took turns with the lead author. Although the project was led by co-researchers, the academic researcher was lead author on the publication (Walmsley and Central England People First History Project Team 2014).

In student projects where co-researchers are part of the research team, student researchers need to ensure that they do enough of the research that it is their own project in order to fulfil the requirements of their qualification (Björnsdóttir and Svensdóttir 2008; Morgan et al. 2014; Dorozenko et al. 2016). The use of co-researchers in student research is another clear example of an ad hoc adaptive approach to inclusion which needs to meet the needs of both academic researchers working within their own world and to meaningfully involve co-researchers.

Box 3 Example of inclusive research in mental health
A self-reflexive account of this research is written up in Case et al. (2014). This article was co-written with the co-researchers.

Research aims: This research aimed to understand the experiences of people with mental ill-health using a particular mental health service. The organization which commissioned the research wanted this qualitative, experiential data to augment existing satisfaction surveys.

Co-researchers: People with a lived experience of mental ill-health. They were people already receiving services through the organization being studied. The co-researchers were employed through a competitive application process and were employed because of commitment to the project goals.

Methods used: 14 focus groups were conducted with 101 people with mental ill-health attending the mental health service. Qualitative analysis of transcripts.

Inclusion: The research derived from interests of an academic researcher and a service provider. Co-researchers were employed in order to carry out the data collection. Co-researchers were given training in research methods over 9 months. Co-researchers facilitated and took notes during the focus groups. An academic researcher also assisted with all focus groups by providing mainly logistical support. Each transcript was analyzed by one co-researcher and one academic researcher. The co-researchers presented the findings. The whole team collaboratively wrote the report.

Impact of co-research: The research effected this desired outcome for the organization of creating organizational change and validating consumer knowledge (see below). The co-researcher input into the language used in the focus groups helped to make the process and questions relevant to the participants. Different types of knowledge were able to be derived from the focus groups when facilitated by co-researchers because the respondents were more forthcoming. The co-researchers contextualized the data through analyzing it within the context of their own lived experience. This provided an insight into the data which were not available to the academic researchers. The collaboration has spawned further inclusive research.

Comments on the process: For the organization, this inclusive research was a natural extension of the collaborative approach taken by the organization. It represented their values of inclusion in practice. It was also a tool for the organization to "elevate the status" of consumers – their inclusion as researchers validated the experiential knowledge they held. Training was provided in an ad hoc fashion where the academic researchers "facilitated" the knowledge of the co-researchers. For the co-researchers, the process was "terrifying" at times (particularly presenting), but overall they felt "valued" and "honored" to be part of the research. They felt that the remuneration was important but also appreciated having something to fill their time.

8 Conclusion and Future Directions

What this chapter aimed to do was to provide an introduction to inclusive research, how it is done, what it means for the people involved, and what to watch out for when creating an inclusive research project. The conclusion aims to suggest how researchers might best put in place initial steps to beginning an inclusive research project. Taken as a whole, this chapter shows there are two essential elements that need to be put in place prior to the research starting: (1) communication and (2) timely preparation.

The first thing needed is to start having conversations with people with a lived experience of disability to find out what is important to them, how a researcher might work in with work that is already being done, or offer skills to groups of consumers who are interested in doing research. From this point, inclusive research will be developed as a collaborative process. Inclusion of co-researchers from the start will help to ameliorate many of the problems identified above and make sure that they are fully included in a project, rather than as an add-on. Their needs are more likely to be considered and their skills used in the best way possible if they are involved from the start.

It is also important to have time to prepare a project over a long period so that issues can be thought through before essential elements such as funding and ethics approval are gained as these processes will usually set in place a methodological approach which is hard to move away from later. While an ad hoc and flexible approach may be the easiest to implement, ad hoc does not equate to a lack of planning. A long lead-time is needed in order to map out, preferably in a collaborative manner with co-researchers, the path ahead, identify potential risks, and develop strategies to mitigate them.

With a focus on choice and control and person-centered care for people with a disability, a central tenet of major policy frameworks in many countries including the UK (e.g., https://hee.nhs.uk/our-work/person-centred-care) and Australia (e.g., National Disability Insurance Scheme Act 2013), it is important that those services which are developed meet the needs of people with a disability or choice and control is meaningless. Including people with a disability in the creation of research which evaluates services or develops new practices will be essential to ensuring that practice can be assessed in relation to questions which are relevant to people with a disability. Choice and control is also meaningless if there is no trust in the voice of people with a disability. Inclusive research helps to validate these voices and change practices which have only viewed the experiences of people with a disability through a lens which is not focused on their needs.

Acknowledgments The research in this chapter developed out of work commissioned by National Disability Services https://www.nds.org.au/ Short sections of the text published here have been released in reports related to this research.

References

Abma TA, Nierse CJ, Widdershoven GA. Patients as partners in responsive research: methodological notions for collaborations in mixed research teams. Qual Health Res. 2009;19(3):401–15.

Anderson LM, Adeney KL, Shinn C, Safranek S, Buckner-Brown J, Krause KL. Community coalition-driven interventions to reduce health disparities among racial and ethnic minority populations. Cochrane Database Syst Rev. 2015;(6)

Arthur B, Knifton L, Park M, Doherty E. 'Cutting the dash' – experiences of mental health and employment. J Public Ment Health. 2008;7(4):51–9.

Azzopardi-Lane C, Callus AM. Constructing sexual identities: people with intellectual disability talking about sexuality. Br J Learn Disabil. 2015;43(1):32–7.

Baart ILMA, Abma TA. Patient participation in fundamental psychiatric genomics research: a Dutch case study. Health Expect. 2011;14(3):240–9.

Bell P, Mortimer A. Involving service users in an inclusive research project. Learn Disabil Pract. 2013;16(4):28–30.

Bigby C, Frawley P. Reflections on doing inclusive research in the "making life good in the community" study. J Intellect Dev Disabil. 2010;35(2):53–61.

Bigby C, Frawley P, Ramcharan P. Conceptualizing inclusive research with people with intellectual disability. J Appl Res Intellect Disabil. 2014;27(1):3–12.

Björnsdóttir K, Svensdóttir AS. Gambling for capital: learning disability, inclusive research and collaborative life histories. Br J Learn Disabil. 2008;36(4):263–70.

Boland M, Daly L, Staines A. Methodological issues in inclusive intellectual disability research: a health promotion needs assessment of people attending Irish disability services. J Appl Res Intellect Disabil. 2008;21(3):199–209.

Brookes I, Archibald S, McInnes K, Cross B, Daniel B, Johnson F. Finding the words to work together: developing a research design to explore risk and adult protection in co-produced research. Br J Learn Disabil. 2012;40(2):143–51.

Buettgen A, Richardson J, Beckham K, Richardson K, Ward M, Riemer M. We did it together: a participatory action research study on poverty and disability. Disabil Soc. 2012;27(5):603–16.

Caldwell J, Hauss S, Stark B. Participation of individuals with developmental disabilities and families on advisory boards and committees. J Disabil Policy Stud. 2009;20(2):101–9.

Carey E, Salmon N, Higgins A. Service users' views of the Research Active Programme. Learn Disabil Pract. 2014;17(4):22–8.

Case AD, Byrd R, Claggett E, DeVeaux S, Perkins R, Huang C, . . . Kaufman JS. Stakeholders' perspectives on community-based participatory research to enhance mental health services. Am J Community Psychol. 2014;54(3–4):397–408.

Chappell P, Rule P, Dlamini M, Nkala N. Troubling power dynamics: youth with disabilities as co-researchers in sexuality research in South Africa. Childhood. 2014;21(3):385–99.

Conder J, Milner P, Mirfin-Veitch B. Reflections on a participatory project: the rewards and challenges for the lead researchers. J Intellect Dev Disabil. 2011;36(1):39–48.

Cook T, Inglis P. Participatory research with men with learning disability: informed consent. Tizard Learn Disabil Rev. 2012;17(2):92–101.

Daniel B, Cross B, Sherwood-Johnson F, Paton D. Risk and decision making in adult support and protection practice: user views from participant research. Br J Soc Work. 2014;44(5):1233–50.

Davidson L, Shaw J, Welborn S, Mahon B, Sirota M, Gilbo P, . . . Pelletier J. 'I don't know how to find my way in the world': contributions of user-led research to transforming mental health practice. Psychiatry. 2010;73(2):101–13.

de Wolff A. The creation of 'We Are Neighbours': participatory research and recovery. Can J Commun Ment Health. 2009;28(2):61–72.

Dorozenko KP, Bishop BJ, Roberts LD. Fumblings and faux pas: reflections on attempting to engage in participatory research with people with an intellectual disability. J Intellect Dev Disabil. 2016;41(3):197–208.

Flood S, Bennett D, Melsome M, Northway R. Becoming a researcher. Br J Learn Disabil. 2013;41(4):288–95.

Garcia Iriarte E, O'Brien P, McConkey R, Wolfe M, O'Doherty S. Identifying the key concerns of Irish persons with intellectual disability. J Appl Res Intellect Disabil. 2014; 27(6):564–75.

Garcia-Iriarte E, Kramer JC, Kramer JM, Hammel J. 'Who did what?': a participatory action research project to increase group capacity for advocacy. J Appl Res Intellect Disabil. 2009;22(1):10–22.

Gillard S, Turner K, Lovell K, Norton K, Clarke T, Addicott R, ... Ferlie E. "Staying native": coproduction in mental health services research. Int J Public Sect Manage. 2010;23(6):567–77.

Gillard S, Simons L, Turner K, Lucock M, Edwards C. Patient and public involvement in the coproduction of knowledge: reflection on the analysis of qualitative data in a mental health study. Qual Health Res. 2012;22(8):1126–37.

Grayson T, Tsang YH, Jolly D, Karban K, Lomax P, Midgley C, ... Williams P. Include me in: user involvement in research and evaluation. Ment Health Soc Incl. 2013;17(1):35–42.

Hancock N, Bundy A, Tamsett S, McMahon M. Participation of mental health consumers in research: training addressed and reliability assessed. Aust Occup Ther J. 2012;59(3):218–24.

Hassouneh D, Alcala-Moss A, McNeff E. Practical strategies for promoting full inclusion of individuals with disabilities in community-based participatory intervention research. Res Nurs Health. 2011;34(3):253–65.

Horsfall J, Cleary M, Walter G, Malins G. Challenging conventional practice: placing consumers at the centre of the research enterprise. Issues Ment Health Nurs. 2007;28(11):1201–13.

Hreinsdóttir EE, Stefánsdóttir G. Collaborative life history: different experiences of spending time in an institution in Iceland. Br J Learn Disabil. 2010;38(2):103–9.

Hutchinson A, Lovell A. Participatory action research: moving beyond the mental health 'service user' identity. J Psychiatr Ment Health Nurs. 2013;20(7):641–9.

Jivraj J, Sacrey LA, Newton A, Nicholas D, Zwaigenbaum L. Assessing the influence of researcher-partner involvement on the process and outcomes of participatory research in autism spectrum disorder and neurodevelopmental disorders: a scoping review. Autism. 2014;18(7):782–93.

Karban K, Paley C, Willcock K. Towards support: evaluating a move to independent living. Hous Care Support. 2013;16(2):85–94.

Kidd J, Edwards G. Doing it together: a story from the co-production field. Qual Res J. 2016;16(3):274–87.

Kramer J, Barth Y, Curtis K, Livingston K, O'Neil M, Smith Z, ... Wolfe A. Involving youth with disabilities in the development and evaluation of a new advocacy training: Project TEAM. Disabil Rehabil. 2013;35(7):614–22.

Letts L, Wilkins S, Law M, Stewart D, Bosch J, Westmorland M. Critical review form – qualitative studies (version 2.0). Hamilton: McMaster University; 2007.

Lewis A, Parsons S, Robertson C, Feiler A, Tarleton B, Watson D, ... Marvin C. Reference, or advisory, groups involving disabled people: reflections from three contrasting research projects. Br J Spec Educ. 2008;35(2):78–84.

Lincoln AK, Borg R, Delman J. Developing a community-based participatory research model to engage transition age youth using mental health service in research. Fam Community Health. 2015;38(1):87–97.

Linz S, Hanrahan NP, DeCesaris M, Petros R, Solomon P. Clinical use of an autovideography intervention to support recovery in individuals with severe mental illness. J Psychosoc Nurs Ment Health Serv. 2016;54(5):33–40.

Makdisi L, Blank A, Bryant W, Andrews C, Franco L, Parsonage J. Facilitators and barriers to living with psychosis: an exploratory collaborative study of the perspectives of mental health service users. Br J Occup Ther. 2013;76(9):418–26.

Martin JA. Research with adults with Asperger's syndrome – participatory or emancipatory research? Qual Soc Work. 2015;14(2):209–23.

McClimens A. This is my truth, tell me yours: exploring the internal tensions within collaborative learning disability research. Br J Learn Disabil. 2008;36(4):271–6.

Morgan MF, Cuskelly M, Moni KB. Unanticipated ethical issues in a participatory research project with individuals with intellectual disability. Disabil Soc. 2014;29(8):1305–18.

NHMRC. Methods for rating the quality of evidence. 2009. Retrieved from Canberra https://www.nhmrc.gov.au/guidelines-publications/information-guideline-developers/resources-guideline-developers

National Disability Insurance Scheme Act (2013). Australian Government.

Nicolaidis C, Raymaker D, McDonald K, Dern S, Ashkenazy E, Boisclair C, ... Baggs A. Collaboration strategies in nontraditional community-based participatory research partnerships: lessons from an academic-community partnership with autistic self-advocates. Prog Community Health Partnersh. 2011;5(2):143–50.

Nicolaidis C, Raymaker D, Katz M, Oschwald M, Goe R, Leotti S, ... The Partnering with People with Disabilities to Address Violence Consortium. Community-based participatory research to adapt health measures for use by people with developmental disabilities. Prog Community Health Partnersh. 2015;9(2):157–70.

Nilsen SE, Myrhaug TH, Johansen M, Oliver S, Oxman AD. Methods of consumer involvement in developing healthcare policy and research clinical practice guidelines and patient information material. Cochrane Database Syst Rev. 2013;(2). https://doi.org/10.1002/14651858.CD004563.pub2.

Nind M, Vinha H. Doing research inclusively: bridges to multiple possibilities in inclusive research. Br J Learn Disabil. 2014;42(2):102–9.

Nind M, Chapman R, Seale J, Tilley L. The conundrum of training and capacity building for people with learning disabilities doing research. J Appl Res Intellect Disabil. 2016;29(6):542–51.

Northway R, Melsome M, Flood S, Bennett D, Howarth J, Thomas B. How do people with intellectual disabilities view abuse and abusers? J Intellect Disabil. 2013;17(4):361–75.

O'Brien P, McConkey R, Garcia-Iriarte E. Co-researching with people who have intellectual disabilities: insights from a national survey. J Appl Res Intellect Disabil. 2014;27(1):65–75.

Ollerton J, Horsfall D. Rights to research: utilising the convention on the rights of persons with disabilities as an inclusive participatory action research tool. Disabil Soc. 2013;28(5):616–30.

Oschwald M, Leotti S, Raymaker D, Katz M, Goe R, Harviston M, ... Powers LE. Development of an audio-computer assisted self-interview to investigate violence and health in the lives of adults with developmental disabilities. Disabil Health J. 2014;7(3):292–301.

Rome A, Hardy J, Richardson J, Shenton F. Exploring transitions with disabled young people: our experiences, our rights and our views. Child Care Pract. 2015;21(3):287–94.

Sherwood-Johnson F, Cross B, Daniel B. The experience of being protected. J Adult Prot. 2013;15(3):115–26.

Stevenson M. Flexible and responsive research: developing rights-based emancipatory disability research methodology in collaboration with young adults with down syndrome. Aust Soc Work. 2010;63(1):35–50.

Strnadova I, Cumming TM, Knox M, Parmenter T. Building an inclusive research team: the importance of team building and skills training. J Appl Res Intellect Disabil. 2014;27(1):13–22.

Strnadová I, Cumming TM, Knox M, Parmenter TR, Lee HM. Perspectives on life, wellbeing, and ageing by older women with intellectual disability. J Intellect Dev Disabil. 2015;40(3):275–85.

Tilly L. Being researchers for the first time: reflections on the development of an inclusive research group. Br J Learn Disabil. 2015;43(2):121–7.

Tuffrey-Wijne I, Butler G. Co-researching with people with learning disabilities: an experience of involvement in qualitative data analysis. Health Expect. 2010;13(2):174–84.

Walmsley J, Central England People First History Project Team. Telling the history of self-advocacy: a challenge for inclusive research. J Appl Res Intellect Disabil. 2014;27(1):34–43.

White EL, Morgan MF. Yes! I am a researcher: the research story of a young adult with Down syndrome. Br J Learn Disabil. 2012;40(2):101–8.

Williams V, Ponting L, Ford K. A platform for change?: inclusive research about 'choice and control'. Br J Learn Disabil. 2015;43(2):106–13.

Woelders S, Abma T, Visser T, Schipper K. The power of difference in inclusive research. Disabil Soc. 2015;30(4):528–42.

# Understanding Sexuality and Disability: Using Interpretive Hermeneutic Phenomenological Approaches

112

Tinashe Dune and Elias Mpofu

Contents

1	Introduction	1954
2	Understanding Sexuality	1955
3	Sexuality and Disability	1956
4	Interpretive and Hermeneutic Phenomenology	1958
5	Interpreting Insider Perspectives	1958
6	Interpretive Hermeneutic Phenomenology Approach Procedures	1959
	6.1 Individuals Are Agents	1959
	6.2 Hone in on Participant Perspectives	1960
	6.3 Participate in the Participants' Daily Life	1961
	6.4 Appreciate that the Researched Are also Researchers	1963
7	Trustworthiness Within an Interpretive Hermeneutic Phenomenology Approach	1963
	7.1 Step 4a: Assessing Credibility	1964
	7.2 Step 4b: Assessing Transferability	1964
	7.3 Step 4c and 4d: Assessing Dependability and Confirmability	1966
8	Reflections on IHPA for Our Study and Its Findings	1968
9	Strengths and Limitations to Applying IHPA	1969
10	Conclusion and Future Directions	1970
References		1971

T. Dune (✉)
Western Sydney University, Sydney, NSW, Australia
e-mail: t.dune@westernsydney.edu.au

E. Mpofu
University of Sydney, Lidcombe, NSW, Australia

Educational Psychology and Inclusive Education, University of Johannesburg, Johannesburg, South Africa
e-mail: elias.mpofu@sydney.edu.au

© Springer Nature Singapore Pte Ltd. 2019
P. Liamputtong (ed.), *Handbook of Research Methods in Health Social Sciences*,
https://doi.org/10.1007/978-981-10-5251-4_130

Abstract

Disability, and those who live with disability, has been researched widely by scholars across a number of fields. However, there has been relatively little research on how people with cerebral palsy (CP) construct their own sexuality and the importance of the sexual scripts involved in this process. Given that sexuality is a fundamental human right with links to identity, health, and belonging, it is important for researchers in this area to engage deeply with understandings of how people with CP construct, understand, and experience their sexuality. This chapter introduces readers to researching constructions, understandings, and experiences of sexuality by applying an Interpretive Hermeneutic Phenomenological Approach (IHPA) with people with moderate to severe CP. It discusses the rational and processes for applying IHPA to engage participants in these sensitive and complex discussions on their lived experiences of understandings of sexuality. The chapter also provides procedural guidelines for applying IHPA to studying sexuality with CP in addition to the strengths and limitations of this approach. IHPA provides a unique advantage to studying heath issues with hidden populations or socially sensitive topics with the general population.

Keywords

Physical disability · Sexuality · Hermeneutic phenomenology · Qualitative methodology · Trustworthiness

1   Introduction

Alex (not real name) and I (Tinashe) were watching television in his relatively accessible, technologically-enhanced and fully mechanized dorm room. As we chatted over the buzzing of the television, our attention was suddenly drawn to a heated and passionate sex scene. The characters had only just met and seemed to be having the best sex of their lives. "Do people really have sex like that?" Alex asked. "I don't think so... No... Of course not," I replied. Alex and I were silenced as both characters simultaneously reached orgasm in a very complicated physical configuration. "Would you want to have sex like that?," I asked. "You must be joking" Alex replied. Alex, who has severe spastic quadriplegic cerebral palsy, explained to me that sex (for him) was about working with what you had. With a wink, he assured me that no one had ever complained.

Cerebral palsy is neurological condition characterized by poor muscle coordination in carrying out voluntary movements (ataxia), with stiff or tight muscles and exaggerated reflexes (spasticity) which is also associated with muscle tone that is too stiff or with loss of tone (Dune 2011). It also might show with additional symptoms like foot or leg dragging. People with CP tend to have a range of secondary conditions such as seizures, communication and hearing disorders, impaired vision, bladder and bowel control issues, pain and abnormal sensations, as well as mental health conditions such as depression and anxiety (Dune 2011). It affects 1 in 50 live births and presents at birth or soon after with prevalence at around 17 million people globally (Oskoui et al. 2013). With increasing medical support, therefore life expectancy more people with CP are living well into adulthood globally. Given that CP is becoming a prevalent health condition in adulthood the need to address personal and social care needs in this population including sexuality (Linton and Rueda 2015).

This chapter introduces readers to researching constructions, understandings, and experiences of sexuality by applying an Interpretive Hermeneutic Phenomenological Approach (IHPA) with people with moderate to severe CP. It discusses the rational and processes for applying IHPA to engage participants in these sensitive and complex discussions on their lived experiences of understandings of sexuality. The chapter also provides procedural guidelines for applying IHPA to studying sexuality with CP in addition to the strengths and limitations of this approach. IHPA provides a unique advantage to studying heath issues with hidden populations or socially sensitive topics with the general population.

2 Understanding Sexuality

Before researchers can engage with this complex topic with a marginalized population, like people with CP, a deeper understanding of sexuality is required. In order to further appreciate how sexuality may influence the human experience, it is important to note the pathways by which it is constructed. Drawing on theory by Simon and Gagnon (1986, 1987, 2003, 2011), it has been found that human sexuality is constructed via public, interactional, and private sexual scripts. This means that sexuality is expressed in all aspects of human life – what we see in the media, to how we interact with other, to what we see in our dreams.

For example, public sexual scripts are created, influenced, and reinforced by attitudes and interpretations presented in popular culture and media (Dune 2015). We absorb these through consuming and engaging with social media, magazines, movies, or television and the internet. Public events, like cultural or religious events, may impact the expression of sexuality as they expose people to images of what is sexually desirable and/or "appropriate." As such, people may be encouraged to engage in sexual activity only with those who are publically prescribed as appropriate (Simon and Gagnon 1986). These public scripts then influence the way in which we actually express our sexuality. Based on public scripts, we are influenced to engage in sexual interactions with people who are "sexy" and the media rarely portrays disability as desirable. These interactional events are characterized by the ways in which we negotiate intimate sexual relationships (i.e., flirting, courting, and dating). From these public and interactional scripts and experiences, we develop private inner dialogues about sexuality (Emerson 1983). These private mental processes influence the way individuals internalize sexual scripts and consolidate perceptions and constructions of sexuality – a mental picture of what is sexy, who is sexy, and what we should and should not be – that guides them in their understanding of their sexuality and that of others (Dune and Shuttleworth 2009). In this way, privatizations of sexual expectations, behavior, and constructions are bound to public and interactional social scripts (Dune 2014a). Figure 1

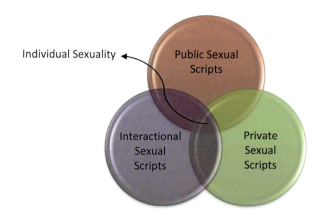

Fig. 1 Influence of public, interactional, and private sexual scripts on the construction of individual sexuality

illustrates that an individual's sexuality (in the center) is therefore a product of public, interactional, and private sexual scripts (as presented in Dune and Shuttleworth 2009).

3 Sexuality and Disability

The ways in which people think about disability and sexuality is linked to how it is publically represented (i.e., cultural norms), how people with disability experience interactional encounters (i.e., intimate relationships), and how they perceive of themselves as sexual beings (i.e., sexual self-concept) (Dune 2012a). Given that people with CP are not often present in popular media or represented as being desirable sexual partners, people may internalize the myth that people with CP are not sexual or not sexy enough to engage with (Linton and Rueda 2015). The invisibility of disability is important to expressions of sexuality as individuals may choose potential sexual partners based on their perceptions of what is socially acceptable.

Notably, research indicates that when people with a disability do engage in sexual relations, contradictions between their experiences and internalized scripts of idealized "appropriate" sexual relationships may occur. If social expectations of sexuality are internalized, then sexual difficulties may result. When expectations are not met, people may experience a negative impact on their sexuality. The influence of idealized sexual scripts is important to consider for people with CP because expectations of normative function, bodily movement, and the body beautiful can make access to sexual relationships difficult for many people in this group (Shuttleworth 2000, 2007).

To successfully engage in sexual activity, an intimate understanding of these domains in which sexuality is defined and appropriated is required. This is important to consider as sexuality as normal for people with CP as it is for their typically developing peers. However, there may be fewer opportunities for people with CP to engage with and express their sexuality due to physical and developmental restrictions which mean that normative movement and functioning is impaired (Dune 2015). For instance, a person with severe spastic CP may find it difficult to gently caress their partner's exposed skin with an open hand therefore changing the way foreplay is assumed to proceed. With little practice comes difficulty in securing opportunities to develop the psychosocial skills required to manage and sustain intimate sexual relationships (Esmail et al. 2010). For instance, being excluded from childhood games like spin-the-bottle or being asked out on a date can result in a limited ability to flirt, approach, consent (or not) potential sexual partners. Individuals, therefore, rely on public, interactional, and private sexual scripts to understanding and expressing their sexuality. However, these understandings and expressions are grounded in the practical realities of opportunity for sexual expression.

Although the consequences of exclusion from sexual opportunities are often perceived of as secondary to other health and medical concerns, being denied this integral aspect of health and well-being can lead to major mental health and physical health issues. As noted, mental concerns include depression, anxiety, and suicidal ideation (Jones et al. 2015). Exclusion from opportunities for sexual development including sexual education can result in people with CP having limited opportunities to practice consent and being susceptible to increased rates of sexually transmitted infections and unwanted pregnancy because of limited knowledge of sexual health and sexual well-being (Alvarelhão and Lopes 2016). Of course, the range of opportunities and/or exclusions exists on a spectrum with some people with severe forms of CP having many sexual opportunities while others with mild to moderate forms of CP having very few. As such, the sexual health and sexual well-being outcomes for a person with CP would logically depend on how opportunities for sexuality are interpreted by the individual in the context of present and evolving social situations. This would include how the individual interprets themselves in the context of public sexual scripts, access to interactional opportunities, and also the person's subjective sense of sexuality and sexual fulfilment.

Sexuality is, therefore, an interpretive experience (Edley 2001; Jackson and Scott 2007). As such, employing a hermeneutic phenomenological approach can assist researchers in gaining a better understanding of how people with CP interpret sexual scripts and construct their sexuality (Heidegger 1962; Husserl 1952, 1980). Research which engages hermeneutic phenomenology is interested in interpretive structures (e.g., sexual scripts and sexual constructions) of experience, how people engage with and understand these experiences in relation to ourselves and others (Ramberg 2008). With its foundations in philosophical studies, hermeneutic phenomenology is useful for examining how public, interactional, and private meanings are deposited and mediated through myth (i.e., myth of sexual spontaneity), religion (i.e., constructions of heterosexuality), art (i.e., erotica), and language (i. e., popular culture and interactional discourse). Ultimately, interpretive hermeneutic phenomenology aims to answer questions about how people interpret the meaning of being, the self and self-identity (Van Manen 2002). The following sections describe this approach and how it can support research related to sexuality with people with disabilities.

4 Interpretive and Hermeneutic Phenomenology

To fully appreciate and begin to understand the role of sexuality in the lives of people with disability, research investigations can benefit from inductively interpreting both the lived experience and the context of the participants and their social interactions (McKiernan and McCarthy 2010). For instance, Koch (1995) emphasized that our cultural and social history informs understandings of one's background or positionality in the world. Interpretive approaches, therefore, conceptualize people's experience as constructed by social and organizational realities (e.g., sexual scripts) (Heracleous 2006). According to Heracleous (2006), interpretive approaches to understanding the human experience acknowledge individuals, in this case research participants with CP, as actors with human agency. Hermeneutic phenomenology also emphasizes the importance of the individual's participation in his or her own creation (Laverty 2003). Heidegger (1962) believed that understanding is a basic form of human existence in that understanding not only helps us know the world, but also makes us the way we are (Polkinghorne 1983). This means that participants are agents who engage in their lives versus simply being along for the ride. This is not to say that life always happens according to their actions or plans but that they, and all people, influence and change their environments and experiences as a result of engaging with them and/or perceiving what those experiences may mean.

In this way, hermeneutic phenomenological inquiry helped to facilitate the aims of the research as it acknowledges that people and the world are indissolubly related through cultural, social, and historical contexts. From this standpoint, the ways in which the participants communicate about their sexuality-related experiences demonstrates how they interpret and shape their own or others' understandings (e.g., the researcher, the researched, sexual partners, friends and family and society) (see also ▶ Chap. 11, "Hermeneutics: A Boon for Cross-Disciplinary Research").

5 Interpreting Insider Perspectives

Qualitative methodology facilitated the application of interpretive hermeneutic phenomenology as it employs a naturalistic approach to comprehending phenomena in "real world settings [where] the researcher does not attempt to manipulate the phenomenon of interest" (Patton 2002, p. 39). Generally, qualitative methodology is broadly defined as "any kind of research that produces findings not arrived at by means of statistical procedures or other means of quantification" (Corbin and Strauss 1990, p. 17). Instead, it allows the "phenomenon of interest [to] unfold naturally" (Patton 2002, p. 39) and supports interpretations of insider perspectives (see also ▶ Chap. 63, "Mind Maps in Qualitative Research").

Insider perspectives refer to individual participant interpretations of a construct or phenomena (Liamputtong 2010; Silverman 2009). An insider perspective allows for the exploration of concepts from the participants own perspective(s). Rich descriptions result from qualitative exploration of insider perspectives (Liamputtong 2010). In turn, rich descriptions allow for interpretative conceptualizations important for comprehensive understandings (Mayoux 2006). In our research, we were able to engage with these perspective and in-depth descriptions through a series of interviews with participants. This helped us to understand the ways that people with CP construct and experience sexuality in light of normative cultural sexual expectations. The next section describes how we operationalized interpretive hermeneutic phenomenology using qualitative methodology in our research.

6 Interpretive Hermeneutic Phenomenology Approach Procedures

IHPA is a way of engaging in research with marginalized populations that requires researchers to develop projects supported by the following foundational tenets:

1. Individuals are agents
2. Hone in on participant perspectives
3. Participate in the participants' daily life
4. Appreciate that the researched are also researchers

6.1 Individuals Are Agents

This first step acknowledges that individuals, including those in marginalized identities, are agents within their social environments. This does not mean that participants control all aspects of their lives, but that they can and do interpret social phenomena in ways that change the nature of those phenomena. Research with this focus seeks to explore a social phenomenon through this agentic lens where participants' perspectives, voices, and interpretations lead the development of understanding and a conceptual framework for a given social phenomenon. This focus should be reflected in the research objectives and methodologies. In our research, this focus led us to center the perspectives of people with CP with the aim of developing a conceptual model based on their constructions of sexuality. An example of how research objectives and methodology can be used to acknowledge that *individuals are agents* is presented below.

Step 1: Example from *Making Sense of Sex with People with Cerebral Palsy*	
Research objectives	Research methodology
1. Use IHPA to understand people with cerebral palsy, construct the concept of sexual spontaneity, how they construct their individual sexuality, and what sexual scripts (public, interactional, private) are prioritized within that construction	Qualitative methodology aims to explore phenomena in its nuanced richness was appropriate for understanding *insider perspectives* of constructions of sexuality by people with cerebral palsy
2. Construct a conceptual model of sexuality which takes into account the interpretive hermeneutic phenomenological experience of people with cerebral palsy	In the context of IHPA, *qualitative methodology* enables access to social phenomena by prioritizing participants' interpretations of often subconscious experiences and personal learning journeys

6.2 Hone in on Participant Perspectives

The bulk of information available both academically and in popular culture presents a deficit model of disability and highlights the social and sexual limitations experienced by this group. While these limitations and experiences are important to recognize and explore IHPAs, focus on individuals as agents can help researchers to engage with participants' agentic constructions, interpretations, and behaviors. In doing so, researchers can understand more of what people with disabilities, and other marginalized identities, are thinking and doing to understand and manage (where necessary) a range of social phenomena. In our research, this allowed us to develop an in-depth interview guide that allowed participants to bring the researchers into their interpretive processes and behavioral responses to sexuality as they knew it. An example of the questions we piloted with participants demonstrates how data collection processes can *hone in on participant experiences* without focusing on impairment, disability, or deficits.

Step 2: Example from *Making Sense of Sex with People with Cerebral Palsy*	
Interview guide	Questions/discussion
Introduction	Some people would say that a person's sexuality is made up of a variety of different things. I'd like to ask you a few questions about some of the ways that people think about sex
Grand tour	Tell me about your views about sexuality
	Tell me what comes to mind
	There is no importance to what comes first
	(a) If participant generates concepts, probe concepts
	(b) If they feel differently ask them why
Private	Generally, sexuality and sexual activities are considered private things. What do you think about that?
	(a) If participant generates concepts, probe concepts
	(b) If they feel differently ask them why
Interactional	When people think about sex, they may often think about having sexual experiences with another person or people. What do you think that means?
	(a) If participant generates concepts, probe concepts
	(b) If they feel differently ask them why
Public	The media presents a lot of information about sex. How do you think that could influence sexuality?
	(a) If participant generates concepts, probe concepts
	(b) If they feel differently ask them why

Using this interview guide the researchers collected data with participants across two or three interactive sessions. The interviews were conducted at a time, place, and format (i.e., face-to-face, telephone, email) of the participant's choosing. During these sessions, the researcher engaged with participants in whatever they were doing at that time (e.g., shopping, eating at a restaurant, relaxing at home, watching TV, and so on) while simultaneously discussing the interview questions. For some participants who had communication difficulties or became stressed or fatigued, the interviews were extended over more than two sessions. Participants who responded via email participated in two or more sessions in order to allow the researcher to seek (via face-to-face, email, or telephone) clarification if needed. All oral interviews were audio-recorded digitally and then transcribed in full. In line with IHPA, the data were analyzed thematically (see Dune and Mpofu 2015) which provided a clearer picture of participants' daily life and lived experience.

6.3 Participate in the Participants' Daily Life

This can include collecting peer-reviewed and gray literature which describes the experiences of individuals like the participants. Perhaps more importantly, researchers should participate in disability awareness, social, and advocacy events. One does not have to be an advocate in these settings, but should authentically participate in as many social experiences, disability-focused or in the mainstream, related to the ways in which participants experience a range of aspects in their daily life. This should be done before, during, and after the research study. In the present study, this engagement facilitated the development, refinement, piloting, and assessing for trustworthiness (described in Step 4 below), analyzing the data and disseminating the findings in impactful and authentic ways. An example of how researchers can *participate in the participants' daily life* before, during, and after the research is presented below.

Step 3: Example from *Making Sense of Sex with People with Cerebral Palsy*			
Engagement	Before	During	After
Academic engagement	Systematized review and analysis of literature on disability and sexuality with particular focus on the public, interactional, and private sexual scripts	Writing on findings from literature review and analysis	Analyzing the findings in line with IHPA
	Exploring CP from medical, historical, and cultural perspectives to better understand how the condition and those who live with it have been constructed	Developing a theoretical perspective around sexual spontaneity with disability	Engaging with literature to compare and contrast the findings
		Piloting of the research with people with CP	Disseminating the findings in academic journals, books, book chapters, and conference presentations

(*continued*)

Step 3: Example from *Making Sense of Sex with People with Cerebral Palsy*

Engagement	Before	During	After
		Engaging with academic literature to scrutinize IHPA with other methodological frameworks	Cohosting events with people with CP to debate and discuss society's role in sexuality and disability
		Refining IHPA for the main study with people with CP	
		Collecting data in line with IHPA with people with CP	
Social/cultural engagement	Exploring popular and social media to examine content, messages, and constructions of sexuality and disability	*Authentically* developing friendships with people with disabilities through the research, other friends, social events, online, and employment opportunities	*Authentically* maintaining friendships with people with disabilities
	Authentically engaging socially with people with disabilities through friends, social events, online, and employment opportunities	Attending social events with research participants during data collection to better understand aspects of their life that they were willing to share	Hosting and participating in mainstream social events (physically and online) related to sexuality
	Participating in mainstream social events (physically and online) related to sexuality (e.g., *Sexpo*)	Participating in mainstream social events (physically and online) related to sexuality	Hosting and participating in disability-specific social events (physically and online) related to sexuality and disability awareness
	Participating in disability-specific social events (physically and online) related to sexuality and disability awareness (e.g., *#deliciouslydisabled*)	Participating in disability-specific social events (physically and online) related to sexuality and disability awareness	
Environmental/ structural engagement	Trying to understand the lived experience of disability and sexuality through role plays with key informants			
	Spending time in a wheelchair and/or navigating infrastructure without using routes employed by typically bodied people			
	When meeting with a friend, student, or colleague with a disability ensuring that the meeting location is accessible (including eating areas and toilets)			

(*continued*)

Step 3: Example from *Making Sense of Sex with People with Cerebral Palsy*

Engagement	Before	During	After
Political/ economic engagement	Becoming familiar with the political debates in the area of disability (and sexuality) by attending stakeholder meetings and workshops	Giving participants space to discuss and demonstrate the role of political and economic structures in their lives and sexualities	Using the research findings to support and advocate for changes in portrayals of disability in popular culture, policy, and legislative documents
	Interrogating the role of political and economic structures in the health and well-being of people with disabilities by reading/engaging with policy and legislation		Revising policy documents in line with IHPA outcomes from the research

6.4 Appreciate that the Researched Are also Researchers

Given that participants are agents in their social environments, their role as co-researchers is important to IHPA. This entails allowing participants to contribute to the flow and outcomes of the research. In line with qualitative methodology, this can be done by ensuring the trustworthiness of research and the data emerging from in-depth insider perspectives. Using the current study as an example, the following section will describe how to engage participants towards assessing and ensuring trustworthiness.

7 Trustworthiness Within an Interpretive Hermeneutic Phenomenology Approach

As noted in Dune and Mpofu (2015), trustworthiness is a factor which every qualitative researcher should concern themselves with while designing a study, analyzing results, and judging the quality of the study. In qualitative paradigms, the term "trustworthiness" encompasses "credibility," "transferability," "dependability," and "confirmability" (Dune and Mpofu 2015). Credibility is the evaluation of how well the study's findings represent a sound conceptual interpretation of the data which comes from participant data. Transferability is the potential of the study's findings to transfer to other settings. Dependability is the assessment of the quality of the collective processes which include data collection, analysis, and theory production. Finally, confirmability is a gauge of the level to which the study's findings are supported by the collected data. These qualities are important in qualitative inquiry as they affirm the value and utility of one's research findings (Dune 2011). In assessing trustworthiness, researchers using IHPA can assure readers that the findings are accurate presentations of participants' interpretive experiences and constructions that are worthy of attention. To assess trustworthiness within our study, we scrutinized the pilot interview guide (described in *Step 2: Hone in on participant experiences*) and processes to ensure they were well aligned with an IHPA. This process as it applies to IHPA is described below.

7.1   Step 4a: Assessing Credibility

Our study was premised upon the assumption that people with CP, in their general understanding of sexuality, would describe it in terms that parallel or mirror private, interactional, and public sexual constructs. As such, a goal of the pilot study was to determine the credibility of these constructs. To tackle the issue of credibility, "member checking" (Liamputtong 2010; Lincoln and Guba 1985) with three key informants was conducted. In the process of member checking, each of the research participants was asked questions via telephone (see Step 2 above) to ascertain whether or not the use of private, interactional, and public conceptualizations of sexuality were credible constructs for this study. In order to ensure interpretive accuracy of the emerging data, both the participants and the researchers identified themes related to sexual scripts within the transcribed interviews. This supported an appreciation of the researched as researchers.

All participants made comments that directly connected private, interactional, and public sexual constructs to one or more personal experiences they had in the past (see Table 1). For example, participants felt that sexuality was made up of many different factors. They further confirmed that while spaces for experiences of sexuality are commonly considered "private" (i.e., in the privacy of one's home or "behind closed doors") the initiation of sexual encounters began in their minds. In addition, participants felt that a satisfactory sex life meant being able to share ones' sexuality with others. This included exploring sexual options and activities with different people and settings in order to discover what they really wanted from intimate relationships. Participants also mentioned that they had fantasized about a celebrity or a friend who resembled a media personality when they engaged in sexual activities with themselves and sometimes with others. The participants explained that the media tells people how to deal with sex as well as who and what is sexy. This feedback suggests that private, public, and interactional sexual constructs may be natural to the discourse of people with cerebral palsy in understanding their sexuality.

7.2   Step 4b: Assessing Transferability

Transferability pertains to the potential for a study's findings to transfer to other settings and/or populations. Transferability was achieved as the findings from the pilot study mirror those of similar nature in other studies (McCabe and Taleporos 2003; Taleporos and McCabe 2001, 2002a, b, 2003) which explored constructions of sexuality (i.e., sexual esteem, sexual satisfaction, and sexual ideation) by and within

Table 1 IHPA example of key informant themes related to private, interactional, and public constructs of sexuality

Key informant	Private construct of sexuality	Interactional construct of sexuality	Public construct of sexuality
1	"…sex, like intercourse, isn't something that I'd do with anybody, anywhere…I mean, like I'd do it at home. Like, it doesn't have to be in the bedroom or in the bed for that matter but in my own space…I'd feel more comfortable"	"Sex is intimate, like not just with yourself, I mean it is in that respect, but I mean like it is connecting…connective…with someone else…Not that having 'sex' with yourself isn't fun or satisfying but doing it with someone else is even better. Like, it reinforces that you are sexy, deserve sex or like you are what or who someone thinks is worthy of 'gettin' jiggy with'"	"I mean who doesn't like a good looking body or person. I mean there are so many good looking people, like sometimes it doesn't have anything to do with disability or anything but when they have the features you know? Those things that make someone sexy…I mean it's hard I'm sure for anyone to resist that"
2	"I mean look, I don't necessarily 'get it on' on a daily basis, physically I mean, but I don't need that either…Look, sometimes it's nice just to think about it, fantasize without anything physical happening. It gets you going, you know?"	"Part of me, my sexuality I mean, is made up of trying, the ability, the desire you know, to try out different things and different people. Like you don't have to have 'sex-sex' but being with different people sensually kind of is like exploring my sexual side…figuring it out you know?"	"Jennifer Hawkins, that's right everybody knows who she is…But I guess I'm a red-blooded male. Any pretty face that smiles at you…Particularly if they're long legged and absolutely gorgeous. You know Donald Trump even seems to think that the sunlight shines out of her every orifice, which it does. And look, to add to the mix, with someone like that, it's really the first time you fall in love although you don't really know it"
3	"I feel like, when I'm having fun, like having sexual type fun with myself or someone else it's just more fun when you think of something or the other person as everything you want…why not right? I mean it's my head, I can think of whatever I want"	"The best part of sex, or doing sexy stuff, with other people is the look on their face. I mean like, you know that look when you know someone is having a good time? Like for me, I suspect for everyone that is like the goal, like to please, satisfy the person you're doing sexy stuff with"	"I mean if you have experiences that aren't what people think is right then that can put a damper on things a bit…I mean it's ruined by what other people think is supposed to happen for men or women or whatever…I mean in that respect things can get confusing because you not doing what 'you're supposed to' even though it makes you happy"

people with cerebral palsy and other physical disabilities. Further, public, interactional, and private constructions of sexuality have been indicated in other relevant research (Shakespeare 1999, 2006, 2013) as salient to experiences of sexuality with disability.

Although this connection was drawn by the researchers, it may be irrelevant if people with CP do not interpret or perceive of the findings in that way. Given that IHPA is concerned with participants' agentic interpretations of sexuality in the context of disability, the key informants (pilot study) and additional participants (main study) were asked if the findings collated by the researchers were in fact transferable to other people with CP that they knew, heard, or read about. All participants agreed that the findings provided a variety of views and experiences related to sexuality they were representative of the experiences of a range of people with CP with varying developmental experiences and/or qualities (Dune 2011, 2012a, 2014a; Dune and Mpofu 2015; Dune 2012b, 2013). Of particular importance to IHPA was participants' affirmation that the findings portrayed people with CP as agents of their own sexuality and sexual experiences. This support of the research's transferability reinforced Steps 1 and 2 of IHPA as described above.

Transferability of this study's protocol was also achieved through making these documents available upon request. While IHPA is focused primarily on participants, a clear description, with examples, of its processes can assist other researchers in their development of IHPA-focused research. As such, providing evidence of the analytic process allows other researchers to repeat, as closely as possible the procedures of this project. In addition, other researchers can utilize IHPA procedures to analyze data from research which explores similar concepts. In this way, transferability can be further accessed and assessed by other research and researchers.

7.3 Step 4c and 4d: Assessing Dependability and Confirmability

Within IHPA, reviewing the dependability of one's research requires assessment of the quality of data collection, analysis, and theory production (Dune and Mpofu 2015). Confirmability relies on dependability and measures the level to which the study's findings are supported by the collected data (Dune and Mpofu 2015). To enhance the dependability and confirmability of the study, an independent audit of the research by a competent peer, familiar with the experiences of sexuality with disability, was completed. This required the independent auditor (who played no part in the study) to review all documents related to the research project, the guidelines and application of IHPA, the study findings, and conclusions (Liamputtong 2013).

The auditor for this project is a practicing occupational therapist with 10 years of experience and has created many training and "how-to" resources to support people with disabilities in enjoying their sexuality. At the time this study was being conducted, she was also conducting research using qualitative research methodologies but had not used hermeneutic phenomenology or the IHPA model developed from our research.

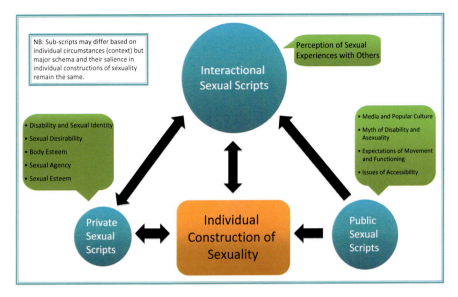

Fig. 2 Model for the construction of individual sexuality in people with cerebral palsy (Dune 2013)

The independent auditor then made an assessment of whether or not the study and its conclusions followed a logical progression through the chain of evidence. Part of this chain may include initial notes on the research question(s), interview schedule, digital audio files, pilot data transcripts, thematic coding and analyses, draft reports and final. To assess the links between the findings and conclusions of the research, the auditor reviewed the emergent conceptual model that depicts the role of public, interactional, and private sexual constructs in the lives of people with CP (see Fig. 2).

The auditors assessment resulted in a written report supporting the links between the research, IHPA, the research findings, and conclusions (full letter available upon request) (Dune and Mpofu 2015). An excerpt of this report is provided below.

Step 4c and 4d: Example from *Making Sense of Sex with People with Cerebral Palsy*		
Independent auditor comments		
Research design	**Data analysis**	**IHPA related**
It is the auditor's opinion that the focus of the study remained consistent with the proposed focus. Ms Dune did, however, slightly reframe her research objectives and questions as it became apparent to her that she was not adequately addressing constructions of sexuality by people with Cerebral Palsy. This revision is to be expected	During the audit meeting with Ms. Dune, I noted the accuracy of the transcription of interviews. I listened to four digital files; two of which were transcribed by Ms. Dune and two which were transcribed by RapidType transcription services. I then noted upon the transcriptions inaccuracies in the text. I listened to at least five pages	It appears that the trustworthiness of the study can be established in that the findings seem to be clearly grounded in the data. The researcher carefully designed her project and employed a number of verification strategies (including the assessment of credibility, transferability, dependability and confirmability) to ensure

(continued)

since qualitative research is an emerging process and initially research questions are tentatively posed. In addition, data collection and verification strategies followed those proposed in the methods section of the proposal. Data analysis procedures slightly changed in that Ms Dune also utilized a manual line-by-line analysis instead of using Nvivo as initially planned. She was also much clearer at the end of the process than at the proposal stage regarding how she approached data analysis. Initially, I had some difficulty making sense of her notes about the evolution of her dissertation included in the audit trail and her manual colored, cut and paste and Nvivo coding system. However, when Ms Dune and I met she walked me through the process and materials pertaining to analysis and specifically outlined the steps she used. The suggestions I made about increasing clarity and brevity within the dissertation were attended to	of each interview (a total of 22 pages) and noted as few as three errors and as many as 20 errors per 5–6 page set. Overall, transcription was exceptionally accurate. The inaccuracies identified were very minor, including single words that were omitted or mistranscribed and/or short phrases that were omitted. In my estimation, the minor inaccuracies did not impact the overall content of the transcriptions	the accuracy of the data. The data were presented in detail in chapter 7 using extensive informants' quotes and descriptive language. Having reviewed all the materials provided to me it appears that the data accurately represents informants' perspectives. The conclusions flow logically from the data presented in results. Comparing those chapters to transcripts, documents, and analytic methods, the conclusions of the study seem warranted

8   Reflections on IHPA for Our Study and Its Findings

The findings from our research acknowledge that people with CP (and potentially other disabilities) are at risk for stigmatization and exclusion from social activities and sexual participation. This is because normative sexual scripts construct people with disabilities as undesirable and/or asexual (Shuttleworth 2000). This perspective may make it difficult for people with disabilities to fulfil the requirements of normative functioning, hegemonic sexual expectations, and romantic expectations.

The current deficit model of disability often reiterated in academic outlets, and popular media ignores how people with disabilities conceptualize sexuality generally and construct their sexuality specifically. IHPA does, however, acknowledge that sexual expectations and their role in an individual's sexual expressions and activities are intrinsically linked to one's social environment. Even so, our research highlights that this approach allows researchers and their participants to engage with people with disabilities as sexual beings and agents who construct sexuality through public, interactional, and private scripts. IHPA allowed for both the researchers and the participants to determine which social, sexual, and cultural factors influence individual constructions of sexuality and served as a building block towards understanding sexuality as something humans actively engage with and influence rather than simply being influenced by constructions of it.

9 Strengths and Limitations to Applying IHPA

Employing IHPA to research on sensitive and complex issues, like sexuality, with people with CP supports the role of participants as agents of social phenomena, research processes, and research outcomes. However, there are methodological challenges which researchers should be aware of when attempting to engage people with disabilities using IHPA. These are described and addressed below.

First, IHPA is best applied to marginalized or hidden population. Such populations can be characterized as those "involved in activities that are considered deviant, such as drug taking, or they may be vulnerable, such as the stigmatised in society, making them reluctant to take part in more formalised studies using traditional research methods" (Atkinson and Flint 2001, p. 3). People with CP, or other disabilities, are often considered a hidden population as they are not only stigmatized but may also require daily and intimate care from a range of providers. These providers, like disability support organizations or other carers, may be the first point of contact between people with disabilities and other parts of the community (Dune 2014b). Providers can, therefore, act as gatekeepers who seek to ensure the health, safety, and well-being of people with disabilities (Dune 2015; Shuttleworth et al. 2010). If the gatekeeper deems the research and its objectives to be inappropriate, they may discourage people with disabilities from participating. Simultaneously, it may be the carers or gatekeepers who encourage people with disabilities to participate in research to engage them in the issues and discussions outside of their regular circle of contact (Dune 2011, 2012b). In line with IHPA, it is paramount that researchers develop deep and nonopportunistic trust with gatekeepers, disability organizations, and people with disabilities (Dune 2011). This will assist in developing a collaborative research relationship with participants and opportunities to maximize recruitment capability. Trust can also be developed through snowballing where referrals are made by acquaintances or peers of those who have participated in the research rather than other more formal methods of identification (Dune 2011). As noted in the IHPA procedures, establishing connections with disability networks through piloting of your recruitment strategies and data collection methods can provide important opportunities for member-checking which assists in establishing the trustworthiness of phenomenological qualitative research and the maintenance of participant engagement over the course of the project.

Second, given that some people with disabilities are characterized as existing within hidden populations the trust required for recruitment and data collection may take the researcher and participant into a gray area of personal and professional space. For instance, when building rapport, the researcher is required to give part of their personal selves within the professional space of the research environment (Emmel et al. 2007). This dissolution of the researcher-participant binary allows the participant to provide robust data and express themselves with little social filtering. Given that IHPA encourages researchers to enter and navigate this gray space, ways for managing these relationships is necessary. As such, this gray space may present issues for both parties when distressing or traumatic experiences are divulged and/or if the researcher-participant relationship ends (Eide and Kahn 2008). In such situations, the line between professional and personal interaction may need to be drawn, bent, or repositioned in order for both parties to process distressing information and/or adapt to the end of the relationship (Robards 2013).

Finally, the analysis and interpretation of qualitative data emerging IHPA for exploring social phenomena with marginalized population should be in line with the principles of trustworthiness. As mentioned, the term "trustworthiness" encompasses "credibility," "transferability," "dependability," and "confirmability." To ensure credibility, the study's findings should represent a sound conceptual interpretation of the data which comes from participant data. Transferability is the potential of the study's findings to transfer to other settings. Dependability is the assessment of the quality of the collective processes which include data collection, analysis, and theory production. Finally, confirmability is a gauge of the level to which the study's findings are supported by the collected data. These qualities are important in qualitative inquiry as they affirm the value and utility of one's research findings (Liamputtong 2013). Failure to engage with these principles when using IHPA for qualitative research reduces it rigor and contextual applicability.

10 Conclusion and Future Directions

This chapter discussed IHPA and its application to research on the lived experiences of people with cerebral palsy and their constructions of sexuality. It is a framework which values participants' perspectives and their agency in both their lives and their ability to articulate and explain their experiences. Given that sexuality is a fundamental human right with links to identity, health, and belonging, it is important for researchers in this area to engage deeply with understandings of how people with disabilities construct, understand, and experience their sexuality. To do this in ways which acknowledge, empower, and value the voices of people with disabilities, better links with this hidden, and stigmatized, population need to be encouraged. Further, sexual and social scripts must be reconstructed to include those who do not seem to meet unattainable expectations. Within this larger call for social change, researchers interested in the lives of people with disabilities should prioritize their voices in the development, engagement, and dissemination of research on their lived experiences – a task which is well supported by an interpretive hermeneutic phenomenological approach.

References

Alvarelhão J, Lopes D. A Guttman scale to assess knowledge about sexually transmitted diseases in adults with cerebral palsy. Sex Disabil. 2016;34(4):485–493.

Atkinson R, Flint J. Accessing hidden and hard-to-reach populations: snowball research strategies. Soc Res Updat. 2001;33(1):1–4.

Corbin J, Strauss A. Grounded theory research: procedures, canons and evaluative criteria. Z Soziol. 1990;19(6):418–27.

Dune T. Making sense of sex with people with cerebral palsy. Unpublished PhD thesis, University of Sydney, Sydney. 2011.

Dune T. Understanding experiences of sexuality with cerebral palsy through sexual script theory. Int J Soc Sci Stud. 2012a;1(1):1–12.

Dune TM. Sexuality and physical disability: exploring the barriers and solutions in healthcare. Sex Disabil. 2012b;30(2):247–55. https://doi.org/10.1007/s11195-012-9262-8.

Dune TM. Re/developing models for understanding sexuality with disability within rehabilitation counselling. Electron J Hum Sex. 2013;16(April). http://www.ejhs.org/volume16/Disability.html.

Dune T. Conceptualizing sex with cerebral palsy: a phenomenological exploration of private constructions of sexuality using sexual script theory. Int J Soc Sci Stud. 2014a;2(2):20–40.

Dune T. Sexuality in chronic illness and disability. In: Johnson A, Chang E, editors. Chronic illness and disability: principles for nursing practice. 2nd ed. Milton: Wiley; 2014b. p. 115–32.

Dune T. "You just don't see us": the influence of public schema on constructions of sexuality by people with cerebral palsy. In: Romaniuk SN, Marlin M, editors. Development and the politics of human rights. Roca Raton: CRC Press – Taylor and Francis Group; 2015. p. 223–48.

Dune T, Mpofu E. Evaluating person-oriented measures to understand sexuality with cerebral palsy: procedures and applications. Int J Soc Sci Stud. 2015;3(4):145–56.

Dune TM, Shuttleworth RP. "It's just supposed to happen": the myth of sexual spontaneity and the sexually marginalized. Sex Disabil. 2009;27(2):97–108. https://doi.org/10.1007/s11195-009-9119-y.

Edley N. Analysing masculinity: interpretive repertoire, subject positions and ideological dilemmas. Discourses as data: a guide for analysis. London: Sage; 2001.

Eide P, Kahn D. Ethical issues in the qualitative researcher – participant relationship. Nurs Ethics. 2008;15(2):199–207.

Emerson C. The outer word and inner speech: Bakhtin, Vygotsky, and the internalization of language. Crit Inq. 1983;10(2):245–64.

Emmel N, Hughes K, Greenhalgh J, Sales A. Accessing socially excluded people – trust and the gatekeeper in the researcher-participant relationship. Sociol Res Online. 2007;12(2). https://doi.org/10.5153/sro.1512., http://www.socresonline.org.uk/12/2/emmel.html.

Esmail S, Huang J, Lee I, Maruska T. Couple's experiences when men are diagnosed with multiple sclerosis in the context of their sexual relationship. Sex Disabil. 2010;28(1):15–27.

Heidegger M. Being and time (trans: Macquarrie J, Robinson E). New York: Harper & Row; 1962.

Heracleous L. Discourse, interpretation, organization. Cambridge: Cambridge University Press; 2006.

Husserl E. Ideas pertaining to a pure phenomenology and to a phenomenological philosophy. Second Book. 1952.

Husserl E. Phantasie, bildbewusstsein, erinnerung zur phänomenologie der anschaulichen vergegenwärtigungen. 1980.

Jackson S, Scott S. Faking like a woman? Towards an interpretive theorization of sexual pleasure. Body Soc. 2007;13(2):95–116.

Jones K, Wilson B, Weedon D, Bilder D. Care of adults with intellectual and developmental disabilities: cerebral palsy. FP Essent. 2015;439:26–30.

Koch T. Interpretive approaches in nursing research: The influence of Husserl and Heidegger. J Adv Nursing. 1995;21(5):827–836.

Laverty SM. Hermeneutic phenomenology and phenomenology: A comparison of historical and methodological considerations. International journal of qualitative methods. 2003;2(3):21–35.

Liamputtong P. Performing qualitative cross-cultural research. Cambridge: Cambridge University Press; 2010.

Liamputtong P. Qualitative research methods. 4th ed. Melbourne: Oxford University Press; 2013.

Lincoln YS, Guba EG. Naturalistic inquiry, vol. 75. New York: Sage; 1985.

Linton KF, Rueda HA. Dating and sexuality among minority adolescents with disabilities: an application of sociocultural theory. J Hum Behav Soc Environ. 2015;25(2):77–89.

Mayoux L. Quantitative, qualitative or participatory? Which method, for what and when. Doing Dev Res. 2006:115–29.

McCabe MP, Taleporos G. Sexual esteem, sexual satisfaction, and sexual behavior among people with physical disability. Arch Sex Behav. 2003;32(4):359–69.

McKiernan M, McCarthy G. Family members' lived experience in the intensive care unit: a phemenological study. Intensive Crit Care Nurs. 2010;26(5):254–61.

Oskoui M, Coutinho F, Dykeman J, Jetté N, Pringsheim T. An update on the prevalence of cerebral palsy: a systematic review and meta-analysis. Dev Med Child Neurol. 2013;55(6):509–19.

Patton MQ. Two decades of developments in qualitative inquiry: a personal, experiential perspective. Qual Soc Work. 2002;1(3):261–83.

Polkinghorne D. Methodology for the human sciences: systems of inquiry. New York: Suny Press; 1983.

Ramberg B. Hermeneutics. In: The Stanford encyclopedia of philosophy. Stanford, California: Center for the Study of Language and Information (CSLI), Stanford University, Stanford, California; 2008. http://plato.stanford.edu/archives/sum2008/entries/hermeneutics.

Robards B. Friending participants: managing the researcher–participant relationship on social network sites. Young. 2013;21(3):217–35.

Shakespeare T. 'Losing the plot'? Medical and activist discourses of contemporary genetics and disability. Sociol Health Illn. 1999;21(5):669–88.

Shakespeare T. The social model of disability. Disabil Stud Read. 2006;2:197–204.

Shakespeare T. Disability rights and wrongs revisited. New York: Routledge; 2013.

Shuttleworth RP. The search for sexual intimacy for men with cerebral palsy. Sex Disabil. 2000;18(4):263–82.

Shuttleworth R. Critical research and policy debates in disability and sexuality studies. Sex Res Soc Policy. 2007;4(1):1–14.

Shuttleworth R, Russell C, Weerakoon P, Dune T. Sexuality in residential aged care: a survey of perceptions and policies in Australian nursing homes. Sex Disabil. 2010;28(3):187–94. https://doi.org/10.1007/s11195-010-9164-6.

Silverman D. Doing qualitative research. New York: Sage; 2009.

Simon W, Gagnon JH. Sexual scripts: permanence and change. Arch Sex Behav. 1986;15(2):97–120.

Simon W, Gagnon JH. A sexual scripts approach. In: Geer JH, O'Donohue WT, editors. Theories of human sexuality. New York: Plenum; 1987. p. 363–83.

Simon W, Gagnon JH. Sexual scripts: origins, influences and changes. Qual Sociol. 2003;26(4):491–7.

Simon W, Gagnon JH. Sexual conduct: the social sources of human sexuality. Piscataway: Transaction Publishers; 2011.

Taleporos G, McCabe MP. Physical disability and sexual esteem. Sex Disabil. 2001;19(2):131–48.

Taleporos G, McCabe MP. Body image and physical disability – personal perspectives. Soc Sci Med. 2002a;54(6):971–80.

Taleporos G, McCabe MP. The impact of sexual esteem, body esteem, and sexual satisfaction on psychological well-being in people with physical disability. Sex Disabil. 2002b;20(3):177–83.

Taleporos G, McCabe MP. Relationships, sexuality and adjustment among people with physical disability. Sex Relatsh Ther. 2003;18(1):25–43.

Van Manen M. Researching the experience of pedagogy. Educ Can. 2002;42(4):24–39.

Ethics and Practice of Research with People Who Use Drugs

113

Julaine Allan

Contents

1	Introduction	1974
2	Ethical Conduct of Research	1974
3	Recruiting Participants	1975
4	Anonymity	1978
5	Surveys Versus Interviews	1979
6	Intoxication	1981
7	Dealing with Distress	1981
8	Research Participants in Substance Treatment Settings	1983
9	Conclusion and Future Directions	1986
References		1986

Abstract

Global harm-reduction strategies aim to prevent or reduce the severity of problems associated with nonmedical use of dependence-causing drugs including alcohol. However, harm reduction strategies have to fit the personal, social, and environmental context of people using drugs to be effective. The best way to develop strategies that fit is to research and understand drug use practices including how, why, and when drugs are used. This chapter discusses a number of ethical and practical factors to consider when planning and conducting research with people who use drugs. Data collection challenges include recruitment of a marginalized and hidden population, gaining consent, ensuring anonymity and responding to harm and distress. Examples are drawn from the author's research on alcohol and other drug use in rural Australian settings including farming and fishing workplaces, on illicit fentanyl use, and with people in treatment.

J. Allan (✉)
Lyndon, Orange, NSW, Australia
e-mail: jallan@lyndon.org.au

© Springer Nature Singapore Pte Ltd. 2019
P. Liamputtong (ed.), *Handbook of Research Methods in Health Social Sciences*,
https://doi.org/10.1007/978-981-10-5251-4_143

Keywords

Drug use · Data collection · Ethics · Intoxication · Harm reduction · Cognitive impairment

1 Introduction

Research on people's nonmedical use of drugs provides vital information for healthcare policy and practice. Problem drug use, particularly alcohol, is one of the most significant health-care challenges globally. Alcohol causes 3.3 million deaths annually, 15.3 million people have drug use disorders, and most countries are experiencing increasing deaths from injecting drug use as well as a burden of disease from injecting-related blood borne viruses (World Health Organization 2016).

The principle of harm reduction was proposed by the World Health Organization in 1973 as an alternative to drug control, which had proven unsuccessful in preventing illicit drug use (Ball 2007). The twentieth World Health Organization (WHO) Expert Committee on Drug Dependence defined harm reduction as preventing or reducing the severity of problems associated with the nonmedical use of dependence-producing drugs (WHO 1974). Harm reduction approaches expanded in the 1980s with attempts to control the HIV-AIDS epidemic particularly within injecting drug user populations (Ball 2007). Understanding why, when, how, and what drugs people use is fundamental to developing strategies to reduce harms.

Research with people who use drugs is the best way to develop understanding of their personal, social, cultural, and environmental context so harm reduction strategies can be effectively targeted (Ruefli and Rogers 2004). However, data collection from a stigmatized, possibly vulnerable, and potentially intoxicated population who may be engaged in illegal activity poses practical challenges. Research should be approved by a properly constituted human research ethics committee before commencing but most committees do not provide detailed advice on ways to be ethical. This chapter examines ethical data collection processes and suggests some strategies that can be used to conduct research with people who use drugs.

The ways we understand people's drug use experiences and preferences has shaped the ways drug and alcohol treatment and harm reduction is provided. This is particularly so at Lyndon, the not-for-profit drug treatment, research and training organization conducting the studies described in this chapter.

2 Ethical Conduct of Research

Research ethics aim to safeguard study participants rights by ensuring they understand what they are being asked to do, and what might happen to them as a result of participating in a specific study (Aldridge and Charles 2008; see also ► Chap. 106, "Ethics and Research with Indigenous Peoples"). Typically, the details of a study are set out in an information sheet that is given to a potential participant or details appear on a screen before information can be collected electronically. Once the study has been explained, people are asked to consent by signing a form, checking a box on a computer screen or by recording their verbal agreement on a digital recorder.

The consent process is straightforward, but the ways you get the information sheet in front of someone assess their capacity to understand it, predict potential harms, and can be challenging. These are the things that have to be explained in an ethics application. For example, in a study examining the impact of different types of heroin on injecting practices, the research team asked for permission to watch the study participant preparing and injecting heroin, recorded the injection on video, interviewed the participant, and then pH tested any drug residue to assess its acidity (Ciccarone and Harris 2015). Numerous ethical issues are highlighted in the research topic and the methods of data collection in the heroin study.

Ethical approval for the study is explained in one of the resulting publications by the following statement;

> Approval for this research was obtained by the institutional review boards of both University of California San Francisco and London School of Hygiene and Tropical Medicine. Given the sensitive nature of the research and vulnerability of the participants, a waiver of written informed consent was obtained. Participants verbally consented to each part of the study separately and could refuse any part. No identifiers were obtained so anonymity could be protected. In the US, a Federal Certificate of Confidentiality was obtained from the National Institutes of Health. (Ciccarone and Harris 2015, p. 1104)

The statement indicates the complexity of approval processes across two countries and a number of ethical review groups including the need for some legal protection for the research team in the form of a government certificate of confidentiality that allows them to refuse to disclose participant's names or identifying details to law enforcement authorities (e.g., https://humansubjects.nih.gov/coc/back ground). But, the example does not explain how you might go about conducting research in such confronting circumstances. Practical strategies for participant recruitment, anonymity, state of intoxication, and engagement in treatment or health care are critical to conduct ethical research with people who use drugs.

3   Recruiting Participants

Prior to asking for consent, potential participants fitting the study criteria have to be found. Studies of recreational drug use by young people have recruited participants by setting up booths at music festivals (Lim 2013) and advertising research on social media sites (Butler et al. 2017). Identifying the most likely sites for contact is critical in participant recruitment (see also ▶ Chap. 5, "Recruitment of Research Participants").

One of the easiest ways to find people who need health care related to drug use is to ask health workers to distribute information about the study or to refer their eligible patients or clients to the researcher. The first challenge to this strategy is health workers perceiving they need to protect people they have a professional duty of care to. While a reasonable concern, it can also mean the health worker is making a judgment about a person's capacity to participate and/or the relevance of the study.

People in positions of power such as health workers can be effective gatekeepers, preventing research participation because of the ostensibly good motive of protecting someone from harm. Gatekeeping has been a significant barrier to the participation in research with people who have intellectual disabilities (McDonald and Keys 2008; Lotan and Ells 2010), mental illness (Welie and Berghmans 2006), and dementia (Olde Rikkert et al. 1995). The consumer advocacy movement in mental health and some substance use settings has resulted in consumer researchers influencing the ways research is conducted and the consumer's participation in all facets of the research process (Woltmann and Whitley 2010; NAOMI Patients Association and Boyd 2012). However, not all projects include a consumer advocate.

Consumer advocacy is uncommon in Australian drug and alcohol services or for people in community health settings who have problem drug use. In my study of perceptions and experiences of health and health care from socioeconomically disadvantaged rural residents (Allan et al. 2010), the participant recruitment strategy primarily involved distribution of flyers with project information and contact numbers of the researchers in the doctor's practice center, community health center, and drug treatment center in a small rural town. When few people were contacting the researchers, health workers were contacted to ask if they had distributed the information or put up the flyers. We found that health workers were reluctant to pass the information on to potential participants because of concern about the client's fragile mental or emotional state and the harm the research interview could cause. They were also concerned that clients may report a negative picture of the health worker's service delivery.

There were two strategies developed to promote the research project and reduce gatekeeping. The first was meetings with the health workers to discuss the project and answer any questions about potential harms to participants or to the reputation of the health service. It became clear that while health workers were concerned about harm to their patients, in the isolated rural location, they perceived it as likely to be their responsibility to address any distress the research caused. Key information subsequently provided by the researchers included ways that would be responded to by connection to telephone support and others outside of the town. Providing information about how many people had become distressed and required support in previous studies reassured them that it was unlikely but could be managed if it occurred. Describing ways by which the location of the health service and the town would be kept out of publications of the results was critical in allaying concerns about their own or their employer's reputation as a health-care provider.

The second strategy to recruit hard–to-reach participants was the use of snowball sampling or respondent-driven sampling (RDS) (Liamputtong 2007; Gile et al. 2015). Once the flyers for the project were circulated and potential participants began to make contact for interviews, it was much easier to explain the project directly and ask callers if they had any friends or family who met the eligibility criteria. In studies of drug use, snowball sampling is the most effective way to infiltrate a hidden network of people, many of whom may not use health services or disclose their drug use to a health worker (Gyarmathy et al. 2014; Wohl et al. 2017).

Snowball sampling via a peer network was used to recruit participants in our study of people who injected fentanyl (Allan et al. 2015). Fentanyl is a synthetic opioid with powerful pain killing and tranquillizing properties which is determined to be approximately 100 times stronger than morphine (Lofton and Phillip 2005). Recreational or nonmedical use of fentanyl has contributed to increased overdose deaths in Australia and globally (Hempstead and Yildirim 2014). In some countries, fentanyl is also produced illegally, made into tablets called carfentanil and added to other opioid drugs to increase their potency resulting in a huge increase in overdose and deaths (Middleton et al. 2016).

In Australia, the increase in deaths is related to increased availability of prescription opioids and reduced availability of heroin (Horyniak et al. 2013). A coronial investigation into 136 fentanyl-related overdose deaths recorded between 2000 and 2011 in Australia found that around one-third (34%) were due to fentanyl poisoning, half of the people (54%) had a history of injecting drug use and two-thirds (64%) had not been prescribed the drug that killed them (Roxburgh et al. 2013). Rural areas are over-represented in the overall number of Australian deaths (Allan et al. 2015).

To recruit participants into the study of fentanyl use in rural Australia, flyers were distributed to rural needle and syringe and opioid treatment programs and a residential withdrawal program in three regional centers in rural NSW and Victoria. Recruitment sites included places where people would be actively using fentanyl and treatment centers for those seeking support to stop. Snowball sampling was used to share information about the project through peer networks with flyers including a clear statement of eligibility criteria.

To be eligible to participate in the study, people had to be over 18, speak English, usually reside in a rural location, and have used fentanyl for nonmedical purposes more than once. The flyers included a contact phone number of the researcher conducting the interviews and prospective participants were asked to call for more information and to arrange an interview time if they wanted to proceed. Both men and women were sought for interviews with the aim of achieving 30% female participation to represent the ratio of deaths from fentanyl overdoses, which was 30/70 female to male at the time (Roxburgh et al. 2013).

The flyers and information sheets offered a fifty dollar supermarket voucher for participating in an interview. It was important to explain in the ethics application that the payment reimbursed people for their time and any travel costs but was not enough to be considered an inducement to participate. Concerns about payment to study participants include that money may be used to buy drugs and that people may not want to participate, or be harmed in some way but consent because they want the payment. Researchers have to consider whether a payment somehow influences the representativeness of their sample. Studies into paying people who use drugs or have been in treatment for research participation have found that concerns are mostly unfounded (e.g., Festinger et al. 2008). Payments improve participation and follow-up rates and are typically used by the recipient for essential purchases (Festinger et al. 2008; Dugosh et al. 2014). Further, payment to a study participant is respectful of the time spent and knowledge shared with the researchers (Liamputtong 2007).

4   Anonymity

Ensuring anonymity was paramount in the rural fentanyl study (Allan et al. 2015). Participants were engaged in illegal activity by using and sometimes selling fentanyl for nonmedical purposes. Only two people in the study had been prescribed fentanyl for a medical reason and both described doctor-shopping – attending multiple medical centers reporting made up symptoms – to obtain the drug. Describing how anonymity would be protected, and its limits, was an important first step to gaining verbal consent to participate.

At the first meeting, the research project was explained, and the potential participant was given an information sheet that had more details than the project flyer. Participants had the opportunity to ask questions about the project and have them answered. Possible risks of participation were described by the interviewer, and strategies for dealing with these were discussed. The risks we had anticipated included being identified and linked to illegal activity including drug trafficking, causing harm to themselves or others because of distress caused during the interviews, and risks to children in the care of the participant because of the participant's drug use. Because the research involved describing illegal activity, consent was verbal, no personal information was collected and pseudonyms were allocated to each participant to link demographic details to their interview data. In contrast to the heroin injecting study described earlier, there were no government certificates of confidentiality for the study.

Because there were no legal safeguards to protect anonymity of the participants, the researchers had to explain the limits of anonymity to warn participants when information would have to be passed to authorities. To avoid collecting reportable information, we did not ask direct questions or seek details about crimes. The information sheet included the following statement;

> **What about my privacy?**
> Your privacy is assured, and no personal information will be collected. Information about children being abused, threatening to kill yourself or others, and drug trafficking is reportable to the police. We will not ask you questions about those things. We will write a report about what we find out. You will be given a different name in the study report so you cannot be recognized by any person or organization.

The warnings about the limits of anonymity were clear enough for the researchers and participants to be able to discuss drug use practices and experiences without requiring detailed descriptions of people, places, or dates. For example, discussing how they obtained fentanyl, the participants used generalizations and talked about what they believed others to have done without going into details. For example, when talking about procuring fentanyl, participant statements included;

> I've always used it through people who've got it themselves. I haven't got it personally, but I know it's easy to get.
> **Q: How would you usually get a hold of it?**
> A: Buy it off the street just through dealers that get it wherever they get it from...they buy it off other people that are prescribed it, that's what I'm aware of anyway. There's a big market for it, yeah, a big market.

The benefits of the study for people who use drugs were perceived by the research team and the research ethics committee to outweigh the individual risks for participants even though potential harms could be identified (Aldridge and Charles 2008). Participant experiences and explanations were critical to the development of strategies for reducing the risk of overdose death from fentanyl and suggesting ways to disseminate information about health risks associated with fentanyl use. Participants also experienced gains in knowledge and understanding about fentanyl via sharing information during the interview with a researcher with good knowledge of substance treatment. Some sought referrals to drug treatment and health care as a result of the interview. For example, few participants were aware of internet sites that shared harm reduction and user experience information with others;

> They just said, "This is how you do it," and I just done it myself... That's all I've heard, yeah. Yeah, that's all, and these guys have been on it the whole time, so no one has ever told me any different.

The researchers provided study participants with information about useful websites and sources of harm reduction information to ensure that the benefits of the study were not just to the research team in the short term.

5 Surveys Versus Interviews

Anonymity was offered to participants in another project that investigated farming and fishing workers' use of drugs and alcohol (Allan et al. 2012a). When providing information about the project, the research team explained that participants could choose a pseudonym to disguise their identity if they were concerned about personal information being revealed in any project reports or publications.

Potential key informants were initially identified from telephone books and local newspapers. Key informants were community members with roles in health, primary industries and business, local government representatives, publicans, police and members of civic groups such as the Country Women's Association. Farm worker and fisher participants were identified through these local industry groups and networks by snowball sampling. A total of 145 farm and fishing workers/contractors, partners of workers, and community leaders across six research sites completed interviews and surveys between November 2010 and May 2011. The farming industry was represented in three sites with a total of 77 participants (53.2%), 46 of whom completed a survey. The fishing industry was represented in three sites with a total of 68 participants (46.8%), 25 of whom completed a survey.

Only two people from 145 participants chose to use a pseudonym for the interviews indicating most participants were either satisfied with assurances of confidentiality or not concerned with being identified in later reports of the study. However, a comparison of the survey and interview results found different types and patterns of drug use between the two data collection methods. In the semistructured interviews we asked;

> This project is about drug and alcohol use. Can you tell me about any drugs or alcohol that you use?

The interview prompts included how often, how much, with whom, and when drugs and/or alcohol were used.

In the survey, we used the Alcohol Use Disorders Identification Test [AUDIT] (Saunders et al. 1993), as a measure of self report drinking levels, and the Alcohol, Smoking and Substance Involvement Screening Test [ASSIST] (Humeniak et al. 2010), to identify drug use other than alcohol. We found that many participants reported more illicit drug use and use of more types of drugs in the survey compared to the interviews. For example, one participant stated in his interview;

> I don't smoke cigarettes but I've smoked marijuana since I was 18. Yeah, I smoke that flat out as a fucking chicken unfortunately.

He made no reference to other illicit drugs in the interview. However, his survey data reported use of amphetamines, marijuana, cocaine, and ecstasy in the previous 12 months. In another example, two participants reported use of marijuana and amphetamines in the past 12 months in their survey responses, yet in the interviews, both talked about marijuana use in the past;

> We were only into pot and a bit of marijuana and beer

> Yeah. I used to smoke a bit of marijuana, but found it did no good for me. I wouldn't go out of my way to buy it.

While interviews and surveys have different research approaches and aims, it is likely that some information is easier to reveal in a survey rather than face-to-face in an interview with a stranger. Surveys offer greater anonymity, particularly for collecting sensitive information where a research participant may perceive they will be judged on the basis of their answer if the data collection is face-to-face. The restriction of a survey is the inability to explore reasons for an answer or the person's experience of an event, including drug use (see also ▶ Chap. 32, "Traditional Survey and Questionnaire Platforms").

Technological developments are offering new opportunities for anonymous data collection because technology can mitigate the risk of disclosing sensitive issues (Lee et al. 2003). A project investigating young people's use of drugs and alcohol and their experiences of health-care interventions in relation to substance use is currently being planned. To correctly identify potential participants we will ask questions about their substance use on an IPad during a standardized data collection for the youth health service they are attending. The IPad tool has demonstrated acceptability among young people and when trialed found that young people accessing the service were two to ten times more likely to disclose sensitive issues including substance use (Bradford et al. 2014; Bradford and Rickwood 2015). When planning a study, data collection methods that offer ways of preserving anonymity and easing disclosure of personal information are important considerations for investigating drug use.

6   Intoxication

Assessing a person's degree of intoxication is important when explaining a research project and seeking consent to participate in a face-to-face data collection from people who may have recently used a drug. Being intoxicated affects a person's ability to understand risks or harms arising from participation in research (Donroe and Tetrault 2017). If an individual with no usual communication difficulties cannot talk to the interviewer about the project and ask relevant questions, they should be assumed to be unable to consent. People who are heavily intoxicated will be unable to consent at the time of intoxication but will be able to later even if the effects of the substance are still experienced. It is important for the researcher to have a way of evaluating the impact of intoxication on the person's ability to participate.

Intoxication is best assessed by having a conversation and identifying if the potential participant understands what is being said to and asked of them (Aldridge and Charles 2008). Those who are slurring their speech, sleeping, or stumbling when walking will be assumed to be too intoxicated to participate in research. However, the type of drug causing intoxication will affect the length of time taken for heavy intoxication to wear off. In the fentanyl study prospective participants were asked if they had consumed any fentanyl within the previous 3 h. If they said yes, the interview was delayed until the effects of intoxication had subsided. This took approximately 2 h and could be assessed via pupil size and orientation to day and time. In this example, it is the degree of intoxication that is assessed rather than any state of intoxication limiting participation in the study.

Intoxication is not just a problem for gaining consent. Alcohol affected people can be aggressive and unpredictable causing a risk of harm to the researcher. Intoxication also affects people's responses to interviews or survey questions. Intoxication is frequently implicated in suicide suggesting that people can be in a negative or vulnerable emotional state at the time of intoxication (Arias et al. 2016). Alcohol use in particular is a risk factor in suicide (Currier et al. 2016). Distress related to intoxication needs to be addressed during data collection.

7   Dealing with Distress

Planning ways to respond to distress is important in preparing to conduct research with people who use drugs. A greater proportion of Australians who drink at risky or high risk levels experience high or very high psychological distress (Allan et al. 2012a). In our farming and fishing study, identifying local support services prior to data collection was important because in general rural Australian men have a higher incidence of depression than men living in cities (Currier et al. 2016). They are also more likely than their city counterparts to have experienced a mental disorder associated with substance use throughout their lifetime (AIHW 2010). Furthermore, farmers and farm workers have higher suicide rates than the national male population (Neufeld et al. 2015). Research also indicates that alcohol misuse may be associated with the pressures experienced by farmers as a result of recurrent drought conditions

(Fragar et al. 2011). Asking questions about why farmers and fishers used drugs and alcohol was expected to provoke discussions of problems experienced by research participants in our study.

Harm or discomfort should be anticipated to come from discussions of the problems associated with substance use and the reasons why people use drugs. Participants may become distressed if, for example, they discuss problematic family relationships, trauma, or mental distress. The research interview does not cause the problems and can provide an opportunity for supportive intervention to be offered if required. However, intoxication can make distress worse and people's responses unpredictable as the effects of a drug increase or decrease. For example, the following two quotes from people using fentanyl demonstrate the challenges people face when dependent on a drug;

> Opiate users are desperate, and desperate human beings go to desperate measures to seek what they desperately need.
> **Q: So how many people do you reckon you've seen overdose?**
> A: Well I know I've seen three or maybe four people in the last 12 or so months, one person twice. I actually had one mate die at just at Christmas time. That was my best mate.
> **Q: How many people do you reckon you've known have died from using the patches?**
> A: My missus, two cousins... about five of them.

Researchers should be alert to distress expressed by tears, shaky voice, averted gaze, and descriptions of difficulty in daily tasks or relationships with no strategy for dealing with the situation, expressions of hopelessness or helplessness where substance use is concerned, and/or examples of violence or victimization. It is important that interviews are conducted by a researcher experienced in responding to distress and making appropriate referrals when necessary. Participants should be warned in the information sheet and in preparation for the recording of the possibility of the interview causing distress before seeking consent to proceed or delaying the consent and data collection if the person is heavily intoxicated.

Researchers require information and contacts for local and accessible support services when discussing intoxication and drug use with study participants. In the farming and fishing study, interviews raised problems associated with loss and grief;

> Ever since he tried to take his own life, he's had so many doubts and so many down days, even that weren't alcohol induced, that I think because – I couldn't manage him; basically I've been his carer. He's an able bodied man, and he's physically fit, but mentally, unfit, unable to make decisions, personally, in his life ... I have to manage every aspect of his personal life, to a degree.
> ... he went to the boat to sleep intoxicated and fell in the bait tank and died, drowned. Now, that wouldn't have happened if he hadn't been intoxicated.

In these examples, it was important for the researcher to follow-up with questions unrelated to the study; asking the interviewee if they needed support for themselves in relation to the incident or situation they described. The research team ensured that information was left with the interviewee with contact numbers for follow-up, even if they said no to wanting to contact someone for support at the time of the interview.

The research team on the farming and fishing study also developed information for support and advice for research participants who were dealing with intoxication and distress in other ways. For example, a farming employer stated;

> Well, we have got a local alcoholic, who did have to have a beer at afternoon tea time, and I didn't sack him because we were desperate but normally if there's any alcohol ingestion during the day, they would be asked to leave or not drink alcohol, or they'd get the boot.

It became clear to the research team early in the study that work place practices were putting employers and employees at risk because intoxication was being ignored;

> Question: It's still happening, farm owners allowing guys under the influence to actually work in the farm?
> Answer: Yes. It's ignored.
> Question: Why do you think it is happening?
> Answer: Because they want the hours done, I suppose they don't care because they want the crops off. They'll see the rain and they don't care who comes.

Health-care workers have clear policies and practices for dealing with intoxication in patient presentations, yet still find it challenging (Donroe and Tetrault 2017). In farming and fishing workplaces, there are no guidelines and the additional disadvantage of rural and remote populations experiencing more limited access to prevention, assessment, and treatment compounds the problem (Allan et al. 2012a). In response to the issues raised in early interviews, the research team developed a factsheet with a list of web links to workplace guidelines on recognizing and responding to intoxication and workplace policies on substance use. The factsheet was given to everyone who participated in the study. It was an important research practice to have information available in the short term because the project recommendations that would address the problems raised in the interviews were more than a year away.

8 Research Participants in Substance Treatment Settings

Substance treatment centers offer an easy way to access potential research participants who have used drugs. However, two challenges face the researcher in drawing a sample from treatment centers. The first is ensuring that participants will not perceive their continuing access to treatment or way they are treated by staff is affected by whether they agree to participate in a study, particularly if they have a cognitive impairment. The second challenge is ensuring that data from a sample of people in treatment adequately meet the aims of the study because of loss to follow-up.

People in drug treatment settings are vulnerable because of the circumstances that result in them needing treatment but also because of the unequal power relationship between the treatment provider and the patient or client (Banks 2016). Potential participants could perceive they are being coerced to join a study. This is particularly the case in residential treatment settings where the withdrawal of treatment could result in increased likelihood of relapse to drug dependence and homelessness. Conducting research in the treatment setting requires a clearly defined assurance that treatment will not be affected by participation or nonparticipation in research. The assurance is usually provided in the information sheet provided to potential participants, but it should also be integrated into the ways data collection is organized and managed.

In our study of the treatment experiences of people with cognitive impairment in a residential substance treatment program, two strategies were employed to assure study participants that the research process was separate to the treatment provided (Collings et al. 2017). Firstly, a project worker gave eligible residents the participant information and consent form and provided a verbal explanation of the research aim and methods. The project worker was not directly involved in the treatment program but visited the program site several times a week to meet potential participants and to conduct cognitive screening with people who had consented to participate in the study. This allowed her to discuss the merits and implications of participation in an impartial way. To address any perceived coercion, the project worker asked eligible individuals to discuss their participation with a family member or friend before deciding to consent or not. The project worker then checked back 2 or 3 days later to find out if the person wanted to participate or not.

The project worker explained to potential participants that participation involved sharing routinely collected individual outcome data using three standardized instruments: cognitive screening results; routinely collected personal demographic information; and the option of taking part in an individual interview when they reached the tenth week of treatment. She also explained how someone could withdraw from participation at any time. All of these meetings and data collection processes were separate from any staff or activities that were part of the treatment program.

Secondly, the researchers remained at arm's length to recruitment by not having any indirect or direct contact with eligible individuals until after they consented to participate. This ensured that the university conducting the research did not exercise any coercive role upon potential participants' decision-making. In reality, the researchers did not contact the participant until just prior to their tenth week in the treatment program which required an additional consent process in reminding people what they had agreed to and asking if they still wished to participate. The preliminary consent was sought shortly after people started in the treatment program and the baseline data collection was being done. This was typically around seven days after entering the program. Nine weeks later, most people had forgotten about the research project and the project worker need to explain again what was required and reminded them what they had consented to. She also explained that they could withdraw their consent if they had changed their minds about participating.

Conducting research with people who had cognitive impairments and used drugs required additional attention to consent processes as described above. However, most research with people who use drugs is highly likely to involve people with cognitive impairments because of the strong association between substance misuse and cognitive problems. Most substances of dependence impair attention, learning and memory, visual-spatial abilities, and executive functioning (Allan et al. 2012b). Traumatic brain injury (TBI) is also highly prevalent in substance treatment populations, leading to significant complexities in the process of treatment and research with this population (Sacks et al. 2009). If research is to be conducted within a treatment setting, then the prevalence rate of cognitive impairment is likely to be around 50% of the people in treatment (Allan et al. 2012b).

Conducting a number of studies in substance treatment settings raises the potential bias of a sample recruited in treatment. Bias is possible in both the participants' characteristics and their perceptions of substance treatment. This is sometime referred to as the treatment halo effect which is recognized in health-care settings but not specifically in research methods (Kerger et al. 2016). The positive halo refers to those who are most successful in treatment and therefore the easiest to recruit, and also to the descriptions of benefits and impact of treatment by people who are still immersed in the treatment setting and may be overly optimistic about how the treatment will affect them in the future (Holbrook 1983; Baltes and Parker 2000).

Research participants drawn from treatment programs are likely to be those who have the most success with the program in that they are able to adhere to it and find it helpful. It is important to check that those who participate in a study in a treatment setting are not significantly different in age, sex, ethnicity, socioeconomic status, or drug use history from those who decline to participate or who leave treatment early. This is particularly important if the aim of the study is to evaluate the effectiveness or acceptability of the particular treatment (Hser et al. 2004; Zhang et al. 2008). In the study described above of people with cognitive impairment in residential rehabilitation, the project worker spent a significant amount of time contacting people who had consented to participate but left the program prior to the 10 week interview.

In preparing for the study, the high rate of drop out from treatment by people with cognitive impairment was identified as a problem in the final sample of those interviewed and also in the 3 month follow-up rates for substance dependence and well-being measures (Brorson et al. 2013). To minimize the loss to follow-up, the consent forms included participants agreeing to three different ways of contact once they had left treatment. Contact methods listed were the person's own mobile phone including recording multiple numbers if they changed over time; one or two family members or friends telephone numbers and addresses; and contact details of any health or welfare worker that the person had been in regular contact with prior to treatment and who they identified as continuing to support them. Family members, friends, and support workers were sent information about the research so they were aware of the person's consent to participate and that they may be contacted to provide information about the research participant's whereabouts or how to get in touch with them. The follow-up strategy resulted in completion of 12 interviews and eight people agreeing to undertake 3-month follow-up measures from a total of 33 people who initially consented to participate in the study; just under half of whom left the treatment program prior to completion.

9   Conclusion and Future Directions

Substance misuse has been identified as the predominant problem for people who are homeless, experiencing domestic violence or engaged with the criminal justice or child protection systems (Butler et al. 2016). These are all key arenas of work for health and social care workers. However, education and training on ways of working with and addressing substance misuse is a significant gap in professional and practice development in the health and social care sector (van Boekel et al. 2013). For example, strategies for working with substance users have been identified as a key training need of social workers post-graduation (Hall et al. 2000).

Research with people who use drugs is critical to inform and develop health and social care practice to effectively address substance misuse. As a result of research conducted with people who use drugs in rural Australia and those seeking treatment in our programs, Lyndon has developed engagement strategies (Allan and Campbell 2011), peer delivered harm reduction strategies (Allan et al. 2015), assessment and referral pathways (Allan and Kemp 2011, 2014), and a treatment program for people with cognitive impairments (Allan et al. 2012b; Collings et al. 2017).

People report they participate in research because they want to help others and influence policy and practice around drug-related laws and treatment (Barratt et al. 2007). Everyone can use human rights principles to support people's involvement in research and facilitate ways for them to participate as safely as possible (McDonald and Raymaker 2013). Overall, there is an incredible generosity and willingness from study participants to share their experiences, their insights into drug use, and their own harm reduction strategies. It is vitally important for researchers and practitioners to acknowledge study participants' contributions and protect them with practical strategies that are ethical.

References

Aldridge J, Charles V. Researching the intoxicated: informed consent implications for alcohol and drug research. Drug Alcohol Depend. 2008;93(3):191–6.

Allan J, Campbell M. Improving access to hard-to-reach services: a soft entry approach to drug and alcohol services for rural Australian aboriginal communities. Soc Work Health Care. 2011;50(6):443–65.

Allan J, Kemp M. Aboriginal and non aboriginal women in new South Wales non government organisation (NGO) drug and alcohol treatment and the implications for social work: who starts, who finishes, and where do they come from? Aust Soc Work. 2011;64(1):68–83.

Allan J, Kemp M. The prevalence and characteristics of homelessness in the NSW substance treatment population: implications for practice. Soc Work Health Care. 2014;53(2):183–98.

Allan J, Ball P, Alston M. What is health anyway? Perceptions and experiences of health and health care from socio-economically disadvantaged rural residents. Rural Soc. 2010;20(1):85–97.

Allan J, Clifford A, Ball P, Alston M, Meister P. 'You're less complete if you haven't got a can in your hand': alcohol consumption and related harmful effects in rural Australia: the role and influence of cultural capital. Alcohol Alcohol. 2012a;47(5):624–9.

Allan J, Kemp M, Golden A. The prevalence of cognitive impairment in a rural in-patient substance misuse treatment programme. Ment Health Subst Use. 2012b;5(4):303–13.

Allan J, Meister P, Clifford A, Whittenbury K. Drug and alcohol use by farming and fishing workers. Canberra: Rural Industries Research and Development Corporation; 2012c. Retrieved 25 Apr 2017 from https://rirdc.infoservices.com.au/items/12-061.

Allan J, Herridge N, Griffiths P, Fisher A, Clarke I, et al. Illicit fentanyl use in rural Australia – an exploratory study. J Alcohol Drug Depend. 2015;3:196. https://doi.org/10.4172/2329-6488.1000196.

Arias SA, Dumas O, Sullivan AF, Boudreaux ED, Miller I, Camargo CA Jr. Substance use as a mediator of the association between demographics, suicide attempt history, and future suicide attempts in emergency department patients. Crisis. 2016;37(5):385–91.

Ball AL. HIV, injecting drug use and harm reduction: a public health response. Addiction. 2007;102(5):684–90.

Baltes BB, Parker CP. Reducing the effects of performance expectations on behavioral ratings. Organ Behav Hum Decis Process. 2000;82(2):237–67.

Banks S. Everyday ethics in professional life: social work as ethics work. Ethics Soc Welf. 2016;10(1):35–52.

Barratt MJ, Norman JS, Fry CL. Positive and negative aspects of participation in illicit drug research: implications for recruitment and ethical conduct. Int J Drug Policy. 2007;18(3):235–8.

Bradford S, Rickwood D. Acceptability and utility of an electronic psychosocial assessment (myAssessment) to increase self-disclosure in youth mental healthcare: a quasi-experimental study. BMC Psychiatry. 2015;15:305. https://doi.org/10.1186/s12888-015-0694-4.

Bradford S, Rickwood D, Boer D. Health professionals attitudes towards electronic psychosocial assessments in youth mental healthcare. Health. 2014;6:1822–33. Published Online July 2014 in SciRes. http://www.scirp.org/journal/health, https://doi.org/10.4236/health.2014.61421.

Brorson H, AJo Arnevik E, Rand-Hendriksen K, Duckert F. Drop-out from addiction treatment: a systematic review of risk factors. Clin Psychol Rev. 2013;33(8):1010–24.

Butler L, Burns L, Breen C. Social media as a recruitment tool for drug use surveys. 2017. Retrieved on 25 Apr 2017 from National Drug and Alcohol Research Centre, UNSW Australia https://ndarc.med.unsw.edu.au/sites/default/files/ndarc/resources/EDRSapril2017_FINAL.PDF.

Ciccarone D, Harris M. Fire in the vein: heroin acidity and its proximal effect on users' health. Int J Drug Policy. 2015;26(11):1103–10.

Collings S, Dew A, Dowse L, Cooney E. Evaluation of project RE-PIN. Receive, encode, process and INtegrate drug and alcohol treatment strategies for people with cognitive impairment: final report. 2017. Retrieved on 25 Apr 2017 from www.lyndon.org.au/research/.

Currier D, Spittal MJ, Patton G, Pirkis J. Life stress and suicidal ideation in Australian men – cross-sectional analysis of the Australian longitudinal study on male health baseline data. BMC Public Health. 2016;16(Suppl 3):1031. https://doi.org/10.1186/s12889-016-3702-9.

Donroe JH, Tetrault JM. Recognizing and caring for the intoxicated patient in an outpatient clinic. Med Clin N Am. 2017;101(3):573–86.

Dugosh KL, Festinger D, Cacciola JS. Examining perceived coercion among incarcerated substance abusers participating in research. Drug Alcohol Depend. 2014;140:e52. https://doi.org/10.1016/j.drugalcdep.2014.02.161.

Festinger DS, Marlowe DB, Dugosh KL, Croft JR, Arabia PL. Higher magnitude cash payments improve research follow-up rates without increasing drug use or perceived coercion. Drug Alcohol Depend. 2008;96(1–2):128–35.

Fragar L, Depczynski J, Lower T. Mortality patterns of Australian male farmers and farm managers. Aust J Rural Health. 2011;19(4):179–84.

Gile KJ, Johnston LG, Salganik MJ. Diagnostics for respondent-driven sampling. J R Stat Soc Ser A. 2015;178(1):241–69.

Gyarmathy VA, Johnston LG, Caplinskiene I, Caplinskas S, Latkin CA. A simulative comparison of respondent driven sampling with incentivized snowball sampling-the "strudel effect". Drug Alcohol Depend. 2014;135:71–7.

Hall MN, Amodeo M, Shaffer HJ, Vander Bilt J. Social workers employed in substance abuse treatment agencies: a training needs assessment. Soc Work. 2000;45(2):141–55.

Hempstead K, Yildirim EO. Supply-side response to declining heroin purity: fentanyl overdose episode in New Jersey. Health Econ. 2014;23(6):688–705.

Holbrook MB. Using a structural model of halo effect to assess perceptual distortion due to affective overtones. J Consum Res. 1983;10(2):247–52.

Horyniak D, Higgs P, Jenkinson R, Degenhardt L, Stoove M, Kerr T, Dietze P. Establishing the Melbourne injecting drug user cohort study (MIX): rationale, methods, and baseline and twelve-month follow-up results. Harm Reduct J. 2013;10:11. https://doi.org/10.1186/1477-7517-10-11. http://www.harmreductionjournal.com/content/10/1/1.

Hser Y, Evans E, Huang D, Anglin DM. Relationship between drug treatment services, retention, and outcomes. Psychiatr Serv. 2004;55(7):767–74.

Humeniuk R. Validation of the alcohol, smoking and substance involvement screening test (ASSIST) and pilot brief intervention: A technical report of phase II findings of the WHO ASSIST Project. 2010. Retrieved May 29, 2015 from http://www.who.int/substance_abuse/activities/assist_technicalreport_phase2_final.pdf.

Kerger BD, Bernal A, Paustenbach DJ, Huntley-Fenner G. Halo and spillover effect illustrations for selected beneficial medical devices and drugs. BMC Public Health. 2016;16:979. https://doi.org/10.1186/s12889-016-3595-7.

Liamputtong P. Researching the Vulnerable: A Guide to Sensitive Research Methods. Sage Publications, London; 2007.

Lim R. Drug use and other risky behaviours in young people attending a music festival. In: Paper presented 15, October, 2013 at the national drug trends conference, Melbourne. 2013. https://ndarc.med.unsw.edu.au/event/2013-national-drug-trends-conference.

Lee FSL, Vogel D, Limayem M. Virtual Community Informatics: A Review and Research Agenda, Journal of Information Technology Theory and Application (JITTA). 2003; 5:1, Article 5. Retrieved on 23 April, 2017 from http://aisel.aisnet.org/jitta/vol5/iss1/5.

Lofton A, Phillip W. Encyclopedia of Toxicology 2005. New York: Elsevier.

Lotan G, Ells C. Adults with intellectual and developmental disabilities and participation in decision kaking: ethical considerations for professional–client practice. Intellect Dev Disabil. 2010;48(2):112–25.

McDonald KE, Keys CB. How the powerful decide: access to research participation by those at the margins. Am J Community Psychol. 2008;42(1–2):79–93.

McDonald KE, Raymaker DM. Paradigm shifts in disability and health: toward more ethical public health research. Am J Public Health. 2013;103(12):2165–73.

Middleton J, McGrail S, Stringer K. Drug related deaths in England and Wales. BMJ. 2016; 355:i5259. https://doi.org/10.1136/bmj.i5259. (Published 17 October 2016).

NAOMI Patients Association and Boyd. NAOMI research survivors: experiences and recommendations. North American opiate medication initiative. Vancouver: British Columbia. 2012. Retrieved on 2 Nov 2015 from http://drugpolicy.ca/wp-content/uploads/2012/03/NPAreport March5-12.pdf.

Neufeld E, Hirdes JP, Perlman CM, Rabinowitz T. A longitudinal examination of rural status and suicide risk. Healthc Manage Forum. 2015;28(4):129–33.

Olde Rikkert MG, Verweij MF, Hoefnagels WH. Informed consent and mental competence of the elderly in medical-scientific studies. Tijdschr Gerontol Geriatr 1995;26(4):152–62

Roxburgh A, Burns L, Drummer OH, Pilgrim J, Farrell M, Degenhardt L. Trends in fentanyl prescriptions and fentanyl-related mortality in Australia. Drug Alcohol Rev. 2013;32(3): 269–75.

Ruefli T, Rogers SJ. How do drug users define their progress in harm reduction programs? Qualitative research to develop user-generated outcomes. Harm Reduct J. 2004;1(1):8. https://doi.org/10.1186/1477-7517-1-8.

Sacks A, Fenske C, Gordon W, Hibbard M, Perez K, Braundau S. Co-morbidity of substance abuse and traumatic brain injury. J Dual Diagn. 2009;5(304):404–17.

Saunders JB, Aasland OG, Babor TF, De La Fuente JR, Grant M. Development of the Alcohol Use Disorders Identification Test (AUDIT): WHO collaborative project on early detection of persons with harmful alcohol consumption-II. Addiction 1993;88(6):791–804.

van Boekel LC, Brouwers E, van Weeghel J, Garretsen H. Stigma among health professionals towards patients with substance use disorders and its consequences for healthcare delivery: systematic review. Drug Alcohol Depend. 2013;131:23–35.

Welie SP, Berghmans RL. Inclusion of patients with severe mental illness in clinical trials: issues and recommendations surrounding informed consent. CNS Drugs. 2006;20(1):67–83.

Wohl AR, Ludwig-Barron N, Dierst-Davies R, Kulkarni S, Bendetson J, Jordan W, Perez MJ. Project engage: snowball sampling and direct recruitment to identify and link hard-to-reach HIV-infected persons who are out of care. J Acquir Immune Defic Syndr. 2017;75(2):190–7.

Woltmann EM, Whitley R. Shared decision making in public mental health care: Perspectives from consumers living with severe mental illness. Psychiatric Rehabilitation Journal. 2010;34(1):29–36.

World Health Organisation [WHO]. WHO Expert Committee on Drug Dependence. Twentieth report. World Health Organ Technical Report Series; 1974. p.1–89.

World Health Organisation [WHO]. WHO expert committee on drug dependence. 2016. Retrieved on 25 Apr 2017 from http://www.who.int/medicines/access/controlled-substances/ECDD_38th_Report_Unedited_version_13032017.pdf?ua=1.

Zhang Z, Gerstein DR, Friedmann PD. Patient satisfaction and sustained outcomes of drug abuse treatment. J Health Psychol. 2008;13(3):388–400.

Researching with People with Dementia

Jane McKeown

Contents

1 Background 1992
2 Involving People with Dementia Throughout the Research Process 1993
3 Involving People with Dementia as Research Participants 1994
   3.1 Recruitment 1994
   3.2 Consent 1996
   3.3 Research Methods and Approaches 1999
4 Considerations in Involving People with Dementia in Research 2000
5 Methodological Considerations 2001
6 Conclusion and Future Directions 2002
References 2003

Abstract

Undertaking research with people with dementia has historically been perceived as problematic, especially as the condition advances and where there are concerns over capacity to make decisions and give informed consent. Much research that does include people with dementia as participants tends to focus on people early on in the condition. By not involving people with dementia across the trajectory of the condition, we are failing to develop important understandings from the perspectives of people living with the condition across a range of research topics. Furthermore, by not involving people with dementia throughout the research process, we may not be exploring the most relevant research topics or not considering the most relevant methods and approaches to capture their experiences. There is an increased understanding on the importance of involving people

---

J. McKeown (✉)
School of Nursing and Midwifery, The University of Sheffield, Sheffield, UK
e-mail: j.mckeown@sheffield.ac.uk

with dementia in all aspects of the research process, but the challenge remains on how to do so in a meaningful and ethical way. Researchers are beginning to explore these challenges in the literature. This chapter will draw upon the evidence base as well as personal experience in the UK to consider approaches to consent and a range of diverse approaches and methods that seek to include rather than exclude people living with dementia throughout the research process.

Keywords

Dementia · Research · Involvement · Ethics · Consent · Qualitative

1   Background

Before the 1990s, it was rare to see the inclusion of people with dementia in research accounts and consequently their perspectives tended not to be reported (Hubbard et al. 2003). At that time, research undertaken emphasized a medical model focus on symptoms (Downs 1997), with studies largely focussed on cognitive function and decline (Hubbard et al. 2003). Cottrell and Schultz (1993) suggest that persons with dementia are being described as a "disease entity," who are seen as unable to share their experience of the condition. Reflecting on the past, a person with dementia notes: "What a hugely missed opportunity it would be if people with Alzheimer's were excluded from the very thing that could be used to gain a fuller understanding of their disease" (Robinson 2002, p. 104).

The effective exclusion of people with dementia from studies has implications for dementia research specifically (McKeown et al. 2010) but also for research more broadly. Investigations about issues affecting older people more generally will not reflect the specific needs of older people with dementia in their findings if these people are not invited to participate. Taylor et al. (2012) discovered that over a quarter of studies reported in an international geriatric medicine journal over a 2-year period explicitly excluded people with cognitive impairment from participating. Including people with dementia in research will surely improve the quality of research about dementia and about older people.

Since the 1990s, perceptions of dementia have started to broaden and person-centered understandings of dementia have emerged initially from the work of Tom Kitwood (1997). These understandings have contributed to a growing acknowledgment that people with dementia are citizens and have rights, including the right for their experiences to be explored through research (Downs 1997). Interest in psychological and biographical aspects of the life experiences of people with dementia has been important in this change (Hubbard et al. 2003). Increasingly, there is consensus that people with dementia should be included in research as active participants, not purely as subjects (Cottrell and Schultz 1993; Downs 1997; Dewing 2002; Hubbard et al. 2003; Hellstrom et al. 2007). However, challenges in including people with dementia in research remain (Hughes and Castro Romero 2015). This chapter will explore the challenges and offer insights in how such challenges are being overcome.

2   Involving People with Dementia Throughout the Research Process

Involving people in research has in the past been perceived purely as recruiting people as subjects or participants. However, a broader understanding of the ways that people with dementia can be engaged across the research process is emerging.

The framework offered by the National Institute for Health Research (2014, p. 14) identified the aspects of the research process where "patients and public" might take an active role. These opportunities for involvement include: Identifying and prioritizing, design, development of the grant proposal, undertaking and managing, analyzing and interpreting, dissemination, implementation, and monitoring and evaluation. Reports of how researchers have involved people with dementia across the research process remain scarce.

One example of where people with dementia have been involved in the development of research ideas is the UK James Linde Alliance research priority setting exercise. A consensus approach was used with people with dementia, family carers, and others interested in dementia to identify the top ten unanswered questions and dementia research priorities. While such an approach is to be applauded, the process is worthy of closer inspection. Kelly et al. (2015) report that 4.1% of the 1563 completed surveys were known to be from people with dementia, and in the final prioritization exercise, 2 of the 18 people involved had dementia. This reflects the challenges in seeking views of people with dementia in their own right and that the views and opinions of family carers and others can prevail over those of the person with dementia.

The need for public and patient engagement in research as a prerequisite for research funding has led to the formation of dementia-specific research advisory groups. Organizations such as the Alzheimer's Society offer peer review of research proposals by volunteers. This is a highly valuable process and can provide researchers into insights they may never have considered. However, questions do need to be asked about the constitution of these groups: Is it people with dementia or is it carers who have the dominant voice? Clearly, carers need a voice and their perspectives are essential but more work is required so that people with dementia have their views heard and separated out from those who care for them.

In the UK, groups of people with dementia, for example, the Scottish Dementia Working Group (SWDG), have started to ask these questions. Indeed, this group published core guiding principles for researchers to involve people with dementia in research (SWDG 2014). A network aimed at bringing groups of people with dementia together, known as the Dementia Engagement and Empowerment Project (DEEP), has been formed. A summary of the work of DEEP is reported by Williamson (2012), and the network encompasses over 50 groups. The DEEP network includes groups interested in informing research in a variety of ways. This emerging dementia "activist" movement promises the potential for a different and exciting future landscape for people with dementia in research.

Moving forward, a critical examination of the involvement of people with dementia within the entire research process deserves far more attention. The remainder of the chapter, however, will explore more specifically the involvement of people with dementia as participants within research studies.

3 Involving People with Dementia as Research Participants

As highlighted at the beginning of the chapter, a number of challenges may arise in the inclusion of people with dementia as research participants. These challenges will be explored through the context of the research process.

3.1 Recruitment

An initial challenge in recruiting people with dementia to research studies is how to access people in order to invite them to join the study. To avoid any suggested coercion by researchers to participate, ethics guidance advises to recruit through "services" or third sector organizations. It is here that researchers may face what Sherratt et al. (2007) term the "gatekeeper challenge."

Gatekeepers come in many forms and may include care home staff, family members, third sector organizations such as Alzheimer's Societies, specialist dementia services, and ethics committees. Several layers of gatekeepers may need to be negotiated with in order to access older people with dementia (Hellstrom et al. 2007). While their intentions are usually good, gatekeepers may sometimes be informed by negative assumptions about the ability of people with dementia to speak for themselves. Hughes and Castro Romera (2015, p. 224) report concerns from family carers that the person with dementia "won't understand," and consequently, they did not want the researcher to make contact. Witham et al. (2015) explore the challenges of recruiting people with dementia when healthcare professionals identify patients as "vulnerable." Even with all the necessary ethical and legal permissions on the part of the researchers, healthcare providers may deny access believing research will add an additional "burden" to an already vulnerable individual (Witham et al. 2015). Historically, there has been a practice of paternalism where people with dementia were deemed as needing "protecting," compared to a more recent change of focus in learning disabilities research with an emphasis on "rights rather than protection" (Hughes and Castro Romera 2015, p. 231).

Such gatekeeping has perhaps withheld opportunities for people with dementia to have participated as fully in all aspects of research (Bartlett and Martin 2002). Equally, it may be that people with dementia "chosen" by services to be advised of research opportunities may not be representative of people with dementia more widely. Published research studies do appear to focus on people in the earlier stages of dementia able to communicate verbally and quite often having support of family carers. It stands to reason that their experiences are likely to be very different from a person with dementia who is more impaired, with communication challenges and living alone for example.

Organizations are developing ways for researchers to more easily make contact with people with dementia. Join Dementia Research in the UK is one such example (see https://www.joindementiaresearch.nihr.ac.uk/). Funded by the National Institute for Health Research, the organization maintains a database of people with dementia and family carers who are interested in participating in research. Researchers can then request the details of people who meet their inclusion criteria.

Building and maintaining positive relationships with gatekeepers is essential. It is important that researchers demonstrate their credibility so that they are not perceived to be "using" people with dementia to meet own research needs. Hellstrom et al. (2007) conclude that there are no easy solutions to the gatekeeper dilemma, and researchers must accept that some form of protection is necessary and desirable. King et al. (2016) emphasize that coercion should be avoided and gatekeepers are likely to remain important. It can be the worker at the local Alzheimer's Society or the Memory Service that is aware when a person with dementia may be overwhelmed by the opportunities for involvement being presented to them. In the broader involvement context, people with dementia themselves identify how tiring such activities can be and that people can feel "over-used" by others who want to involve people with dementia in their work (Litherland 2015).

While accepting that there can be negative personal consequences of involvement for the person with dementia, the potential personal benefits need to also be recognized. Researchers perhaps need to understand and be able to articulate the potential benefits for the person with dementia in participating in research. Data collection can, in itself, be therapeutic for the person with dementia and researchers describe how participants were pleased to be listened to (Barnett 2000; Clarke and Keady 2002). One person with dementia recognized that his experiences of living with dementia may benefit others with the condition (McKillop 2002). A person with dementia I worked with described how participating in forums for sharing his experience made him feel valued as a person and "less like an in-growing toe nail." People with dementia report that participating in research and contributing to the development of new knowledge can help the person to experience value and positive satisfaction.

In addition to gatekeeper issues, recruiting a diverse range of people living with dementia to research studies appears a challenge. Carmody et al. (2014) identify the additional difficulties in recruiting people from culturally and linguistically diverse communities (see also ▶ Chaps. 107, "Conducting Ethical Research with People from Asylum Seeker and Refugee Backgrounds," and ▶ 5, "Recruitment of Research Participants"). Researchers have started to recognize the differences in experience that may result from gender, the type of dementia a person has, their sexual orientation, whether they live alone or with family, whether they are physical fit or frail, and whether they live in the community or in a residential facility or are in hospital. The diversity of people with dementia is currently not widely reflected in participants in reported research studies. In their *Call for Change* in research with people with dementia, Carmody et al. (2014) urge researchers to strengthen the relevance of their research by avoiding the application of overly restrictive exclusion criteria.

3.2 Consent

There can be assumptions by the very nature of a person having a diagnosis of dementia that they are unable to consent. Gaining consent from people with dementia to participate in research is a complex issue. Usual informed consent processes can exclude people with dementia (Dewing 2008). Also, if a person is assessed as unable to consent, there can be assumptions that they cannot take part in the research. These assumptions will be explored in more detail in this section and possible solutions to the challenges discussed.

In England and Wales, the Mental Capacity Act (DH 2005) states that people are deemed capable unless there is evidence to the contrary. The Act (DH 2005) stresses that a diagnosis of dementia does not necessarily mean a person is unable to consent and decisions are situational. If doubts are raised, then the Act asks for a thorough assessment around the following questions: Does the person have impairment of mind or brain? If so, does that impairment mean that the person is unable to make the decision in question at the time it needs to be made? If there are doubts over capacity, then an assessment needs to be made of the person's ability to have: a general understanding of what decision needs to be made and why, a general understanding of the likely consequences of making (or not) the decision, an ability to understand, retain, use, and weigh up the information, and an ability to communicate this.

Maintaining a clear record of conversations and decisions is essential to maintain an audit trail of the consent process. Hughes and Castro Romero (2015) reported comments that participants in their study had made which the researchers perceived to indicate participant understanding of the research they were being invited to take part in. Consent was subsequently recorded verbally on an audio device. To account for the fluctuating nature of consent and challenges for some people with dementia in remembering the previous research conversations, Hughes and Castro Romera (2015) revisited consent upon each research encounter. This process approach to consent is increasingly being reported as good practice when researching with people with dementia.

Dewing (2002) and Hubbard et al. (2003) suggest that a "one-off" act of attaining consent is inadequate for people with dementia, particularly in qualitative research, where consent is viewed as a continual, ongoing process between the researcher and the participant. "One-off" consent process is seen to place the person with dementia in a less powerful position than the person seeking the consent (McCormack 2002). A number of terms have been used to describe ongoing consent. These include: process consent (Usher and Arthur 1998; Reid et al. 2001; Dewing 2002, 2007), ongoing negotiated consent (Crossan and McColgan 1999), and narrative-based approach to consent (McCormack 2002). A framework incorporating key elements of process consent models is proposed by Dewing (2008) where consent is seen as a process running through the entire research project. These elements include:

preparation and background, establishing basis for consent, initial consent, ongoing consent monitoring, and feedback and support. McKeown et al. (2010) report how they utilized this framework within their study on the use of life story work with people with dementia. Researchers may choose to consider this as a pathway to undertaking research with people with dementia in addition to more traditional approaches.

In their review of strategies to maximize the involvement of people with dementia in research, Murphy et al. (2015) propose the CORTE guidelines. Reported in relation to undertaking interviews with people with dementia, the guideline is characterized by attentions to: **CO**nsent, maximizing **R**esponses, **T**elling the story, and **E**nding on a high. **CO**nsent is understood through face-to-face encounters where an importance is placed on getting to know the person. Research assistants were trained to attend to verbal, nonverbal, and behavioral responses of people with dementia during the consent process. Such approaches are thought to reduce people being coerced into research participation. Maximizing **R**esponses necessitates adaptations to the research interview to enable the person to participate as fully as possible. Again, there is an emphasis on getting to know the person and to conduct interviews at a pace, time, and location to suit the individual with dementia. **T**elling the story is a consideration of alternative approaches such as the use of prompts or diaries to help the person share their unique experiences. An important aspect is to enable the voice of the person to be heard as separate from that of their family carer. **E**nding on a high emphasizes the importance of leaving the person with a sense of achievement from the research encounter. This may require building in time after the research interview to spend talking with the person.

As dementia progresses, however, it is accepted that abilities of comprehension, making judgements, reasoning, communicating, and remembering may become increasingly impaired (Hubbard et al. 2002). Thereby, capacity to informed consent may be affected. There is also the possibility that some people may have capacity to consent at the start of a research project but may lose capacity during the project. Some people with dementia have fluctuating levels of capacity to consent. The perceived associated complexity and additional ethical approvals may mean that researchers choose to exclude people with dementia who are assessed as not having capacity to consent. This would appear to be the case as research accounts of people with dementia lacking capacity are scarce.

In the past, proxy consent was often used in place of informed consent from the person with dementia. This approach identified a 'proxy'; most often a relative or close supporter, who knew the person before they developed dementia. This person is then asked, based on the past wishes and values of the person with dementia, to make a decision about whether they should or should not participate. A weakness in this approach is that often the person with dementia and their carer will have different views. The potential for conflict between people with dementia and their proxies has been highlighted by researchers (Sachs et al. 1994; Stocking et al. 2006). Communication with carers of people with dementia suggests that they are often surprised at the choices their relatives make with regard to diet or participation in activities, compared to the past, so it is difficult to ensure that other values and preferences remain the same with the experience of dementia. Being a proxy decision-maker may prove a burdensome activity for some care givers, and Bartlett and Martin (2002) draw attention to the lack of practical guidance on the best way to involve carers in the process. A further concern regarding the use of a proxy is that the person with dementia is not always meaningfully included in the process as the attention is on the researchers and proxy's responsibilities (Dewing 2002), consequently disempowering a person who has made their own decisions throughout life.

Approaches to the involvement of people lacking capacity to consent vary between nations. It is, therefore, incumbent for researchers to be aware of the ethical frameworks relevant to the country in which the research is to be undertaken. In England and Wales, the Mental Capacity Act (DH 2005) provides clear guidance to researcher on including people with dementia who are not able to consent. Additionally, the British Psychological Society (Dobson 2008) developed further helpful guidance, based on the MCA and the Incapacity Act 2000 in Scotland, to support researchers through the process. Case studies are used to exemplify the process.

When a person is assessed as lacking capacity, the Mental Capacity Act advises the appointment of a "consultee." This person does not "consent" on behalf of their relative; rather, they advise whether they think their relative would want to participate based on past wishes and "best interests." The researcher, however, is obliged to ensure that involvement is in participants' best interests and is in keeping with their wishes. Assent from the person with dementia is required even when the proxy has given consent. Assent is defined as "a verbal agreement to participate based on less than full understanding" (Keyserlingk et al. 1995, p. 340). McKillop, who has dementia (McKillop and Wilkinson 2004), urges researchers to seek permission from the person with dementia before interviewing but believes the involvement of family is important. He suggests differing views to participation between the carer and the person with dementia can be negotiated with the researcher to prevent any confrontation and that "if anything goes wrong they (the carer) are left to pick up the pieces" (p. 119). Similarly, advice from the SWDG (2015) is that researchers should communicate with people who know the person with dementia well. This becomes especially important when there has been a gap between research encounters, to find out whether the circumstances of the person with dementia have changed in any way. Carmody et al. (2014) highlight that there remains no consensus among researchers on the application of assent and dissent in research with people with dementia. Perhaps, a broadening of the debate beyond capacity to consent or not would over time lead to more inclusive engagement of people with a wider range of experience of dementia into research participation.

Dewing (2008) and McCormack (2003) argue that researchers need to present a range of alternative approaches to ethics committees, and consent must move towards methods that fully engage the person lacking in capacity in the research process. Ethical processes do remain challenging for researchers. King et al. (2016, p. 26) report that "stubborn determination" was required to ensure that simplified information sheets remained understandable to people with dementia while also meeting the needs of the ethics committee. While accepting the need for enhanced ethical procedures, Henwood et al. (2015) draw attention to the strain it can place on research budgets and timescales. Similarly, Holland and Kydd (2015) outline the learning that emerges about ethical issues during the course of a research study and the necessary amendments that may need to be made. As the ethical debate broadens in the literature, then researchers can learn from one other and avoid replicating mistakes.

3.3 Research Methods and Approaches

A further consideration needed in research with people with dementia is the choice of research methods to ensure they are an appropriate approach meaningfully gather the experiences of the participant. This section explores some of the necessary adaptations required to usual methods and identifies the emergence of more creative approaches.

Interviews remain the most reported research data collection with people with dementia. The literature draws attention to the adaptations needed to enable people with dementia to participate in a meaningful way. The development of rapport between the researcher and the person with dementia before the interview is important. This may mean a "preinterview conversation" (Digby et al. 2016) or "chit-chat" (Murphy et al. 2015) on a nonresearch-related topic to convey a genuine interest in the person. This is consistent with the ethical frameworks and guidance discussed earlier in the chapter. A flexible approach to gathering data is preferable to a rigid structure. In reflecting on their research design, Hubbard et al. (2003) concluded that they did not provide the flexibility required to respond to the needs of individual participants. They give the example of how a rigid research protocol prevented them from being able to respond to a participant who was more communicative on a day the researcher happened to be on the ward to see another participant. Flexible approaches are supported by McKillop and Wilkinson (2004), for example, arranging a further visit if a person is becoming tired. It is helpful to establish the "best time of day" for the person with dementia to participate and to avoid noisy environments which can be distracting and make it harder for the person to concentrate. The SWDG (2015) identify the need to feel safe and secure as paramount for any meaningful research activity with people with dementia. Research methods need to instil this sense of security in people with dementia.

The empirical evidence-base to support the use of focus groups as a research method with people with dementia appears to be lacking. Reports from the use of focus groups in service development initiatives identify the method as relevant for people with dementia. A particular reported advantage is the peer support that is offered and that people with dementia can take cues from the responses of others if they lose track of the topic. Again, adaptations are advised. For example, it is best to focus on just one topic for discussion, and it can help if the group is held in an environment that is the focus of the conversation and if people are grouped by topic interest (Bamford and Bruce 2002; Savitch et al. 2006). Bamford and Bruce (2002) report storytelling by people with dementia as a potential hindrance to gathering data relevant to the research topic. Skilled facilitation is required to enable people to tell their stories while also bringing them back to the research topic and can also help ensure no one person dominates the group discussion (Savitch et al. 2006). Focus groups become more difficult to use with people with more advanced dementia and communication difficulties and with people who find it hard to remain seated for very long (Bamford and Bruce 2002).

Traditional methods of data collection may be less meaningful for people with dementia as their condition advances. It is incumbent on researchers to develop more innovative and creative approaches to enable the views of people with dementia to be heard in research (Alzheimer's Europe 2011). Killick (2001), in discussing how best to gain the views of people with dementia, believes that direct questioning can lead to anxiety and increased confusion. He suggests that time and encouragement is needed to "tease out" their perspectives; he often represents people's views in poetry or narrative. In the context of service evaluation, Murphy (2007) notes that people with dementia can perceive an interview as a "test" and feel under pressure. He urges evaluators to prioritize the relationship with the person over the asking of questions. If discrepancies are evident in a conversation with the person, it is important these are not "thrown in the person's face" but dealt with through sensitive questioning (McKillop and Wilkinson 2004).

Researchers are rising to the challenge and starting to explore alternative approaches to research and or service evaluation with people with dementia. For example, photographic storyboards were used to explore care transitions for people with dementia by Parke et al. (2015). However, their innovative approach to hearing the voice of people with dementia identified challenges as the dementia progressed. Talking Mats© have been advocated as a visual way of seeking the views of people with dementia (Murphy et al. 2010). Bartlett (2012) shares her experience of using video diaries and conversations arising out of other activities, such as walking, hand massage, and singing, have been found to be both enjoyable and providing a promising approach to gaining the views of people with dementia (Allan 2001). Participatory video as a research method is detailed by Ludwin and Capstick (2011). Over the next few years, it is likely that these and other methods will become the subject of empirical examination and do suggest a future where broader range "dementia-specific" research methods can be drawn upon.

4 Considerations in Involving People with Dementia in Research

It was previously suggested that, on the whole, people with dementia do benefit from sharing their views and experiences in research encounters. However, there are some considerations that researchers should be aware of.

Carmody et al. (2014) highlight that some people with dementia may not fully understand the boundaries of research conversations and may share experiences that are more private. This is reported by McKeown et al. (2010), giving the example of a participant who shared very personal information as part of a research study. In this case, the researchers excluded the personal details from the research study. It could be construed that the person with dementia had lost capacity, but it may be that the relationships formed as part of an inclusive methodology blur the boundaries for some people between a research conversation and a friendly chat. This highlights the responsibilities that researcher need to accept before working with this group of people.

Carmody et al. (2015) identify a further challenge that researchers may become aware of inappropriate carer/partner behavior as part of the research encounter that warrants notification to relevant authorities.

In contrast to the previously reported experience of Bamford and Bruce (2002) of stories being a challenge within focus groups, Digby et al. (2016) describe "tangential stories" as a means for people with dementia to take some control over the research encounter. Such stories are often used if a person is uncertain of what is being asked or what to say. Stories about the past may also be interpreted as reflecting a sense of how the person is feeling in the present (Digby et al. 2016). The challenge for the researcher is to value personal stories and know how to interpret them and sensitively steer the conversation back to the research topic in a respectful way.

Concern has been raised over ensuring that terms used to describe dementia are in keeping with the individual understanding of the person with dementia about their diagnosis. The term "dementia" was not used by Hellstrom et al. (2007) unless it was introduced by the person or their family; the term "memory problem" being used instead. Bartlett and Martin (2002) ask whether fully informed consent is only possible when the person is fully aware of their diagnosis, but concomitantly, appreciating the harm and distress that may be evoked by a researcher unwittingly giving the person a diagnosis. It appears important to meet the participants on their own terms and not insist on them admitting that they have dementia.

5 Methodological Considerations

The choice of methodology is an important consideration for undertaking research with people with dementia. It is suggested that more traditional research paradigms perhaps do not take into consideration some of the more participatory approaches that may be helpful in dementia research (Swarbrick 2015). Ontological and epistemological positions need to be considered and made transparent in methodological choices (see also ▶ Chap. 6, "Ontology and Epistemology").

Power is an important issue and involves more than the different status of researchers and the researched, particularly when the factors of age and disability are considered. Furthermore, the different status attributed to health professionals and researchers compared to service users must be acknowledged. It must be questioned whether a nonhierarchical researcher/participant relationship is ever possible or even desirable, and Miller (1998) underlines the dangers of participants divulging more then they may have wished if they believe the relationship to be reciprocal.

As person-centered practices develop then so too should person-centered research. Such approaches are characterized by a "sustained commitment to participants to ensure the value of the person is held central" (McCormack 2003, p. 182). This especially is relevant for people with dementia and their carers to ensure that they do not feel "used." Reflecting on the McKillop's experience as a person with dementia, McKillop and Wilkinson (2004) suggest that if the researcher is not authentic, warm, and genuine, then the person with dementia may pick up on this and be uncomfortable in an interview. Prompting a person-centered approach to research, McCormack (2003) urges researchers to avoid the "hit and run" approach and consider what should be offered to participants following the research. This might be a copy of their recorded interview, sharing findings, ongoing supervision, or training for staff. Participating in research and having their views and experiences heard and valued was perceived by Cowdell (2008) as "nourishing" the personhood of participants with dementia. Such person-centered approaches privilege the needs of the person with dementia over the research encounter itself.

McCormack (2003) argues that person-centered research involves researchers being sensitive and prepared for the variety of unpredictable challenges that may arise in the practice setting. In the prevailing culture of person-centered care and research, perhaps researchers without these skills should not be undertaking such research. An emphatic statement by the SDWG (2014) is that researchers need to have training before being "let loose" with people with dementia. There does need to be an understanding of dementia and sensitivity as to how it affects people on an individual basis (McKeown et al. 2010), certainly if person-centred principles are to be adhered to.

6   Conclusion and Future Directions

The chapter has summarized the range of ways that people with dementia can be "involved" in research more generally, before specifically focussing on the involvement of people with dementia as participants in research studies.

A number of challenges have been identified, ranging from assumptions of others about dementia, ethical and legal aspects along with practical issues to involving people. These challenges, while often there to protect the person with dementia, can result in excluding people from taking part in research. While this is a tragedy for the quality of research, it is also disappointing that people with dementia cannot benefit from feeling valued through sharing their experiences and knowing that they are contributing to a greater understanding of dementia.

Reflecting on the literature over the past 25 years has demonstrated advances, in that people with dementia are participating more in research studies and are viewed far more respectfully than just "disease entities." This has perhaps benefited research that involves people with dementia with capacity to consent and people in the earlier stages who are able to communicate and participate using conventional research data collection methods. The next few years may see further advancements in understanding how to best involve people lacking in capacity along with a broader range of data collection approaches to gather experiences in a meaningful way. Ethical processes must continue to protect people with dementia, especially those who may not have capacity to consent, but at the same time must not be so onerous that they deter researchers from including people with dementia in studies.

References

Allan K. Communication and consultation: exploring ways for staff to involve people with dementia in developing services. Bristol: The Policy Press; 2001.

Alzheimer Europe. The ethics of dementia research. Alzheimer Europe Report. 2011. http://www.alzheimer-europe.org/EN/Ethics/Ethical-issues-in-practice/2011-Ethics-of-dementia-research. Accessed 27 Oct 2016.

Bamford C, Bruce E. Successes and challenges in using focus groups with older people with dementia. In: Wilkinson H, editor. The perspectives of people with dementia: research methods and motivations. London: Jessica Kingsley; 2002. p. 139–64.

Barnett, E. Involving people with dementia in designing and delivering care: 'I need to be me!' London: Jessica Kingsley Publishers; 2000.

Bartlett R. Modifying the diary interview method to research the lives of people with dementia. Qual Health Res. 2012;22(12):1717–26.

Bartlett H, Martin W. Ethical issues in dementia care research. In: Wilkinson H, editor. The perspectives of people with dementia: research methods and motivations. London: Jessica Kingsley Publishers; 2002. p. 47–62.

Carmody J, Traynor E, Marchetti E. Barriers to qualitative dementia research: the elephant in the room. Qual Health Res. 2014;25(7):1013–9.

Clarke C, Keady J. Getting down to brass tacks: a discussion of data collection. In: Wilkinson H, editor. The perspectives of people with dementia: research methods and motivations. London: Jessica Kingsley Publishers; 2002. p. 25–46.

Cottrell V, Schultz R. The perspective of the patient with Alzheimer's disease: a neglected dimension of dementia research. The Gerontologist. 1993;33(2):205–11.

Cowdell F. Engaging older people with dementia in research: myth or possibility? Int J Nurs Older People. 2008;3(1):29–34.

Crossan B, McColgan G. Informed consent: old issues re-examined with reference to research involving people with dementia. Paper presented at the British Sociological Association annual conference, Glasgow; 1999.

Department of Health. Mental Capacity Act. London: HMSO; 2005.

Dewing J. From ritual to relationship: a person-centred approach to consent in qualitative research with older people who have dementia. Dementia: Int J Soc Res Pract. 2002;1(2):157–71.

Dewing J. Participatory research: a method for process consent with persons who have dementia. Dementia: Int J Soc Res Pract. 2007;6(1):11–25.

Dewing J. Process consent and research with older persons living with dementia. Res Ethics Rev. 2008;4(2):59–64.

Digby R, Lee S, Williams A. Interviewing people with dementia in hospital: recommendations for researchers. J Clin Nurs. 2016;25(7–8):1156–65.

Dobson C. Conducting research with people not having the capacity to consent to their participation: a practical guide for researchers. Leicester: British Psychological Society; 2008.

Downs M. The emergence of the person in dementia research. Ageing Soc. 1997;17(5):597–607.

Hellstrom I, Nolan M, Nordenfelt L, Lundh U. Ethical and methodological issues in interviewing persons with dementia. Nurs Ethics. 2007;14(5):608–19.

Henwood T, Baguley C, Neville C. Achieving ethics approval in residential aged care research: a protective process or barrier. Australas J Ageing. 2015;34(3):201–2.

Holland S, Kydd A. Ethical issues when involving people newly diagnosed with dementia in research. Nurs Res. 2015;22(4):25–9.

Hubbard G, Downs M. Tester S. Including the perspectives of older people in institutional care during the consent process. In: Wilkinson H, editor. The perspectives of people with dementia: research methods and motivations. London: Jessica Kingsley; 2002. p. 63–82.

Hubbard G, Downs M, Tester S. Including older people with dementia in research: challenges and strategies. Aging Ment Health. 2003;7(5):351–62.

Hughes T, Castro Romero M. A processual consent methodology with people diagnosed with dementia. Qual Ageing Older Adults. 2015;16(4):222–34.

Kelly S, Lafortune L, Hart N, Cowan K, Fenton M, Brayne C. Dementia priority setting partnership with the James Lind Alliance: using patient and public involvement and the evidence base to inform the research agenda. Age Ageing. 2015;44(6):985–93.

Keyserlingk E, Glass K, Kogan S, Gauthier S. Proposed guidelines for the participation of persons with dementia as research subjects. Perspect Biol Med. 1995;38(2):319–61.

Killick J. "The best way to improve this place": gathering views informally. In: Murphy C, Killick J, Allan K, editors. Hearing the user's voice: encouraging people with dementia to reflect on their experiences of services. Stirling: Dementia Services Development Centre; 2001. p. 6–9.

King A, Hopkinson J, Milton R. Reflections of a team approach to involving people with dementia in research. Int J Palliat Nurs. 2016;22(1):22–7.

Kitwood T. Dementia reconsidered: the person comes first. Buckingham: Open University Press; 1997.

Litherland R. Developing a national user movement of people with dementia: learning from the Dementia Engagement and Empowerment Project (DEEP). Joseph Rowntree Foundation. 2015. https://www.jrf.org.uk/report/developing-national-user-movement-people-dementia. Accessed 27 Oct 2016.

Ludwin K, Capstick A. Using participatory video to understand diversity among people with dementia in long-term care. J Psychol Issues Organ Cult. 2011;5(4):30–8.

McCormack B. The person of the voice: narrative identities in informed consent. Nurs Philos. 2002;3(2):114–9.

McCormack B. Researching nursing practice: does person-centredness matter? Nurs Philos. 2003;4(3):179–88.

McKeown J, Clarke A, Ingleton C, Repper J. Actively involving people with dementia in qualitative research. J Clin Nurs. 2010;19(13–14):1935–43.

McKillop J. Did research alter anything? In: Wilkinson H, editor. The perspectives of people with dementia: research methods and motivations. London: Jessica Kingsley; 2002. p. 109–14.

McKillop J, Wilkinson H. Make it easy on yourself! Advice to researchers from someone with dementia on being interviewed. Dementia: Int J Soc Res Pract. 2004;3(2):117–25.

Miller T. Shifting layers of professional, lay and personal narratives. In: Ribbens J, Edwards R, editors. Feminist dilemmas in qualitative research. London: Sage; 1998. p. 58–71.

Murphy C. User involvement in evaluations. In: Innes A, McCabe L, editors. Evaluation in dementia care. London: Jessica Kingsley Publishers; 2007. p. 214–29.

Murphy J, Gray C, Wyke S, Cox S, van Achterberg T. The effectiveness of the talking mats framework in helping people with dementia to express their views on well-being. Dementia: Int J Soc Res Pract. 2010;9(4):454–72.

Murphy K, Jordan F, Hunter A, Cooney A, Casey D. Articulating the strategies for maximising the inclusion of people with dementia in qualitative research studies. Dementia: Int J Soc Res Pract. 2015;14(6):800–24.

National Institute for Health Research. Patient and public involvement in health and social research: a handbook for researchers. 2014. http://www.rds.nihr.ac.uk/wp-content/uploads/RDS-PPI-Handbook-2014-v8-FINAL.pdf. Accessed 27 Oct 2016.

Parke B, Hunter K, Marck P. A novel visual method for studying complex health transitions for older people living with dementia. Int J Qual Methods. 2015;14(4):1–11.

Reid D, Ryan T, Enderby P. What does it mean to listen to people with dementia? Disab Soc. 2001;16(3):377–92.

Robinson E. Should people with Alzheimer's disease take part in research? In: Wilkinson H, editor. The perspectives of people with dementia: research methods and motivations. London: Jessica Kingsley Publishers; 2002. p. 101–7.

Sachs G, Stocking C, Stern R, Cox D, Hougham G, Sachs R. Ethical aspects of dementia research: informed consent and proxy consent. Clin Res. 1994;42(3):403–12.

Savitch N, Zaphiris P, Smith M, Litherland R, Aggarwal N, Potier E. Involving people with dementia in the development of a discussion forum: a community-centred approach. In: Clarkson J, Langdon P, Robinson P, editors. Designing accessible technology. London: Springer; 2006. p. 237–47.

Scottish Dementia Working Group Research Sub-group. Core principles for involving people with dementia in research: innovative practice. Dementia: Int J Soc Res Pract. 2014;13(5):680–5.

Sherratt C, Soteriou T, Evans S. Ethical issues in social research involving people with dementia. Dementia: Int J Soc Res Pract. 2007;6(4):463–79.

Stocking C, Hougham G, Danner D, Patterson M, Whitehouse P, Sachs G. Speaking of research advance directives: planning for future research participation. Neurology. 2006;66(9):1361–6.

Swarbrick C. The quest for a new methodology for dementia care research. Dementia: Int J Soc Res Pract. 2015;14(6):713–5.

Taylor J, DeMers S, Vig E, Borson S. The disappearing subject: exclusion of people with cognitive impairment and dementia from geriatrics research. J Am Geriatr Soc. 2012;60(3):413–9.

Usher K, Arthur D. Process consent: a model for enhancing informed consent in mental health nursing. J Adv Nurs. 1998;27(4):692–7.

Williamson T. A stronger collective voice for people with dementia. Joseph Rowntree Foundation. 2012. http://www.jrf.org.uk/publications/stronger-collective-voice. Accessed 8 Aug 2015.

Witham G, Beddow A, Haigh C. Reflections on access: too vulnerable to research? J Res Nurs. 2015;20(1):28–37.

Researching with Children

Graciela Tonon, Lia Rodriguez de la Vega, and Denise Benatuil

Contents

1 A New Approach in Research with Children	2008
2 Sociopolitical Dimension	2009
3 Cultural Dimension	2010
4 Psychological Dimension	2012
5 Different Methodologies	2013
5.1 Quantitative Methods	2013
5.2 Qualitative Methods	2015
5.3 Mixed Methods	2016
6 Ethics and Research with Children	2016
7 Conclusions and Future Directions	2018
References	2019

Abstract

Research with children is a vast and complex field, as it is influenced by the conceptions of childhood prevalent in each historical period, each particular culture, and each research team conception. In addition, research with children encompasses different points of view: social, political, cultural, and

---

G. Tonon (✉)
Master Program in Social Sciences and CICS-UP, Universidad de Palermo, Buenos Aires, Argentina

UNICOM- Universidad Nacional de Lomas de Zamora, Buenos Aires, Argentina
e-mail: gtonon1@palermo.edu

L. Rodriguez de la Vega
Ciudad Autónoma de Buenos Aires, University of Palermo, Buenos Aires, Argentina
e-mail: liadelavega@yahoo.com

D. Benatuil
Master Program in Social Sciences and CICS, Universidad de Palermo, Buenos Aires, Argentina
e-mail: dbenatuil@iname.com

© Springer Nature Singapore Pte Ltd. 2019
P. Liamputtong (ed.), *Handbook of Research Methods in Health Social Sciences*,
https://doi.org/10.1007/978-981-10-5251-4_123

psychological. This chapter provides a review of the current state and the recent developments in each of these fields. Research with children presents researchers with the challenge of finding methods that are well-suited to children and that recognize the importance of children's experience and agency. Such methods should promote a respectful approach based on ethics. We conceive children as the real protagonists, and thus believe they need to be addressed directly. In the same way, children's self-expression, understanding, and empowerment should be promoted through the use of different techniques. For the purposes stated above, this chapter explores the possibility of using quantitative, qualitative, and mixed methods and the emerging of new proposals such as the inclusion of technologies and arts-based methods that present significant future perspectives.

Keywords

Research methods · Children · Sociopolitical · Culture · Psychology · Agency

1   A New Approach in Research with Children

Children represent a particular population group that presents characteristics in relation to the historical time in which they live. They show their perceptions and opinions about reality in a free and creative way. Therefore, to work with them, researchers need to do it in a participative form. This chapter proposes a reflection on research with children from different points of view: social, cultural, psychological, and political, on the basis of different quantitative and qualitative methodologies.

In the 1990s, Qvortrup stated that childhood was a specific and distinct social structure, a permanent social category that is exposed to the same forces that affect adulthood but in a different manner, subject to both paternalism and marginalization. Over the decades that followed such contribution, children's everyday life has been considered to develop in three levels: a material level consisting of the economic, work-related, and technological dimensions; a level of social relations including the family and the community; and a cultural level comprising values and opinions (Gaitán Muñoz 1999).

Childhood, as a social category, has been traditionally defined in a disqualifying manner, as everything that children could not yet be or do, or by comparing their current roles with those they might perform in the future, when they grew up, disregarding what they could do in the present time. In addition, it was a widespread practice in the social sciences to consider children as not competent enough to provide information about their personal and social experiences, as if they were passive observers of the processes they are part of and that take place in their lives (García and Hecht 2009).

Studies concerning children's lives have historically focused on asking adults. In this respect, Hirschfeld (2002) considers that the absence of children as research participants is due to an impoverished view of cultural learning – a view that overestimates the role of adults and underestimates the contribution of children in cultural reproduction – and to the general disregard for the scope and force of children's culture.

In this chapter, we aim to place children as true protagonists of this study; hence our decision to ask them directly. Thinking about children as protagonists of research implies exceeding the traditional model that is based on the idea that the researcher is the only bearer of knowledge and the person investigated is the passive object that the researcher deals with. It is important to go beyond the idea that boys and girls are the objects of the research and consider them the subjects of it. According to Sen (2000), they are perceived as active agents of change instead of passive recipients of benefits.

This change in the approach of childhood researchers implies a reconfiguration of the researcher-person bond that leads to a new situation according to which both are social and political subjects, bearers of a biographical situation.

Taking into account the theoretical considerations above, the focus of this chapter is to move forward in the perception of children as agents and protagonists. Nowadays, in the research field of social sciences, the condensation and presentation of data on a topic is not sufficient; it is necessary to consider how representations and dialogues have originated and which is the perspective of the social actors (Appadurai 2001).

Research with children presents a challenge that, firstly, leads us to revise our practice as researchers to go on building a scenario that allows the true prominence of the person.

2 Sociopolitical Dimension

We agree with Gaitán Muñoz (1999) that children construct the social reality they live in, transferring experiences to those who will follow them in time, recreating the reality they have been given and developing their own culture. This protagonism strengthens children's abilities and demands, as well as their independent and influential role in society (Liebel 2007).

Children also perform a political role if we define politics according to Rabello de Castro (2007), as the activity that brings people together for them to be able to interpret their own existence and the world that surrounds them. In this sense, politics is understood as a space to build our common sense and collective action. This notion is rooted in Ancient Greece, where interest in children was not for what they were but for what they would become in the future as adults in charge of the *polis* (Kohan 2003). In current societies and according to Kohan (2003), children represent something that is guarded; they are one of the strongest symbols of lack of freedom and power. This is a reason why the subject of emancipation becomes all the more interesting.

The prominence of children and the development of their freedom does not only refer to the autonomy or independence they have, it is also based on their active relationship with the world around them, which is related to the social structure where they develop their lives and which gives them the possibility of having an active role in society (Liebel 2007).

It is important to point out that "participation implies considering subjects as the protagonists of the decisions; in this way, participation is more than acting together, it is about making decisions together" (Tonon 2012, p. 15). At the same time, the possibilities of participation depend on the nature of social institutions and, in this sense, the State and society play a leading role in terms of responsibility (Sen 2000). The reason for this is that the construction of any democratic society requires the participation of citizens since childhood.

3 Cultural Dimension

From a symbolic perspective, we may define culture as a "universe of senses" (Giménez 2005), and identity as a group of cultural repertories that have been internalized by social actors, allowing them to differentiate themselves by symbolically creating social frontiers (Giménez 2000). Against this background, studies have been conducted that focus on children in different cultural environments, evidencing the way those environments contribute to creating specificities, that need to be addressed when a research process develops.

Fossheim (2013) sustains that research with children faces different challenges and suggests that there is the need to reflect about three principles: respect, beneficence, and justice of which every culture has different ways to express.

Meanwhile, Boddy (2013) notes that when considering, for example, ethnicity, it should be noted that it covers many different aspects and adds that it is necessary to recognize the intersectionality, i.e., social class, ethnicity, race, sexuality, and gender, while the identity of study participants as children is an element of additional intersection.

In considering this issue, adult research has helped to develop some ethical principles that in his consideration can be extended to studies with children, such as: ensuring given freely fully informed consent and the right to withdraw from research participation.

Other studies analyze the difference between the so-called individualistic and collectivist cultures. Within this framework, Hanson (1992) contends that in individualistic cultures, caregivers encourage children to develop a behavior that will allow them to act independently as early as possible. Kibria (1993), on the other hand, states that in collectivist cultures children are encouraged to ask adults for help instead of solving their problems on their own. Such behavior, according to Kibria, fosters a greater confidence in the other person and, potentially, leads to greater cohesion within the group.

In the context of other studies that consider the cultural specificities of children participating in them, we can mention studies that have reviewed cultural differences among different groups in relation to language. Clark (2000), for example, found socioeconomic differences in parent-child dyads: professional parents tend to talk more to their children than working-class parents and these, in turn, talk more to their children than parents in poverty. These particularities were strongly connected with the vocabulary used by 3-year-old children.

Hymes (1967), on the other hand, argues that every culture develops its own concept of communicative competence. Moreno (1997), for example, highlighted the differences between American-European and Latin mother-child dyads in connection with the tying of shoelaces. The study focused on questions mothers ask their children aged between 3 and a half and 5: although the questions were similar, generally, American and European mothers asked questions requiring an answer based on the immediate field of perception, while Latin mothers usually made questions about mental representations that were beyond the child's immediate field of perception.

In considering different cultural groups, an obvious reference should be made to children that grow up in bilingual or plurilingual environments. In this respect, different studies (Jackson-Maldonado et al. 1993; Junker and Stockman 2002) have concluded that no delay or particularity in language development was observed in children in bilingual homes. However, Clark (2000), among others, states that it may be very detrimental for a child to learn a language without being able to use it later on because the environment does not present opportunities to put it into practice. Fernandez (2007) refers to different studies that consider that second-language learning contributes to the development of met linguistic awareness (necessary for reading) in that it expands a child's idiomatic experience (Yelland et al. 1993; Liddicoat 2001).

In respect to resources related to cultural sensitivity, Roer-Strier and Rosenthal (2001) state that every parent has an image of their child that guides their childrearing and socialization practices. This image is so fundamental that parents carry it with them even when they immigrate to another context, where it may even limit the child, but it is so deeply held that it is not questioned by the parents.

There are other differences and similarities among different cultural groups. For example, Fu et al. (2007) studied moral understanding in individual and collective-oriented groups from an analysis of truth and lies with children aged 9–11 from Canada and China. This analysis was made through the reading of stories with characters that face moral dilemmas about whether to lie or tell the truth to help a group but harm an individual or vice versa. After reading, the children were required to do certain activities connected to those stories. The major cultural differences lay in choices and moral evaluations. Chinese children chose lying to help a collective but harm an individual, and they rated it less negatively than lying with opposite consequences (they also rated truth telling to help an individual but harm a group less positively than the alternative). Canadian children did the opposite. According to the authors, the major findings obtained were the following: (a) few cross-cultural differences were found in children's categorizations of truths and lies; (b) the cultural environment in which children are socialized plays a significant role in their decisions about whether they might lie and the moral evaluations of lying and truth telling; and (c) there is an interaction between age and children's choices and moral judgments of lying and truth telling (as age increased, Chinese children's choices and moral evaluations increasingly favored the interests of a group over truthfulness and Canadian children became less stringent in their insistence on being truthful and were more inclined to protect the individual). These findings suggest that enculturation processes may play a relevant role in children's development of moral distinctions between truthful and untruthful communications.

4 Psychological Dimension

The perception of childhood has changed over time and these changes have had an impact on all spheres, including psychology and education. From being seen as unimportant in society, children became kings worthy of unconditional love (Aries 1962), and hold at present an intermediate position in which they receive affection and have limits set on them. They also have rights and responsibilities, interests, needs, concerns, and fears. They are regarded as research participants. The new outlook on childhood recognizes children's potential agency, normalizing them as individuals who take part in research in relation to adults (Christensen and James 2000; Mieles Barrera and Tonon 2015).

Research with children in the field of psychology is vast and has undergone different stages. Traditionally, the focus was placed on pathologies, risks, child mortality, use and abuse of substances, violent and risky behavior. These variables were correlated to, for instance, parental styles and family features (Kwan and Ip 2009), cultural differences (Szapocznik and Kurtines 1993; Berry 1997; Lau et al. 2005), socioeconomic variables such as poverty (ECLAC/UNICEF TACRO 2010; Espíndola Advis and Rico 2013; Bornstein and Bradley 2014), and sociodemographic variables such as the level of education (WHO 2012).

The study of salutogenic aspects, such as children's characteristics and possibilities at different stages of their lives, was overlooked for quite a long time. Most common measures of early childhood development pertain to deficiencies in achievements, problem behaviors, and negative circumstances. The absence of problems or failures, however, does not necessarily indicate proper growth and success (Moore et al. 2004; Ben-Arieh 2005). Measures of risk factors or negative behaviors are not the same as measures that gauge the presence of protective factors or positive behaviors (Aber and Jones 1997). In recent years, psychological research has moved from a focus on human distress and psychopathology to happiness, and life satisfaction (Ben-Arieh et al. 2013).

This new theoretical outlook oriented to working on potentialities rather than deficiencies. Thus, placing great value on children's accomplishments, strengths, and values provides a fairer look, and does not pathologize childhood or focus on what children lack in comparison with adults, on what still has not been accomplished.

This paradigm, described as positive psychology or salutogenic approach, has yielded a plethora of research and developments on different aspects. In the field of psychology, well-being is probably the most widely used approach. It has prompted much quantitative and qualitative research, and various instruments have been developed to measure it, with some of them having been used in different languages and countries (Lyubomirsky and Lepper 1999; Ben Arieh 2000).

Extensive research reveals the factors identified by children as central to the development of well-being: love, care, attention, support, security, and the company of their parents; family economic and labor stability; time for playing and sharing with their friends; obtaining high grades at school and taking part in cultural, sports, or artistic activities; having access to technology; not being ill; being satisfied with their physical appearance; experiencing values such as respect, sharing, responsibility, and helping others.

Dissatisfaction is associated with the following: quarrels among parents on issues related to child support, among others; the death of someone close; insecurity, slovenliness, disorder, and traffic problems in the cities where children live; problematic situations undergone by children in the city or bullying; parents' economic and labor difficulties; unfavorable conditions at school, reduced spaces; being unable to participate in community groups or activities; feeling sick (Bradshaw et al. 2007; Mieles-Barrera and Tonon 2015).

In addition to well-being, the salutogenic approach also includes other aspects such as happiness, resilience, children's personal life skills such as self-esteem, assertiveness, work capacity, safety, physical status, children's engagement in work, play, social interactions (Lyubomirsky and Lepper 1999; Li et al. 2011; Sanders et al. 2012; Ager 2013; Peterson 2013).

This shift in approach has been fundamental, as it places a growing emphasis on child well-being rather than just on survival, and on enhancing positive outcomes rather than confronting negative impacts, as well as on the voices of children rather than only on adult perspectives (Kamerman 2010).

5 Different Methodologies

The selection of the appropriate research method depends not only on the fact that we are working with children, but also, and fundamentally, on the social, political, cultural, and economic context in which children's lives develop. We, thus, delineate three possible options: the quantitative method, the qualitative method, and mixed methods.

5.1 Quantitative Methods

The field of research with children usually employs quantitative, qualitative, and mixed methods. Each of them presents its own strengths and weaknesses, which must be thoroughly understood for an appropriate selection of methods in each case. The quantitative approach is extensively used for general research purposes and research with children is no exception.

Quantitative methods offer certain advantages, and they prove extremely useful when large samples need to be analyzed, as their level of standardization allows collecting data more easily and quickly, and the results are less prone to researcher bias (Sampieri et al. 1996; Pita Fernández and Pértegas Díaz 2002; Ben-Arieh 2008; see also ► Chap. 63, "Mind Maps in Qualitative Research").

One of the advantages of standardized instruments such as indices, surveys, and questionnaires is that they can be used in similar conditions with different populations and contexts. This facilitates comparisons among different groups, cultures, and nations. These methods are important and necessary as they allow obtaining, among others, epidemiologic data and nationwide indices, which will provide a basis for the design of public policies, allocation of resources in social and health areas, and so on (Cook and Reichardt 1986; Sampieri et al. 1996; Binda and Balbastre-Benavent 2013).

Traditionally used in the field of research with children, quantitative research allows comparing groups and obtaining indices, such as poverty and well-being. Standardized instruments are used to compare variations among countries and cultures, and to measure changes over time. Examples of this type of research include: the UNICEF Annual State of the World's Children Reports; Child well-being Innocenti Report Card; Doing better for children – OECD (Chapple and Richardson 2009).

Today, the use of social indicators is widely accepted and recognized as an important tool in shaping social policies. The questions asked in this connection concern the type and quality of the indicators used. Furthermore, when we do collect data and information on the state of our children the question should be asked: What do we measure and by what means? (Ben Arieh 2000).

Regardless of the research method to be used, the researcher must ask himself or herself what he or she is measuring, whether it is appropriate for his or her culture, and the actual reality of the subjects that participate in the research. It is also critical that the researcher questions the viability of the technique to be used for data collection. Before using indices or questionnaires, the researcher must critically assess the participants' possibilities of understanding the instrument used, the need for language adaptation, the use of group data collection, the extent of the instrument and its viability taking into account the participants' age – due consideration should be taken in this respect of the fact that children's concentration span is more limited than adults' – and the convenience of implementing other techniques (such as graphs and visual methods) as part of the process (Punch 2002; see also ► Chaps. 116, "Optimizing Interviews with Children and Youth with Disability," ► 117, "Participant-Generated Visual Timelines and Street-Involved Youth Who Have Experienced Violent Victimization," ► 102, "Understanding Refugee Children's Perceptions of Their Well-Being in Australia Using Computer-Assisted Interviews," ► 99, "Visual Methods in Research with Migrant and Refugee Children and Young People," and ► 100, "Participatory and Visual Research with Roma Youth").

One of the major disadvantages of quantitative methods is that, as much focus is placed on large samples and the comparison of populations, the singularity of the participants is usually overlooked. The quantitative approach tends to provide standardized measurements, with the consequent loss of singularity of both the participant(s) and the researcher's creativity. Therefore, in order to choose the most appropriate method, the researcher should first clearly define the purpose of the research.

5.2 Qualitative Methods

The social reality is constructed through social processes that develop at the same time in a material particular area and other subjective and symbolic. In this context, social actors develop their action within frameworks of certain conditions involving a social world and a natural world, demarcating the borders of their social practices.

The main purpose of using qualitative methods then is to understand the meaning held by the participants regarding the events, situations, and actions in which they are involved, the context in which they act and its influence on their actions, and the process in which actions take place, which at the same time enable the identification and generation of new theoretical understanding about the lives of the participants (Maxwell 1996; Liamputtong 2013; Bryman 2016; see also ► Chap. 63, "Mind Maps in Qualitative Research").

It should likewise be considered that the qualitative approach implies gaining access to the world of the research participants (in this case, children), and involves a report on their cognitive and emotional aspects (Gilbert 2000), which inevitably brings up this topic from the very start, not to mention that any particularly sensitive question regarding the participants' accounts may not only be touching to them but also to the researchers. Moreover, Collins and Cooper (2014) focus on emotional intelligence, regarding it as "a capacity for recognizing our own feelings and those in others for the purpose of motivating and managing our relationships and ourselves" and, considering the complexities of human interaction as well as the complexities of research work – which resorts to interaction as a data collection method – they further point out that emotional intelligence is an innovative alternative to the learning and development of qualitative research techniques. Furthermore, they believe that the emotional intelligence framework includes two main areas: (a) personal competence and (b) social competence, which respectively apply to self-management and social awareness, thus enhancing qualitative research with a more flexible role and, subsequently, more interesting findings.

Considering specifically children, Rodriguez Pascual (2006) notes that there has been a finding that the assumptions about the social life of children underestimated their ability to function as active social agents, interpreting and influencing social situations. This has had both theoretical and methodological effects and although he alludes to sociology, we think it can be considered in general.

For that reason, even if today, we can say that children have been constituted as study subjects per se, they realize their own experiences with their own voices, they are subjects of study from a present dimension and not only as future adults, and all this occurs in the context of the characterization of childhood as a structural and cultural component of societies (James and Prout 1997; Rodriguez Pascual, 2006), we have to remember what Fuhs (1999) points out about the fact that adults occupy a status that is established on the basis of an asymmetrical relationship that children know. That is, in research with children, intergenerational axis is key, with different scopes in research (for example, research is controlled by adults).

5.3 Mixed Methods

When we make reference to mixed research methods, we should clarify that such methods must be distinguished from methodological triangulation or collaboration.

Triangulation was defined by Denzin (1978) as the combination of methodologies for the study of the same phenomena or process. However, in practice, this approach has been dominated by quantitative methods to the detriment of qualitative ones, and it is difficult to find studies which give both methods an equally important weight.

According to Coffey and Atkinson (1996, p. 19), the combination or juxtaposition of different research techniques does not reduce the complexity of our understanding, given that the more we examine our data from different viewpoints, the more we can reveal – or, in fact, construct – about its subjectivity.

The mixed methods approach is a type of research in which the researcher or team of researchers combines qualitative and quantitative elements in order to gain depth of understanding and corroboration (Johnson et al. 2007; see also ▶ Chaps. 4, "The Nature of Mixed Methods Research," and ▶ 40, "The Use of Mixed Methods in Research").

In the case of research with children, the use of mixed methods allows the researcher to understand in a deeper sense what children are thinking and feeling. For example, in an analysis of the quality of life of children it is interesting to use a mixed methods approach in which the use of quantitative methods and qualitative methods can facilitate the comprehension of the opinions children have about their own quality of life, instead of asking the adults about the children's opinions.

The decision of using mixed methods are considered to be an approach to knowledge that integrates theory and practice from multiple viewpoints, perspectives, positions, and standpoints both on a qualitative and quantitative basis (Johnson et al. 2007). In this way, the importance of the use of mixed methods resides in the first question that leads the research project (Tonon 2015).

When we refer to mixed methods research, we need to explain that we do not understand those methods to be the sum of the results obtained from the use of quantitative and qualitative methods. Rather, mixed methods result in an integration, which is greater than the mere sum of them and allows the construction of a new identity.

6 Ethics and Research with Children

Research with children poses a series of ethical challenges that concern us both in our capacity as researchers and as individuals (see also ▶ Chap. 106, "Ethics and Research with Indigenous Peoples"). In this respect, Fossheim (2013) identifies three basic principles: respect for the persons, doing good (on the part of the researcher), and justice. All these, in addition to the evident complexities of the researcher's personal reflections, highlight the awareness of the working context as critical to assess such notions from the perspective of the participants. That is to say, not only is knowledge coconstrued, but also a "field of interaction," which modulates the different ethical approaches in the respect to the different, the learning process, and the reaction of surprise/amazement (i.e., how should written informed consent be handled in certain contexts where the main weight of culture is oral?). Thus, research with children, on children, their views, perceptions, emotions, ideas, and so on calls for a larger contribution, that of the exercise of real interaction with the recognition of children as subjects of law.

Researchers must approach children with curiosity, attention, sensitivity and simplicity, and above all, with the conviction that children have a lot to say. There is no one more qualified than them to speak about what they are like, what happens to them, and what they need.

Against this background, any cleavages of power present in the study must be thoroughly identified and analyzed. This implies an assessment of the researcher's principles and context (i.e., are such principles relevant to the rules of ethical research proposed in the context of the researcher's residence and in the context in which the research is being conducted?) as well as the scope of the relationship with children – a matter generally discussed in the context of qualitative research. In other words, an ethical, dialogical, and flexible space must be created on the basis of the ethics of interaction. As discussed by Abebe (2009, p. 463, cited by Kjørholt 2013):

> [Ethics] entails a moral consideration grounded in respect for local, gendered and socio-spatial constructions of childhood, as well as the need to go beyond acknowledging such complexities to ask how moral and ethical spaces are (re) produced and who they actually serve.

Besides that, Boddy (2013) points out that there are discussions of research ethics, especially when considering children, between protection and participation. On this issue, Powell et al. (2011) sustain that there is no essential conflict between the right to have a voice and the right to be protected but it is a question of balance.

In recent decades, there can be identified various ways of working with children in research. One of them is the one that reflects the traditional research model with an asymmetrical power relationship between researchers and researched. In this way of involving children in research, the researcher is attributed expert status and we can talk about "research on children." Another way is that in which the research is still directed by adult researchers but involving children in some or all parts of the research. We talk then about "research with children." Another way is the one of "research by children," in which children initiate, develop, analyze, and disseminate research (with the necessary skills to do that, of course, having being taught on them) (Backe-Hansen 2013).

In any case, we ought to carry on with our research work, providing critical considerations on new conceptual and methodological contributions in a context of a dynamics of power derived from the identities engaged in dialogue; paying special attention to temporality and spatiality – the support variables of any social relation – with firm belief that the production of knowledge is not external to social construction; and, above all, privileging human relationships.

7 Conclusions and Future Directions

Children represent a particular population group that presents characteristics in relation to the historical time in which they live. They show their perceptions and opinions about reality in a free and creative way; therefore, to work with them, researchers need to do it in a participative form (See also ▶ Chap. 116, "Optimizing Interviews with Children and Youth with Disability," and ▶ 117, "Participant-Generated Visual Timelines and Street-Involved Youth Who Have Experienced Violent Victimization").

Research with children must take account of the sociopolitical, cultural, and psychological dimensions. This is so given that "[i]n working with children, we are gaining access not only to their knowledge and subjective experiences, but also to the whole complex of their culture, family life, beliefs, and the social collective imaginary" (Glokner Fagetti 2007, p.75). From a political standpoint, we agree with Kohan (2003, p. 279) who argues that "those who deny children the ability to think do so because they have previously created an authoritarian and hierarchical image of thought, an image that excludes that which will be then branded as incapable."

With respect to children's different cultural contexts, we can state that such contexts seem to contribute to the creation of specificities that translate into life experiences in different spheres (Hanson 1992; Edwards 2005; Fu et al. 2007). Along the same lines, we argue that in recent years, psychology has moved away from the classical focus on children's pathologies and risks towards more salutogenic aspects (Ben-Arieh et al. 2013; Peterson 2013).

With respect to the methodological aspects, significant developments have taken place in recent years that have led to new perspectives. This chapter provided an analysis of quantitative and qualitative methods and explored the relevance of using mixed methods. Mention should also be made of new emerging proposals – such as the inclusion of technologies and arts-based methods – that present significant future perspectives. The inclusion of technologies such as the internet and the use of email – as is the case with email-based surveys – facilitates, among others, large-scale surveys (Scott 2000). In addition, image-making technologies – providing children with digital video cameras, participative video – (Cochran-Smith and Lytle 2011) and other technology-supported creative productions such as blogs bring the researcher closer to the language and forms of expression used by children and young people.

Another significant line of research is provided by the arts-based methods: the self-portrait, graphic elicitation, mapping, timelines (Bagnoli 2009; Boydell et al. 2012). These methods present the advantage of allowing the researcher to come closer to children, as the use of visual and graphic language – which children find more comfortable – provides them with empowerment and a sense of agency (Bagnoli 2009). Children often feel more confident in creating drawings, photographs, and videos than words. Additionally, children's visual culture is central in childhood studies (Prosser and Burke 2008). These methods enable the adult researcher to gain insight into the children's world, while at the same time respect their language and capture different experiences. They are also comprehensive, offer a variety of choices, and can be used at different stages of the research. They open up future lines that can lead to new perspectives, which should not lose sight of the ethical aspects and the role of children as agents.

All of the above considerations lead to the conclusion that our theoretical-methodological approach to research on, for, and with children will be guided by the way in which we perceive children. In this sense, we view children as protagonists. This challenges us to reflect on our research practices, the creation of ethical spaces, and who they are really useful for. The reason for this is that we understand the production of knowledge to be part of the construction of society, and underscore once again the nonneutrality of the knowledge thus produced and the researcher's commitment to his or her work.

It is likely that many researchers will involve children in their research. We hope that what we have discussed in this chapter will provide some theoretical standpoints that researchers can adopt in their research with children around the globe.

References

Aber LJ, Jones S. Indicators of positive development in early childhood: Improving concepts and measures. In: Hauser RM, Brown BV, Prosser WR, editors. Indicators of children's wellbeing, pp. 395–408. New York: Russell Sage Foundation.

Ager A. Annual research review: resilience and child well-being–public policy implications. J Child Psychol Psychiatry. 2013;54(4):488–500.

Appadurai A. La modernidad desbordada: dimensiones culturales de la globalización. Montevideo: Ediciones Trilce; 2001.

Aries P. Centuries of childhood: a social history of family life. New York: Vintage Books; 1962.

Backe-Hansen E. Between participation and protection. Involving children in child protection research. In: Fossheim H, editor. Cross-cultural child research ethical issues. Norway: The Norwegian National Research Ethics Committees; 2013. Retrieved from https://www.etikkom.no/globalassets/documents/publikasjoner-som-pdf/cross-cultural-child-research-webutgave.pdf. Accessed 27 Jun 2015.

Bagnoli A. Beyond the standard interview: the use of graphic elicitation and arts-based methods. Qual Res. 2009;9(5):547–70.

Ben-Arieh A. Beyond welfare: measuring and monitoring the state of children – new trends and domains. Soc Indic Res. 2000;52(3):235–57.

Ben-Arieh A. Where are the children? Children's role in measuring and monitoring their wellbeing. Social Indicators. 2005;74(3):573–596.

Ben-Arieh A. The child indicators movement: past, present, and future. Child Indic Res. 2008;1(1):3–16.

Ben-Arieh A, Kaufman NH, Andrews AB, George RM, Lee BJ, Aber LJ. Measuring and monitoring children's well-being, vol. 7. Dordreth: Springer; 2013.

Berry JW. Immigration, acculturation, and adaptation. Appl Psychol. 1997;46(1):5–34.

Binda NU, Balbastre-Benavent F. Investigación cuantitativa e investigación cualitativa: buscando las ventajas de las diferentes metodologías de investigación. Rev Cienc Econ. 2013;31(2):179–87.

Boddy J. Ethics tensions in research with children across cultures, within countries. A UK Perspective. In: Fossheim H, editor. Cross-cultural child research ethical issues. Norway: The Norwegian National Research Ethics Committees; 2013. Retrieved from https://www.etikkom.no/globalassets/documents/publikasjoner-som-pdf/cross-cultural-child-research-webutgave.pdf. Accessed 27 Jun 2015.

Bornstein MH, Bradley RH, editors. Socioeconomic status, parenting, and child development. Oxon: Routledge; 2014.
Boydell KM, Gladstone BM, Volpe T, Allemang B, Stasiulis E. The production and dissemination of knowledge: a scoping review of arts-based health research. Forum Qualitat Social Res. 2012;13(1):Art. 32. http://www.qualitative-research.net/index.php/fqs/article/view/1711/3328
Bradshaw J, Hoelscher P, Richardson D. An index of child well-being in the European Union. Soc Indic Res. 2007;80(1):133–77.
Bryman A. Social research methods. 5th ed. Oxford: Oxford University Press; 2016.
Chapple S, Richardson D. Doing better for children. OECD; 2009.
Christensen P, James A, editors. Research with children: perspectives and practices. London: Falmer Press; 2000.
Clark E. The proceedings of the thirtieth annual child language research forum. Stanford: Center for Study of Language and Information; 2000.
Cochran-Smith M, Lytle SL. Commentary – changing perspectives on practitioner research inquiry: perspectives, processes and possibilities. Learning Landscape. 2011; 4(2):7–30.
Coffey A, Atkinson P. Making sense of qualitative data: complementary research strategies. Thousand Oaks: Sage; 1996.
Collins CS, Cooper JE. Emotional intelligence and the qualitative researcher. Int J Qual Methods. 2014;13(1):88–103. http://ijq.sagepub.com/content/13/1/88.full.pdf+html. Accessed 25 Jun 2015
Cook TD, Reichardt CS. Métodos cualitativos y cuantitativos en investigación evaluativa. Madrid: Morata; 1986.
Denzin NK. The research act: a theoretical introduction to sociological methods. New York: Praeger; 1978.
ECLAC/UNICEF TACRO (The Americas and the Caribbean Regional Office). Pobreza infantil en América Latina y el Caribe, LC/R.2168. Santiago: ECLAC; 2010.
Edwards CP. Children's play in cross-cultural perspective: a new look at the six cultures study. Faculty Publications, Department of Child, Youth, and Family Studies. Paper 1. 2005. http://digitalcommons.unl.edu/cgi/viewcontent.cgi?article=1000&context=famconfacpub. Accessed 9 Oct 2015.
Espíndola Advis E, Rico MN. Child poverty in Latin America: multiple deprivation and monetary measures combined. In: Minujin A, Nandy S, editors. Global child poverty and well-being: measurement, concepts, policy and action Bristol. UK: The Policy Press.pp; 2013. p. 379–418.
Fernandez S. Promoting the benefits of language learning. Report to the department of education and training. Melbourne: Research unit for multilingualism and cross cultural communication at the university of Melbourne. 2007. Retrieved from https://www.eduweb.vic.gov.au/edulibrary/public/teachlearn/student/promobenefitslanglearning.pdf.
Fossheim H. Introduction. In: Fossheim H, editor. Cross-cultural child research: Ethical issues. Oslo: Norwegian National Research Ethics Committees; 2013. p. 9–16. https://www.etikkom.no/globalassets/documents/publikasjoner-som-pdf/cross-cultural-child-research-webutgave.pdf. Accessed 27 Jun 2015.
Fu G, Xu F, Cameron CA, Heyman G, Lee K. Cross-cultural differences in children's choices, categorizations, and evaluations of truths and lies. Dev Psychol. 2007;43(2):278–93. http://www.ncbi.nlm.nih.gov/pmc/articles/PMC2581463/
Fuhs B. Die Generationenproblematik in der Kindheitsdforshcung /The Problem of Generations in the Research of Childhood. In Honig, 1999.
Gaitan Muñoz L. El espacio social de la infancia. Madrid: Comunidad de Madrid Conserjería de Sanidad y Servicios Sociales; 1999.
García M, Hecht AC. Los niños como interlocutores en la investigación antropológica. Consideraciones a partir de un taller de memoria con niños y niñas indígenas. Telluso. 2009;7:163–86.
Gilbert K, editor. The emotional nature of qualitative research. Boca Raton: CRC Press; 2000.
Giménez G. Identidades en globalización. Espiral, Vll. 2000;19:27–48.

Giménez G. La concepción simbólica de la cultura. In: Giménez G, editor. Teoría y análisis de la cultura. México: CONACULTA e Instituto Coahuilense de Cultura; 2005. p. 67–88.

Glokner Fagetti V. Infancia y representación. Hacia una participación activa de los niños en las investigaciones en Ciencias Sociales. Revista Tramas, subjetividad y procesos sociales. N° 28. Diciembre. México DF: Universidad Autónoma Metropolitana; 2007. p. 67–83.

Hanson MJ. Families with Anglo-European roots. In: Lynch EW, Hanson MJ, editors. Developing crosscultural competence: a guide for working with young children and their families. Baltimore: Brookes; 1992. p. 65–87.

Hirschfeld L. Why don't anthropologists like children? Am Anthropol. 2002;104(2):611–27.

Hymes D. Why linguistics needs the sociologist. Soc Res. 1967;34:632–47.

Jackson-Maldonado D, Thal D, Marchman V, Bates E, Gutierrez-Clellen V. Early lexical development in Spanish-speaking infants and toddlers. J Child Lang. 1993;20:523–49.

James A, Prout A, editors. Constructing and reconstructing childhood: contemporary issues in the sociological study of childhood. London: Falmer Press; 1997.

Johnson R, Onwuegbuzie B, Turner L. Toward a definition of mixed methods research. J Mixed Methods Res. 2007;1:112–33.

Junker DA, Stockman IJ. Expressive vocabulary of German-English bilingual toddlers. Am J Speech Lang Pathol. 2002;11(4):381–94.

Kamerman S. Preface. In: Kamerman S, Phipps S, Ben-Arieh A, editors. From child welfare to child well-being. Dordrecht: Springer; 2010.

Kibria N. The changing lives of Vietnamese-Americans. Princeton: Princeton University Press; 1993.

Kjørholt AM. "Childhood studies" and the ethics of an encounter: reflections on research with children in different cultural contexts. In: Fossheim H, editor. Cross-cultural child research: ethical issues. Oslo: Norwegian National Research Ethics Committees; 2013. p. 17–44. https://www.etikkom.no/globalassets/documents/publikasjoner-som-pdf/cross-cultural-child-research-webutgave.pdf. Accessed 27 Jun 2015.

Kohan W. Infancia entre educación y filosofía. Barcelona: Ed. Laertes; 2003.

Kwan YK, Ip WC. Life satisfaction, perceived health, violent and altruistic behaviour of Hong Kong Chinese adolescents: only children versus children with siblings. Child Indic Res. 2009;2(4):375–89.

Lau AS, McCabe KM, Yeh M, Garland AF, Wood PA, Hough RL. The acculturation gap-distress hypothesis among high-risk Mexican American families. J Fam Psychol. 2005;19(3):367.

Li H, Martin AJ, Armstrong D, Walker R. Risk, protection, and resilience in Chinese adolescents: a psycho-social study. Asian J Soc Psychol. 2011;14(4):269–82.

Liamputtong P. Qualitative research methods, 4th edn. Melbourne: Oxford University Press; 2013.

Liddicoat A. Learning a language, learning about language, learning to be literate. Babel. 2001;35(3):12–5.

Liebel M. Paternalismo, participación y protagonismo infantil. In: Corona Caraveo Y, Linares Pontón ME, editors. Participación infantil y juvenil en América Latina. México: UAM; 2007.

Lyubomirsky S, Lepper S. A measure of subjective happiness: preliminary reliability and construct validation. Soc Indic Res. 1999;46:137–55.

Maxwell J. Qualitative research design: an interactive approach. New York: Sage; 1996.

Mieles Barrera M, Tonon G. Children's quality of life in the Caribbean: a qualitative study. In: Tonon G, editor. Qualitative studies in quality of life methodology and practice, Social indicators research series, vol. 55. Switzerland: Springer; 2015. p. 121–48.

Moreno R. Everyday instruction: a comparison of Mexican American and Anglo mothers and their preschool children. Hisp J Behav Sci. 1997;19:527–39.

Moore KA, Lippman L, Brown B. Indicators of child well-being: The promise for positive youth development. The annals of the american academy of political and social science. 2004;591(1):125–145.

Peterson C. The strengths revolution: a positive psychology perspective. Reclaiming Child Youth. 2013;21(4):7–14.

Pita Fernández S, Pértegas Díaz S. Investigación cuantitativa y cualitativa. Cad Aten Primaria. 2002;9:76–8. http://www.fisterra.com/mbe/investiga/cuanti_cuali/cuanti_cuali.asp

Powell MA, Graham A, Taylor NJ, Newell S, Fitzgerald R. Building capacity for ethical research with children and young people: an international research project to examine the ethical issues and challenges in undertaking research with and for children in different majority and minority world contexts – report prepared for the Childwatch International Research Network. Oslo, Norway. 2011. Retrieved from http://epubs.scu.edu.au/cgi/viewcontent.cgi?article=1033&context=ccyp_pubs. Accessed 26 Jun 2014.

Prosser J, Burke C. Image-based educational research: childlike perspectives. In: Knowles JG, Cole A, editors. Handbook of the arts in qualitative research: perspectives, methodologies, examples and issues. London: Sage; 2008. p. 407–20.

Punch S. Research with children: the same or different from research with adults? Childhood. 2002;9(3):321–41.

Rabello de Castro L. Participación política en el contexto escolar: experiencias de jóvenes en acción colectiva. In: Corona Caraveo Y, Linares Pontón ME, editors. Participación infantil y juvenil en América Latina. México: Universidad Autónoma Metropolitana; 2007. p. 17–45.

Rodriguez Pascual I. Redefiniendo el trabajo metodológico cualitativo con niños: el uso de la entrevista de grupo aplicada al estudio de la tecnología EMPIRIA. Rev Metodología Cienc Soc. 2006;12:65–88. Retrieved from http://www.redalyc.org/articulo.oa?id=297124008003. Accessed 23 Jul 2015

Roer-Strier D, Rosenthal MK. Socialization in changing cultural contexts: a search for images of the "adaptive adult". Soc Work. 2001;46(3):215–28.

Sampieri RH, Collado CF, Lucio PB. Metodología de la investigación. México: Edición McGraw-Hill; 1996.

Sanders J, Munford R, Liebenberg L. Young people, their families and social supports: understanding resilience with complexity theory. In: Ungar M, editor. The social ecology of resilience. New York: Springer; 2012. p. 233–43.

Scott J. Children as respondents: the challenge for quantitavive methods. In: Christensen P, James A, editors. Research with children: perspectives and practices. London: Routledge; 2000. p. 98–119.

Sen A. Desarrollo y libertad. Bogota: Ed. Planeta; 2000.

Szapocznik J, Kurtines WM. Family psychology and cultural diversity: opportunities for theory, research, and application. Am Psychol. 1993;48:400–7.

Tonon G. Young people's quality of life and construction of citizenship, Series SpringerBriefs in well-being and quality of life research. Dordrecht: Springer; 2012.

Tonon G, editor. Qualitative studies in quality of life methodology and practice, Social indicators research series, vol. 55. Cham: Springer; 2015.

WHO. Social determinants of health and well-being among young people. Copenhagen: World Health Organization Regional Office for Europe; 2012.

Yelland G, Pollard J, Mercuri A. The metalinguistic benefits of limited contact with a second language. Appl Psycholinguist. 1993;14:423–44.

Optimizing Interviews with Children and Youth with Disability

116

Gail Teachman

Contents

1 Introduction	2024
2 Framing Research with Children	2025
3 Assembling Multiple Customizable Interview Methods	2026
3.1 Setting the Stage	2026
3.2 Role Play with Character Dolls or Puppets	2028
3.3 Cartoon Captioning	2030
3.4 Photo-Elicitation	2032
3.5 Vignettes	2033
3.6 Sentence Starters	2034
4 Partnering with Parents	2035
5 Child–Researcher Power Relations	2036
6 Conclusion and Future Directions	2038
References	2038

Abstract

While there is a growing body of literature explicitly outlining methods for interview studies with children, few have focused on engaging children with disabilities. This chapter describes innovative techniques, strategies, and methods for engaging disabled children and youth in qualitative interviews. A child interview methodological approach is described with an emphasis on three key elements: assembling a range of customizable interview methods; partnering with parents; and consideration of the power differential inherent in child–researcher

G. Teachman (✉)
McGill University, Montreal, QC, Canada
e-mail: gail.teachman@mail.mcgill.ca

interactions. Drawn from the author's research, examples are used to illustrate the methods and discuss how they were adjusted as the research unfolded. The methods and strategies discussed in the chapter might equally inform interview research with participants of all ages and abilities.

Keywords

Child · Disability · Interviews · Qualitative research · Youth

1 Introduction

Researchers undertaking interviews with disabled children might ask: Is there a need for specialized methods? How can existing methods be adapted to support participation of disabled children in the research process generally, and in interviews specifically? If time constraints limit a child's participation in research to a single interview, how can data generation be optimized? This chapter addresses these questions and outlines an innovative methodological approach to combining techniques, strategies, and methods for optimizing interviews conducted with disabled children and youth (see also Teachman and Gibson 2013). A note on terminology: acknowledging debate on the matter, the term "disabled child" and other variants are used in line with critical disability scholarship which emphasizes that disability is not an individual trait. Rather, it is produced through social relations (Barnes 1995; Morris 2001).

Many articles and books have focused on the process of engaging children in interviews (e.g., Docherty and Sandelowski 1999; Morrow 2001; Kortesluoma et al. 2003; Christensen 2004; Irwin and Johnson 2005; Kirk 2007; Christensen and James 2008; Danby et al. 2011). Most often, the methods described involve multiple interviews or prolonged engagement with children in participatory ethnographic designs. Some exemplary work is available to guide studies involving more extended engagement in children's health contexts (Drew et al. 2010; MacDonald et al. 2011; Nicholas et al. 2011). These relatively intensive methods are generally desirable to produce in-depth, situated data. However, researchers might design studies involving single interviews with child participants in a variety of circumstances; for example, when there are limits on the time available with participants and their families, when multiple interviews might create an unreasonable economic burden for families, or when access to participants is limited by distance.

Children with disabilities accumulate a great deal of expertise specific to growing up with impairments and through early immersion in the worlds of medicine, rehabilitation, and special education. Their perspectives are vital to informing understandings of the impact of these institutions and services on the lives of disabled people. Research that involves a single child interview can sometimes be more readily accommodated by parents of disabled children because of the time constraints and busy schedules their families typically experience. Less information is available to guide single interview methods with children and youth with disabilities.

Following a brief review of principles framing research with children, the chapter highlights three key elements of a methodological approach for doing research with disabled children. These include: (a) assembling multiple, readily adaptable interview methods; (b) partnering with parents prior to interviews; and (c) reflexive consideration of the power relations inherent in child–researcher interactions. Empirical examples are used to illustrate the methods (see Box 1 below). The potential benefits and limitations of these approaches are discussed, along with description of the iterative process involved in modifying and adapting interviewing methods as research unfolds. The methods described in this chapter could be applied in many types of research, but will be particularly useful in studies where opportunities to become familiar with child participants prior to an interview are limited, and where data generation is limited to one or two interviews.

> **Box 1 Study Context**
> All the examples described in this chapter are drawn from a study conducted in Ontario, Canada, that explored the beliefs, assumptions, and experiences of disabled children, their parents, and clinicians regarding the importance of walking (Gibson et al. 2012; Gibson and Teachman 2012). The study was framed within a critical social science perspective (Eakin et al. 1996) with explicit intentions to explore ways that dominant normative discourses about walking are reproduced, reformulated, and resisted, and to (re)consider taken-for-granted assumptions in the field of children's rehabilitation. Six pairs of participants (each pair consisted of a child or youth with cerebral palsy and one of his or her parents) were interviewed for a total of 12 study participants. The children and youth who participated were between 7 and 18 years of age. All had been involved in some type of walking therapy and used an assistive device (e.g., walker, wheelchair) for at least some of their mobility needs. Separate one-on-one interviews with parents and their children were conducted in participants' homes in all but one instance. The study will be referred to as "the walk study" throughout this chapter.

2 Framing Research with Children

In research that is informed by the *new social studies of childhood*, children are positioned as active, competent, and expert research participants whose perspectives on issues that affect their daily lives in the here and now are important and likely to differ from those of their parents (James and Prout 1997; Christensen 2004; see also "Research with Children & Participant-Generated Visual Timelines with Street-Involved Youth Who Have Experienced Violent Victimization"). This view contrasts with previously dominant approaches wherein children were viewed as adults-in-the-making, incapable of understanding and sharing their own experiences. Those views suggested that children's lives should be understood and interpreted through a more "mature" adult lens. Shifts in how childhood and children are conceptualized have challenged researchers to reconsider taken-for-granted assumptions inherent in many methods developed specifically for children (Christensen and Prout 2002;

Punch 2002; Christensen 2004; Irwin and Johnson 2005; Kirk 2007). For example, Christensen (2004) noted a weakening of conventional research approaches in which researchers do things "to" children, and a move toward researchers researching "with" children. The approach challenges the assumption that special methods are needed for research, that a different set of ethical standards is required, or that the problems faced during research process are unique to working with children (p. 165). It has been argued, for example, that novel techniques for use in qualitative research with children might be equally helpful when interviewing adults (Punch 2002; Kirk 2007). Many visual and arts-based methods described in research with children were originally developed and continue to evolve in research with adults (Miles 1990; Catterall and Ibbotson 2000).

This chapter is aligned with childhood scholars who recommend that research practices be reflective of individual children's experiences, interests, values, and routines, and that researchers consider the unique ways that children routinely express and represent themselves (Punch 2002; Kortesluoma et al. 2003; Christensen 2004). This recommendation is particularly relevant when undertaking research with disabled children where planning and adaptation might be necessary to align the research methods with each participant's abilities, preferences, and communication styles. In what follows, three key elements of methodological approach used in research with disabled children are illustrated with examples drawn from the author's research.

3 Assembling Multiple Customizable Interview Methods

3.1 Setting the Stage

It can be helpful to begin a child interview by introducing an activity that helps to set the stage. This type of activity can help set up the interview as a discussion rather than a more formal exchange of questions and answers, and establish there are no "right" answers. This type of activity is often referred to as a "warm-up." In the walk study, an "interviewer card game" was incorporated. The activity was similar to a method termed "talking cards" by Moore et al. (2008), who reported it was highly rated by child participants when they were asked to reflect on interview methods that they would recommend. In the walk study, the method was used to: (a) increase comfort and reduce anxiety on the part of participants; (b) aid the interviewer in gauging the communication style of the participant; (c) diminish the power differentials by giving explicit permission for the child to question the interviewer and by reinforcing that there were no "right" answers; and (d) provide early cues to guide the interviewer's choices of strategies, techniques, and methods from the interview toolkit best suited for that participant.

The game involved the participant and interviewer taking turns to choose from a deck of question cards, and then play the role of interviewer by asking the question printed on the card. Two identical sets of three cards were prepared: one set for the participant and an identical set for the interviewer. The child was given the opportunity to go first by selecting one of her or his cards, and then reading the question to the interviewer, who would answer the question.

> **Box 2 Interview Card Game**
> **Questions and Example Responses**
> Can you tell me a little bit about yourself?
> What is your best talent?
> If you could have any super power, which super power would you choose?
>
> **Sample responses from an interview with a 13-year-old girl**
> Interviewer (I): If you could have any super power, what would you choose?
> Child (C): Make my wheelchair have wings and fly.
> I: That sounds like a lot of fun. Where would you go?
> C: Around the world.
> I: What color would it be?
> C: Pink.

The girl who contributed the responses in the above example was sitting in her pink power wheelchair as she made these comments, which provided revealing insight into the ways she had incorporated her wheelchair into her sense of herself, even extending into an imaginary world of possibilities. In another instance, warm-up discussion about imagined super powers opened opportunities for a later discussion about how the participant envisioned "ideal" mobility devices. Thus, while the interview activity card questions did not relate to the research focus (concerning the value of walking and walking therapies), some participants' responses provided rich insights into disabled children's identities, and their views about their bodies, mobility options, and assistive devices. One older participant, an 18-year-old youth who had just begun his first year of university, was not engaged by the activity and elected to play the game as follows:

Interviewer (I): You can pick any one of these, and ask me a question first, then, I'll ask you one.
Youth (Y): I don't want to ask you anything.
I: Can I ask you one of these?
Y: Sure.
I: I'm going to start with this one: What is your best talent?

Although the youth went along with this one-sided version of the game, his cues were judged to indicate that he was more comfortable participating in ways that he perceived as "adult." Given these early cues from the participant, the interviewer presented only some of the assembled methods (vignettes and photographs to elicit discussion of preferred mobility choices) and did not present cartoon captioning or role play methods, judging that the youth was likely to view these as childish. This example demonstrates how, in some cases, older teens and young adults who are keen to establish recognition of their advancing maturity might interpret game-like activities as being childish. However, activities such as games, vignettes, and role play might be very engaging for other participants, including adults (Guillemin 2004).

The use of an activity to set the stage for the interview served as a reminder to refrain from making assumptions about participants' developmental skills and abilities based on age or diagnosis. When a 10-year-old participant was unable to read the cards, the researchers reflected on the potential for the activity to contribute to a sense of failure and anxiety for some participants at the very onset of the interview. In all subsequent interviews, participants were offered a choice to read the cards or ask the interviewer to read the cards. In a related example, as the research unfolded, a pattern was noted in instances where the participants tended to repeat the interviewer's responses to the game card questions. This highlighted the possibility that child participants in the study judged the researcher's responses to be appropriate and safe answers. This points to at least some susceptibility for study participants to respond in ways that they perceived to be desirable, or "what researchers want to hear."

3.2 Role Play with Character Dolls or Puppets

Dolls or puppets can be used in qualitative interviews with children to allow them to discuss potentially sensitive issues from the imagined perspective of a doll or puppet through role play (Jager and Ryan 2007; Epstein et al. 2008; Aldiss 2009). Researchers have noted that in studies with disabled children, the doll or puppet's gender, ethnicity, and physical appearance influenced participants' conduct during the interview, suggesting that children might relate more easily to puppets that appeared more "like" them (Epstein et al. 2008). In other research (Teachman 2006), when disabled children were offered a choice of boy or girl puppets, every participant chose his or her same-gender puppet to act out a role play activity. Character dolls were outfitted with splints which the participants appeared to find highly engaging, since many had worn similar splints. In the walk study, participants were offered a choice from among five dolls representative of different genders and ethnicities (different hair and skin color) with arm or leg splints. They could also choose to position the doll with either a doll-sized walker or wheelchair (see Fig. 1).

The following reflective memo describes the participant cues that were used to guide the researchers' decision on whether to include the puppet methods during an interview with a 13-year-old girl:

> The girl seemed interested in the puppets as I prepared for the interview. My initial impressions led me to decide to include this role-play activity in the interview, based on my impression of her level of maturity and sense of playfulness. She did quite easily relate to that activity, choosing the blonde, girl doll and placing her in a wheelchair.

Use of the dolls or puppets can be introduced by saying, for example: "We are going to pretend," or "Let's act out a story." Then, a storyline can be set up. In the walk study, this involved explaining that the participant could elect to play the role of a child who had cerebral palsy, while the interviewer would use another doll to play the mother character. One of the dolls was slightly larger than the others and was

Fig. 1 Character dolls/puppets and mobility aids

dressed to appear more like an adult. The participant was instructed to pretend his or her character was at home watching a favorite television show when their mother interrupts to say that it is time to go to therapy. The interviewer then voiced the role of the mother, animating the doll to emphasize what was being said. Several potential scenarios were prepared with a loosely outlined script that allowed the interviewer to respond with probes during the role play, depending on the emergent dialogue (see Box 3).

> **Box 3 Sample Role Play Script and Probes**
> Mother (interviewer): It's time to get ready to go to therapy. I know it's your favorite show, but we need to get going.
> Yes, I know you don't want to stop but we'll be late. Don't you want to keep working on your walking?
> Let's talk about why you're going to therapy every week.
> If you could choose by yourself, whether to go to therapy or not, what would you like to do?
> What would happen if you didn't go to therapy?

This scenario was designed to elicit children's views about attending therapy, its role and importance, and the value of walking. Participants' dialogue in the role play was interpreted as reflective of their understandings of the everyday discourses within their home and school life.

The dolls or puppets were not used with all children in the walk study. The researcher made this decision during the interview depending on the child's cues. A 10-year-old boy in the study showed considerable interest in the puppets, asking about them when he entered the room at the onset of the interview. He engaged immediately with one of the boy-puppets and talked directly to the puppet as well as "through the puppet" within the context of role play. But, the researcher elected not to present this method during the interview with a similarly aged boy of 12. That participant appeared comfortable throughout the interview and engaged with the vignette, photograph, and cartoon methods. But, he repeatedly responded to probes by saying, "I dunno [don't know]" and periodically leaned back in his chair while yawning in an exaggerated way. He denied being tired and appeared more engaged after being given the opportunity to take a break or end the interview (consistent with the researcher's ethical commitment to observe for and address indications of possible fatigue). This participant also frequently ended thoughtful responses by saying, "And blah, blah, blah, blah," as if detached from his own comments. These behaviors can be interpreted as indications that the boy was establishing an identity as a "cool" detached teenager.

Participants and researchers alike can be thought of as performing particular identities during interviews. Fernqvist (2010, p. 1321) similarly described the work of "doing age" where, for example, youth might challenge an interviewer by actively calling on internalized age-related behaviors. This is not to suggest these behaviors are planned, rather they reflect internalized strategies for managing the interview. For example, frequent "I don't know" responses from a young participant who is otherwise sharing rich insightful commentary might be a signal that the topic is uncomfortable. To avoid further discussion, Fernqvist suggests the participant reverts to "doing child"; that is, playing the role of a child who does not have knowledge of importance. In the walk study, the 12-year-old boy's frequent yawning and repetition of the phrase "blah, blah, blah" were interpreted as indications that he was asserting himself by "doing teen," and strategies to deflect dialogue about feelings and topics that were less comfortable. The young participant shared rich, detailed descriptions about walking, mobility devices, and therapies despite his frequent "doing teen" responses and behavior.

3.3 Cartoon Captioning

Cartoon captioning is a projective technique developed with adult participants in market research (Broeckelmann 2010; Doherty and Nelson 2010) and in educational research with children and adults (Warburton 1998; Catterall and Ibbotson 2000). In this method, participants are presented with a cartoon drawing and asked to fill in empty "speech bubbles." This method is especially engaging because of its novelty because it generates curiosity, while being conducive to interaction (Catterall and Ibbotson 2000). Participants must imagine how others might respond to the situation portrayed in the cartoon, in much the same way as vignette methods which are described later in this chapter.

In the walk study, the cartoon captioning method was engaging and elicited perspectives that might otherwise have been missed. However, some children in the study struggled to compose captions. They responded with one or two words, and were unable to say more when probed. It could be that some participants thought their responses needed to be humorous and found this intimidating. As well, the cartoon images were not amenable to adaption so, for some participants, the cartoon scenarios might have been unfamiliar experiences, making it more difficult for them to respond meaningfully. An example cartoon-completion activity from the walk study is shown in Box 4.

Box 4 Example Cartoon Completion

In the example in Box 4, two youths of indeterminate age are illustrated on a sidewalk. One, a girl, is seated on a power scooter; the other is a boy holding a skateboard. The speech bubbles above their heads were either partially completed or left blank. The activity was included toward the end of the interview, and modified depending on the sophistication of language and other cues from the participant. Participants were to comment on how they might complete the cartoon. A 9-year-old boy in the study responded this way:

Scooter girl (G):	Let's have a race.
Skateboard boy (B):	No, I don't want to race.
G:	Why not?
B:	Because I hate racing you because you go faster than me.

3.4 Photo-Elicitation

Photographs brought into an interview by a researcher can help scaffold and elicit dialogue with children (Morrow 2001; Kirk 2007). In the walk study, a series of photographs of everyday environments were selected to represent spaces that children and youth might navigate (See Box 5). The series included several variations of school classrooms, a school gymnasium, an outdoor playground, a home kitchen, a sidewalk, and various park or natural environments. These were presented in sequence as a slide show presentation on a laptop computer to elicit discussions about participants' mobility challenges and preferences. Youth were asked a series of questions related to ways they would choose to mobilize in each environment. The photographs proved to be effective in prompting rich discussions with all participants, and further, the concrete nature of the activity seemed to provide a break for some participants from more challenging discussion of abstract feelings. For example, the photo-elicitation method seemed to reduce anxieties evidenced by a 10-year-old boy who appeared to struggle with some of the concepts discussed in the interview. He was quick to say, "I don't know." However, he relaxed and appeared very engaged when using the laptop computer to scroll through and comment on the photographs. The images contextualized the discussion in a very concrete way that helped him to feel more confident in responding.

Box 5 Photo-Elicitation Examples

In another instance, a 12-year-old boy immediately appeared more engaged when presented with the photographs, as described in an extract from the researcher's interview notes:

> The boy, who had been acting bored and frequently yawning, became very engaged by the photographs. I used language like: "If you were right inside this picture, how you want to move about?" or "How would you like to get around if you were right here?", because he was actually touching the screen, asking, "Where am I in the picture?" He seemed to be imagining himself in the picture so I tried to match my prompts to his cues.

3.5 Vignettes

Vignettes are short scenarios that set up a specific context and situation that requires some type of response or resolution by imagined characters. They help elicit participants' values, meanings, and beliefs (Finch 1987) about the topic of inquiry. Vignettes have been described as a method that allows interviewers to combine a "systematic structured approach with the expression of 'emic' or personal meanings" (Miles 1990, p. 38). The method was developed for use with adult research participants who were asked to write a series of vignettes based on their own experiences to record and reflect on specific situations and experiences. Researchers noted that an advantage of the methods was that participants' vignettes reflected common experiences which when shared, conveyed to the reader a sense that "I am not alone" (Miles, p. 41). The method has since been modified so that researchers provide prewritten vignettes during an interview and ask participants how they might resolve the described situation (Barter and Renold 1999, 2000; Wilks 2004). Vignettes should be constructed to resonate with participants' experiences but remain ambiguous enough to prompt judgments, decision making, and descriptions of the views or experiences that influence participants (Barter and Renold 2000).

In the walk study, vignettes were paired with an image of a child who resembled the participant (in relation to gender, age, and mobility device used). The purpose of this pairing was to provide a visual representation that could contextualize the vignette and help engage the child participant. Vignette details were modified to reflect each participant's situation and experiences, based on prior discussion with a parent. A sample vignette appears in Box 6.

Box 6 Sample Vignette

I want to tell you a story about Andrew. This is Andrew in the picture. (Show generic photo of boy using wheelchair.) He is in Grade 3 and he has CP [cerebral palsy] – like you do. He does some things at school differently than the other kids in his class but mostly he tries to be the same as the other kids. One thing that is different for Andrew is his walking – he uses a walker and he needs extra time to get around his school. Sometimes, he has a hard time keeping up with the other kids. Lots of times, he gets tired and must take a rest. He gets upset about that. One day last week, his teacher said, "I think that you should use your wheelchair when we go out for recess."

Probes included:

What do you think Andrew will say?
How do you think he will feel about it?
What does he decide to do?
Do you have a story like that?

In the walk study, participants all acted engaged and responded easily to the vignettes. The method evoked descriptions of children's similar personal experiences along their views and feelings about those experiences, shifting with ease between sharing their views about the vignette and their views of their own experiences. The method was especially effective in eliciting discussion about ways that children in the study viewed themselves and wished for others to view them, as illustrated by an example in Box 7.

> **Box 7 Sample Vignette Response**
> A 12-year-old girl commented on the importance of using her walker and standing tall, which she identified as a highly valued aspect of being "like everyone else." She shared that when using a walker, she felt people saw her as being the same as everybody else who walks. She contrasted this with concerns regarding how others view her when she is using her wheelchair:
> Child (C): I like that I get exercise, 'cause everybody else walks. I don't want to be like a person that doesn't walk at all.
> Interviewer (I): Is it important to try to be like everybody else when you can?
> C: Yeah, yeah, it's pretty important to me.
> I: What do you think it would be like for somebody like the girl in the story if she used the wheelchair?
> C: Well, she knows that she can walk; it's just that people see that she maybe can't walk. She keeps having to tell people that, like, she can walk. It's just that she doesn't.
> GT: What about the walker – do you think that about the walker?
> C: No, because they see that I just use it for, like support, and not for, just using it. Do you have a story like that?

3.6 Sentence Starters

Sentence starters are used as an interview method when a participant is having difficulty initiating responses. Using this technique, researchers can give children permission to talk about sensitive topics or to express views that they might otherwise deem "inappropriate" to express to an adult. The method also tends to decrease the sense that the participant is under pressure to identify and share his or her own views, because the interviewer is responsible for beginning the sentence and the sentence can be framed as a general rather than personal statement.

> **Box 8 Example Sentence Starters**
> In the walk study, instead of asking participants to specify whether they liked or disliked elements of their own therapy, participants were asked to complete sentences such as:
> "The best thing about therapy is..."
> "The worst thing about therapy is..."

The method might prompt concerns that research participants will be led toward specific responses. However, in the walk study, children and youth responded to the sentence starters in a variety of ways, including deflection of the topic area. The following excerpt from an interview with a 12-year-old girl demonstrates that the technique does not necessarily lead participants toward a predictable response:

Interviewer (I):	I think you've also tried some Botox? [a drug injected to treat muscle spasticity]
Child (C):	Yeah.
I:	Okay. So, if I said, "I like Botox because…," Can you fill in that blank? "I like Botox because…".
C:	Ah, no, I can't say anything.
I:	Okay, that's fine. How about this one: "I didn't like Botox because…"
C:	Because it's, um, painful, and it's like, for me it doesn't really help anyways … I just have them because sometimes it helps, but sometimes it doesn't.

This sequence of responses, generated using sentence starters, helped elicit more in-depth discussion where the girl explained her feeling about this treatment and her motivation for complying, even though she was uncertain of its effectiveness. Her responses provided insights into her perspective about the relative merits of the treatment that might not have otherwise been discussed in the interview.

The methods set out here could be thought of as a toolkit that, once assembled, can be used to optimize data generation with disabled children through individualized, flexible interviews, dynamic interactions, and rich information exchanges. When combined, the methods provide opportunities to shifting the focus of an interview from the participant's experiences and beliefs to a projected "other" whose imagined views and actions are informed by the participant's own experiences, beliefs, and understandings. In doing so, the methods collectively function to add aspects of fun, and help create a comfortable space for children to share their views.

4 Partnering with Parents

Prior to involving children in an interview, it is helpful to have some contextual knowledge of the participant's life and circumstances. This can help the researcher identify specific areas that might have resonance for that individual child. Working with parents prior to an interview to learn about the preferences of the child can also be important to identify strategies that might contribute to the child's comfort in the interview and therefore, optimize the data generated (Irwin and Johnson 2005). The importance of partnering with parents in research with children has also been noted by Barter and Renold (2000) who, in a study of peer violence among in residential children's homes, reported that study participants were better able to engage in discussion when they had experience with the situations depicted in the vignettes.

Initially, in the walk study, limited demographic information about study participants was elicited through a phone conversation with parents prior to the interviews (age, diagnosis, and walking ability). Reflecting on the interview process, it became apparent that to increase the depth and quality of the data being generated, the interviewer needed more contextual information about the child's current and recent therapies, their school placement, their typical modes of mobility, and the assistive devices they used. In subsequent interviews, this information was elicited in conversation with a parent prior to conducting the child interview. The background information helped the interviewer more readily adapt elements of the methods described earlier in this chapter. For example, role-play and vignette activities were individualized to include familiar types of therapy, similar school settings, and narratives that were likely to resonate for the child or youth participant. The following extract from the interviewer's notes describes an example from the study:

> I had followed through on the idea of getting a little more information ahead of time from the parent to allow me to customize the role-play and other vignette activities, to assist the child to draw from their own experiences and project these onto the characters. This appeared to be very helpful today, as the young girl commented: "Just like me!" on hearing the vignette and when asked, "Do you have any stories like that?" she readily described similar situations that she had experienced.

Parents can be a valuable resource and contribute toward establishing a supportive and comfortable interview frame that acknowledges the interdependence of family members. It is important to differentiate this strategy from one where a parent's comments are interpreted as representations or proxies of the child's viewpoints. Rather, information shared by parents serves as a point of departure for later engagement with the child participant and helped optimize conditions where they were supported represent their own views about the topic of inquiry.

5 Child–Researcher Power Relations

Conducting interviews in a child's home is desirable as it can aid in addressing power differentials and creating rapport. It also has the advantage of maximizing opportunities to observe participants in their preferred personal space. These advantages come with some degree of unpredictability (MacDonald and Greggans 2008). For example, a child who might struggle to sit in one place throughout an interview will be more likely and able to move about in the familiar home environment than would be possible in a stark, institutional space. This can lead to what Irwin and Johnson termed "kinetic conversations" (2005, p. 826).

In the walk study, whenever feasible, interviews were conducted in the child's home and participants were invited to select a location within their home for the interview. One 10-year-old boy chose his hockey-themed bedroom; the location helped establish a sense of intimacy, comfort, and privacy, where he readily took charge of aspects of the interaction. During the interview, he frequently pointed to furnishings, photographs, and other favorite possessions to illustrate a point he was making. The setting and the artifacts it contained elicited novel dialogues, and contributed rich descriptive data through interviewer observations. Effectively, the setting allowed the participant to host the interviewer, sharing power, at least to some extent, within the interaction. In another situation, an interview was conducted with a 13-year-old girl while the family home was being renovated and in conditions that involved significant noise and interruptions. Still, the home setting provided contextual grounding and opportunities for discussion that another location would not have elicited. The participant's excitement about the renovations led to discussion about an elevator that was being installed to make her home more accessible. This opened further dialog about walking and mobility issues, which was the focus of the research.

Research with disabled children necessitates ongoing reflexivity on the part of the interviewer in relation to the power differential between researcher and child participant. Many novice interviewers wonder how they should present themselves and their research aims. Christensen suggested that child interviewers "take on and perform themselves as an unusual type of adult, one who is seriously interested in understanding how the social world looks from children's perspectives but without making a dubious attempt to be a child" (2004, p. 174). It is also helpful to aim for rapport that connotes a working relationship between the child and the interviewer, as opposed to a friendship with the child (Irwin and Johnson 2005). Being mindful that children are actively encouraged not to talk to strangers, interviewers might expect to encounter some difficulty in building rapport with young children, especially during a first or single interview. However, disabled children and youth, being accustomed to the attentions and interventions of numerable adult professionals, might be more inclined to go along with professionals even – perhaps especially – when it is uncomfortable, because they have internalized understandings that it is "for their own good." Boundaries around privacy and trust might be blurred. Therefore, researchers should reflect on whether this group of young people might be more trusting of interviewers and less guarded about sharing intimate, personal insights into their worldviews than other groups of youth. Interviewers should take extra steps to explain and model ways that children can assert their rights as research participants.

By providing a range of methods within a single interview, researchers can be better prepared with options to reduce a child's anxiety about providing "right" answers and facilitate opportunities for comfortable dialogue – sometimes with the interviewer and sometimes with imagined characters or peers. Researchers typically review parameters for participating or withdrawing from research with participants. But, interviewers can go further by explicitly rehearsing with participants options for signaling that they want to stop, take a break, or refrain from answering a question. Children who have learned to defer to adults might not easily act on their option to withdraw. In the walk study, this was addressed through detailed scripted examples (see also Kirk 2007), such as, "If you don't want to talk about something, you can just say, 'I don't want to talk about that,' and it will be okay." The interviewer explained that she would interrupt and ask specific questions if the youth showed signs of fatigue or seemed anxious. These reminders served to demonstrate that the participant and interviewer would share responsibility within the interview.

6　Conclusion and Future Directions

This chapter has introduced a methodological approach for combining and adapting interview methods to optimize the quality of data generated during single interviews with disabled children and youth. The importance of partnering with parents early in the process has been highlighted. Parents can provide important contextual information that will aid researchers in building rapport and providing a supportive, safe interview frame wherein children and youth can tell their own stories and share their views. Just as Kortesluoma et al. (2003, p. 440) emphasized, "children certainly know more about what they know than interviewers do", researchers will do well to acknowledge that parents know far more about their child than interviewers do. Multiple sources of information, when used reflexively, can enhance data collection and interpretation. Partnering with parents prior to an interview can get the interviewer "into the same ballpark," so that they are introducing topics and asking questions that likely resonate with individual children's experiences. This approach might be equally helpful in some research with adults, for example, when adult participants have difficulties with communication.

The methodological approaches reported in this chapter reflect learning from a small pilot study with children with disabilities across a relatively wide age range. Even though individual children and youth of a certain age might tend to be more comfortable with certain methods, it was optimal to be prepared to select and combine methods within an interview based more on the participant as an individual than on his or her age or diagnosis. The assembling of multiple interview methods allows the interviewer to take on an improvisational approach during interviews that is responsive to participants' unique experiences, contexts, abilities, and ways of communicating, as well as the evolving interactions within each interview.

It is important to conclude the chapter by noting that the quality of data generated through interviews with children (and adults), regardless of method, is always reliant on the skills of the interviewer; the methods are an adjunct to those skills. The approach to interviews outlined in this chapter could be used in interviews with participants of varying ages and abilities. These methods and approaches have since been used , and continue to be refined, in several subsequent studies involving disabled children in interviews about various aspects of children's rehabilitation.

References

Aldiss S. What is important to young children who have cancer while in hospital? Child Soc. 2009;23:85–98.

Barnes C. Disability, cultural representation and language. Crit Public Health. 1995;6(2):9–20.
Barter C, Renold E. The use of vignettes in qualitative research. Soc Res Update. 1999;25(9):1–6. Retrieved from http://sru.soc.surrey.ac.uk/SRU25.html
Barter C, Renold E. 'I wanna tell you a story': exploring the application of vignettes in qualitative research with children and young people. Int J Soc Res Methodol. 2000;3(4):307–23.
Broeckelmann P. Exploring consumers' reactions towards innovative mobile services. Qual Mark Res Int J. 2010;13(4):414–29.
Catterall M, Ibbotson P. Using projective techniques in education research. Br Educ Res J. 2000;26(2):245–56.
Christensen P, Prout A. Working with ethical symmetry in social research with children. Childhood. 2002;9(4):477–97.
Christensen PH. Children's participation in ethnographic research: issues of power and representation. Child Soc. 2004;18(2):165–76.
Christensen P, Monrad, James A. Research with children: perspectives and practices. New York, NY: Routledge; 2008.
Danby S, Ewing L, Thorpe K. The novice researcher: interviewing young children. Qual Inq. 2011;17(1):74–84.
Docherty S, Sandelowski M. Focus on qualitative methods: interviewing children. Res Nurs Health. 1999;22(2):177–85.
Doherty S, Nelson R. Using projective techniques to tap into consumers' feelings, perceptions and attitudes...getting an honest opinion. Int J Consum Stud. 2010;34(4):400–4.
Drew S, Duncan R, Sawyer S. Visual storytelling: a beneficial but challenging method for health research with young people. Qual Health Res. 2010;20:1677–88.
Eakin J, Robertson A, Poland B, Coburn D, Edwards R. Towards a critical social science perspective on health promotion research. Health Promot Int. 1996;11(2):157–65.
Epstein I, Stevens B, McKeever P, Baruchel S, Jones H. Using puppetry to elicit children's talk for research. Nurs Inq. 2008;15(1):49–56.
Fernqvist S. (inter)active interviewing in childhood research: on children's identity work in interviews. Qual Rep. 2010;15(6):1309–27.
Finch J. The vignette technique in survey research. Sociology. 1987;21(1):105–14.
Gibson BE, Teachman G. Critical approaches in physical therapy research: investigating the symbolic value of walking. Physiother Theory Pract. 2012;28(6):474–84.
Gibson BE, Teachman G, Wright V, Fehlings D, Young NL, McKeever P. Children's and parents' beliefs regarding the value of walking: rehabilitation implications for children with cerebral palsy. Child Care Health Dev. 2012;38(1):61–9.
Guillemin M. Understanding illness: using drawings as a research method. Qual Health Res. 2004;14:272–89.
Irwin LG, Johnson J. Interviewing young children: explicating our practices and dilemmas. Qual Health Res. 2005;15:821–31.
Jager J, Ryan V. Evaluating clinical practice: using play-based techniques to elicit children's views of therapy. Clin Child Psychol Psychiatry. 2007;12(2):437–50.
James A, Prout A. A new paradigm for the sociology of childhood? Provenance, promise and problems. In: James A, Prout A, editors. Constructing and reconstructing childhood: contemporary issues in the sociological study of childhood. London: Falmer Press; 1997. p. 7–33.
Kirk S. Methodological and ethical issues in conducting qualitative research with children and young people: a literature review. Int J Nurs Stud. 2007;44:1250–60.
Kortesluoma R, Hentinen M, Nikkonen M. Conducting a qualitative child interview: methodological considerations. Methodol Issues Nurs Res. 2003;42(5):434–41.
MacDonald J, Gagnon A, Mitchell C, Di Meglio G, Rennick J, Cox J. Include them and they will tell you: learnings from a participatory process with youth. Qual Health Res. 2011;21:1127–35.
MacDonald K, Greggans A. Dealing with chaos and complexity: the reality of interviewing children and families in their own homes. J Clin Nurs. 2008;17(23):3123–30.
Miles MB. New methods for qualitative data collection and analysis: vignettes and pre-structured cases. Int J Qual Stud Educ. 1990;3(1):37–51.

Moore T, McArthur M, Noble-Carr D. Little voices and big ideas: lessons learned from children about research. Int J Qual Methods. 2008;7(2):77–91. Retrieved from http://ejournals.library.ualberta.ca/index.php/IJQM/article/view/1941/1362

Morris J. Impairment and disability: constructing an ethics of care that promotes human rights. Hypatia. 2001;16(4):1–13. https://doi.org/10.1111/j.1527-2001.2001.tb00750.x.

Morrow V. Using qualitative methods to elicit young people's perspectives on their environments: some ideas for community health initiatives. Health Educ Res. 2001;16(3):255–68.

Nicholas D, Picone G, Selkirk E. The lived experiences of children and adolescents with end-stage renal disease. Qual Health Res. 2011;21:162–73.

Punch S. Research with children: the same or different from research with adults? Childhood. 2002;9(3):321–41.

Teachman, G. (2006). Becoming a writer: Social constructions of writing from children with cerebral palsy (Masters dissertation). Retrieved from ProQuest Dissertations and Theses database. (AATMR2114).

Teachman G, Gibson BE. Children and youth with disabilities: innovative methods for single qualitative interviews. Qual Health Res. 2013;23(2):264–74.

Warburton T. Cartoons and teachers: mediated visual images as data. In: Prosser J, editor. Image-based research: a sourcebook for qualitative researchers. London: Falmer; 1998. p. 235–54.

Wilks T. The use of vignettes in qualitative research into social work values. Qual Soc Work. 2004;3(1):78–87.

Participant-Generated Visual Timelines and Street-Involved Youth Who Have Experienced Violent Victimization

117

Kat Kolar and Farah Ahmad

Contents

1 Introduction	2042
2 The Study: Methodological Considerations	2044
2.1 Study Setting and Design	2044
2.2 Data Collection	2045
2.3 Data Analysis	2046
2.4 Participant Demographics	2047
3 Timeline Implementation: Introducing Timeline Mapping to Participants	2047
4 Timeline Styles	2048
5 Thematic Findings	2049
5.1 Theme 1: Rapport Building	2050
5.2 Theme 2: Participant as Navigators	2052
5.3 Theme 3: Therapeutic Moments and Positive Closure	2053
6 What We Have Learnt	2054
7 Conclusion and Future Directions	2057
References	2058

Abstract

Despite growing interest in the use of visual methods as a way to engage with issues of representation, meaning, and power relations in qualitative research, only limited literature is available on the use of participant-generated imagery in guiding or supplementing semi-structured or open-ended interviewing methods in the health and social science disciplines, or in navigating issues of interviewing vulnerable persons who have experienced trauma. We draw from a study

K. Kolar (✉)
Department of Sociology, University of Toronto, Toronto, ON, Canada
e-mail: kat.kolar@mail.utoronto.ca

F. Ahmad
School of Health Policy and Management, York University, Toronto, ON, Canada
e-mail: farahmad@yorku.ca

© Springer Nature Singapore Pte Ltd. 2019
P. Liamputtong (ed.), *Handbook of Research Methods in Health Social Sciences*,
https://doi.org/10.1007/978-981-10-5251-4_125

exploring resilience among street-involved youth to investigate how participant-created visual timelines inform verbal semi-structured interviewing with persons who have experienced personal victimization in the form of violence, as well as structural marginalization. To guide future research efforts, the process of timeline implementation is discussed in depth. Analysis of timelines was conducted through a critical emancipatory research lens. Three overarching themes developed through analysis of timelines are explored here: (a) rapport building, (b) participants as navigators, and (c) therapeutic moments and positive closure. In the discussion, we engage with the potential of visual timelines to supplement and situate semi-structured interviewing and illustrate how the framing of research is central to whether that research facilitates increased participant authority in the research process, enhances trust, and ensures meaningful, accountable engagement.

Keywords

Resilience · Timeline · Visual methods · Street-involved youth · Qualitative interviews

1 Introduction

In qualitative research, visual methods encompass analysis of a wide variety of mediums, from found images and visual objects to the use of participant-generated imagery (i.e., imagery that was produced by participants specifically within the context of a research study) (Rose 2001; Guillemin and Drew 2010; Pauwels 2010; Jackson 2012). Despite growing interest in the use of visual methods as a way to engage with issues of representation, meaning, and power relations in qualitative research, only limited literature is available on the use of participant-generated imagery in guiding or supplementing semi-structured or open-ended interviewing methods in the health and social science disciplines, or in navigating issues of interviewing vulnerable persons who have experienced trauma (e.g., Guillemin 2004; Goodrum and Keys 2007; Umoquit et al. 2008; Bagnoli 2009; Horsfall and Titchen 2009; Berends 2011; Patterson et al. 2012). This research has contributed to the understanding of potential uses and strengths of visual methods, including the building of rapport, enhanced contextualization of narratives, and nonverbal communication as a way to access "othered" ways of knowing. However, detailed exploration of the implementation and use of visual methods with marginalized groups is still required. Also essential is critical engagement with the ways in which visual methods may inform or pose new concerns for the researcher-participant relationship.

This chapter contributes to the growing literature on visual methods and participant-generated imagery by providing an analysis of the implementation and findings of a study using participant-created visual timelines and semi-structured interviewing to explore resilience among street-involved youth in Canada's Greater Toronto Area [GTA]. Timelines are a visual, arts-based data collection method, derived from a broader framework of graphic elicitation designs (Umoquit et al.

2008; Bagnoli 2009; Sheridan et al. 2011). Timelines are created from a participant's life events, placed in some sort of chronological arrangement, with visual indication of the significance or meaning attached to highlighted events (Berends 2011; Patterson et al. 2012). The aims of this chapter are to examine the potential of participant-generated visual timelines to supplement and situate semi-structured interviewing with marginalized groups, as well as to provide guidance for health and social science researchers regarding implementation of timeline mapping (see also ▶ Chaps. 67, "Timeline Drawing Methods," ▶ 70, "Body Mapping in Research," and ▶ 71, "Self-portraits and Maps as a Window on Participants' Worlds").

The use of in-depth narrative interviews on sensitive topics or with marginalized groups not only rouses concerns regarding potentially exploitative research relationships, but also involves issues regarding development of rapport (here understood as accountable, meaningful engagement with participants (see Holland 2007; Liamputtong 2007; Nicholls 2009). We recognize that interviews may elicit anxiety as participants reflect on and share potentially traumatic or difficult experiences (Hollway and Jefferson 1997). Researchers must work to ensure that any potential distress which may be caused by study involvement be minimized for participants. To address these concerns, selection of research methods must involve critical consideration of how these methods structure power dynamics between the researcher and participant. In addition, by prioritizing reciprocal engagement of participants such that they have say in how the research proceeds, researchers can facilitate increased participant authority in the research process and enhance trust and meaningful, accountable engagement (Holland 2007; Karnieli-Miller et al. 2009; Nicholls 2009).

Qualitative methods can be integrated to make data collection situations more amenable to participants who have experienced marginalization, as well as to allow diversified exploration and representation of participant life experiences (Kesby 2000; Umoquit et al. 2008; Patterson et al. 2012; Liamputtong 2013). The combination of graphic elicitation methods such as visual timelines with verbal interviewing provides one such possibility to address these issues. Timelines have been used to explore a wide variety of issues, including the trajectory of substance abuse and treatment (Berends 2011), the impact of financial incentives on clinical behaviors (Umoquit et al. 2008), and barriers to health of people experiencing homelessness (Patterson et al. 2012). The available visual methods literature suggests that use of timelines in tandem with in-depth narrative interviews may enhance the data collection experience and data quality, particularly when researching sensitive topics or marginalized groups (Harper 2003; Berends 2011; Sheridan et al. 2011).

Although this literature has begun to assess some of the strengths and limitations of visual timeline implementation, much remains to be elaborated on how sources of data, topics of investigation, and epistemological approaches all inform timeline interview processes and outcomes. Academic articles on timelines focus largely on the content of timelines at the expense of what their form contributes to an understanding of various social phenomena (exceptions include Bagnoli 2009). Some researchers note that they could not include individual timelines due to ethical issues (e.g., Berends 2011), could only provide researcher-created timelines after interviews had been completed (e.g., Patterson et al. 2012), or were limited to the use of prestructured diagrams in the interview (e.g., Umoquit et al. 2008). Berends (2011) suggests that the increased availability of research utilizing visual methods is necessary because analyses of individual timelines in combination with supporting interview text will facilitate more holistic understandings of data by readers. Further, considering the overall lack of participant-generated timeline implementations with groups who experience social structural marginalization, it is clear that research on use of timelines across more varied populations is still required to understand differences in uptake and response to this method. To speak to these gaps in the timeline literature, here we provide an in-depth critical discussion of our implementation of timeline mapping with street-involved youth. We also conducted a parallel study using timeline mapping with South Asian immigrant women who were survivors of partner violence; the details are presented elsewhere given the focus on street-involved youth for this chapter (see Ahmad et al. 2013; Kolar et al. 2015).

2 The Study: Methodological Considerations

2.1 Study Setting and Design

This study was conducted in the city of Toronto, in collaboration with a community agency serving street-involved youth in 2010. Dr. Patricia Erickson, a researcher on the study team, had done considerable prior research on street youth issues with this community agency. She met with agency staff at an early stage of study planning in order to explain the project and gain their participation, and also committed to returning and presenting the findings when data collection was complete (Kolar et al. 2012).

A critical emancipatory and feminist lens guided the implementation and analysis of this research project (Kolar et al. 2012; Ahmad et al. 2013). Such a lens values power-conscious epistemology where the interview is approached as an active, co-constructive process between the participant and researcher. This perspective moves a researcher away from conventional approaches that treat interviews as pipelines between the research "subject," positioned as the passive conveyor or object of knowledge, and the researcher who is the source of objective authority eliciting information (Smith 1990; Kesby 2000; Nicholls 2009; Gringeri et al. 2010; see also ▶ Chaps. 118, "Capturing the Research Journey: A Feminist Application of Bakhtin to Examine Eating Disorders and Child Sexual Abuse," and ▶ 119, "Feminist Dilemmas in Researching Women's Violence: Issues of Allegiance, Representation, Ambivalence, and Compromise"). Emphasis is placed on participant narratives by asking simple, open-ended questions and active listening by the interviewer. This lens requires that reflexivity be a central practice, whereby researchers critically engage with how the very production of knowledge and interaction with participants is situated in social relations and power inequalities. For our analysis, this involved looking to how the experiences of street-involved youth – experiences and social locations which are conventionally marginalized – could be given voice in ways that addressed the concerns and interests of participants. Our study sought to understand how timelines in in-depth interviews could provide a venue for participants to tell their stories and how timelines could be used to recognize and legitimize participants' understandings of resiliency (Maxwell 1992; White and Klein 2008).

For this study, a qualitative design of face-to-face, in-depth interviews was employed. The interview guide was developed through review of literature addressing resilience and concerns specific to street-involved youth, from available literature on timeline implementation (particularly Bagnoli 2009), and through consultation with the collaborating agency serving street-involved youth in Toronto. The interviews were shaped around a semi-structured interview guide with open-ended questions on resilience, defined as resources (internal or external) that may assist individuals in their engagement with and navigation of adversity (PreVAiL 2010; Kolar 2011). Farah Ahmad, as a research team member, conceptualized the inclusion of timeline mapping as meeting the critical emancipatory and feminist perspectives of the research. This investigation of resilience was intended to disrupt benchmarks of positive adaptation that frequently reflect values of White, middle-class families (Ungar 2004). Such benchmarks are particularly problematic for those who have lived in resource-limited and volatile contexts, such as street-involved youth (Ungar 2004; Kolar et al. 2012). Interview questions were accordingly intended to encourage participants to identify what constituted a "resilience resource." For example, interview questions included: "Can you draw a timeline depicting events that were important in your life?" and "Can you tell me if there were other supports that you wish you had when you were going through this difficult time?" A short sociodemographic survey was conducted with participants prior to commencing the in-depth interviews. The collaborating agency reviewed the study protocol and provided feedback. Research ethics approval was obtained from the University of Toronto.

2.2 Data Collection

This study was designed to be exploratory, aiming to develop a contextually appropriate understanding of resilience in relation to the specific challenges faced by street-involved youth. Thus, sampling was not aimed at identifying a "representative sample" of street-involved youth in Toronto, but rather at inductively exploring variation in experiences of resilience and generating new insights through in-depth investigation. For exploratory studies, a small number of cases are recommended (i.e., fewer than 20) (Kuzel 1999; Crouch and McKenzie 2006). Youth between 18 and 26 years of age were eligible to participate if they had experienced street-involvement, had experienced violence since becoming street-involved and/or had experienced childhood maltreatment, and viewed themselves as having made "positive changes" in their lives. It is recognized here that what constitutes contextually appropriate indications of positive development will be different for street-involved youth in comparison to the general population due to the volatile and often dangerous social environments street-involved youth occupy, and structural barriers that they face. Through collaborating discussions between researchers and agency counselors on identifying contextually appropriate understanding of resilience for street-involved youth, "positive changes" were determined to mean that participants have been engaging in activities that promote their mental health, well-being, and coping, including: (a) addressing addictions and past trauma, (b) establishing more supportive relationships, and (c) pursuing goals such as education, stable housing, or employment. Counselors at the collaborating agency identified participants meeting these criteria and provided potential participants with preliminary information about the study in the form of flyers. Interested participants contacted the study coordinators for further details (for more information on sampling, see Kolar et al. 2012).

Interviews with street-involved youth were held in a private space provided by the collaborating agency and were conducted by one team member (Kat Kolar). Prior to proceeding with interviews, the interviewer discussed the in-depth interview process with each participant, provided them with a copy of the consent information sheet, and then gave a verbal overview of the consent sheet. The interviewer addressed any questions or concerns of participants before obtaining verbal consent to proceed. The verbal consent process was used in order to reinforce the anonymity of participants, as no record of participant names was kept post interview in any file. Participants also received an honorarium of 30 dollars. One interview was conducted with each participant, with interviews lasting 90 min on average. Interviews were tape recorded and transcribed, and visual timelines were kept by the interviewer.

Upon completion of each interview, the interviewer was asked to write a research memo wherein they reflected on her experience of the interview. This involved thinking about the tone or mood of interaction between the interviewer and participant, the impact of the study focus of resiliency on the interview process, the impact of the use of timeline mapping on the interview process, and consideration of how participants engaged in the timeline mapping activity, as well as of interviewer thoughts or feelings on timeline mapping. These reflection notes were instrumental to the reflexive engagement of researchers with the process of knowledge production in this study.

2.3 Data Analysis

The study data included transcribed narratives of the participants, the timeline maps, and interviewer reflection notes. These were thematically coded (King and Horrocks 2010). The thematic open coding framework was developed to investigate how timelines could assist researchers in better understanding the experiences discussed by participants and how timelines shaped the data collection process. Coding involved analysis of both *content* and *form* of data using a two-stage team-based approach and was conducted by hand. Primary open coding was conducted by one team member (Linda Chan) who was not involved in data collection. This initial set of codes was then reviewed and refined by all team members over the course of several group meetings before consensus on codes was reached. All participant names provided in this manuscript are pseudonyms.

In the following sections, several salient issues that emerged from our study will be discussed. These include: a description of timeline implementation, an overview of timeline styles, and the themes identified in exploring the impact of timeline mapping on interview dynamics.

2.4 Participant Demographics

It is essential to briefly describe the demographic characteristics of the participants in our study. All participants described themselves as survivors of violence who had experienced homelessness. Participants were 19–26 years old and reported diverse ethnic origins (self-identifying as White, Black, Aboriginal, East Indian, or Latino), with half of them being born outside of Canada. Most had high school education and an annual income of less than 10 thousand dollars. Eight of the ten participants reported having access to housing at the time of the interviews. Detailed descriptions of participant demographics and experiences of resilience have been reported elsewhere (Kolar et al. 2012).

3 Timeline Implementation: Introducing Timeline Mapping to Participants

To introduce the timeline method to participants, the interviewer began by providing a brief description of the timeline as a tool that is intended to assist researchers in better understanding the important life experiences of participants, particularly by showing a more holistic picture of life events than can often be captured in one-time verbal interviews. The research team anticipated that some participants may be confused by verbal instructions to draw a timeline or may be uncomfortable with this request due to feeling that they have poor drawing or writing skills. In order to make participants more comfortable with the potentially unfamiliar task of drawing a timeline, participants were then shown sample timelines created by the interviewer. These sample timelines were intended to help stimulate creative engagement by participants and to provide them with a sense of flexibility for creating their own timelines, while simultaneously reassuring them that there was no "wrong way" to create a timeline and that spelling was not a concern. The interviewer emphasized that timelines did not have to be done in any specific way and so were not prestructured. Sample timelines took a variety of forms, including simple straight lines, text-heavy lists, and nonlinear representations such as swirls and undulating lines. These instructions were intentionally broad because of the exploratory nature of this study, aiming to identify resilience processes and resources among street-involved youth that cannot be captured by prestructured resilience scales, as these scales fail to account for the context-specific struggles and successes of marginalized groups (Kolar et al. 2012).

Upon explaining the timeline method and answering any preliminary questions on how to proceed, the interviewer gave the participant a pen and large sheet of paper and asked the participant to begin the timeline when they felt ready to do so. Participants created their own timelines. Direct involvement of the interviewer in timeline creation was minimal. The interviewer sometimes contributed to timeline creation when requests were made for direct assistance (e.g., spelling) and also provided reassurance when participants expressed confusion about how to create or organize their timelines.

In this study, timelines were created in two stages. In the first stage, the interviewer introduced the interview format by asking participants to draw the timeline while thinking about indicating the:

> ...important events that stick out in your mind when you reflect on your life. It can be the first time you slept on the street, it can be a time that you were really hopeful or satisfied...

Once participants indicated that they had completed their timeline, the interviewer then probed about several life events that participants had identified in their timeline map in order to build contextual detail during the remainder of the interview. The timelines thus continued to play a central role throughout the interviewing process, with participants often adding contextual details to their timelines as the interview proceeded.

Towards the end of the interview, participants were asked to add another section to their existing timelines which indicated their goals and expectations for the future and then were asked to discuss these with the interviewer. This "future timeline" was the second stage of the timeline mapping process implemented in this study.

4   Timeline Styles

Two distinct timeline mapping styles were prominent from the variety of timelines created by the participants: the list-like timeline (Fig. 1) and the continuous-line timeline (Figs. 2 and 3). Square brackets in the timeline figures indicate information that has been anonymized in order to preserve participant confidentiality.

List-like timelines were text-heavy and tended to describe life events chronologically (see Fig. 1). These timelines consisted of columns with brief notes (e.g., short phrases and keywords) with or without dates (e.g., according to participant ages, the year an event occurred, or time periods as indicated by holidays or seasons). Some participants separated positive and negative life events into different columns or used signs and symbols (such as positive (+) or negative (−) signs or emoticons (such as "smiley faces")) to indicate how they felt about these events or how they perceived these events to have impacted them. Some participants also used dots or Xs to mark important events on the timeline. Some of the timelines provided below have been previously published (see Kolar et al. 2015).

Several participants created continuous-line timelines: they drew a line and used spikes, dips, angles, waves, and curves to represent positive or negative dimensions of their experiences (see Fig. 2). Most of these continuous timelines were constructed horizontally. Similar to the list-like timelines, some participants used dashes, dots, or Xs to indicate events significant to them. Often, this timeline was complemented with varying types of notes and dates, either below, above, or beside the continuous line.

Other timeline styles included a unique hybrid of the continuous-line and list-like timeline styles (see Fig. 3). For instance, one timeline (not shown in this chapter)

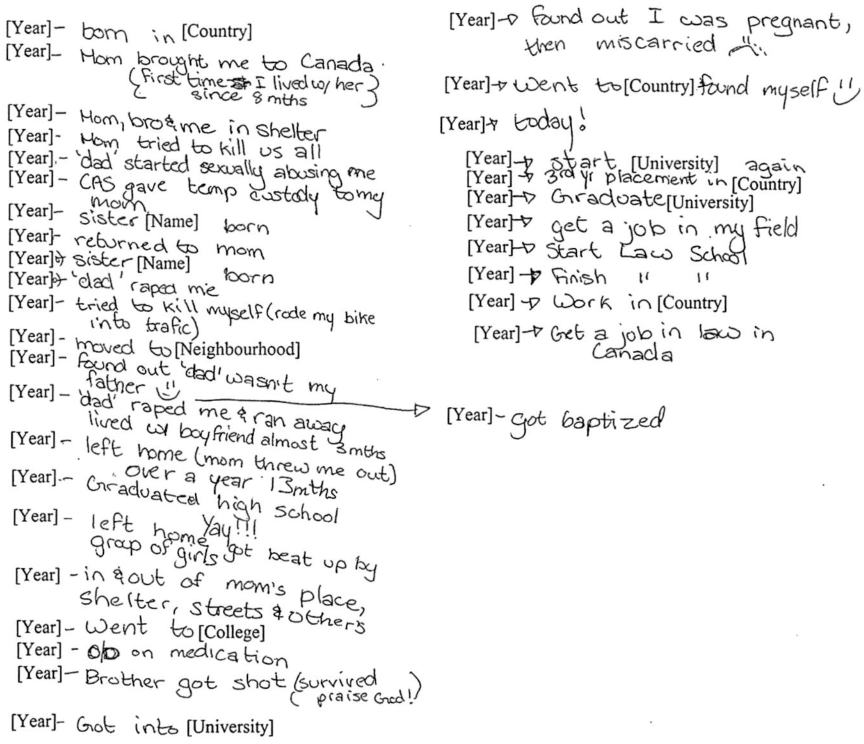

[Year] - born in [Country]
[Year] - Mom brought me to Canada. (First time I lived w/ her since 8 mths)
[Year] - Mom broke me in shelter
[Year] - Mom tried to kill us all
[Year] - 'dad' started sexually abusing me
[Year] - CAS gave temp custody to my mom
[Year] - sister [Name] born
[Year] - returned to mom
[Year] - sister [Name] born
[Year] - 'dad' raped me
[Year] - tried to kill myself (rode my bike into trafic)
[Year] - moved to [Neighbourhood]
[Year] - found out 'dad' wasn't my father ☺
[Year] - 'dad' raped me & ran away lived w/ boyfriend almost 3 mths
[Year] - left home (mom threw me out)
[Year] - Graduated high school over a year 13 mths Yay!!!
[Year] - left home grop of girls got beat up by
[Year] - in & out of mom's place, shelter, streets & others
[Year] - Went to [College]
[Year] - O/D on medication
[Year] - Brother got shot (survived praise God!)
[Year] - Got into [University]

[Year] → found out I was pregnant, then miscarried 🙁
[Year] → Went to [Country] found myself ☺
[Year] → today!
[Year] → start [University] again
[Year] → 3rd yr placement in [Country]
[Year] → Graduate [University]
[Year] → get a job in my field
[Year] → Start Law School
[Year] → Finish " "
[Year] → Work in [Country]
[Year] → Get a job in law in Canada

[Year] - got baptized

Fig. 1 List-like Timeline

entailed a continuous line with a list of years from the participant's birth to the present. In this timeline, the participant expanded upon the difficult or tumultuous years he experienced by adding a list of significant events underneath the year these events corresponded with. This participant also drew several pictures underneath these tumultuous years. The participant described his timeline as "messy" and explained that this messiness depicted the complexity of many of his life experiences.

5 Thematic Findings

Here, we discuss three themes that were identified in our coding of timelines, interviews, and interviewer reflection memos. First, the use of timelines encouraged *rapport building* by reducing traditional hierarchies of a research interview. Second, timeline mapping allowed the *participant to navigate* the interview space through reflection and boundary setting around their experiences. Finally, the use of timelines facilitated *positive closure* to many of the interviews by providing participants with an opportunity to envision future timelines in light of their survivorship and resiliency.

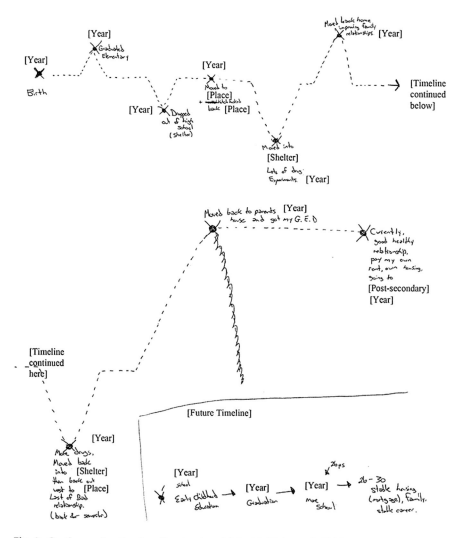

Fig. 2 Continuous-line timeline (Previously published in Kolar et al. 2015)

5.1 Theme 1: Rapport Building

Analysis of timeline narratives reveals that rapport building between the interviewer and participants was mediated by a *life-story* approach taken by participants, purposeful use of *topic-shift* and *self-disclosure* by the interviewers, and the *interactive* nature of the timeline. The rapport-building theme was supported by interviewer reflections as well.

Participants took a life-story approach when mapping their timelines. This open-ended approach allowed them to share significant life events unobtrusively while the interviewer became an involved listener. This style of conversation

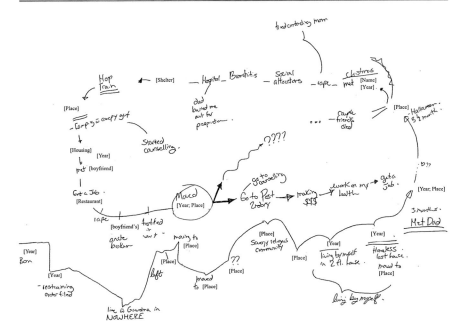

Fig. 3 Hybrid timeline (Previously published in Kolar et al. 2015)

facilitated more participant control over the structure and direction of the conversation by shifting the interactional dynamic away from the ask-and-answer of conventional, interviewer-directed verbal interviews. For example, Rachel created a list-like timeline, where she used two columns to represent positive or negative life events that corresponded with the year these events occurred. Aligning with the way Rachel created a timeline framed by chronological dates, the interviewer was able to ask resilience-related questions regarding the time and experience of, for instance, Rachel's move to Canada, or that of her move into stable housing following a period of homelessness. By structuring her own responses through the timeline, Rachel could more directly exercise control over the direction of the interview. At the same time, the interviewer was able to use the timeline to probe for eliciting richer descriptions.

The use of topic-shift was another device that allowed the interviewer to build rapport with participants. The interviewer used the timelines to point to certain important events, both positive and negative, and asked about resources and strategies that helped participants to cope with experiences of marginalization, discrimination, and violence. This style of questioning allowed participants to respond without providing details of the difficult event or its gravity, instead guiding participants towards discussing their sources of support. If participants became visibly distressed or preoccupied with reflecting on traumatic events, the interviewer also utilized the timeline to topic-shift by subtly redirecting participants away from highly distressing reflections towards strategies and resources that participants indicated were important in enhancing their lives, as was the intended focus of the study.

As noted above, the introduction of the timeline by the interviewer included several examples of how one may draw a timeline. The interviewer also briefly shared her personal timeline with participants and used humor to describe the appearance of her own timeline in order to convey a relaxed attitude towards what a timeline could look like. Interviewer self-disclosure and use of humor led to a more comfortable start to the interviews and so enhanced rapport.

The *interactivity* of the timelines facilitated a sense of participant comfort and momentum in the interviews. The interviewer encouraged participants to reflect on major life events in creating the timeline. For example, the interviewer assured participants that "you don't have to worry about missing anything, because you can always add stuff in [to the timeline] while we're talking" (to Jordan). Likewise, the interviewer supported participants in generating timelines by giving suggestions during descriptions of their life stories. For instance, Jade iteratively constructed her timeline, filling in details on the timeline regarding how her experience of homelessness changed over the course of several years while she verbally described these experiences to the interviewer who probed for contextual details:

> Jade: [After being in the hospital] I went to [a shelter] and it was an interesting place. It was just for women ... I was the only person who wasn't like, smoking crack in the building.
> Interviewer: When was that? Was that in [year], then?
> Jade: Yeah, Christmas would have been [year]. [Fills in timeline]. ... Bronchitis, hospital, [shelter name] [Adds these to timeline]. Ok ... at the same time, I suppose like a couple of my friends died ... [by] overdose. Yeah. Like I've never done heroin, but apparently it's like a fickle drug. [Fills in timeline].

Reflection notes written by the interviewer lend support to the rapport-building capacity of timeline mapping. The interviewer noted:

> The timeline provided a middle ground, which the participant constructed ... This allowed the researcher and participant to reach some sort of working consensus on the ground to be covered in the interview, and acted as a reminder to both of how the interview would proceed. ... I personally used the timeline to jog my memory of previously mentioned items in the interview to be expanded on.

The interviewer found that the timeline acted as an external mediating object through which the interviewer and participant could interact. The physical presence of the timeline decreased the interviewer's "need for note taking... [and] increased [her] capacity to pay direct attention to the participant and to prevent from distracting them."

5.2 Theme 2: Participant as Navigators

The creation of timelines facilitated active engagement of participants through *reflection* on major life events and through *visual aspects* by drawing these events on paper. Reflecting on and recalling life events through timeline mapping allowed participants to create a sense of direction in terms of what they felt comfortable or able to share when asked the interview questions. For example, during an interview, a participant used the timeline to create boundaries around what she could contribute to the interview:

> Interviewer: How long were you living with your dad? When did you move out with your grandparents?
> Mona: [I was] too young to remember the year and too young to remember anything, so, I wouldn't be able to help with anything around here. [Points to flat line at beginning of timeline].

In this way, the visual aspects of the timeline acted as a navigational filter for participants. To begin with, participants themselves identified which events were significant enough to be added to the timeline. Then, they were able to separate positive and negative events using spikes or dips in the continuous-line timeline, and columns and dates in the list-like timeline. Some participants also expressed emotions by adding emoticons and small diagrams for particular life events. Through such visual distinction of life events, participants were able to preemptively choose what they were comfortable talking about, and with this had the opportunity to exercise more control in directing the interview. This finding was also supported by interviewer reflection notes (e.g., see the reflection note provided above in the section on Theme 1).

5.3 Theme 3: Therapeutic Moments and Positive Closure

The timelines created opportunities for participants to experience therapeutic moments in the interview. Such moments were facilitated by the *dual focus* of timelines on positive and negative events, as well as by the addition of a *future timeline* by participants towards the end of the interview.

The dual focus of timelines on both positive and negative events highlighted by participants helped to maintain participant emotional comfort and calmness within the interview by reducing stimulation of distress about traumatic experiences. For example, in referring to his timeline, one participant described the simultaneity of positive and negative experiences:

> This year [was] up and down [pointing to timeline]. Like good, but [I also experienced the] biggest low [because I was a victim of] violence. Because I had [created an art] exhibition on the hate and the violence on the street. And [one week later] I was [a] victim of same violence... I do an [art] exhibition... but at the same time that [violence] happen to you. (Angelo)

This dual focus of the timelines, particularly in combination with the topic-shift strategy described above (see Theme 1), allowed for the interviewer to redirect the participant toward reflections on more positive experiences if a participant seemed "stuck" on discussing difficult events, thereby minimizing emotional distress.

It is important to note that the topic of resiliency in this study was fundamental to minimizing potentially distressing emotional impacts of the timeline mapping process and interview upon participants. Regarding the role of a resilience focus in this research project with marginalized groups, the interviewer commented:

> The timeline represents a picture of the lifecourse with which [participants are] confronted at the end of the interview. If the focus of the interview had solely been on negative experiences, such a representation of trauma and setbacks, this would do little to provide positive closure in the interview ... when using the timeline, especially with vulnerable populations, it may be unethical and cause undue harm if the tool is used to explore only negative experiences.

In the second stage of timeline mapping, participants were asked to think about their aspirations and goals for the future. Some of them added a future timeline and this generally facilitated positive closure of the interview. For example, in one future timeline, a participant identified her immediate goals followed by her long-term dream:

> So, I think that I've just come to terms with accepting being in school and that a career takes long ... I'm just taking it a year at a time. But I know by the time I'm thirty I'd like to have things like my own home. (Leanne)

The interviewer noted that participants appeared to find the process of "concretely lay[ing] out their future plans on a piece of paper to visualize their future" to be therapeutic, and this "created an uplifting emotional shift that provided a sense of closure for both interviewer and participant."

At the same time, the interviewer found that participant's use of future timelines was related to their fears of experiencing a future characterized by ongoing (or escalating) violence and socioeconomic marginality. Some participants found the future timeline difficult to envision because they did not want to "think too far ahead ... and be set up for failure" (Christopher), or they wanted to focus on the "most simple things" in the present (Angelo). The interviewer thus noted that the implementation of a future timeline component in the mapping process may produce "feelings of ambivalence, uncertainty, and fears of failure" for some participants and indicated that future timelines need to be used with sensitivity.

6 What We Have Learnt

Methodological literature which explicitly addresses participant-generated imagery in the form of diagramming methods (e.g., drawings, sketches, or outlines) remains in short supply (Jackson 2012). This study contributes to the limited literature on participant-generated timeline mapping through its unique approach of examining multiple sources of data (i.e., timelines, verbal interviews, interviewer reflection notes, and a short socio-demographic survey) and a team-based approach wherein team members discussed and came to agreement on a thematic coding scheme and analysis. Timelines have been found to strengthen data by enhancing interviewer-participant rapport, mutual understanding, and reflexivity through interactive and supportive engagement with a time-lined representation of a participant's life (Berends 2011; Sheridan et al. 2011; Jackson 2012). Although interviews that involve only one point of research contact with each participant (i.e., one-time interviews) limit researcher engagement with participants and so may pose difficulties for building rapport, such an approach may be necessary as a result of issues of confidentiality (i.e., preserving anonymity), with maintaining contact with mobile or transient populations (e.g., street-involved youth), or because of limited study finances. Our findings indicate that timeline implementation need not be limited to repeat interviews in order to be effective: self-disclosure and interactivity through timelines created a common ground between participants and interviewers, increasing the comfort of participants, and thereby enhanced rapport.

The study findings show that timelines help to focus a participant's attention on the interview by acting as both a memory aid and a visual guide for how the interview will progress, as well as to situate responses within personal and structural contexts while highlighting important events in an individual's life story. That is, the timeline creates a visual middle ground between interviewer and participant from which both can draw to iteratively inform interview questions and responses. In this study, strategies such as the topic-shift were used by the interviewer to ensure that the interview could stay focused on the aim of the study, to move the interview along to meet time constraints, and to ensure minimal emotional distress by redirecting focus when participants appeared to feel distressed. Participants' navigation of the interview was facilitated through their reflection on life events, which encouraged a focus on time-lined event points and allowed participants to walk the interviewer down the path they had illustrated. The timeline also provided a flexible, creative, and collaborative space for communication of meaning, struggle, emotions, and experience through visual aspects. Others have recognized the timeline as a tool to organize and accumulate data, helping to place the research construct in the context of a participant's life events (Berends 2011). Complementary aspects of timelines for enhancement of verbal interviewing are also confirmed by other research utilizing visual methods (Bagnoli 2009; Patterson et al. 2012; Sheridan et al. 2011).

Navigating trauma and emotional distress should be a primary concern for researchers who explore emotionally sensitive topics, particularly when interviewing people who have experienced extensive marginalization. To allow participants to share difficult experiences in a manner that does not cause prolonged emotional distress, an interview approach that fosters a safe and supportive space is necessary for participants to disclose sensitive stories (Goodrum and Keys 2007). Therapeutic aspects built in to the data collection method can help participants address the stress of discussing a difficult or traumatic experience (Horsfall and Titchen 2009; Osei-Kofi 2013). To create a supportive interview space, we relied on the timelines to maintain a focus on resiliency in a context of marginalization and to guide participants to discussing other topics if they appeared to become distressed. Adding a future timeline provided a point of projection that allowed the researchers to consider the possibilities that participants saw for themselves, at the same time that it operated as a goal-setting tool to facilitate positive closure. However, it is important to note that taking resilience as the topic of investigation allowed the timeline to be an overall reflection of achievements and coping strategies, rather than a confrontation of the participant with their perceived failures and negative experiences; that is, on their own, timelines do not ensure positive closure. Thus, in the current study, the use of visual methods is not assumed to be emancipatory simply because the mode of expression and communication is expanded to non-verbal data, but because such methods should arise as the result of a reflexive research process that maintains awareness and critical engagement with issues of power and representation (e.g., Horsfall and Titchen 2009; Mason and Davies 2009; Osei-Kofi 2013).

In this exploration of marginalization and resilience, timelines can be interpreted as providing lifecourse imagery that is representational of intersections of social structure and individual experience. Although a detailed analysis of this kind is beyond the scope of this chapter, a constructivist approach to lifecourse analysis is conducive to exploring how lifecourse imagery is itself used by participants to construct their experiences in developmental terms, and how this imagery provides insight into the ways that participants make sense of, and position the relative importance of, various events in navigating the struggles they face arising from marginalization and violence (Holstein and Gubrium 2000). The different temporal logics evidenced by the timelines in this study (e.g., use of lists, directions of lines, and the use of spirals or diverging lines to illustrate multiple potential future trajectories) are one example of how constructionist approaches to lifecourse analysis provide an avenue for researchers to better grasp the ways that participants establish the relevance and meaning of particular events for how they experience the present and understand possibilities for the future.

Other contributions of participant-generated timeline mapping include capacity for triangulation of data. The sequencing of events in the form of timelines can aid participants in recollecting personal events and can thereby facilitate more accurate and holistic researcher understanding of a participant's life story – an understanding that may be particularly difficult to develop through one-time ask-and-answer verbal interviews (Bagnoli 2009; Jackson 2012). Further, because participants exercised direct control over the visual representation of important life experiences in the form of timeline mapping and then verbally explained these maps to the interviewer, this decreased the risk of misinterpretation of participants' stories and so allowed for enhanced trustworthiness and validity of the analysis (Karnieli-Miller et al. 2009). In addition, analysis of interviewer reflection notes acted as a component of data triangulation because these notes allowed for insight into the reflexive process of evaluating the implementation of timelines. Interviewer reflection notes also provided insight into how timelines and verbal interviews inform one another and made visible the research process through the eyes of the interviewer.

The critical emancipatory and feminist approach of this project guided our interest in using the timeline to act as a middle ground between the interviewer and participant, giving marginalized participants a voice through nonconventional forms of communication, as well as through their increased control in directing the interview. This research approach, in combination with a focus on resilience, resulted in highlighting the strengths of street-involved youth, as well as in exploring the multifaceted coping strategies they engage in, thereby moving beyond much extant research which remains restricted to documenting the marginalization, victimization, and deviance of street-involved persons. Under a critical emancipatory and feminist approach, it was important for the research team to give back to the community of participants and to debrief participants to ensure no harm. As such, access to counseling was provided post interview to ensure that any prolonged distress that may have arisen as a result of the interview process could be addressed. Further, participants were encouraged to debrief with counselors or the co-coordinators of the collaborating agency serving street-involved youth if they wished to do so, and to contact researchers if they had any questions regarding the project outcomes. Upon research completion, researchers disseminated results in a newsletter designed for street-involved youth and met with agency employees to provide a question and answer session regarding the research process and outcomes (for more information on post-research engagement with participants and the collaborating agency, see Kolar et al. 2012). In order to encourage the timeline being seen as a creative and self-exploratory project by participants, we suggest that researchers create a copy of the timelines for analysis and provide original timelines to participants. However, this is only possible if the research is conducted with persons who will be seen on more than one occasion, or in the case of one-time interviews such as our own, where copying or scanning facilities are available.

7 Conclusion and Future Directions

It cannot be overemphasized how central the topics of investigation and selection of research approaches are for informing the implementation and analysis of timeline mapping. On their own, timelines do not inherently provide a more equitable research medium. Rather, as illustrated in this chapter, timelines hold this potential when mobilized with particular goals in mind, including managing and maintaining awareness of power relationships between interviewers and participants, and engaging in issues of representation and voice for marginalized groups. Similarly, the resilience focus in the overarching project facilitated the expression and documentation of experiences of nonconventional coping and achievement for participants through the use of timelines and verbal interviews; without this resilience focus, the use of timelines does not necessarily provide positive closure. When combined with power-conscious epistemologies and a research focus that facilitates critical engagement with the representation of experiences of coping and success of marginalized groups, timeline methods may greatly supplement investigation of complex constructs through a life-story approach, the use of visual aspects, and increased participant control of the interview.

Two limitations need be mentioned here. First, we have made clear that this study is limited to application in one-time interviews; experiences of and issues with implementing timelines in repeat interviewing may differ. Second, this study examined the integration of timelines with qualitative interviews specifically with street-involved youth. Readers should not assume that the advantages and issues explored in this chapter apply equally to all groups (e.g., see Kolar et al. 2015 and Ahmad et al. 2013 on differences in timeline implementation and participant experiences between street-involved youth and South Asian immigrant women who have experienced domestic violence). This study was not intended to be automatically transferable to diverse groups, but to inform readers and researchers who are considering using timelines in their own work of issues for consideration. For instance, more involved coaching of participants through their mapping of timelines may be required for participants who are unfamiliar or uncomfortable with visual modes of expression in order to decrease their discomfort and uncertainty, and thereby enhance engagement. The implementation of timelines should thus include consideration of participants' levels of familiarity with visual methods, as some groups may require more involved coaching in order to more effectively engage with timeline methods.

References

Ahmad F, Rai N, Petrovic B, Erickson P, Stewart D. South Asian immigrant women and resilience as survivors of partner violence. J Immigr Minor Health. 2013;15:1057–64.

Bagnoli A. Beyond the standard interview: the use of graphic elicitation and arts-based methods. Qual Res. 2009;9:547–70.

Berends L. Embracing the visual: using timelines with in-depth interviews on substance use and treatment. Qual Report. 2011;16:1–9.

Crouch M, McKenzie H. The logic of small samples in interview-based qualitative research. Soc Sci Inf. 2006;45:483–99.

Goodrum S, Keys J. Reflections on two studies of emotionally sensitive topics: bereavement from murder and abortion. Int J Soc Res Methodol. 2007;10:249–58.

Gringeri C, Wahab S, Anderson-Nathe B. What makes it feminist? Mapping the landscape of feminist social work research. Affilia: J Women Soc Work. 2010;25:390–405.

Guillemin M. Understanding illness: Using drawings as a research method. Qual Health Res. 2004;14(2):272–89.

Guillemin M, Drew S. Questions of process in participant-generated visual methodologies. Visual Stud. 2010;25(2):175–88.

Harper D. Reimagining visual methods: Galileo to Neuromancer. In: Denzin NK, Lincoln YS, editors. Collecting and interpreting qualitative materials. 2nd ed. Thousand Oaks: Sage; 2003. p. 176–98.

Holland J. Emotions and research. Int J Soc Res Methodol. 2007;10:195–209.

Hollway W, Jefferson T. Eliciting narrative through the in-depth interview. Qual Inq. 1997;3:53–70.

Holstein JA, Gubrium JF. Constructing the life course. Dix Hills: Rowman & Littlefield; 2000.

Horsfall D, Titchen A. Disrupting edges – opening spaces: pursuing democracy and human flourishing through creative methodologies. Int J Soc Res Methodol. 2009;12:147–60.

Jackson K. Participatory diagramming in social work research: utilizing visual timelines to interpret the complexities of the lived multiracial experience. Qual Soc Work. 2012;12(4):414–32.

Karnieli-Miller O, Strier R, Pessach L. Power relations in qualitative research. Qual Health Res. 2009;19(2):279–89.

Kesby M. Participatory diagramming: deploying qualitative methods through an action research epistemology. Area. 2000;32(4):423–35.

King N, Horrocks C. Interviews in qualitative research. Thousand Oaks: Sage; 2010.

Kolar K. Resilience: revisiting the concept and its utility for social research. Int J Ment Health Addict. 2011;9:421–33.

Kolar K, Erickson PG, Stewart D. Coping strategies of street-involved youth: exploring contexts of resilience. J Youth Stud. 2012;15:744–60.

Kolar K, Ahmad F, Chan L, Erickson PG. Timeline mapping in qualitative interviews: a study of resilience with marginalized groups. Int J Qual Methods. 2015;14(3):13–32.

Kuzel AJ. Sampling in qualitative inquiry. In: Crabtree B, Miller W, editors. Doing qualitative research. Thousand Oaks: Sage; 1999. p. 33–45.

Liamputtong P. Researching the vulnerable: a guide to sensitive research methods. London: Sage; 2007.

Liamputtong P. Qualitative research methods. 4th ed. Melbourne: Oxford University Press; 2013.

Mason J, Davies K. Coming to our senses? A critical approach to sensory methodology. Qual Res. 2009;9:587–603.

Maxwell J. Understanding and validity in qualitative research. Harv Educ Rev. 1992;62:279–301.

Nicholls R. Research and indigenous participation: critical reflexive methods. Int J Soc Res Methodol. 2009;12:117–26.

Osei-Kofi N. The emancipatory potential of arts-based research for social justice. Equity Excell Educ. 2013;46:135–49.

Patterson M, Markey M, Somers J. Multiple paths to just ends: using narrative interviews and timelines to explore health equity and homelessness. Int J Qual Methods. 2012;11:132–51.

Pauwels L. Visual sociology reframed: an analytical synthesis and discussion of visual methods in social and cultural research. Sociol Methods Res. 2010;38(4):545–81.

Preventing Violence Across the Lifespan [PreVAiL] Research Network. Theme 2: Resilience. 2010. Retrieved from: http://prevail.fims.uwo.ca/theme2.html

Rose G. Visual methodologies: an introduction to the interpretation of visual methods. London: Sage; 2001.

Sheridan J, Chamberlain K, Dupuis A. Timelining: visualizing experience. Qual Res. 2011;11:552–69.

Smith D. The conceptual practices of power: a feminist sociology of knowledge. Toronto: University of Toronto Press; 1990.

Umoquit M, Dobrow M, Lemieux-Charles L, Ritvo P, Urbach D, Wodchis W. The efficiency and effectiveness of utilizing diagrams in interviews: an assessment of participatory diagramming and graphic elicitation. BMC Med Res Methodol. 2008;8:53–65.

Ungar M. Nurturing hidden resilience in troubled youth. Toronto: University of Toronto Press; 2004.

White J, Klein D. The feminist framework and poststructuralism. In: White J, Klein D, editors. Family theories. 3rd ed. Thousand Oaks: Sage; 2008. p. 205–40.

Capturing the Research Journey: A Feminist Application of Bakhtin to Examine Eating Disorders and Child Sexual Abuse

118

Lisa Hodge

Contents

1 Introduction	2062
2 A Feminist Approach to Bakhtin's Dialogism	2063
3 Bakhtin's Authoritative and Internally Persuasive Discourse	2065
4 The Study: Situating the Context	2066
5 Narrative Bio-sketches	2066
5.1 Analiese	2067
5.2 Ollie	2067
6 Dialogical Semistructured Interviews	2068
7 Poetry as Part of a Layered Account	2071
8 Visual Methods as Another Layer	2072
9 Conclusion and Future Directions	2075
References	2076

Abstract

Important links have been established between eating disorders and child sexual abuse. These medical and positive studies, however, have causally quantified the link, and analysis has remained within the parameters of individual psychology. Thus, women's perspectives and experiences are ignored. In this chapter, I argue for a feminist application of Mikhail Bakhtin's sociological linguistics when examining women's experiences of eating disorders and child sexual abuse and the links between them. Bakhtin's theoretical constructs – authoritative and internally persuasive discourse – can enable researchers to expose the seemingly objective *truths* that overshadow alternative discourses competing for expression. I also argue for the use of a *layered account*, a technique that enables me to incorporate artistic expression. This chapter begins with an overview of how

L. Hodge (✉)
College of Health and Biomedicine, Victoria University, Melbourne, VIC, Australia
e-mail: Lisa.hodge@vu.edu.au

Bakhtin's theoretical constructs can promote a feminist paradigm for analyzing women's understandings. Drawing from my research that examined the nature of the relationship between women's experiences of child sexual abuse and eating disorders, I demonstrate how Bakhtin's theoretical constructs can expose the hidden mechanisms of control found in these gender-based oppressive practices. I also illustrate how drawing and poetry, when used in qualitative research methodologies, can create space for interactional discovery and give voice to the unspeakable.

Keywords

Bakhtin · Sociological linguistics · Feminist paradigm · Women · Eating disorder · Child sexual abuse · Drawing · Poetry

1   Introduction

My previous work as a social worker at the Eating Disorders Association of South Australia provided the impetus for the doctoral research upon which this chapter is based. I use the blanket term *eating disorders* throughout this chapter because I have found, as have others, that, despite common assumptions, anorexia and bulimia do not necessarily exist independently of each other, as oppositional categories (e.g., see Burns 2004; Warin 2004). Anorexia and bulimia are practices commonly engaged in either sequentially or simultaneously. Like Burns (2004), I push for a notion of fluidity between the eating disorders and the importance of challenging the way in which we have become accustomed to thinking about the categories of anorexia and bulimia that privileges anorexia above other types of eating distress.

I was driven to examine the link between eating disorders and child sexual abuse, because I found it to be an under-explored area in feminist research. Feminist scholars had acknowledged the significance of child sexual abuse in the backgrounds of women with eating disorders (see Wooley 1994; Malson 1998; Bordo 2003; Warin 2010). Yet, feminist research had not focused on women who had experienced both. Medical and positivist studies had causally quantified the link between eating disorders and child sexual abuse (Smolak and Muren 2002; Sanci et al. 2008; Chen et al. 2010). Although significant in naming the link, these studies portrayed women in pathologizing ways and ignored women's perspectives and experiences. The problem with quantitative approaches is that pathologized accounts of women frame them as victims; they are stigmatized and silenced, and the problem becomes worse.

It was *not* the aim of my research to suggest that all eating disorders in women are a direct result of sexual trauma. However, child sexual abuse is frequently an influencing factor and can no longer be ignored when understanding eating disorders and how women recover. Central to the study was the question of gender, as child sexual abuse and eating disorders are argued to be two highly gendered phenomena. For example, the majority of individuals affected by eating disorders are women (Malson and Burns 2009), and this is also the case with child sexual abuse victims

(Warner 2009; Stoltenborgh et al. 2011). The vast majority of child sexual abuse perpetrators are male with females comprising only 1–4% of all sexual offenders (Peter 2009). Despite child sexual abuse being a criminal offense, perpetrators are often portrayed as normal men who have been seduced by a precocious child or forced into the abusive behavior by an inadequate wife/mother (Hooper 1992; Bolen 2001). Moreover, child sexual abuse is conceptualized in medical terms as resulting in permanent damage to the personality (Warner 2009). Within this view, women are depicted as being made ill through their experience. This is also common in eating disorder theorizations, which is one example of how these experiences are similarly gendered in some respects.

As medicine has "owned" the study of the body and its disorders since the seventeenth century (Lafrance and McKenzie-Mohr 2013), dominant medical discourse portrays eating disorders as personal and internal problems of biochemical irregularities or internal psychological maladjustment (Malson and Burns 2009). Rather than positioning eating disorders within a framework in which analysis remains within the parameters of individual psychology, this chapter shifts the focus from "a damaged personality" to an eating disorder being "an understandable response" to sexual trauma. Indeed, the women who participated in my study articulated and understood eating disorders in a multiplicity of ways. Eating disorders were constructed as self-punishing and self-destructive, yet simultaneously self-producing. In response, I examined the social and structural constraints on the women and how their eating disorders could be viewed as an extreme way to "cope" with child sexual abuse.

As women feel and think about the social world through the body, a body which is not independent of social relations, but continuously constituted by them (Kleinman et al. 1997; Yang et al. 2007), emotions can illuminate aspects of women's experiences and add power in understanding, analysis, and interpretation. Thus, emotions, which are understood in this chapter to be products of both the body and discourse, offer valuable insights into social relations and social dynamics (Probyn 2005). Yet, not everything can be realized in language. The chapter also demonstrates how artistic processes can give participants a voice to resist with. Artistic forms can provide a different window into the lives of participants (Clarke et al. 2005) and enable the expression of powerful emotions that might not always be easily expressed in a clear or linear fashion (Furman 2006a). In this way, the theoretical and methodological framework for this research project grew out of the desire to counter silence and make women's "invisible inner thought, vision, or experience visible" (Ramm 2005, p. 66).

2 A Feminist Approach to Bakhtin's Dialogism

As language is inevitably distilled with culturally ingrained, authoritative discourses, it was important to understand *how* meaning production is bound by the larger relevant discourses in which meanings are produced. As such, I turned, in part, to Russian philosopher Mikhail Bakhtin's (1895–1975) sociological linguistics, for a theoretical framework. Bakhtin (1981) developed, over a prolific career of some 50 years, a theory known as *dialogism*, in which he conceived of all discourse as dialogical. Bakhtin argues that there is no one interpretation, no single harmonious worldview, and no single truth (Bakhtin 1981). Communicative acts only have meaning in particular situations or contexts. Bakhtin (1981, p. 276) writes:

> Any concrete discourse (utterance) finds the object at which it was directed already as it were overlain with qualifications, open to dispute, charged with value, already enveloped in an obscuring mist-or, on the contrary, by the 'light' of alien words that have already been spoken about it. It is entangled, shot through with shared thoughts, points of view, alien value judgements and accents. The word, directed toward its object, enters a dialogically agitated and tension-filled environment of alien words, value judgements and accents, weaves in and out of complex interrelationships, merges with some, recoils from others, intersects with yet a third group: and all this may crucially shape discourse.

What Bakhtin is suggesting here is that we do not learn words from a dictionary, we acquire them from hearing or reading the words of others and, therefore, they are marked with the voices of those prior contexts. (Bakhtin 1986)

I used Bakhtin's rich body of theories as both as a methodological tool when conducting interviews and as a theoretical lens when analyzing the interview transcripts. In adopting a Bakhtinian perspective toward interviewing, I understood that no voice exists in isolation but is shaped by, relates to, and competes with other voices. In this sense, the dialogical interviews were regarded as a joint project where meanings were mutually constructed between the women and me. Communication was seen as an interactive process where the women were encouraged to be active agents through reflective processes. This included the use of specific questions that instigated self-reflections (Frank 2005) and me showing each of the women the analysis of her interview. In this way, dominant discourses that held binding authority within the women's narratives could be overturned together, and other discourses that were competing for expression could be exposed and analyzed within subsequent interviews. By inviting the women to participate in the deconstruction of their own experience as acts of engagement with me, the roles of researcher and participant were destabilized, and my own voice, perspectives, narrative, and knowledge were intertwined with theirs (see also ▶ Chap. 119, "Feminist Dilemmas in Researching Women's Violence: Issues of Allegiance, Representation, Ambivalence, and Compromise").

Yet, like almost all literary critics in the first half of the last century, Bakhtin does not include women as authors or speakers in his discussion of literature. Bakhtin's socially and historically grounded concept of language does, however, lend itself to feminist criticism because he recognizes that literature exists in a political context by taking into account the various determining social and historical factors of language. A gender-aware application of Bakhtin can disrupt patriarchal language and explore marginal voices within dominant discourses (Bauer and McKinstry 1991). As such, a Bakhtinian theoretical framework enabled me to illuminate multiple meanings by providing a strategy for resisting hierarchical and normalizing discourses. Through a feminist application of Bakhtin, I was able to consider the contradictions within discourses and the multiple ways in which women negotiate *how* eating disorders, as a practice, work "across multiple planes" (Probyn 1987, p. 210). Interrogating the dominant discourses that emerged in the women's narratives gave voice to their rarely discussed experiences and provided the opportunity for powerful counter-narratives to established hegemonies to emerge and psychiatric and biomedical discourses to be challenged.

3   Bakhtin's Authoritative and Internally Persuasive Discourse

To understand the embodiment of women's experiences of child sexual abuse and eating disorders, I used Bakhtin's key concepts, *authoritative discourse* and *internally persuasive discourse*. These two theoretical constructs described in Bakhtin's *Discourse in the Novel* inform my analysis and promote a feminist agenda (1981). Bakhtin contends that consciousness is a process of interaction among authoritative and internally persuasive discourses; the self is, therefore, an event of language experience. Moreover, the body is defined by those with an authoritative voice of what is the "truth" of human nature. Through the use of Bakhtin's theoretical concepts, I am able to remain attuned to how power works in the process of communicating with the women I interviewed. Understanding the social, political, and historical implications of words passing through a speaker on their way to a listener is critical to understanding how different discourses compete with a single utterance (Alcoff 2008). I drew on Bakhtin's notion of the authoritative and internally persuasive discourse to expose seemingly objective truths in the women's narratives that overshadowed alternative discourses competing for expression. An authoritative discourse has such binding authority that it seems untouchable, inspiring only adoration and respect, and it maintains the status quo. In contrast, an internally persuasive discourse is denied all privilege, as it is "frequently not even acknowledged in society" (Bakhtin 1981, p. 342). Where an authoritative discourse is weighted with authority and appears to remain within a single language system, an internally persuasive discourse is open and unfinished (Bakhtin 1981).

Bakhtin (1981) argues that when reading a text authoritatively, a seemingly objective truth about the meaning of that text is established. From this position, authoritative discourses are complicit with patriarchal ideology. This is a dangerous prospect for women, as male truths are spoken with such power that their assumptions are no longer questioned. A prominent example of an authoritative discourse is the medical discourse of eating disorders. Medical discourses pathologize the experiences of women and, as such, minimize attention to possibilities for recovery (Becker 2010). Searching for alternative discourses in the women's narratives offered the opportunity to incite struggle and encourage creativity. As meaning is produced (or policed) through discourse, rather than revealed through it (Burr 2007), these concepts enabled me to examine how gendered and disciplinary discourses created and sustained the women's eating disorders.

4 The Study: Situating the Context

I conducted multiple semistructured qualitative interviews with seven women who had an eating disorder and had experienced child sexual abuse; my aim was to create a "living dialogue" in which authoritative discourses could be teased out and disrupted (Francis 2012, p. 5). To recruit participants, colored flyers asking participants to volunteer for the study were placed around the University where I undertook my PhD in South Australia. The flyers asked women to volunteer for the study who were 18 years or older and who self-identified as having an eating disorder and unwanted sexual experiences in their past. The flyers also stated that participants were required to participate in multiple interviews. As child sexual abuse is considered by radical feminists to be a manifestation of the oppression of women, inherent in a patriarchal society, both the causes and consequences can be markedly different depending on one's gender (Reavey and Gough 2000). Thus, I chose to focus on experiences of women, and therefore only women were recruited. Seven women were recruited and each participant was interviewed up to five times. The point at which the data reached saturation point determined the number of interviews conducted with each woman. I planned to recruit more participants if I had found this not to be the case when analyzing the data; however, no further recruitment was required.

Potential participants who had seen a flyer and chose to contact me were given a verbal description of the study as well as a written information sheet. The information sheet contained details about the study's aims, processes, and confidentiality, and it outlined how potential participants could participate and provided contact details of the researcher and the university's Human Research Ethics Committee. Participants were provided with a second information sheet and verbal description if they wanted to include their artwork, poetry, and journals, which clearly explained the use, collection, storage, dissemination, and future use of the artwork, poetry, and journals. The women signed consent forms before participating in this research: one to participate in the interviews and a second form prior to their artwork, poetry, and journals being collected.

This research involved individual interviews averaging 90 min each in duration, which were digitally audiotaped and then transcribed *verbatim*. Interview participants were asked to self-select fictitious names at the start of the first interview. During transcription of interviews, any identifying features were deleted and fictitious details inserted in order to ensure confidentiality. These fictitious details were sustained through data analysis and reporting of outcomes. I asked the women to talk in detail about their understanding of their experience.

5 Narrative Bio-sketches

This next section introduces the reader to two of the seven women who constituted the sample of this research by providing a compressed narrative bio-sketch (as appropriated by Berger 2006) for each woman. Following Berger's (2006) approach in her study of HIV-infected women drug users, the narrative bio-sketches presented here lay the groundwork for understanding the context of the women's lives. They show snippets of how the women talked about their experience of an eating disorder and their experience of child sexual abuse.

5.1 Analiese

Analiese was a 24-year-old woman who had an older brother who lived interstate and parents who were professionals. She saw her "mum a fair bit" but did not have a lot to do with her father. Her parents would frequently "split up and get back together and split up." Analiese told her mother that she would move out if her father came back home again, and so, with 24 h' notice, Analiese found herself "homeless for about 12 months." Analiese said that she had "dissociative identity disorder" which was diagnosed "about 4 or 5 years ago," and she considered this to be "a post-trauma related condition where basically different personalities develop." Analiese had had "ongoing contact with the Mental Health Services" since she was about 12 or 13 years old, "not really knowing what's going on." She spent most of her high school years "in and out of hospital" and that this was for "the eating disorder stuff and depression and suicide attempts."

Analiese was diagnosed with anorexia or what was described "as anorexia in the DSM." She "had big issues with vomiting," so she could never vomit up her food but that she "would exercise." She said "there was a lot of sexual abuse" in her childhood and that she "basically grew up in the child sex industry" being "used in the production of pornography" from "2 years old" until she was a teenager, when she "rebelled against everything and left and ran away from home." Very quickly, Analiese realized that the only way she really knew how to survive was to work in the "sex industry."

Analiese "did a lot of stuff as a kid," like "self-harming" and "took a lot of drugs, did a lot of the normal feral kid stuff but probably a lot more mass extreme." Part of her eating disorder was "probably related to that." When she was first hospitalized for anorexia at 14 or 15, when they were trying to get her to eat, she felt she did not "deserve food." She had to battle the negative effects of the "psych meds" that made her keep on the weight and that her "obsession" with numbers and getting to a certain weight started after that. Analiese said her parents had "a lot of denial that there was an issue" and that they would not allow her to "see a psychiatrist or psychologist," so she accessed services at "Second Storey" because she could access them "without parental consent." When her father found her medication in her room, he became extremely angry and threw her "across the room." Analiese said her mum is very supportive now and that "she knows what DID [dissociative identity disorder] is, where it comes from, and she still won't ask anything." Analiese is no longer "skinny enough to be diagnosed with anorexia" but that she still has "a lot of the behaviors."

5.2 Ollie

When I first interviewed Ollie, she was a 26-year-old woman who had two older brothers and was raised in a practicing Christian family where she "had to be a Christian" and "had to be well behaved" and act like a "little 50s housewife." She "was molested by a babysitter at 3." She said she "was raped at 4" and that she understood this to have been "my father" for an unknown period of time and was then raped at 17 by a boyfriend. Ollie always believed that she was "never going to be as good as anybody else at any given thing"; she was "never going to be as beautiful" or was "never going to be as valued as anybody else." She understood that her "worth was so much less" because she had "been ruined by something." Ollie had no memory of her childhood – it was "blank." She started having flashbacks or memories at different stages although she had "always just put them down to nightmares or day mares." At the age of 19, Ollie "started asking questions"; she started experiencing memories that were "almost like body memories" saying that she could "almost feel it happening at times." Ollie was depressed until she "started having flashbacks." Now, instead of feeling really depressed, she felt "shame" that it was her "fault" and that she was "disgusting" and "no-one's ever going to want that." At age 4, Ollie woke up one morning to find that she had a "double inguinal hernia" which she explained was "caused by severe trauma." She had to have surgery to stitch all her "muscles back together."

Ollie described having the eating disorder "since probably 9 or 10 years old" and went on to say that her "negative attitude towards food" tied "into a really negative body image and self-image," and the two were "linked somehow to the abuse." Ollie said the eating disorder "wasn't so much a cry for help"; it was more about if she was treated badly, it was her "weapon" as it made her feel like she "had some power too."

Her father "used to bait" her "in front of the family," and it would get to the point where she avoided him as she "couldn't trust him" and "did not feel safe." When he came within a meter of her, she "already felt violated in some way" not because she "knew that something had happened before" but because he made her "feel dirty." At about 14 or 15, she started self-harming as a way of "releasing stuff" that was happening in her family and because she believed that she "deserved to feel pain on a regular basis." Ollie said self-harming was a "coping mechanism" that she used to help her cope with emotions or anything that made her "feel out of control" and that it was "a punishment too." Ollie also said "the limitation of food was another coping mechanism" for her and went on to say that, by starving, she could "get to control" of one area of her life and that she was rewarded for it by "getting thinner." Ollie always felt guilty about eating, no matter how much she ate, and would always go to bed feeling "dirty and heavy." She said she felt she was the issue and that "the abuse left its dirty paw prints" all over her, and if she made herself smaller through starving, she was "making the mess smaller." It was also important for Ollie to have structure and order and that she needed to "have a plan," because having a plan was "another aspect of the control." She said that "without a plan all the things" that she had been running from would "finally catch up."

6 Dialogical Semistructured Interviews

Following a dialogical method, which considers verbal responses to be "themselves built on responses to historic utterances made by ourselves and others" (Francis 2012, p. 4), I conducted multiple, individual, semistructured, face-to-face interviews with Analiese and Ollie. Conducting more than one interview with each woman enabled discourses to be built on or resisted. A distinguishing feature of dialogic research is including oneself as an active participant in the narrative and its interpretation (Riessman 2008; see also ▶ Chap. 24, "Narrative Research"). In these dialogical interviews, my communication with the women was not approached from the point of view of transmission of information but was seen as an interactive process in which both speaker and listener played an active role (Lahteenmaki 1998).

The interviews were conducted in a place of the women's choosing, where they felt most comfortable and at a time that suited them. I conducted the interviews over a 12-month period to allow time to transcribe *verbatim* and analyze each interview before conducting subsequent interviews with each woman. After each interview, I analyzed the transcripts using a thematic analysis (Patton 2015; see also ▶ Chap. 48, "Thematic Analysis"). After each interview with each woman, and before conducting further interviews with her, a copy of her interview transcript was offered to her.

When undertaking the multiple interviews with each participant, the research process became a dialogue between the researcher and participant. In a similar way to Lather and Smithies (1997, p. xvi), I wanted to position "the reader as thinker, willing to trouble easily the understood and the taken-for-granted." Thus, the more room that I made "for the voices, language, and moving stories of the women's lives," I hoped, the more engaged the readers would be in the analysis (Abma 2002, p. 25). This is because, if the text refers to the reader's own lifeworld and they can connect their experiences with the text, additional meaning can be gained from it (Abma 2002). As such, I present large amounts of the women's interview transcripts.

I used Bakhtin's understandings to tease out the discourses that inform both the speaking and hearing integral to the construction of each theme. As Bakhtin (1981, p. 282) insists, "understanding comes to fruition only in the response." Prior to conducting further interviews with each woman, I showed her my written analysis containing the emerging themes and her quotes that were used to determine these themes. New interview questions were designed, and subsequent interviews were conducted after each woman had the opportunity to comment on my interpretations of her interview. This process was repeated for each interview. As a dialogical researcher, I was more than an observer of the women's lives outside the interview; I participated in acts of engagement with them in understanding that their stories were the site of struggles permeated by multiple voices, how each voice was the site of multiple voices, and what was the contest among these voices.

Figure 1 provides an overview of the dialogical process between the women and me. Figure 1 shows, at a glance, the first layer in the analysis.

There was potential for conflict between my feminist dialogic frame of reference and the women's interpretations, as I tried to explain their lives without violating their reality. However, from a Bakhtinian view, meanings are negotiated. Each woman and I did our best to arrive at a compromise regarding our interpretations. Lahteenmaki (1998, p. 91) suggests

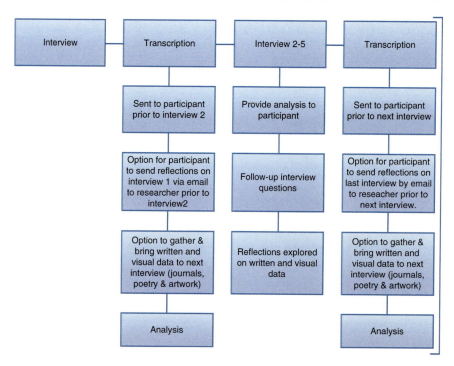

Fig. 1 Creating layered accounts through the dialogical interview process

> Communication always includes ... an element of struggle over meanings ... it must be borne in mind that absolute disagreement is also impossible in the sense that, in order to disagree on something with someone, there naturally has to be a common ground mutually shared by both parties.

Thus, the dialogical process opens up quite different assumptions about the creation and meaning of truth and knowledge (Riessman 2008). This is particularly important for research on child sexual abuse and eating disorders where hegemonic discourses can conceal the potential for other understandings and thereby the potential for resisting silenced trauma.

As I did not want to passively record where each woman was in her life, I used specific questions that were developed from the emerging themes to instigate self-reflections with each woman. Thus, the women actively coauthored, or co-constructed, in this dialogical process. Multiple meanings were negotiated, and workings of power were acknowledged and analyzed together. This method of conveying information fitted with my feminist agenda, as it allowed the women to determine at least some of the terms by which they would be known.

The women's written and visual data was also used to help engage the women in talk. Using an adaption of Rose's (2008) framework as a guide (see Hodge 2014), I asked Analiese to explain the meaning of her drawing and the metaphors Ollie used in her poem. When the interviews were transcribed and analyzed, I could go back to the women with their poetry and drawing and use my analysis to instigate further questions. Ollie and Analiese then had the opportunity to critically reflect and expand on the meanings of their poem and drawing after discursive themes had emerged in my analysis. Thus, the interviews were fluid, and their structure was determined to a large extent by the women.

7   Poetry as Part of a Layered Account

Creative writing is used as an avenue for expressing feelings and emotions, which is considered easier than through talk. Poems are a powerful narrative tool which can further empathize and understand, as poems speak "in a unique way from the interior of human experience in a way that science can never" do (Shapiro 2004, p. 172). Thus, poems have been argued to better represent the speaker and are compelling research revelations in their own right (Lahman et al. 2010). The use of poetry in this study offered a different window into the lives of participants (Clarke et al. 2005), because what was included in the poem was just as important as what was left out of it (Poindexter 2003). Poetry opened up possibilities for discovering new meanings as it allowed the women to critically reflect on experience, with at least some of the habitual meanings held in abeyance (Oiler 1983). This critical reflection offered space to voice-silenced traumas.

Poetry, as a source of data, fitted well with the dialogical interviews. This is because it is less linear than other texts (Gannon 2001). It is in the pauses and gaps and motion that readers can insert their own lived experience and their "various selves to create embodied knowledges" (Gannon 2001, p. 791). Thus, the women's poetry was used in a reciprocal relationship in which insights and meanings were developed and shared (Furman 2006b). One way to mitigate the richness of differing meanings within a single poem is to encourage full inclusion of the text, allowing readers "to act as triangulated investigators" (Shapiro 2004, p. 175). As such readers can vicariously enter into Ollie's lived experience, in a way that is not available in other modes of writing/research.

Ollie said that using poetry to talk about her experiences of child sexual abuse and an eating disorder was a positive experience, as it offered another layer of understanding and a way to give voice to unarticulated experiences. Child sexual abuse is often difficult to speak of directly, and for Ollie, poetry was a means to effectively convey powerful and complex emotion. As Furnam (2004, p. 163) contends, poems transcend the limits of language and provide in-depth understanding of lived experience "operating on the level of image, the poem resounds in the mind." Unlike other forms of expression, for Ollie, words, space, and sound merged, and this was critical to the articulation of meaning. Poetry opened a space for Ollie that was "attentive to multiple meanings" and more easily enabled the accessing of subjugated perspectives (Leavy 2009, p. 64). As Leavy (2009) points out, listeners and readers tend to be moved by the power of poetry, and the impact of a metaphor was more powerful than if Ollie had merely described her experience. Ollie used poetry to represent the essence of what she wanted to say in a vivid, sensory way.

In *Ollie's Poem*, she has used forceful words such as "raped," "miracle," and "freedom" and phrases including "in my normal voice," "when I cut myself," and "I want to kill myself" to express the extent of her emotional turmoil and her feelings, in which she was metaphorically dead and alive at the same time.

Ollie's Poem I want a love that knows no fear. I want to stop ruining what I have. I want a love that even time will lie down and be still for. I want to FEEL. I want to know that I'm worth more than what I weigh. I want to be able to say I'm good at something. And believe it. I want to know why I feel like I've been raped. I want it to be OK when I cut myself, I want scars to be acceptable. I want to be able to say "I love you" in my normal voice. I want a miracle! I want some help! I want to know who I am. I want my own space and freedom. I want to enjoy eating instead of hating it. I want to kill myself, instead of always talking and thinking about it. I want more romance and less sex. I want to want sex.

The repetition of "I want" is strong in how it eludes to a sense of powerlessness. It also creates space for Ollie's own voice in the poem to say what she could not say out loud and in public. In the phrases "I want to stop ruining what I have" and "I want to know that I'm worth more than I weigh," Ollie connected her feelings of worthlessness to her eating disorder. Ollie's poetry symbolically expressed her fear of being alone in her struggles and her confusion around sexual relationships. Although eating disorders have been portrayed in numerous ways – as a means of avoiding sexual contacts and feelings; as an expression of anger, inflicted upon the woman herself as a form of punishment; as a coping strategy for self-hatred and powerlessness; as posttraumatic stress symptoms related to severe boundary violations; and as a way to make oneself sexually unappealing – for Ollie, an eating disorder was a metaphor for a symbolic relieving of abusive experiences. Ollie's body was the place where knowledge, memory, and the pain of sexual trauma were stored. As the body is integral to the sense of self, discourses of femininity that Ollie had to negotiate rendered the body a trap. Ollie's struggle with her embodied feelings of worthlessness was primarily regulated through shame. An eating disorder was understood as a process of transformation of this emotion through the physical body. Her shame became contained through disordered eating practices and thus made safe. As Lawrence (1984) points out, an eating disorder can affect the relationship between oneself and one's body, and for Ollie an eating disorder was a solution to a problem that she found impossible to deal with any other way. It was painfully difficult to choose to give up, which shows the complexity of her agency. Being able to exercise her agency, I am arguing, provided Ollie with a sense of safety, and thereby was used to negotiate the complexity and ambiguity of sexual relationships in the aftermath of child sexual abuse.

8 Visual Methods as Another Layer

The ability of drawing, to surface unspoken thoughts and feelings, has long been accepted by art therapists (Malchiodi 2007). Therapists have used this tool for many decades to facilitate expressions that are not easily put into words (Malchiodi 2007).

Within psychology, images have traditionally been given the status of a more simplistic form of communication and confined to use with children or those deemed less able to communicate thoughts and feelings (Reavey 2011). Drawing was used in this study as a visual product and a process, offering a way of exploring the multiplicity and complexity that is the base of much social research interested in human experience (Guillemin 2004; see also ▶ Chap. 101, "Drawing Method and Infant Feeding Practices Among Refugee Women"). Specifically, drawings were used to encourage Analiese to talk about her experiences, affording her every opportunity to frame her own experiences, unencumbered by my biases. This enabled me to collaborate with Analiese, which fitted well with my Bakhtinian dialogic feminist methodological approach.

Some methods for analyzing visual data are suggested to be more methodical than others. They lay down very precise criteria for analysis and a step-by-step procedure. My aim in using visual data was not to obtain an understanding *of* the participant's drawings. I wanted to understand with the drawings about the life of the participant in this study. In order to *look at* the image and *look behind it,* I drew on Rose's (2008) critical visual framework. According to Rose (2008), the interpretation of visual images exhibit three sites at which meanings of an image are made: the site of production of an image, the site of the image itself, and the site of the audience. I followed Guillemin's (2004) adaption of Rose's framework, where the data comprised both the visual images and the participant's verbal descriptions of the image. It was important to ask Analiese to describe her drawing as an essential part of the method, as it elicited the nature of the drawing and why she chose to draw that particular image. This process instigated critical reflections on how Analiese's artwork represented her experience of child sexual abuse and an eating disorder, and the significance of what she had drawn in relation to her previous statements made during the interview. The events, experiences, and interactions that preceded the drawing all worked to produce the understandings that were embedded in the drawing (Guillemin 2004).

I also left room for readers to interpret the pictures themselves. Radley (2010, p. 268) suggests "that what pictures portray and what stories narrate are better thought of as versions of our experience of the world than as constructions of the world that we experience." Although the images needed to be contextualized to some degree by words, Rose (2008) argues the actual image itself can be used as evidence to develop and support, or to supplement, research findings. Pink (2001) suggests exploring the relationship between the visual and the social and cultural contexts of knowledge production by using a reflexive approach toward the collation and analysis of visual data, which does not depend on translating visual evidence into verbal knowledge. Thus, the participant's drawings were used to present things that words could not, as "not everything that can be realised in language can also be realised by means of images or vice versa" (Kress and van Leeuwen 1996, p. 17).

Figure 2, a black-and-white drawing, was drawn by Analiese about 7 years ago, and it was one of the first that was specifically about the eating disorder. Here, Analiese's skeletal body is stripped of any facial features, hair, fingers, or clothes suggesting a total lack of power over her life – a facelessness and defenselessness.

Images of demons dominate the picture. Analiese has portrayed herself as running but barely able to stay outside of the grasp of the multiple demons that are attempting to envelope her, further emphasizing an overwhelming sense of powerlessness. As such, this drawing is also full of movement, marked by a moving body running away from the demons. The demons are also in motion, close behind the running body. In this way, Analiese's body is in perpetual motion, her sense of self always on the move, trying to get away from the demons so that they do not overcome her and become part of her own self. In this sense, the embodied self is not static or passive – it is active, even though it responds to the trauma it cannot name easily, or the emotional labor which goes into staying ahead, staying in control. The drawing, I am suggesting, names the emotional effort, the bodily labor, put into staying in control, all of which cannot be named with words but is nevertheless perceived visually.

When I asked Analiese for her reading of Fig. 2, she said it was about controlling her food as a way of running away from her demons and everything that was behind her. Having demons following her was a common theme in her drawings. This was because they represented "the magnitude of emotion" and the inner demons that she developed as a result of feelings of worthlessness. This was significant because later

Fig. 2 Analiese's self-portrait

in the interview, Analiese said she was beaten as a child if she expressed emotion. When I asked Analiese if she connected this picture to the abuse, she said that her "circumstances" were "still very strongly dictated by the abuse." Even though she tried to distance herself from the experience, there were "triggers" that took her back:

> There are probably elements of trying to get away from that and some of the stuff that I'm dealing with now as well as it is about actually accepting that some of my circumstances are still very strongly dictated by the abuse and the main perpetrator of the abuse is still a very powerful person in my life. So I guess it's almost that bind of you trying to run away from it but they keep up with you. So things like the fact that the perpetrator of the abuse or the main perpetrator is still a very strong person in my life is kind of one of those things that you can't escape them constantly, they're following you and there's all those triggers that kind of take you back. So there's that and then there's other things like being in places and stuff that kind of trigger things and all of a sudden you remember stuff and yeah, so everything kicks off. (Analiese)

Bakhtin (1981, p. 401) contends that "all direct meanings and direct expressions are false." This means that when we seek to understand a word, it is not the direct meaning the word gives to objects and emotions that matters but rather the self-interested use to which this meaning is put and the way it is expressed by the speaker, a use determined by the speaker's position. Taking Bakhtin's point into consideration, both Analiese's narrative and drawing suggest the enormity of her enduring struggle in "trying to run away" from the emotional "stuff" she was trying to deal with. However, she was powerless in the sense that her emotional state was destabilizing and unsettling and constantly evading her attempts to govern it.

9 Conclusion and Future Directions

This chapter has shown that through the internally dialogic quality of discourse, it allows Ollie and Analiese to make connections about their experiences of child sexual abuse and possible influences in later life. The chapter has also shown how the women made sense of their experiences through drawings and poetry, which flowed smoothly as a continuation of the narrative. By incorporating poetry as data, the "commonsense" assumptions of authoritative discourses that "center" women can be ruptured, and internally persuasive discourses about the female body emerge. The women spoke of a loss of control, and they attributed this loss to experiences of child sexual abuse and sexual trauma. They used their bodies to negotiate internally persuasive discourses – that is, alternative positions and voices – to challenge or contradict authoritative discourses of the female body. Drawing and poetry facilitated recall of painful memories, which otherwise might have been difficult to share with others. Controlling their bodies through an eating disorder gave the women a sense of control over their lives, and this was directly related to the control that was lost to them in child sexual abuse.

Trying to be perfect and in control are common in eating disorder theorizations. However, from a Bakhtinian position, there is a different story waiting to be told. The women's narratives articulated a controlled body that was about more than perfectionism; it was about trying to cope with a loss of control of the body and emotions and thereby a loss of self that was experienced as part of child sexual abuse. The women's preoccupation starvation functioned as a powerful "normalizing" strategy. As such, getting thin was not about getting thin, per se, but rather, staying in control of their bodies, emotions, memories, and thoughts about the abuse. Put differently, not eating was an assertion of individual control where food and body weight featured as the only arenas in which control was possible. In this sense, as Malson (1998, p. 123) describes, the controlled body quite explicitly signifies total control and "the positive subjectivity signified by the thin body here is accompanied by a subtextual subjectivity of 'failure' in all other aspects of life." Yet, as Bakhtin (1981, p. 348) asserts, "one's own discourse and one's own voice... will sooner or later begin to liberate themselves from the authority of the other's discourse." From this position, disordered eating served as an attempt by the women to stave off the painful effects from their experience of abuse and to manage and expunge feelings and emotions, memories, and thoughts while providing a sense of control through the body. Starving signified having a sense of voice about something they could not talk about openly or freely.

I want to conclude by advocating for two things. First the link between child sexual abuse and eating disorders has implications for the treatment and prevention of eating disorders. Ultimately, eating disorder treatment depends on changing the social conditions that underlie their etiology, and not simply on individual healing. This is because, as this chapter has argued, we need to consider ways of relating to human distress that go beyond *illness* and *pathology*. Indeed, the prevention of eating disorders depends on women's access to cultural, political, social, and sexual justice. Secondly, we must reconsider how we approach the topics of child sexual abuse and eating disorders. This is not a matter of selecting or fine-tuning one approach over another. Rather, it is about rethinking how our approaches and methods frame the basis through which we understand trauma, silence, violation, emotions, and the body in the context of child sexual abuse and eating disorders. It is about rethinking how our approaches frame human beings' capacity and agency to respond to, and make sense of, and resist trauma while living a livable life. My hope is that this chapter has made readers more aware of the rich promise of utilizing dialogic theory. Although none of us can be freed from ideological biases, I hope we can be freed from a kind of monologism that turns all ideologies into falsehoods. I also hope that the capacity of creative arts to convey meaning in alternative ways is realized, and that drawing and poetry will be incorporated into more research methodologies, especially when exploring facets of human experience too difficult to describe in words, yet too important to ignore.

References

Abma TA. Emerging narrative forms of knowledge representation in the health sciences: two texts in a postmodern context. Qual Health Res. 2002;12(1):5–27.

Alcoff LM. The problem of speaking for others. In: Jackson AY, Mazzei LA, editors. Voice in qualitative inquiry: challenging conventional, interpretive, and critical conceptions in qualitative research. Hoboken: Routledge; 2008. p. 117–35.

Bakhtin MM. Discourse in the novel. In: Holquist M, editor. The dialogic imagination: four essays (trans: Emerson C, Holquist M). Austin: University of Texas Press; 1981. p. 259–422.

Bakhtin MM. The problem of speech genres. In: Emerson C, Holquist M, editors. Speech genres and other late essays (trans: McGee VW). Austin: University of Texas Press; 1986. p. 60–102. (Original work published 1979).

Bauer DM, McKinstry SJ, editors. Feminism, Bakhtin, and the dialogic. Albany: State University of New York Press; 1991.

Becker D. Women's work and the societal discourse of stress. Feminism Psychol. 2010;20(1):36–52.

Berger M. Workable sisterhood: the political journey of stigmatized women with HIV/AIDs. Princeton: Princeton University Press; 2006.

Bolen RM. Child sexual abuse: its scope and our failure. New York: Springer; 2001.

Bordo S. Unbearable weight: feminism, western culture and the body. Berkeley: University of California Press; 2003.

Burns M. Eating like an ox: femininity and dualistic constructions of bulimia and anorexia. Feminism Psychol. 2004;14(2):269–95.

Burr V. Social constructionism. 2nd ed. New York: Routledge; 2007.

Chen LP, Murad MH, Paras ML, Colbenson KM, Sattler AL, Goranson EN, Zirakzadeh A. Sexual abuse and lifetime diagnosis of psychiatric disorders: systematic review and meta-analysis. Mayo Clin Proc. 2010;85(7):618–29.

Clarke J, Febbraro A, Hatzipantelis M, Nelson G. Poetry and prose: telling the stories of formerly homeless mentally ill people. Qual Inq. 2005;11(6):913–32.

Francis D. Gender monoglossia, gender heteroglossia: the potential of Bakhtin's work for re-conceptualising gender. J Gend Stud. 2012;21(1):1–15.

Frank AW. What is dialogical research and why should we do it? Qual Health Res. 2005;15:964–74.

Furman R. Poetic forms and structures in qualitative health research. Qual Health Res. 2006a;16(4):560–6.

Furman R. Poetry as research: advancing scholarship and the development of poetry therapy as a profession. J Poet Ther. 2006b;19(3):133–45.

Furnam R. Using poetry and narrative as qualitative data: exploring a father's cancer through poetry. Fam Syst Health. 2004;22(2):162–70.

Gannon S. (Re) presenting the collective girl: a poetic approach to a methodological dilemma. Qual Inq. 2001;7:787–800.

Guillemin M. Understanding illness: using drawings as a research method. Qual Health Res. 2004;14(2):272–89.

Hodge L. Lost for words: drawing as a visual product and process to give voice to silenced experiences. In: Chonody JM, editor. Community art: creative approaches to practice. Champaign: Common Ground Publishing; 2014. p. 60–76.

Hooper CA. Mothers surviving child sexual abuse. London: Routledge; 1992.

Kleinman A, Das V, Lock MM, editors. Social suffering. Berkeley: University of California Press; 1997.

Kress GR, van Leeuwen T. Reading images: the grammar of visual design. New York: Routledge; 1996.

Lafrance MN, McKenzie-Mohr S. The DSM and its lure of legitimacy. Feminism Psychol. 2013;21(1):119–40.

Lahman MKE, Rodriguez KL, Richard VM, Geist MR, Schendel RK, Graglia PE. (Re)forming research poetry. Qual Inq. 2010;17(9):887–96.

Lahteenmaki M. On meaning and understanding: a dialogical approach. Dialogism. 1998;1:74–91.

Lather P, Smithies C. Troubling the angels: women living with HIV/AIDS. Boulder: Westview; 1997.

Lawrence M. The anorexic experience. London: The Women's Press; 1984.
Leavy P. Method meets art: arts-based research practice. New York: Guilford Press; 2009.
Malchiodi CA. The art therapy source book. New York: McGraw-Hill; 2007.
Malson H. The thin woman: feminism, post-structuralism and the social psychology of anorexia nervosa. New York: Routledge; 1998.
Malson H, Burns M. Re-theorising the slash of dis/order: an introduction to critical feminist approaches to eating dis/orders. In: Malson H, Burns M, editors. Critical feminist approaches to eating dis/orders. London: Routledge; 2009. p. 1–6.
Oiler C. Nursing reality as reflected in nurses' poetry. Perspect Psychiatr Care. 1983;21(3):81–9.
Patton MQ. Qualitative research and evaluation methods. 4th ed. Thousand Oaks: Sage; 2015.
Peter T. Exploring taboos: comparing male- and female-perpetrated child sexual abuse. J Interpers Violence. 2009;24(7):1111–28.
Pink S. Doing ethnography: images, media and representation in research. London: Sage; 2001.
Poindexter CC. Research as poetry: a couple experiences HIV. Qual Inq. 2003;8(6):707–14.
Probyn E. The anorexia body. In: Kroker A, Kroker M, editors. Body invaders: panic sex in America. Montreal: New World Perspectives; 1987. p. 201–12.
Probyn E. Blush: faces of shame. Minneapolis: University of Minnesota Press; 2005.
Radley A. What people do with pictures. Vis Stud. 2010;25(3):268–79.
Ramm A. What is drawing? Bringing the art into art therapy. Int J Art Ther Formerly Inscape. 2005;12(2):63–7.
Reavey P, editor. Visual methods in psychology: using and interpreting images in qualitative research. New York: Routledge; 2011.
Reavey P, Gough B. Dis/locating blame: survivors' constructions of self and sexual abuse. Sexualities. 2000;3(3):325–46.
Riessman CK. Narrative methods for the human sciences. Thousand Oaks: Sage; 2008.
Rose G. Visual methodologies: an introduction to the interpretation of visual materials. 2nd ed. London: Sage; 2008.
Sanci L, Coffey C, Olsson C, Reid S, Carlin J, Patton G. Childhood sexual abuse and eating disorders in females: findings from the Victorian adolescent health cohort study. Arch Pediatr Adolesc Med. 2008;162(3):161–7.
Shapiro J. Can poetry be data? Potential relationships between poetry and research. Fam Syst Health. 2004;22(2):171–7.
Smolak L, Muren SK. A meta-analytic examination of the relationship between child sexual abuse and eating disorders. Int J Eat Disord. 2002;31:136–50.
Stoltenborgh M, van Jzendoorn MH, Euser EM, Bakermans-Kranenburg MJ. A global perspective on child sexual abuse: meta-analysis of prevalence around the world. Child Maltreat. 2011;16(2):79–101.
Warin M. Primitivising anorexia: the irresistible spectacle of not eating. Aust J Anthropol. 2004;15(1):95–104.
Warin M. Abject relations: everyday worlds of anorexia. New Brunswick: Rutgers University Press; 2010.
Warner S. Understanding the effects of child sexual abuse: feminist revolutions in theory, research and practice. New York: Routledge; 2009.
Wooley S. Sexual abuse and eating disorders: the concealed debate. In: Fallon P, Katzman MA, Wooley S, editors. Feminist perspectives on eating disorders. New York: Guilford Press; 1994. p. 171–211.
Yang LH, Kleinman A, Link BG, Phelan JG, Lee S, Good B. Culture and stigma: adding moral experience to stigma theory. Soc Sci Med. 2007;64(7):1524–35.

Feminist Dilemmas in Researching Women's Violence: Issues of Allegiance, Representation, Ambivalence, and Compromise

119

Lizzie Seal

Contents

1. Introduction 2080
2. Researching Women's Violence 2080
3. Questions of Researcher Allegiance 2083
4. Questions of Representation 2085
5. Questions of Ambivalence and Compromise 2087
6. Conclusion and Future Directions 2089
References 2090

Abstract

This chapter explores feminist dilemmas in researching women's violence. It suggests that women's use of violence is a sensitive topic for feminist researchers because feminists have sought to delineate the role of male violence in continuing women's subordination. Highlighting women's violence potentially detracts from this. Feminists also wish to avoid lending credence to misogynistic and antifeminist stereotypes, which inaccurately claim that women are equally as violent as men. Researching women's violence using feminist methodologies, which place value on creating knowledge from women's experiences, hearing marginalized voices, and democratizing the research process, raises dilemmas. The chapter considers these dilemmas across three areas – questions of allegiance, questions of representation, and questions of ambivalence and compromise. Allegiance refers whether researchers are "on the side" of their research participants, which can be a complex issue if their research participants have harmed others. The politics of representation are significant to how data are interpreted and how research participants and their actions are portrayed when writing up

L. Seal (✉)
University of Sussex, Brighton, UK
e-mail: e.c.seal@sussex.ac.uk

sensitive data. Researchers may experience feelings of ambivalence when they find their research participants difficult to empathize with and this can compromise researchers by making them in some sense vulnerable. The chapter discusses a range of examples in order to highlight these issues. Due to the limited methodological literature specifically on women's violence, it also draws on insights from other relevant feminist and criminological studies.

Keywords

Women's violence · Feminist methodology · Allegiance · Politics of representation · Ambivalence

1 Introduction

Researching violence by women from a feminist perspective poses some substantial dilemmas. For feminists, this is a sensitive topic entailing thorny ethical and political issues. The concept of "sensitive research" is well-established and refers to research which has the potential to cause harm (particularly emotional harm) to the participants and researcher (Dickson-Swift et al. 2007; Liamputtong 2007). Violence is a sensitive topic as it involves behavior that is manifestly harmful to others, and which also lays the perpetrator open to state-sanctioned punishment (Lee and Renzetti 1993). More than this, researching violence "raise[s] wider issues related to the ethics, politics and legal aspects of research" (Stanko and Lee 2003, p. 3).

This chapter examines methodological issues for feminist researchers engaged in researching violent women. In particular, it addresses questions of allegiance – whether researchers are "on the side" of their research participants, the politics of representation entailed in writing up sensitive data, the feelings of ambivalence researchers experience when they find their research participants difficult to empathize with and the ways in which this may compromise them. Before considering these issues, the chapter explicates why women's violence is a sensitive topic for feminist researchers and reviews feminist studies in this area.

2 Researching Women's Violence

Feminist methodologies have been employed to build a significant and indispensable body of research on gender violence. These methodologies emphasize democratizing the research process, enabling marginalized voices to be heard, taking a reflexive approach, and creating knowledge that will lead to political change (Hesse-Biber 2012). Feminist research is not distinguished by use of a particular method or approach, but rather by its grounding in feminist theory and ethics (Ramazanoglu and Holland 2002). Violence against women has been a crucial strand of feminist research and "is one of the most sensitive areas of research that feminists are engaged in" (Skinner et al. 2005, p. 10). In order to challenge women's subordinate position in patriarchal societies, it is necessary to both fully understand and combat violence against women.

Violence *by* women is less well researched by feminists and it is necessary to ask whether it is a suitable topic for feminist research at all. As Burman et al. (2003, p. 74) assert, "feminists have traditionally ignored female violence, fearing potentially negative political and social costs for the feminist movement more generally." This avoidance is explicable in the context of women's comparatively low use of violence, particularly in its serious and lethal forms (DeKeseredy 2011; Hester 2013). Ristock (2002, p. x) identifies that "[s]ecrets are sometimes kept for strategic reasons within liberatory movements such as feminism that are trying to eradicate the globally pervasive phenomenon of male violence against women."

There is a danger that attention to violent women confirms or lends credence to misogynistic fears and misperceptions, which lead to exaggerations about the prevalence of violence of by women (DeKeseredy 2011). Women who use violence are frequently demonized in ways which replay wider, sexist stereotypes (Seal 2010). Renzetti (1999, p. 51) admits that she was initially reluctant to research women's use of violence "for fear that my work will be used against women." She identifies the belief that "women are as violent as men but are not held accountable for their violence" as a supporting principle of antifeminism (p. 42). However, as she persuasively argues, far from this being a reason that feminists should avoid researching violent women, it is essential that they do so from a feminist perspective and via use of feminist methodologies.

Where violence by women, even fatal violence, can be contextualized as a response to violent abuse from male partners, it more easily fits into feminist frameworks (Seal 2010). However, there are also many feminist studies of more "atypical" violence by women, which is not in response to male violence. Child abuse and sex offending by women have been analyzed from a feminist perspective (Fitzroy 2001; Matravers 2008), as has women's interpersonal violence in same-sex relationships (Ristock 2002; Hester and Donovan 2009). There are feminist studies on women who kill their own children (Wilczynski 1997; Oberman and Meyer 2008). Seal (2010) examines cases of women who commit "unusual" murders – that is, where they have killed someone other than a male partner or their own child. This encompasses a wide array of these rare cases, from the murder of a female relative to "serial killings."

Bringing feminist perspectives to studying women's violence enables it to be contextualized, particularly as women who commit violence are portrayed in the media and elsewhere as transgressive, having violated norms of nurturance and care. They may be described as unnatural or even as "monsters" (Morrissey 2003). Feminist research can challenge sexist and antifeminist stereotypes about violent women by situating their actions in relation to socioeconomic factors, past experiences of abuse, and social isolation. Oberman and Meyer's (2008) interviews with women imprisoned for killing their children reveal common themes of poverty, abusive relationships (with partners and with parents), and gaps in social welfare provision. Matravers (2008, p. 305) explores how the preconviction lives of female sex offenders in her research "were characterised by social isolation and disadvantage." Many had histories of substance abuse and depression. By raising the significance of gendered social inequalities to women's violence, feminist research helps to contextualize it in relation to the realities of women's everyday lives (Burman et al. 2001).

Kruttschnitt and Carbone-Lopez's (2006, p. 345) analysis of women's narratives of their violence highlights reasons for violent behavior that range beyond prior victimization and social deprivation to include "the desire for money, respect and reparation." Giving prominence to women's narratives of their own violent behavior fulfils the feminist principle of taking women's experiences as the basis for research (Ramazanoglu and Holland 2002) and characterizes several feminist studies of women's interpersonal violence (Kruttschnitt and Carbone-Lopez 2006) and of women who kill (Oberman and Meyer 2008). These studies also seek to highlight women's agency in using violence, without resorting to stereotypical or anti-feminist explanations (see also Morrissey 2003).

Another important aspect of feminist research into women's violence is to assess their treatment by the criminal justice system. Studies demonstrate that policies which encourage or instruct the police to make arrests in reported incidents of domestic violence have led to the disproportionate arresting of women, despite their lower likelihood, in comparison with men, of using physical violence, threats, or harassment (Hester 2013). Ballinger's (2000) study of the 15 women executed in twentieth-century England revealed that although overall women were more likely to be reprieved from the death penalty than men, where they had killed adults rather than children they were actually statistically less likely to be reprieved. Ballinger (2011, p. 112) emphasizes the need for feminists to create alternative knowledge to that generated by state-mandated punishment in order to "reconstruct new configurations of 'truth' which allows hitherto silenced groups to speak for themselves."

Feminist perspectives must be applied to women whose "deeds fall between the cracks of the normative representation women" (Frigon 2006, p. 4) in order to advance feminist aims of "challenging stereotypes and restrictive gender norms" (Seal 2010, p. 3). Attention to the gender representation of violent women helps to illuminate how the norms of "appropriate femininity" are socially and culturally maintained. In her study of 28 Canadian women found guilty of murdering their husbands, 1866–1954, Frigon (2006) argues that judgments of women's performance of their wife and mother roles were particularly significant. Such judgments can have implications for women's sentencing in the criminal justice system (Wilczynski 1997), but also serve "the interests of both state and patriarchal power by reproducing discourses that circumscribe women's roles" (Seal 2010, p. 7). Women's violence threatens the gender order, and social, legal, and cultural responses to it reflect "more general, inchoate anxieties about the potential for the feminine character to disrupt the order of things" (Heberle 2001, p. 55).

There is a strong case to be made for feminist research on women's violence and there is an increasingly well-developed existing literature on this. However, this does not diminish the fact that substantial dilemmas attach to researching violent women using feminist methodologies. These relate not to the selection of methods, but to the ethical and political issues that the research entails. Feminist methodologies are rooted in a political commitment to ending women's subordination, but also to bringing feminist principles to the research process. As discussed above, these include undertaking research from a position of solidarity, eroding power differences as far as possible, and enabling marginalized voices to be heard and with the aim of effecting positive change.

Some of these principles are complicated, or even potentially attenuated, when researching women who have harmed others. While there is feminist work to be done in terms of contextualizing women's violence and challenging derogatory and limiting stereotypes, the researcher is also faced with the reality that the women they research have caused the suffering of others. This may limit the empathy that researchers can feel with them and can induce feelings of ambivalence (Seal 2012). The rest of the chapter discusses these ethical and political issues for feminist research on women's violence. As there is only a very small methodological literature on research in this area, it draws on relevant wider literature from criminological and feminist research in order to do so.

3 Questions of Researcher Allegiance

In order to unpack issues of allegiance as they relate to criminological research, Liebling (2001) returns to Becker's (1967) famous question to sociological researchers – "whose side are we on?". She asserts "any social research is also a human process and it can therefore be fraught with personal dilemmas" (p. 481). Research is also necessarily political "because it involves wielding power" (p. 481). Liebling (2001) reflects that criminologists who firmly place their allegiance with the imprisoned or those classed as "offenders," and oppose attempts to see the perspective of the "superordinate" such as prison governors and police officers, have moral courage and a strong sense of what is right – but this inflexibility can seem troubling. The complexities of the social world and the tensions between different values are not addressed if the researcher firmly pledges allegiance before conducting the research. Liebling (2001) argues that to acknowledge this does not remove the validity of a political position, or the desire to advance social justice through research. Rather, researchers need to recognize the contingency of their values.

For feminist researchers, the answer to "whose side are we on?" would be those who are disadvantaged by gender power imbalances. This frequently means women, but also includes men. Due to their social positioning, women who use violence do usually fall within this category – as discussed above, they are vilified as evil, masculinized, or dismissed as not in control of their actions. However, in many circumstances, they have also misused their power against others – especially in "atypical" cases. To straightforwardly declare to be "on their side" would erase the suffering of their victims. The choice of allegiance is uncertain and can be troubling (Liebling and Stanko 2001).

Much of the time, social science researchers do not put themselves in the position of researching people with whom they cannot form an empathetic relationship

(Godfrey 2003). Allegiance can be relatively straightforward in studies of people who are socially marginalized or excluded, but less so when they are "morally marginal" and without "victim-status," and where "more ambiguous moral and empathetical positions" are involved (p. 57). Criminological research involves venturing into this ambiguous territory when it focuses on convicted murderers, pedophiles, and other serious sex offenders. The challenge for researchers becomes one of how to deal with exploring morally unacceptable behavior "whilst still attempting to understand the context of their subjects' lives" (Godfrey 2003, p. 58).

Sollund (2008) faced this challenge in her interviews with minority ethnic men in Norway who had been convicted of crimes against women. She focused on two interviews with men who had killed women, one of whom who also raped his victim. Sollund argues that researchers often omit their feelings from their published accounts of their research, whereas she came to understand her feelings toward these interviewees as significant to the data she collected. One of the men, John, had stabbed a woman to death while he was severely mentally ill. The crime had been reported widely in the media, and Sollund was shocked, when she interviewed him, to realize that he was the perpetrator of this crime. However, she felt sympathy for John and believed "he was a genuinely good person" (p. 188) as he was quiet and respectful, and expressed remorse. Yusuf, on the other hand, "had an offensive attitude in the interview" and came up with excuses for his actions (p. 188). Sollund felt that he was dangerous and hoped that he would not be released from prison. Both men had come to Norway as refugees and Sollund reflects that she could be "accused of taking advantage of the vulnerable situation of male refugees with an uncertain future" (p. 184). She also acknowledges that she found it easier to appreciate the wider context and mitigating factors of John's situation because of his demonstration of shame and remorse than Yusuf's, even though she knew that Yusuf had been a forced solider before fleeing to Norway.

Cowburn's life history research with incarcerated male sex offenders (Cowburn 2007) was undertaken from a profeminist standpoint as his aim was to help change dominant and harmful forms of male behavior. Taking a reflexive approach, Cowburn (2007, p. 284) analyzes his own role in the interviews and relates how he found it "impossible to listen passively to justifications of abuse." He provides the example of one particular interview in which the participant described the sexual abuse of his daughter as resulting from "excessive love" (p. 285). Cowburn directed the conversation in a way which enabled him to challenge this man's beliefs, but acknowledges that he was able to do so because of the power and authority he had as the interviewer. Asserting researcher power over the participant in an interview situation breaks the principle of feminist methodology of democratizing the research encounter. However, not to have challenged the man's justifications could have seemed like an endorsement of them. Cowburn (2007, p. 286) concludes that "allegiances are not always simple."

Similar issues were faced by Blagdon and Pemberton (2010), who also interviewed convicted male sex offenders. Such men have of course mistreated and caused the suffering of others, but their marginalized and stigmatized position within both prison and the wider community means that they are also vulnerable, which needs to be recognized by researchers. Building rapport with participants was important but meant that Blagdon and Pemberton (2010, p. 272) "had to reconcile their own moral positions" with this. This could, on the one hand, be because they found that they liked a particular interviewee, or on the other could be because they disliked them. Like Cowburn (2007), they also faced the dilemma of how not to seem like they were colluding with interviewees' justifications of abusive behavior without directly challenging them. In answer to whether researchers must take sides, Blagdon and Pemberton (2010, p. 278) argue that they must "see the merit in both sides" – in their case, their participants' perspective and also that of the prison regime, which was specifically designed for sex offenders – while appreciating the tensions that this produces.

In my archival research from case files on women accused of murder in mid-twentieth-century England and Wales (Seal 2012, p. 695), I encountered cases where the woman was not easy to sympathize with "as the circumstances of the killing did not exemplify feminist concerns about gendered power differences" and questions of allegiance were vexed. One of these was a 54-year-old woman named Renee Hargreaves, who poisoned Ernest Massey, a 78-year-old family friend who had come to live with Renee after his wife died, by putting weed killer in his tea. She was found guilty of involuntary manslaughter on the basis that she had only intended to make him ill, rather than to kill him, and she received an 18-month prison sentence. Circumstantially, it appeared from the case file material that Renee may have intended to kill Ernest as the pathologist believed that he had been poisoned over a period of 2 or 3 days. She also stood to inherit money from Ernest.

In this work, I note that "Renee's case poses the dilemma of carrying out feminist research without strong feelings of empathy" for the woman in question (p. 695). Renee killed someone with relatively less social power than she had and she was not treated harshly by the criminal justice system. I acknowledge that in my study I could not "place [my]self 'on the side' of the women accused of murder as many of them had done terrible things, and to people who were less powerful" than they were (p. 695). This raises the question of "what to do with emotional reactions" that do not contribute to feminist politics (p. 695).

4 Questions of Representation

Blackman (Blackman 2007) addresses the issue of "hidden ethnography," which is where the researcher's emotional reactions to their participants are absent from completed, published accounts of the research. This could be because this data is controversial or because revealing such emotions poses a threat to the researcher or their participants. The necessity of not only recognizing that research is an emotional process, but also of incorporating reflection on researchers' emotional responses into data analysis, is increasingly accepted (Holland 2007; Liamputtong 2007; Dickson-Swift et al. 2008; see also ► Chap. 123, "Emotion and Sensitive Research"). However, this becomes complicated when negative emotions are entailed that do not seem to contribute to the aims of the research.

Such dilemmas arise from the complexity of the politics of representation. Kirsch (1999) examines how the politics of interpretation and representation are significant to feminist research. Feminists may face "interpretive conflicts" when interviewing women who do not share their values. There is no easy solution to how to represent this in published accounts and how to avoid the unequal power dynamics at play between the researcher and the researched. Writing up is where the researcher exercises the most power over their research participants as they present the definitive version of their findings and of the research process itself.

The dilemmas of carrying out feminist research with "unsympathetic" women are explored by Luff (1999) in relation to interviews she conducted with women in the "moral lobby," which has generally been characterized as "antifeminist." As a woman from a white, middle-class background that was similar to that of her participants, Luff could establish rapport with them and could also experience feelings of warmth toward some of the women. In general, however, they expressed views that contrasted sharply with her own. Adhering to the principles of feminist research, such as equalizing power relations, was by no means straightforward. The interviewees were relatively socially powerful women, but as research participants Luff (1999) exercised a degree of power over them in terms how she represented them in published accounts. Luff (1999, p. 699) felt an ethical commitment not to include in her writing the views participants expressed on feminism where such material could potentially compromise confidentiality and "might place them in a potentially vulnerable or marginalised position within their organisations." This was a dilemma as it meant collusion, on the part of a feminist researcher, with the women remaining in antifeminist organizations.

One frequent way of challenging othering and marginalization is to offer more positive imagery that counters derogatory stereotypes. However, it must be acknowledged that representations of disadvantaged and subordinate populations cannot always be positive (Hall 1996). Hester and Donovan (2009) reflect on feminist discomfort in relation to researching domestic violence in same-sex relationships. They state that "there were strong tendencies to minimize, hide, and deny the existence of such abuse" due to the fear of exacerbating homophobia or seeming to offer support to conservative discourses of "traditional family values" (p. 162). These questions of how to represent "troublesome" data and research participants are questions of ethics and morality, and do not have straightforward answers (Preissle and Han 2012).

The notion of "hidden" data applies to documentary methods, as well as to fieldwork. In some ways, it has even greater relevance as documentary research frequently entails sifting through hundreds of pages of data, only a tiny fraction of which can be quoted in published research. Unlike interview transcripts or ethnographic fieldnotes, the chance to consult subjects on their feelings about what should be included is usually not available. In my research on women accused of murder (Seal 2012), I found indications of prior violence toward their victims on the part of the some of the women. One of these was Marilyn Bain, a woman in her twenties who in 1962 stabbed her female flatmate, Jan, in the ribs. Jan died 3 days later after her wound became infected. Marilyn's statement to the police referred to "quarrels" between her and Jan and also explained that when fighting, she always punched Jan in the ribs rather than the face so that she would leave no clearly visible marks. To me, this account seemed to bear strong similarities with the actions of male perpetrators of violence against women, raising the danger of confirming antifeminist arguments about the prevalence and nature of women's use of violence (see also Renzetti 1999).

In terms of the politics of representation, data such as this demands the feminist researcher's recognition of women's interpersonal violence toward other women, and its parallels with men's violence against women. This does not mean succumbing to antifeminist arguments that overstate women's violence and underplay men's, but it does entail the "dangers of knowledge" (Ristock 2002). My other published work on Marilyn Bain's case does not include discussion of Marilyn's violence toward Jan before the night of the stabbing (Seal 2009a, 2010). This is because my research questions concerned gender representation in the criminal justice system and Marilyn's account of punching Jan did not feature in portrayals of her femininity. However, reading file material such as this created ambivalence in terms of how I felt about the women I researched and complicated the issue of allegiance (Seal 2012).

A different publication from my research (Seal 2009b) examines the public sympathy that was expressed for Edith Chubb, a woman who in 1958 killed her sister-in-law, Lilian, by strangulation. She was convicted of involuntary manslaughter and sentenced to 4 years in prison. Although violent women have frequently been vilified in both historical and contemporary contexts, I explore how Edith's portrayal as a respectable, hard-working woman and "put upon" mother facilitated positive responses to her case in the form of a newspaper-led campaign to release her from prison. This campaign involved establishing a dichotomy between Edith as an overworked mother and Lilian as a "lazy" spinster (Lilian was in full-time paid work). In terms of the politics of representation, I highlight how derogatory stereotypes of unmarried womanhood constructed Lilian as a "deserving" victim. Cases where women receive public sympathy can be used to reinscribe antifeminist arguments about undue leniency toward violent women. It is, therefore, necessary to pay close attention to the complexities of representations of femininity, particularly how certain "positive" portrayals may be constructed in relation to other negative stereotypes of womanhood.

5 Questions of Ambivalence and Compromise

Reflexivity, the "self-critical sympathetic introspection and the self-conscious *analytical* scrutiny of the self as researcher" (England 1994, p. 82, italics in original), is central to feminist approaches. This principle demands that the ambivalence entailed by researching violent women is admitted to and discussed. Ristock (2002, p. 39) conceptualizes reflexivity as the willingness to look at "my own meaning-making processes." She describes her research into interpersonal violence in lesbian relationships as exposing her to the "dangers of knowledge," especially as she had extensive experience as an activist in relation to men's violence against women (p. 41). In their research on child abuse, Jackson et al. (2013, p. 9) noticed that "female-perpetrated abuse was experienced as particularly problematic to grasp." Employing "emotional reflexivity," they concluded that this particular difficulty in comprehending abuse by women resulted from the "gendered norms of care" held by some members of the research team (p. 9). Therefore, reflexivity is essential to unpacking the influence of the researcher's own anxieties, assumptions, and feelings in relation to violent women.

Mixed feelings about research participants are, to at least some degree, likely to be an element of most research, feminist or otherwise, that takes an in depth look at human beings. Tensions and contradictions are part of the research process and must be navigated. Researchers must live with "ambiguity, difficult decisions, and a certain openness to change in the world" (Plummer 2001, p. 228). Where topics are sensitive or controversial, feelings of ambiguity and ambivalence often come to the fore. These must be negotiated in order to achieve reflexivity (Hollway and Jefferson 2012). Ordinary life is characterized by "complexity, paradox, provisionality, changeability and unpredictability," which cannot be erased from the research process – and may also be necessary in order to achieve understanding of what is being researched (p. 165).

Campbell (2002, p. 28) recommends that researchers ask, "What do we feel conflicted about and why?" in order to gain insight into moments of discomfort. This issue is considered by Reeves (2010), who conducted ethnographic research in a probation hostel for sex offenders. Once she had left the hostel, she sometimes encountered former participants in the street. This made her uneasy as the men were classified as high-risk offenders, but also ashamed of this unease because the purpose of her research was to shed light on their little understood experiences of the criminal justice system. This "ethical and emotional conflict" was only resolved by the lessened of chance meetings as participants were removed from the area, and then by Reeves' own move away for unrelated reasons (Reeves 2010, p. 328). She reflects that her unease arose from the changed context that chance meetings entailed – the protection of the hostel institution was no longer there.

Friedman's (1991, p. 109) exploration of the emotional facets of qualitative fieldwork notes that researchers can become "too close" to their participants, but can also experience "negative emotions that threaten to interfere with the development of an in-depth understanding of the people whose lives are being studied." Friedman's research was with female police officers and after spending much time with them, she felt close to them. However, this entailed recognizing aspects of their behavior that she would prefer not to have seen. She saw that "some of the women police officers whom I had grown to like were overtly, unapologetically racist" (Friedman 1991, p. 115). Whether to include this material in written versions of the research was a dilemma in terms of the politics of representation as she had an "impulse to portray the women consistently in the most flattering light possible" (p. 116).

Presser (2005) makes the case for carrying out feminist research with violent men. Such research needs to help identify why men perpetrate violent crimes but also, more controversially, needs to pay attention to "the humanity of the men" (p. 2067). She argues that feminist criminologists situate women's offending in relation to social marginalization and inequality, but "are not 'doing feminist methodology' when it comes to studying violent men" (p. 2068). In order to remedy this, she asserts that feminist researchers should employ "strong reflexivity," which incorporates the wider social, economic, and political contexts of men's actions. In identifying violent men, researchers rely on the power of the state to provide access to incarcerated participants (or in keeping records that can be accessed by documentary researchers).

This power imbalance between the researched and researcher should be acknowledged, even where the researched in question are men who have committed violent crimes against women. Presser's (2005) interviews involved being sexualized by some of the men, but she also felt empathy for them, particularly for one participant who was on death row. She articulates how ambivalence in research is not restricted to personal emotions but also to political positions. Her feminism meant that she wanted to challenge male violence but also to highlight the social positioning of marginalized men and the state's control over them.

In an article which considers Presser's (2005) account among others, Griffin (2012) develops the notion of the "compromised researcher." She applies this in particular to feminist researchers and explores examples of where they have been made vulnerable or in some way contaminated by their research, whether through choice of topic or methodology. She does not use the term "compromised" to allege unethical or unsatisfactory research practice. Rather, Griffin argues that researchers are compromised because of certain assumptions and conventions that attach to feminist research. One assumption is that in feminist research "you are what you do" and that for it to be otherwise inevitably compromises the research (p. 337). This means that there is a convention for feminists to research "the same" – in other words, other women – and to choose topics that are "close to their heart" and of which they have personal experience (p. 339). Remaining within these conventions entails its own compromises and vulnerabilities for the researcher, but stepping outside of them to research the "different" and "not experienced" raises questions about the researcher's ability to fully and ethically represent their subjects.

6 Conclusion and Future Directions

Unlike researchers such as Presser who have researched violent men, researchers who are both female and feminist who study violent women are researching "the same" in terms of shared gender (although their subjects may well be different in terms of social class position, ethnicity, sexuality, and age). As discussed, this potentially entails compromise in terms of the politics of representation by highlighting behavior that feeds into misogynistic and antifeminist portrayals of women. Despite being women researching other women, developing empathy and allegiance with violent women may be difficult or even undesirable for feminist researchers. Feminist research on violent women may seek to deconstruct derogatory stereotypes, to place women's violence "within wider contexts of social inequality and marginalisation" (Seal 2012, p. 698) or to highlight previously overlooked forms of victimization. However, it would not be ethical to uncritically pledge allegiance with violent women as research participants on the basis that they may have been negatively portrayed and/or "ended up as state defined 'offenders'" (Seal 2012, p. 698).

The complexity and ambivalence of these cases must be analyzed and lived with via a reflexive approach, with all the compromise that this may engender. Although there is a thriving feminist literature on different aspects of women's violence, there remains a paucity of methodological discussion within this. Future research in this area should provide further consideration of reflexivity, emotions, allegiance, and ambivalence as they relate to feminist studies of women's violence.

References

Ballinger A. Dead woman walking: executed women in England and Wales 1900–1965. Aldershot: Ashgate; 2000.
Ballinger A. Feminist research, state power and executed women. In: Farrall S, Sparks R, Maruna S, Hough M, editors. Escape routes: contemporary perspectives on life after punishment. Abingdon: Routledge; 2011. p. 107–33.
Becker HS. Whose side are we on? Soc Probl. 1967;14(3):239–47.
Blackman SJ. "Hidden ethnography": crossing emotional borders in qualitative accounts of young people's lives. Sociology. 2007;41(4):699–716.
Blagdon N, Pemberton S. The challenge of conducting qualitative research with convicted sex offenders. Howard J. 2010;49(3):269–81.
Burman M, Batchelor SA, Brown JA. Researching girls and violence: facing the dilemmas of fieldwork. Br J Criminol. 2001;41(3):443–59.
Burman M, Brown J, Batchelor S. Taking it to heart: girls and the meaning of violence. In: Stanko EA, editor. The meanings of violence. London: Routledge; 2003. p. 71–89.
Campbell R. Emotionally involved: the impact of researching rape. London: Routledge; 2002.
Cowburn M. Men researching men in prison: the challenges for profeminist research. Howard J. 2007;46(3):276–88.
DeKeseredy WS. Violence against women: myths, facts, controversies. Toronto: University of Toronto Press; 2011.
Dickson-Swift V, James EL, Kippen S, Liamputtong P. Doing sensitive research: what challenges do qualitative researchers face? Qual Res. 2007;7(3):327–53.
Dickson-Swift V, James EL, Liamputtong P. Undertaking sensitive research in the health and social sciences: managing boundaries, emotions and risks. Cambridge: Cambridge University Press; 2008.
England KVL. Getting personal: reflexivity, positionality, and feminist research. Prof Geogr. 1994;46(1):80–9.
Fitzroy L. Violent women: questions for feminist theory, practice and policy. Crit Soc Policy. 2001;21(1):7–34.
Friedman T. Feeling. In: Ely M, Anzul M, editors. Doing qualitative research: circles within circles. London: Routledge; 1991. p. 107–38.
Frigon S. Mapping scripts and narratives of women who kill their husbands in Canada 1866–1954: inscribing the everyday. In: Burfoot A, editor. Killing women: the visual culture of gender violence. Waterloo: Wilfrid Laurier Press; 2006. p. 3–20.

Godfrey B. "Dear reader I killed him": ethical and emotional issues in researching convicted murderers through the analysis of interview transcripts. Oral Hist. 2003;31(1):54–64.

Griffin G. The compromised researcher: issues in feminist research methodologies. Sociol Forskn. 2012;49(4):333–47.

Hall S. New ethnicities. In: Morely D, Chen K-H, editors. Stuart Hall: critical dialogues in cultural studies. London: Routledge; 1996. p. 442–51.

Heberle R. Law's violence and the challenge of the feminine. Stud Law Polit Soc. 2001;22:49–73.

Hesse-Biber SN. Feminist research: exploring, interrogating, and transforming the interconnections of epistemology, methodology, and method. In: Hesse-Biber SN, editor. Handbook of feminist research: theory and praxis. 2nd ed. Thousand Oaks: Sage; 2012. p. 2–26.

Hester M. Who does what to whom? Gender and domestic violence perpetrators in English police records. Eur J Criminol. 2013;10(5):623–37.

Hester M, Donovan C. Researching domestic violence in same-sex relationships – a feminist epistemological approach to survey development. J Lesbian Stud. 2009;13(2):161–73.

Holland J. Emotions and research. Int J Soc Res. 2007;10(3):195–209.

Hollway W, Jefferson T. Doing qualitative research differently. 2nd ed. London: Sage; 2012.

Jackson S, Backett-Milburn K, Newall E. Researching distressing topics: emotional reflexivity and emotional labour in the secondary analysis of children and young people's narratives of abuse. Sage Open, April–June, 2013. p. 1–12.

Kirsch GE. Ethical dilemmas in feminist research: the politics of location, interpretation and publication. Albany: State University of New York Press; 1999.

Kruttschnitt C, Carbone-Lopez K. Moving beyond the stereotypes: women's subjective accounts of their violent crime. Criminology. 2006;44(2):321–51.

Lee RM, Renzetti CM. The problems of researching sensitive topics: an overview and introduction. In: Lee RM, Stanko EA, editors. Researching sensitive topics. London: Sage; 1993. p. 49–65.

Liamputtong P. Researching the vulnerable: a guide to sensitive research methods. London: Sage; 2007.

Liebling A. Whose side are we on? Theory, practice and allegiances in prisons research. Br J Criminol. 2001;41(3):472–84.

Liebling A, Stanko B. Allegiance and ambivalence: some dilemmas in researching disorder and violence. Br J Criminol. 2001;41(3):421–30.

Luff D. Dialogue across the divides: 'moments of rapport' and power in feminist research with anti-feminist women. Sociology. 1999;33(4):687–703.

Matravers A. Understanding women who commit sex offences. In: Letherby G, Williams K, Birch P, Cain M, editors. Sex as crime? London: Routledge; 2008. p. 299–320.

Morrissey B. When women kill: questions of agency and subjectivity. London: Routledge; 2003.

Oberman M, Meyer C. When mothers kill: interviews from prison. New York: New York University Press; 2008.

Plummer K. Documents of life 2: an invitation to a critical humanism. London: Sage; 2001.

Preissle J, Han Y. Feminist research ethics. In: Hesse-Biber SN, editor. Handbook of feminist research: theory and praxis. 2nd ed. Thousand Oaks: Sage; 2012. p. 583–605.

Presser L. Negotiating power and narrative in research: implications for feminist methodology. Signs. 2005;30(4):2067–90.

Ramazanoglu C, Holland J. Feminist methodology: challenges and choices. London: Sage; 2002.

Reeves CL. A difficult negotiation: fieldwork relations with gatekeepers. Qual Res. 2010;10(3): 315–31.

Renzetti C. The challenge to feminism posed by women's use of violence in intimate relationships. In: Lamb S, editor. New versions of victims: feminists struggle with the concept. New York: New York University Press; 1999. p. 42–56.

Ristock JL. No more secrets: violence in lesbian relationships. London: Routledge; 2002.

Seal L. Issues of gender and class in the Mirror newspapers' campaign for the release of Edith Chubb. Crime Media Cult. 2009a;5(1):57–78.

Seal L. Discourses of single women accused of murder: mid twentieth-century constructions of 'lesbians' and 'spinsters'. Women's Stud Int Forum. 2009b;32(3):209–18.

Seal L. Women, murder and femininity: gender representations of women who kill. Basingstoke: Palgrave; 2010.

Seal L. Emotion and allegiance in researching four mid-20th-century cases of women accused of murder. Qual Res. 2012;12(6):686–701.

Skinner T, Hester M, Malos E. Methodology, feminism and gender violence. In: Skinner T, Hester M, Malos E, editors. Researching gender violence. Cullompton: Willan; 2005. p. 1–22.

Sollund R. Tested neutrality: emotional challenges in qualitative interviews on homicide and rape. J Scand Stud Criminol Crime Prev. 2008;9(2):181–201.

Stanko EA, Lee RM. Introduction: methodological reflections. In: Lee RM, Stanko EA, editors. Researching violence. London: Routeldge; 2003. p. 1–11.

Wilczynski A. Mad or bad?: child killers, gender and the courts. Br J Criminol. 1997;37(3):419–24.

Animating Like Crazy: Researching in the Animated Visual Arts and Mental Welfare Fields

120

Andi Spark

Contents

1 Introduction 2094
2 Background to the Project 2094
3 Methodological Approach: In Detail 2098
   3.1 Identifying a Triangulated Heuristic Framework 2098
   3.2 Identifying Symptoms of the Health Condition 2099
4 The Creative Structural Process 2100
   4.1 Snapshots and Thumbnails 2100
   4.2 Fragments and Vignettes 2101
   4.3 Incorporating Humor 2103
5 Affordances of Animation 2104
   5.1 Visualizing the Invisible 2105
   5.2 Penetration 2106
6 Conclusion and Future Directions 2107
References 2108

Abstract

This research straddles the divide between creative arts practice and social science methodologies, with a focus on outlining a practical approach to developing short-form mixed-media format animated projects that address serious issues such as postnatal depression. Using a multimethod approach that incorporates elements of heuristic, practice-based, autoethnographic, and narrative enquiry methodologies, this chapter describes how animation can be utilized for both communicative, informative, and entertaining purposes. Explaining how animation works in emphasizing symbols and metaphors to elicit empathetic responses, and using examples from various independent creative practitioner's approach to

A. Spark (✉)
Griffith Film School, Queensland College of Art, Griffith University, Brisbane, QLD, Australia
e-mail: a.spark@griffith.edu.au

their work, the focus here is on delivering authentic and authoritative projects for a targeted audience.

Keywords

Animation · Mental health · Postnatal depression · Mothering · Creative arts practice · Fragmented narrative communication

1 Introduction

The practical aspects of researching within the visual arts are readily covered through a practice-based or practice-led methodology (Hamilton and Jaaniste 2009; Biggs and Henrik 2010). This approach is also common in animation production research, whereby the practitioner develops screen-based artworks through a process of writing, image making, and compiling a creative output in a critical feedback loop cycle (Brown and Sorensen 2008; Barrett and Barbara 2010). Researchers in the health and social science area nominally follow a structured gathered-data based approach, being one example among many other common methodologies (Denscombe 2010). This chapter looks at a multimethod approach as a convergence of enquiry into ways animation can effectively be used to convey social health and welfare issues in a nondidactic format.

My research explores the context and practice of creating an animated project specifically responding to the issue of postnatal depression (PND). The study takes an exploratory self-reflective practice approach wherein I examine my own and others' experiences and responses to mental health issues surrounding childbirth. I correlate these to associated themes of the representation of adult women, social constructs and expectations of women as mothers, concepts of taboo and abjection, along with ideas of embodiment, memory, and fragmented storytelling. Coupled with this, I employ an analytical data-based review of scientific and medical knowledge of the postnatal depression field and connect these aspects of information (creative, sociological, and medical) into script development for animation. Further iterative development occurs in a heuristic mode, taking into account my own knowledge-based and intuitive practice (including use of technology, storytelling methods, and use of visual symbols and metaphors) to create new ways of expressively conveying information. The project is realized as a transmedia style interactive web-based series of animated vignettes, entitled *Coming Through*.

2 Background to the Project

Some background to this project is essential in understanding the methodological approach. I developed the *Coming Through* project by directly responding to my own experiences as well as anecdotal evidence of others' experiences and through reviewing the list of recognized PND symptoms that have been identified through clinical practice evidence. The methodology comprised a distinctive heuristic approach (Moustakas 1990), documenting human experience through including auto-ethnographic (Orbe and Boylorn 2014; see also ▶ Chap. 30, "Autoethnography") and grounded theory research (Buchanan and Bryman 2011; see also ▶ Chap. 18, "Grounded Theory Methodology: Principles and Practices") along with information analysis combined with action-based practice (Pink 2012). It also considered elements of documentary and pondered notions of authenticity and truth within the fictionalized genre of animation. A focus on gathering information and data on specific mental states via clinical practice evidence affords an element of authenticity and epistemological truth, unclouded by sentiment or cultural bias. However, my intention for this project is to highlight the deeply personal lived experience of people encountering this illness. This methodology is common to expressive documentary treatments, used in both collateral therapy (Carlisle et al. 2009) and narrative/informational contexts. A striking example of animators working with mental health patients in therapeutic and creative ways are Swiss filmmakers Nag and Gisele Ansorge's 20-year project (1962–81), where they gathered stories and testimony from psychiatric patients at the Cery Hospital Psychiatric Clinic of the University of Lausanne, Switzerland, from which they created film vignettes (for example, "The Poet and the Unicorn" [1963], "Seven Nights of Siberia" [1967]). As a filmmaker, Ansorge was deeply affected by the exposure to the patients, which also translated into highly engaging audience encounters. He states:

> The way in which the patients analyzed things and made judgments allowed me to learn to know myself better. I discovered a world profoundly human, which had to work with very great suffering, the intensity of which is hard to imagine....These films should not arouse compassion, but rather an interest and a sense of active conscience in the face of the mystery of mental illness. (Ansorge 1998, p. 38)

In this same way, I draw attention to a general audience's consciousness about PND and reexamine my own notions of maternal mental health and well-being in the framework of identifying as an animatrix (essentially, an animation practitioner with a woman-centered focus).

The "story" of PND is complex and multifarious; each affected woman experiences different symptoms in different ways (Buist et al. 2009). Thus, I endeavored to accommodate this complexity in my approach to developing the project. My aim is for the work to remain multifaceted, affording a space for a variety of voices. Therefore, the project is designed as a series of animated vignettes, each responding to one or more of the identified symptoms, such as feeling inadequate or worthless as a mother, being exhausted, empty, sad, tearful, or being unable to think clearly or make decisions. Some of these overlap or involve nuances and subtleties; for example, "empty, sad, and tearful" may evoke images of crying, but this symptom also correlates to "not being able to cry." The differences between being "sad" and "empty" can also be interpreted in varying ways. The vignettes or episodes are intended to work as a memory jog, providing a moment of recognition in either the visual or thematic content. Although aimed at an ideal audience of postnatal mothers, it is not intended to provide a therapeutic function, moreover offering an engaging entertainment. However, with further development and engagement with health professionals and organizations, this may become possible. Humor is another essential factor in this project. Rather than a verbal gag or pun, the kind of humor here operates on the shock of an unexpected visual twist or an ironic juxtaposition of elements. Many of the vignettes are inconclusive; they do not rely on a punchline, climax, or denouement. Some aspects of the humor rely on popular culture references, such as a "used-car salesman" characterization. However, these are in almost universal contemporary use so as to be recognizable across most modern cultures and age groups. Rather than focus on a political or overtly didactic approach, I am more concerned with the idea of communal connection in a similar way to Vera Neubauer's films (Neubauer n.d.). In an interview and profile piece about her animated career, Leslie Felperin (2012, p. 71) states:

> In contrast to the worthy agit-prop of some feminist animators, Neubauer's films refuse to preach and seem to prefer to problematise through abstraction and personalise through the use of autobiography that declines to claim for itself any universal application, while still inviting the viewer's identification.

The project is constructed to also contribute to an entertaining, information-based website where the vignettes can be accessed individually or reviewed as a whole series. The vignettes are accessible in random order, in much the same way that experiencing the illness does not necessarily occur in any particular order. Also, more importantly, not all symptoms present in all people, so the site is intended to be searchable to enable viewers to find an episode that particularly relates to them. It is also possible to view an assembled edit of a number of vignettes to be watched in the form of a short film.

In the first iteration of what is expected to be a long-term and larger-scale project, the works were collated and screened in looping sequences on very small screens (iPad minis) displayed concurrently. One defining aspect of creating transmedia works, in that they can be effectively displayed across a variety of distribution modes, for example, from small-screen mobile devices to gallery-based projections. The design of the project allowing for interactive selection encourages further stories to be potentially collated from viewers and readers who may contribute another anecdote of an event or situation that is a variant descriptor of a particular symptom. These can be translated into illustrations or animations. In the long term, for example, there may be five or six (or more) different ways of showing the concept of "appetite change" or "extreme lethargy" which can be added to the site creating that polymorphous "voice" noted previously.

It is imperative that each episode is short, as women suffering through this condition have very little available time, and most of it is in short bursts. They also have a considerably lower cognitive load due to the demands of multitasking in caring for a baby. Using simple sketch-like images adds to the sense of immediacy and of a story being told by "everywoman" and stories shared between friends and told in the "now." Again, this channels Neubauer's approach of using a strong graphic style "which gives a sense of swift and urgent execution, does not fetishise technical perfection" (Felperin 2012, p. 70). Essential to this stylistic treatment is the consideration that these microstories or vignettes are a small part of a larger, more comprehensive, voice that will encompass multiple viewpoints from various women, even from individuals with multiple ways of recounting their experiences. As Felperin (2012, p. 70) continues,

> ...Neubauer's films attempt to tell stories through montages of striking images and fragmentary scenes which refuse to pull the wool of linearity over the spectator's eyes. Instead, the time of her narratives is fractured, the 'plots' cut up and reassembled on the editing table, evoking the feeling of stories half-remembered, narrated by someone perhaps with... digressive tendencies....., or perhaps with unreliable memory, or perhaps just someone trying to say not 'this and then this' but everything at once.

The sketch drawings in *Coming Through* are easy to read, being uncluttered and with an almost naif/naïve quality that does not require sophisticated renderings of perspective or depth of field. In his book *Understanding Comics*, artist and theorist Scott McCloud (1994, p. 30) argues that the abstracted cartoon image emphasizes specific details and that by "stripping down an image to its essential 'meaning', an artist can amplify that meaning in a way that realistic art can't." Similarly, the character design and visual vernacular mimics a simplified cartoon style. Honess Roe (2013, p. 110) emphasizes the way that we perceive words and images differently and that our understanding of visual language predates our formal education in verbal and textual language, whereby images can reach the emotions before they are cognitively understood. This is a deliberate contrast to most other works in this genre (i.e., information or communication videos about maternal health issues) that use live-action testimonials. In this way, a cartoonified character can potentially appeal to a greater demographic, in that engaging the audience is not restricted to their personal identification with the actor or persona on screen (e.g., race, status, age), which prevents them from "making judgements based on appearance" (Honess Roe 2013, p. 114). Honess Roe argues that "non-indexical media may be the most vibrant and evocative way of remembering the past" and, as I also argue, can evoke an internal feeling of recognition. Through the ability of the animated character to transform or transmogrify, possibilities for self-identification are enhanced. In this project, I use a lead character based on a rough caricature that represents myself.

The embodied self as cartoon character becomes one step removed from the real me but also from the real viewer. Although the character may appear as a heteronormative nondysmorphic white female, she is intended to represent the pantheon of various bodies and selves. The caricature displays as a simplified female form, echoing McClouds' construct of "stripping down... to essential meaning" in the same way we understand the Male/Female symbols denoting public toilets. In this way, metamorphosis is also an essential tool to quickly represent various body states, locations, positions, and mental images (Wells 1998, p. 69). Likewise, the use of signs and symbols is essential to rapid recognition in a short space of screen time. Many symbols can effectively take the place of dialogue or voice-over, which is an important consideration. The episodes are designed to be watched quickly and quietly (so as not to disturb a sleeping baby) and intended to be available across cultures and language barriers. Visual communication tools including physical cues (postures and gestures), facial expressions (and eye movement), and the staging and spatial relationship between figures convey a more universal meaning than dialogue (Pease 1981) Despite this intention, there are also some episodes that incorporate text on screen as speech bubbles or thought bubbles as a short-cut to amplify the performance acting of the characters. In this way, instead of needing more screen time to fully act out the way the character may be thinking, the text can contribute to an explanation.

3 Methodological Approach: In Detail

The multimethods approach is necessary in this type of research to enable both the creative impetus to effectively couple with the scientific evidence. An initial heuristic approach applies to the creative development phase.

3.1 Identifying a Triangulated Heuristic Framework

According to Moustakas (1990), an early key theorist of heuristic research, there are six stages of heuristic research. The first stage, initial engagement, involves the researcher identifying an issue they are passionate about with personal and social implications. From there, a question will be formed. The second stage, immersion, implies that the researcher is immersed and deeply involved in all things related to their question. At the third stage, the incubation phase, the researcher takes a step back from the intensity of the immersion stage and allows their acquired tacit knowledge and intuition to clarify their understanding. The fourth stage, illumination, involves a breakthrough in understanding and awareness. The fifth stage, the explication phase, requires the researcher to examine their newfound understanding and to identify new constituents and themes arising from the illumination process. Creative synthesis is the final phase. This involves the expression of the themes and findings into a creative form.

In my specific multimethod approach, there is a distinctive break between the third and fourth phase, that extends to data examination and analytical review of the specific subject matter, from a scientific, medical, structural, and formal content basis. In this case, gathering information on the established symptoms and clinical presentations of postnatal depression patients from a wide variety of data sources. This is then synthesized to clarify the key analogous factors across different clinical data reports and a table or list is developed which then identifies and correlates the functional knowledge with the initial creative emotional response to the recollected symptoms or incidents. Further data gathering takes place at this stage, in collecting additional anecdotal responses from other sources, in this case including both verbal and written personally related stories. These are also tracked to the recognized symptoms. From here, the "script" will be developed, incorporating the initial immersive impressions, established factual data, and extended experiential reports. Essentially, this leads to the fifth phase describes by Moustakas, integrating new constituent knowledge. However, this interim additional stage creates the relationship between a purely personal illuminative and expressive process and one that confers a clinical connection with medical and social science factual data.

3.2 Identifying Symptoms of the Health Condition

Briefly noting the facts associated with diagnosing PND (AIPC 2015) will help to understand the way the creative elements are incorporated into the project. Clinical and psychological research into this condition over the past 20 years, particularly that led by the late Professor Sherryl Pope and a team of researchers from the Women and Infants Research Foundation at the King Edward Memorial Hospital in Western Australia, has clarified a list of identifiable symptoms (Pope et al. 2000). These are categorized into three nonhierarchical areas: feelings, actions, and thoughts. They are listed below.

Feelings:
- having a very low mood
- feeling inadequate and a failure as a mother
- having a sense of hopelessness about the future
- feeling exhausted, empty, sad, tearful
- feeling guilty, ashamed or worthless
- experiencing anxiety or panic
- experiencing fear for and of the baby
- experiencing fear of being alone or going out

Actions:
- a lack of interest or pleasure in usual activities (including sex)
- insomnia or excessive sleep, nightmares
- appetite changes (not eating or overeating)
- decreased energy and motivation
- withdrawal from social contact
- not looking after personal hygiene
- inability to cope with daily routines

Thoughts:
- being unable to think clearly or make decisions
- experiencing a lack of concentration and/or poor memory
- thinking ideas about suicide
- thinking about running away from everything
- worrying about partner leaving
- worrying about harm or death occurring to partner or baby (Pope et al. 2000)

Diagnosed PND can occur within weeks or sometimes months after giving birth (Beyondblue n.d.; Black Dog Institute n.d.). It may not affect a woman until she weans her baby or until her second or third child is born. As LeBlanc (1999, p. 162) observes:

> Postnatal depression does not normally strike suddenly on any particular day but takes hold insidiously in such a way that its presence may never be recognised or acknowledged by the mother, her spouse, the clinic sister or even her medical practitioner. Instead, her sense of wellbeing is steadily eaten away day by day until it seems to the mother, and everyone around her, that this is her 'normal self'. The longer the depression goes undiagnosed, the harder it becomes for a woman to admit that all is not well in her world.

PND can last from several weeks to months or even years, with up to 50% of women either diagnosed late or still reporting symptoms after 12 months (PANDA n.d.). Furthermore, "recovery has the sense of being two steps forward, one step backwards" (LeBlanc 1999, p. 162). The prevalence of PND is cited at around one in seven women, or approximately 15% of births (Cox et al. 1987). It occurs across social, cultural, and economic boundaries, and symptoms are universally comparable. However, it has been established that lower socio-economic status combined with limited maternal support are contributing factors to higher percentages (Buist and Bilszta 2006; Westall and Liamputtong 2011).

4 The Creative Structural Process

My working process straddles the two worlds of clinical-based communicative structures and an independently based intuitive model. The communicative structural approach focuses on preproduction and planning wherein every element of what will finally be seen on screen is meticulously preformulated. However, this research approach requires an allowance for spontaneous changes and unintentional additions or "happy accidents" in the execution, particularly because of the intimacy of the ideas and story. I commenced articulating my ideas with single image sketches as a kind of cartoon or comic picture, much like a snapshot at the core of the emotional and physical sense recalled from a period of depression. These initial sketches form keystone conceptual design images for the project and stimulated the animation process. Julie Roy (2012, p. 38), from the National Film Board of Canada (NFB), notes a similar modus operandi in Michele Cournoyers's filmmaking approach:

> This importance accorded to the raw material of the unconscious is expressed in her process. She animates the initial fragment, then a second, a third, and so on. These fragments are then stuck on the wall in her studio. A storyboard gradually develops.

4.1 Snapshots and Thumbnails

In an interview in 2001 with the NFB, which provided support for the creation of her works, Cournoyer herself talks about the development of her work: "I never knew what was going to happen. I was communicating with my unconscious, the demons, angels and everything else inside me. I was in a state of need. It was utterly compulsive" (Roy 2012, p. 38). Following the initial concept sketches, "thumbnail images" are created to plot out the action or scenario for each scene. In fact, many of the shots do not require a lot of "animating," as the essence of the emotion can be read from a simple single image. At the time of creating this work, there is no sense of a linear narrative story, just fragments of memory and raw feelings. As Cournoyer reflects in her interview essay with animation critic, biographer, and director of the Ottawa Animation Festival, Chris Robinson, "It [i.e., the process] became more and more liberating, I worked in a primitive, direct communication with my devils and found the story in the execution" (Robinson 2005, p. 95). Pierre Hebert, a contemporary of Cournoyer and fellow Canadian, also reflects on the intensity of emotion Cournoyer expresses in her work: "I realised that she's seeking a kind of welling up of material, where she gives expression to something that is beyond her control" (quoted in Roy 2012, p. 32). In the same way, channelling emotions is a way to revisit the raw intensity of an experience, that may translate to the screen more effectively that clinical or numerical data and textual facts.

My first raw early concept sketches include many "crying" images – overflowing tanks of water, volcanoes of tears, drowning – as well as graphical pop culture references, such as a snakes-and-ladders board representing losing the plot and the idea of taking one step forward and then fifteen steps back; a "Used Baby Yard," replete with "snake-oil salesman" type character to trade in babies who cannot be adequately looked after by their mothers; and stereotypical supermarket sales banners and signs changing to threatening slogans of devouring babies (see Fig. 1 "Tears for No Reason," Fig. 2 "Dodgy City," and Fig. 3 "Desolator"). These sketches are then redeveloped into concept art works which serve as a signpost to the way the project could be envisioned, as well as the tone and treatment of the story, using visual and textual puns in the titles, such as "Tears for No Reason," "Rain Hat," "Dodgy City," "The Desolator," "Nutcase," and "Taedium Vitae" (see Fig. 4 "Tired of Life").

4.2 Fragments and Vignettes

Most animation productions, including short films, are predicated on a finished script. This is the format recognized by organizations that fund projects, although as noted, visual treatment, including character design and concept art, is also fundamental to understanding the range and scope of any project. However, in this case, the story does not play out in a sequential manner but emerges in fragments. The quest is to form these snippets of vision and memory into some kind of cohesive scenario. Similarly, Cournoyer's artistry lies in the relationships between the scenarios: "It is in the process of taking these fragments, each of which has its own meaning, and linking them together so that an emotion, a guiding line, maybe even a narrative, emerges" (Roy 2012, p. 38). This approach is risky because it does not conform to the expected three-act story formula nor can it rely on simply disseminating factual information. It presupposes that the visuals may be strong enough to get the ideas across. It also enhances the sense of the project being a collection of disparate ideas loosely strung together by a connecting theme or style, similar to an anthology of cartoons. One film that employs this "fragment" approach is Marjut

Fig. 1 Spark, Andi 2012 "Tears for No Reason." Digital paint

Fig. 2 Spark, Andi 2012 "Dodgy City." Digital paint

Rimminen's *I'm Not a Feminist, But...* (1986), which is an animated interpretation of drawings from a book of the same name by illustrator Christine Roche. The drawing style is varied and the vignettes do not follow a strictly narrative pattern, although there are linking elements through some of the character scenarios and the soundtrack, which includes a selection of song motifs. As curator Ruth Lingford (n.d.) notes, "they are mostly not laugh-out-loud funny, but Rimminen gives the film an entertaining pace and a lightness of touch."

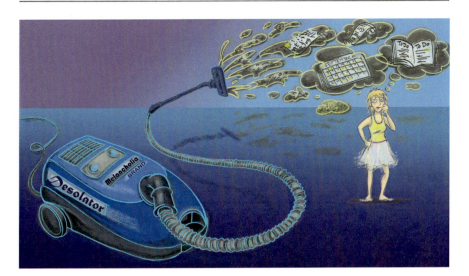

Fig. 3 Spark, Andi 2012 "Desolator." Digital paint

Fig. 4 Spark, Andi 2012 "Tired of Life." Digital paint

4.3 Incorporating Humor

Accordingly, the vignettes in *Coming Through* vary in style of humor; some elicit a wry chortle, and others a surprised laugh. In discussing comics and graphic novels, theorist Elisabeth El Rafaie (2012, p. 70) states that "psychoanalytic approaches to humour propose that jokes can function as a welcome release from the constant need to repress our socially unacceptable desires." Likewise, I uphold that using humor in this context, particularly in discussing contentious subjects like feminism, gender disparity, disability, and mental illness, offers comic relief for an audience who may be identifying with similar socially unacceptable thoughts or feelings (Furniss 1994).

Animation evokes nuances of comic timing and visual hilarity, although many of the projects of this ilk, dealing with serious issues including alienation, illnesses, depression, sexual abuse, or sexual fantasy have an unmistakable seriousness about them (e.g., Alison de Vere's *The Black Dog* [1987]; Michèle Cournoyer's *Le Chapeau* [1999]; Suzan Pitt's *Asparagus* [1979]) as well as an overtly feminist tone. Alternately, Marjane Satrapi's graphic novel *Persepolis* (2004) and the subsequent animated film of the same name (2007) deals with all of these issues with an entirely different tone. It resonates with this research approach particularly the quasi-autobiographical nature of the story appearing to be from a singular voice but created through an amalgam of many disparate incidents and the fact that the work was many years in the making. Hernandez (2010, p. 80) writes about independent animation created by women and comments on the *Persepolis* film:

> Despite Satrapi's personal involvement with the story, the passing of time has helped her to reach a historical perspective, a reflection opposed to anger – the seed of extremism and fanaticism. Instead of being solemn and tragic, she wanted to appeal to all audiences through her own experiences, irony and a sense of play... Thus, humour became a key weapon for exposing the absurdity of everyday situation...

Elements of this double entendre humor and linguistic and visual conflation is common to contemporary internet memes. Contemporary comical slice-of-life strips, such as "Cyanide and Happiness" (2004–) with its nihilistic social commentary, Allie Brosh's (Brosh 2011) episodes in depression at *Hyperbole and a Half* (2009–13) and Nick Seluk's *The Awkward Yeti* (2011–), are closest to the kind of self-referential humor that suited this subject matter. A combination of humor and poignancy place a different spin on the way social science content may be visually articulated.

5 Affordances of Animation

Animation is a way of visualizing the invisible. It is regarded as an effective communication tool, being able to provide complicated information in a short amount of time. As Wells (1998) notes, animation uses many narrative strategies; for example, condensation and ellipsis (compressed or edited continuity); synecdoche (a small image or idea that represents the whole or more complex picture); symbolism or metaphor (recognizable signs); sound and choreography (movement, pattern, and rhythm); and penetration (ability to "see inside" and depict inner workings). As it is not reliant on linguistic comprehension or ability, animation can have narratives that are neither linear nor rational. It also proves effective in communicating information to audiences through associative recognition of appealing characters that represent "every person" (Wells 1998, p. 122). In her introduction to *Animating the Unconscious: Desire, Sexuality and Animation*, Jayne Pilling (2012, p. 12) echoes Wells's list of affordances:

> Animation can make a unique contribution to the exploration and expression of states of mind, unconscious impulses, sexuality and sensory experience. Unrestricted by the dictates of photographic realism and traditional narrative, animation can make such experience palpable via visual imagination, metaphor, metamorphosis and highly creative use of sound.

5.1 Visualizing the Invisible

Animation is useful in showing elements of psychological vulnerability; as Wells (1998, p. 184) comments, "animation has become a vehicle by which inarticulable emotions and experiences may be expressed." It can also portray issues that cannot be shown as real life images, whether elements of stories or testimony too painful or impossible to photograph or real life events that have not been caught on film and are incapable of adequately being recreated in live action. The quality of penetration is an important aspect of animated biographies, autobiographies, and documentaries; in particular, those that focus on mental states of mind (Mosaic Films 2015). Referencing artist Paul Klee's famous quote: "Art is not to reproduce what we can already see, but to make visible what we cannot" (Deleuze and Guattari 1987, p. 48), I qualify the term "animation" in this context as an artist's tool that communicates ideas.

In this spirit, using animation to tell a quasi-documentary style story has many benefits, as evidenced by Orly Yadin and Sylvie Bringas's animated short film *Silence* (1998), which tells the story of a young holocaust survivor. In an essay published from the 2003 "Holocaust and the Moving Image" conference and subsequent book of the same name, Yadin (2005) lists several key points that demonstrate how animation can facilitate authentic communication, and particularly its compatibility with the documentary format. She writes: "Animation can be the most honest form of filmmaking . . . [This honesty] lies in the fact that the filmmaker is completely upfront about his or her intervention with the subject" (2005, p. 169). With animation, it is obvious that the images are constructed, and therefore, "if we believe the film to be true it is because we believe the intention was true" (Yadin 2005, p. 169). As she continues, "animation is less exploitative of its subjects" than live-action films because "there is no danger of being uncomfortably voyeuristic." Moreover, she argues, using animation is "a gesture of respect towards [sensitive] subjects" (Yadin 2005, p. 169). Importantly, "animation can take the viewer to locations unreachable through conventional photography . . . [it] is very useful for saying a lot in very few frames, and saying it ambiguously enough for the audience to bring its own interpretation and experience to the screen . . . [Therefore] animated characters can seem more real than actors" (Yadin 2005, p. 170). As Yadin (2005, p. 170) observes, when we accept that we are entering an animated world, "we tend to suspend disbelief, and the animation acquires a verisimilitude that drama-documentaries hardly ever achieve." Utilizing animation in the social sciences field has the potential to penetrate the cognitive consciousness of the target audience.

5.2 Penetration

As noted, much of the historical literature and even contemporary audiovisual information about the issue of PND and aspects of care and support are not only extremely dry but often quite depressing themselves. Through quantitative research, I note that many of the video pieces take a documentary form, including first-person interviews and reflective anecdotes. Although these work in one sense, when watching videos, I invariably experienced a lack of connection with the person appearing on screen. Some of this was due to ethno-sociographic factors (i.e., the woman was of a different age, had a different economic status, or possessed different cultural attitudes from me), but I also found the format very boring (Brunsdon 1986). Likewise, reading through long testimonials of other people's encounters becomes tedious and repetitive. I wanted to balance this by using the affordances of animation, such as penetration, metamorphosis, synecdoche, and other visual or symbolic cues (Wells 1998) so as to be able get *inside* the mind, beyond the corporeal, and be able to show undiluted thoughts.

The quality of "penetration," enabling the illustrated moving image to represent what cannot be physically seen, is an important aspect of defining animated documentaries, and specifically animations that approach the issue of mental health. As part of the creative structural phase of this research approach, a contextual practice review is essential. This takes into account creative practice works, particularly animated works of a similar form and genre, to comparatively analyze in terms of content, theme, form, and function. One contemporary seminal series of works in this realm is *Animated Minds (2003)* commissioned by Mosaic Films and developed by Dr. Andy Glynne in consultation and collaboration with people who suffer from various diagnosed mental health conditions (Glynne 2003). The series of short films is based on narration of real life experiences including bipolar disorder, obsessive compulsive behavior, agoraphobia, and autism spectrum disorder. A comparative analysis content review revealed some key similarities in the use of symbol, metaphor, and form to portray myriad states of mental disruption. The *Animated Minds* series of films employ a variety of commonly used "penetrative" elements as follows: multiple layered images; deconstructed images; flickering/sped-up images; repetitive images; disconnected metamorphoses; being alone in a wide space/room/corner; images of falling, drowning, crouching (holding head); lying on a bed; being held or dangled on a hook or hanger; uncontrolled movement; hole in body/penetration beneath skin; symbols and diagrams; and words in text. Through this connective research, many of these symbols and metaphors are also employed in the *Coming Through* series as a way of shortcutting the additional visual vocabulary required to read the project. Using these techniques provides an opportunity to enhance the dissemination of extended knowledge into medical and scientific areas and the social sciences fields.

6 Conclusion and Future Directions

In developing original creative responses to the experience of PND, the work contributes to the body of knowledge in this field, both from a social service viewpoint and as an expressive model for cross-disciplinary visual and screen arts practice associated with the animation medium. On review of the "multi-methods" methodology used throughout this research, I also endeavor to contribute a framework for approaching visual art-based research in the humanities or social health and social sciences fields (see also ▶ Chap. 64, "Creative Insight Method Through Arts-Based Research"). The studio component is essentially practice-based, contributing my own work to the field of knowledge, which developed from a heuristic approach evolving from deep immersion and engagement with the subject matter and developing a response using tacit knowledge and intuition. This also would be considered autoethnographic in the way I tell my own story (see also ▶ Chap. 30, "Autoethnography"). However, it also crosses into the area of narrative enquiry in terms of "re-storying" by using anecdotes and interpretations of others' similar experiences of PND (see also ▶ Chap. 24, "Narrative Research"). An action research methodology also contributes a key factor in this approach through a systematic feedback loop on analysis of the practice-based work in connection with an analysis of medical and scientific data or knowledge – in this case, tracking the content of my animated vignettes to the recognized and acknowledged list of symptoms of PND. Action research also includes the reflective feedback loop embodied in the critical reflection and iterative production (see also ▶ Chap.17, "Community-Based Participatory Action Research"). This also incorporates a phenomenological methodology, not only through the subjective, experiential, self-descriptive content but also through the approach to the actual animating process utilizing concepts of flow and risk in the "happy accident". This research approach includes practice-led methodologies, discussing the nature of artistic practice which also includes content analysis, process analysis and discourse analysis surrounding other key practitioners works while considering this singular overarching practice as a case study to exemplify this methodological approach. A multimethod approach combining elements across the creative and sciences fields affords possibilities for deeper understanding of the contexts and content in research projects.

Further developments utilizing this approach may include a more defined structure in establishing clinical or diagnostic evidence, incorporating a stronger feedback loop from medical practitioners in the initial stages of the creative practice. It is also desirable to include further examples and anecdotes from patients or associated people experiencing the targeted condition. Prospective projects in this genre remain focused on the lived-experience of an illness or situation. Some examples are about the experience of being diagnosed with cancer and breast or reproductive organ cancer specifically. Highlighting this area maintains the focus on the underrepresented demographic of the adult female, particularly in how mature women are shown in animation forms.

References

Ansorge N. Animated films in psychiatry: the psychiatric clinic of the University of Lausanne Cery Hospital, 1962–1981. Animation World. 1998;3(2):36–41. http://www.awn.com/animationworld/animated-films-psychiatry-psychiatric-clinic-university-lausanne-cery-hospital-1962

Australian Institute of Professional Counsellors (AIPC) Postnatal depression: onset, prevalence and consequences. 2015. 7 May. http://www.aipc.net.au/articles/postnatal-depression-onset-prevalence-and-consequences/

Barrett E, Barbara B, editors. Practice as research: approaches to creative arts enquiry. London: I B Tauris; 2010.

Beyondblue. n.d. https://www.beyondblue.org.au/

Biggs M, Henrik K, editors. The Routledge companion to research in the creative arts. New York: Routledge; 2010.

Black Dog Institute. n.d. http://www.blackdoginstitute.org.au/

Brosh A. Adventures in depression. Hyperbole and a Half (blog), 27 Oct. 2011. http://hyperboleandahalf.blogspot.com.au/2011/10/adventures-in-depression.html

Brown AR, Sorensen A. Integrating creative practice and research in the digital media arts. In: Smith H, Dean R, editors. Practice-led research, research-led practice in the creative arts. Edinburgh: Edinburgh University Press; 2008. p. 153–65.

Brunsdon C, editor. Films for women. Essex: British Film Institute Publishing; 1986.

Buchanan DA, Bryman A, editors. The Sage handbook of organizational research methods. Thousand Oaks: Sage; 2011.

Buist A, Bilszta J. The Beyondblue national postnatal depression program: prevention and early intervention 2001–2005. Final report volume I: national screening program. 2006. https://www.beyondblue.org.au/docs/default-source/8.-perinatal-documents/bw0075-report-beyondblue-national-research-program-vol2.pdf?sfvrsn=2

Buist A, Bilszta J, Eriksen J, Milgrom J. Women's experience of postnatal depression: beliefs and attitudes as barriers to care. Aust J Adv Nurs. 2009;27(3):11.

Carlisle S, Henderson G, Hanlon PW. 'Welling': a collateral causality of modernity? Soc Sci Med. 2009;69(10):1556–60.

Cox JL, Holden JM, Sagovsky R. The Edinburgh postnatal depression scale. Br J Psychiatry. 1987;150:782–6.

Cyanide & Happiness. 2004–. Explosm. http://explosm.net/

Deleuze G, Guattari F. A thousand plateaus: capitalism and schizophrenia. Translated by Massumi B. Minneapolis: University of Minnesota Press; 1987.

Denscombe M. The good research guide: for small scale social research projects. Maidenhead: McGraw Hill/Open University Press; 2010.

Felperin L. On Vera Neubauer. In: Pilling J, editor. Animating the unconscious: desire, sexuality and animation. New York: Columbia University Press; 2012. p. 58–63.

Furniss M. What's so funny about cheese? and other dilemmas: the Nickelodeon television network and its female animation producers. Animation J. 1994;Spring:4–22.

Glynne A. Animated minds. Mosaic Films. http://mosaicfilms.com/animated-minds/. 2003.

Hamilton JG, Jaaniste LO. The effective and the evocative: reflecting on practice-led research approaches in art and design. In Woodrow, Ross (Ed.) Interventions in the Public Domain, ACUADS Publishing, Queensland College of Art, Griffith University, Brisbane, Queensland; 2009.

Hernández ML. A film of one's own: the animated self-portraits of young contemporary female animators. Animation. 2010;5(1):73–90.

LeBlanc W. Naked motherhood: shattering illusions and sharing truths. Sydney: Random House; 1999.

Lingford R. The BFI and animation. n.d.. http://www.screenonline.org.uk/film/id/529975/

McCloud S. Understanding comics: the invisible art. New York: Harper Perennial; 1994.

Mosaic Films (2015) Animated minds: stories of postnatal depression. Mosaic Films. http://mosaicfilms.com/animated-minds-postnatal-depression/
Moustakas C. Heuristic research: design, methodology, and applications. Newbury Park: Sage; 1990.
Neubauer V. Adult animation. n.d. http://www.veraneubauer.com/adultanimation.html
Orbe MP, Boylorn RM, editors. Critical autoethnography: intersecting cultural identities in everyday life in writing lives. Walnut Creek: Left Coast Press; 2014.
Pease A. Body language: how to read others thoughts by their gestures. Sydney: Camel Publishing Company; 1981.
Perinatal Anxiety and Depression Australia (PANDA). n.d. http://www.panda.org.au/
Pilling J, editor. Animating the unconscious: desire, sexuality and animation. New York: Columbia University Press; 2012.
Pink S, editor. Advances in visual methodology. London: Sage; 2012.
Pope S, Watts J, Evans S, McDonald S, Henderson J. Postnatal depression: a systematic review of published scientific literature to 1999. Canberra: National Health and Medical Research Council; 2000. https://www.nhmrc.gov.au/_files_nhmrc/publications/attachments/wh29_a.pdf
Rafaie EE. Autobiographical comics: life writing in pictures. Jackson: University Press of Mississippi; 2012.
Rimminen M. I'm not a feminist, but... Marjut Rimminen Animation in Association with Channel 4. 8 minutes; 1986.
Robinson C. Unsung heroes of animation. London: John Libbey Publishing; 2005.
Roe AH. Animated documentary. London: Palgrave Macmillan; 2013.
Roy J. The body and the unconcious as creative elements in the world of Michèle Cournoyer. In: Pilling J, editor. Animating the unconcious: desire, sexuality and animation. New York: Columbia University Press; 2012. p. 19–30.
Satrapi M, Paronnaud V. Persepolis. Sony Pictures. 95 minutes; 2007.
Satrapi M. Persepolis I, II. New York: Pantheon; 2003–2004.
The Awkward Yeti: Featuring heart and brain. 2011–. http://theawkwardyeti.com/
Wells P. Understanding animation. London: Routledge; 1998.
Westall C, Liamputtong P. Motherhood and postnatal depression: narratives of women and their partners. Springer: Dordrecht; 2011.
Yadin O. But is it documentary? In: Haggith T, Newman J, editors. The Holocaust and the moving image: representations in film and television since 1933. Columbia: Wallflower Press; 2005. p. 168–81.
Yadin O, Bringas S. Silence. Yadin Productions. 11 minutes; 1998.

Researching Underage Sex Work: Dynamic Risk, Responding Sensitively, and Protecting Participants and Researchers

121

Natalie Thorburn

Contents

1. Introduction 2112
   1.1 What Constitutes "Sensitive" Research? 2112
   1.2 Researching Sensitively: Using Feminist Principles 2112
   1.3 The Dynamic Nature of Risk 2113
2. Case Study: Researching Survivors of Sexual Exploitation 2114
   2.1 Overview of the Research Design 2114
   2.2 Agency and Consent 2115
   2.3 Involving Gatekeepers 2115
   2.4 Building the Research Relationship 2116
   2.5 Maintaining Confidentiality Despite Illegality 2119
   2.6 Considering Context: Personal, Social, and Political 2119
   2.7 Isolation and Responding to Danger 2119
   2.8 Trust-Building and Researcher Vulnerability 2120
   2.9 Managing Postproject Interest 2121
3. Strategies for Engaging with Vulnerable Populations 2122
   3.1 Agency Support and Cooperation 2122
   3.2 Importance of Interpersonal Approaches 2122
   3.3 Knowledge of Trauma 2122
   3.4 Risk Management in Advance 2123
4. Conclusion and Future Directions 2123
References 2124

Abstract

Navigating ethical issues solely involving wholly competent adults who are physically safe and who play agentive roles in negotiating their social environments is often rife with difficulty. Navigating these ethical issues with human participants who are underage; or whose cognitive-emotional processes are potentially impacted by

N. Thorburn (✉)
The University of Auckland, Auckland, New Zealand
e-mail: N.Thorburn@auckland.ac.nz

previous trauma; or who lack autonomy, freedom, or basic safety makes the preparation and execution stages of research exponentially more difficult. This chapter recounts the challenges inherent in the author's experience of conducting research with adolescent sex workers, many of whom have complex trauma histories. It goes on to discuss the dynamic nature of risk as it relates to research with vulnerable populations, particularly in regard to physical safety, emotional and psychological safety, consent, confidentiality, and interpersonal power within the research relationship. Finally, methods of identifying and managing these risks are discussed.

Keywords

Exploitation · Feminist · Methodology · Prostitution · Sensitive research · Victims · Violence · Vulnerability

1 Introduction

1.1 What Constitutes "Sensitive" Research?

Whether an interview topic is considered "sensitive" depends on how it is regarded within social and cultural norms (McCosker et al. 2001). Lee (1993) argues that the three topics that are typically considered sensitive are stressful or personal events such as sexual abuse, identity issues that are stigmatizing, and topics with political overtones that may be invite controversy. The involvement of adolescents in sex work, with its nuances of sexual violence, stigmatization, and political positioning, can be seen to contain all three elements. However, topics considered "sensitive" to researchers might not necessarily be viewed as such by prospective participants, as many report positive outcomes resulting from their opportunities to speak freely and in a safe environment about these personal experiences (McCosker 1995). Sieber and Stanley (1988) propose an alternative to the conventional conceptualization of "sensitive" studies, arguing that sensitive research is instead research that poses potential adverse complications for the target group, either through the act of participating in research or through the outcomes arising from the research product. This is echoed by Lee's (1993) argument that research sensitivity is determined by the level of threat inherent in participating, and notes that this threat does not apply only to participants, but to anyone directly or indirectly involved in the research (see also Liamputtong 2007). Violence against women, however, fits naturally into the category of sensitive research (Fontes 2004), as it contains threats to both physical and emotional safety, and also contains themes of intimacy, sexual integrity, and personal power (Lee and Renzetti 1990; Liamputtong 2007).

1.2 Researching Sensitively: Using Feminist Principles

Given that the use of underage people in sex work clearly represents a type of gender-based violence, with perpetrators being typically male and victims typically female and/or children, the use of a feminist research methodology is fitting. Feminist research is centered on unmasking the lived experiences of women, and in particular, women whose expressions of their experiences have historically been silenced (Ezzy 2002; Liamputtong 2007; Olesen 2000; Sands 2004; see also ▶ Chap. 119, "Feminist Dilemmas in Researching Women's Violence: Issues of Allegiance, Representation, Ambivalence, and Compromise"). Contiguous goals include the consideration of women in the context of the entirety of their social lives and individual situations, with reference to the influence of economic status, class, sexuality, suppression of identity, and most of all, gender (Darlington 1993; Olesen 2000). Furthermore, it requires the researcher's commitment to furthering the interests of women involved with the study, taking into account the wider policy context of both the stated problem and the research findings, and acknowledging their subjective positioning with regard to their background, worldview, racial, sexual, and gender identity, and social influences (Alston and Bowles 2003).

A wealth of literature supports the conclusion that, if done well, participants are likely to find their participation in sensitive research satisfying and beneficial (Martin et al. 2001; Rabenhorst 2006; Campbell et al. 2009; Edwards et al. 2009). Four issues central to doing sensitive research benevolently are identified by McCosker et al. (2001); namely, accurately defining the topic, accessing participants, avoiding concealment and distrust within the researcher-participant relationship, and actively promoting safety. While literature concerning the practical issues of safety in research into experiences of violence is scarce (McCosker et al. 2001), some studies focus on the psychological safety of the researcher and/or participants (see Burr 1996; Young 1997; Rowling 1999; Dickson-Swift et al. 2008; Hom and Woods 2013). Strategies suggested to mitigate the risks of psychological harm and triggering of painful emotions for participants include building rapport prior to the interview, taking steps to preserve confidentiality and anonymity if there is potential danger, ensuring researchers have appropriate professional backgrounds that enable them to manage sensitive disclosures, providing information about options for crisis counselling and on-going support, and using debriefing to assess immediate impacts (McCosker et al. 2001; Liamputtong 2007). This case study will, therefore, set out some of the interview methods, contact processes, and risk management techniques used with the adolescent sex workers who participated in the research.

1.3 The Dynamic Nature of Risk

Mulla and Hlavka (2011) express their discomfort with the generalization of all victims of violence into a particular "class" of vulnerable participants, suggesting that victimhood, or victim status, is a fluid state and does not invariably distort levels of safety or risk involved in interviewing survivors of violence. The concept of "risk" to participants should be considered within the contexts of social position, culture, and environment; for example, a participant whose daily life is saturated by "risk" may view the potential for psychological distress to occur during an interview as minor in comparison (Miller et al. 2006). The concept of vulnerability must, therefore, be fluid, and assessed on a case-by-case basis, with care taken not to impute preconceived notions of victimhood or vulnerability onto prospective participants, thereby further disempowering them

(Alcoff and Gray 1993; Lamb 1999). Mulla and Hlavka (2011) thus position participants, like researchers, as "moral agents," who should be given sufficient information and options to be able to make decisions in a manner similar to the decision-making processes used in consent decisions as medical or legal subjects, arguing that to assume a level of incapacity following victimization may in itself be harmful both to the participant and to the researcher. Moreover, this principle of adhering to self-defined notions of competence and consent is not solely restricted to the recruitment and consent stages; rather, personal narratives should also be treated with integrity throughout the data analysis and discussion stages, to mitigate the potential for secondary violence that "arises when victims, subjected to epistemic categories and understandings that do not fully encapsulate their experiences, do not recognize themselves in the research product" (Mulla 2008, p. 1514). Accordingly, Clark and Walker (2011) argue that researchers should address risk considerations at each stage of research, from explication of the research purpose to decisions about publication. At the same time, research with any supposed "vulnerable" group must be both intentional and reflexive, and privilege both process and outcome (Fonow and Cook 1991; Stanley and Wise 1993).

2 Case Study: Researching Survivors of Sexual Exploitation

2.1 Overview of the Research Design

The aim of the research was to explore participants' experiences of having been involved in sex work as children or young people under the age of 18. Specifically, it sought to find out what the early life experiences and current social situations were of adolescent sex workers, how they understood and organized their interactions with their clients, and how they experienced interventions from formal services.

While initially sourcing ten participants, two interviews were terminated soon after beginning due to explicit risk to participants. Consequently, only eight participants completed their interviews. While more would have been ideal, eight is still a robust number given the depth of the each interview and the challenges inherent in accessing a population that is transient, typically shrouded in secrecy, and often suspicious of any professional involvement. The aim was not to generalize findings to all adolescents involved with underage sex work, but rather to explore and increase understanding of the experiences of the sample and the policy and practice implications for young people who might share these experiences. The eight participants were all aged 16–20, and had been involved with USW prior to turning 18, with all participants beginning formal sex work between the ages of 12 and 16.

The selection of participants aged 20 and under only, despite the limitation this imposed on sample size, was enforced so that the experiences analyzed would be relevant to current policy frameworks and practitioner initiatives. In addition, due to the reformulation and recontextualization that retrospective remembering of experiences invariably involves, the perspectives of adolescents were considered more likely to provide insight into the meanings they attributed to their experiences within a shorter time-frame of these occurring.

2.2 Agency and Consent

While emotionality throughout the interviews was by no means a unique experience, it did not indubitably indicate lack of agency or impaired ability to consent to being part of the research. However, this constituted a principal consideration, and being accustomed to working with survivors of trauma enabled me to differentiate between emotional states that demonstrated healthy responses to the recall of traumatic events, and those that appeared to threaten participants' emotional equilibrium beyond a level conducive to giving continuous and meaningful consent. Threats to consent were characterized by blankness, inability to recall, unquestioning compliance, and needing to prompt responses. These signs were useful in letting me know that the person may be moving into unhealthy remembering, and, accordingly, that it was time to phase out of interviewing and into a supportive role. In addition, on starting the study, a key concern was whether participants' autonomy may be inhibited by abusive or controlling situations; however, while several disclosed this occurring in the past, only two (one of whom did not complete the interview) had dominant themes of being controlled in the present throughout her interview. On reflection, this is likely to be associated with the method of recruitment; the act of responding to a recruitment flyer and taking the initiative to speak to a researcher for a small voucher incentive (and, as it was later discovered, the chance to tell an attentive listener the full story) indicates a level of agency and choice that is unlikely to occur within a context of control.

2.3 Involving Gatekeepers

As previously discussed, the question of whether to involve parents to give consent on behalf of adolescents, who may have experiences of trauma unknown to parents or caregivers, poses a significant methodological dilemma (Campbell et al. 2014). Limiting samples to only those whose caregivers are aware of their histories of victimization excludes an important subgroup of prospective participants. Campbell et al. (2014) propose a prospective design, where invitations to participate are extended by a third-party organization providing services. However, this too has risks: young people may be unintentionally coerced into participation if they are recruited directly by somebody within a service they are receiving help from (Campbell and Dienemann 2001). To this end, Edleson and Bible (2001) suggest developing relationships between advocates and researchers to enable ethical access to participants. Researchers seeking participants in violence against women research may also solicit participation through recruitment flyers at helping services, counselling clinics, Laundromats, childcare centers, and agency waiting rooms to reduce the potential for any coercion, as this method requires participants to initiate contact rather than the reverse (Fontes 2004).

Accordingly, the first step in the recruitment of participants involved approaching agencies known to have contact with the population group and requesting that they display recruitment flyers at their premises. However, the relationships between potential professional intermediaries or gatekeepers for the population being researched are not always straightforward. There was a less than enthusiastic response than I had hoped for from the agencies I approached to display my flyers. Many did not respond to any attempts to contact them, and the Family Planning Association, from which I had anticipated a positive response, declined my request to have recruitment flyers displayed in their reception areas, citing differences in values and implied judgment of clients evident in the short brief I gave them. This was apparently due to the explanation included in the brief of the legal framework surrounding USW, which included the clarification that according to New Zealand legislation, "clients" are legally offenders when they purchase sexual services from adolescents under 18.

However, certain organizations and practitioners helped not only to facilitate access but also to develop lines of questioning (or appropriate terminology) to promote communication between researcher and participant. The New Zealand Prostitutes' Collective (NZPC) and Social Workers in Schools (SWiS) were invaluable for these purposes. The professionals who did become involved in some way with the project, whether to offer advice or assistance, were typically those who had been individually contacted and met with in person – fulfilling, in short, the "gatekeeper" role judiciously. Conversely, those who were sent the standard email appeared to be more averse to the idea of being seen to endorse research – and a researcher – with which they were unfamiliar.

2.4 Building the Research Relationship

Another chief concern was the extent to which participants would feel able to talk freely to someone who was effectively an untrusted stranger. Paradoxically, given the dominant themes of reluctance to engage with professionals that featured in all of the interviews, this problem did not eventuate. However, strategies designed to mitigate the potential for this to occur centered on methods of contact – I considered it imperative to have the opportunity to exchange text messages with, talk on the phone to, or have a "warm-up" period with participants, in order to allay any concerns they may have about being identified, being judged, or being emotionally triggered by the interviews. Correspondingly, while I had several topics on my interview agenda, the interviews were largely participant-directed, as this was felt to increase their level of control. Given the relative likelihood of them having experienced multiple incidents of having control stripped from them by abusive others in the past, I wanted to minimize the risk of replicating the feelings associated with those experiences. While the reasons behind the ease with which the interviews progressed are difficult to isolate, it is likely that maximizing participants' agency throughout the process is associated with their willingness to engage.

As suggested by McCosker et al. (2001) and Dworski-Riggs and Langhout (2010), the relational strategies used with participants are instrumental in facilitating openness, comfort, and a sense of safety. There were three (often extremely simple in practice) primary interpersonal considerations I used while conducting interviews with adolescent sex workers – participant pacing, nonjudgmental questioning, and maintaining a therapeutic presence. The first consideration, participant pacing, has three intersecting objectives: it allows raw data to emerge spontaneously based on participants' perceptions of what is important, thereby privileging their narratives over those of the researcher and preserving the integrity of data; it allows participants to be in control of the process, affording them additional agency; and it avoids retraumatization by consistently presenting disclosure of abuse or marginalizing experiences as optional. It was not entirely participant-led, as I offered guidance and asked questions at times, but the depth and breadth of answers was left up to participants. For example:

Natalie: "We can start wherever you like – perhaps how you came to be involved in sex work?"
Participant: "Do you want just the sex stuff, or all the stuff?"
Natalie: "It's up to you – however much you'd like to tell me."
Participant: Okay. Well, I . . . it's hard to know where to start really. Maybe I'll start by telling you a bit about the sex industry and then give you the background later, like why I did it"

Natalie: "You've mentioned escape a couple of times, can you tell me a bit more about that?"
Participant: "Uhhh yup . . . I'll try. I just. . ... There were times when . . . oh, it's so hard to explain, sorry!"
Natalie: "You're doing great, take your time."
Participant: "There was . . . there were times when I just felt like if I stayed still too long it would be . . . too hard to deal with. The feelings I was having of just not being, not being . . . good enough I guess? Not being cared for, not having family to go back to"

The second consideration, nonjudgmental questioning, was a strategy I saw as essential to avoid imputing judgment through my participation in the conversation. Literature into underage sex work highlights both push and pull factors; I needed to avoid texturing their experiences for them by assuming they had experienced any particular event negatively. This was also integral to avoiding shaming or embarrassing participants. My approach, therefore, was to follow the lead of participants – if they used negative language to denote an experience, so did I.

Participant: ". . .it was pretty nasty shit."
Natalie: "You mentioned the nasty shit that happened. How'd you cope with that?"
Participant: "I don't know, sometimes. Like, sometimes I wonder why."
Natalie: "Why what?"

Participant:	"Like, why I did it . . . like, I see girls talking about it, like I was with a couple of people and we were driving through town one night, and they were yelling out to prostitutes. . . like, wow, I used to do that. . ."
Natalie:	"So you saw other people doing what you used to do? How did that feel?"
Participant:	"I don't know, like . . . some people are so judgmental!"
Natalie:	"And when you do ask yourself why, what kind of answers do you come up with?"
Participant:	"Like, my big answer for it is just because I was lonely. I felt unwanted. And all my social group were doing it, and I felt wanted."
Natalie:	"When you cut yourself, what did that do for you?"
Participant:	"It, it sounds crazy, but it helped. It made me feel something and I was so switched off. I also hated my body. It was kind of like punishing myself, cos for such a long time I felt like it was my fault. If I had been different, or done something different, it wouldn't have happened."
Natalie:	"Really?"
Participant:	"Ummm, I don't know. Maybe? Um, I . . . I should've known better. Or not been so dumb. Or, like, told someone, then maybe I wouldn't have held it all on my own in my mind for so long."
Natalie:	"What was it like keeping it secret?"
Participant:	"Um, shitty. It was really hard but I just couldn't voice it. I was too scared people would see it in my face, and would know."

The final consideration – maintaining a therapeutic presence – necessitated therapeutic microskills such as reflecting back emotions, validating experiences, and gently challenging narratives of self-blame. My reasons for this were two-fold: these techniques engendered an honesty and depth that may not have been otherwise possible and which added a raw richness to the data, but, more importantly, because not to do so in the face of emotional vulnerability of young people would have felt unacceptably divergent to my professional training and values.

Participant:	"So when I had those feelings, or when I was around guys who, umm, if they, you know, there was something about them that made me feel unsafe, I would leave. Or if I felt like someone was going to reject me or not like me, I would leave. That way I wouldn't get hurt. Like hurt emotionally."
Natalie:	"That sounds really rough and really scary. Would you have liked someone to intervene?"
Participant:	"Yes! But I couldn't ask them to. I thought if I told someone how I was feeling, or like what I'd been doing, they'd just be disgusted, and not like me."
Natalie:	"It seems like you coped really well in a pretty hard situation."

2.5 Maintaining Confidentiality Despite Illegality

Given that the participants in the study were legally the victims of crime and were disclosing information that could have the propensity to identify their offenders or the nature of the offending, steps were taken to minimize the risk to participants, and to a lesser extent, the risks to me as the researcher. These included encouraging the participant to use an alias, and not collecting any identifying demographic information from participants. The rationale behind this stemmed from two concerns: firstly, the concern that if transcripts or consent forms were ever compromised, participants' safety was less likely to be at risk; and secondly, if I was later asked questions about disclosures of illegal activity, I could answer truthfully without breaching participants' anonymity. This turned out to be propitious, as subsequent media attention regarding the alleged involvement of police officers with underage sex workers in the study prompted investigation by the Independent Police Complaints Authority. The combination of participant-led interview data, the use of aliases, and not collecting demographic information meant that I could truthfully answer Police questions without having to make decisions about either withholding information or compromising researcher ethics pertaining to confidentiality.

2.6 Considering Context: Personal, Social, and Political

New Zealand does have comprehensive laws prohibiting the use of children and young people under 18 years of age in sex work; however, the extent to which these are actively enforced and accompanied by supportive measures for survivors is debatable. Further, the laws did not, at that time, recognize the sale of children or young people for sexual purposes occurring within a context of organized crime as trafficking, and participants displayed a predictable lack of faith in statutory agencies to protect and support them when considering whether to report violent crimes. This limited the options available to participants, as existing services – many of which they were already familiar with – were unlikely to meet their range of needs, and most had already developed negative outcome expectations from their interactions with statutory agencies. Several expressed the belief that if there had been an appropriate service that they perceived as supportive rather than punitive, and did not pose a threat to their continued autonomy, they would have willingly engaged with it. Unfortunately, in the absence of political will, professional knowledge, or local resources for such an issue, avenues for external assistance for participants were extremely limited, which largely negated the role separation of "researcher" and "practitioner."

2.7 Isolation and Responding to Danger

It has been acknowledged that an immediate response by specialist services and/or law enforcement following victim-survivors' disclosures of being harmed in a context of sexual exploitation is imperative to limit further harm (Pearce 2009; Jordan et al. 2013). However, in lieu of formal services set up especially for this population group, or consent to contact Police when a participant disclosed being at imminent risk of harm, the person receiving the disclosure is, by necessity, cast into the position of needing to act to promote the victim-survivor's safety. In some situations, this leads to an on-the-spot decision about whether to reject the helping role in favor of assuring safety (both the participant's and the researcher's), and acting as a practitioner, albeit informally, to address immediate psychological and physical needs and to facilitate access to services – knowing that with some participants' complex, transient backgrounds, the opportunity to facilitate intervention is unlikely to reoccur quickly.

Two participants disclosed that they were significantly at risk of imminent harm during interviews. One of these risks was to the participant's physical integrity; the other to their emotional well-being as talking (although directed by her) elicited traumatic memories that appeared to prompt traumatic dissociation and, accordingly, required both immediate support and provisions for longer-term help. While both had already given immensely rich data pertaining to serious elements of other-controlled underage sex work not captured by other participants, the seriousness of their risks necessitated data collection becoming secondary to safety planning. Both interviews were, therefore, stopped and the recordings deleted. For the first situation of risk, Police were engaged to secure safety; for the second, psychological first aid was used to stabilize her immediate crisis state before accessing specialist services for follow up. Unfortunately, because of the agreement that contact between participants and I would only be initiated by them for safety reasons, I was unable to follow up on either of their situations for my own peace of mind. However, both situations gave me valuable experience in recognizing situations where risk intersects with participation in research, and managing this risk by prioritizing the wellness of participants.

2.8 Trust-Building and Researcher Vulnerability

Gaining the trust of victim-survivors often necessitates some level of self-disclosure (Ahrens et al. 2007; Campbell et al. 2010), and if this is deemed appropriate by the researcher, this may be premised on shared experiences; for example, addiction, sexual assault history, mental health challenges, parenting, or domestic violence (Ceglowski 2000; Kanuha 2000; Hayman et al. 2012; Campbell et al. 2014; Reddy et al. 2006). At the time, some reciprocal self-disclosure seemed imperative, in order to acknowledge the disclosures of the victim-survivor. However, in some instances, this information gave power to third parties who were secondarily privy to these disclosures, which was potentially dangerous given the contentious nature of the research and the possible ramifications of exposure for anyone profiting from sexual exploitation. After the discovery of the resulting safety threats, I then learned to steer such self-disclosure toward innocuous experiences (for instance, anxiety prior to public speaking, or similarly "safe" vulnerabilities) that could not be subsequently exploited for the secondary listener's gain.

2.9 Managing Postproject Interest

Most studies propose techniques to maximize safety and focus on the researcher's stance and approach within interviews, the methods employed to coordinate meetings while safeguarding both the researcher and participant from harm, and strategies to increase participants' perceptions of their safety during their involvement (see May 2001; Shuy 2002; Northcutt and McCoy 2004; Sullivan and Cain 2004; Baker 2005; Stringer and Simmons 2014). However, such studies rarely expound on researchers' strategies to mitigate the risks and vulnerabilities arising from informal contact following publication, or on the enmeshment between researchers' academic, professional, and personal lives as they become involved with inherently dangerous phenomena. Following my research into underage sex work and subsequent media coverage, I received unsolicited contact from a range of people, including sex workers, people who had been sexually exploited during childhood, and people who spoke of being controlled by crime organizations. Given the number of messages via social media being received, and the range of motivations behind them, it was difficult to identify which of these individuals might be motivated to make contact with me because of their own experiences, and which were simply bystanders interested in the research findings. In some cases, young women who initiated contact had social media accounts indicating that they were feminist activists; in other words, whose public presence was consistent with that of other interested parties who contacted me with a view to engaging in academic debate. Some later revealed their status as victim-survivors of sexual exploitation, in either opportunistic or organized contexts. In other instances, the perception of me as an apparently trusted advocate as a result of my previous research results in girls obtaining my contact details and giving me information. Both situations resulted in ethical dilemmas about what and when to report to Police that extended well beyond the length of the initial project. The informal nature of this contact exacerbated the complexities of ethical decision-making; I was no longer protected by the limitations of the "researcher role" and the accompanying luxury of having clearly defined criteria and procedures for responding to crises, threats, and disclosures of harm. Further, as a registered social worker, I had a mandated obligation to report situations where children are at risk of harm, meaning that any stories implying the abuse of anyone under the age of 18 presented an immediate dilemma of whether or not to report. The implications of this mandate were far from straightforward; often the allusions to underage involvement of victims were vague and lacking in any kind of detail that could be realistically reported to statutory agencies. These decisions consequently end up being almost entirely subjective, without the safeguards inherent in obtaining secondary input. In this project, these dilemmas were managed by facilitating relationships with Police and passing on information that involved anyone currently underage, while also making efforts to ensure that they understood the complexities of situations and the need to proceed sensitively.

3 Strategies for Engaging with Vulnerable Populations

3.1 Agency Support and Cooperation

As demonstrated, gatekeepers are instrumental in enabling access to participants, and in guiding the researcher to communicate with a population group that may be unfamiliar or hard to reach. In the event that a participant requires help, the easiest way to facilitate this is to have already established avenues for assistance. By creating relationships with helping professions and with police, the search for appropriate services may be expedited, offering peace of mind and additional safety when confronted with complex risks. This may be particularly beneficial once the research is completed, but unsolicited contact, information, and reports of harm that require external help continue.

3.2 Importance of Interpersonal Approaches

Talking about trauma has the potential to be retraumatizing, and participants in research projects focusing on marginalization or interpersonal violence have often lived through events they experienced as traumatic. Strategies to mediate the risk of reinforcing feelings associated with this trauma are suggested in the literature, such as building rapport through initial phone calls and emails, demonstrating empathy throughout engagement, using appropriate self-disclosure, and using open-ended but sensitively phrased questions (Enosh and Buchbinder 2005; Peters et al. 2008). The risk of exacerbating distress through discussion about sensitive experiences will also be partly mitigated by allowing participants to pace the interviews.

Victims of interpersonal violence have often been disempowered, silenced, and controlled. Repositioning the balance of power to lie with the participant, rather than researcher, helps to avoid re-experiencing dynamics of control as a result of the research (Northcutt and McCoy 2004). Baker (2005) suggests also paying attention to issues such as dress style, speech, and professionalism as potential barriers to engagement. Participants should be afforded the right to tell their stories in their own words without this process being dominated by the interviewer (May 2001; Shuy 2002). Appropriate self-disclosure can also be used, ideally centering on vulnerabilities that will not compromise either party if they are passed on.

3.3 Knowledge of Trauma

It became apparent when reviewing similar previous studies that participants in this study were likely to have multiple and intersecting experiences of trauma. This has implications for interviewing: participants should have control over what experiences they choose to disclose, and control the pace of such disclosures, to avoid being overwhelmed or feeling coerced into sharing intensely personal experiences.

3.4 Risk Management in Advance

Obviously, it is impossible to plan for every contingency – participants, like all individuals, are complex beings and present multifaceted and interweaving issues and experiences. However, some of the basic (and likely) safety needs can be prepared for – for this group, some of the anticipated needs included being at risk from boyfriends, gang members, or pimps, both in general and as a result of them discovering the participants' involvement in the study; retraumatization through discussion of physical and sexual violence histories; possible familial violence or homelessness; and the risk of disclosure involving children and young people being harmed, necessitating police notification. Accordingly, step-by-step plans were made for these situations, although they needed to be adapted, updated, and reformulated in conjunction with the participant at the time of risk disclosure, in recognition of the participant's right to agency (in other words, a say in what safety measures should be taken), and complicating factors (for instance, traumatic dissociation). As recommended by Clark and Walker (2011) and Duong (2015), these risk management plans should encompass potential risks arising at each stage of the research process, from explanation of the project, actual contact, and representation of the findings. This risk management plan should incorporate managing the aftermath of the research, as subsequent public interest in research into controversial social conditions is unlikely to be a unique experience.

4 Conclusion and Future Directions

Ultimately, researching violence against women, and especially violence against women that occurs within a context of organized crime, is fraught with risk. In my experience, this level of risk was elevated by the absence of public recognition and the consequent lack of adequate service provision for this group, leading to immense pressure to blur professional, personal, and researcher boundaries. It is therefore critical to put a number of safeguards in place and to plan for contingencies that may occur during or following face-to-face contact with participants or prospective participants, including obvious or physical risks, the potential for interviewing style to mitigate or heighten participants' personal vulnerabilities, the possibility of being positioned as a first responder to disclosures of imminent risks of harm, and the prospect of on-going or additional contact after closure of the project. However, these risks can be principally managed by due consideration of the likelihood of prior experiences and corresponding adjustment to methods of contact, relationship building, and interviewing, and by the institution of a robust plan to manage threats of safety at each stage. Finally, the paramountcy of facilitating a researcher-participant relationship that induces participant trust, comfort, and nonjudgment cannot be overstated. This relationship fulfils dual functions in sensitive research projects: it establishes the infrastructure from which rich and sincere data can emerge, and it fulfils an ethical imperative in the quest to gain such data while causing minimal harm and maximum benefit to potentially vulnerable participants.

The increasingly diverse methods of participant recruitment offered by social media and corresponding potential for participants to use both anonymity and deception present a range of unprecedented ethical issues. The discussion in this chapter is therefore hoped to provide some insight into the potential issues inherent in new methods of contact and unorthodox research contexts when researching sensitive topics, and provide some strategies for ethical engagement that safeguard both the researcher and participant.

References

Ahrens CE, Campbell R, Ternier-Thames NK, Wasco SM, Sefl T. Deciding whom to tell: expectations and outcomes of rape survivors' first disclosures. Psychol Women Q. 2007;31:38–49.

Alcoff L, Gray L. Survivor discourse: transgression or recuperation? Signs. 1993;18(2):260–90.

Alston M, Bowles A. Research for social workers: an introduction to methods. Crows Nest: Allen and Unwin Ltd.; 2003.

Baker LM. Ethical issues for researchers: interviewing victims of trauma. In: Yamanashi J, Milokevic I, editors. Researching identity, diversity and education. Teneriffe: Post Pressed; 2005. p. 105–21.

Burr G. Unfinished business: interviewing family members of critically ill patients. Nurs Inq. 1996;3:172–7.

Campbell JC, Dienemann JD. Ethical issues in research on violence against women. In: Renzetti CM, Edleson JL, Bergen RK, editors. Sourcebook on violence against women. Newbury Park: Sage; 2001. p. 57–72.

Campbell R, Adams AE, Wasco SM, Ahrens CE, Sefl T. Training interviewers for research on sexual violence: a qualitative study of rape survivors' recommendations for interview practice. Violence Against Women. 2009;15:595–615. https://doi.org/10.1177/1077801208331248.

Campbell R, Adams AE, Wasco SM, Ahrens CE, Sefl T. "What has it been like for you to talk with me today?": the impact of participating in interview research on rape survivors. Violence Against Women. 2010;16(1):60–83.

Campbell R, Greeson MR, Fehler-Cabral G. Developing recruitment methods for vulnerable, traumatised adolescents: a feminist evaluation approach. Am J Eval. 2014;35(1):73–86.

Ceglowski D. Research as relationship. Qual Inq. 2000;6:88–103.

Clark JJ, Walker R. Research ethics in victimization studies: widening the lens. Violence Against Women. 2011;17(12):1489–508.

Darlington Y. The experiences of childhood sexual abuse: perspectives of adult women who were sexually abused in childhood. Department of social work. Brisbane: University of Queensland; 1993.

Dickson-Swift V, James EL, Kippen S, Liamputtong P. Risk to researchers in qualitative research on sensitive topics: issues and strategies. Qual Health Res. 2008;18(1):133–44.

Duong KA. Doing human trafficking research: reflections on ethical challenges. J Res Gend Stud. 2015;5(2):171–90.

Dworski-Riggs D, Langhout RD. Elucidating the power in empowerment and the participation in participatory action research: a story about research team and elementary school change. Am J Community Psychol. 2010;45:215–30.

Edleson JL, Bible AL. Collaborating for women's safety: partnerships between research and practice. In: Renzetti CM, Edleson JL, Bergen RK, editors. Sourcebook on violence against women. Thousand Oaks: Sage; 2001. p. 73–95.

Edwards KM, Kearns MC, Calhoun KS, Gidycz CA. College women's reactions to sexual assault research participation: is it distressing? Psychol Women Q. 2009;33:225–34.

Enosh G, Buchbinder E. The interactive construction of narrative styles in sensitive interviews: the case of domestic violence research. Qual Inq. 2005;11(4):588–617.

Ezzy D. Qualitative analysis: practice and innovation. Allen and Unwin: Crows Nest; 2002.

Fonow MM, Cook JA, editors. Beyond methodology: feminist scholarship as lived research. Indiana: Indiana University Press; 1991.

Fontes L. Ethics in violence against women research: the sensitive, the dangerous and the overlooked. Ethics Behav. 2004;14(2):141–74.

Hayman B, Wilkes L, Jackson D, Halcomb E. Exchange and equality during data collection: Relationships through story sharing with lesbian mothers. Nurse Res. 2012;19(4):6–10.

Hom KA, Woods SJ. Trauma and its aftermath for commercially sexually exploited women as told by front-line service providers. Issues Ment Health Nurs. 2013;74:75–81.

Jordan J, Patel B, Rapp L. Domestic minor sex trafficking: a social work perspective on misidentification, victims, buyers, traffickers, treatment, and reform of current practice. J Health Behav Soc Environ. 2013;23(3):356–67.

Kanuha VK. 'Being' native versus 'going native': conducting social work research as an insider. Soc Work. 2000;45(5):439–47.

Lamb S. New versions of victims: feminists struggle with the concept. New York: New York University Press; 1999.

Lee RM. Doing research on sensitive topics. Thousand Oaks: Sage; 1993.

Lee RM, Renzetti CM. The problems of researching sensitive topics: an overview and introduction. CRVAW Faculty Journal Articles, 36; 1990. Retrieved from http://uknowledge.uky.edu/crvaw_facpub/36

Liamputtong P. Researching the vulnerable: a guide to sensitive research methods. London: Sage; 2007.

Martin JL, Perrott K, Morris EM, Romans SE. Participation in retrospective child sexual abuse research: beneficial or harmful? What women think six years later. In: Williams LM, Banyard VL, editors. Trauma and memory. Thousand Oaks: Sage; 2001. p. 149–59.

May T. Social research: issues, methods and process. Buckingham: Open University Press; 2001.

McCosker H. Women's conceptions of domestic violence during the childbearing years. Masters of Nursing thesis. Brisbane: Queensland University of Technology; 1995.

McCosker H, Barnard A, Gerber R. Undertaking sensitive research: issues and strategies for meeting the safety needs of all participants. Qual Soc Res. 2001;2(1):327–53.

Miller RL, Forte D, Wilson BD, Greene GJ. Protecting sexual minority youth from research risks: conflicting perspectives. Am J Community Psychol. 2006;37(304):341–8.

Mulla S. There is no place like home: the body as the scene of the crime in sexual assault intervention. Home Cult. 2008;5(3):301–26.

Mulla S, Hlavka H. Gendered violence and the ethics of social science research. Violence Against Women. 2011;17(12):1509–20.

Northcutt N, McCoy D. Interactive qualitative analysis: a systems method for qualitative research. Thousand Oaks: Sage; 2004.

Olesen VL. Feminisms and qualitative research at and into the millennium. In: Denzin N, Lincoln Y, editors. Handbook of qualitative research, 2nd edn. London: Sage; 2000. p. 215–56.

Pearce J. Young people and sexual exploitation: It's not hidden, you just aren't looking. New York: Routledge-Cavendish; 2009.

Peters K, Jackson D, Rudge T. Research on couples: are feminist approaches useful? J Adv Nurs. 2008;62(3):373–80.

Rabenhorst MM. Sexual assault survivors' reactions to a thought suppression paradigm. Violence and Victims. 2006;21(4):473–81. https://doi.org/10.1891/vivi.21.4.473.

Reddy M, Fleming MT, Howells NL, Rabenhorst M, Casselman R, Rosenbaum A. Effects of method on participants and disclosure rates in research on sensitive topics. Violence Vict. 2006;21(4):499–506.

Rowling L. Being in, being out, being with: affect and the role of the qualitative researcher in loss and grief research. Mortality. 1999;4(2):167–82.

Sands RA. Narrative analysis: a feminist approach. In: Padgett D, editor. The qualitative research experience. Belmont: Thoms; 2004. p. 48–78.

Shuy RW. In-person versus telephone interviewing. In: Gubrium JF, Holstein JA, editors. Handbook of interview research: context and method. Thousand Oaks: Sage; 2002. p. 537–55.

Sieber JE, Stanley B. Ethical and professional dimensions of socially sensitive research. Am Psychol. 1988;43(1):49–55.

Stanley L, Wise S. Breaking out again. London: Routledge; 1993.

Stringer C, Simmons G. Stepping through the looking glass: researching slavery in New Zealand's fishing industry. J Manag Inq. 2014;24(3):253–63.

Sullivan C, Cain D. Ethical and safety considerations when obtaining information from or about battered women for research purposes. J Interpers Violence. 2004;7(5):603–18.

Young M. Bearing witness to the unspeakable. Women Ther. 1997;20(1):23–5.

The Internet and Research Methods in the Study of Sex Research: Investigating the Good, the Bad, and the (Un)ethical

122

Lauren Rosewarne

Contents

1   Introduction 2128
2   The Ethics of Studying Sex 2128
3   Secondary Research: The Internet and Desk Research 2130
4   Primary Research: Online Materials 2132
5   Primary Research: The Internet and Participant Recruitment 2133
6   Primary Research: The Internet and Sexual Subculture Participant Recruitment 2135
7   Primary Research: Online Ethnography 2137
8   The Internet and Deceptive Responses 2138
9   The Internet and Truthful Responses 2140
10 Conclusion and Future Directions 2140
References 2141

Abstract

The Internet has thoroughly revolutionized sex. On an individual level, the technology has become a key source in exploring sexuality, researching sexual interests, and participating in erotic activity, both vicariously and potentially even physically. For scholars, the Internet has given effortless access to academic databases and archives, to social media sites and public diaries, and notably to a world of possible research participants, in turn dramatically altering the ways sex gets studied. This chapter outlines, analyzes, and problematizes the use of the Internet in sex research, drawing on a wide range of literature on research ethics as well as my own background as a sex researcher, an author of a range of recent material specifically about the Internet, a supervisor of several dissertations on new media, and a long-time member of my university's human ethics committee.

L. Rosewarne (✉)
School of Social and Political Sciences, University of Melbourne, Melbourne, VIC, Australia
e-mail: lrose@unimelb.edu.au

Keywords

Internet · Sexuality · Ethics · New media · Technology · Sexology

1 Introduction

The Internet has thoroughly revolutionized sex. On an individual level, the technology has become a key source in exploring sexuality, researching sexual interests, and participating in erotic activity, both vicariously and potentially even physically (Rosewarne 2011, 2015, 2016a). For scholars, the Internet has given effortless access to academic databases and archives, to social media sites and public diaries, and notably to a world of possible research participants, in turn dramatically altering the ways sex gets studied.

This chapter outlines, analyzes, and problematizes the use of the Internet in sex research. I begin with a brief discussion of the ethics of researching sex and, more specifically, the role of the Internet in this endeavor. I follow with an examination of the Internet as a tool in secondary source data collection. I explore the technology's use in recruiting research participants: both in general via the utilization of an easy means to broadcast requests, and then, more specifically in targeting the hard-to-reach, notably members of sexual subcultures. Lastly, the role of the Internet as shaping research participation is examined: both its usefulness in concealing identity – and thus potentially fostering enhanced honesty – as well as the deception potential that such anonymity fosters.

This chapter draws on a wide range of literature on research ethics as well as my own experience as a sex researcher, an author of a range of recent material specifically about the Internet (Rosewarne 2016a, b, c), a supervisor of several dissertations on new media, and a member of my university's human ethics committee since 2010.

2 The Ethics of Studying Sex

Sex researcher Leonore Tiefer wrote a 1991 essay criticizing the persistent call from within the discipline for "rigor" in sexology. Tiefer considers that this as, at least partly, a ham-fisted response to the struggle sexology has had in securing legitimacy, and is a call designed to somehow counter the tireless "yes, it's interesting, but it isn't science" criticisms of sex research (Tiefer 1991, p. 596). Numerous theorists in fact have spotlighted the struggle that sexology has had in being recognized as legitimate in a relatively conservative academic environment (Irvine 1990; Waynberg 2009). Many of the underpinnings of this struggle – i.e., suspicions about the prurient interests of researchers (Rosewarne 2011; Thomas 2016), and the widespread belief that sex is a private matter and a topic in bad taste to casually discuss (Rosewarne 2013) – are the very reasons why such research is often considered problematic:

asking people about their sex lives, fantasies, and attitudes has long been considered sleazy and invasive, if not also *low-brow*, in the academy.

Research – particularly the kind that involves humans – is under permanent pressure to be ethical. Ethics committees at universities and hospitals go to great lengths to ensure that safeguards are in place to protect both participants and researchers (see also ▶ Chap. 106, "Ethics and Research with Indigenous Peoples"). Such committees exist to thwart troublesome or unworkable research and to provide guidance to scholars on how to improve research design. The geographer Clare Madge (2007) discussed the conduct of human research online and summarized the five key areas prioritized in research ethics policies: informed consent, confidentiality, privacy, debriefing, and netiquette. While these concerns are relevant for any human research project, each factor – notably consent, confidentiality, and privacy – have additional relevance in sex research. For all those reasons that sex is considered private, embarrassing and difficult to talk about are the very reasons that special care needs to be taken when conducting sex research, notably when consent, confidentiality, and privacy have heightened relevance. In a world where judgment, marginalization, and criminalization often occur as a result of exposed sexual interests, research into sex necessitates that effort goes into both comprehensively informing project participants and – as far as possible – ensuring their confidentiality. It should be noted that achieving these things has additional burdens online. Social researchers Jesse Bach and Jennifer Dohy (2015, p. 319), for example, identified the troubles they encountered in establishing consent while using the Internet to study human trafficking: "informed consent, is exceedingly difficult when researching online commercial sex advertisements due to the clandestine nature of the crime and the environments that host it."

It has been contended that scholars undertaking online sex research are not uniquely burdened in regard to ethics, but just need to be mindful of the ethical demands placed on human research of *any* kind and the necessity to keep abreast of best practice around sex research (Wagner et al. 2004; Dewey and Zheng 2013; see ▶ Chap. 106, "Ethics and Research with Indigenous Peoples"). It is, however, worth questioning whether conducting such research online creates any additional ethical quandaries, a topic addressed by numerous scholars (Binik et al. 1999; Madge 2007). Madge (2007, p. 656), for example, summarizes the existing literature, identifying:

> It has been suggested that online research ethics raise many interesting debates as the computer stands 'betwixt and between' categories of alive/not alive, public/private, published/non-published, writing/speech, interpersonal/mass communication and identified/anonymous.

Here, Madge spotlights the complexity of online interactions whereby ideas about geography, privacy, and identity have different meanings online than off. Should the identities, for example, encountered online be considered "real"? Should statements made in social media or in a blog be treated as on the public record? Are exchanges made in chatrooms considered private or public conversations? As Madge highlights, medium specificities need to be kept in mind. Studying sex online also necessitates a rethink of some unique practical and theoretical factors relating to ethics. Binik et al. (1999, p. 82) for example, question: "Are paper and electronic consent forms interchangeable? Can we promise anonymity and confidentiality on the Internet?" The authors extend their concerns to the determination of age: "The researcher probably cannot use the Internet to verify the minor's real circumstances and responses to the research (e.g., whether they are actually safe from harm as a result of participation)" (Binik et al. 1999, pp. 84–85). While Binik and colleagues posed these questions in 1999 – in the earliest days of mainstream Internet use – they nonetheless remain concerns relevant to researchers today.

Discussed later in this chapter is virtual ethnography: of relocating fieldwork to an online space. While indeed, such research boasts appeals, the idea of a researcher "lurking" in online spaces, without making their presence known, conflicts with a range of ethical principles established by research bodies. Binik et al. (1999, p. 83), for example, reference the American Psychological Association whose guidelines note that psychologists should "describe themselves and their activities and should avoid deceptive statements and inappropriate or excessive inducements." While these guidelines are applicable to on- and offline research, there is heightened applicability in cyberspace whereby a website user might be conducting activity which they may realize is not quite "private" in a literal sense, but nonetheless neither is it an activity they want documented in a research publication. Mentioned earlier was privacy and confidentiality. A further concern noted by Binik et al. (1999, p. 86) is the inability to completely guarantee data security: "Promises of anonymity on the Internet can rarely, if ever, be given with 100% certainty, since a persistent hacker or an official with a court order may be able to discover the identity of research participants." Hacking indeed remains a concern, however, given that most scholars today would use Internet-connected computers to store sensitive data collected in *offline* settings anyway, hacking is not uniquely or additionally pronounced for online research.

While obvious risks – reputational, emotional, and psychological – exist for participants in sex research, numerous risks also exist for researchers. Doing online sex research potentially exposes a researcher to illegal sexual activity and prohibited sexual images; situations which could place a scholar in a legal tangle and which are circumstances unique to online sex research.

3 Secondary Research: The Internet and Desk Research

Most research projects will begin with a desk research stage whereby readily available, on-topic material is assembled without fieldwork. Secondary research materials – work that has already been published such as books and reports and journal and newspaper articles – are reviewed to gauge what is already known about a topic, to ascertain what areas remain to be investigated and to determine whether fieldwork is necessary (see also ▶ Chap. 29, "Unobtrusive Methods"). If fieldwork remains desirable, the desk research stage helps brief a scholar about what research methods have used previously – both to positive and negative outcomes – in order to design suitably ethical projects and to expand a field of research. The Internet has completely revolutionized this process.

The *desk* in *desk research* was once used relatively loosely whereby this process could be executed at the scholar's own desk but also at desks and on surfaces in a range of libraries and archives. The Internet, however, has made *desk review* much more literal whereby the entirety of the desk review process can be conducted from one's own desk, in a fixed location, via utilization of an Internet connection. Doing so saves enormous amounts of time navigating through documents and commuting between libraries and other locations. Interned-aided desk research is also notably cost effective: not only is money saved on travel, but if the scholar is affiliated with an education institute, they will likely have access to full-text scholarly databases.

Aside from cost-saving and convenience, the Internet broadens the range of secondary materials available for analysis. On a cursory level, it means that a newspaper archive search can easily be conducted from one's own desk as opposed to sitting behind a microfiche machine in a library. Equally, the (not uncontroversial) Google Books library project has resulted in some 30 million books being scanned and (in varying degrees) made accessible to scholars no matter their location (Wu 2015). Amazon's "look inside" feature accomplishes something similar.

While the usefulness of the Internet in conducting a desk review is undeniable, it would be naïve to ignore that very worthwhile research projects that have transpired in its entirety without a single interview or survey being conducted. Tiefer (1991), for example, surveys a range of sexuality studies which use advertisements or cinema as a dataset; my own research is also heavily reliant on the analysis of popular media to tell the story of our relationship with sexuality. That said, it is necessary to acknowledge the limitations of desk research. Such problematizing is done not to devalue the importance of briefing oneself on the field, but simply to acknowledge that this stage does not exist without shortcomings.

Desk research in general and, more specifically in the context of sex research, is perceived as having some notable limitations; limitations which, incidentally, motivate many scholars to undertake *primary* research (as addressed later in this chapter). First, the very nature of desk research necessitates that only materials already in existence are explored. Such research is perceived not merely as frequently uninteresting, but as lacking in the innovation, newness and gravitas that new results would have. This is particularly important in the context of publication. Many journals prefer to publish new and seemingly innovative research as opposed to review articles without primary data. This frequently skews the research that scholars elect to produce. Marketing scholars Paul Hague et al. (2013, p. 41), in their book on research methods, identify a range of other shortcomings:

> It may be that we are suspicious of the secondary sources because we had no involvement in their compilation. It may be that the data we are looking for are not in quite the form we require. It could be that we have not searched long enough or dug deep enough to see if this information is already available. Sometimes, desk research seems too easy. A big decision surely needs a lot of money spending on it and merits an original piece of research.

While Hague et al. were discussing desk reviews in the field of marketing, these concerns also plague sex research. Tiefer's acknowledgement of the criticism of, for example, feminist analyzes of advertisements or films, is part of a bias toward "scientific" studies as opposed to secondary data analysis; the latter which might still teach us about society and the sexual mores and scripts produced within but which does not utilize primary data.

4 Primary Research: Online Materials

Autobiographies, memoirs, diaries, and correspondence are all part of the deluge of materials considered as primary sources. While, as discussed earlier, the Internet has made a range of materials – books and archives and articles, for example – accessible, the Internet itself has also been crucial in the *creation* of such material. People have, of course, been writing and publishing diaries and memoirs for centuries. Whereas the diaries that got studied historically were commonly documents written privately and studied or published posthumously – even if, as writer and literary critique Thomas Mallon (1984) speculated, many writers secretly imagined an audience for their musings – blogs take a different form. As described by media theorist Geert Lovink (2008, p. 6): "Blogs experiment with the public diary format, a term that expresses the productive contradiction between public and private in which bloggers find themselves." Blogs are a very good illustration of Madge's betwixt/ between categories as an example of writing online that is both simultaneously private and public (see also ▶ Chap. 77, "Blogs in Social Research").

Blogging, a practice dating back to the earliest days of the Web, allows users to write and instantly publish entries. While the concept of a deliberately public diary raises concerns about performance and authenticity (Lomborg 2014; Whitehead 2015; van Nuenen 2016), the very same concerns also plague more traditional diaries (Mallon 1984), and as expanded on throughout this chapter, the idea of their only being one true self and one true expression is fraught (Rosewarne 2016d). Private diaries as well as more public ones like blogs provide a fascinating resource for scholars, particularly in relation to sex.

The writer and memoirist, Kerry Cohen (2013, p. 12), identifies that "[words] help us see who we are in our darkest, most private places. There are few memoirs this is truer for than sex memoirs, for nothing elicits vulnerability quite the way sex can." Certainly for sex researchers, the sex memoir has served as useful research source material. While books like Frank Harris's volumes *My Life and Loves* (1922–1927) or Ingeborg Day's *Nine and a Half Weeks* (1978) have historically proven illuminating for scholars, the number of book-length sex memoirs pale in comparison to sex blogs, material predominantly produced by women (Attwood 2009) and which often produce a level of immediacy and explicitness absent from offline publications. While the veracity of sex blogs and whether considering them as akin to diaries are debates had elsewhere (Cardell 2014), the Internet nonetheless has facilitated the creation of a new source of primary source material where first-person sexual confessions are made available online, providing valuable individual testimonies, obtained without having to apply for permission from an ethics committee.

While sex blogs are a good example of a new source of public data available for analysis, they also provide insights into some very medium-specific ways of being sexual: from insight into netporn-aided sexual fantasy (Muise 2011) to the exhibitionist titillation that comes from public sexual confessions, of the caliber that only the Internet can so easily deliver (Wood 2008; Rosewarne 2011, 2014, 2016a; Fullwood et al. 2013).

This section has focused on the Internet's role in the creation of public diaries and thus the creation of new sources of data for investigations into human sexuality. Blogs, however, are not the only primary source research material available on the Internet. The study of the contents of sexual interactions in chatrooms (Koch et al. 2005) and in online games (Marteya et al. 2014), the presentation of the sexual self in dating profiles (Rosewarne 2016a, b), in vlogs (Biel and Gatica-Perez 2013), and in amateur porn (Paasonen 2010), and analysis into the sexual interactions between members of social network sites such as Fetlife (Fay et al. 2015), along with the activity logs of shopping sites (Coulson 2015), each provide new and notably *medium-specific* materials for scholars to mine for insights into human sexuality.

5 Primary Research: The Internet and Participant Recruitment

While the Internet has provided sex researchers vast quantities of new information, the technology has also had a major role in reconceptualizing a very traditional area of research: participant recruitment. Primary research is about the conjuring of new questions and the obtaining of new answers to research quandaries that either have not yet been posed, or – at the very least – not posed in the way that a scholar intends to. In order to undertake this kind of research, a constant supply of people willing to fill in surveys, answer questions, and sit in laboratories are needed. The Internet helps enormously in this regard. Psychologists Danielle Murray and Jeffrey Fisher (2002) summarized the range of reasons that the Internet has become so indispensable in soliciting research participation including that doing so is efficient and cost effective, that scholars are able to get samples up to four times larger than those organized in-person, and that much money is saved on paper and other stationery costs by relocating operations online. The Internet enables research participants to be found anywhere in the world and for a potentially larger sample of primary data to be collected. From Interviews using Skype (Deakin and Wakefield 2014) or questionnaires using Survey Monkey (Waclawski 2012; see also ▶ Chap. 76, "Web-Based Survey Methodology"), the Internet has meant that some of the traditional impediments to primary research such as recruitment time, costs, and geography are rendered less important. Scholars have also noted other advantages to online primary research such as physical distance from researchers encouraging participants to self-disclose in ways less likely to transpire in a face-to-face environment (White and Thomson 1995; Reid and Reid 2005). Such a method also helps a scholar to circumvent "gatekeeper" concerns (whereby access to participants in a particular organization or clinical setting needs to be granted to a researcher) (Coulson 2015).

Conducting research online also allows participants greater flexibility with the completion of, for example, surveys, enabling them to do so at a time, and in a place, convenient for them (Coulson 2015).

Using the Internet for participant recruitment boasts a range of obvious positives. Worth acknowledging, however, are some of the concerns; concerns which do not devalue the use of the technology in this regard, but nonetheless force deeper thought on research design.

Scholarly work on online methodologies frequently (although increasingly less so) spotlight limitations such as not everyone having a computer or Internet access, thus in essence excluding some likely already-marginalized people. While these factors of course are becoming decreasingly relevant in a world of smartphones, wifi, and public libraries with Internet access, they nonetheless remain relevant in countries where Internet use is not ubiquitous. Equally, early online research concerns, such as fears about those white, wealthier, educated men who once dominated Internet use and, in turn, datasets, have become dramatically less relevant 20 years on where online activity is near universal.

More pressing, however, are issues created by some of the benefits of the technology. In any research project, concerns are raised about representativeness and the degree to which the sample reflects the broader community. While the Internet facilitates the advertising of projects and the ability to locate and easily target individuals and communities, ultimately the sample still ends up being self-selected: these are individuals who have *volunteered* to participate. While the lack of representation in a self-selected sample will exist regardless of where participants are recruited and is not an Internet-specific problem, nonetheless, if the Internet is selected as the exclusive recruitment tool over other sampling methods such as *stratified random sampling* or *opportunity sampling* (Liamputtong 2013; Patton 2015), then this lack of representativeness may be more pronounced.

Mentioned earlier was the convenience factor of research participants being able to complete surveys or answer online questions whenever it is most convenient. A downside of this, however, is that the researcher has little control over the setting in which participation transpires, something problematized by criminologist Lynne Roberts (2007, p. 23): "This means [researchers] cannot tailor instructions to an appropriate level for an individual, clear up any misunderstandings (unless contacted by e-mail) or ensure the survey or measure is completed in an environment free from distractions." A connected concern centers on response quality: studies indicate that written responses tend to be briefer and less comprehensive than verbal ones (Burton and Bruening 2003); thus without a researcher asking questions, follow-ups, explanations, and expanded answers are thwarted.

Noted earlier was the ability to engage in global recruitment of research participants. This, of course, creates its own challenges. In nursing scholars Eun-Ok Im and Wonshik Chee's work (2004) on online methodologies, they discussed a range of projects where the Internet was used for recruitment. If documents, for example, are only made available in English, the ability to globally recruit means in practice that only people with relatively strong English skills can participate. Even then, mastery of English does not fully account for uses of, for example, idioms which may seem unimportant in a survey design but may result in uninterpretable answers. Im and Chee (2004, p. 295) reflected on their study and identified: "Since data were collected using only English, the validity of foreign terms identified to be used in cancer pain descriptions from the nine countries may be threatened by language problems of the participants as well as the translation process."

Another concern, and one specifically pertaining to ethics and sexuality, is anonymity. While anonymity online raises issues pertaining to identity and honesty, and while sometimes it may encourage participants who might be reluctant to expose their identity, there are some notable shortcomings. The capacity for online participants to create a false identity – in line with the identity play that the Internet is renowned for (Rosewarne 2016a, b) – means that factors frequently essential in data collection, like the gender of the participant, may not be accurately gleaned online.

Like any research method or sampling technique, using the Internet has a series of deficits. Such factors, however, have not diminished the desirability of using the technology, particularly given that doing so gives scholars not just a way to target a lot of possible participants efficiently and effectively, but enables specific kinds of participants to be targeted, something with pronounced relevance to sex research.

6   Primary Research: The Internet and Sexual Subculture Participant Recruitment

The Internet has completely revolutionized how individuals experience sexuality, it turn altering expressions of intimacy and becoming a key source of informal education (Rosewarne 2011, 2015, 2016a). Resultantly, the Internet has become a one-stop shop for sex researchers: there they can observe and collate, and as discussed, *recruit* research participants. Social researchers Wendy Bostwick and Amy Hequembourg (2013, p. 658), for example, discussed the use of the Internet in specifically targeting bisexuals, identifying that the technology "has opened up innumerable avenues to conduct targeted recruitment and research. Bisexual-specific listservs are not new, but the proliferation of social media venues (e.g., Facebook and Meetup groups) has made Internet recruitment of bisexual participants particularly appealing for both face-to-face studies and Internet-based survey research." Highlighted here is the use of the technology to investigate populations notoriously difficult to target, in turn helping to create truly representative research samples. Psychologists Danielle Murray and Jeffrey Fisher (2002, p. 6) discuss some of the limitations of social science research which tends to rely on university undergraduate participants because "this population is an easy-to-access, convenient, and inexpensive group of participants…." Such studies invariably exclude "hard-to-reach" subjects such as those who live in rural areas, who are not out, who are transgender or intersex, who eschew sexuality labels, who are sex workers, who have been trafficked or are any of these populations in association with other factors such as drug-use, homelessness, and mental illness (McDermott and Roen

2012; McCormack 2014; Bach and Dohy 2015; Barros et al. 2015). Murray and Fisher (2002, p. 7) propose that the Internet is an answer in accessing these populations: "Use of the Internet for data collection has already shown an increase in diversity over that of college student populations, and as computers become more accessible to the general public, the diversity of potential samples will increase dramatically."

Use of the Internet in attracting research participants boasts a broad number of advantages in sex research. Firstly, while many reasons explain the "hard-to-reach" nature of certain minority sexual populations, a central reason for this is shame: these populations are hidden because there is a cost (real or perceived) to identity revelation (Rosewarne 2011). Participating in research online – with the ability to take advantage of anonymity and not having to engage in face-to-face contact – is considered one way to attract research participants who, otherwise, would not be inclined to participate and thus may not get their voices included in research projects (Liamputtong 2007, 2013).

While accessing hard-to-reach populations has been made substantially easier through use of the Internet, it is necessary to identify the shortcomings of using the technology in this regard. While on one hand, the Internet provides many ways to target groups, it should not be taken for granted that doing so is effortless. In psychologists Ilan Meyer and Patrick Wilson's work (2009) on sampling in lesbian, gay, and bisexual populations, the authors spotlight that many websites, "particularly those that are sexually explicit or deal with provocative subject matters," actually *prohibit* online solicitation of study respondents. On Fetlife, for example, it is very common for profiles to include a statement objecting to the information posted being used in research projects. Meyer and Wilson similarly identify that while advertising for research participants online might get an advertisement seen by many people, there is little correlation between the number of eyes on an ad and the number of respondents.

Another hard-to-reach population for the purposes of sex research is children, something that Binik et al. (1999, p. 84) address:

> In the past, researchers did not have easy or direct access to minors or patients independent of their parents, caretakers, schools, or some third party or institution. This insured that legal third parties were at least minimally aware of the research and were involved in giving consent. Now, many thousands of users under the age of 12 and legally defined as children use the Internet every day... Potentially important research with minors (e.g., relating to surviving sexual abuse or childhood sexuality practices) might be effectively carried out over the Internet, possibly even more effectively than face-to-face interviewing....

While the ethics pertaining to conducting research with children – specifically around issues of sexuality – is addressed elsewhere (Flanagan 2012; Sparrman 2014), as outlined by Binik et al., the Internet creates both a motivation to undertake research with children – as a result of things such as access and their exposure to erotic content – but also creates the capacity to reach children in ways outside of traditional education settings often blocked by gatekeepers.

7   Primary Research: Online Ethnography

An interesting hybrid between primary and secondary is online ethnography whereby fieldwork gets undertaken from the comfort of a desk. Discussed earlier was the capacity for an enormous amount of research to be conducted from one's own desk. While commonly desk work is associated with secondary source data collection, the Internet has also revolutionized how *primary* research is conducted.

Thinking of the Internet as a place is well established (Rosewarne 2016a, b, c). Thinking of it as a somewhere that people can go, or be from, underpins popular perceptions of it as a badlands or Wild West at one end of the spectrum and as a new frontier at the other. As related to fieldwork, the idea of the Internet as its own geographic site(s) creates the capacity for a scholar – without leaving their desk – to observe social interactions; a research method known as *virtual ethnography* (Hine 2008). In practice, online ethnographies have been conducted on communities as diverse as Brazilian migrants (Schrooten 2012) to software developers (Cordoba-Pachon and Loureiro-Koechlin 2015), but for the purposes of my chapter, it is ethnographies in the realm of sexual communities that make this research method particularly useful. Discussed earlier was the Internet offering the capacity to analyze a range of online activities such as blog posts, chatroom interactions, and amateur pornography. In fact, the virtual possibilities for ethnography are much broader, as outlined by Binik et al. (1999, p. 82):

> The growth and popularity of personal Internet services allow for novel investigations of sexuality at home, in the absence of physical presence, and under conditions of relative anonymity. By making use of existing or experimental on-line sex therapists and sexual self-help or entertainment groups, researchers can study topics such as interpersonal attraction, flirting, sexual language, sexual self-help, sexual writing, role playing, and therapeutic relationships.

Such ethnographies have been conducted widely in the study of sex. Social researcher Faracy Grouse (2012), for example, conducted an ethnography of the sexual behaviors of Second Life players using an avatar to investigate how intimacy gets transformed without physical contact. Criminal justice scholars Kristie Blevins and Thomas Halt (2009) used similar techniques to study the attitudes of male clients of sex workers as exhibited in Web forums. Sociologist Joy Hightower (2015) conducted a virtual ethnography which observed the interactions of women on a lesbian dating website to examine gendered bodily presentation, label use, and peer perceptions.

Just as the Internet has dramatically benefited the desk review stage of research, it has also overhauled ethnography (see also ► Chap. 26, "Ethnographic Method"). While many scholars will, of course, still want to undertake in-person observations in places like gay bars (Johnson 2005), pride marches (Ammaturo 2016), and swingers conventions (Kimberly 2016), the Internet opens up scope for conducting this kind of research without leaving one's desk. This can make work substantially cheaper, more convenient, and able to overcome geographic boundaries, but it also taps into a reality that the Internet does not just create new ways and places to study sex, but rather the technology has completely overhauled the way sex itself is experienced and thus, in turn, generates its own, medium-specific data. In my book *Intimacy on the Internet: Media Representations of Online Connections* (Rosewarne 2016a), I examined the range of ways that the Internet has revolutionized the experience of sexuality, from changing how we meet partners, maintain relationships, self-stimulate, fantasize and obsess, have "sex," hook up, cheat and betray, experience our interests vicariously, dabble in subversive sexual practices, expand our networks, and how we feel less alone. In each of these areas, the Internet plays a crucial role and one that necessitates research methods that observe these practices in situ. Virtual ethnography is one method that facilitates this. This technique also has advantages of being less intrusive than interviews, focus groups, or physical observations and it can be viewed as (comparatively) more authentic than had the researcher orchestrated a space for interactions to transpire.

Like each of the methods discussed in this chapter, virtual ethnography is not without criticism. Scholars have problematized this method as so fundamentally dissimilar to traditional fieldwork that it is inappropriate to dub it as such (Lenihan and Kelly-Holmes 2016). An extension of this is that by just observing, for example, a lesbian's interactions in a dating website, does not provide data on her life in its entirety (although, arguably, even offline methodologies would struggle with gleaning such information). Limitations similarly exist in extrapolating data from sexual spaces online and assuming that this provides insight into offline activity, although again, it is hard to imagine that any methodology could truly encapsulate every aspect of identity. Worth noting, the same concerns about authenticity that plague other digital research methods plague ethnography. Communication is somewhat compromised if visual cues cannot be monitored. Similarly, ethical concerns exist about researchers "lurking without consent" (Roller and Lavrakas 2015, p. 190). While these limitations should not be considered as deal breakers – consent can be obtained and a broader definition of "authenticity" can be utilized (Rosewarne 2015, 2016b) – nonetheless it is worth identifying that this method does have unique elements that need to be accounted for.

8 The Internet and Deceptive Responses

In this section, the role of the Internet in obtaining more deceptive data, and alternatively *more truthful* data is examined. While deception is a possibility in any human research, this concern is more pronounced in the context of work done online. Since its inception, the Internet has been framed as a kind of badlands (Rosewarne 2016a, b). A key underpinning of this perception is the capacity for the concealment of identity and, in turn, the concealment of potentially duplicitous intent. While this can partly be explained by techno- and cyberphobia whereby things that are new are both feared and perceived as malevolent (Rosewarne 2016c), the reality is that some people *are* frequently deceptive online. The ability to conceal identity, to play with identity, to don the guise of another gender, and to exaggerate appearance are all behaviors fostered by the default-anonymity of the Internet and what have come to be construed as one the many gameplay-like behaviors executed online; something legal scholar Chris Ashford (2009, p. 298) discusses:

> These sites enable the formation of a "virtual identity" which may be regarded as "false"... This is particularly relevant in the construction of age. This may take the form of a 40-something male becoming a 30-something for the purposes of a networking site, or may attract the attention of the law where this is seen as representing a "subversive" motive, for instance in the "grooming of children."

The possibility of this kind of deception has indeed long haunted online research. Sociologist Christine Hine (2008, p. 263), for example, discussed the issue of duplicity in her discussion on virtual ethnography:

> Online ethnography, and indeed all research using data collected online, has been dogged by the question of authenticity... Such was the association between the Internet and identity play in the early days that considerable doubt was expressed whether enough trust could be placed in what people said online for their words to constitute grounds for any sound conclusions to be drawn.

While the capacity for deception must be factored into research design, this should be considered as less an impediment and more so as a complicating factor, and one that ultimately can yield fascinating findings in new areas. Ashford (2009, p. 302), for example, discusses the complexity of researching online sexual identities and the expanded scope created for unique kinds of analysis:

> The Internet enables the identity of the researcher to be recast as a fluid, relentlessly shifting construct. Those researchers who maintain Facebook and other Web 2.0 accounts, project one self; another self may be projected on a dating or hook-up site, another in the classroom and a further self at a conference presentation. No single self can, or should, be regarded as 'true' in any absolute sense.

Outside of identity play, it is worth noting other kinds of deception transpiring as a result of the anonymity fostered by the Internet. The social scientist Ian Greener (2011, p. 52), for example, spotlights that the ability to be anonymous creates the capacity to be deceptive without consequence:

> People taking part in research can behave in remarkably dishonest ways when the assurance of anonymity is in place. Research participants have been shown to be more likely to steal and to lie about test results they have taken, for example, when they believe they are anonymous.

While this might be interesting if a research project is about deception, it is often perceived as less helpful if scholars are seeking personal insight. While scholars have put efforts into refining techniques to validate identity and to verify things like age and signature, the reality is that truth and identity are concepts needing to be reconceived in the design of research projects using online methodologies.

9 The Internet and Truthful Responses

At the same time that we have questioned the truth of online responses and the *real* of online identities, there has been another set of conversations transpiring about the Internet's capacity to aid *greater honesty*: that anonymity can facilitate truth-telling because a person is not self-censoring in fear of judgment; i.e., serving as an online illustration of Oscar Wilde's famous "Give a man a mask and he'll tell you the truth." As applied to social research, Greener (2011, pp. 51–52) also sees capacity for positives to come from anonymity: "Offering anonymity will lead to respondents being more honest, and feeling that they can say what they believe without being concerned whether their answers will in some way be used against them."

Numerous scholars have spotlighted a capacity for heightened self-revelation online. Social researcher Mark Griffiths (2010, p. 9), for example, discusses the disinhibiting effects that have contributed to this: "For populations discussing sensitivities issues like addiction, this may lead to increased levels of honesty and therefore higher validity in the case of self-report." In the context of sex research, disinhibition can lead to heightened self-revelation of a caliber that sometimes struggles to emerge in offline research environments.

While as noted earlier, the identity concealing or identity fabricating possibilities of the Internet need to be acknowledged, so too do the very reasons why researchers have so enthusiastically embraced online spaces: research participants sometimes open up and self-disclose in ways that offline environments might never permit.

10 Conclusion and Future Directions

The Internet has completely revolutionized every aspect of sex: from overhauling our fantasies, supplementing our masturbation, helping us connect, hook-up, cheat, rinse, and repeat. It is no surprise, therefore, that with all these changes transpiring online, that the interest of scholars would be piqued. Scholars have delved into every aspect of sex online, probing the how, the where, and the why and putting under the microscope each element of this new private life accoutrement. Of course, for scholars, the role of the Internet in sex research is much more than just a new set of stuff to study. As discussed throughout this chapter, the Internet is *itself* something to study, but it is also a tool to explore sexual behavior occurring online as well as offline: research on online dating sites and hookup apps invariably involves online-instigated activity that often ends up playing out in real life.

While the Internet serves researchers as a tool and a source of data, it is important to recognize the complexity of this. Going online to do research creates an enormous array of benefits but also a range of methodological shortcomings and ethical concerns. Such factors do not devalue online research, but nonetheless create pause for thought for scholars and a necessity to think very carefully about research design.

A key challenge for researchers going forward – and, notably, a key issue for university ethics committees – is keeping abreast of new online tools that aid with recruiting participants and also yield new materials of the kind sex researchers might be keen to study. In my experience, for example, students were navigating the use of Tweets and Facebook posts in research long before ethics committees settled on best practice as related to such material. Just as governments are challenged with needing to write legislation that meets constant technological change, scholars are charged with the same burden in regard to ensuring the ethics of their online methodologies.

References

Ammaturo FR. Spaces of pride: a visual ethnography of gay pride parades in Italy and the United Kingdom. Soc Mov Stud. 2016;15(1):19–40.

Ashford C. Queer theory, cyber-ethnographies and researching online sex environments. Inf Commun Technol Law. 2009;18(3):297–314.

Attwood F. Intimate adventures: sex blogs, sex 'blooks' and women's sexual narration. Eur J Cult Stud. 2009;12(1):5–20.

Bach J, Dohy J. Ethical and legal considerations for crafting rigorous online sex trafficking research methodology. Sex Res Soc Policy. 2015;12:317–22.

Barros AB, Dias SF, Martins MO. Hard-to-reach populations of men who have sex with men and sex workers: a systematic review on sampling methods. System Rev. 2015;4:141–51.

Biel J, Gatica-Perez D. The YouTube lens: crowdsourced personality impressions and audiovisual analysis of vlogs. IEEE Trans Multimedia. 2013;15(1):41–55.

Binik YM, Kenneth M, Kiesler S. Ethical issues in conducting sex research on the internet. J Sex Res. 1999;36(1):82–90.

Blevins KR, Holt T. Examining the virtual subculture of Johns. Contemp Ethnogr. 2009;38(5):619–48.

Bostwick W, Hequembourg AL. Minding the noise: conducting health research among bisexual populations and beyond. J Homosex. 2013;60:655–61.

Burton LJ, Bruening JE. Technology and method intersect in the online focus group. Quest. 2003;55(4):315–27.

Cardell K. Dear world: contemporary uses of the diary. Madison: The University of Wisconsin Press; 2014.

Cohen K. Sex memoirs. Am Book Rev. 2013;34(6):12–3.

Cordoba-Pachon JR, Loureiro-Koechlin C. Online ethnography: a study of software developers and software development. Balt J Manag. 2015;10(2):188–202.

Coulson N. Online research methods for psychologists. New York: Palgrave Macmillan; 2015.

Deakin H, Wakefield K. Skype interviewing: reflections of two PhD researchers. Qual Res. 2014;14(5):603–16.

Dewey S, Zheng T. Ethical research with sex workers: anthropological approaches. New York: Springer; 2013.

Fay D, Haddadi H, Seto MC, Wang H, Kling C. An exploration of fetish social networks and communities. Lect Notes Comput Sci. 2015:195–204. https://arxiv.org/pdf/1511.01436v1.pdf. Accessed 14 June 2016.

Flanagan P. Ethical review and reflexivity in research of children's sexuality. Sex Educ. 2012;12(5):535–44.

Fullwood C, Melrose K, Morris N, Floyd S. Sex, blogs, and baring your soul: factors influencing UK blogging strategies. J Am Soc Inf Sci Technol. 2013;64(2):345–55.

Greener I. Designing social research: a guide for the bewildered. London: Sage; 2011.

Griffiths MD. The use of online methodologies in data collection for gambling and gaming addictions. Int J Ment Heal Addict. 2010;8(1):8–20.

Grouse F. Becoming Mireila: a virtual ethnography through the eyes of an avatar. In: Brabazon T, editor. Digital dialogues and community 2.0: after avatars, trolls and puppets. Oxford: Chandos Publishing; 2012. p. 105–20.

Hague P, Hague N, Morgan C. Market research in practice: how to get greater insight from your market. Philadelphia: Kogan Page; 2013.

Hightower J. Producing desirable bodies: boundary work in a lesbian niche dating dite. Sexualities. 2015;18(1/2):20–36.

Hine C. Virtual ethnography: modes, varieties, affordances. In: Fielding N, Lee RM, Blank G, editors. The Sage handbook of online research methods. London: Sage; 2008. p. 257–70.

Im E, Wonshik C. Recruitment of research participants through the Internet. Comput Inf Nurs. 2004;22(5):289–97.

Irvine J. Disorders of desire: sex and gender in modern American sexology. Philadelphia: Temple University Press; 1990.

Johnson CW. 'The first step is the two-step': hegemonic masculinity and dancing in a country-western gay bar. Int J Qual Stud Educ. 2005;18(4):445–64.

Kimberly C. Permission to cheat: ethnography of a swingers' convention. Sex Conv. 2016;20(1):56–68.

Koch SC, Mueller B, Kruse L, Zumbach J. Constructing gender in chat rooms. Sex Roles. 2005;53(1–2):29–42.

Lenihan A, Kelly-Holmes. Virtual ethnography. In: Hua Z, editor. Research methods in intercultural communication: a practical guide. Malden: Wiley; 2016. p. 255–67.

Liamputtong P. Researching the vulnerable. London: SAGE; 2007.

Liamputtong P. Qualitative research methods. 4th ed. Melbourne: Oxford University Press; 2013.

Lomborg S. Social media, social genres: making sense of the ordinary. New York: Routledge; 2014.

Lovink G. Zero comments: blogging and critical internet culture. New York: Routledge; 2008.

Madge C. Developing a geographers' agenda for online research ethics. Prog Hum Geogr. 2007;31(5):654–74.

Mallon T. A book of one's own: people and their diaries. New York: Ticknow & Fields; 1984.

Marteya RM, Stromer-Galleyb J, Banksc J, Wud J, Consalvoe M. The strategic female: gender-switching and player behavior in online games. Inf Commun Soc. 2014;17(3):286–300.

McCormack M. Innovative sampling and participant recruitment in sexuality research. J Soc Pers Relat. 2014;31(4):475–81.

McDermott E, Roen K. Youth on the 'virtual' edge: researching marginalized sexualities and genders online. Qual Health Res. 2012;22(4):560–70.

Meyer IH, Wilson PA. Sampling lesbian, gay, and bisexual populations. J Couns Psychol. 2009;56(1):23–31.

Muise A. Women's sex blogs: challenging dominant discourses of heterosexual desire. Fem Psychol. 2011;21(3):411–9.

Murray DM, Fisher JD. The Internet: a virtually untapped tool for research. J Technol Hum Serv. 2002;19(2–3):5–18.

van Nuenen T. Here I am: authenticity and self-branding on travel blogs. Tour Stud. 2016;16(2):192–212.

Paasonen S. Labors of love: Netporn, Web 2.0 and the meanings of amateurism. New Media Soc. 2010;12(8):1297–312.

Patton MQ. Qualitative research and evaluation methods. 4th ed. Thousand Oaks: Sage; 2015.

Reid DJ, Reid FM. Online focus groups: an in-depth comparison of computer mediated and conventional focus group discussions. Int J Mark Res. 2005;47:131–62.

Roberts L. Opportunities and constraints of electronic research. In: Reynolds RA, Woods R, Baker JD, editors. Handbook of research on electronic surveys and measurements. Hershey: Idea Group Reference; 2007. p. 19–27.

Roller M, Lavrakas PJ. Applied qualitative research design: a total quality framework approach. New York: The Guilford Press; 2015.

Rosewarne L. Part-time perverts: sex, pop culture and kink management. Santa Barbara: Praeger; 2011.

Rosewarne L. American taboo: the forbidden words, unspoken rules, and secret morality of popular culture. Santa Barbara: Praeger; 2013.

Rosewarne L. Masturbation in pop culture: screen, society, self. Lanham: Lexington Books; 2014.

Rosewarne L. School of shock: film, television and anal education. Sex Educ. 2015;15(4):553–65.

Rosewarne L. Intimacy on the internet: media representations of online connections. New York: Routledge; 2016a.

Rosewarne L. Cyberbullies, cyberactivists, eyberpredators: film, TV, and Internet stereotypes. Santa Barbara: Praeger; 2016b.

Rosewarne L. Cinema and cyberphobia: Internet clichés in film and television. Aust J Telecommun Digit Econ. 2016c;4(1):36–53.

Rosewarne L. Choose your own (miss) adventure: single ladyhood in 2016. Meanjin. 2016d;75(3):32–40.

Schrooten M. Moving ethnography online: researching Brazilian migrants' online togetherness. Ethnic Racial Stud. 2012;35(1):1794–809.

Sparrman A. Access and gatekeeping in researching children's sexuality: mess in ethics and methods. Sex Cult. 2014;18:291–309.

Thomas J. Getting off on sex research: a methodological commentary on the sexual desires of sex researchers. Sexualities. 2016;19(1):83–97.

Tiefer L. New perspectives in sexology: from rigor (mortis) to richness. J Sex Res. 1991;28(4):593–602.

Waclawski E. How I use it: Survey Monkey. Occup Med. 2012;62:477.

Wagner G, Bondil P, Dabees K, Dean J, Fourcroy J, Gingell C, Kingsberg S, Kothari P, Rubio-Aurioles E, Ugarte F, Navarrete RV. Ethical aspects of sexual medicine. J Sex Med. 2004;2(1):163–8.

Waynberg J. 1908–2008: a century of sexology and still no legitimacy? Theol Sex. 2009;18(1):1–3.

White GE, Thomson AN. Anonymized focus groups as a research tool for health professionals. Qual Health Res. 1995;5:256–61.

Whitehead G. The evidence of things unseen: authenticity and fraud in the Christian mommy blogosphere. J Am Acad Relig. 2015;83(1):120–50.

Wood EA. Consciousness-raising 2.0: sex blogging and the creation of a feminist sex commons. Fem Psychol. 2008;18(4):480–7.

Wu T. What ever happened to Google Books? The New Yorker, September 11. 2015. http://www.newyorker.com/business/currency/what-ever-happened-to-google-books. Accessed 10 June 2016.

Emotion and Sensitive Research

123

Virginia Dickson-Swift

Contents

1 Introduction	2146
2 Research Work as "Emotion Work"	2146
3 The Research Process	2148
3.1 Ethics	2148
3.2 Data Collection	2149
3.3 Data Analysis	2150
4 Issues for Other Research Team Members	2151
4.1 Transcriptionists	2151
4.2 Research Assistants	2152
5 Working with Secondary Data	2153
6 Sources of Researcher Support	2154
6.1 Informal Support	2154
6.2 Protocols and Guidelines	2155
6.3 Guidelines for Transcriptionists	2157
7 Conclusion and Future Directions	2158
References	2159

Abstract

Qualitative research on sensitive topics is often an emotional journey, not only for the participants but for others that may be involved along the way. It is now more than 20 years since Raymond Lee authored the seminal works *Doing Research on Sensitive Topics, Researching Sensitive Topics*, and *Dangerous Fieldwork* that raised the awareness of the challenges that researchers can face. More recently, Lee and Lee (2012) warned that the emotional challenges that researchers face when doing fieldwork are now difficult to ignore. Given this warning and the

---

V. Dickson-Swift (✉)
LaTrobe Rural Health School, College of Science, Health and Engineering, LaTrobe University, Bendigo, VIC, Australia
e-mail: v.dickson-swift@latrobe.edu.au

© Springer Nature Singapore Pte Ltd. 2019
P. Liamputtong (ed.), *Handbook of Research Methods in Health Social Sciences*,
https://doi.org/10.1007/978-981-10-5251-4_141

growing numbers of reports from researchers, both empirically and in reflective accounts, an examination of the issues is timely for both novice and experienced researchers. Drawing on earlier empirical work with researchers in Australia (Dickson-Swift 2005) and published accounts, this chapter provides an overview of the emotional challenges inherent in this type of research. Suggestions for researchers, research supervisors, and others involved in the research team are presented. These can be adopted by academic or research institutions to ensure that researchers have the necessary support to carryout this important research.

Keywords
Sensitive research · Qualitative · Emotions · Emotional labor · Ethics · Training

1 Introduction

Undertaking qualitative research is often an emotional journey, not only for the participants but for others that may be involved along the way. There is growing evidence that researchers, research supervisors, transcriptionists, and research assistants face a number of emotional challenges while participating in qualitative research, particularly when that research focuses on sensitive topics and/or vulnerable populations (see Liamputtong 2007). Interest in the issues researchers face when researching sensitive topics is not new. More than 20 years ago, Ray Lee and Claire Renzetti authored a number of seminal texts outlining some of the key challenges researchers may face (see for example Lee and Renzetti 1993; Lee 1995; Renzetti and Lee 1993). The focus of their work was on both physical and emotional risks for researchers and participants across the spectrum of research methods. More recently, Lee and Lee (2012) warned that the emotional challenges researchers face when doing fieldwork cannot be ignored. Given this warning, and the growing numbers of empirical and reflective reports from researchers, an examination and consideration of the issues researchers face is timely. While much of the information in this chapter is drawn from accounts from qualitative projects, the key messages are also applicable to those researchers working on quantitative or mixed method studies. In this chapter, I draw on my earlier empirical work with researchers in Australia (Dickson-Swift et al. 2005) and a range of published accounts from across the globe, drawn from a range of disciplines to provide an overview of the emotional challenges faced by researchers. I provide some suggestions for researchers, research supervisors, and others involved in the research team. These can be adopted by academic or research institutions to ensure that researchers have the necessary support they may need.

2 Research Work as "Emotion Work"

Emotion work theory can provide a framework for understanding researchers' experiences throughout the research process. The concept of "emotion work" was initially developed by Arlie Hochschild in her now classic study *The Managed Heart* (1983), which explored the experiences of flight attendants and how they managed their emotions on a day-to-day basis on the job. The terms "emotional labor" and "emotion work" are often used interchangeably in the literature. Initially, these two concepts were developed by Hochschild (1983) to mean different things. "Emotional labor" was used to refer to emotional management during work done for a wage, and "emotion work" was used to refer to the work involved with dealing with other people's emotions (James 1989). In her definition of emotion work, Hochschild (1983, p. 7) states that it is "management of feeling to create a publically observable facial and bodily display whereby people work on managing their own and the other people's feelings to comply with a set of 'feeling rules'" that direct the type, intensity, and duration of the emotion (Stets and Turner 2014). The concept of emotion work and the resultant consequences of undertaking it have been documented in a range of occupations including airline staff (Hochschild 1983), front line service (including retail assistants (Van Maanen 1990; Ashforth and Humphrey 1993; Grandey et al. 2012), nurses and caring staff (Pisaniello et al. 2012; Bailey et al. 2015; Lovatt et al. 2015), physicians (Larson and Yao 2005), beauty therapists (Sharma and Black 2001), call center staff (Mulholland 2002), teachers (including academics) (Bellas 1999; Ogbonna and Harris 2004; Isenbarger and Zembylas 2006), barristers and legal staff (Harris 2002; Anleu and Mack 2005), clergy (Grauel 2002; Cotton et al. 2003), and sex workers (Sanders 2004), models (Mears and Finlay 2005). While there is not yet an extensive body of work focusing on emotion work done by researchers, interest in the other members of research teams in this area is growing (see for example, Campbell 2002; Dickson-Swift et al. 2009; Carroll 2013; Fitzpatrick and Olson 2015).

For the purposes of this chapter, I use the phrase emotion work when referring to any effort made by those participating in research (participants, researchers, or other members of the research team) to manage emotion and emotional displays. In relation to research, emotion work can take the form of what has been termed "surface acting" and "deep acting" (refer to Goffman 1959 for an in-depth discussion of these concepts). Surface acting takes place when individuals manage the observable emotional expressions by controlling verbal or facial expressions (e.g., holding in tears) as well as other gestures and/or bodily displays (e.g., altering facial expressions) (Hochschild 1979, 1983). Deep acting, on the other hand, is thought to be a more complex activity requiring utilization of the techniques of cognitive emotion work. This can include attempting to alter the emotion by changing the thoughts associated with that emotion. For example, bodily emotion work includes attempts to change bodily processes to alter the emotional responses and expressive emotion work includes altering outward expressions to shape underlying emotions (Hochschild 1983). More recently, authors interested in emotion work in research have also utilized the concept of "habitus" to explain and extend the theory to include the unconscious, embodied, and habitual aspects of emotion work (Fitzpatrick and Olson 2015).

While not all research into researcher emotions uses the concept of emotion work to describe the work undertaken, there is a growing body of research that refers directly to researcher emotion. Examples from the UK and Europe (Bloor et al. 2007, 2010; Mitchell and Irvine 2008; Parker and O'Reilly 2013; Benoot and Bilsen 2015), USA and Canada (Campbell 2002; Carroll 2013), and Australia and New Zealand (McCosker et al. 2001; Dickson-Swift et al. 2009; Johnson 2009; Bahn 2012; Bowtell et al. 2013; Mckenzie et al. 2016) provide both empirical and reflective accounts across a wide range of health and social science disciplines (including health, sports psychology, urban studies, geography, marketing, and occupational research). While this list is not exhaustive, it is illustrative of the breadth and depth of the documented accounts of the emotional challenges researchers might face.

3 The Research Process

In this section, I draw a range of published material to outline some key challenges documented by researchers throughout the research process. These will be provided in sections relating to ethics and risk, data collection, and data analysis and include some discussion of the impacts on transcriptionists and research assistants.

3.1 Ethics

Ethics guidelines are primarily focused on nonmaleficence and beneficence with regard to research participants which is the principal focus of ethical review (see also ► Chap. 106, "Ethics and Research with Indigenous Peoples"). As part of the review process, committee members consider emotional and physical risks to research participants and ensure that a number of strategies are in place to mitigate any harm (National Health & Medical Research Council 2007; Social Research Association 2006). Researchers completing ethics applications are well versed in the need to protect participants from physical harm but are not required to systematically consider any emotional harms that they may be exposed to. In her reflections on a 6-year social research project focused on a community street soccer program, Emma Sherry (2013, p. 280) shares her experiences:

> Each aspect of the paperwork required significant levels of detail and thought, to ensure no harm was done to my vulnerable research participants. But what was missing was a section that led me to consider any potential harm to myself as the researcher, let alone an opportunity to debrief.

Institutional review boards (IRBs) are well versed in protecting participants, but the protection of researchers is often not considered in deliberations (Dickson-Swift et al. 2005). Many of the existing safety protocols designed to protect researchers focus narrowly on physical risk which are often considered part of an organizational duty of care under Occupational Health and Safety legislation (Noblet 2003; Social Research Association 2006; National Health & Medical Research Council 2007; Kennedy et al. 2012). Ethics committees and IRBs have procedures and policies in place to manage physical risks to researchers when undertaking fieldwork; however, guidelines and policies for managing and mitigating emotional are still largely absent. Frustration and lack of preparedness for the emotionality of some fieldwork encounters are not uncommon;

> Nothing that I read in planning this study prepared me for the emotionality of the research process. I read recommendations about how I should address confidentiality, harm, deception and privacy, but there was not much written on such things as the impact on the researcher of listening to people talk about their grief, their fears and anxieties, sometimes being expressed for the first time and times of crisis. (Rowling 1999, p. 175)

Susanne Bahn's (2012) paper outlining the issues in relation to risks in fieldwork highlights that risk reduction policies for researchers often fail as they mostly rely on Heads of School/Department, PhD supervisors, or Chief Investigators to decide on the level of risk. Since early 2000, there have been a number of publications relating to risk in research questioning whose responsibility it is to ensure that researchers are safe in the field (see for example, Kenyon and Hawker 2000; Johnson and Clarke 2003; Dickson-Swift et al. 2005; Bloor et al. 2007).

In a study focused on researcher safety, Kenyon and Hawker (2000) set up an online discussion board which attracted 46 participants from the UK, Australia, USA, Finland, Norway, Sweden, Italy, and Canada. Only one of the 46 participants had ever been issued with a safety code of practice that outlined key issues for researchers and how to handle them. Similarly, in an Australian study completed in 2005, Dickson-Swift and colleagues investigated 37 university ethics application forms to determine the number that addressed the safety of the researcher. They found that in 78% of cases, there was no reference to researcher safety across the domains of physical, emotional, and psychological safety (Dickson-Swift et al. 2005). While there has been no follow-up work in this area within the Australian context, there has been some similar research undertaken in the UK. In 2007 Bloor and colleagues undertook a Commissioned Inquiry into the Risk to Wellbeing of Researchers in Qualitative Research (Bloor et al. 2007) conducted by Qualiti (Qualitative Research in the Social Sciences: Innovation, Integration and Impact; a node of the Economic and Social Research Centre's National Centre for Research Methods). As a follow up to this study 83 PhD students were invited to post their stories onto a website and then 13 participated in an in-depth interview to elaborate on the practices in place to protect researchers in the field (Bloor et al. 2010). The recommendations from the inquiry, the study report and subsequent publication included providing safety in the curricula, health and safety audits for all university departments, and specific questions in ethics applications that addressed contextual safety issues (Bloor et al. 2010, p. 52).

While we have not seen the widespread adoption of this approach to identifying and managing risk for researchers, a number of guidelines have been suggested which will be presented later in this chapter.

3.2 Data Collection

Researchers have provided many accounts of emotionally charged situations during data collection. Rebecca Olson (Fitzpatrick and Olson 2015, p. 51) reflects on her own experiences of research with caregivers to explore their responses to a cancer diagnosis of a spouse;

At times, I felt sadness and frustration in response to his stories, but I didn't cry. I was aware of my facial expressions and actively tried to sustain an "open" and "active listening" expression: lips closed, gaze fixed on either Joe or my tea cup. I left the interview feeling tired but indebted to Joe for his honesty. Several days later, while watching television with a group of friends, as a character learned his father was dying from cancer, I felt the sadness of Joe's and other interviewees' stories and cried for half an hour.

Similarly, Rebecca Campbell (2002, p. 6) shares her reflections on interviewing rape victims,

As I listened to her preface her story with the information about her rape I realized this was something more than just a research interview. I was not just the project director, not just a researcher, I was the first person she was going to tell to trust with this information. I was being given something very fragile, and yet very strong. It was a sobering responsibility. We both knew what was coming was going to be hard for both of us.

There are also many accounts of researchers reporting physical and sometimes emotional exhaustion from undertaking interviews. Considering the nature of the interviews, this is not surprising.

...by the time I got home I was just like exhausted, just emotionally exhausted. I was just interviewing all day, I would have done more than five interviews – I was just interviewing all day, by myself and I was really buggered, yeah I was had it. I found it really emotionally draining ... it got to the point where I had to allocate time ... I can go and do the interview in the morning but then I will have to block out the rest of the day because after that I need to go home and I need to digest what has happened ... go through it in my head. (Dickson-Swift et al. 2009, p. 71)

Researchers also highlight that it is often difficult not to get drawn into the emotion, especially when face-to-face with another person who is experiencing emotion (Dickson-Swift et al. 2009; Jafari et al. 2013). Some researchers do not attempt to hold back or manage their emotions during the interview process, instead preferring to become part of the experience themselves. Dickson-Swift et al. (2009, p. 64) reported occasions where researchers were emotionally overwhelmed during the research, stating that this was often directly attributable to the participants becoming emotional.

I cried pretty much the whole way through it because once she got upset it was impossible not to be upset you know.... seven no its eight – eight out of ten people I interview cry and they cry sometimes uncontrollably, it's a very sad thing to talk about ... and how can you, as a person not get caught up in those feelings of sadness. (Dickson-Swift et al. 2009, p. 64)

3.3 Data Analysis

It is important to recognize that the emotional impact can extend well beyond the data collection phase as the researcher moves into data analysis and writing up of the findings. The effects on the researcher can be cumulative, resulting in emotional exhaustion (Woodby et al. 2011). Data analysis often becomes a time for reflection on emotions for researchers,

> The complex emotions I experienced while collecting the data were relived and recounted through the passing months and years when I conducted manual data analysis. On reflection, I realise that during both the data collection and analysis phases, I engaged in an amount of emotional management. Storing away sadness and fear during the exhilarating fieldwork phase only to be unexpectedly revisited by it during the long and lonely phases of data analysis. (Jafari et al. 2013, p. 1189)

Based on a qualitative study focusing on elder neglect, Band-Winterstein et al. (2014, p. 536) reflect on some of the challenges related to the emotion connections that became evident during the analysis process.

> They "got to me", I was touched, they managed to break through my walls. At first, I tried to keep my distance and let the person talk. After some time, I found myself shamelessly asking questions about their personal lives. And I felt that I was connecting with them. They had waited for me, which was very touching and I don't know how to explain it, only that "I am taking them with me."

Charlotte Benoot (Benoot and Bilsen 2015, p. 5), a novice researcher undertaking interviews with people caring for cancer patients, reflects on her feelings in the coding and analysis phase of the research.

> The act of coding and transcribing made me relive the emotions I suppressed while conducting an interview, without the necessity of controlling them. This was because I was alone at that time, which gave me more freedom to express my feelings, and also because I had had more time to take the story in. Moreover, the repeated listening, replaying, and typing intensified the emotional responses to sensitive materials.

4 Issues for Other Research Team Members

In addition to considering ethics and risk to the researcher, there has been an increase in acknowledgment of the possibility of physical and/or psychological harm being extended beyond participants to include transcriptionists and research assistants and those working with secondary data.

4.1 Transcriptionists

While there are many documented examples of emotional risks and emotion management for researchers, transcriptionists working with data from interviews, particularly on sensitive topics, are often left out. There is now evidence that listening to audio recordings and transforming those audios for analysis the potential to have an emotional impact on the transcriptionist (Gilbert 2001; McCosker et al. 2001;

Darlington and Scott 2002; Gair 2002; Warr 2004; Lalor et al. 2006; Etherington 2007; Sherry 2013; Kiyimba and O'Reilly 2015). Often considered a clerical role, the act of transcription can impact on those that undertake it.

> It is definitely important to know what you are getting into, not to think of it as a typing job, think of it, um, as something that will have repercussions that you will wake up in the night and think about it on a level that you just wouldn't expect from that sort of, that level of clerical work really (Kiyimba and O'Reilly 2015, p. 102).

People who undertake transcription of interview data are one of the few people (besides the researcher) to hear the actual voices and expressions of emotion of those people participating in the interview (Kiyimba and O'Reilly 2015). Concern for those that undertake transcription work has been documented for almost 20 years with Gregory et al. (1997) and McCosker et al. (2001) examining the role of the transcriber as a vulnerable person. A transcriptionist must listen and re-listen to the data to undertake the process of capturing the spoken word which can result in the data becoming embedded within their consciousness. They are often not considered in the ethics process except for the signing of a confidentiality agreement. This poses a problem when the data refer to disturbing events or traumatic life experiences (Gregory et al. 1997). There are a number of accounts of transcriptionists becoming emotional when listening to the data (Gregory et al. 1997; McCosker et al. 2001; Wilkes et al. 2014; Kiyimba and O'Reilly 2016). One recent example draws on research with 12 transcriptionists in Australia and New Zealand (Wilkes et al. 2014). Some transcriptionists reported finding the process of transcription overwhelming resulting in them deciding not to take on any more work related to some research projects. Participants also highlighted that they faced a number of challenges during the process including negative emotional effects (e.g., anger and sadness) and negative physical effects (sleeplessness, vomiting, headaches, and stress) and that they used a number of personal strategies to manage these effects (including debriefing, support from family and friends, and taking time out).

Evidence has shown that there is emotional risk for transcriptionists which are heightened if opportunities are not provided for support that includes debriefing and training including opportunities to talk about the emotional impacts of the work that they do (Etherington 2007; Kiyimba and O'Reilly 2015). A recent study undertaken across the UK involving 9 transcriptionists showed that emotional distress was perceived as a threat to the emotional welfare of those involved (Kiyimba and O'Reilly 2015). In some studies, the process of transcription has been considered as a form of emotion work which has may lead to burnout (Hochschild 1983; Dickson-Swift et al. 2009; Kiyimba and O'Reilly 2015). In order to ameliorate the risks to transcriptionists, they also need to be provided with practical and emotional support throughout the research process.

4.2 Research Assistants

While there are some considerations for researchers and issues for transcriptionists are gaining more attention, research assistants are another group that potentially face emotional challenges related to the work that they do (Bahn and Weatherill 2012; Benoot and Bilsen 2015; Mckenzie et al. 2016).

> I knew we had something set up for the participants. But suddenly I wondered if we had it set up for the research assistants. If I carried this story home in my head for a couple of days, what might be the impact on a more junior team member? I must make a point of meeting up with <> for a chat. It will be useful for me too. Self awareness can be such a draining thing, ha, ha? (Bahn and Weatherill 2012, p. 8)

Many researchers report reaching a point of emotional saturation from undertaking emotion work (Sherry 2013). This sense of emotional saturation was also reported in the study by Dickson-Swift et al. (2009, p.72).

> ... it's not just about saturation of when you don't get new themes...it's about your saturation as well – how much you can actually take and I could not, could nothave fronted for another one of those interviews.

Emotional saturation may be accompanied by feelings of exhaustion, burnout, guilt, tiredness, sleeping difficulties, anxiety, and gastrointestinal upsets (Dickson-Swift et al. 2009; Benoot and Bilsen 2015). In the study by Benoot and Bilsen (2015) Charlotte Benoot reflects on her own experiences in undertaking research with cancer patients that lived alone:

> As a consequence, I started to internalize the bodily complaints my patients had, which means that I often literally could feel the pain symptoms or nausea from the patient I was interviewing at that time. Another consequence was that the fear of getting cancer myself grew with every interview I took (Benoot and Bilsen 2015, p. 5).

5 Working with Secondary Data

Social research utilizing secondary data is growing in popularity in health and social sciences. There is increasing acknowledgment that undertaking this type of research may also pose emotional risks to those undertaking it (Moran-Ellis 1996; Fincham et al. 2008; see also ▶ Chap. 119, "Feminist Dilemmas in Researching Women's Violence: Issues of Allegiance, Representation, Ambivalence, and Compromise"). Secondary data in this context refer to data that have been assembled by another person rather than data collected originally by a researcher. Data sources for this type of research are often documentary but could also include a range of other types (e.g., digital, visual, or aural) (Fincham et al. 2008). Fincham and colleagues undertook a study focusing on the review of coronial files of people who had suicided. Their research highlighted that the contents of such files can have a profound emotional impact on those who review them and that researchers undertaking this type of work need to consider their own self-care and have a range of support systems in place to ameliorate any risks to researchers or other members of the team. In an earlier study, Moran Ellis (1997) refers to the possibility of emotional risk for researchers as feeling "pain by proxy" providing examples from a study on child sexual abuse. She reports,

> I felt appalled by what I was finding out, and I felt much pain by proxy for the children who had been subjected to what amounts to physical as well as emotional and sexual assault. I could barely contemplate the pain they had felt...And yet I found I couldn't not think about it. (Moran-Ellis 1996, p. 181)

6 Sources of Researcher Support

In the following section, I outline some of the supports that may be useful for researchers.

6.1 Informal Support

A number of researchers have reported using informal support networks of colleagues, trusted friends, and family members for support and debriefing throughout the research (Hubbard et al. 2001; Dickson-Swift et al. 2009; Fahie 2014). This informal peer support is very important for researchers particularly as the much of the discussion about emotions in research and how the researcher actually "feels" in the process is often done informally at the photocopier, coffee machine, or in the corridors (Dickson-Swift et al. 2009). In a study exploring researcher trauma for researchers working on sexual violence research, Jan Coles et al. (2014, p. 106) reported a number of instances of how informal support was used.

> At the time, I did not realize how vitally important it was to protect myself. The organization I worked for did not provide an embedded support system for its staff. Our support came from each other as colleagues and friends, and to an extent, this enabled me to survive mentally, but it did not help me eradicate the root causes. I became adept at burying the emotional stress but, of course, it continues to surface in a number of guises.

Mick Bloor et al. (2007, p. 34) sum up the problems associated a reliance on using informal networks for this type of support,

> Whilst it is inevitable to a certain extent that there will be off-loading at home, the formal exploitation of informal networks – for example, building them into research designs – is not deemed appropriate, and such strategies do not absolve research funders and institutions of their responsibilities to researchers.

While it is clear that undertaking qualitative research can be emotionally challenging for many researchers, it is important to note that some disciplines may better prepare researchers to deal with the emotional challenges through their postgraduate programs. But this may not be the case for all researchers (novice or otherwise). Most postgraduate students have access to regular supervision within the university setting from their immediate supervisors. However, other more experienced researchers attached to universities or large research centers may not necessarily have access to regular formalized supervision (Campbell 2002; Carroll 2013). If we are to truly create a space for researchers to explore the emotional nature of the work that they do, then we need to ensure that appropriate support is offered, both institutionally and individually for researchers to do that (Benoot and Bilsen 2015).

6.2 Protocols and Guidelines

Most university and research organizations pay attention to the issue of physical safety for researchers, and there are risk management policies and protocols designed to mitigate any harm. Physical risks tend to be more easily recognizable (Paterson et al. 1999; Bowtell et al. 2013), and recommendations to ameliorate these types of risk include identifying possible threats and developing written safety protocols, not volunteering personal information and avoiding interviews in private homes (among other things). Sampson et al. (2008) determined that researchers actually perceive emotional harm as more prevalent in the field that physical harm. This highlights that the development and implementation of emotional safety guidelines for all member of the research team are just as important as the well-used physical safety protocols. However, emotional safety does not receive the same recognition within research guidelines leaving many researchers vulnerable to emotional harms. Almost 20 years ago, Martin Schwartz (1997, p. x) pointed out some the emotional challenges faced by researchers:

> Most academic research programs can provide advice on when logistic regression is a better tool than discriminate function analysis, but few have mentors who can talk about how to handle your uncontrollable tears late at night after a day of conducting interviews with victimized women.

Despite the many calls to introduce standardized protocols to protect researchers and research team members from emotional harm, it appears that few institutions and research centers have formalized them to date (Bloor et al. 2007; Dickson-Swift et al. 2008; Kiyimba and O'Reilly 2016). Questions have been raised about whose responsibility it would be to ensure that any protocols are implemented. Arguably, the safety (both emotional and physical) is as much a responsibility of ethics committees as participants (Dickson-Swift et al. 2008; Bowtell et al. 2013). Researchers have been encouraged to consider the ethical mantra of "do no harm" in relation to themselves as well as their research participants (Lee-Treweek and Linkogle 2000).

In the absence of any standard formalized recommendations, researchers have outlined their own approaches to emotional and physical risk management. Declan Fahie (2014), a social researcher from the UK, recommends the following personal safety tips for researchers (see Box 1).

> **Box 1 Personal Safety Tips for Researchers**
> - Undertake a risk assessment as part of the design
> - Do not disclose personal details (home address or phone numbers)
> - Ensure that your interviews are carried out in public places as much as possible (libraries, public meeting rooms, community houses, etc.)
> - Inform your supervisor of your location and carry a mobile device and SMS when you arrive and leave
> - Get a dedicated SIM card or voice mail box just for the research (these can be destroyed later)
> - Monitor carefully the interview to assess the emotional impact and response of the interviewee
> - Monitor your own response and have a plan for debriefing
> - Make sure you have regular sessions with your supervisor (or mentor) so you can talk through the research process and any effects on you
> - Don't be afraid to write, talk, and discuss your responses and seek help when it is needed
>
> Adapted from Fahie (2014).

In Australia, Bowtell et al. (2013) have called safety protocols to include training to teach researchers about being aware of how they feel in relation to emotion saturation and tips for recognizing emotional exhaustion and strategies to reduce researcher fatigue (see Box 2). They recommend that the assessment of emotional safety risks "predate or at least be developed concurrently with the ethics application for any research project" (Bowtell et al. 2013, p. 659).

> **Box 2 Ten Emotional Safety Tips for Researchers**
> 1. Researchers should acknowledge that an emotional impact is inherent to qualitative research.
> 2. Before applying to an HREC, the researcher and supervisory team should undertake a detailed assessment of potential risks to the emotional safety of the researcher, as well as to the emotional safety of research participants.
> 3. The research team should discuss the boundaries that lie between the researcher and participants and how individual researchers might maintain these while establishing rapport.
> 4. Research supervision should regularly include reviewing the emotional impact of the research process on the researcher through both discussion and review of field notes.
> 5. Supervisors should proactively arrange emotional support for researchers from a suitably qualified professional who understands the nature of the

(continued)

> **Box 2** (continued)
> research. Regular debriefing should take place from the start of the study, rather than in response to an incident or event. To promote honest communication and uphold privacy, this person should not be a member of the research team.
> 6. Researchers should be encouraged to regularly practice mindfulness and engage in emotional auditing via memos or research diaries after each interview and when reviewing recordings and transcripts.
> 7. Researchers need to create boundaries between home and work. Having a balanced lifestyle reduces the risk of burnout.
> 8. Research supervisors should support the researcher to briefly "step away" from the intensity of the research process as an appropriate response to promoting emotional safety when major challenges arise.
> 9. Regular departmental meetings (e.g., seminar series) should encourage discussion of the emotional impact of the research on the researcher.
> 10. The challenging moments of research should be shared within the qualitative research community so that others can learn.
>
> Note: HREC = human research ethics committee.
> Reproduced with permission Bowtell et al. 2013.

Researchers may be well advised to participate in clinical supervision to deal with any issues that may arise throughout the research process that cannot be discussed openly with supervisors. This type of formalized supervision is now a requirement of many of the codes of ethics of professional associations across the world (Australian Guidance and Counselling Association 1997; Australian Psychological Society 2002; Australian Nursing Council 2003; International Federation of Social Workers 2006; Social Research Association 2006). It has been recommended that researchers and research institutions should also be encouraged to have arrangements for such supervision formalized within their guidelines (Bloor et al. 2007; Bowtell et al. 2013). There are a number of good examples of how supervision and support for researchers can be built into research projects. The Social Policy Research Unit (SPRU) at the University of York used a model of group psychotherapy to support researchers who were interviewing recently bereaved parents (Corden et al. 2005). Similarly, Natalie Wray et al. (2007), in their study of gynecological cancers, reported using both the university counselor for debriefing after emotionally distressing interviews and utilizing a fee-for-service psychotherapist to assist in dealing with researcher distress.

6.3 Guidelines for Transcriptionists

Transcriptionists can also be emotionally affected by the work that they do. Gregory et al. (1997, p. 297) propose the following tips for those undertaking transcription work.

- Be included in the ethics process.
- Be encouraged to have a process for self-care.
- Be fully informed of the nature of research and type of data.
- Be altered prior to the transcription of potentially 'challenging' or 'difficult' interviews.
- Have regular scheduled debriefing sessions.
- Have prompt access to an appropriate person for crisis counseling.
- Have a clearly documented process for the termination after transcription is completed that includes a resolution of personal issues that may have arisen as a consequence of the work.
- Be encouraged to journal thoughts and feelings which may then become a part of the fieldwork notes in some studies.

A discussion of these prior to undertaking any work would ensure that transcriptionists are fully prepared for the work that that they do.

7 Conclusion and Future Directions

As researchers, research assistants, supervisors, and transcriptionists we need to take emotions within research seriously into account. If we do then we open a space within which we can explore practical strategies to work with our emotional responses. Like Fitzpatrick and Olson (2015), in this chapter, I have demonstrated that rather than viewing researchers' emotions as risks to be avoided, researchers, ethics committees, supervisors, transcriptionists, and colleagues should value emotions as integral to human life and respond accordingly by encouraging researchers to reflect on their own emotions. In this chapter, I have provided a range of examples from researchers that should encourage others to consider these risks. In doing this, we bring to light aspects of our experience that may be particularly problematic for novice researchers or those researching sensitive topics. I have outlined the need to put in place process and policies to protect **all** members of the research team from both physical and emotional harm and provided a number of tips to assist. By doing this, we create an environment where both researchers and supervisors feel able to talk about challenging issues and devise strategies to ameliorate any risks. As part of this, it is important for us to incorporate discussions of psychological well-being and emotional risk into professional development and research coursework programs and to demand that formalized guidelines are developed that address all harms that researchers may face.

Future directions for research in this area should include more empirical work that documents the experiences of researchers and others in the research team involved in a range of qualitative projects. With further evidence of the risks inherent in this research, a comprehensive and universal standard of minimum policy requirements can be developed. This type of policy could be implemented by researchers, ethics committees, research institutes and universities to ensure the emotional and physical safety of those involved in research is protected.

References

Anleu S, Mack K. Magistrates' everyday work and emotional labour. J Law Soc. 2005;32(4): 590–614.
Ashforth B, Humphrey RH. Emotional labour in service roles: the influence of identity. Acad Manag Rev. 1993;18:88–115.
Australian Guidance and Counselling Association. Code of ethics. Canberra: Australian Guidance and Couselling Association; 1997.
Australian Nursing Council. Code of conduct for nurses in Australia. Canberra: Australian Nursing Council; 2003.
Australian Psychological Society. Code of ethics. Melbourne: Australian Psychological Society; 2002.
Bahn S. Keeping academic field researchers safe: ethical safeguards. J Acad Ethics. 2012;10(2): 83–91.
Bahn S, Weatherill P. Qualitative social research: a risky business when it comes to collecting 'sensitive' data. Qual Res. 2012; https://doi.org/10.1177/1468794112439016.
Bailey S, Scales K, Llyod J, Schneider J, Jones R. The emotional labour of health-care assistants in inpatient dementia care. Ageing Soc. 2015;35(02):246–69.
Band-Winterstein T, Doron I, Naim S. 'I take them with me' – reflexivity in sensitive research. Reflective Pract. 2014;15(4):530–9.
Bellas ML. Emotional labor in academia: the case of professors. Ann Am Acad Pol Soc Sci. 1999;561:91–110.
Benoot C, Bilsen J. An auto-ethnographic study of the disembodied experience of a novice researcher doing qualitative cancer research. Qual Health Res. 2015; https://doi.org/10.1177/1049732315616625.
Bloor M, Fincham B, Sampson H. Qualti (NCRM) commissioned inquiry into the risk to well-being of researchers in qualitative research. Cardiff ESRC National Centre for Research Methods. Cardiff University, Wales; 2007.
Bloor M, Fincham B, Sampson H. Unprepared for the worst: risks of harm for qualitative researchers. Methodol Innov. 2010;5(1):45–55.
Bowtell EC, Sawyer SM, Aroni RA, Green JB, Duncan RE. "Should I send a condolence card?" Promoting emotional safety in qualitative health research through reflexivity and ethical mindfulness. Qual Inq. 2013;19(9):652–63.
Campbell R. Emotionally involved: the impact of researching rape. New York: Routledge; 2002.
Carroll K. Infertile? The emotional labour of sensitive and feminist research methodologies. Qual Res. 2013;13(5):546–61.
Coles J, Astbury J, Dartnall E, Limjerwala S. A qualitative exploration of researcher trauma and researchers' responses to investigating sexual violence. Violence Against Women. 2014;20(1):95–117.
Corden A, Sainsbury R, Sloper R, Ward B. Using a model of group psychotherapy to support social research on sensitive topics. Int J Soc Res Methodol. 2005;8(2):151–60.
Cotton S, Dollard M, de Jonge J, Whetham P. Clergy in crisis. In: Dollard M, Winefield AH, Winefield HR, editors. Occupational stress in the service professionals. London: Taylor and Francis; 2003.
Darlington Y, Scott D. Qualitative research in practice: stories from the field. St. Leonards: Allen and Unwin; 2002.
Dickson-Swift V, James E, Kippen S. Do university ethics committees adequately protect public health researchers? Aust N Z J Public Health. 2005;29(6):576–82.
Dickson-Swift V, James EL, Kippen S, Liamputtong P. Risk to researchers in qualitative research on sensitive topics: issues and strategies. Qual Health Res. 2008;18(1):133–44.
Dickson-Swift V, James BL, Kippen S, Liamputtong P. Researching sensitive topics: qualitative research as emotion work. Qual Res. 2009;9(1):61–79.

Etherington K. Working with traumatic stories: from transcriber to witness. Int J Soc Res Methodol. 2007;10(2):85–97.

Fahie D. Doing sensitive research sensitively: ethical and methodological issues in researching workplace bullying. Int J Qual Methods. 2014;13:17.

Fincham B, Scourfield J, Langer S. The impact of working with disturbing secondary data: reading suicide files in a coroner's office. Qual Health Res. 2008;18(6):853–62.

Fitzpatrick P, Olson RE. A rough road map to reflexivity in qualitative research into emotions. Emot Rev. 2015;7(1):49–54.

Gair S. In the thick of it: a reflective tale from an Australian social worker/qualitative researcher. Qual Health Res. 2002;12(1):130–9.

Gilbert KR. Collatoral damage? Indirect exposure of staff members to the emotions of qualitative research. In: Gilbert KR, editor. The emotional nature of qualitative research. London: CRC; 2001. p. 147–61.

Goffman E. The presentation of self in everyday life. Harmondsworth: Penguin; 1959.

Grandey A, Foo S, Groth M, Goodwin R. Free to be you and me: a climate of authenticity alleviates burnout from emotional labor. J Occup Health Psychol. 2012;17(1):1–14. https://doi.org/10.1037/a0025102.

Grauel T. Overseeing the overseers: supervision of christian clergy in Australia. In: McMahon M, Patton W, editors. Supervision in the helping professions: a practical approach. Sydney: Prentice Hall; 2002. p. 261–72.

Gregory D, Russell CK, Phillips LR. Beyond textual perfection: transcribers as vulnerable persons. Qual Health Res. 1997;7(2):294–300.

Harris LC. The emotional labour of barristers: an exploration of emotional labour by status professionals. J Manag Stud. 2002;39(4):553–84.

Hochschild A. Emotion work, feeling rules and social structure. Am J Sociol. 1979;85:551–7.

Hochschild A. The managed heart: the commercialization of human feeling. Berkeley: University of California Press; 1983.

Hubbard G, Backett-Milburn K, Kemmer D. Working with emotions Issues for the researcher in fieldwork and teamwork. Int J Soc Res Methodol. 2001;4(2):119–37.

International Federation of Social Workers. Ethics in social work: statement of principles. http://www.ifsw.org/. (2006).

Isenbarger L, Zembylas M. The emotional labour of caring in teaching. Teach Teach Educ. 2006;22(1):120–34.

Jafari A, Dunnett S, Hamilton K, Downey H. Exploring researcher vulnerability: contexts, complications, and conceptualisation. J Mark Manag. 2013;29(9–10):1182–200.

James N. Emotional labour: skill and work in the social regulation of feelings. Sociol Rev. 1989;37:15–42.

Johnson N. The role of self and emotion within qualitative sensitive research: a reflective account. ENQUIRE. 2009;4:23.

Johnson B, Clarke J. Collecting sensitive data: the impact on the researchers. Qual Health Res. 2003;13(3):421–34.

Kennedy F, Hicks B, Yarker J. Challenges and work experiences of oncology researchers. Psycho-Oncology. 2012;21:14.

Kenyon E, Hawker S. "Once would be enough": some reflections on the issue of safety for lone researhers. Int J Soc Res Methodol. 2000;2(4):313–27.

Kiyimba N, O'Reilly M. The risk of secondary traumatic stress in the qualitative transcription process: a research note. Qual Res. 2015;60:1–9. https://doi.org/10.1177/1468794115577013.

Kiyimba N, O'Reilly M. An exploration of the possibility for secondary traumatic stress among transcriptionists: a grounded theory approach. Qual Res Psychol. 2016;13(1):92–108.

Lalor JG, Begley CM, Devane D. Exploring painful experiences: impact of emotional narratives on members of a qualitative research team. J Adv Nurs. 2006;56(6):607–16.

Larson EB, Yao X. Clinical empathy as emotional labor in the patient-physician relationship. JAMA. 2005;293(9):1100–6.

Lee RM. Dangerous fieldwork. Thousand Oaks: SAGE; 1995.

Lee R, Lee Y. Methodological research on "sensitive" Topics: a decade review. Bull Sociol Methodol. 2012;114(1):35–49.

Lee R, Renzetti C. The problems of researching sensitive topics. In: Renzetti CM, Lee RM, editors. Researching sensitive topics. Newbury Park: SAGE; 1993. p. 3–12.

Lee-Treweek G, Linkogle S, editors. Danger in the field: risk and ethics in social research. London: Routledge; 2000.

Liamputtong P. Researching the vulnerable: a guide to sensitive research methods. London: SAGE; 2007.

Lovatt M, Nanton V, Roberts J, Ingleton C, Noble B, Pitt E, Seers K, Munday D. The provision of emotional labour by health care assistants caring for dying cancer patients in the community: a qualitative study into the experiences of health care assistants and bereaved family carers. Int J Nurs Stud. 2015;52(1):271–9.

McCosker H, Barnard A, Gerber R. Understanding sensitive research: issues and strategies for meeting the safety needs for all participants. Forum Qual Soc Res. 2001;2(1):1.

Mckenzie SK, Li C, Jenkin G, Collings S. Ethical considerations in sensitive suicide research reliant on non-clinical researchers. Res Ethics. 2016; https://doi.org/10.1177/1747016116649996.

Mears A, Finlay W. Not just a paper doll: how models manage bodily capital and why they perform emotional labor. J Contemp Ethnogr. 2005;34(3):317–43.

Mitchell W, Irvine A. I'm okay, you're okay?: reflections on the well-being and ethical requirements of researchers and research participants in conducting qualitative fieldwork interviews. Int J Qual Methods. 2008;7(4):31–44.

Moran-Ellis J. Close to home: the experience of researching child sexual abuse. In: Hester M, Kelly L, Radford J, editors. Women, violence and male power. Buckingham: Open University; 1996. p. 176–87.

Mulholland K. Gender, emotional labour and teamworking in a call centre. Pers Rev. 2002;31(3): 283–303.

National Health & Medical Research Council. National statement on ethical conduct in human research. Canberra: National Health and Medical Research Council and Australian Vice Chancellors Committee; 2007.

Noblet A. Building health promoting work settings: identifying the relationship between work characteristics and occupational stress in Australia. Health Promot Int. 2003;18(4):351–7.

Ogbonna E, Harris L. Work intensification and emotional labour among UK university lecturers: an exploratory study. Organ Stud. 2004;25(7):1185–203.

Parker N, O'Reilly M. "We are alone in the house": a case study addressing researcher safety and risk. Qual Res Psychol. 2013;10(4):341–54.

Paterson BL, Gregory D, Thorne S. A protocol for researcher safety. Qual Health Res. 1999;9(2): 259–69.

Pisaniello SL, Winefield HR, Delfabbro PH. The influence of emotional labour and emotional work on the occupational health and wellbeing of South Australian hospital nurses. J Vocat Behav. 2012;80(3):579–91.

Renzetti C, Lee RM, editors. Researching sensitive topics. Newbury Park: SAGE; 1993.

Rowling L. Being in, being out, being with: affect and the role of the qualitative researcher in loss and grief research. Mortality. 1999;4(2):167–81.

Sampson H, Bloor M, Fincham B. A price worth paying? Sociology. 2008;42(5):919–33.

Sanders T. Controllable laughter: managing sex work through humour. Sociology. 2004;38(2): 273–91.

Schwartz MD, editor. Researching sexual violence against women: methodological and personal perspectives. Thousand Oaks: Sage; 1997.

Sharma U, Black P. Look good, feel better: beauty therapy as emotional labour. Sociology. 2001;35(4):913–9.

Sherry E. The vulnerable researcher: facing the challenges of sensitive research. Qual Res J. 2013;13(3):278–88.

Social Research Association. A code of practice for the safety of social researchers. 2006. Available at http://www.the-sra.org.uk/documents/word/safety_code_of_practice.doc. Accessed 19 Sept 2007.

Stets JE, Turner JH. Handbook of the sociology of emotions. Dordrecht: Springer; 2014.

Van Maanen J. The smile factory work at Disneyland. In: Frost PJ, Louis LF, Lundberg CC, Martin J, editors. Reframing organizational culture. Newbury Park: SAGE; 1990. p. 58–76.

Warr D. Stories in the flesh and voices in the head: reflections on the context and impact of research with disadvantaged populations. Qual Health Res. 2004;14(4):578–87.

Wilkes L, Cummings J, Haigh C. Transcriptionist saturation: knowing too much about sensitive health and social data. J Adv Nurs. 2014; https://doi.org/10.1111/jan.12510.

Woodby L, Williams B, Wittich A, Burgio K. Expanding the notion of researcher distress: the cumulative effects of coding. Qual Health Res. 2011;21(6):830–8.

Wray N, Markovic M, Manderson L. "Researcher saturation": the impact of data triangulation and intensive-research practices on the researcher and qualitative research process. Qual Health Res. 2007;17(10):1392–402.

Doing Reflectively Engaged, Face-to-Face Research in Prisons: Contexts and Sensitivities

124

James E. Sutton

Contents

1 Introduction	2164
2 Contextualization of Research with Prisoners in Prisons	2166
2.1 The Nature of Prisons	2166
2.2 The Nature of Prisoners' Lives	2167
2.3 Sensitive Contexts for Researchers	2168
3 Ethical Considerations	2170
3.1 Formal Review Boards	2171
3.2 Beyond Formal Review Boards	2172
4 Doing Reflectively Engaged Prisoner Research	2173
4.1 Managing Boundaries and the Presentation of Self	2174
4.2 Emotional Management	2175
5 Conclusion and Future Directions	2175
References	2176

Abstract

This chapter begins by establishing an historical trajectory of face-to-face research with prisoners. It goes on to identify fundamental features of prisons and prisoners' lives that make them sensitive settings and populations for researchers to study, and it then presents ethical considerations that researchers must be mindful of when carrying out this kind of research. Ethical concerns both within and beyond the scope of formal Institutional Review Boards are outlined and explored, as are researcher strategies for managing boundaries and emotions when doing prison research. To the extent that ethics, emotions, researcher presentation of self, and similar themes have been written about within the context of prison research, they have primarily been framed as considerations

J. E. Sutton (✉)
Department of Anthropology and Sociology, Hobart and William Smith Colleges, Geneva, NY, USA
e-mail: jsutton@hws.edu

© Springer Nature Singapore Pte Ltd. 2019
P. Liamputtong (ed.), *Handbook of Research Methods in Health Social Sciences*,
https://doi.org/10.1007/978-981-10-5251-4_137

for qualitative field researchers. By way of contrast, an underlying assumption of this chapter is that those who do other forms of face-to-face research with prisoners, including quantitative self-report surveys, mixed-method approaches, and focus groups, should similarly engage with these themes. This chapter ultimately endorses being reflectively engaged with the setting, the research process, and oneself when doing face-to-face research in prisons, regardless of the substantive goals of one's study or the particular research methods one employs. Accordingly, the issues raised in this chapter will be relevant to a range of health and social science researchers who enter prisons to study prisoners.

Keywords

Prison · Corrections · Research ethics · Sensitive research · Reflexivity · Total institution

1 Introduction

Systematic, face-to-face observation within prisons dates back to when John Howard visited over 200 European prisons in the 1700s and then reported on their abhorrent conditions (Howard 1780). The first prison-based research projects in the United States began to emerge in the early-to-mid 1900s, including Joseph Fishman's expose of sexual behavior and abuse in prison (Fishman 1934), Donald Clemmer's ethnography of prison culture (Clemmer 1940), and Gresham Sykes' examination of how prisoners respond to deprivations and totalitarian power (Sykes 1958). A recurring theme in these early prison studies was the illumination of harmful prison conditions, often for the purpose of promoting penal reform.

US prisoners were studied by ethnographers and others more frequently in the 1960s and 1970s. However, there was subsequently little ethnographic prison research being conducted when the US prisoner population was growing exponentially in the 1990s and early 2000s (Wacquant 2002; Reiter 2014). This was in large part due to more formal restrictions being placed on prison research beginning in the 1970s (Wakai et al. 2009; Cislo and Trestman 2013). Most recently, the past decade has seen a resurgence of critically engaged prison ethnography, with much of the contemporary work occurring in Europe (Reiter 2014; Drake et al. 2015).

The world's prisoner population has been growing rapidly and significantly, with an unprecedented 10.35 million people now incarcerated worldwide (Walmsley 2013). The United States imprisons approximately 2.2 million of these prisoners (Kaeble et al. 2015), making it the world's leader in the use of incarceration. Despite the notable growth in incarceration in recent years, we know less about the lives of prisoners than we do about people in free society (Cislo and Trestman 2013). Hence, there remains a current need for more research in prisons given the number of people who now experience incarceration firsthand. Similar to the early prison studies that were referenced in the opening paragraph, further research can be used to highlight harmful conditions in modern prisons and to ultimately support reform efforts.

Prison research can also help to better understand afflictions that are common among prisoners (Brewer-Smyth 2008) and to inform responses to other problems that extend beyond prison walls. The fact that most prisoners will ultimately be released is typically an afterthought, yet the sheer number of individuals who now cycle in and out of correctional facilities puts myriad strains on the community (Petersilia 2009). For instance, returning prisoners bring home high rates of infectious disease upon release (Bick 2007; Brewer-Smyth 2008) and have ongoing struggles with addiction and drug-related offending (Hser et al. 2007). Insights gained from doing research with prisoners can potentially guide efforts to alleviate these kinds of problems.

In many jurisdictions, crime policy has increasingly become more interlaced with social policy. Accordingly, whereas prison researchers have historically been criminologists and sociologists, the current nature of incarceration now brings researchers from a broader range of health and social sciences into prisons to conduct research. Whether the purpose is to expose oppressive prison conditions (Ross and Richards 2003), to conduct epidemiological research (Johnson et al. 2015), to examine romantic relationships between prisoners and those on the outside (Comfort 2008), or to pursue some other objective, this is a dynamic time for prison-based research. Moreover, although ethnography has traditionally been the conventional research strategy used in prison environments, other face-to-face methods including surveys (Sutton 2011), mixed method approaches (Harvey 2007; Jenness 2010), and focus groups (Naylor 2015) are now also being used more frequently (see also ► Chaps. 4, "The Nature of Mixed Methods Research," ► 27, "Institutional Ethnography," and ► 33, "Epidemiology,").

Simply put, the backgrounds, objectives, and methods of prison researchers today are more diverse than they were in the past, and recent incarceration trends are likely to sustain the current resurgence of prison research into the future. Toward this end, this chapter provides an overview of common issues that emerge when doing face-to-face research with prisoners in prisons. As has typically been the case within the literature, the focus here is primarily on issues that have been pertinent to prison research conducted in the USA and other English-speaking nations (see Cunha 2014, for an examination of prison ethnography in a broader global context).

A number of ethical and practical considerations are introduced, and a case for reflectively engaging in the research process is put forth. One of the reasons for writing reflectively about carrying out research is that it can potentially be helpful for future researchers (Sutton 2011; Jewkes 2012). The intended audience for this chapter is, therefore, novice researchers from a range of health and social sciences, although more seasoned prison researchers will likely benefit from this overview as well. Although ethnographic themes are prevalent, this chapter is ultimately geared toward those who engage in any form of face-to-face research. A guiding premise is that the unique and sensitive nature of studying prisoners has implications for all researchers, irrespective of the particular methods they might use. I turn now to outlining fundamental features of prison contexts that researchers should be mindful of.

2 Contextualization of Research with Prisoners in Prisons

Prisons are complex environments, and prisoners' lives are complicated. Each of these dynamics poses unique research challenges in its own right, and when they converge the difficulties are further compounded. Downing et al. (2013, p. 493) propose "distinguishing between foreground and background dynamics. Foreground dynamics involve the physical, practical elements of a research space…and background components involve the lived realities of respondents." Put another way, those doing face-to-face research with prisoners must take the natures of prisons and prisoners' lives into account when developing their research designs, collecting data, and contextualizing their findings.

2.1 The Nature of Prisons

Researchers working in prisons need to be cognizant of forces that operate on the organizational and meso levels within these environments. Prisons are total institutions, which can "be defined as a place of residence and work where a large number of like-situated individuals, cut off from the wider society for an appreciable period of time, together lead an enclosed, formally administered round of life" (Goffman 1961, p. xiii). Aside from prisons, other common examples of total institutions include mental hospitals and immigrant detention centers. Regardless of the specific organizational context, inmates in these kinds of settings similarly become subservient to the routines and external social controls that structure their daily lives.

Prisons are ultimately bureaucratic organizations that prescribe positive behaviors, proscribe negative conduct, and comprise webs of interactions based on formal statuses and roles (Hart 1995). Accordingly, as is the case with other bureaucratic organizations, prisons are often experienced as impersonal, disempowering entities by the individuals who are found within them. The organizational practices and routine activities of those who work in total institutions reinforce us/them divisions between staff and prisoners, which often lead to inmates being treated as though they were objects rather than humans (Goffman 1961). Within this context, imprisonment and detachment from free society result in pain and deprivation, as inmates lose the privacy, agency, security, and intimacy with others that are often taken for granted in the broader community (Sykes 1958).

These preceding observations establish a conceptual foundation from which incarceration can be better understood. There are also a number of more specific stressors commonly found in prisons that vary in magnitude from one institutional setting to the next. For instance, prisons can be barren, noisy, and musty places, which can be arduous for prisoners and staff to endure over time. Poor confinement conditions, such as inadequate ventilation and overcrowding, result in further discomforts and in some cases can even facilitate the spread of disease (Hoge et al. 1994; Coninx et al. 2000). When compared to other kinds of research settings, prisons feature more pronounced effects of the research environment on the researched given that they are closed systems that restrict movement (Johnson 2015).

The most challenging stressor for many prisoners is navigating the inmate social order. Clemmer (1940) and Sykes (1958) were the first scholars to outline the central tenets of the inmate code, which tend to be organized around toughness, antagonism toward prison staff, and the enforcement of the unofficial normative systems developed by prisoners. Violence, victimization, gang activity, and other forms of deviance are now often intertwined with the inmate social order, and hence they pervade the public spaces within many correctional facilities (Johnson 2002). Inmates learn how to survive in the face of these complex forces and are otherwise socialized into the prison culture through the process of "prisonization" (Clemmer 1940). The nature of prisons and the prisonized ways in which inmates adapt can, therefore, add further complication to prisoners' lives.

2.2 The Nature of Prisoners' Lives

Prisoners are diverse, and they end up in prison for a variety of reasons. Nonetheless, for prisoners as a group, there are a number of recurring themes in their lives that put them at greater risk for involvement with the criminal justice system and related problems. Those who do prison research should, therefore, be equipped with a working sense of what prisoners' lives entail.

Generally speaking, cumulative social disadvantage is prominent within the prison population. Prisoners as a group feature high likelihoods of coming from impoverished communities and families, have low levels of educational attainment, face poor job prospects, experience stigma in the community, tend to come from disadvantaged class and race backgrounds, and are often low in cultural capital (Western and Pettit 2010). Many have also been physically and/or sexually abused when they were children (Brewer-Smyth 2008), suffer from serious mental illnesses (Fazel and Seewald 2012), and contend with physical illnesses and diseases (Bick 2007; Johnson et al. 2015). Substance abuse is a common coping strategy for prisoners (Brewer-Smyth 2008). It is, therefore, unsurprising that a systematic review of prior studies found that 10–48% of male prisoners had drug problems upon coming to prison, which is a higher rate than is found among the general population (Fazel et al. 2006).

Several foreground and background characteristics that are common in prisons and prisoners' lives have now been presented. It is important to recognize that, in practice, these dynamics are often mutually reinforcing. For instance, a preexisting mental illness might be aggravated by the routine procedures of total institutions. Or, new mental anguish may emerge upon experiencing the deprivations of confinement for the first time or after having been victimized by other inmates. Similarly, racial or other tensions might be triggered by the stresses of living in overcrowded facilities or by adherence to precepts of the inmate social order. The nature of these dynamics and intricacies of their interconnections illuminate some of the reasons why prisoners can be sensitive populations to study and prisons can be sensitive settings in which to study them.

2.3 Sensitive Contexts for Researchers

Whether unnoticed or blatant, the patterned interactions, power hierarchies, objectification, and other features of total institutions are ubiquitous when doing face-to-face research with prisoners in prisons. Those who have yet to do this kind of research may have trouble fully appreciating the ways in which these dynamics come into play when collecting data. Several examples from both my own work and the works of other researchers are, therefore, provided in this section to further underscore the uniqueness of doing research in these settings.

As formal bureaucratic organizations, prisons prioritize their own organizational goals over the objectives of researchers (Hart 1995; Sutton 2011; Sloan and Wright 2015; see also ▶ Chap. 125, "Police Research and Public Health"). Preserving institutional security is the primary mission for those charged with operating prisons (Hart 1995; Wakai et al. 2009; Beyens et al. 2015), and toward this end schedules, rules, and routines are regularly adhered to in order to maintain order. This can have unintended consequences for researchers who are trying to complete crucial steps in the research process. A few examples of how organizational practices can obstruct meetings with interviewees are instructive.

When doing research in US prisons, I frequently encountered challenges upon my arrival due to the front desk staff not having received the correct paperwork to grant me admittance (Sutton 2011). I had little recourse in these instances. Staff members at the entrance were similarly powerless, even though in many instances these same staff members knew what I was doing, knew that my project had been approved by the prison system, and had allowed me in just the day before and on dozens of other occasions. Another researcher found that "inmates are reluctant to participate [in research] if it means they will miss the opportunity to participate in work or educational programming for the day...this can result in refusal to participate altogether" (Downing et al. 2013, p. 488).

These two examples demonstrate how research is secondary to and often trumped by organizational routines and procedures. Prisons differ from other sites in that the researcher is dependent on the organization when carrying out the research (Newman 1958). While one might correctly observe that researchers who do field work in other settings lack control over the situations that emerge during the research process, researchers in prisons and other total institutions are unique in that they actually cede control to a more powerful actor in order to accomplish integral research tasks.

Sometimes, the inherent power disadvantage between researchers and staff manifests itself in surprising ways. When doing research in a prison, the individuals within the organization that are familiar with the researcher's project and who grant approval are usually administrative staff who are removed from the prison's daily activities. The prison employees that researchers usually interact with on a regular basis typically have little understanding of the researcher's protocol or research procedures in general. This can produce unique and unexpected challenges.

For instance, staff members might take the initiative to arrange meetings for the researcher with prisoners that the researcher had not been planning to talk with

(Naylor 2015). In my own research, I had a few times when staff members suggested that I interview prisoners that they thought could be interesting. On other occasions, they arranged to have multiple people come to meet with me at the same time to help with efficiency, despite the fact that my project required me to meet with interviewees individually. These kinds of seemingly innocuous incidents can result in tensions if staff members who are genuinely trying to help feel that their efforts are unappreciated or if prisoners feel disrespected due to their routines being unnecessarily disrupted. These examples underscore the inherent lack of control that researchers have over the research process when they cede power to the institution, irrespective of the intentions of the individuals who then enact that power.

Given the emphasis on order in prison, an implication for researchers would be to take efforts to fit into the routine. However, this can be complicated in practice due to us/them dichotomies between staff and inmates that are inherent in total institutions (Goffman 1961; Sutton 2011). Within a closed, formally structured system, where do researchers who are neither "us" nor "them" fit in? As previously noted, researchers are dependent on staff to facilitate access to prisoners and assist with other logistics. Researchers, therefore, need to maintain positive relations with staff, although this can potentially be detrimental to the research if prisoners perceive the researcher to be on the side of the prison. At the same time, researchers need to establish trust with inmates that they are collecting information from, which can hinder working relationships with staff who view the researcher as being on the side of prisoners.

Balancing loyalties while maneuvering within the power structure of a total institution can be trying for those who do face-to-face research in prison. However, the most difficult part of collecting data in these settings for many researchers is being exposed to prisoners' lived realities firsthand. I have previously written about my own experiences of feeling powerless when seeing inmates who were isolated from free society being dehumanized within a total institution (Sutton 2011). Being unable to provide direct assistance to people who are in need can also be distressing, as Bosworth and Kellezi (2016) experienced when doing research in an immigrant detention facility in Britain.

Those who are preparing to do research with prisoners should expect to encounter a diverse group. Whereas some prisoners will not exhibit any visible indicators of pain, there will be others whose problems are immediately clear. Demeanor, facial expressions, and disposition often hint at foreground stressors associated with incarceration. I have also done interviews with a handful of prisoners who had bruises, self-cutting scars, and other embodiments of foreground and background trauma. Additional inmate challenges such as language barriers, being illiterate, being low in cultural capital, and being under the influence of psychotropic drugs tend to become apparent through interaction. In sum, the cumulative disadvantage commonly found among the prisoner population is often easy to see when doing face-to-face research in prison.

An issue that has not yet been addressed is the possibility of danger to researchers in prison. Brewer-Smyth (2008) notes that three potential harms that researchers might encounter in prison are disease, physical violence, and psychological distress when being exposed to prisoners' pain. She concludes that the likelihood of experiencing psychological distress is much higher than the chances of contracting a disease or being assaulted. There are certainly many dangerous people in prison, although violence toward researchers is not among the prevalent themes in the literature. But with this said, Lonnie Athens was once intentionally locked in a cell with a dangerous prisoner who was not on his interview list (Rhodes 1999, p. 51). Athens was set up by staff in this particular instance, which is an extreme example of how researchers can encounter problems when navigating us/them divisions in prison.

As should now be clear, doing face-to-face research in prison is a visceral experience (Jewkes 2012). The selected examples that have been provided up to this point are in no way exhaustive, but they are nonetheless indicative of the types of dynamics that emerge when doing research in a complex environment with people who have complicated lives. Furthermore, it must be noted that many of these dynamics occur simultaneously. It is, therefore, crucial for researchers to have a sense of these themes prior to carrying out their research. Researchers should also have an understanding of common ethical dilemmas that emerge when engaging with the sensitivities associated with prisons and prisoners' lives.

3 Ethical Considerations

Prisoners are vulnerable to coercion given their inherent powerlessness within total institutions. There is a long history of prisoners being exploited by researchers and used in experiments that would now be considered unethical by today's standards, including high-profile abuses that were carried out under the Nazi regime (Wakai et al. 2009; Cislo and Trestman 2013). Prisoners have often served as a convenient pool of research subjects over the years. For instance, in the mid-1900s prisoners were used in the majority of trials for new drugs leading up into the 1970s (Cislo and Trestman 2013).

The 1970s marked a pivotal time for research, as formal guidelines and review boards increasingly emerged to protect the rights of prisoners and other research subjects (see also ▶ Chap. 106, "Ethics and Research with Indigenous Peoples"). Some have argued that prisoners have now actually become "overprotected" and "understudied" in more recent years as a result (Wakai et al. 2009; Cislo and Trestman 2013; Johnson et al. 2015). The U.S. Department of Health and Human Services released the Belmont Report in 1979, which is a document that specifies ethical principles and guidelines for protecting human subjects when doing research (US Department of Health and Human Services 1979). In spirit, the Belmont Report is consistent with the Nuremberg Code that was prepared during the time of the Nuremberg trials (Cislo and Trestman 2013). In practice, these two documents feature several overlapping recommendations for conducting ethical research.

The Belmont Report stipulates that individuals who are being studied must be made aware of the potential risks and benefits of their research participation. They must ultimately give informed consent before participating, and their participation must be voluntary. Moreover, consent must be free of undue influences or rewards that could potentially compel participant acquiescence, and participants must additionally be given the right to withdraw from participating at any time. The Belmont Report recognizes that the ability to agree to participate in research may be affected by factors such as age, developmental maturity, and incarceration, and it, therefore, calls for special efforts to be taken towards protecting prisoners, children, and other vulnerable populations. Formal institutional review boards now routinely screen research proposals to ensure that they comply with the stipulations outlined in the Belmont Report and other ethical standards (see also ▶ Chap. 106, "Ethics and Research with Indigenous Peoples").

3.1 Formal Review Boards

Ethical standards for research in the USA and many of its peer nations in the English-speaking world are typically maintained by formal review boards that operate within universities and similar settings in which researchers work. Institutional review boards in the USA are required to have a prisoner or prisoner representative participate when proposals involving prison research are reviewed (Brewer-Smyth 2008; Wakai et al. 2009). University researchers who work with human subjects are likely accustomed to securing review board approval from their places of employment. However, some may not realize that when doing prison research, there is typically a second formal review board administered by the prison system that will also need to grant approval before research can be conducted (Wakai et al. 2009). Prison researchers must, therefore, satisfy multiple gatekeepers in order to conduct research in prison (Naylor 2015).

Moreover, some funding agencies may require additional steps before research can commence. For instance, researchers doing studies supported by the U. S. Department of Health and Human Services (DHHS) are required to get a DHHS prison certificate that verifies that the project adheres to DHHS standards (Brewer-Smyth 2008). Along these lines, US researchers can also apply to the National Institutes of Health for a Certificate of Confidentiality (Wolf et al. 2004; Brewer-Smyth 2008). Certificates of Confidentiality legally protect researchers from having to provide identifying information about their research participants to authorities, and researchers' applications for Certificates of Confidentiality are rarely rejected in practice (Wolf et al. 2004). In my own research, I have found that having a Certificate of Confidentiality helps prisoners feel more comfortable about their participation.

A dilemma that often comes up in prison research is whether or not to compensate participants. This issue raises concerns about undue influence as spelled out in the Belmont Report. Monetary compensation is generally not permitted, nor are prisoners allowed to use their research participation to garner favor in parole hearings (Wakai et al. 2009; Waldram 2009). In my own research, I had hoped to be able to provide participants with a candy bar or similar item from the prison vending machine to thank them for their time. However, this request was not allowed by the prison's review board. As Brewer-Smyth (2008, p. 123) correctly observes, shampoo and other "low-cost items that can be purchased for less than $5 per inmate may seem to be so great to the inmate that his or her ability to weigh the associated risks becomes impaired because of the restricted prison environment." Setting concerns about undue influence aside, providing compensation to participants might also cause management problems for prisons. For instance, institutional security could potentially be threatened if prisoners who are not invited to participate in research perceive that other inmates are receiving special treatment.

Whereas formal review boards help researchers ensure that their projects are in compliance with ethical standards, their demands may inadvertently introduce methodological challenges to research designs. For instance, review boards may ultimately determine, and thereby limit, the specific institutions that researchers visit or the actual prisoners that researchers interact with. For quantitative researchers, this can hinder probability sampling and, therefore, jeopardize generalizability. Qualitative researchers face their own challenges. For instance, review boards often require that researchers submit a list of questions that will be asked before a project can be approved. However, this demand is problematic for phenomenological researchers who rarely know the exact direction that their research will take prior to entering the field (Easterling and Johnson 2015).

3.2 Beyond Formal Review Boards

The formalized standards of review boards are reified through their enforcement and the compliance that they command. One result of this is that newer researchers may assume that ethical considerations have been satisfied once review board standards have been met. This is an erroneous assumption. When doing face-to-face research with prisoners in prisons, a number of additional, unanticipated ethical challenges often emerge that fall beyond the purview of the review board process.

For instance, it was previously noted that some prison staff may take it upon themselves to help researchers without first consulting with them. This potentially raises an ethical dilemma given that these staff members are unlikely to follow proper researcher protocol when attempting to help with research tasks. It was also previously noted that prisoners as a group are low in cultural capital. One might ponder whether this poses an ethical dilemma if participants have a limited understanding of the research process. Bosworth and Kellezi (2016) struggled with this issue when realizing that the detained immigrants that they studied had "unrealistic expectations" about the extent to which the researchers could help them.

I have previously written about the possibility that loneliness might constitute an undue influence when doing prison research (Sutton 2011). Prisons, and therefore prisoners, are often located far from prisoners' home communities, which can result in prisoners having few or no visitors from the outside world (Sutton 2011; Downing et al. 2013). Prisons are also dull and monotonous places (Martin 2000). As has already been shown, it is assumed that items like candy bars and small bottles of shampoo can influence one's ability to make a free decision to participate in research within the restrictive prison environment. It, therefore, seems plausible that opportunities to interact with researchers could similarly compel participation from prisoners who are regularly deprived of stimulation and contact with individuals who are not of the prison.

These select examples are intended to demonstrate that the ethics of doing research with prisoners in prisons are not always clear, nor can every ethical consideration be addressed prior to actually carrying out the research. Moreover, good ethical practice in nonprison settings does not always serve researchers and participants well within total institutions. For instance, administering written consent forms to participants is a standard practice that is typically required when gaining informed consent, with the researcher and participant then each retaining a copy of the signed consent form for their records (see ▶ Chap. 106, "Ethics and Research with Indigenous Peoples"). However, the detained immigrants who participated in Bosworth and Kellezi's research were often uncomfortable signing their names (Bosworth and Kellezi 2016). Naylor (2015) encountered similar challenges and observed that having a signed consent form can potentially put one at risk within the prison environment.

It is imperative that prison researchers think critically about the ethics of what they do on an ongoing basis, and they should take special efforts to consider how the natures of prisons and prisoners' lives pose unique ethical challenges. Formal guidelines and review boards help with addressing many crucial concerns. Yet, they do not anticipate every ethical dilemma that will emerge when doing research. Moreover, in some cases, they may even have problematic applications when their stipulations are applied to face-to-face research with prisoners in prisons. For these reasons, it is crucial for those who do this kind of research to be flexible and reflectively engaged in their work.

4 Doing Reflectively Engaged Prisoner Research

As the previous sections convey, conducting face-to-face research with prisoners is challenging and unpredictable. A number of sensitive issues arise when doing this kind of research, including unanticipated ethical dilemmas. It is beyond the scope of this chapter to present a comprehensive list of suggestions that addresses all of the considerations one must account for when doing prisoner research (see King and Liebling 2007 and Sutton 2011 for thorough lists of practical recommendations). This chapter instead advocates more broadly for adopting a reflectively engaged approach when doing research in prisons.

To the extent that ethics, researcher presentation of self, emotions, and similar themes have been examined in the prison research literature, they have primarily been framed as considerations for ethnographers. Reflexivity, which "involves developing a consciousness of one's self in the process of research" (Drake et al. 2015, p. 11) similarly tends to connote qualitative field research. Yet, in practice, those who do other forms of face-to-face research with prisoners, including quantitative self-report surveys, mixed-method projects, and focus groups, also deal with these themes (Sutton 2011). Being reflectively engaged with the setting, the research process, and oneself can, therefore, benefit any researcher who studies prisoners in prisons, regardless of the substantive goals of his or her study or the particular research methods that he or she employs. The following explorations of strategies for managing boundaries and emotions model themes of reflective engagement when doing prison research.

4.1 Managing Boundaries and the Presentation of Self

The question of where researchers who are neither "us" nor "them" fit into us/them dichotomies was posed earlier in this chapter. The answer will likely depend on factors such as the goal of the project, if one is studying prisoners, staff, or both, and whether one is doing ethnography or having one brief interaction with interviewees. Setting these kinds of considerations aside, some helpful insights can be derived from examining how other researchers have balanced this tension.

One strategy for managing these boundaries is to take turns focusing on and connecting with each side (Liebling 2001; Beyens et al. 2015). This approach might entail embracing each side as situations dictate (Nielsen 2010), or a researcher could start with one side and then shift to the other (Beyens et al. 2015). An advantage of managing boundaries by taking both sides at various points would be having an opportunity to gain multiple perspectives, while downsides would include the effort this would entail and the potential for backlash from the first side after transitioning to the second.

Another strategy is to embrace an outsider status (King 2000). I have taken this approach in my own research, and doing so allowed me to gain invaluable insights (Sutton 2011). I took the advice of other researchers and emphasized the fact that I was not connected to the prison system (Newman 1958). For instance, I made sure to engage in acts such as waiting for escorts to ensure that my outsider status was clear (King 2000; Johnson 2015).

I also emphasized the fact that I was there to learn rather than to advocate for either prisoners or staff (King 2000). Toward this end, I introduced myself as a researcher rather than a criminologist because I feared that the word criminologist might have a negative connotation. When researching detained immigrants, Bosworth and Kellezi (2016) encountered respondents that were put off by the fact that they were being studied by criminologists despite not having done anything wrong. The criminologist title implies law enforcement to some and prisoner advocate to others, so I found that embracing the university researcher label worked best. Others have similarly noted that being connected to a university increases the likelihood of being taken seriously by prisoners and staff in prison settings (Newman 1958; Jacobs 1977; King 2000; Martin 2000).

Within a closed system that is structured around power hierarchies and formal roles, researchers will inevitably be categorized by those who inhabit the system. Being reflectively engaged in the research process can, therefore, enable researchers to exercise more agency over how they are ultimately placed. Moreover, being reflectively engaged can also help ensure that researchers select the most fitting presentation of self for their purposes given that what constitutes best practices in this line of research is often situational. Having to manage these kinds of boundaries is among the reasons why establishing connections to those being researched in prisons can be especially difficult when compared to doing research in other environments (Brewer-Smyth 2008; Waldram 2009).

4.2 Emotional Management

Along with managing boundaries and impressions, those who do prison research also need to engage in emotion management when they are in the research setting and potentially when they are thinking about their research at other times (Nielsen 2010; Sutton 2011). In recent years, there have been several written accounts of the emotional experiences of field researchers who work in prisons. This is a positive trend given that future researchers can learn from reading about the emotional experiences of those who have come before them (Liamputtong 2007; Sutton 2011; Jewkes 2012; see also ▶ Chap. 123, "Emotion and Sensitive Research").

Spending time in a prison can result in intense emotional experiences (Wacquant 2002; Jewkes 2012). It is, therefore, crucial for researchers to remain reflectively engaged in what they are doing because intense emotion can cause distress when researchers are exposed to trauma and other unsettling realities found in prison (Brewer-Smyth 2008). Moreover, researchers who react to their emotions when doing face-to-face research in total institutions also run the risk of making their interviewees uncomfortable. For instance, an intense reaction to something shared may cause the interviewee who shared it to stop talking (Bosworth and Kellezi 2016). An effective strategy for reflective engagement on an emotional level is to have regular conversations with close colleagues who can provide support, give advice, and identify instances where the researcher has become too immersed (Sutton 2011; Liebling 2014; Beyens et al. 2015). On a final note, researchers should ultimately keep in mind that their exposures to these settings are temporary and voluntary and come with the option to withdraw at any time, which is not the case for prisoners and staff (Drake et al. 2015).

5 Conclusion and Future Directions

This chapter's primary goal has been to make the process of doing research with prisoners in prisons accessible for a general health and social sciences audience. It has provided overviews of the nature of prisons and prisoners' lives while highlighting research sensitivities and exploring ethical implications. A reflectively engaged orientation has ultimately been endorsed for the purposes of helping researchers identify, manage, and resolve the complex and complicated issues that regularly emerge when doing this kind of research.

Although prisons and prisoners are diverse, there are recurring themes when researching prisoners in prisons. For instance, prisons are emotionally charged places and prisoners' lives tend to feature cumulative foreground and background disadvantage. Moreover, structural power dynamics and ethical dilemmas are inherent in research in prisons and other total institutions, and the research process is often unpredictable. Accordingly, aspiring prison researchers should learn about other researchers' experiences as part of a broader effort to be as prepared as possible (Sloan and Wright 2015). Prison researchers should also plan for the unexpected (Sutton 2011), and toward this end, they should be dynamic, spontaneous, and flexible (Martin 2000; Johnson 2015). Finally, and most basically, those who do research with prisoners in prisons should "plan to learn more than [they] envisioned" (Sutton 2011, p. 58).

Two general conclusions about future directions for prison research can now be derived from the preceding sections. First, there will likely be an uptick in research done in prisons given that the World's prisoner population has been growing. If incarceration continues to expand, there will inevitably be more people subjected to the problems associated with imprisonment. This should in turn bring researchers from an increasingly broad range of disciplines into prisons as they work to develop solutions.

Second, and also relative to the past, future prison research will be more diverse in terms of researchers' backgrounds, objectives, and methods. Research in prison has its origins in the ethnographic tradition and in fields such as criminology and criminal justice. These methodological and disciplinary boundaries will need to become more encompassing in order to better illuminate harmful conditions and facilitate penal reform. This brings me to a final conclusion, which is that researchers who study prisoners in prison will need to increasingly focus on identifying common challenges, refining best practices, and engaging in scholarly dialogue as they carry out this work.

References

Beyens K, Kennes P, Snacken S, Tournel H. The craft of doing qualitative research in prisons. Int J Crime, Justice Social Democr. 2015;4(1):66–78.
Bick JA. Infection control in jails and prisons. Clin Infect Dis. 2007;45(8):1047–55.
Bosworth M, Kellezi B. Doing research in immigration removal centres: ethics, emotions and impact. Criminol Crim Justice. 2016:1–17. https://doi.org/10.1177/1748895816646151.
Brewer-Smyth K. Ethical, regulatory, and investigator considerations in prison research. Adv Nurs Sci. 2008;31(2):119–27.
Cislo AM, Trestman R. Challenges and solutions for conducting research in correctional settings: the U.S. experience. Int J Law Psychiatry. 2013;36:304–10.
Clemmer D. The prison community. Boston: Christopher; 1940.
Comfort M. Doing time together: love and family in the shadow of the prison. Chicago: University of Chicago Press; 2008.
Coninx R, Maher D, Reyes H, Grzemska M. Tuberculosis in prisons in countries with high prevalence. Br Med J. 2000;320(7232):440–2.
Cunha M. The ethnography of prisons and penal confinement. Annu Rev Anthropol. 2014;43:217–33.
Downing S, Polzer K, Levan K. Space, time, and reflexive interviewing: implications for qualitative research with active, incarcerated, and former criminal offenders. Int J Qual Methods. 2013;12:478–97.

Drake DH, Earle R, Sloan J. General introduction: what ethnography tells us about prisons and what prisons tell us about ethnography. In: Drake DH, Earle R, Sloan J, editors. The Palgrave handbook of prison ethnography. New York: Palgrave MacMillan; 2015. p. 1–16.

Easterling BA, Johnson EI. Conducting qualitative research on parental incarceration: personal reflections on challenges and contributions. Qual Rep. 2015;20(10):1550–67.

Fazel S, Seewald K. Severe mental illness in 33,588 prisoners worldwide: a systematic review and meta-regression analysis. Br J Psychiatry. 2012;200(5):364–73.

Fazel S, Bains P, Doll H. Substance abuse and dependence in prisoners: a systematic review. Addiction. 2006;101:181–91.

Fishman JF. Sex in prison: revealing sex conditions in American prisons. New York: National Library Press; 1934.

Goffman E. Asylums: essays on the social situation of mental patients and other inmates. Garden City: Anchor Books; 1961.

Hart CB. A primer in prison research. J Contemp Crim Just. 1995;11:165–75.

Harvey J. An embedded multimethod approach to prison research. In: King RD, Wincup E, editors. Doing research on crime and justice. 2nd ed. Oxford: Oxford University Press; 2007. p. 487–99.

Hoge CW, Reichler MR, Dominguez EA, Bremer JC, Mastro TD, Hendricks KA, Breiman RF. An epidemic of pneumococcal disease in an overcrowded, inadequately ventilated jail. N Engl J Med. 1994;331(10):643–8.

Howard J. The state of prisons in England and Wales. Warrington: William Eyres; 1780.

Hser Y, Longshore D, Anglin MD. The life course perspective on drug use: a conceptual framework for understanding drug use trajectories. Eval Rev. 2007;31(6):515–47.

Jacobs JB. Stateville: the penitentiary in mass society. Chicago: University of Chicago Press; 1977.

Jenness V. From policy to prisoners to people: a "soft-mixed methods" approach to studying transgender prisoners. J Contemp Ethnogr. 2010;39(5):517–53.

Jewkes Y. Autoethnography and emotion as intellectual resources doing prison research differently. Qual Inq. 2012;18(1):63–75.

Johnson R. Hard time: understanding and reforming the prison. 3rd ed. Belmont: Wadsworth; 2002.

Johnson J. Researching in prison: first hand encounters of a first-time prison researcher. Te Awatea Rev. 2015;12(1):15–20.

Johnson ME, Kondo KK, Brems C, Eldridge GD. HIV/AIDS research in correctional settings: a difficult task made even harder? J Correct Health Care. 2015;21(2):101–11.

Kaeble D, Glaze L, Tsoutis A, Minton T. Correctional populations in the United States, 2014. Washington, DC: Bureau of Justice Statistics (BJS), US Department of Justice, & Office of Justice Programs; 2015.

King RD. Doing research in prisons. In: King R, Wincup E, editors. Doing research on crime and justice. New York: Oxford University Press; 2000. p. 285–312.

King RD, Liebling A. Doing research in prisons. In: King R, Wincup E, editors. Doing research on crime and justice. 2nd ed. New York: Oxford University Press; 2007. p. 431–51.

Liamputtong P. Researching the vulnerable: a guide to sensitive research methods. London: Sage; 2007.

Liebling A. Whose side are we on?: theory, practice and allegiances in prisons research. Br J Criminol. 2001;41:472–84.

Liebling A. Postscript integrity and emotion in prisons research. Qual Inq. 2014;20(4):481–6.

Martin C. Doing research in a prison setting. In: Jupp V, Davies P, Francis P, editors. Doing criminological research. Thousand Oaks: Sage; 2000. p. 215–33.

Naylor B. Researching human rights in prisons. Int J Crime Justice Soc Democr. 2015;4(1):79–95.

Newman DJ. Research interviewing in prison. J Crim Law Criminol Police Sci. 1958;49(2):127–32.

Nielsen MM. Pains and possibilities in prison: on the use of emotions and positioning in ethnographic research. Acta Sociol. 2010;53(4):307–21.

Petersilia J. When prisoners come home: parole and prisoner reentry. New York: Oxford University Press; 2009.

Reiter K. Making windows in walls: strategies for prison research. Qual Inq. 2014;20(4):417–28.

Rhodes R. Why they kill: the discoveries of a maverick criminologist. New York: Vintage Books; 1999.

Ross JI, Richards SC. Introduction: what is the new school of convict criminology? In: Ross JI, Richards SC, editors. Convict criminology. Belmont: Wadsworth; 2003. p. 1–14.

Sloan J, Wright S. Going in green: reflections on the challenges of "getting in, getting on, and getting out" for doctoral prisons researchers. In: Drake DH, Earle R, Sloan J, editors. The Palgrave handbook of prison ethnography. New York: Palgrave MacMillan; 2015. p. 143–63.

Sutton J. An ethnographic account of doing survey research in prison: descriptions, reflections, and suggestions from the field. Qual Sociol Rev. 2011;7(2):45–63.

Sykes G. The society of captives. Princeton: Princeton University Press; 1958.

US Department of Health and Human Services. The Belmont report: ethical principles and guidelines for the protection of human subjects of research. Washington, DC: US Department of Health and Human Services; 1979.

Wacquant L. The curious eclipse of prison ethnography in the age of mass incarceration. Ethnography. 2002;3(4):371–97.

Wakai S, Shelton D, Tretman RL, Kesten K. Conducting research in corrections: challenges and solutions. Behav Sci Law. 2009;27:743–52.

Waldram JB. Challenges of prison ethnography. Anthropol Newsl. 2009;50(1):4–5.

Walmsley R. World prison population list. 10th ed. London: King's College London International Centre for Prison Studies; 2013.

Western B, Pettit B. Incarceration & social inequality. Daedalus. 2010;139(3):8–19.

Wolf LE, Zandecki J, Bernard L. The certificate of confidentiality application: a view from the NIH institutes. IRB Ethics Hum Res. 2004;36(1):14–8.

Police Research and Public Health

Jyoti Belur

Contents

1 Introduction 2180
2 Researching Public Health in Policing 2181
3 Unpicking the "Hard-to-Reach" Nature of Police Organizations: Barriers to Police Research 2183
   3.1 Cultural Barriers 2183
   3.2 Divergent Aims 2184
   3.3 Lack of Trust and Cross-Cultural Communication 2184
4 "Who Is the Researcher" and "What Is Being Researched"? Important Questions in Police Research 2185
5 Moving from the Personal to the General: Lessons Learnt from Doing Police Research 2187
   5.1 Negotiating Access 2189
   5.2 Agreement on Research Objectives 2190
   5.3 Establishing Trust 2190
   5.4 Knowledge Exchange and Dissemination 2191
6 Conclusion and Future Directions 2192
References 2193

Abstract

This chapter examines the challenges involved in conducting face-to-face research with police officers with respect to their work in sensitive areas or with hard-to-reach populations. The intersection between law enforcement and public health is drawing greater academic attention through the concept of harm reduction policing (Ratcliffe 2015). The refocusing of police attention from concentrating purely on crime to reducing harmful effects on individuals and the community has created an emerging social space that needs further exposition.

J. Belur (✉)
Department of Security and Crime Science, University College London, London, UK
e-mail: j.belur@ucl.ac.uk

© Springer Nature Singapore Pte Ltd. 2019
P. Liamputtong (ed.), *Handbook of Research Methods in Health Social Sciences*,
https://doi.org/10.1007/978-981-10-5251-4_138

The changing nature of policing and a rise in societal demands for security has increased the overlap between law enforcement and public health. This chapter lays down some basic guidelines to aid researchers operating in this space based on the author's personal experience and drawing upon the research experience of others working in this area. It begins by identifying fundamental features of policing culture that make it a sensitive organization to access and research. It then discusses the difficulties in approaching gatekeepers and negotiating access to police data and individual officers. The discussion then focuses on the experience and implications of researching the police organization as an insider and/or as an outsider (Brown 1996). Finally, ethical and practical considerations involved in carrying out this kind of research are discussed.

Keywords

Public health research · LEPH · Police research · Access negotiation · Research aims · Researcher status

1   Introduction

This chapter examines the challenges involved in conducting research in areas where policing and public health protection intersect. The overlap between law enforcement and public health is drawing greater academic attention through the concept of harm reduction policing (Ratcliffe 2015). The refocusing of police attention from concentrating purely on crime to reducing harmful effects on individuals and the community has created an emerging social space that needs further exposition. The changing nature of policing and a rise in societal demands for security have increased the overlap between law enforcement and public health. The domains of policing and public health intersect at the point where the state is concerned with the promotion of healthy lifestyles, protection of the vulnerable, and injury prevention. Securing certain aspects of the public health agenda depends on co-operation from the police and other arms of law enforcement. The police, in turn, need public health systems and structures to deal with complex social issues involving prevention of violence, injury, and social harms. The key elements uniting both disciplines are the shared objectives to protect the public and promote harm reduction.

However, this interdependence has not always been recognized by the police or public health authorities. Traditionally, the police have concerned themselves with patrol, investigation and detection of crime, maintenance of law and order, and public security. Protecting and promoting public health has not been high on their agenda of priorities or their self-identity (Van Dijk and Crofts 2016), even though they are responsible for preventing harm and securing well-being of populations, which, among other things, is squarely within the purview of public health (WHO 1948). Recently, there has been a spate of arguments for the two fields to recognize their interdependence in protecting the population's well-being and there are calls for both communities to work together as frontline organizations in providing direct interventions to securing health and ameliorating risky behaviors (see Van Dijk and Crofts 2016 for a summary). The police mission has expanded to include crime control and order maintenance with social service (Millie 2013). Concurrently, increasing focus on "victims" and "vulnerable people" for law enforcement has brought a whole range of public health related issues such as domestic violence, violence in public places, human trafficking, prostitution, and mental health under the purview of policing policy and practice (Van Dijk and Crofts 2016). Conceivably, the concept of vulnerability will bridge the existing gap between health and police sciences (Bartkowiak-Théron and Asquith 2016; see also ▶ Chap. 110, ""With Us and About Us": Participatory Methods in Research with "Vulnerable" or Marginalized Groups").

Evidence indicates that the police can help or hinder access to resources and efforts made by public health organizations and NGOs working, for example, in the area of HIV prevention with sex workers and drug addicts to promote public health addicts (Open Society Foundation 2014). Using a recent example of research on the relationship between public health objectives and policing conduct with respect to drug taking behavior, Hayashi et al. (2013) suggest that the police treatment of drug addicts and drug enforcement policies can affect whether the public health agenda of reducing drug misuse is achieved or not. More often than not, public health and law enforcement agencies work in silos and do not share conceptual frameworks and practice synergies that arise from the expertise, skills, and knowledge of each other's personnel (Bartkowiak-Théron and Asquith 2016).

The intention here is not to reiterate the case for why law enforcement and public health agencies must collaborate, but to suggest that the two fields intersect and, therefore, research at these points of conjunction needs to be interdisciplinary. However, as with practitioners, researchers in both sciences remain experts in their particular domain and, therefore, find it difficult to research the interconnections. This chapter explores challenges for public health researchers working with law enforcement.

2 Researching Public Health in Policing

The crossover between law enforcement and public health, termed as LEPH, is attracting academic and practitioner interest in recent years (Punch and James 2016) but remains a fledgling field of research. More traditional public health-related research in policing mainly explores how police misconduct or negligence actually can cause public health hazards, for example, police shootings (excess use of force), custodial violence, or negligence. Further, a substantial section of public health research on policing issues explores the victim's perspective on police response or behavior focusing on provision of police service to victims of drug abuse (Maher and Dixon 1999; Cooper et al. 2004) or rape (Jameel 2010). There are a few examples of public health research done with police officers exploring police perspectives on public health-related issues which include evaluations of interventions, for example, in cases of domestic violence (Hovell et al. 2006). Even these projects did not generally involve primary research with the police, but working with police records and open-source crime data. Similarly, other public health-oriented research is focused on how the police can aid or mar efforts to improve public health and safeguard community security. The benefits of such cross-disciplinary research seem to be focused on how police actions are currently detrimental to public health (Maher and Dixon 1999; Kerr et al. 2005), and policy implications are discussed in terms of what the police should be doing in order to improve public health.

The benefits of such research for public health are clear. Perhaps this might be the direct result of the fact that public health scholars researching the police are clearly focused on public health outcomes of their research. However, the other side of the equation remains largely neglected, i.e., how can this research be beneficial for the police? This is not to assert that improving public health is not in the interests of the police, but perhaps the dearth of police researchers getting involved in research that has public health implications has meant that these issues are not viewed from the perspective of the police.

It follows that most public health research on the police is in the "critical research tradition," which "prides itself on its detachment and independence from the police as subject matter, and, almost always, manages to find fault with rather than celebrate the role and activities of the public police. It does not seek to directly change the police but to contribute through its expert voice and publications to the thinking of governments and legislators" (Bradley and Nixon 2009, p. 423). Most of this research is "on" the police rather than "with" the police (Cockbain and Knutsson 2015). Unsurprisingly, police leaders are deeply unhappy and nervous about the "spirit and tenor" of such critical research (Young 1991).

The police are traditionally considered to be a closed organization and difficult to access (Skolnick 1975; Punch 1989). Access to organizational knowledge can be of two kinds: documentary data, in the form of official statistics; and documents or experiential data, embedded in surveys and interviews. While on the one hand, the police are heavily bound by restrictions of data protection legislation in most Western democratic countries, limiting both their ability and desire to give access to sensitive data. On the other hand, access to experiential data, in the form of ethnographies or interviews with police officers, is equally difficult. Negotiations have to be conducted not only with gatekeepers in the form of senior management and bureaucracy, but repeatedly with individual participants and place managers in the organization (Punch 1989) (see also ▶ Chaps. 126, "Researching Among Elites," and ▶ 124, "Doing Reflectively Engaged, Face-to-Face Research in Prisons: Contexts and Sensitivities").

Further, there are some natural barriers between academics and the police that make it difficult for them to work together unless they "negotiate," "communicate," and "stand in the other person's shoes" (Fleming 2010) in order to foster better relations. The first section of this chapter discusses some barriers between academia and policing in more general terms. The second section then focuses on different kinds of police-related research and types of researchers that exist to contextualize how this might affect researcher access to the police organization. The third section discusses challenges involved in researching law enforcement based on my personal experience of conducting two police research projects which had public health implications. Finally, I draw out some conclusions for the best way forward to foster an ethical and mutually beneficial police – public health research relationship.

3 Unpicking the "Hard-to-Reach" Nature of Police Organizations: Barriers to Police Research

The research relationship between police and academics is fraught with difficulties. Three main barriers are discussed here: differences in subculture; intended aims; trust and communication.

3.1 Cultural Barriers

The police are traditionally a closed, hierarchical organization and very protective of their patch. Relative resistance to research, a closed culture (Punch 1993; Reiner 2000), and the sensitive nature of many of their operations lend itself to a culture of secrecy, making the police a difficult organization to access for outsiders (Kennedy 2015).

Part of the distrust and suspicion is rooted in ignorance of the true purpose and objective of the research and what is perceived to be "academic waffle" and armchair theorizing by academics with no experience of real-life policing. The urgent nature of their daily work means police priorities are shifting constantly and results are demanded instantaneously. The police like solutions that are unambiguous and decisive (Lum et al. 2012; Rojek et al. 2015). On the other hand, the research community would like to be cautious and careful about their conclusions and are comfortable with uncertainty (Strang 2012). The police are constantly accountable to various stakeholders (the public, politicians, and the press) and are rewarded for delivering concrete actions (Rojek et al. 2015). They, thus, value quick and dirty research results that help them achieve their goals. On the other hand, academics are rewarded for scholarly productivity in the form of academic publications. Prestigious publications are dependent on high-quality research of great methodological rigor and conclusions that are the result of slow and reflective deliberation throughout and after the research, and are necessarily a time-consuming process. Thus, not only are the impetus for conducting research divergent between the two communities, but the mismatch of timescales involved also affects research relevance for practitioners (Weisburd and Neyroud 2011).

Academics and researchers, therefore, need to find common research ground to facilitate an entry route, and continued cooperation requires academics to constantly negotiate with gatekeepers and individuals at entry, as well as throughout the research process. The police are aware that researchers need them more than they need researchers to do their work (Engel and Whalen 2010). Hence, it becomes incumbent upon researchers to take that extra step to understand and appreciate police culture and operational pressures and adapt their research aims accordingly to accommodate police expectations without compromising on the integrity of the research process.

3.2 Divergent Aims

For the police, research would be useful only if it is of direct operational or tactical relevance to them. As discussed earlier, results need to be delivered in the short run and in areas that are priority for the organization or senior leaders (Lum et al. 2012). Thus, the police are more likely to be supportive of research that is action and/or policy oriented and helps them achieve their targets. Research, on the other hand, can be purely for the purpose of developing and testing theory (Madensen and Sousa 2015). Both communities might, therefore, hold different views on what kind of knowledge is valuable and worth investing in (Buerger 2010).

Police research can be a "mirror" that reflects the complexities and dilemmas of police operations and processes with the aim of understanding them better; or it could act as a "motor" to impel change and reform in the organization (Innes 2010). Thus, of the two types of research traditions, critical police research and policy police research (Rojek et al. 2015), researchers involved in the latter type of research, are more inclined to work closely with the police to provide theories, ideas, and evidence for the purpose of improving police practice (Bradley and Nixon 2009).

Past experience indicates that often the association between researchers and practitioners can be unidirectional – researchers have been known to just "take" data without actually giving anything useful back in return to the police (Tompson et al. 2017). "Data robbers" are rarely tolerated by the organization and one unhappy experience with such a researcher can "foul the nest" making it difficult for future researchers to access the organization (Punch 1989).

Finally, while the police are sensitive to negative press coverage, academics, on the other hand, are impelled by the need to be transparent and open in disseminating research findings (Tompson et al. 2017). The police would not want any procedural lacunae or policy shortcomings to be highlighted and consequently make them vulnerable to public criticism or their reputation damaged. Conversely, the wider the public discussion of their research, the more beneficial it is for the researchers' academic reputation. This makes for an uneasy relationship between the two communities since their goals can be divergent.

3.3 Lack of Trust and Cross-Cultural Communication

Some of the subcultural factors and the divergence of goals can be responsible for a lack of trust between police and researchers, unless care is taken to build that trust at the beginning and throughout the research process. The experience of the 'critical police research' era has left the police organization suspicious of researchers. Trust can be gained by ensuring that the research is credible and has clear practical applicability, and the resulting policy or operational recommendations are amenable to implementation (Engel and Whalen 2010), especially in the short run. By acknowledging and respecting police tradecraft and individual expertise, as well as the constraints they operate under, researchers can begin to bridge the gap in trust (Stanko 2007; Brown 2015).

One of the main reasons for lack of trust is due to poor communication between police and researchers. Often, it is said communication between the research community and police practitioners is like the "dialogue of the deaf" (MacDonald 1986) with academics being accused of using language that is arcane and inaccessible to practitioners. Failure to regularly report research progress to practitioners can also give rise to suspicion and impact upon trust. Ultimately, police practitioners prefer short, unambiguous reports with clear action directives which will make their job easier (Rojek et al. 2015). On the other hand, academics tend to use language that is cautious and draw conclusions that are more tentative and circumspect than the police would prefer (Madensen and Sousa 2015). Thus, an appropriate communication strategy, where academics have the freedom to disseminate research findings through appropriate academic outputs and also produce reports for practitioners which use simple, succinct, and direct language, becomes essential for engendering trust between the two communities.

4 "Who Is the Researcher" and "What Is Being Researched"? Important Questions in Police Research

The twin pillars supporting a good research project are the credentials (and abilities) of the researcher and the quality of the project itself, judged by its aims and proposed impact. Thus, understanding what kind of research is valued by, and what type of researcher is more acceptable to, the police organization becomes important for any potential researcher.

Police research is of four kinds: research by the police, research for the police, research on the police, and research with the police (Innes 2010). Research "by" the police is conducted by serving police officers or civilian staff and often remains internal to the organization. Much of this research is rarely, if ever, published for external consumption. Research "for" the police is contracted research conducted by academics or think tanks on specific topics commissioned by the funding police organization – this means researchers are given access to those parts of the organization that are of interest to the police. Research "on" the police, their processes and procedures, that possibly involve irregularities (e.g., research on police violence, deviance, abuse, investigations, and soon) are usually topics that are of primary interest to the researcher. Such research usually identifies lacunae and problem areas in policing. The organization can become difficult to access for this kind of research. Conceivably, public health research on topics involving police response to vulnerable individuals or police actions that can contribute to increasing vulnerability, are more likely to be research "on" the police, giving rise to access issues. Finally, research "with" the police is usually conducted in areas that are of policy and operational importance to the police organization and forms part of the larger impetus toward becoming more evidence-based (e.g., research on what works in crime prevention, efficient resource allocation, and so on). They focus on topics that are of interest to, and therefore, more likely to be supported by, the police. Collaborative research, with the police, gets the best results in terms of genuine knowledge exchange and getting access to aspects of policing otherwise unavailable. Existing examples of successful collaborations of public health-related research done with the police rather than on the police (Reid and Walton 2015) indicates that such projects can be very productive. However, collaboration would presume that the police are equally interested in the research topic and are willing to cooperate not only in terms of giving access, but also possibly by allocating resources.

The challenges involved in getting access to the police have been documented by several researchers (see Reiner and Newburn 2008) and often who the researcher is affects whether and what kind of access they would get. This affects the extent of cooperation a research project is likely to receive from the police organization. Brown (1996) talks about four kinds of researchers, who ally most closely with Innes' different kinds of police research discussed above:

- Inside insiders (research by the police) – internal or in-house research conducted by police officers
- Inside outsiders (research for the police) – professional researchers who work in internal research departments of the police force or professional researchers commissioned by the police to conduct research
- Outside insiders (research on or with the police) – former or retired police officers or staff who become academics and conduct research on the police – (I fall in this category!)
- Outside outsiders (research on the police) – academics or professional researchers who conduct independent research on the police from the outside on behalf of academic institutions or grant funding bodies

The degree of acceptance of researchers by police organizations clearly depends on whether they perceive that the researcher understands policing culture, appreciates the pressures and constraints under which the police operate, and respects practitioner knowledge and experience. Thus, the degree of association a researcher has had with or in the police impacts upon whether they will get easy access to the organization. Further, research reputation and word of mouth testimonials based on previous experience with research have a bearing on whether access will be granted. Whether this access is maintained would depend on the conduct of the researcher and the degree of trust he is able to forge with the practitioners he works with. Whether researchers need in-depth access to and, therefore, greater cooperation from the police would depend on the nature of the research project and research question. It is, therefore, politic for them to consider whether they can move from being total outside outsiders to becoming inside outsiders by co-opting police research priorities for getting greater police cooperation.

Research "on" the police conducted by academic researchers is often the most difficult since there is little reason for the police to grant access. Furthermore, the era of critical research on the police ensured that the police feel under attack or become defensive when they are being researched. Most research on the police, therefore, remains confined to using police data, either open source data or police recorded data which requires minimal police co-operation.

Furthermore, the research approach can either be overt or covert research, or even research that is overt but with covert aims. These approaches carry their own baggage of ethical issues, not only in the initial approach and presentation to gatekeepers and during the conduct of the research, but also whether and how the research outputs will be disseminated, especially if they require approval of the gatekeepers before publication. However, over the years, instances of covert research projects are becoming fewer and further apart given the strong ethics regime that is in place in most reputed universities across the world. Although the aim of my research on use of deadly force was focused on "extrajudicial killings," I framed it to cover police use of deadly force in all situations, including public order situations and riots. Thus, while the research aim was not covert or deceitful by any standards, its focus on "extrajudicial killings" was not highlighted while negotiating access.

5 Moving from the Personal to the General: Lessons Learnt from Doing Police Research

Other challenges in researching the police include issues of ethics; questions of validity and reliability; informed consent; confidentiality and anonymity; sampling; access, social context, and personal security (Brewer 1990). Drawing from personal experience of conducting several police-related research projects, some of which had retrospective relevance to public health and some with a specific public health angle, there are a few lessons that can be drawn for researchers desirous of conducting LEPH research. In the process of writing this chapter and reflecting on my research projects with the police, I realized retrospectively that most of my police-related research projects had a public health aspect to them, which were never articulated or explored. Researching police violence – shootings of organized criminals (Belur 2010) or in the course of counter terrorism operations (Belur 2013) as well as police responses to domestic violence (Belur 2008; Belur et al. 2014a) – has implications for public health in ways that were not always at the forefront of the research agenda. It was only during the course of conducting research on a multidisciplinary project looking at medicolegal responses to women as victims of burns in India (Belur et al. 2014b; Daruwalla et al. 2014) that the clear implications of the cross over between policing practice and its impact on public health became glaringly obvious.

It is often the case that police researchers are so focused on crime or deviance that they seldom recognize the public health implications of their research. On the other hand, public health researchers tend to focus on violence-related police data and the victims' perspective and neglect the policing aspect of the problem. Disciplinary expertise can often restrict the overlap of policing studies with public health research. Multidisciplinary studies are recommended but not always possible. My experience of conducting two police research projects on public health-related topics not only reiterated some of the challenges already identified in the literature, but also highlighted some additional nuances of conducting research in countries with a lesser tradition of police research, such as India, where the research was situated. Both projects were qualitative explorations to increase understanding of police processes and perceptions and involved in-depth interviews with police officers and access to their data. Both of these were researcher-led projects, and involved an "outside insider" approach. As a former Indian police officer, I was an "insider," but I was researching in police forces that were not the one I served (see also ▶ Chaps. 91, "Space, Place, Common Wounds and Boundaries: Insider/Outsider Debates in Research with Black Women and Deaf Women," and ▶ 92, "Researcher Positionality in Cross-Cultural and Sensitive Research"). Furthermore, I was conducting the research under the aegis of an independent academic institution and was not in active service, hence brought in an outside perspective. Both research projects were "on" the police – not the most recommended model – but tried to incorporate police participation in some of the decision-making process as explained below.

The first project was on understanding police justifications for use of deadly force in India. The research explored a peculiar feature of Indian policing called *encounters*, which is portrayed as a spontaneous unplanned "shoot-out" between the police and alleged criminals in which the criminal is usually killed, but there are no police injuries (Belur 2010). These killings were suspected to be abuse of deadly force, but it was an open secret that the police carried out these extrajudicial killings of suspected criminals and had a degree of social acceptability and approval (Belur 2009). The criminological study involved conducting in-depth interviews with police officers of various ranks in the city of Mumbai to understand officers' perceptions and accounts of *encounters* and how they explained, excused, or justified this conduct to themselves and to their various audiences. Officers interviewed had either themselves exercised deadly force, or supervised its use, or facilitated the process and management of the aftermath. It was a highly sensitive topic which required careful negotiation of practical considerations and officer expectations and trust in an ethical manner, while maintaining research integrity and transparency.

The second project was conceived as a multidisciplinary approach to understanding the medicolegal response to women as victims of burns, especially cases of dowry deaths in two cities in India. The research, thus, focused on the response of health professionals and police to women victims of burns. Part of the research was centered exclusively on understanding the police response to women victims of burns, especially in cases of dowry-related burns. In India, special legislation (Section 304B of the Indian Penal Code provides that if the death of a woman is caused by burns or physical injury or occurs in doubtful circumstances within 7 years of her marriage, and there is evidence to show that before her death, she was subjected to cruelty by her husband or his relative for demand of dowry, then the husband or the relative shall be deemed to have caused her death) covers the unnatural death of a woman within 7 years of marriage, as such deaths are suspected to be suicides or homicides resulting from harassment by the husband or in-laws for dowry. Historically, such dowry-related deaths are most commonly associated with burning, usually presented as kitchen accidents. The study involved conducting in-depth interviews with police officers who investigated or supervised investigations of dowry-related burn deaths in the cities of Mumbai and Delhi. This was again a sensitive topic, and involved understanding of the nuances of police investigative and decision-making processes. Given the explicit focus on the public health aspect of this research meant that the findings were shaped and policy implications were framed and disseminated within a public health perspective (Daruwalla et al. 2014).

I encountered several challenges during the course of conducting these and other projects with the police. Some challenges faced were specific to the context of research in India or the research projects themselves, but others are more common to researching the police organization as such and echo findings in the literature. I focus on the challenges identified earlier which specifically relate to accessing the police for research purposes which may be of relevance for public health research.

5.1 Negotiating Access

Presenting oneself to gatekeepers in police organizations is extremely important when researching a bureaucratic and hierarchical organization like the police since first impressions matter (Warren and Rasmussen 1977). Other researchers have also paid a great deal of attention to how researchers position themselves when approaching senior leaders and individuals within the police organization during the course of the research and the difficulties they had in accessing the organization (see Brunger et al. 2016). Factors such as gender (Huggins and Glebbeek 2003; Belur 2013), status (Reiner 2000), ethnicity, age, and class (Manderson et al. 2006) have an influence on whether and how much access will be granted (see also ► Chap. 92, "Researcher Positionality in Cross-Cultural and Sensitive Research" and ► 126, "Researching Among Elites"). It is important to be aware of this while approaching the police to negotiate access.

My experience was relatively straightforward. I usually got a meeting with senior officers as an ex-senior police officer. However, the experience of getting official permission to conduct research was different during the first study (when I was still a serving officer on study leave) as compared to the second, when I was no longer serving, but was an academic working in a "foreign" university was slightly different. The first project was more sensitive for the police, but I was given unconditional access. In the second case also I received permission, but more hesitantly and there were murmurings about getting permission from the Ministry to allow access to "outsiders." However, once access was granted, there was complete cooperation. At this stage, presenting the research proposal and its proposed outputs as being beneficial to the organization is very important, even if it is research "on" the police. Personal integrity of the researcher and the institution they represent also matters a great deal at this stage. Unless senior officers are convinced that the research will help the organization or have faith that the researcher will, at the very least, not damage their reputation, chances of success in gaining access are limited.

Further, access negotiation does not end with senior leaders, it has to be renegotiated with each individual during the course of the research. As a hierarchical organization, officers of lower ranks may feel compelled to participate in the research on orders from above, but actually getting them to communicate and answer questions or supply data fully and honestly is a challenge as my experience revealed. Other researchers have found that negotiating this "secondary access" can be quite difficult (Punch 1989). In my experience, officers would agree to be interviewed, but would either say they did not know the answer or give monosyllabic replies, reducing the interview to mere formality. Although they are unhelpful as sources of data to answer the main research question, such interview experiences nevertheless provide food for thought in terms of – was the interviewee uncooperative because the researcher did not manage to establish trust (and why) or were the nonanswers saying something about the topic or the interviewees themselves that is revealing of the underlying phenomenon. However, in order for researchers to appreciate the nuances of this research, it is important for them to have an in-depth understanding of police subculture and tacit organizational rules in order to overcome some of the cultural barriers identified.

5.2 Agreement on Research Objectives

An effective way of getting cooperation from the police while doing research is to move the focus away from research on the police to research with the police (Innes 2010). However, this implies that both researchers and police practitioners involved have similar priorities and compatible goals and the research output is of practical value to the police. Since both research projects were researcher-led and, therefore, were on topics that interested me personally, it was important to include the aspects of research that might also be of interest to the participating police organization. I took care to spell out the benefits of conducting that piece research for the police in the initial request to access the organization.

I found that police leaders in India had little or no prior experience of research (the Indian police are severely under researched) but were fairly open to giving access based on personal credentials. However, it always helped that the research was framed so that it had practical value for the organization. In order to be more inclusive in setting the research agenda, I always asked whether leaders had any particular aspect of the problem that they would like the research to include. Although I had a methodologically sound sampling framework, I invariably asked the leaders for suggestions on areas and field locations that would be most suitable for the research and whom to interview within those requirements. This was partly to give them the opportunity to contribute to the research but mainly in recognition of their expert knowledge of their area and personnel. To their credit, police leaders were very supportive in suggesting names of officers who had the most experience on the topic or police stations that were particularly prone to the issue under consideration. This more often than not resonated with choices I might have made based on my study of the available evidence.

5.3 Establishing Trust

The police in countries with a history of police research are suspicious of researchers as a legacy of the critical police research tradition era. In countries such as India, which has been relatively under-researched, the relative lack of exposure to police research, as well as a cultural disposition to maintain secrecy, means that researchers have to establish personal rapport with senior officers and gatekeepers in order to gain their trust. It is again vital at this stage that there is clear understanding regarding what the outputs of the research will be and how it would be fed back to the organization.

However, researching the police has in the past led to uncovering corrupt or deviant practices, consequently straining the ethical boundaries of research protocol (Punch 1989). This indeed places the researcher in a difficult situation while deciding on a response to deviance uncovered during the course of the research. On the one hand, the principles of ethical practice demand that the interests of the interviewee be preserved as much as possible, and on the other, turning a blind eye to deviance can be morally unacceptable. Further, whether the researcher should or should not report corrupt or deviant behavior has implications on the establishment of trust and whether the research project can be successfully completed. My experience of the research on police use of deadly force showed me that officers revealed a number of malpractices, which had serious criminal justice consequences, in faith and trust. I did suffer serious pangs of conscience about potential whistleblowing behavior, if I were to write up the findings. Finally, I resolved the dilemma by protecting my interviewees' identity and criminally incriminating personal information, but at the same time reporting that deviant behavior was occurring and how it was being protected by the organization. This may not have been a morally acceptable path for some researchers, but for the sake of protecting the interests of my interviewees and consolidating a relationship of trust between academic researchers and the police, I decided it was not my place to reveal past misdemeanors. Ultimately, the researcher has to take moral and ethical decisions based on their individual conscience (van Maanen 2008).

An important part of establishing trust is recognizing practitioner knowledge and cultural capital and respecting police processes and operating constraints (Stanko 2007). I encouraged officers to tell me their perspective and was consciously respectful and objective in responding to their opinions. I found that this approach was very useful especially when I was presented with opinions that I would have instinctively disagreed with. Distancing myself from the research and trying to view the problem from the participant's perspective not only made me a more objective researcher, but also enhanced the quality of the research.

5.4 Knowledge Exchange and Dissemination

Knowledge exchange is conceived as a two-way street, with researchers expected to feedback research findings to the police, but at the same time involves learning from practitioners and incorporating their experiential knowledge into the research. Being receptive to practitioner views, carefully observing, listening, asking pertinent objective questions, and understanding police processes and culture (Brown 2015) are all part of the knowledge exchange agenda and also contribute to building trust and confidence.

As discussed above, the end aims of research for the police and researchers are sometimes divergent. Academics are interested in academic outputs in the form of journal articles or monographs and practitioners are interested in reports that provide them with concise policy and operational suggestions. For researchers to cement an ongoing research relationship with the police, it is important that research findings are communicated to practitioners appropriately at various levels of the organization. Conveying the real-life application of research knowledge to police practitioners has distinct advantages (Fleming 2010) and augurs well for future collaborations.

It is also important that any communication plans are discussed at the initial stages to avoid confusion or misunderstanding once the research is complete. While the police would like to deflect critical media scrutiny, academic communication aims to reach the widest audiences (Fleming 2010). To avoid a conflict, the two communities must negotiate the terms on which research findings will be disseminated at the very beginning and clarify what levels of vetting, if any, the police might have on publication. However, researchers have a duty of care toward their participants and should protect the organization and individuals from suffering adverse publicity or other reputational or professional damage as a result of cooperating with the researcher. This will have a negative impact on future research collaborations and the relationship of trust.

It was my experience that the time lag between the time the research was conducted and when findings were analyzed and finally written up could be from a few months to a few years, especially in the case of ethnographic studies. In that period, officers who had given access to the organization had moved on and there was no follow up on presenting research findings. It was also the case that incumbent officers were not interested in the findings, once they were actually written up. In forces where the link between the police and academic research is not well established, this can often be the case. But, it should be every researcher's endeavor to disseminate headline research findings to the organization that supported the research immediately following completion of the research.

6 Conclusion and Future Directions

It is indisputable that policing has a definite impact on certain aspects of public health protection. Thus, related public health research would benefit from greater access to and better understanding of the policing of a variety of issues including gender-based violence, domestic violence, drug abuse, child and elder abuse as well as treatment of suspects in custody, police violence and police deviance. However, the pursuit of the ideal of collaborative research, or research "with" the police where the researcher and the police work together to achieve a particular aim would be possible mainly in cases where the outcome of the research directly benefits the police in the short run and is of interest to them. Thus, this kind of research requires the two communities to find a research topic that is of mutual interest and a research question that is of mutual benefit. Personal experience of conducting police research in public health-related topics revealed that, while the starting point of this process can be complicated enough (i.e., finding a research topic and question that is of mutual interest), the process of collaborative research itself can be quite demanding and often unrealistic. It requires building of trust and confidence, understanding of each other's culture, and should be based on open and frequent communication in easily understandable language (Tompson et al. 2017). In reality, most academic research happens to be research "on" the police, strengthening the perception that the police are hard to reach. However, this chapter highlights some steps that a researcher needs to consider when approaching the police for research access. It is important in order to get access to and researching "with" the police collaboratively, research aims should coincide with police priorities and have policy application in the short and medium term. In the current economic and social climate, it is important that there are close ties and cooperation between researchers and practitioners in order for research to retain its intellectual and to have practical and policy implications that are of practical relevance to the police (Innes 2010).

In the days to come, a growing need to support vulnerable populations, especially the mentally ill, provides a strong impetus for greater LEPH cooperation; simultaneously, economic cutbacks and political imperatives for the police to only focus on cutting crime has meant severe withdrawal of police resources for what is perceived to be "social work" by politicians (Punch and James 2016). Given this juxtaposition, there is a greater requirement to adopt an evidence-based approach to understand "what works" as also "what doesn't work" in order to efficiently allocate limited law enforcement resources (Punch and James 2016). Calls for the police to reimagine themselves as "guardians" instead of "warriors" (Rahr and Rice 2015) creates the possibility of a new wave of research to further the guardianship agenda (Wood and Watson 2016). To use the simple example of enhancing the role of police foot patrol, Wood et al. (2015, p. 211) suggest that "the way forward for theory, policy, and practice" for LEPH devises creative ways of dealing with urban vulnerability through multiagency management of "places" instead of "cases," rather than the traditional approach of changing officer attitudes and behavior. There is a call for academics and practitioners in the LEPH field to make a concerted effort "to integrating theories and methods, combined with a cross-system commitment to reducing fragmentation" (Wood and Watson 2016, p. 8). Thus, the future of LEPH research rests in closer cooperation between researchers and practitioners, transcending disciplinary boundaries, sharing knowledge and expertise, and working toward mutually beneficial goals.

References

Bartkowiak-Théron I, Asquith N. Conceptual divides and practice synergies in law enforcement and public health: some lessons from policing vulnerability in Australia. Polic Soc. 2016; https://doi.org/10.1080/10439463.2016.1216553.

Belur J. Is policing domestic violence institutionally racist? Polic Soc. 2008;18(4):426–44.

Belur J. Police use of deadly force: police perception of a culture of approval. J Contemp Crim Justice. 2009;25(2):235–52.

Belur J. Permission to shoot? Police use of deadly force in democracies. New York: Springer; 2010.

Belur J. Status, gender and geography: power negotiations in police research. Qual Res. 2013;14(2):184–200.

Belur J, Tilley N, Daruwalla N, Kumar M, Tiwari V, Osrin D. The social construction of dowry deaths. Soc Sci Med. 2014a;119:1–9. https://doi.org/10.1016/j.socscimed.2014.07.044.

Belur J, Tilley N, Osrin D, Daruwalla N, Kumar M, Tiwari V. Police investigations: discretion denied yet undeniably exercised. Polic Soc. 2014b;25(5):439–62.

Bradley D, Nixon C. Ending the "dialogue of the deaf": evidence and policing policies and practices. An Australian case study. Police Pract Res. 2009;10(5):423–35.

Brewer JD. Sensitivity as a problem in field research: A study of routine policing in Northern Ireland. American behavioral scientist. 1990;33(5):578–593.

Brown J. Police research, some critical issues. In: Leishman F, Loveday B, Savage S, editors. Core issues in policing. London: Longman; 1996.

Brown R. Tip-toeing through the credibility mine field: gaining social acceptance in policing research. In: Cockbain E, Knutsson J, editors. Applied police research. Abingdon/Oxon: Routledge; 2015.

Brunger M, Tong S, Martin D, editors. Introduction to policing research. Oxon: Routledge; 2016.

Buerger ME. Police and research: two cultures separated by an almost-common language. Police Pract Res Int J. 2010;11(2):135–43.

Cockbain E, Knutsson J. Introduction. In: Cockbain E, Knutsson J, editors. Applied police research. Abingdon/Oxon: Routledge; 2015.

Cooper H, Moore L, Gruskin S, Krieger N. Characterizing perceived police violence: implications for public health. Am J Public Health. 2004;94(7):1109–18.

Daruwalla N, Belur J, Kumar M, Tiwari V, Sarabhai S, Tilley N, Osrin T. A qualitative study of the background and in-hospital medicolegal response to female burn injuries in India. J Women's Health. 2014;14:142. http://www.biomedcentral.com/1472-6874/14/142

Engel R, Whalen J. Police-academic partnerships: ending the dialogue of the deaf. The Cincinnati experience. Police Pract Res. 2010;11(2):105–16.

Fleming J. Learning to work together: police and academics. Polic J Policy Prac. 2010;4(2):139–45.

Hayashi K, Small W, Csete J, Hattirat S, Kerr T. Experiences with policing among people who inject drugs in Bangkok, Thailand: a qualitative study. PLoS Med. 2013;10(12):e1001570. https://doi.org/10.1371/journal.pmed.1001570.

Hovell M, Seid A, Liles S. Evaluation of a police and social services domestic violence program: empirical evidence needed to inform public health policies. Violence Against Women. 2006;12(2):137–59.

Huggins M, Glebbeek M. Women studying violence male institutions: cross gendered dynamics in police research on secrecy and danger. Theor Criminol. 2003;7(3):363–87.

Innes M. A "mirror" and a "motor": researching and reforming policing in an age of austerity. Polic J Policy Prac. 2010;4(2):127–34.

Jameel J. Researching the provision of service to rape victims by specially trained police officers: the influence of gender– an exploratory study. New Crim Law Rev. 2010;13(4):688–709.

Kennedy D. Working in the field: police research in theory and in practice. In: Cockbain E, Knutsson J, editors. Applied police research. Abingdon/Oxon: Routledge; 2015.

Kerr T, Small W, Wood E. The public health and social impacts of drug market enforcement: a review of the evidence. Int J Drug Policy. 2005;16(4):210–20.

Lum C, Telep CW, Koper CS, Grieco J. Receptivity to research in policing. Justice Res Policy. 2012;14(1):61–96.

MacDonald B. Research and action in the context of policing: an analysis of the problem and a programme proposal. 1986, Unpublished document of the Police Foundation of England and Wales.

Manderson L, Bennett E, Andajani-Sutjahjo S. The social dynamics of the interview: age, class, and gender. Qualitative health research. 2006;16(10):1317–1334.

Madensen T, Sousa W. Practical academics: positive outcomes of police-researcher collaborations. In: Cockbain E, Knutsson J, editors. Applied police research. Abingdon/Oxon: Routledge; 2015.

Maher L, Dixon D. Policing and public health: law enforcement and harm minimization in a street-level drug market. Br J Criminol. 1999;39(4):488–512.

Millie A. The policing task and the expansion (and contraction) of British policing. Criminol Crim Just. 2013;13(2):143–60.

Open Society Foundation. To protect and to serve: how police, sex workers and people who use drugs are joining forces to help improve health and human rights. New York: Open Society Foundation; 2014.

Punch M. Researching police deviance: a personal encounter with the limitations and liabilities of field-work. Br J Sociol. 1989;40(2):177–204.

Punch M. Observation and the police: the research experience. In: Hammersley M, editor. Social research: philosophy, politics and practice. London: Sage; 1993. p. 181–99.

Punch M, James S. Researching law enforcement and public health. Polic Soc. 2016; https://doi.org/10.1080/10439463.2016.1205066.

Rahr S, Rice S. From warriors to guardians: recommitting American police culture to democratic ideals. Washington, DC: Department of Justice; 2015.

Ratcliffe J. "Harm reduction policing". Ideas in American policing, no 15. Sept 2015. Police Foundation. 2015. Accessible at http://www.policefoundation.org/publication/harm-focused-policing/

Reid J, Walton S. Protecting people and promoting healthy lives in the West Midlands. London: Public Health England; 2015.

Reiner R. The politics of the police. Oxford: Oxford University Press; 2000.

Reiner R, Newburn T. Police research. In: King R, Wincup E, editors. Doing research on crime and justice. 2nd ed. Oxford: Oxford University Press; 2008.

Rojek J, Martin P, Alpert GP. Developing and maintaining police-researcher partnerships to facilitate research use: a comparative analysis. New York: Springer; 2015.

Skolnick J. Justice without trial. 2nd ed. New York: Wiley; 1975.

Stanko B. From academia to policy making: changing police responses to violence against women. Theor Criminol. 2007;11(2):209–20.

Strang H. Coalitions for a common purpose: managing relationships in experiments. J Exp Criminol. 2012;8(3):211–25.

Tompson L, Belur J, Morrison J, Tuffin R. How to make police-researcher partnerships mutually effective. In: Knutsson J, Tompson L, editors. Advances in evidence based policing, Crime science series. Abingdon: Routledge; 2017.

Van Dijk A, Crofts N. Law enforcement and public health as an emerging field. Polic Soc. 2016; https://doi.org/10.1080/10439463.2016.

Van Maanen J (2008) The moral fix: on the ethics of fieldwork. In: Pogrebin M (ed.) Qualitative Approaches to Criminal Justice. Thousand Oaks, CA: Sage, pp. 363–376.

Warren C, Rasmussen P. Sex and gender in field research. J Contemp Ethnogr. 1977;6(3):349–69.

Weisburd D, Neyroud P. New perspectives in policing: police science – toward a new paradigm. Harvard executive session on policing and public safety. Washington, DC: National Institute of Justice; 2011.

Wood J, Watson A. Improving police interventions during mental health-related encounters: past, present and future. Polic Soc. 2016; https://doi.org/10.1080/10439463.2016.1219734.

Wood J, Taylor C, Groff E, Ratcliffe J. Aligning policing and public health promotion: insights from the world of foot patrol. Police Pract Res. 2015;16(3):211–23.

World Health Organization. Definition of health. In: Preamble to the constitution of the world health organization as adopted by the international health conference, 19–22 June 1946. New York. 1948, Entered into force on 7 Apr 1948.

Young M. An Inside Job, Oxford: Clarendon Press; 1991.

Researching Among Elites

126

Neil Stephens and Rebecca Dimond

Contents

1	Introduction	2198
2	Planning Data Collection	2200
	2.1 Ethical Issues and Health Elites	2200
	2.2 Sampling and Access	2201
3	Doing Data Collection	2202
	3.1 Interviewing Health Elites	2202
	3.2 Observing Health Work: The Clinic and Laboratory	2203
	3.3 Conferences as Sites for Research Among Health Elites	2205
	3.4 The Documentary Cultures of Health Elites	2206
4	Other Health Elites	2206
	4.1 Patients as Health Elites	2207
	4.2 Social Scientists as Health Elites	2208
	4.3 Health Elites as Social Scientists	2208
5	Health Elites as Collaborators: Researching-About, and Researching-With, Health Elites	2209
6	Conclusion and Future Directions	2210
References		2210

Abstract

Health elites are powerful actors in the medical domain and it is essential that social scientists engage with their work. However, there is a specific set of methodological challenges to conducting this research. This chapter articulates key issues to consider before undertaking research with health elites by drawing

---

N. Stephens (✉)
Social Science, Media and Communication, Brunel University London, Uxbridge, UK
e-mail: Neil.Stephens@Brunel.ac.uk

R. Dimond
School of Social Sciences, Cardiff University, Cardiff, UK
e-mail: DimondR1@Cardiff.ac.uk

© Springer Nature Singapore Pte Ltd. 2019
P. Liamputtong (ed.), *Handbook of Research Methods in Health Social Sciences*,
https://doi.org/10.1007/978-981-10-5251-4_135

upon examples from the authors' own research practice. It starts by identifying the ambiguities in defining exactly what constitutes a health elite by drawing upon important literature on the topic. Section 2 discusses ethical issues in health elite research, including providing a sample consent form. It then articulates sampling and access issues with elites, for example, the benefits of purposive and snowball sampling. Section 3 articulates key challenges firstly in interviews and secondly in observational work with health elites (in clinics and laboratories), by stressing the need for flexibility in approach. This is followed by a discussion of conferences as sites for research among health elites, and the resources of elites' documentary cultures. Section 4 reflects upon the increasing significance of patients and social scientists as health elites, and instances of health elites as social scientists. Section 5 considers health elites as collaborators by discussing the rewards and challenges of collaborating with fellow researchers active in generating new knowledge about health. The chapter closes by pointing forward toward the continued need for qualitative social science to engage with health elites, and for researchers to be informed by a methodological awareness of the challenges and rewards of doing so.

Keywords

Elites · Ethnography · Health · Medical sociology · Qualitative methods · STS

1   Introduction

Health research frequently involves contact with, or researching among, individuals in elite positions, and with that comes a set of key issues to reflect upon. Social science research often focuses on the vulnerable in society. The researcher is considered to hold the power in the research relationship, and traditional scholarship on research methods often provides advice for the researcher on the assumption of these power relations. Elite research is less common, although no less important, and requires some reformulation of common ideas in research methodology. Elites in research are associated with the idea of "studying up" and of researching those who are more powerful than the researcher. Where there is power, there are elites, and there is power everywhere. This means that defining exactly what does and does not constitute elite is difficult, and perhaps an unproductive process. We could understand elites as those who are powerful relative to society, in general, or relative to others in their social space, or, as "studying up" implies, relative to the social scientist. Odendahl and Shaw (2002) provide a detailed account of multiple typologies of elites, and show why frequently these do not capture the full diversity. Health elites could be divided into clinical, research, policy, and commercial elites, but it remains likely that this typology too fails to capture important components.

Moving beyond divisions between professional domains, Zuckerman (1996) uses the term "ultra-elite" to describe the most powerful, or revered, among an elite group in her study of Nobel laureates. Payne and Payne (2004) opt for the term "expert witnesses" to describe elites because the term highlights the valuable contribution to research they can make due to their privileged information, their role as representative of a particular institution or culture, their potential to play a gatekeeping role for others in their group, or their ability to make recommendations for the direction of the project.

However elites are defined, and whether it is because of their public profile or the position they hold in their organization, their elite status in society raises particular issues for the researcher. Research should involve sensitivity to hierarchies in professional settings. In one encounter, Neil Stephens was aware that he was interviewing the director of an organization straight after conducting an interview with an employee. Each party was suspicious about what the other was disclosing in the interview and directly asked Neil to disclose this information. On another occasion, members of an oversight committee refused to take part in an interview, and it was only later that Neil realized it was because they were making staff redundant and did not want this knowledge shared with the researcher or with other staff (Stephens and Dimond 2015a).

When researching among health elites, it can be beneficial to adopt the approach of Aldridge (1993) in his work on Anglican clergy-men and women that stresses the importance of recognizing and making the best of both the differences and commonalities between the researcher and the researched (see also Stephens 2007 for another illustration of this approach). As this chapter will demonstrate, social scientists are themselves in some regard powered and can share common ground with elites in the health sector.

We have conducted nine projects collectively and individually with health elites and we draw on these as case studies to illustrate the points made in this chapter. Rebecca Dimond is a medical sociologist primarily focusing on clinician/patient/family relationships and genetic disease. Neil Stephens is a Science and Technology Studies scholar primarily focusing on innovative biomedical technologies. Both operate with ethnographic and qualitative methods. This chapter uses our work to illustrate key methodological points. It draws upon: (i) Rebecca's study of family/patient/clinician interactions in 22q11 deletion syndrome; a rare genetic disease (Dimond 2014a, b, c), (ii) Rebecca and Neils' interview and observational study of mitochondria disease; a rare disease for which an IVF-based preventative intervention was developed and then legalized in the UK (Dimond 2013, 2015a, b), (iii) Rebecca and Neils' interview and observational study of an anonymous biobank; an institution that holds human biological material for use in biomedical research (Stephens and Dimond, 2015a, b), (iv) Neils' ethnography of the UK Stem Cell Bank; the first institution in the world to hold and regulate national human embryonic stem cell use (Stephens Atkinson and Glasner 2008a, b, 2011, Stephens, Lewis and Atkinson 2013), and (v) Neil's ethnography and interview studies of research active tissue engineers (Stephens 2010, 2013; O'Riordan et al. 2016).

This chapter articulates some of the key issues and challenges in researching among health elites. It focuses on qualitative methods, primarily interviews, and ethnographic or observational work. It is ordered loosely following the format a research project might take. This starts with *planning data collection* (Sect. 2), with subsections on ethics and sampling and access, respectively. Next is *doing data collection* (Sect. 3), with subsections on interviewing elites, clinic observations, conferences as research sites, and documentary cultures, respectively. This is followed by other groups who might be considered elite, such as patients and social scientists, and researching-with health elites. The chapter closes with conclusion and future directions (Sect. 6).

2 Planning Data Collection

Research design and methodology are a key part of any research project. This section focuses on some key issues specific to researching among health elites, and compliments other chapters in the volume which focus on the more general components of good research practice found in other health research settings.

2.1 Ethical Issues and Health Elites

Most research projects will require ethical approval before carrying out any research. The key element of ethical approval is to ensure the safety of participants, and in health research, this often means ensuring that patients are protected (Stark and Hedgecoe 2010; Israel 2015; see also "Ethics and Research"). Although the "elite" might not appear vulnerable at first, their protection should nonetheless be central in planning any research project.

The elite can be vulnerable in different ways to other respondents. They might be well known, and might feel their public image is at risk if they or their words are represented in a particular way. The most important aspect when researching elites is to be as open as possible about the intentions of the project (and see below for a discussion about the problems and pitfalls of collaborating with elite participants). The consent form template (see Fig. 1) is an example where some issues can be clarified from the start. There are two main aspects which are included here which may not appear on other consent forms:

- Observational notes and interview quotations will generally appear without formal attribution (i.e., anonymously). If I request formal attribution, the research team may identify me if it does not undermine the anonymity of other participants.
- I recognize that the specialist nature of [*respondent specialism*] means some readers may identify me even if interview quotations and observational notes are used without attribution.

Both of these points refer to the fact that because of the unique positon held by the elite, anonymization might not be possible.

For research involving cutting edge science, additional issues can be at stake, requiring an extra layer of approval. Some scientists, for example, may be concerned

I, the undersigned, agree to participate in the above project. I understand that:

• Observations are being conducted of my professional practice.

• Observational notes will be written or audio recorded and stored.

• In some instances photographic, audio or video recordings may be made of my professional practice. In these instances I will be made aware and the use of digital materials in the analysis and dissemination of the research will be discussed with [*Researcher name*] who will respect our agreed position on usage.

• Interviews will be conducted, transcribed and stored.

• The observational notes and interview transcripts may be quoted in academic journals, conference papers, and books, and potentially wider dissemination avenues.

• Observational notes and interview quotations will generally appear without formal attribution (i.e. anonymously). If I request formal attribution the research team may identify me if it does not undermine the anonymity of other participants.

• I recognise that the specialist nature of [*respondent specialism*] means some readers may identify me even if interview quotations and observational notes are used without attribution.

• My participation is voluntary and I have the right to withdraw from the research process at any time. I can withdraw by informing [*Researcher name*].

Print name: _____ Signature: _____ Date: _____

Fig. 1 Potential consent statement for interview and observational work with health elites

about disclosure if commercially sensitive information is involved, which has intellectual property implications. This was the case for Neil's research of a tissue engineering laboratory, which was not allowed to begin until he had met with the commercial research office. The result was that Neil was enrolled as a team member, which meant he was not external to the project and thus the participants did not "disclose" information when they shared it with him. Secondly, Neil had to agree to allow the commercial research office to check any reports in advance of dissemination. In practice, no change to Neil's work was required, but without talking to the commercial research office and securing their consent, the research would not have been allowed to take place.

2.2 Sampling and Access

A key aspect of any research project is deciding who you are going to include in your study, a process known as sampling. Researching among health elites poses distinctive sampling benefits and hurdles. Firstly, the elite will generally have a public profile which means a researcher will be able to find the details of those he/she wishes to invite to participate. Secondly, some individuals, such as the only person in the world who has made a laboratory grown hamburger or the only director of a stem cell bank, are globally unique. It is highly likely, therefore, that sampling will involve a purposive sample where the researcher identifies and makes contact with specific key people rather than random sampling to find respondents who represent a broader group (Liamputtong 2013; Patton 2015; see also "The Nature of Qualitative Research"). This means that a project exploring management strategies of a unique stem cell bank or a rare disease clinic will be able to draw on different sampling strategies than one investigating children's experiences of going to the doctor, for example.

But elite status also presents hurdles. Blix and Wettergren (2015), in their research with the Swedish judiciary, recognized that gaining and securing access to an elite group involves extensive emotional labor. Although they might have a public profile and appear publically accessible, in practice, making contact can be difficult, and carefully prepared emails, letters, or phone calls might not actually reach their target. It is always worth thinking about the strategy for contact, speak to the elite's secretary or PA and do not be concerned if you do not receive an immediate reply.

Once contact has been made, or once the research is underway, it is useful to use snowball sampling (Liamputtong 2013; Patton 2015) to make contact with other potential respondents. Your contact can recommend other people in the field to speak to, and might even offer a personal introduction. This approach would mean that the difficulties of making that initial contact can be minimized, but also shows how sampling and access can be deeply entwined practices (see also ▶ Chap. 5, "Recruitment of Research Participants").

3 Doing Data Collection

Once ethics, sampling, and access are secured, the fun work of collecting data and being active in health elite social worlds begins. Data can be collected in multiple sites through multiple methods. This section identifies some key research sites and articulates methods issues related to them, including interviewing health elites, observing clinics, labs and conferences, and documentary analysis.

3.1 Interviewing Health Elites

Interviews are one of the most common approaches to researching among health elites. Interviews are relatively easy to organize, and can be conducted either face to face or via telephone or Skype. Stephens (2007) provides a useful exploration of the advantages and disadvantages associated with each method and their practical implications. One of the advantages of course is that telephone/Skype interviews are cheaper to conduct than face-to-face interviews, which can be an important consideration for a project with limited resources and where the elite in a particular field are based in different countries.

Elite interviewing can raise particular challenges in addition to the general expectations about the role of the researcher when conducting interviews. Our first key message is for the researcher to appear professional and be flexible (Harvey 2011). Arrive early for interviews, and dress according to the space. Morrissey (1970) stresses the need for flexible timetabling to respond to elites' busy diaries, and the potential for interviews to be cut short or be interrupted. It is worth clarifying at the start how much time they can afford, and be prepared to sit and wait while your interviewee takes a phone call or goes out to meet a patient or colleague. The second key message is to prepare for the interview. There is a growing amount of literature on elite interviewing, which frequently stress the specific nature of the elite interview, particularly around power. Ostrander (1993) argues the social scientist should use nonverbal cues and direct questioning to assume a dominant position during interviews. Cassell (1988), Dexter (1970), and Hunter (1993) stress preparing for gate-keeping questions through which the interviewee tests the caliber of the interviewer. Petkov and Kaoullas (2016) suggest using an intermediary who can help a researcher establish trust with the respondent and would be able to intervene if the respondent is "deliberately or unintentionally withholding information" (p. 1). Elites are typically highly engaged in the interview topic and capable of discussing detailed points at length, but they might not agree to take part in a research project if they feel that it would be a waste of their time. Once again, preparation is key – which means potentially learning about both the topic and your participant. One issue is whether the researcher needs what Collins and Evans (2007) call "interactional expertise," that is, the ability to speak the technical language of the expert being interviewed. Mikecz (2012, p. 482) highlights the importance of gaining rapport, claiming "the success of interviewing elites hinges on the researcher's knowledgeability of the interviewee's life history and background."

Although interviewing elites can appear daunting at first, in our experiences, interviews often work very well, are relatively easy to conduct, and frequently are enjoyable and informative encounters. Out of more than a 100 elite interviews conducted by Neil, only one could be described as going badly – in that the interviewee seemed reluctant to engage in the spirit of the interview – yet even this interview produced excellent data. Particularly in the health field, the elite are often aware of the value of evidence-based research and that their participation can contribute to this knowledge culture. We have found that once a slot in their diary has been arranged, health elites are typically affable, articulate, and keen to contribute. Indeed, although some interviews have been cut short, more often than not our participants have been able and willing to rearrange their diaries in order to extend the interview, in extreme instances with interviews lasting 3, 4, or 5 hours. Our final message about interviewing elites is about engaging the participant so that they see the merit in the research and enjoy the opportunity to reflect upon their professional role, activities, and personal perspectives.

3.2 Observing Health Work: The Clinic and Laboratory

The clinic is possibly the first site that health researchers think about when undertaking observational research, as the consultation exemplifies the relationship between health professional and patient. Likewise, when thinking about the work of scientists, most people would think about the laboratory. The power of the health professional or the scientist has been a key theme within the sociology of work and the hierarchy of the clinic and laboratory can, therefore, have important implications for the researcher. Here, we discuss the important role of a gatekeeper, and the tension between wanting to observe mundane everyday work compared with the expectation that research will focus on the extraordinary.

Gaining access to the clinic or laboratory will often depend on forging a good relationship with its director, and this person will then act as an important gatekeeper for the research. One obvious example is that the clinician is in a position to tell the researcher when and where meetings will take place, and can help set up access, e.g., informing the receptionist to expect the researcher to attend. But, it also means that, to a large extent, the researcher can become wholly dependent on the gatekeeper, which might lead to limited access if he or she does not understand or recognize the value of social science research.

There might also be confusion about the role of the ethnographer. Ethnographic work requires attentiveness to the mundane and everyday activity conducted in a given site (see also ▶ "Ethnographic Method"). Yet, it might be assumed by clinic or laboratory staff that the researcher wants to observe specific and special "events." Within a medical space, routine tasks such as team meetings, chats over coffee, or training students might not seem like important "work" and appear mundane to those performing them. But, these activities are rich in detail for the ethnographer (Atkinson 2015). Thus, once access is secured, the researcher might still need to engage in a process of negotiation about what to witness, why, and when. The problem, of course, is that health professionals will no doubt be busy and in-depth conversations about the aims and philosophy of the research are not their top priorities.

The health professional as gatekeeper situation might also mean that the researcher witnesses events far beyond what was envisioned in the planning of the project. For example, on one occasion, Rebecca arrived at a clinic to interview the director, but was instead led into a "viewing" room, where along with a technician, speech therapist, and father, she watched a consultation with a 6-year-old girl and her mother through a one-way window. Rebecca had not been introduced to anyone in the room, but as they were all watching the consultation proceed, she felt unable to explain her presence or ask for their consent. Indeed, at the direction of the clinician, it is possible that many of those who fall under the researcher's gaze, including patients, parents, medical students, receptionists, and members of the clinic team, are not always explicitly informed about their presence. On many occasions such as this, the researcher might feel like a "guest" in someone else's busy work environment and, therefore, relatively powerless to influence proceedings.

It is possible that the key contact will act as a gatekeeper throughout the time when the researcher is observing. It might be the clinician who identifies which patients to observe and it can also be the clinician who decides when note taking is appropriate, even though this might be specifically included as a tick box on the consent form. Indeed, a researcher might be directed in many other ways by a clinician; for example, it is not unusual to be asked to take charge of children during a consultation or to go to the waiting room to call in the next patient. These occasions blur the boundaries of whether the researcher is a participant in the event or a nonparticipant observer. Whatever the task, demonstrating a willingness to get involved might be important for securing future collaboration, as long as it does not ultimately compromise the research.

3.3 Conferences as Sites for Research Among Health Elites

Conferences are very useful sites for a researcher. First of all, they offer an introduction to the field, a chance to see what kinds of information are exchanged, who are the elite (for example, who acts as host or chair but also who are mentioned as success stories), and how individuals relate to each other (Dimond, Bartlett and Lewis 2015; González-Santos and Dimond 2015). One of the advantages of going to a conference is that the researcher will have a wide opportunity to interact with others, and fortunately this means many opportunities for meeting "the elite." In addition, conferences blur the boundaries of informal and formal working practices, and subsequently you might talk to the person sitting next to you during a session, or have a chat and hear gossip over lunch or at a wine reception. Collins (2004) argues these informal opportunities could be even more important than attending the formal presentation sessions. It means that when you go to a conference, do not just focus on attending the presentations but look around, identify key people in the field, and seek out opportunities to engage with them.

One of the advantages of observing conferences is that it is easy for a researcher to make notes. Note taking is an essential part of the ethnographer's tool box, but as with any site, decisions have to be made as to when and how notes are made. Fortunately, the conference provides a natural stage for recording events – welcome packs contain a logoed pen and notepad and the physical location, and rows of chairs facing forward emphasis the conferences as a site of learning. But, the occasions when an ethnographer takes notes might not correspond with the "norms" of the event. This was the case at a parent-led conference when Rebecca realized that no one else around her was writing. Although many attendees, who were mostly parents, held a pen in their hands, they would only jot down an email or website, the details of a medical contact, or the name of a treatment. On another occasion, at a scientific conference, it was the timing of note taking that attracted attention. Rebecca noticed that those sitting around her only made notes during particular presentations. In contrast, Rebecca wanted to take notes through the entire day and this meant during the welcome address, the speaker introductions, and the question and answer sessions. Being visible as a researcher is not a problem, but it means that the researcher needs to think about "impression management" (Hammersley and Atkinson 2007), which includes deciding what to say to others about the research, what to wear, and how to behave.

3.4 The Documentary Cultures of Health Elites

An advantage of researching elites is that it is likely that researchers will be able to access publically available information about the participants and their work. Documentary evidence can be used in combination with other research methods, for example, to help with the preparation for an interview. But, documents also represent a textual footprint that can be used as research data. Prior (2003) highlights how what counts as a "document" is wide ranging, including videos, diaries, paintings, and photographs, all of which can produce a rich insight into a social world (see also ▶ "Unobtrusive Methods").

The example of the value of documents as data given here is based on the mitochondrial donation project. This germ-line technology required a change in the UK law for it to be offered to patients, and this involved a lengthy process of consultations and reports. There were several parliamentary debates which were transcribed and available online. There were also four scientific reviews and several public consultations all of which resulted in publically accessible reports. Some of the evidence submitted by patients, professionals, and publics were also publically accessible. As the legalization of mitochondrial donation was controversial, it also attracted extensive media coverage. Media interest in the techniques involved attention grabbing headlines and overviews of new developments, but what was interesting in the mitochondria debates was that the media was also used by key people in the field as a way of declaring their personal or professional perspectives. Letters were published by high profile individuals or groups; for example, one letter included signatures from five Nobel Prize winners, and another was signed by 40 scientists from 14 different countries. Those directly involved in developing the techniques or were members of oversight committees also published sometimes lengthy opinion pieces. Thus, media at a particular time could be used not just to identify key people in the field but also to explore their public accounts.

The wealth of publically available documents, particularly around a controversial or current debate, can enable researchers to conduct research by accessing information about the activities of elite health professionals. But, the researcher is reliant on the reports being made public, and this is clearly not the case at all times. Openness depends on institutional requirements, culture, and national legislation. In the UK, the law now makes it possible to access information through a Freedom of Information request. This might produce documents which have not been made publically accessible and might reveal different insights or new perspectives.

4 Other Health Elites

The first thoughts that come to mind for many when they think of health elites are professionals: clinicians, scientists, and policy-makers. However, when thinking about the specific methodological issues raised when researching health elite, it is useful to recognize other groups who might be considered, or might consider themselves, as elites.

4.1 Patients as Health Elites

It is productive to categorize some patients as a type of health elite. These are patients engaged in the politics and practice of managing the healthcare of other patients beyond their own, either through, or in conflict with, mainstream health institutions. They typically fall somewhere between patient advocacy and patient representation.

A classic study on patients as both experts and political activists is Epstein's (1996) work on HIV activism and the credibility struggles they engaged in to challenge and reconfigure mainstream scientific opinion in the field. Kent (2003) provides a related example of confrontational patient activism about lay experts and the politics of breast cancer. These are important sites of research, but patient elites are also evident in contexts less antagonistic to mainstream health policy, and can be embedded within the routine operation of healthcare practice.

The mitochondrial donation policy debate featured two related types of patient elites. The first – the Lily Foundation – is a voluntary patient-run patient advocacy group that delivered a powerful voice on behalf of families with mitochondrial disease. The second – Muscular Dystrophy UK – provides advocacy and support for, by, and on behalf of, people with muscular dystrophy through the professionalization of patient representation.

The illustrative case studies on biobanks also involved patient elites in the form of "lay board members" or "patient representatives" on oversight committees (Roth 2011; Mallik 1997). They contribute to discussions about an institution's day-to-day policy and long-term plans, in some instances even being party to decisions to shut institutions down (Stephens and Dimond 2015a; see also Stephens et al. 2008). Some "lay" members feature on multiple oversight boards and make a significant time commitment to decision-making roles. Of course, the term "lay expert" is problematic, as their expertise is not "lay" but unaccredited through formal qualifications (Collins and Evans 2007). Methodologically, the term also points to the hybridity of their status, suggesting a form of "non-elite elites." These are not institute directors or Nobel Prize winners, so their type of eliteness requires consideration.

We can understand people in these positions as patient elites in at least three complementary ways. Firstly, they are elites within their own communities, adopting a position of representation of the voice of others, thus giving them power within their communities and access to other professional power elites. Secondly, they are elites in that their work – in policy debates and on oversight committees – is the work of powered people, patients or not. And thirdly, while we cannot generalize from our own projects, some people in this position transpose elite professional experiences from other domains into their patient advocacy work, deploying the education and cultural capital developed elsewhere into the health field. This said, they are not traditional elites as their patient-ness situates them differently to professionalized power, so we advise remembering the hybridity of the "lay expert" and the importance of being reflexive about research relationships when "studying up" with patient elites.

4.2 Social Scientists as Health Elites

Social science has become an important component of health practice, and as such social scientists often hold elite positions in health settings. The case studies included interviews with bioethicists, lawyers, sociologists, and anthropologists about their work in shaping health and biomedical policy and practice. Health economists and political scientists are just some of the other social science elites that can be encountered. Interviewing people who properly understand the theoretical frameworks that you work with, and maybe have read (or peer reviewed) your papers, can be both enjoyable and challenging. Interviews with participants who closely share your disciplinary perspective sometimes feel entwined with a sense of shifting between knowing and puzzlement as the interviewee recognizes, or tries to recognize, the ideas underpinning your line of questioning. They can also sometimes spontaneously begin reflexive analysis of their own accounts, articulating how their previous utterance might be understood by Foucault, or Latour, or Goffman. This is about establishing rapport as a reflexive operation; both interviewer and interviewee contributing to the smooth running of the interview. It is also about negotiating elite status, and the status similarities and differences between interviewee and interviewer.

While the intrasocial science elite interview may be framed by its intrasocial science form, we do not believe this invalidates or undermines the data collected. This is for two reasons. Firstly, in analysis, it is simply dealt with, recognized, and understood reflexively for the situated account that it is. All interviews are situated, and this is just a distinctive form of that situatedness. Secondly, social science is an important component of contemporary biomedicine and that should not be written out of the account but accommodated within it by acknowledging its role. During Neil's ethnography of the UK Stem Cell Bank, he and fellow social scientists sometimes joked that there were more social scientists that had conducted interviews at the Bank than there were stem cell lines in the Bank itself. Numerically the joke was true, and the humor hinted at an over-supply of social scientists in the field, but actually this should be understood as an indication of the role of social science in biomedicine and should be treated as a topic to research as opposed to an embarrassment to be written out. The same can be true of observational work. In one encounter, Neil stood in a line of three ethnographers all writing fieldnotes about the practices they were observing. He ensured his fieldnotes recorded this fact, as it was an important interaction to analyze. Social science can be constitutive of biomedical and healthcare practice, and we should deal with it reflexively as a resource to improve our analysis.

4.3 Health Elites as Social Scientists

Finally, as well as social scientists becoming health elites, progressively health elites are becoming social scientists, in that medically trained professionals are more and more involved in studying and conducting social research. Neither of us are health elites in this traditional sense, so we are not in a position to draw upon our own experience on being, for example, a medical consultant interviewing other medical consultants. However, work has been published on these relationships, and we direct the reader toward Luam and Sima (2006) for an account.

5 Health Elites as Collaborators: Researching-About, and Researching-With, Health Elites

Social scientists and health elites share a number of professional activities. Shared institutional and cultural foci can include conferences, publications, public engagement interests, and the demand for securing research funding. Together, these can facilitate researching-with, in addition to researching-about, health elites. Examples include copresented talks, coauthored publications, coinvestigator research applications, and coordinated public engagement activities. In the health research sector, this researching-with is typically distinctive to researching among elites, as their elite status is part-premised in the professional cultures that are shared.

Coauthoring or copresenting with health elites brings both challenges and opportunities. It can allow fruitful exchange and mutual learning between coauthors. It can also allow social science research to reach wider audiences and results in a valuable output on end-of-award reports. But, it does require attention to both voice and messaging as the text works to unify divergent interests, methods, and world views. One of our examples (see Stacey and Stephens 2012) takes the unusual step of decoupling the voices by featuring alternating sections from the authors detailing their experience of participating in health elite research.

Similar tensions run through applying for and conducting interdisciplinary research. Joint projects vary from those in which the disciplines are distinct but in parallel (perhaps through a work-package structure) to those in which disciplines genuinely intermingle their practices and outputs; a distinction that Lewis and Bartlett (2013) term individual and collective interdisciplinarity. When researching-with health elites, a full realization of the latter could be a collaborative ethnography in which the participants are active in designing research questions, analysis, and publishing (Lassiter 2005), or a health-based intervention project in which the social science directly impacts upon patient outcomes.

In all of these contexts, the social scientist must be attentive to the power relations active in researching-with health elites and how their own contribution is situated in relation to the groups they study. Researchers should be reflexive about the relationship between critique and legitimacy when entering into researching-with arrangements. Does researching-with undermine the critical capacity of the work? Does researching-with operate to lend legitimacy to the fields being studied? How is the social science research situated within the broader political economy of health? We do not believe there are simple (or indeed right or wrong) answers to these questions. But, we urge the social scientist to be aware of the potential and operate reflexively in doing so.

6 Conclusion and Future Directions

In this chapter, we have highlighted the difficulty of identifying what or who counts as elite. But, whether this includes all or some health professionals, scientists, policy makers, expert patients, and even social scientists, it is clear that research involving the elite creates particular challenges for the researcher. As we have discussed, the elite might have a public profile, which makes them potentially easy to identify, but there are also drawbacks. Being involved in a research project, where they might not be fully in control of the output, might be seen as a risky strategy. We can also assume that elites are particularly busy professionals, which places additional burdens on researchers to be flexible with timing, and be ready to grab opportunities for contact wherever possible. But, we also recognize, that as the experts in their field, they are also highly engaged. If the researcher gets it right, and prepares for the interview or observation, then the elite can be agreeable, supportive, and above all, interested in progressing the research. Thus, despite the challenges, the benefits of researching elite, and potentially being helped to produce good research, make the endeavor worthwhile.

Looking forward, the future for qualitative research with health elites is both demanding and rewarding. Novel medical technologies and new organizational and interactional forms in healthcare mean the ongoing succession of new research topics in health will continue. Innovative technologies arise from the laboratories and the increasing complexity of the globalized world means new socioeconomic contexts give rise to uncertainties that will require exploration. What we can be sure about is that, in many cases, these novel situations will have health elite involvement, and it is important there are social scientists there to analyze what happens.

References

Aldridge A. Negotiating status: social scientists and the Anglican clergy. J Contemp Ethnogr. 1993;22:97–112.

Atkinson P. For ethnography. London: Sage; 2015.

Blix SB, Wettergren Å. The emotional labour of gaining and maintaining access to the field. Qual Res. 2015;15(6):688–704.

Cassell J. The relationship of observer to observed when studying up. In: Burgess RG, editor. Studies in qualitative methodology. London: JAI Press; 1988. p. 89–108.

Collins HM. Gravity's shadow: the search for gravitational waves. Chicago: University of Chicago Press; 2004.

Collins H, Evans R. Rethinking expertise. Chicago: University of Chicago Press; 2007.

Dexter LA. Elite and specialised interviewing. Evanston: Northwestern University Press; 1970.

Dimond R. Patient and family trajectories of mitochondrial disease: diversity, uncertainty and genetic risk. Life Sci, Soc Policy. 2013;9(1):2.

Dimond R. Parent-led conferences as sites of medical work. Health. 2014a;18(6):631–45.

Dimond R. Negotiating blame and responsibility in the context of a 'de novo' mutation. New Genet Soc. 2014b;33(2):149–66.

Dimond R. Negotiating identity at the intersection of paediatric and genetic medicine: the parent as facilitator, narrator and patient. Sociol Health Illn. 2014c;36(1):1–14.

Dimond R. Techniques of donation: 'three parents', anonymity and disclosure. J Med Law Ethics. 2015a;3(3):165–73.

Dimond R. Social and ethical issues in mitochondrial donation. Br Med Bull. 2015b;115(1):173–82.

Dimond R, Bartlett A, Lewis JT. What binds biosociality? The collective effervescence of the parent conference. Soc Sci Med. 2015;126:1–8.

Epstein S. Impure science: AIDS, activism and the politics of knowledge. Berkeley, California: University of California Press; 1996.

González-Santos S, Dimond R. Medical and scientific conferences as sites of sociological interest: a review of the field. Sociol Compass. 2015;9(3):235–45.

Hammersley M, Atkinson P. Ethnography: principles in practice. 3rd ed. London: Routledge; 2007.

Harvey WS. Strategies for conducting elite interviews. Qual Res. 2011;11(4):431–41.

Hunter A. Local knowledge and local power: notes on the ethnography of local community elites. J Contemp Ethnogr. 1993;22:35–58.

Israel M. Research ethics and integrity for social scientists: beyond regulatory compliance. 2nd ed. London: Sage; 2015.

Kent J. Lay experts and the politics of breast implants. Public Underst Sci. 2003;12(4):403–21.

Lassiter L. The Chicago guide to collaborative ethnography. Chicago: University of Chicago Press; 2005.

Lewis J, Bartlett A. Inscribing a discipline: tensions in the field of bioinformatics. New Genet Soc. 2013;32(3):243–63.

Liamputtong P. Qualitative research methods. 4th ed. Melbourne: Oxford University Press; 2013.

Luam C, Sima J. Interviewing one's peers: methodological issues in a study of health professionals. Scand J Prim Health Care. 2006;24(4):251–6.

Mallik M. Patient representatives: a new role in patient advocacy. Br J Nurs. 1997;6(2):108–13.

Mikecz R. Interviewing elites addressing methodological issues. Qual Inq. 2012;18(6):482–93.

Morrissey C. On oral history interviewing. In: Dexter LA, editor. Elite and specialised interviewing. Evanston: Northwestern University Press; 1970. p. 109–18.

Odendahl T, Shaw AM. Interviewing elites. In: Gubrium JF, Holstein JA, editors. Handbook of interview research: context & method. Thousand Oaks: Sage; 2002. p. 299–316.

O'Riordan K, Fotopoulou A, Stephens N. The first bite: imaginaries, promotional publics and the laboratory grown burger. Public Underst Sci. 2016; https://doi.org/10.1177/0963662516639001.

Ostrander SA. Surely you're not in this just to be helpful – access, rapport, and interviews in three studies of elites. J Contemp Ethnogr. 1993;22:7–27.

Patton MQ. Qualitative research and evaluation methods. 4th ed. Thousand Oaks: Sage; 2015.

Payne G, Payne J. Key concepts in social research. Sage key concepts. London: Sage; 2004.

Petkov MP, Kaoullas LG. Overcoming respondent resistance at elite interviews using an intermediary. Qual Res. 2016;16(4):411–29.

Prior L. Using documents in social research. London: Sage; 2003.

Roth D. A third seat at the table: an insider's perspective on patient representatives. Hastings Cent Rep. 2011;41(1):29–31.

Stacey G, Stephens N. Social science in a stem cell laboratory: what happened when social and life sciences met. Regen Med. 2012;7(1):117–26.

Stark L, Hedgecoe A. A practical guide to research ethics. In: Bourgeault I, Dingwall R, De Vries R, editors. The sage handbook of qualitative methods in health research. London: Sage; 2010. p. 589–607.

Stephens N. Collecting data from elites and ultra elites: telephone and face-to-face interviews with macroeconomists. Qual Res. 2007;7(2):203–16.

Stephens N. Growing meat in laboratories: the promise, ontology, and ethical boundary-work of using muscle cells to make food. Configurations. 2013;21(2):159–81.

Stephens N, Dimond R. Unexpected tissue and the biobank that closed: an exploration of value and the momentariness of bio-objectification processes. Life Sci, Soc Policy. 2015a;11(1):14.

Stephens N, Dimond R. Closure of a human tissue biobank: individual, institutional, and field expectations during cycles of promise and disappointment. New Genet Soc. 2015b;34(4):417–36.

Stephens N, Atkinson P, Glasner P. The UK stem cell bank: securing the past, validating the present, protecting the future. Sci Cult. 2008a;17(1):43–56.

Stephens N, Atkinson P, Glasner P. The UK stem cell bank as performative architecture. New Genet Soc. 2008b;27(2):87–98.

Stephens N. In Vitro Meat: Zombies on the Menu? Scripted. 2010;7(2):394–401.

Stephens N, Atkinson P, Glasner P. Documenting the doable and doing the documented: bridging strategies at the UK stem cell bank. Soc Stud Sci. 2011;41(6):791–813.

Stephens N, Atkinson P, Glasner P. Institutional imaginaries of publics in stem cell banking: the cases of the UK and Spain. Sci Cult. 2013a;22(4):497–515.

Stephens N, Lewis J, Atkinson P. Closing the regulatory regress: GMP accreditation in stem cell laboratories. Sociol Health Illn. 2013b;35(3):345–60.

Zuckerman H. Scientific elite. New Brunswick: Transaction; 1996.

Eliciting Expert Practitioner Knowledge Through Pedagogy and Infographics

127

Robert H. Campbell

Contents

1 Introduction	2214
2 Domain-Specific Definitions of Expertise	2214
3 Estimating the Correct Number of Interview Candidates	2216
4 Learning-Based Knowledge Elicitation	2216
5 Cognitive Dissonance Theory	2218
6 Conducting the Research Interviews	2219
7 What the Findings Tell Us	2221
8 Conclusion and Future Direction	2222
References	2223

Abstract

Qualitative research routinely requires expert practitioner knowledge to be elicited. However, effectively eliciting tacit or implicit knowledge can be problematic. This chapter presents a method in which pedagogy and infographics were combined to elicit the knowledge of expert professionals. During interlocutions, using a progressive series of infographics accompanied by explanations, research participants were quickly taught new topics. Then as the learning occurred, they were asked to reflect on their experience using their new knowledge as a lens. Deployed with Information Systems practitioners, the approach was effective, bringing forth 130,000 words of relevant and advanced discourse. Although details of the Information Systems research are presented in the chapter as an illustration, the chapter's foci are the method's underpinning principles and deployment. It is believed that this approach could be easily transferred into a range of qualitative research domains. Given the ambiguity

R. H. Campbell (✉)
The University of Bolton, Bolton, UK
e-mail: R.H.Campbell@bolton.ac.uk

© Springer Nature Singapore Pte Ltd. 2019
P. Liamputtong (ed.), *Handbook of Research Methods in Health Social Sciences*,
https://doi.org/10.1007/978-981-10-5251-4_136

surrounding the term, the concepts of expert and expertise are also discussed along with the challenge of establishing definitions for a given domain.

> **Keywords**
> Expert · Expertise · Expert knowledge elicitation · Pedagogy · Infographics · Information systems practitioners · Information systems research · Qualitative research

1 Introduction

Expert knowledge elicitation, although difficult (Kidd 1987), is a proven empirical technique exploited in a range of disciplines (Hoffman et al. 1995). This chapter presents a method in which knowledge was elicited from expert Information Systems (IS) practitioners on the acceptance of new computer systems. A method that, it is proposed, could easily be adapted for deployment in other domains.

Based on a defendable epistemic assumption that significant understanding and good practice can be found in the knowledge and competencies of expert practitioners, the challenge was to interact with experts in a manner that went beyond superficial conversation, and would in addition to their explicit understanding, elicit their implicit and tacit knowledge. What Leonard and Swap (2005) describe as the special forms of "experience-based expertise" or the "deep smarts" that define an expert. This was achieved through pedagogy and a type of graphic known as an infographic. Accordingly, this chapter pursues three major themes: the challenge of defining expertise; pedagogy as a route to knowledge elicitation and the use of infographics in this process; and for illustration purposes, the Information Systems (IS) research for which this method was originally developed.

This chapter discusses my own experience of research that involved the elicitation of knowledge (on the acceptance of new computer systems) among Information Systems practitioners. The chapter opens with a discussion on expertise and the difficultly of defining experts in a given domain. A short section then considers how many experts need to be interviewed in research of this type. The theory behind the pedagogy-based approach by which knowledge was elicited is then addressed before a description of the interviews themselves. Finally, the effectiveness of the approach is reviewed along with suggestions for future work.

2 Domain-Specific Definitions of Expertise

Although expert knowledge elicitation is an established empirical technique (Hoffman et al. 1995), there is no agreed definition of an "expert" or "expertise" that spans all subject matters (Hoffman et al. 1995; Gobet and Campitelli 2007; Germain and Ruiz 2009). The only real cross domain consensus is that expertise constitutes a blend of domain specific knowledge, skills, and experience (Germain and Ruiz

2009). Qualifying criteria are topic dependent (e.g. Germain 2006). Accordingly, a reasonable place to start any expert interview-based research is to acquire or construct a relevant definition of expert.

Definitively defining expertise in any given subject could prove to be a significant research venture in its own right (Germain 2006; Gobet and Campitelli 2007), but this is not normally necessary. In most cases, all that is required is a definition adequate for candidate selection. Hoffman et al. (1995) surveyed definitions of "expert" and proposed a return to craft guilds terminology. It is a significant observation that, failing to find clear definitions in modern literature, they opted to revive a Middle Ages description. Accordingly, Hoffman et al. present a taxonomy with seven respective categories: naivette, novice, initiate, apprentice, journeyman, expert, and master. At one end of this comprehensive spectrum is the naivette "who is totally ignorant of a domain" (p. 132) with masters being those who are the expert in a subdomain, "whose judgements set the regulations, standards or ideals" (p. 132). Most relevant, however, is their definition of an expert:

> The distinguished or brilliant journeyman, highly regarded by peers, whose judgements are uncommonly accurate and reliable, whose performance shows consummate skill and economy of effort, and who can deal effectively with rare or "tough" cases. Also expert is one who has special skills or knowledge derived from extensive experience with sub domains. (p. 132)

The focus of the IS research in which this method was deployed was the user acceptance of new computer systems. Interview candidates needed to be experts in this field, and unsurprisingly, there was no existing definition. Accordingly, based on the Hoffman et al. (1995) definition, the following was constructed.

IS implementation experts:

- Are highly regarded by their peer group and referred to using distinguishing terminology such as leader, expert, best, or strongest
- Have practitioner experience in excess of 8 years
- Have led the implementation of least three major systems and have participated in many more
- Have a proven track record of dealing effectively with "tough" situations

The numerical values contained in this definition came from my own experience of corporate recruitment, in which minimal experience requirements were set for those entering senior positions. For more general fields of expertise, the criteria set by professional bodies could be of assistance. It is not claimed that this definition is definitive but it was adequate. It was also, for verification purposes, exposed to a selection of practitioners in the field who confirmed it to be a reasonable definition. Participants who met these criteria were then selected from a range of sectors and organizations.

A common belief sometimes even quoted on popular television (e.g., BBC 2015) is that 10,000 h of practice produces an expert, regardless of the individual, practice, or domain. Given its prominence, this merits a special mention. Popularized by the best-selling book "Outliers" (Gladwell 2008), the "10,000-Hour Rule" (Gladwell 2008) is perhaps best described as a misrepresentative simplification of Ericsson et al. (1993). Although not without merit, no credible academic literature supports this simplistic definition.

3 Estimating the Correct Number of Interview Candidates

Estimating the correct number of purposively sampled participants is also known to be problematic (Guest et al. 2006; Onwuegbuzie and Leech 2007). For qualitative research, general guidance is that data gathering should continue until saturation has been reached (Onwuegbuzie and Leech 2007; Liamputtong 2013; see also ▶ Chap. 63, "Mind Maps in Qualitative Research"). Reviewing use of the common term "theoretical saturation," Guest et al. (2006) found that although this was routinely proposed as a milestone for establishing a sample size, the same literature "did a poor job of operationalizing the concept of saturation, providing no description of how saturation might be determined and no practical guidelines for estimating sample sizes for purposively sampled interviews" (p. 60). They go on to review work where interview numbers are suggested, exposing an erratic set of figures. Although many papers suggest small numbers (perhaps only five or six participants) often to be adequate (Guest et al. 2006), it has to be concluded that no one can say how many interviews are enough.

In my research, the interviews were long (about 90 min) and being conducted "expert to expert" were intensive and productive. After a pilot interview, it was thus estimated that saturation would be reached quite quickly. Accordingly, 23 candidates were originally identified, 19 approached, and then only 15 were interviewed. All were convenience sampled from my personal network. On reflection, just the eight strongest interviews would have been sufficient, but it took all of the interviews to identify those eight. Debatably, the final three also need not have occurred as no new major themes emerged. They did, however, serve as an assurance that saturation had occurred and provided additional supporting examples.

4 Learning-Based Knowledge Elicitation

This method is based on the epistemic assumption that significant understanding and good practice can be found in the knowledge and competencies of expert practitioners. Modeling this on the famous four stages of competence model (see Fig. 1), it could be said that expert practitioners have significant unconscious and conscious competence that causes them to recognize, understand, and manage phenomena. Making the same observation through Kolbs (1984) experimental learning theory (see Fig. 2), it could be said that expert practitioners have concrete experience that they may or may not have reflected on or conceptualized.

In my research, the intention was to use learning as a vehicle by which participants would come to reflectively observe and/or abstractly conceptualize their

Fig. 1 The four stages of competence

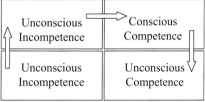

Fig. 2 Kolb's experimental learning cycle (After Kolb 1984)

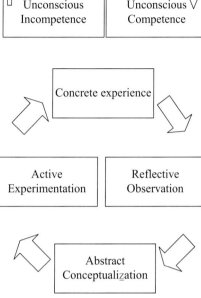

concrete experience. Specifically in this case, the focus was user attitudes during IS implementations. Accordingly, participants were taught notions of attitude change from social and cognitive psychology. As the learning occurred, they were then encouraged to reflect on their experience using this new knowledge as a lens.

The learning was assisted by infographics. Infographics are diagrams (and other graphics) specifically developed to portray information. A famous example is the London Underground map. Infographics seek to provide simple access to just the information that people need for the time that they need it. They need not be memorable, entertaining, or artistic. In the London underground example, information is particularly fleeting and superficial, passengers ignore the complexity of an underground rail system, extracting from the graphic just the information they need at that moment (i.e., which train on which platform).

Participants were provided with a progressive series of infographics accompanied by explanations and guidance that incrementally built their understanding of the relevant theory, sufficient to invoke reflective observation and abstract conceptualization. Their understanding of the theory was neither deep and durable nor precise, but it was adequate. As each interview proceeded, new infographics supported by dialogue incrementally provided additional new knowledge. Then, as the participants learned, they were encouraged to reflect on their practice using their new knowledge as a lens; considering where related phenomena had occurred, the impact, cause, and management responses. The dialogue was then audio recorded and transcribed. In this case, the transcribed data was subjected to a thematic analysis (Braun and Clark 2006; Liamputtong 2013; see also ▶ Chap. 48, "Thematic Analysis").

5 Cognitive Dissonance Theory

The attitude change theories used in this case were Cognitive Dissonance (Festinger 1957), the Elaboration Likelihood Model (Petty and Cacioppo 1986), and an amalgamation of those that have evolved from Tajfel's Social Identity Theory (Tajfel and Turner 1979). Before going onto discuss the interviews, this section introduces Cognitive Dissonance Theory by way of example.

At its most basic, Cognitive Dissonance Theory proposes that people are inclined to change behaviors and attitudes to ensure consistency with belief, values, and perceptions. Failure to acknowledge this consistency of perception causes dissonance. Festinger, and some of those who followed him (e.g., Sakai 1999), actually created formulae to define its impact and scale. The greater the dissonance the greater the motivation to resolve it and the probability of change. Should dissonance be caused by contrasting attitudes, change usually involves weaker attitudes giving way to the stronger. Since the late 1960s, researchers have attempted to understand what motivates dissonance, and three dominant revisions have been proposed with supporting evidence (Harmon-Jones and Harmon-Jones 2007), namely, Self-Consistency Theory (Aronson 1968, 1999), Self-Affirmation Theory (Steel 1988), and "'A New Look at Dissonance Theory" sometimes called Aversive Consequences (Cooper and Fazio 1984):

- Self-Consistency Theory proposes that the self concept (or a violation thereof) is the primary cause of dissonance; people suffer dissonance if they compromise their own self-image.
- Self-Affirmation Theory proposes that people uphold a set of values and thus maintain an overall self-image. Simon et al. (1995) performed significant interpretive work and found that when suffering dissonance, if an individual confirms a value (any value, relevant, or otherwise), then attitude change does not occur, they are effectively distracted from the dissonance and its effect is subdued. The Self-Affirmation premise is that dissonance is caused by the disruption of the holistic self-image, if something then confirms the self-image (anything), consistency is restored.
- A New Look at Dissonance Theory was published after Cooper and Fazio (1984) repeated some of Festinger's original experiments and concluded that dissonance is invoked when an individual feels responsible for possible adverse consequences. When people were paid substantially to tell a lie, they avoided dissonance because the perpetrator of the bribe is the originator but, when the pay is low, the subject has no one else to blame and becomes dissonant.

An early experiment (Aronson and Mills 1959) compared the experiences of women who had undergone an initiation to join a group; some had undergone a severe initiation while others had undergone a mild initiation. Those who underwent the severe initiation were subsequently found to value group membership more. The discomfort of the initiation had invoked dissonance forming attitudes that valued group membership. This is a notion that is now well established and employed for various purposes (e.g., Wicklund and Brehm 1976; Axsom and Cooper 1985; Beauvois and Joule 1996; Draycott and Dabbs 1998). An extreme witness to this would be gang initiations. In another example, Festinger and Carlsmith (1959) paid people to lie, claiming that a tedious task was interesting. When the fee was high, people did not experience dissonance (they did it for the money and felt fine), but when the fee was low, there was significant dissonance. Unable to justify their lies, their attitudes about the job changed and the proclamations thus ceased to be lies and became genuine representations of belief. This can all be aligned with Brehm's (1956) findings that cognitive dissonance is greater when a choice is hard. It stands to reason: if multiple options appeal then choosing can be difficult; conversely, if given the choice between something pleasant and something horrible, the choice is simple.

As well as providing an introduction to cognitive dissonance this section intends to provide some idea of this theory's complexity. The other two attitude change theories employed are equally complex. However, the intention was not that participants should truly understand cognitive dissonance, only that they should learn it a to level where "reflective observation" and "abstract conceptualization" are invoked and it could serve as a lens for reflection on their experience.

6 Conducting the Research Interviews

In my research, each interview began with a confidentiality statement confirming that no information would be published that might enable the participant or any other actor discussed to be identified. Given the nature of the organizations involved, this guarantee was necessary for a significant "warts and all" corpus to be gathered. Given the importance of interviewee comfort (Hair et al. 2000; Babbie 2016), interview locations were decided by participants and accordingly ranged from private offices, to board rooms, coffee shops, and people's homes. Participants were also given breaks as required and in one case the interview was concluded on another day. Once the introductions were complete, participants were shown a simple graphic (see Fig. 3) on an A4 laminated sheet. It might be observed that this is a simple illustration, comparable to what a teacher might quickly draw on a pen board during class.

A simple accompanying explanation was then provided. In each case, it was ad-libbed but went something like this:

> Cognitive dissonance refers to any uncomfortable mental state that could cause attitudes or behaviors to change. For example, anger, disappointment, embarrassment, confusion, shock,

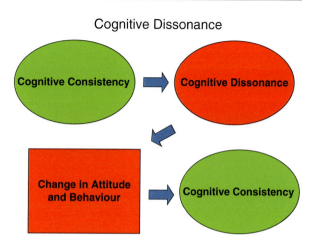

Fig. 3 Infographic used to introduce the concept of cognitive dissonance

moral dilemma and so forth. It is a generic term for uncomfortable cognitive experiences that cause one to reconsider.

As an example, consider a parent who suddenly discovers that an activity is hurting their child. On the diagram, the parent would start in the first green oval "cognitive consistency." Unaware that they are harming their child, their thoughts are consistent and comfortable. Then when they discover that "cake for breakfast everyday" is bad for their child they enter the red uncomfortable area "cognitive dissonance." To get out of this discomfort they need to change their attitudes or behaviors, taking them into the red square. Perhaps their attitude toward cake may change and as a result, they might start to provide something healthier for breakfast, or their attitude toward the source of the information may change causing them to conclude that it is invalid. Either way, with an attitude change complete and perhaps also a change in behavior, they are returned to the comfortable state of cognitive consistency.

Participants were then asked if they felt they had a basic understanding of what cognitive dissonance was. In most cases, they did but if necessary, further clarification and illustrations were provided. They were then guided to reflect on their experiences of cognitive dissonance among users during IS implementations, how it became manifest, its cause, effect, how it was managed and good practice that emerged. In particular, they were encouraged to reflect on real experiences and not imagined scenarios as this is understood to produce more accurate and reliable information (Ericsson and Simon 1993; Cote et al. 2005). As participants concluded a reflection, they were asked if there were any more occasions they could think of. For some participants, this basic understanding of cognitive dissonance kept them reflecting for some time where as others needed additional stimuli after only a few minutes. Accordingly, as required to keep each participant reflecting and conceptualizing, more advanced cognitive dissonance related theory was gradually introduced using additional graphics and explanations. This continued until a substantial amount of discourse had been elicited or it was apparent that no more would be forthcoming. The same process was then undertaken for the second and third attitude change theories.

7 What the Findings Tell Us

This method was effective in eliciting expert knowledge. During the 15 interviews, 160 projects from 111 different organizations were referenced. With an average length of 89 minutes, the interviews produced 22 h and 20 min of relevant discourse, which when transcribed, came to 137,495 words. Table 1 provides a break down of these figures for each participant along with information about their levels of experience.

As well as eliciting a substantive quantity of data, from their facial expressions, postures, pauses, and questions, it was constantly apparent that participants were thinking deeply. Indeed, postinterview some explicitly expressed that this had been the case. Some said that the process had helped them to learn from their own experience, one requested a copy of the audio files for further reflection and another expressed surprise at his own knowledge.

It was also apparent that the infographics had supported the learning process. Without prompting, three participants made positive comments about them. All focused on them and in some way handled them, either tapping, stroking, holding,

Table 1 Basic information about research participants and interlocutions

Participant number	Interview duration (minutes)	Number of organizations that the participant has worked for full time	Years of user acceptance experience	Projects or Systems Referenced	Organizations involved
1	102	1	10	6	2
2	106	1	15	11	11
3	101	6	34	16	11
4	96	2	10	12	3
5	78	1	10	9	8
6	84	6	41	22	20
7	94	1	33	8	5
8	108	1	8	7	1
9	85	8	15	14	8
10	77	1	14	9	1
11	70	6	14	7	8
12	52	3	30	6	2
13	87	18	26	14	14
14	91	1	28	5	5
15	109	4	14	14	12
Total	**1340**	**60**	**302**	**160**	**111**
Average	**89**	**4**	**20**	**11**	**7**

or flicking between them. The effectiveness of each graphic and explanation however differed between participants. Although all participants claimed to understand all of the theories presented, the degree of discourse elicited by each was variable. That which invoked a 40 minute reflection with one participant invoked nothing with another. Given this unpredictability, it was beneficial that a range of theories and infographics were at hand.

With respect to candidate selection and the proposed definition of expertise, as all interviews were effective, this had clearly been adequate. It was apparent, however, that the most impressive insights tended to come from older candidates with over 25 years of experience, suggesting that a status of "elder expert," unrecognized by this method, may have been beneficial. The intensity and momentum of the discourse was also aided by the researcher also being an "expert" who accordingly understood industry trends, history, and shibboleths. This became evident when during the tenth interview a "PDM system" was mentioned, something of which I had no knowledge. At this point, the interview digressed for 439 words while I acquired an understanding adequate to continue. This deviation constituted about 5% of that interview. An apparent limitation, therefore, is that this method can not tolerate many such examples of interviewer ignorance.

A final observation is the breadth of discussion that was invoked. As an inductive investigation, this was beneficial, drawing out events from any number of projects in each participant's career. Care did need to be taken, however, to retain a boundary, in this case ensuring that only Information System implementations were being discussed. Participants easily, for example, digressed onto "attitudes towards cycling helmets" and responses to government legislation. Although such tangents were discouraged, they support the proposition that this method could be reapplied to acceptance in other scenarios.

8 Conclusion and Future Direction

This research has demonstrated that pedagogy supported by graphics can assist in the elicitation of expert knowledge. In this case, the focus was the introduction of new Information Systems, but with little adaption the same method could be applied to elicit expert knowledge on the acceptance of other change, such as changes to organizational structures, quality systems, work-related procedures, regulations, legislation, or corporate strategy. Using different theory and graphics, it is also believed that this method could be adapted to elicit expert knowledge in a range of different domains.

An apparent limitation of the method is the required level of interviewer competency. As it stands, they must have both an expert knowledge of the subject matter and the ability to educate. In this case, as theoretical saturation was reached quickly, it was not an obstacle, but in other applications where participants are more geographically dispersed or numerous, it could be problematic. To overcome this, a more systematic version of the method might be developed. Could the learning for example be facilitated by e-learning and a standard set of follow-up prompts devised? If the process could be simplified or automated, this would enable more interviewers to be involved and/or more participants to be reached, increasing the range of research areas to which it might be applied.

References

Aronson E. Dissonance theory: progress and problems. In: Abelson RP, Aronson E, McGuire WJ, Newcomb TM, Rosenberg MJ, Tannenbaum PH, editors. Theories of cognitive consistency: a sourcebook. Chicago: R& McNally; 1968. p. 5–27.

Aronson E. Dissonance, hypocrisy, and the self-concept. In: Harmon-Jones E, Mills J, editors. Cognitive dissonance: progress on a pivotal theory in social psychology. Washington, DC: American Psychological Association; 1999. p. 103–26.

Aronson E, Mills J. The effect of severity of initiation on liking for a group. J Abnorm Soc Psychol. 1959;59:177–81.

Axsom D, Cooper J. Cognitive dissonance and psychotherapy: the role of effort justification in inducing weight loss. J Exp Soc Psychol. 1985;21:149–60.

Babbie E. The practice of social research. 14th ed. Boston: Cengage Learning; 2016.

BBC. The woman who lived. Dr Who. The British Broadcasting Corporation. 2015.

Beauvois JL, Joule RV. A radical dissonance theory. London: Taylor and Francis; 1996.

Braun V, Clarke V. Using thematic analysis in psychology. Qual Res Psychol. 2006;3:77–101.

Brehm JW. Post decision changes in the desirability of alternatives. J Abnorm Soc Psychol. 1956;52:384–9.

Cooper J, Fazio RH. A new look at dissonance theory. In: Berkowitz L, editor. Advances in experimental social psychology, vol. 17. Orlando: Academic; 1984. p. 229–64.

Cote J, Ericsson KA, Law MP. Tracing the development of athletes using retrospective interview methods: a proposed interview and validation procedure for reported information. J Appl Sport Psychol. 2005;17:1–19.

Draycott S, Dabbs A. Cognitive dissonance: an overview of the literature and its integration into theory and practice of clinical psychology. Br J Clin Psychol. 1998;37:341–53.

Ericsson KA, Simon HA. Protocol analysis: verbal reports as data. Rev ed. Cambridge, MA: The MIT Press; 1993.

Ericsson KA, Krampe RT, Tesch-Romer C. The role of deliberate practice in the acquisition of expert performance. Psychol Rev. 1993;100:393–4.

Festinger L. A theory of cognitive dissonance. Stanford: Stanford University Press; 1957.

Festinger L, Carlsmith JM. Cognitive consequences of forced compliance. The Journal of Abnormal and Social Psychology. 1959;58(2):203–210.

Germain M. Stages of scale development and validation: the example of the generalized expertise measure (GEM). Columbus: Academy of Human Resource Development; 2006.

Germain M, Ruiz CE. Expertise: myth or reality of a cross-national definition? J Eur Ind Train. 2009;33(7):614–34.

Gladwell M. Outliers: the story of success. New York: Little, Brown and Company; 2008.

Gobet F, Campitelli G. The role of domain-specific practice, handedness and starting age in chess. Dev Psychol. 2007;43:159–72.

Guest G, Bunce A, Johnson L. How many interviews are enough? An experiment with data saturation and variability. Field Methods. 2006;18(1):59–82.

Hair JF, Bush RP, Ortinau DJ. Marketing research: a practical approach for the new millennium. Boston: Irwin/McGraw-Hill; 2000.

Harmon-Jones E, Harmon-Jones C. Cognitive dissonance theory after 50 years of development. Z Sozialpsychol. 2007;38:7–16.

Hoffman RR, Shadbolt NR, Burton AM, Klein G. Eliciting knowledge from experts: a methodological analysis. Organ Behav Hum Decis Process. 1995;62(2):129–58.

Kidd AL, editor. Knowledge acquisition for expert systems: a practical handbook. New York: Plenum Press; 1987.

Kolb DA. Experiential learning: experience as the source of learning and development. Englewood Cliffs: Prentice-Hall, Inc.; 1984.

Leonard D, Swap WC. Deep smarts: how to cultivate and transfer enduring business wisdom. Boston: Harvard Business School Publications; 2005.

Liamputtong P. Qualitative research methods. 4th ed. Melbourne: Oxford University Press; 2013.

Onwuegbuzie AJ, Leech NL. A call for qualitative power analyses: quality & quantity. Int J Met. 2007;41:105–21.

Petty RE, Cacioppo JT. Communication and persuasion: central and peripheral routes to attitude change. New York: Springer; 1986.

Sakai H. A multiplicative power-function model of cognitive dissonance: toward an integrated theory of cognition, emotion, and behavior after Leon Festinger. In: Harmon-Jones E, Mills J, editors. Cognitive dissonance: progress on a pivotal theory in social psychology. Washington, DC: American Psychological Association; 1999. p. 267–94.

Simon L, Greenberg J, Brehm JW. Trivialization: the forgotten mode of dissonance reduction. J Pers Soc Psychol. 1995;68:247–60.

Steele CM. The psychology of self-affirmation: sustaining the integrity of the self. In: Berkowitz L, editor. Advances in experimental social psychology, vol. 21. San Diego: Academic; 1988. p. 261–302.

Tajfel H, Turner JC. An integrative theory of intergroup conflict. In: Austin WG, Worchel S, editors. The social psychology of intergroup relations. Monterey: Brooks-Cole; 1979. p. 33–47.

Wicklund RA, Brehm JW. Perspectives on cognitive dissonance. Hillsdale: Erlbaum; 1976.

Index

A
Aboriginal Chronic Disease Support Group, 1574
Aboriginal people, 1713–1716
  equality, 1712
  ethical approval, 1717–1718
  historical research, awareness of, 1711–1712
  National Aboriginal and Torres Strait Islander Suicide Prevention Strategy, 1716
  participatory action research, 1717
  principles of research, 1714–1715
  reciprocity, 1712
  reporting and disseminating findings, 1718–1719
  research participants, 1718
  respect, 1712
  responsibility, 1713
  spirit and integrity, 1712
  suicide in Aboriginal communities, 1709–1711
  survival and protection, 1713
Aboriginal people, in Australia
  cultural protocol, 1567
  data collection, for research, 1568–1569
  focus groups, 1569
  grounded theory, 1567
  Health workers, 1566
  professional identity, 1566
  research findings, 1571–1572
Academic presentation, research findings and public policies, *see* Research findings and public policies
Access negotiation, 2189–2190
Accountability, 274
Adapted alternating treatments design (AATD), 592
Adaptive randomization, 649
Adequate randomization methods, 648
Ad-hoc approach, 1937
Adorno, Theodor W., 136
Advocacy, 54, 56, 57, 59, 1885
Africana womanism, 1582
Africanist Sista-hood in Britain, 1581–1584, 1595–1597
Agency, 2012, 2018
Alcohol Use Disorders Identification Test (AUDIT), 1980
Allegiance, 2083–2085
Allocation concealment, 649
Alternating treatments design (ATD), 590–592
Altruism, 77
Ambivalence, 2087–2089
A Measurement Tool to Assess Systematic Reviews (AMSTAR 2), 1030
American Anthropological Association, 429
American Oral History Association, 430
American Time Use Survey (ATUS), 1226
Analysis, 918, 919
  data management, 922–924
  data selection, 920–922
  goals of, 919–920
Analysis of networks, 770
Analysis of variance (ANOVA) approach, 612–614
Analysis procedures, 863, 870, 873, 878
Analytic reflexivity, 514, 515
Animated visual arts researching, 2100
  background projects, 2094–2098
  fragments and vignettes, 2101–2103
  health condition symptoms, 2099–2100
  humor incorporation, 2103–2104
  invisible visualisation, 2105–2106
  penetration, 2106
  snapshots, 2100, 2101

Animated visual arts researching (*cont.*)
  thumbnails, 2100, 2101
  triangulated heuristic framework, 2098–2099
Animation, *see* Animated visual arts researching
Anonymity, 1358–1360, 1978–1979
Anorexia, 100
Antirealism, 162–163
Aotearoa New Zealand, 1508, 1511
Application programming interfaces (APIs), 501
Applied conversation analysis, 484–485
Appraisal, 1014–1016, 1022, 1023, 1025
  criteria for, 1014
  guidelines, 1024
  qualitative research, 1015–1017
  qualitative studies, 1015
Array factor, 758–761
Arts-based health research, 1304, 1316
Arts-based methods, 1256, 1262
Arts-based research (ABR)
  arts and health movement, 1136
  creative arts therapies, 1136–1138
  creative insight method, 1077
  definition, 1076, 1132
  example, 1138–1139
  methodology in healthcare, 1133–1134
  narrative writing, 1141–1142
  poetry writing, 1142
  theoeretical location, 1134–1136
  visual arts, 1142
Assent, 1998
Asylum seekers, 1812, 1816, 1823, 1825, 1873–1874
Asynchronous email interviewing method, 1084
Attachment, 1898
Attention-deficit hyperactivity, 108, 111
Attention probes, 406
Attributable proportion, 570
Audience(s)
  analysis, 357–359, 362
  and online face-work, 1358–1360
AUDIT, *see* Alcohol Use Disorders Identification Test (AUDIT)
Australia, 1692, 1694, 1695, 1697, 1699, 1702
Autism spectrum disorder (ASD), 1158
Autobiographical narrative, 414–415
Autoethnography, 1504, 1830, 1832–1834, 1836, 1840, 1842, 1843
  analytic, 514–515
  characteristics, 515
  data analysis, 519
  description, 510, 513
  design questions for, 516–518
  methodological parameters, 511
  methodology, 515
  theory and methods, 511
Automethodologies, 511–513

B

Background/introduction (BI) section, 988–989
Bakhtin, Mikhail, 2069, 2075, 2076
  authoritative and internally persuasive discourse, 2065
  dialogism, feminist approach to, 2063
Bayesian meta-analysis, 790
Being
  and becoming, 194–195
  language, 197–198
  meaning, 193
  thereness of, 193–194
  in world of care, 195
Best worst scaling (BWS)
  DCE, 633–634
  objects, 631–632
  profile, 632–633
Bhaskar, Roy, 161, 164, 165
Bias(es), 238, 239, 241, 248, 250, 251, 666, 667
Biological plausibility, 576
Biomedical model, 104–105
Biopsychosocial model, 105–106
Black feminism, 1582, 1583
Black women
  insider/outsider research, 1582, 1583, 1586, 1588, 1590, 1591, 1593, 1595, 1596
  working in public sector, 1581
Blinding, 649–650
Block randomization, 648
Blogs/Blogging, 1083
  and confessional society, 1355
  as documents of life, 1356–1358
Blogs in social research
  data analysis, 1361–1362
  data collection, 1360–1361
  ethics, 1362–1364
  qualitative research, 1354, 1359
Body mapping
  definition, 1238–1239
  ethical aspects of, 1251–1253
  examples, 1241
  history of, 1240
  method, 1080–1081
  planning for study, 1241–1245

process of, 1239–1240
  research using, 1246
  as therapeutic tool, 1240
  uses, 1241
Bottom-up cueing, 1227
Breaching, 277
Breastfeeding, 1765, 1766, 1768, 1770, 1772
BringItOn, 360
Brokerage, 773, 774, 777, 780
Bronfenbrenner's ecological systems theory, 1203
BWS, *see* Best worst scaling (BWS)

C
Calendar methods, 1225
  health-related issues, 1231–1233
  self-administered, 1231
  well-being and surveys, 1224–1227
Canada, Australia, Aotearoa New Zealand and the United States (CANZUS), 1503
Card-based discussion method, 1171–1172
Cartoon captioning, 2030–2031
Case control study, 565
Case study research, 318
  advantages of, 323–324
  definitions, 319–320
  generalization (*see* Generalization)
  limitations of, 321–323
  multiple-case design, 321
  single-case design, 321
Cellphone surveys, 1084
  epidemiological biases
    (*see* Epidemiological biases, in cellphone surveys)
  mobile phone subscriptions, 1404–1405
Central organizing concept, 845, 856
Centre for Digital Storytelling (CDS), 1305
The Centre for Evidence Based Medicine's systematic review appraisal tool, 1030
Centre for Oral History and Digital Storytelling, 430
Channel analysis, 358
Children, 2008–2009
  cultural dimension, 2010–2012
  ethics and research with, 2016–2017
  psychological dimension, 2012–2013
  research methods (*see* Research methods)
  socio-political dimension, 2009–2010
Children's perspectives on health and illness, 1726
Child sexual abuse, 2062, 2065, 2066, 2070–2073, 2075, 2076

Choice of analysis model, 957, 965
Clarification probes, 406
Classic question response model, 1226
Cluster analysis, 950
Cluster randomized trials, 655–656
Cochrane Qualitative and Implementation Methods Group (CQIMG), 787
Code, 847, 848, 855
  deductive orientation, 853
  inductive orientation, 853
  latent, 853
  semantic, 853
Codebook, 847, 849
Coding, 303–306, 918, 920, 922, 923, 926–931, 933, 934
  reliability, 847–849, 857
Cognitive impairment, 1983–1985
Cohort study, 565
Collaboration, 385, 416, 771, 772, 774, 776–778, 1519, 1712, 1713, 1716, 1718, 1719
Collage, 1746, 1747
Collective analysis, 534, 536
Collective memory, 426
Colonization, 255–257, 1692, 1695, 1698, 1699
Coming Through, 2094, 2097, 2103, 2106
Community-based participatory research (CBPR), 1850, 1854, 1863
  community participation, 289–290
  community strengths and dynamics, 292
  definition, 286
  design and conduct research, 293
  disseminate and translate research findings, 293
  feedback and interpret findings, 293
  health priorities, 292–293
  methodological and ethical considerations, 295
  partnership, 292, 294
  power equilibrium, 290
  principles, 288–289
  research process, 291
  social justice and equity, 291
  training, 294
Community case management (CCM), 339
Community engagement, 1567, 1576, 1577
Community-Up approach, 1517
Comparative effectiveness research (CER), 378, 659
Complementary research methods, 1184
Computer-assisted interviews (CAIs), 1778
  accessible and attractive interfaces, 1782
  children's agency and control, 1781–1782

Computer-assisted interviews (CAIs) (*cont.*)
    quantitative and qualitative analyses, 1782
    refugee children's well-being, 1783–1789
Computer-assisted qualitative data
    analysis, 1248
Comte, Auguste, 154, 156, 157
Concept maps, 1745, 1750
Concourse, 754–755
Concurrent triangulation design, 701–702
Confessional society, blogging, 1355
Confidentiality, 1897
Confounding, 572
Connectivity, 772
Consensus, 717–719, 721
    methods, 739, 746
Consent, 1996–1999
Consistency, 576
Consolidated Criteria for Reporting Qualitative
    Health Research (COREQ), 973, 1052
Consolidated Standards of Reporting Trials
    (CONSORT) statement, 972, 987, 1040
Construct elicitation, 1101–1103, 1105
Constructive alternativism, 1096
Constructive empiricism, 163
Constructivism, 58, 60, 834–835, 1098, 1116
Constructivist epistemology, 1096–1097
Constructivist GTM (CGTM), 301
Constructivist paradigm, 1655, 1656
Constructivist research, 300–303, 307–308,
    311–313
Constructivist sensibility, 1135
Constructivist theories, 1097
Construct validity, 670–671
Construing, 1097–1098, 1100, 1103
Consumers, 1936, 1944
Contact varieties, 1622
Content analysis (CA), 178–179, 901
    conventional, 830
    and critical realism (*see* Critical realism)
    definition, 828–829
    directed, 830
    history, 830
    qualitative descriptive studies, 836
    summative, 829–830
Content validity, 670
Continuation probes, 406
Conventional content analysis, 830
Convergent design, 63
Convergent parallel, 701–702, 705
Conversation analysis, 278
    applications, 483–485
    data analysis, 479–483
    data collection, 477–479

description, 472
development, 473–474
empirical ground of analysis, 475
institutional interactions, 475–476
methodological foci, 474
sequential organization, 475
structural organization, 474–475
Conversation Analytic Roleplay Method
    (CARM), 484
Conversion hysteria, 213
COREQ checklist, 975
Co-research, 1947
Cost-benefit analysis, 361
Creative arts practice, 2106, 2107
Creative arts therapies, 1136–1138
Creative research methods, 1188
    See also Innovative research methods
Credibility, 1964
Crisis in social psychology, 119
Criterion validity, 671–672
Critical alternatives, 106–108
Critical appraisal
    definition, 1028
    mixed methods research (*see* Mixed
        methods research)
    observational studies, 1040–1046
    skills, 1028, 1029
    systematic reviews, 1029–1034
Critical Appraisal Skills Programme (CASP)
    checklist, 1030
Critical discourse analysis (CDA), 128–129
Critical ethnography, 511
    description, 225
    HIV, marriage and mining in Papua New
        Guinea, 229–231
    objective-subjective dichotomy, 232–234
    reflexivity, 231–232
Critical interpretive synthesis, 789
Critical psychology, 126
Critical realism, 164–165
    actual reality, 834
    constructivism, 834
    critical theory, 835
    definition, 832
    empirical reality, 834
    knowledge production, 833
    mechanisms, 833
    positivism, 834
    qualitative descriptive research, 836
    real, 834
    social norms, 833
    social structures, influence of, 833
    use of, 837–839

Critical reflexive practice, 262
Critical Skills Appraisal Program (CASP), 1052
Critical theory, 134, 135, 137, 835
   epistemological content of, 136–140
   epistemology, 134
   method of, 140–143
   normative turn, 143–147
   political epistemology, 135
Cross-cultural research, 1499
Cross disciplinary research, 198–201
Cross-language research strategy
   finding capable interpreters, 1641–1643
   social position and subjectivity, of interpreters, 1643–1646
   survey translation, 1646–1647
   in Uganda, 1647, 1649
Crossover designs, 605, 607–609
Crossover studies, 656–657
Cross sectional study, 565
Crystallization, 519
Cultural awareness, 239, 245–246
Cultural dimension, 2010–2012
Culturally and linguistically diverse (CALD) communities, 227
Culturally Deaf women, 1580, 1581, 1590
   interviewing in sign language, 1585, 1587
Cultural materialism, 458
   Marxist, 462
Cultural research on human development, ethical issues, *see* Ethical issues, human development
Cultural responsiveness, 1527, 1539
Cultural safety, 1526, 1527, 1529, 1530
Cultural understanding, 1831, 1832
Culture, 225, 301, 307–310, 313, 314
   as an object of intervention, 227–228
   ethnographic tradition and critique, 226
   structural determinants of disempowerment and health, 228–229
Cumulative duration, 946

D

Dasein, 193
Data analysis, 29, 41, 45–47
Data collection, 416, 991, 1118, 1568–1569, 1975, 1979, 1982, 1984
Data enhancers, 17–18
Data quality, mixed methods research
   analytic adequacy, 1060
   analytic integration rigor, 1060
   data rigor, 1059
   data transparency, 1059
   sampling adequacy, 1059
Data saturation, 19
Decision analysis (DA), 380
Decision-making tools, translational research, *see* Translational research
Decolonization, 1510, 1512, 1521
Decolonizing methodologies, 1544, 1545, 1547, 1549, 1551, 1552, 1555
Decolonizing research, 1852–1855, 1865
Deliberation, 1169
Delphi technique, 716, 731
   applications, 718
   modifications, 724
   qualitative-quantitative design, 730
   question types, 717
   roles, 718
Dementia, 1992
   consent, 1996–1999
   DEEP, 1993
   methodological considerations for research, 2000–2001
   people recruitment, 1994–1996
   perceptions of, 1992
   research methods and approaches, 1999–2000
Dementia Engagement and Empowerment Project (DEEP), 1993
Demonstration designs
   for non-reversible behaviors, 589–590
   for reversible behaviors, 588
Descriptive study, 565
Designing for dissemination and implementation (D4D&I), 368, 372
Design quality, mixed methods research
   design rigor, 1059
   design strength, 1059
   design suitability, 1059
   design transparency, 1058–1059
Development, 1681, 1686, 1688
Dewey, John, 158
Diabetes care, for Aboriginal people, *see* Aboriginal people, in Australia
Dialectics, 134, 136–139
   and intersubjectivity, 143
Digital afterlife, 1314–1315
Digital methods
   advantages, 1082
   asynchronous email interviewing method, 1084
   blogs, 1083
   cell phone survey, 1084

Digital methods (*cont.*)
  netnography, 1083
  phone survey introductions and response rates, 1084
  synchronous text-based online interviewing method, 1084
  web-based survey methodology, 1083
Digital story sharing, 1312–1314
Digital storytelling (DST), 437, 1304–1305, 1879
  codified process, 1305–1306
  defined, 1305
  elements of, 1306
  initiating critical dialogue, 1307–1314
  method, 1081
  negotiating voice, 1310–1312
  voice produce, 1309–1312
  workshop-based practice, 1306–1307
Diglossia, 1622
Directed content analysis, 830
Disability, 1939–1941
  interview methods for child with (*see* Interview methods, disabled children)
  and sexuality, 1956–1957
Discourse analysis (DA), 128, 882
  components of, 887
  and discursive psychology, 882–884
  and FDA, 884–886
  feminist discourse analysis, 887–893
Discrete choice experiments (DCEs), 624, 625, 627, 629, 631, 633, 634
  attributes and levels, 636–638
  cognitive burden, 635
  comparative study, 635
  data collection, 640
  experimental design, 638–639
  health-related research, 627, 635
  reporting, results, 640
  statistical analysis, 640
  UK study of preferences, 628
Discursive practices, 885, 886
Discursive psychology (DP), 129–130, 882–884, 887
Disease, 101
Displacement, 1762, 1773
Display of family life, 1466
Distress, 1981–1983
Documents of life, blogs as, 1356–1358
Dominant western paradigm, 263
Donor-driven prevention programs, 231
Dose-response, 576
Drapetomania, 107

Drawing, 2070, 2072–2076
Drawing method
  adult participants, 1762
  breast milk, 1772, 1773
  cultural beliefs, 1768
  data collection, 1765
  draw and talk/draw and write techniques, 1762
  empathy, 1765
  empowerment, 1767
  encouragement, 1765
  ethnicity, 1763
  freedom in Australia, 1771–1772
  motherhood, 1769–1770
  mother's rejection, 1765
  postmodernism, 1761
  refugee backgrounds, 1762–1763
  "trial-and-error" adjustments, 1764
  vulnerability, 1770–1771
Drug use, 1975, 1976, 1978, 1980, 1982, 1985
DST, *see* Digital storytelling (DST)
Dual frame telephone surveys, 1405–1407
Dubbo Aboriginal Research Team, 1567

E

Eating disorder, 2062, 2065–2068, 2070–2073, 2075, 2076
Ecological study, 566
Ecological systems theory, 1203
Effect size test, 674
Electronic interview via email, 1390–1395
Elites, 1907, 1909, 1914
  health (*see* Health elites)
Email
  as communication and research tool, 1387–1390
  delayed responses and absence of punctuation, 1396
  electronic interview via, 1390–1395
  indirect contact and body language, 1396–1397
  informality, 1397–1398
  as interview tool, 1395–1398
  sampling possibilities and cost-effectiveness, 1398
  validation, 1397
Emancipation, 1741
Emancipatory research, 1938
Embedded design, 702–703
Embodiment research, 1080–1082
Emoticon, 1396

Emotional labor, 2147
Emotions and sensitive research
    data analysis, 2150–2151
    data collection, 2149–2150
    emotion work, 2146–2148
    ethics, 2148–2149
    informal support, 2154–2155
    protocols and guidelines, 2155–2157
    research assistants, 2152–2153
    secondary data, 2153–2154
    transcriptionists, 2151–2152, 2157–2158
Empathic testimonial approach, 1925
Empathy, 238–240, 246
    interdisciplinary, 240–241
    limitations, 250–251
    modes of, 241–243
    practicing, 243–250
    risk of, 250–251
    self-reflection, 248–250
    situational, 243
Empirical studies, 992
Emplacement, 1276
Empower, 1169
Empowerment, 1741
End-of-life care, 828, 836, 837, 839
Endometriosis, 215
Engagement, 1168, 1170
    choreography, 1177
    public, 1169, 1171, 1177, 1179
    rethinking, 1168
Enhancing the Quality and Transparency of Health Research (EQUATOR) Network, 972
Enhancing Transparency in Reporting the Synthesis of Qualitative Research (ENTREQ) statement, 973, 977
Epidemiological biases, in cell phone surveys
    cost-efficiency and feasibility, 1412–1413
    non-coverage bias, 1408–1410
    non response bias, 1410–1411
Epidemiology, 561
    causes, 573–577
    challenges, 577–578
    confounding, 572
    descriptive, 561
    determinants, 561
    experimental study, 564
    history of, 561–563
    measurement error, 573
    observational study, 565–566
    random error, 571–572
    selection bias, 573

Episodic records, 496–497
Epistemology, 102–104, 152, 154, 159–161, 167, 225, 853, 857
    arts-based research, 1138–1139
Equity, 287, 291, 295, 296
Ethical issues, human development
    children's folk-biological reasoning, 1900
    developmental research, 1901
    harm, 1898–1899
    informed consent, 1895–1897
    privacy, 1897–1898
    psychological research, 1900
Ethical standards
    beyond purview of review boards, 2172–2173
    formal review boards, 2171–2172
Ethics, 431, 536–537, 1882–1884, 1936, 1943, 1949, 1974–1975, 1977, 1979, 1994, 1998, 2128–2130, 2135, 2136, 2140, 2141, 2148–2149
    of silence, 435
Ethnographic methods, 232, 444, 511
    advantages, 452
    criticisms of, 453
    future research, 453–454
    innovative forms, 449–452
    reflexivity, 446–447
    traditional approaches, 447–449
Ethnography, 226, 2199, 2208, 2209
    critical challenges to, 468
    definition, 458
    model of, 458
    symbolic interactionism, 176–177
Ethnomethodology, 473
    accountability, 274
    in action, 280–281
    breaching, 277
    and common sense procedures, 276
    and conversational analysis, 276, 278
    data collection techniques, 277–278
    description, 271
    emergence of, 272–273
    and ethnography, 278
    indexicality, 275
    membership categorization analysis, 275–276
    perspective, 271–272
    practicalities, 279–280
    reflexivity, 274
    role of, 270
    strengths and advantages, 279
    weaknesses and limitations, 278–279
Evaluation indicators, 345–347

Evaluation research, 356
    culture/ethnicity, 350
    development of indicators, 347–350
    evaluation indicators, 345–347
    features of, 335
    formative, 357–359
    gender-sensitive indicators, 350–351
    logic models and results based frameworks, 340–342
    PRECEDE-PROCEED model, 342
    process, 359–361
    purpose and phases of, 344–345
    realistic evaluation theory, 340
    study designs and methodologies for, 346
    summative, 361–362
    ToC (*see* Theory of change (ToC))
Evidence-based health care, 3
Excellence in Research for Australia (ERA), 368
Expanded bioecological theory, 1203
Expert(s), 716, 717, 719–722, 724, 1907, 1909, 1915, 2214–2216
Expertise, 2214–2216, 2222
Expert knowledge elicitation, 2214
    cognitive dissonance theory, 2218–2219
    learning based knowledge elicitation, 2216–2218
    research interviews, 2219–2221
    research participants and interlocutions, 2221
Explanatory sequential design, 63–64
Exploitation, sexual, *see* Sexual exploitation
Exploratory sequential design, 64–65
Expressive therapy, 1135
External validity, 373, 375, 383–384
    online survey research, 1344, 1350

F
Facebook, 359
FaceSpace Project, 361
Face-to-face recruitment, 86, 87, 91
Face-to-face research in prisons
    ethics, 2170–2171
    with prisoners, 2166–2167
Face-to-face (F2F) surveys, 548, 554, 555
Facilitator, 741
    orientation, 1800, 1802, 1804
Factorial design, 654–655
Faith-based HIV prevention programs, 229
Feminism and healthcare
    description, 206–207
    politically, historically and culturally embedded healthcare, 207–208

Feminist, 2112–2113, 2121
    methodology(ies), 458, 462, 2080–2083
    paradigm, Bakhtinian theoretical framework (*see* Bakhtin, Mikhail)
Feminist discourse analysis
    aim and method, 887–888
    PMS (*see* Premenstrual syndrome (PMS))
Feminist pragmatist model
    description, 209–210
    radical objectivity, 218
    relational and systems critical approach, 218
    solution-focussed epistemology, 216–217
Fidelity research, 381–383
Fieldwork, 447, 1676, 1679, 1682, 1684, 1687, 1688
Flexible conversation techniques, 1225
Focus group(s), 740, 743, 747, 975, 979
    research, 1800–1802, 1807
Forecasting, 716–718
Forest plots, 821
Formal review boards, 2171–2172
Formative evaluation, 357–359
Foucauldian discourse analysis (FDA), 128–129, 884–887
Four-phase model, timelines, 1197–1198
Fragmented narrative communication, 2094, 2101
Frankfurt School's method, 140
Freelister, 1439
Freelisting, 1432–1443
    advantages, 1433, 1434
    analysis, 1438–1440
    domain analysis, 1436–1438
    method, 1085
    obstacles of using, 1433
    potential shortcoming of, 1434
    recognition of, 1432
    salience analysis, 1438
    systematic data collection, 1441
    with ethnographic interviews, 1440–1443
    written, electronic and oral interviews, 1435–1436
Free recall, 1438
Freund, Alexander, 435
Funnel plots, 822

G
Gender equality, 211–212
Gender sensitive evaluation, 350–351
Generalization
    analytical, 325
    formal, 318, 324

naturalistic, 326–329
nomothetic, 325
statistical, 325
Generative taxonomy, 503
Genuine integration, 686
Go-alongs, *see* Walking interviews
Good reporting of a mixed methods study (GRAMMS) tool, 1062
Grammatical issue, 1623
Graphic elicitation, 1195
Grounded theory methodology (GTM)
  approaches, 301
  characteristics, 302
  coding, 303–306
  constructivist, 301, 307–308
  conventional approaches, 306
  describes, 300
  in health social science, 308–314
  process of, 302
  value of, 301
Grounded theory research, 1655, 1656
Group allocation, 1037
Group discussion, 1169

H

Habermas, Jürgen, 141, 142
Hard-to-reach people, 1907, 1910, 1914
Harm, 1892, 1898–1899
  reduction, 1979
Healing culture, 261
Healthcare personnel, 209, 215–216
Health disparities, 1851–1852
Health elites, 2199, 2209
  clinic and laboratory, 2203–2205
  conferences, 2205
  documentary cultures of, 2206
  and ethical issues, 2200–2201
  interviewing, 2202–2203
  patients, 2207
  sampling and access, 2201–2202
  social scientists, 2208–2209
Health inequities, 1851–1852, 1862
Health information systems
  formative research, 357–359
  process evaluation research, 359–361
  summative evaluation research, 361–362
Health information technologies (HITs), 356
Health promotion, 514, 516, 520, 522, 524, 525, 1544, 1546, 1550, 1552
Health social sciences research
  critical realism in, 165
  netnography in, 1324
  post-positivism in, 160

Heidegger, 191
  See also Hermeneutics
Hermeneutics, 190
  cross disciplinary research, 198–201
  current neglect of, 191–193
  in research, 191
  tenets of, 193–198
Heterogeneity, 808, 812, 815, 819, 821
Heuristic research, 2098–2099
Hidden populations, 1907, 1911
History, content analysis, 828, 830–832
HIV/AIDS, 1152–1158
Home–school communication, 1468–1474
Horkheimer, Max, 136
Hull Floods project, 1450
Human health, 105, 107
Hybrid design, 66–67
Hybrid trial designs, 375–376
Hysteria, 213
Hysterical women, 213–215

I

IMAGINE, 1173
  analyzing, 1179–1180
  application/situation cards, 1174–1175
  card types and choreography, 1174–1176
  challenges, 1169–1170
  discussion groups groups, 1177–1179
  exploring and analyzing the issue, 1172–1174
  future cards, 1175–1176
  issue/context cards, 1175
  story/statement cards, 1174
  validating the card, 1176–1177
Incidence, 567–568
Inclusion, 1170
Inclusive research, 1936
  academic researcher experiences, 1944–1945
  ad-hoc approach, 1937
  characterization, 1937
  components, 1938
  co-researcher experiences, 1942–1944
  co-researcher involvement, 1941–1942
  disability, 1939–1941
  methodological issues, 1945–1948
In-depth interviewing method, 17
In-depth interviews, 393, 396, 402
Indexicality, 275
Indigenist paradigm, 262
Indigenist research, 1545, 1548–1559

Indigenous Australians, 1544, 1546
Indigenous Knowledge, 260–261
Indigenous people(s), 1508, 1511, 1514, 1848, 1854, 1856, 1858, 1861–1863
Indigenous research, 1528, 1531, 1533, 1541
Indigenous statistics, 1697–1698
   Aotearoa New Zealand, Māori families, 1699
   Australia, Indigenous children's education, 1702–1703
   development principles, 1696
   Indigenous data sovereignty, 1696–1697
   social norms, values and racial hierarchy, 1699
Indigenous ways of knowing, 1796, 1797, 1801, 1807
Individualized profiles, 1782
Infant feeding
   cultural perspectives, 1767
   ethnographic studies, 1761
   refugee backgrounds, 1763
   refugee perspectives of, 1768
Inference transferability, mixed methods research, 1061–1062
Infographics, 2214, 2217, 2221
Information computer technology (ICT) surveys, 551
Information systems practitioners, 2214
Information systems research, 2214
Informed consent, 1893, 1895–1897
Innovation, 688–689, 1169
   new perspectives on, 1169
Innovative researcher, 1074–1075, 1087
Innovative research methods
   arts-based and visual methods, 1076–1080
   body and embodiment research, 1080–1082
   classification, 1073
   definition, 1073
   digital methods, 1082–1085
   as emerging research methods, 1074
   innovative researcher, 1074–1075
   mind maps in qualitative research, 1076
   personal construct qualitative methods, 1075
   textual methods of inquiry, 1085–1087
Insider(s), 1609–1610
   perspectives, 1958–1959
Insider/outsider research
   Africanist Sista-hood in Britain, 1581–1584, 1592–1597
   data collection, 1585–1587
   design, 1581
Institutional discourse, 460, 461, 464

Institutional ethnography (IE)
   data collection, 466–467
   described, 458
   framework, 459
   Marxist cultural materialism, 462
   recent works, 459
   work places, 459
Instrumentalism, 163
Integration
   description, 685
   genuine, 686
   multistream visual data, 689
   principles and practices, 686
   researcher role, 687–688
Interactions, 472
   health care, 483–484
   institutional, 476
   See also Conversation analysis
Internal consistency reliability test, 668–669
Internationalization of English, 1621
Internet, 1387
   and deceptive responses, 2138–2139
   desk research, 2130–2132
   online ethnography, 2136–2138
   online materials, 2132–2133
   and participant recruitment, 2133–2135
   and sexual subculture participant recruitment, 2135–2136
   and truthful responses, 2139–2140
Interpretation, 29
Interpreters, 1655, 1657–1658, 1662, 1665–1668, 1671
Interpretive hermeneutic phenomenological approach (IHPA), 1959, 1968
   foundational tenets, 1959–1963
   strengths and limitations, 1969–1970
   trustworthiness, 1963–1968
Interpretive rigor, mixed methods research
   interpretive agreement, 1061
   interpretive bias reduction, 1061
   interpretive consistency, 1061
   interpretive correspondence, 1061
   interpretive distinctiveness, 1061
   interpretive efficacy, 1061
   interpretive transparency, 1061
   theoretical consistency, 1061
Interprofessional relationships, 772
Inter-rater/inter-observer reliability test, 668
Inter-subjectivity, 413
Intervention design, 65–66
Interventionism, 227–228

Intervention research, 584, 587, 589, 598
Interview(s), 173, 975, 979, 980, 1115, 1655, 1657, 1658, 1661, 1663
  data transcription, 1670
  interview guide, 1664–1665
  language use, 1667–1668
  materials, 1394–1395
  participants, contact, 1391–1392
  participants selection, 1391
  questions, 1390–1391, 1393
  second interviews, 1668–1669
  study design, 405
Interviewers, 549–551
Interviewer speech vs. voice/vocal
  characteristics, 1418–1420
  minimizing non response, 1420–1423
Interview methods, disabled children
  cartoon captioning, 2030–2031
  child-researcher power relations, 2036–2038
  partnering with parents, 2035–2036
  photo-elicitation, 2032
  role play, 2029
  sentence-starters, 2034–2035
  stage setting, 2034–2035
  vignettes, 2033–2034
Interview tool, email as, 1396
Intoxication, 1975, 1981, 1983
Investigator, participant interface, 73–74
Involvement of people with dementia, 1993–1995, 1997
Irish Republican Army (IRA), 433

J

Jeffersonian transcription system, 486
Joanna Briggs Institute's critical appraisal tool for systematic reviews, 1030
Journalling, 920, 926
  audit trail, 920, 926

K

Kaupapa Māori health research, 1500
  axiology, 1516–1520
  epistemology, 1514–1515
  methodology, 1515–1516
  ontology, 1512–1513
  paradigm, 1511–1512
kaupapa Māori research, 1531, 1533
Key performance indicators (KPIs), 362
Kinetic conversations, 2036

Knowledge, 3, 4, 6, 1002
  exchange, 777, 780
  research, 1003–1006
  scientific, 1006
  social, 1000
  transfer, 1005, 1007

L

Laddering method, 1102–1104
Ladder of participation' model, 1927
Language, qualitative research, *see* Qualitative research, multiple language groups
Language interpretation, 1640, 1641, 1649
Latent, 853
Latvian Legal Reform Project (LLRP), 1123
Law enforcement and public health (LEPH), 2181, 2187, 2193
Lay-expert divide, 1180
LEPH, *see* Law enforcement and public health (LEPH)
Levels of measurement
  definition, 957
  and descriptive statistics, 959
  interval measurement, 958
  nominal measurement, 957
  ordinal measurement, 957
  ratio measurement, 958
  significance of, 956–957
Lexical issue, 1619, 1628
Lexicographic pattern, translation, *see* Translation
LHC method, *see* Life history calendar (LHC)
Life course, 938–940, 1202
Life course theory, 1203–1204
Life grids, 1186
  timelining method, 1193
Life history calendar (LHC), 1202, 1204–1205
  adaptation of, 1209
  evolution of, 1205–1207
  implementation, 1210–1215
  method, 1079, 1194
  qualitative study, 1207–1210
Life history data, 941, 950
Life history interview, 426, 427, 430, 432, 434
  analyzing and disseminating, 435–438
  described, 426
  preliminary considerations, 428–430
Life story mapping, 1194
Life world, 412
Lingua franca, 1621
Linguistic contact, 1622
Linguistic treatment, 1623–1624

Literature, 920
Lived experience, 414, 450, 453
Living historically, 536, 538
Logical framework approach (LFA), 340, 341
Logical positivism, 157, 160
Longitudinal data, 619
Longitudinal study designs
    advantages and problems, 605–606
    follow-up studies, 606
    repeated measures (*see* Repeated measures)
Luo medical culture, 1727

M

Māori, 1527, 1529–1531
Māori-centred research, 1533
Marcuse, Herbert, 136
Marginalized people, 1909, 1910
Masturbation, 107
Materiality, 1177, 1179
Maternal Early Childhood Sustained Home-visiting (MECSH®) program, 373, 387
MAXQDA, 926
Meandering timeline, 1186
Meaning-making, 417
Measurement, 664
    aetiology, 675–676
    decision-making, 677
    error, 573
    evaluation of interventions, 676
    monitoring, 676–677
Measures of association
    OR, 569
    RR, 568
Media outreach strategies, 91
Medical sociology, 2199
Member checking, 408
Membership categorization analysis, 275–276
Memory work, 528
    analysis, 533–535
    as autoethnography, 529
    as collective method, 529
    ethical considerations, 536–537
    as research approach, 530
    rewriting memories, 535
    social theories, 535–536
    and social transformation, 536
    theme/question, identification of, 531–532
    writing memories, 532–533
Memo writing, 1570

Mental Capacity Act, 1996, 1998
Mental health, 2094, 2106
Mental models, 110–112
Meta-analysis, 807, 811, 814, 820–822, 824
Meta-ethnography, 792, 799, 800
Metaphorical analysis, 1747
Meta-summary, 792, 793
Meta-synthesis, 807, 817, 824
Meta-synthesis of qualitative health research, 786–788
    assessment, 799
    content analysis, 796
    epistemological position, 789, 790
    inclusion/exclusion criteria, 795–796
    keywords/search terms, 795, 797
    objectives, 788, 789
    paradigm dimension, 790–793
    protocol, 797
    quantitative approaches, 789
    question design, 793–795
    reciprocal translation, 799
    scope, 793
    searching and selection/sampling studies, 798
    systematic reviews, 787
    text mining tools, 797
Meta-theory, 792
Methodological approach, translation, *see* Translation
Methodological quality, 1030, 1032–1034, 1047, 1054
Methodology, 1693–1695, 1697–1703, 2113
    life history interview as, 427–430, 438
Methods section, 989–992
Migrant and ethnic minority (MEM) groups, 1740
Migrant and refugee children, visual research, 1728–1734
    See also Visual research
Mind mapping
    assessment, 1120–1123
    case study, 1123–1125
    construction of, 1116
    data analysis, 1118–1119
    data collection, 1118
    definition, 1115
    health and social sciences, 1125–1126
    and mixed methods, 1123
    plan research, 1117–1118
    presenting data, 1119–1120
    theory and methods, 1116–1117
Minds maps, 1076
Minority recruitment, 93

Mixed methods, 684, 696, 697, 919, 921, 922, 933
  data analysis and integration, 707–708
  data collection and sampling, 705–707
  handling conflicting results from, 710–711
  reporting of, 708–709
  study designs, 699–704
Mixed methods appraisal tool (MMAT), 1053–1057
Mixed methods research (MMR), 4, 7, 52, 62, 67, 68, 752–753
  convergent criteria, 1052
  convergent design, 63
  definition, 52–53
  evolution of, 53–54
  explanatory sequential design, 63–64
  exploratory sequential design, 64–65
  feasibility, 1058
  foundational element, 1057
  in health social sciences, 54, 56–57
  hybrid design, 66–67
  individual components approach, 1053
  inference transferability, 1061–1062
  integration, need for, 1064
  interpretive rigor, 1060–1061
  intervention design, 65–66
  MMAT, 1054
  philosophical positions (see Philosophical positions)
  planning transparency, 1058
  quantitative and qualitative components, 1063
  rationale transparency, 1058
  report availability, 1062
  reporting transparency, 1062
  research questions and methodology, 1063
  transparency, need for, 1063
  undertaking research phase, quality criteria for, 1058–1060
  use of, 1064
  yield, 1062
MNL, see Multinomial logit (MNL)
Mobile research method, see Walking interviews
Mobilities, 1272
Modified Delphi, 718, 724
Mothering, 2094, 2095, 2100
Mothers, drawing method, see Drawing method
Motivation, 1901
Multicultural society, 1499
Multi-methodological research, 357
Multinomial logit (MNL), 631
Multiphase design, 704

Multiple-case design, 321
Multivariate approach, 614
Music therapy, 1137

N
Nanomedicine, 1173
Narrating Our World (NOW) project, 1732
Narrative analysis, 863, 866, 868
  and analysis of narratives, 870
  and cognition, 868–869
  findings of, 876
  outcomes of, 875
  techniques, 871–872
Narrative inquiry, 862, 863, 869–571, 874, 877, 878
  in health and social sciences, 865–866
Narrative research, 413
  advantages, 419
  autobiographies, 414
  disadvantages, 420
  formal structural and functional analysis, 416
  health and social sciences research, 418–419
  interpretation, 417–418
  lived, 414
  person-centered therapy, 414
  place, 415
  relational, 413–414
  sociality, 415
  temporality, 415
Narrative synthesis, 789, 795
  CERQual approach, 801
  PICOS approach, 795
Narrative therapy, 1246
National Health and Medical Research Council (NHMRC), 287
  hierarchy, 1029
Naturalistic generalizations, 326–329
The Nature of Qualitative Research, 912
Needs analysis, 357–358
Netiquette, 1396
Netnography, 1083, 1322–1323
  conducting ethical, 1329–1332
  data analysis, 1328–1329
  data collection, 1327–1328
  data set size, 1334
  defined, 1323–1324
  ethical considerations, 1332
  in health social science research, 1324
  lack of face-to-face interaction, 1332
  planning and Entrée, 1325–1327

Netnography (cont.)
  process, 1325
  researcher as participant, 1333–1334
  research representation and evaluation, 1329
Neutral patient model, 209
New media, 2128
The New mobilities paradigm, 1272
New public management (NPM), 216
New Zealand, 1692, 1695, 1699–1702
Nominal group technique (NGT), 738
  advantages and disadvantages, 745–747
  establishing core outcomes, 740–741
  guideline development, 739
  identifying research priorities, 739–740
Non-Indigenous Australians, 1546
Non-reversible behaviors
  AATD, 592
  demonstration designs for, 589
Normativism, 107
Normativity, 134, 135, 138, 141, 144
NVivo, 926, 932

O

Objectivist research, 300–302, 308, 314
Observation, 497–499
Observer bias, 666
Odds ratio (OR), 569
One-off consent process, 1996
Online anonymity, 1358–1360
Online Delphi, 720
Online face-work, audiences and, 1358–1360
Online research methods, 1323
Online survey methods
  advantages and problems associated with, 1341–1342
  critiques of, 1342–1343
  data collection and analysis issues, 1347–1348
  enhancing response rates using, 1346–1347
  measurement issues with, 1345
  platforms and services, 1348–1349
  resources for creating/managing, 1347
  sampling issues and, 1344–1345
Online surveys, 542, 545, 548, 552–554
  merits and limitations of, 553–554
  potential biases, 546–547
  sample sizes for, 548
  social desirability, 549–550
Ontology, 101–102, 161, 164, 165, 167, 192, 199

Operant psychology, 583
Optimal matching (OM), 948
Oral history interviewing, 393
Oral History Society, 430
Oral standard variety/oral cultured norm, 1622
Outsiders, 1609–1610

P

Paired t-test, 674
Panel, 719
Paradigmatic analysis of narratives, 869, 870, 872
Paradigms, 301, 302, 314
Parallel cueing, 1227
Parallel studies, 654
Parenting, 1894, 1899
PAR, *see* Population attributable risk (PAR)
Participant engagement, 77
Participant feedback, 1807
Participant-guided mobile methods, 1081, 1292–1293
  data from diverse participants, 1298–1299
  eliciting different data, 1297
  enabling complex insights, 1299
  limitations and challenges, 1300–1301
  power and control, 1293–1295
  risks to consider with mobility, 1298
  spatially-situated research, 1295–1297
Participant observation, 448–449
Participant voice, 1930
Participation process, 1170
Participatory action research (PAR), 103, 286
  flexibility and interdisciplinarity, 1744
  migration, gender scripts and cultural change, 1748–1752
  Roma youth and drug use, 1744–1748
  with youth, 1741–1742
Participatory analysis method, 1307, 1308, 1313
Participatory justice, 1170
Participatory methods, CBPR, *see* Community based participatory research (CBPR)
Participatory research, 1266, 1576, 1712, 1717
  description, 1920–1922
  models, 1926–1930
  principles, 1930–1932
  visual methods for children and young people, 1726–1728
  vulnerability, 1922–1926
  with children and young people, 1725–1726
Participatory research methodologies, 1875–1876

Participatory video, 1743
Paternalism, 208–210
Paternalistic model, 208–209
PCP, *see* Personal Construct Psychology (PCP)
Pedagogy, 2214, 2222
Peer research, 1742
  migration, gender scripts and cultural change, 1748–1752
  Roma youth and drug use, 1744–1748
Personal construct psychology (PCP), 1075, 1096
  analysis of constructs, 1105
  constructivist epistemology, 1096–1097
  constructs and construing, 1097–1098
  cross-cultural perceptions, 1099–1100
  elicitation constructs in interview, 1099–1100
  features, 1098–1099
  interviewing using elicited constructs, 1101–1102
  laddering, 1102–1104
  pictor technique, 1105–1108
  reflecting on use of, 1110
  self-characterization sketch, 1109–1110
  triadic method of construct elicitation, 1100–1101
Person-centered healthcare (PCH), 215, 217
Person-centered therapy, 414
Perspective-taking, 239, 242–245
Phenomenology, 1831–1832
Philosophical positions
  frameworks, 57–59
    paradigmatic challenges and successes, 59–61
Philosophical stances, 108
Philosophy, 239
Photo-elicitation, 450, 1455, 2032
  methodologies, 1732
Photographic diary method, 1925
Photolanguage, 1878–1879
Photovoice, 1148, 1455, 1732–1733, 1750
  autism, 1158–1161
  data analysis, 1151–1152
  HIV/AIDS, 1152–1158
  method, 1078
  procedures, 1150–1151
  in public health, 1149
  theoretical basis, 1149
  uses, 1148
Pictor technique, 1105–1108
Pilot testing, 639
Pitt County Study, 1232–1233

Placing issues in cross-cultural and sensitive research
  age issues, 1607–1608
  gender issues, 1604–1607
  race, culture and ethnicity issues, 1608–1610
  shared experiences, 1612–1613
  social class issues, 1610–1612
PlayDecide card game, 1171
PMS, *see* Premenstrual syndrome (PMS)
Poetry, 2066, 2071–2072, 2075, 2076
Police research, 2181–2185, 2190
  access negotiation, 2189–2190
  cultural barriers, 2183
  knowledge exchange and dissemination, 2191–2192
  lack of trust and cross-cultural communication, 2184–2185
  research by the police, 2185, 2186
  research for the police, 2185, 2186
  research on the police, 2185, 2186
  research with the police, 2185, 2186
  trust, establishment of, 2190–2191
Political epistemology, 135
  critical theory, 135
Politics of representation, 2085–2087
Polyglossia, 1622
Popper, Karl, 159
Population attributable risk (PAR), 570
Positionality, 1603
Positivism, 153, 154, 834
  health social sciences research, 160
  logical, 157
  post-positivism, 158–160
  social sciences and Comtein classical positivism, 156–157
  western science and positivist notions, emergence of, 155–156
Postal Delphi, 719
Postnatal depression (PND), 2094, 2095, 2098, 2099, 2107
Postpositivism, 58, 60
Power relations, 126
Practical, robust implementation and sustainability model (PRISM), 380
Practice-based research, 525
Pragmatic-explanatory continuum indicator summary (PRECIS) tool, 376
Pragmatic trials, 376–377, 658–659
Pragmatism, 59, 61, 1097
Predisposing, Reinforcing and Enabling Constructs in Educational Diagnosis and Evaluation-Policy, Regulatory, and

Organizational Constructs in Educational and Environmental Development (PRECEDE-PROCEED) model, 342
Preference elicitation, 625–627
Preferences and values, 626
Preferred Reporting Items for Systematic Reviews an Meta-Analyses (PRISMA) statement, 972, 1052
Premenstrual dysphoric disorder (PMDD), 885
Premenstrual syndrome (PMS), 889
   biomedical constructions of, 891
   facilitating agency and self-care, 892–893
   feminist Foucauldian discourse analysis of, 889
Prevalence, 566–567
PRISMA guidelines, 798
Prison
   corrections, 2167
   nature of, 2166–2167
Privacy, 1892, 1897–1898
Process evaluation, 359–362
Program design, 517, 520
Program evaluation, 342
Prostitution, 2116, 2118
Protocol design, 653
Protocol registration, 653–654
Psychiatric art, 1136
Psychological dimension, 2012–2013
Psychology, 239
Public health, 1148, 1149, 1161
   evaluation research (*see* Evaluation research)
   and police research (*see* Police research)
   research, 224, 234
Public policies
   definition, 1002
   and research findings (*see* Research findings and public policies)
Purposive sampling, 18

Q

Q Assessor, 757, 758
Q methodology
   advantage, 762
   bi-person factor analysis, 757
   concourse development, 754–755
   definition, 753
   factor analysis, 753–754
   factor array, 758–761
   limitations, 758, 762–763
   pros and cons of, 763, 764
   P set selection, 755
   Q Assessor analysis, 758
   Q set, 755
   Q sorting, 756–757
   subjectivity, 762
   traditional statistical software, 758
   user participation, 762
Q sorting
   electronic techniques, 757
   manual techniques, 756
   Q Assessor, 757
Qualitative analysis, 1163
Qualitative data
   analysis, 20–21
   collection methods, 17–18
   schema analysis (*see* Schema analysis)
Qualitative data analysis software (QDAS), 919
   case's code, 923
   comparative analyses, 931
   folders, 923
   goals of analysis, 919–921
   matrix coding query, 932
   reviewing and exploring data, 924–926
   selecting data for analysis, 921–922
   sets of codes, 924
   sorting and coding data, 926–931
   users of, 919
Qualitative description, 836
Qualitative inquiry
   analytic strategies, 20–21
   description, 10
   features, 11
   vs. qualitative research, 12–14
   real-world issue examination, 12
Qualitative interviewing, 2045
   advantages, 393–394
   aims, 393
   case studies, 394–399
   description, 393
   participant selection, 400–401
   planning, 402–403
   practice, 404–408
   recruitment, access and ethics, 402
Qualitative, meaning/terminology, 10
Qualitative methodology, 1958, 1959
Qualitative methods, 2199
Qualitative/quantitative research, 109–110
Qualitative research, 4, 5, 300, 301, 476, 482, 1014–1016, 1018, 1020, 1022–1025, 1114, 1184, 1187, 1193, 1239, 1244, 1251, 2146, 2154, 2216
   appraising, 1015–1016
   approach, 11

challenges of, 974
conceptualizing rigor in, 1016–1017
dementia, 1996
guidelines for appraising, 975–977, 1021–1024
inductive approach, 15
mapping stages of, 1118
method, 318, 320, 1877–1879
methodological frameworks, 16–17
and mind mapping, 1115–1116 (*see also* Mind mapping)
proposed strategy for, 1024–1025
research assistants (*see* Research assistants)
saturation, 19
source of data, 18–19 (*see also* Qualitative inquiry)
standards of, 973
and trustworthiness, 19–20
with children, 2026
Qualitative researchers, 14, 15
Qualitative research, multiple language groups
budget, 1671
confidentiality, 1669
constructivist grounded theory, 1655, 1656
co-participants, 1663
cultural competency training, 1658
data analysis, 1669–1670
debriefs, 1667
informed consent, 1662–1663
interpretation method, 1666
interpreters and translators, 1657–1658
interview data, transcription of, 1670
interview guide, 1664–1665
language use, 1667–1668
pre-brief meeting, 1665, 1666
reporting findings, 1670–1671
second interviews, 1668–1669
Qualitative story completion method, 1086
Quality, 973, 974, 982
criteria, 1015, 1021, 1023, 1025
Quality of life (QOL), 546
Quantitative analysis, key measures/ variables for, 991
Quantitative data analysis, 956
bivariate analysis, 962
causal research questions, 960
choice of analysis model, 965
descriptive analysis, 962
descriptive statistics, 963
exploratory research questions, 960
inferential analysis, 963, 965
levels of measurement (*see* Levels of measurement)
multivariate analysis, 962
practice of, 965–968
relational research questions, 960
statistical hypothesis, 964
univariate analysis, 961, 962
Quantitative, meaning/terminology, 10
Quantitative research, 4, 6, 7, 29, 1028, 1047
benefits, 30
bias, 41
defined, 29
development, 36–37
evidence pyramid, 38
experimental, 30
interpreting, 29, 47–48
limitations, 30
methods, 1876–1877
Quasi-experimental study, 564
Questionnaires
data quality, 546
focus group, 555
format and design, 544–545
potential biases, 546–547
validity and reliability, 545–546

R

Racism, 257
Radical objectivity, 218
Random digit dialing (RDD), 1405
Random error, 571–572
Randomization, 648–649
Randomized controlled trials (RCT), 160, 375, 384, 564, 647–648
allocation concealment, 649
blinding, 649–650
cluster, 655–656
comparative effectiveness research, 659
crossover, 656–657
factorial design, 654–655
IIT analysis, 650–651
parallel, 654
pilot studies, 652
pragmatic, 658–659
protocol design, 653
protocol registration, 653–654
randomization, 648–649
reporting of, 657–658
superiority, equivalence and non-inferiority, 657
Random utility theory (RUT), 629
Ranking, 743, 745
Rapid ethnography, 1292

Rapport, 405
Rating scale scores, 111
RCT, *see* Randomized controlled trials (RCT)
Reach, Effectiveness, Adoption, Implementation and Maintenance (RE-AIM) model, 379–380
Realism, 161
 critical, 164–165
 scientific, 161–162
Realistic evaluation theory, 340
Realist research, 300–302, 309, 310
Recruiting Older Adults into Research (ROAR) Toolkit, 90
Recruiting participants
 drug use, 1975
 fentanyl use, 1977
 health care, 1975
 reduce gatekeeping, 1976
 snowball sampling, 1977
Recruitment strategies, 77–79
 adults, 81–90
 barriers and challenges, 74–75
 children and young people, 79–81
 facilitators, 75–77
 older adults, 90–92
Reflexive practice, 260
Reflexive thematic analysis, 848–849
 candidate themes, 854
 constructing themes, 854
 design considerations, 849–852
 familiarization, 852–853
 generating codes, 853–854
 producing the report, 857
 revising and defining themes, 855
Reflexivity, 274, 446–447, 536, 1643
 emotional management, 2175
 managing boundaries, 2174–2175
Refugee(s), 1812, 1816, 1818, 1821, 1824, 1826, 1873–1875, 1884–1885
 camps, 1813, 1816, 1820, 1824
 women, drawing method and infant feeding practices (*see* Drawing method)
Refugee children
 CAIs (*see* Computer-assisted interviews (CAIs))
 respect, 1780
 well-being, 1781
Register, 1621
Relational aesthetics, 1135
Relational map, 1263–1265
Relational narrative, 413–414
Relativism, 1097

Reliability
 data collection, 667
 external factors, 667
 internal consistency, 668–669
 inter-rater/inter-observer, 668
 researcher/observer, 666
 respondent, 667
 sources of errors, 666–673
 test-retest, 668
Repeated measures
 ANOVA, 612–614
 classic repeated measures and growth curve designs, 606–607
 correlation structure, 615–616
 crossover designs, 607–609
 marginal models, 618–619
 mean plots, 610–612
 missing data, 614–615
 mixed model regression approach, 615
 multivariate approach, 614
 panel plots, 609–610
 transition models, 619
Report availability, 1062
Reporting
 bias, 994
 checklists, 987–995
 quantitative findings, 989
 of study design, 990
 sufficient and relevant information, 990
 transparency, 1062
Reporting guidelines, 972–974, 979, 982, 987–995
 for qualitative research, 974–977
Reporting quality, 1062
 definition, 1034
 of experimental studies, 1040
 of studies, 1046
 systematic reviews, 1034–1035
Representativeness, 548–549, 556
Research, 3, 4, 6, 152–154, 160, 165, 1240, 1245–1247
 aims, 2183, 2193
 approach, 4, 5
 design, 54, 56, 60
 ethics, 1854, 1856, 1858, 1859, 1863, 1974–1975
 people with dementia (*see* Dementia)
 with young people, 1725–1726
 See also Body mapping
Research assistants
 in academic and non-academic institutions, 1677
 conceptual and ethical issues, 1686–1688

definition, 1677
interpreters, 1677
large-scale surveys, implementation of, 1677
recruitment of, 1680–1684
researcher, position of, 1678
research timeframe and funding, 1679
role of, 1679
tasks, 1684–1686
Researcher status, 2189
Research findings and public policies
academic research, 1008
action research, 1008
communicational obstacles, 1005
credibility and communication, 1005
decision-making process, 1006
enlightenment model, 1004
epistemic-praxis obstacles, 1005
historical-political obstacles, 1005
influence and legitimacy, 1005
instrumentation research, 1008
interactive model, 1004
knowledge driven-model, 1003
knowledge transfer, 1005
philosophical obstacles, 1005
planning research, 1008
political model, 1004
politics and institutions, 1005
problem solving model, 1004
research quality, 1007
social researchers, 1007
tactical model, 1004
tactic obstacles, 1005
temporary obstacles, 1005
Research methodology
methods section, 989–992
novelty in, 987
Research methods
innovation and use of technologies, 688–689 (*see also* Integration)
mixed methods, 683–684
mixed research, 2016
qualitative, 2015
quantitative, 2013–2014
recommendations for projects, 690
researcher role, 687–688
triangulation, 683
Research with prisoners
prisoners lives, 2167
prisons nature, 2166–2167
reflexivity, 2173–2175
sensitive contexts for, 2168–2170

Resilience, 257–258, 2045, 2047, 2054, 2056, 2057
Respect, 1780, 1783, 1787, 1788, 1790, 1791
Response rates, 1418, 1419, 1425
online survey, 1346–1347
Results based management (RBM), 341, 342
Results section, 992–993
Retrospective data, 940
Reversible behaviors
ATD, 591
demonstration designs for, 588
Rigor, 839, 1014–1017, 1021, 1022, 1945
Risk of bias, 807, 812, 816, 818–820, 822, 824
Risk ratio (RR), 568
Roma youth
peer research on migration, gender scripts and cultural change, 1748–1752
visual and participatory research, 1744–1748
Rorschach inkblot test, 1481
Running records, 495
RUT, *see* Random utility theory (RUT)

S

Salutogenic approach, 2012, 2013
Schema analysis, 898, 900–901
group/meta-schemas, 903–904, 908–911
individual schemas, 901–903, 905–907
interpretation of meta-schemas, 904, 911–912
qualitative analytic practices, 912–913
validity, rigor and trustworthiness of data, 912
Science and Technology Studies (STS), 2199
Scientific antirealism, 162–163
Scientific realism, 161–162
Script writing, 1306
Search strategy, 809, 810, 812, 813
Selection bias, 573
Self-administered freelists, 1435
Self-characterization sketch, 1109–1110
Self-initiated experiential education project (SEEP), 327
Self-portrait(s), 1258–1262
method, 1080
Self-report methods, 496
Semantic, 853
Semi-structured interviews, 393, 402
Semi-structured life history calendar (SSLHC) method, 1202, 1207, 1210, 1215

Sensitive research, 2112–2113, 2124, 2168
  age issues, 1607–1608
  and emotions (*see* Emotions and sensitive research)
  gender issues, 1604–1607
  race, culture and ethnicity issues, 1608–1610
  shared experiences, 1612–1613
  social class issues, 1610–1612
Sensitive research methodology and approach
  animated visual arts and mental welfare fields, 1913
  dementia, people with, 1911
  disability research, 1911
  eating disorders and child sexual abuse, 1912
  elites/experts, 1909
  and emotions, 1914
  ethics and practice of research, 1911
  face-to-face research in prisons, 1914
  hard-to-reach people, 1910, 1914
  Internet in sex research, 1914
  marginalized people, 1908–1910
  participant-generated visual timelines, 1912
  police research and public health, 1914
  underage sex work, 1913
  vulnerable people, 1908–1910
  with children, 1912
  women's violence, 1913
Sensitive topics, 1371, 1380, 1909, 1911, 1913, 1914
Sensitivity
  responsiveness, 673
  sources of risk, 674
  statistical methods, 674–675
Sensory methods, 451–452
Sentence starters, 2034–2035
Sequence analysis, 936–939
Sequence data, 939–941
Sequence probes, 406
Sequential cueing, 1227
Sequential explanatory design, 699–700
Sequential exploratory design, 700–701
Sequential multiple assignment randomized trial (SMART), 609
Sexology, 2128
Sexual exploitation, 2116–2118, 2121–2122
  agency and consent, 2115
  confidentiality, 2119
  gatekeepers, 2115–2116
  isolation and responding to danger, 2119–2120
  personal, social and political, 2119
  trust-building and researcher vulnerability, 2120–2121
Sexuality, 1956–1957, 2128, 2131, 2133, 2135, 2136, 2138
  and disability, 1956–1957
  IHPA (*see* Interpretive hermeneutic phenomenological approach (IHPA))
Shannon's entropy index, 947
Sharing authority, 426
Sick and disabled women, 212–213
Simple randomization, 648
Single-case designs, 321, 322, 326, 582
  AATD, 592
  ABAB design, 588
  ATD, 591
  baseline logic, 585–586
  development of, 583
  graphical presentation of raw data, 593
  individual case, 583–584
  multiple baseline design, 589–590
  procedural fidelity, 598
  repeated measurement, dependent variable(s), 584–585
  repeated, systematic manipulations, independent variable(s), 585
  research questions, 587–588
  response-guided decision-making *vs.* randomization, 598–600
  selection of dependent measures, 597–598
  visual analysis guidelines, 593–597
Single-subject designs, *see* Single-case designs
Situational mapping, 306, 307, 313
Snowball sampling, 85
Social and emotional well-being, 1708, 1710, 1717, 1719
Social constructionism, 118, 863
  discursive psychology, 129–130
  essentialism, 122–123
  Foucauldian/critical discourse analysis, 128–129
  historical and cultural specificity, 121
  illness, 126–127
  knowledge, social processes and social action, 121–122
  language, 124
  objectivity and universalism, 124–125
  origins and influences, 118–120
  power and politics, 125–126
Social constructivism, 163
Social desirability, 542, 548–550, 555, 556
Social determinants of health (SDoH), 1873, 1876, 1886
Social exclusion, 1931, 1932

Social justice, 286–288, 291, 295, 296
Social mechanisms, 463
Social media, 359, 362, 1387
Social network analysis (SNA), 771–772
   and health care research, 776–780
Social network, concepts, 771–772
Social networking sites (SNSs), 1326, 1327, 1332
Social participation, 1565, 1576, 1577
Social pathways, 1203, 1204
Social position, of interpreters, 1641, 1643–1646
Social science(s), 153, 154, 1238
   and Comtein classical positivism, 156–157
   critical realism in, 165
   Dewey's approach to, 158
   post-positivism in, 160
   research assistants (*see* Research assistants)
Social transformation, 529, 536
Social vulnerability, 1908
Social work practitioners, 1102–1104
Sociological linguistics, 2063
Socio-political dimension, 2009–2010
Software, qualitative data analysis, *see* Qualitative data analysis software (QDAS)
Solicited diary methods, 1085, 1448–1449
   addressing the limitations of, 1459–1460
   approaches to, 1451–1454
   capture technologies, 1454–1455
   data analyzing, 1456–1457
   examples, 1460–1461
   methodological underpinnings, 1449–1451
   qualitative, 1448
   quantitative, 1448
   strengths and limitations of, 1457–1458
   typology of research questions, 1452–1454
Space and time consideration, 1628
Specificity, 576
SPSS, *see* Statistical Package for Social Sciences (SPSS)
SRM, *see* Standardized response mean (SRM)
Stability, 727, 729, 732
Stakeholder capture effect, 1170
Standardized response mean (SRM), 675
Standards for Reporting Diagnostic Accuracy Studies (STARD), 973
Standpoint theory, 462–463
   institutional ethnographers, 458
Statistical Package for Social Sciences (SPSS), 956, 958, 965, 966
   statistical software, 758

Statistics, Indigenous peoples, *see* Indigenous statistics
Steering probes, 406
Story analyst, 869
Story completion (SC), 1480
   advantages, 1483
   analysis approaches to avoid, 1493
   benefits of, 1482–1484
   completion instructions for, 1487
   data analyzing, 1491–1493
   as data collection method, 1482–1484
   flexibility of, 1484
   frequency counts, 1493
   story maps, 1491–1492
   theoretical lens, 1481–1482
Story completion (SC) research
   asking additional questions in, 1488
   data collection and ethical considerations in, 1489–1490
   ethics in, 1490
   examples, 1484
   piloting in, 1490
   sampling in, 1488–1489
   stem and study design, 1484–1486
   topics and questions, 1484
Story map, 1491–1492
   template, 1186
Storytellers, 413, 869
Stratified randomization, 648
Street-involved youth, 2044, 2045, 2047, 2057
Strengthening the Reporting of Observational Studies in Epidemiology (STROBE) Statement, 972, 987
Strengths-based research, 1716, 1719
Strengths, weaknesses, opportunities, and threats (SWOT) analysis, 358
Structured vignette analysis, 1831, 1836–1839
STS, *see* Science and Technology Studies (STS)
Study participants, 78, 91
Subjective idealism, 789
Subjectivity, of interpreters, 1641, 1643–1646
Substance treatment
   cognitive impairment, 1984
   drug treatment, 1983
   treatment program, 1984
Suicide research, Aboriginal people, *see* Aboriginal people
Summative content analysis, 829–830
Summative evaluation, 361–362
SurveyMonkey®, 1390

Surveys, 175–176
  definition, 542
  traditional (*see* Traditional surveys)
Survival analysis, 619
Sustainable developmental goal (SDG), 555
Symbolic interactionism, 170
  content analysis, 178–179
  ethnographies, 176–177
  experimental methods, 179–180
  interview methods, 173–174
  methodological divergence, 172–173
  surveys, 175–176
  tenets and propositions of, 170–171
Synchronous text-based online interviewing method, 1084, 1370, 1373, 1378–1381
  development, 1371–1373
  interviewer effect, 1371–1372
  potential interviewees, 1372
  preference method, 1375–1378
  usage, 1373–1375
Systematic analysis, 480
Systematic error, 571, 573
Systematic reviews, 807, 809, 811
  protocols, 811–812
  stages of, 808

T
Target population, 73, 77, 81
Team-working, 899, 903
Technology, 2128, 2133, 2135, 2138
Teddy diaries, 1086, 1466–1467
  bear qualities, 1472–1474
  collective voice development, 1469–1470
  comparative analytical context, 1468–1474
  home–school communication, 1468–1474
  interpretations of, 1468
  significant routines, 1470–1471
  socially saturated naturally occurring data, 1467–1468
Telephone interviewing practice recommendations, 1425–1428
Telephone surveys, 1420
  interviewers, 1418–1420
  nonresponse to, 1418
Temporality, 575
Test-retest reliability, 668
Textual analysis, 464, 1788
Textual method of inquiry, 1085
Thematic analysis (TA), 901
  codebook, 849
  coding reliability, 847–848
  domain summaries *vs.* shared-meaning-based-patterns, 845–847
  history, 844–845
  method, 20
  reflexive (*see* Reflexive thematic analysis)
Thematic map, 855, 856
Theme, 845–849
  candidate, 854–855
  constructing, 854
  revising and defining, 855–857
Theoretical appropriateness, 1779, 1790
Theoretical questions, 1618
Theoretical sensitivity, 688
Theory based evaluation, 335–336
  realistic evaluation theory, 340
  ToC, 336–340
Theory of change (ToC), 336, 337
  CCM, in Indonesia, 339
  and classic change theories, 339–340
  elements, 338
  mapping, 338
Thick description, 1788
Time diary methods
  conversational survey technique, 1226
  cueing properties, 1227
  National Time Accounting framework, 1229–1230
  well-being and surveys, 1224–1227
Timeline(s), 1184, 1185, 1191, 1192, 1197
  concept inclusion, 2045
  critical emancipatory and feminist approach, 2056
  data analysis, 2046–2047
  diagramming methods, 2054
  drawing method, 1079
  implementation, 2055
  mapping to participants, 2047–2048
  organize and accumulate data, 2055
  participant as navigators, 2052–2053
  rapport building, 2050–2052
  sequencing of events, 2056
  styles, 2048–2049
  therapeutic moments and positive closure, 2053–2054
  visual, 2046
  See also Timeline drawings
Timeline drawings
  advantages, 1187–1189
  challenges, 1189–1191
  definition, 1185
  four-phase model, 1197–1198
  review, 1191–1197

Time use research, 1221
  description, 1222
  well-being, 1223–1224
Top-down cuing, 1227
Total institution, 2166
Traces, 492–494
Traditional surveys
  data quality, 546
  ethics, 554–555
  F2F surveys, 555
  and ICT based surveys, 551–553
  interviewer bias, 550–551
  merits and limitations of, 553–554
  pilot test, 555
  potential biases, 546
  primacy and recency effects, 551
  questionnaire format and design, 544–545
  reliability and validity, 545–546
  representativeness, 548–549
  social desirability and acquiescence bias, 549–550
  sustainable developmental goal, 555
Training, 2152, 2156
Transferability, 1964
Transformative design, 703–704
Translation, 1622–1623, 1640, 1642, 1646–1647, 1657, 1659, 1662, 1666
  competence, 1629–1630
  cultural issues, 1623
  difficulties, strategies and rules, 1626–1629
  ethical issues, 1620–1621, 1630–1632
  grammatical issues, 1623
  lexical issues, 1619
  linguistic treatment, 1623–1624
  pragmatic issues, 1623
  rhetorical issues, 1623
  semantic issues, 1620
  syntactic issues, 1620, 1623
  task of, 1624–1626
Translational research
  business research methods, 381–382
  collaboration and community engagement, 385
  decision analysis, 380
  definition, 369
  external validity, 383–384
  fidelity research, 382–383
  health social science research, 372–373
  hexagon tool, 378–379
  humans, 375
  hybrid research models, 375–376
  non-reflexive and reflexive clinical research, 372
  pragmatic trials, 376–377
  PRISM, 380
  RE-AIM model, 379–380
  theoretical research, 374
  types, 369–370
Trauma aware practice, 256
Triadic method, 1100–1101
Triangulation, 683
Trustworthiness, 19–20, 1016, 1963–1968
Turbulence, 947
Twenty statements test (TST), 175

U
Unmotivated looking, 480
Unobtrusive methods
  episodic records, 496–497
  and ethics, 499–500
  generative problem, 502–503
  online methods, 500–501
  running records, 495
  traces, 492–494
  triangulation, 502
User led research, 1937

V
Validity
  action inference, 672
  cellphone surveys, 1408
  construct, 670–671
  content, 670
  criterion, 671–672
  interpretive inference, 672
Verstehen, 196–197
Victims, 2112, 2114, 2119, 2121, 2122
Vignettes, 1831, 1832, 1834, 1836, 1837, 1841–1842, 2033–2034
Violence, 2112, 2113, 2115, 2120, 2122, 2123
Visual analysis, 1814, 1818, 1823, 1826, 1827
Visualization, graphical
  index plot, 942–944
  modal plot, 945
  state distribution plot/chronogram, 944–945
Visualizing, 919
Visual maps, 1125
Visual methodology, 1240, 1251
Visual methods, 449–451, 2055, 2056
  maps, 1262–1265
  self-portraits, 1258–1262
Visual narrative inquiry, 1195
Visual representations, 1813, 1817, 1822, 1826

Visual research, 1077, 1724
  calendar and time diary methods, 1079, 1080
  card-based discussion method, 1078, 1079
  maps and drawings, 1728–1732
  methods, 1148
  participatory methods, 1726–1728
  photographs and videos, 1732–1734
  photovoice method, 1078
  semi-structured life history calendar method, 1079
  timeline drawing method, 1079
  with marginalized youth, 1742–1744
Voice
  negotiating, 1310
  producing, 1309
Volunteer Family Connect (VFC) program, 373, 387
Vulnerability, 2113, 2118, 2120–2123
Vulnerable groups, 1923
Vulnerable people, 1908, 1911, 1913

W

Walking interviews, 1081, 1292, 1296
  case studies, 1283–1287
  conducting, 1281–1283
  considerations and limitations, 1279–1281
  description, 1270, 1271
  ethical considerations, 1282
  strengths and characteristics, 1272–1279
War metaphors, 212
Web-based survey
  data collection and analysis issues, 1347–1348
  platforms and services, 1348–1349
Web-based survey methodology, 1083
Well-being, 106, 108, 114
  calendar interviewing methodology, 1231–1233 (*see also* Calendar methods)
  flexible conversation techniques, 1225
  National Time Accounting framework, 1229–1230 (*see also* Time diary methods)
White guilt, 262
Womanism, 1582
Women, 2062
  Bakhtin's authoritative and internally persuasive discourse, 2065
  Bakhtin's dialogism, feminist approach to, 2063–2065
  child sexual abuse and eating disorders, experiences of, 2067–2068
  dialogical semistructured interviews, 2068–2071
  oppression of, 2066
  poetry, 2071–2072
  visual methods, 2072–2075
Women's violence
  allegiance, 2083–2085
  ambivalence and compromise, 2087–2089
  feminist methodologies, 2080–2083
  politics of representation, 2085–2087
Writing strategies
  author's original research on, 995
  BI section, 989
  journal articles, 987
  quantitative research, 996
  results section, 992–993
  scientific paper, 996
Written standard variety/written normative language, 1622

Y

Yarning, 1575
Youth, 2027, 2035
  interview methods for disabled (*see* Interview methods, disabled children)
  mental health research, 1307